Basic Mathematics

Basic Mathematics

10th EDITION

Marvin L. Bittinger

Indiana University Purdue University Indianapolis

PEARSON

Addison
Wesley

Boston   San Francisco   New York
London   Toronto   Sydney   Tokyo   Singapore   Madrid
Mexico City   Munich   Paris   Cape Town   Hong Kong   Montreal

3/09

Publisher	Greg Tobin
Editor in Chief	Maureen O'Connor
Executive Editor	Jennifer Crum
Executive Project Manager	Kari Heen
Project Editor	Katie Nopper
Editorial Assistants	Elizabeth Bernardi, Alison Macdonald
Managing Editor	Ron Hampton
Senior Designer	Dennis Schaefer
Cover Designer	Dennis Schaefer
Supplements Supervisor	Emily Portwood
Media Producers	Sharon Smith, Ceci Fleming
Software Development	Mary Durnwald, TestGen; Rebecca Williams, MathXL
Marketing Manager	Jay Jenkins
Marketing Coordinator	Tracy Rabinowitz
Senior Author Support/ Technology Specialist	Joseph K. Vetere
Senior Prepress Supervisor	Caroline Fell
Rights and Permissions	Dana Weightman, Ellen Keohane
Senior Manufacturing Buyer	Evelyn M. Beaton
Art and Design Services	Geri Davis/The Davis Group, Inc.
Production Coordination	Kathy Diamond
Composition	Beacon Publishing Services
Illustrations	William Melvin, Network Graphics
Cover Photo	© Lindsey P. Martin/Corbis

About the cover: The aurora borealis, or northern lights, as photographed in the skies over Alaska.

Many of the designations used by manufacturers and sellers to distinguish their products are claimed as trademarks. Where those designations appear in this book, and Addison-Wesley was aware of a trademark claim, the designations have been printed in initial caps or all caps.

Photo credits appear on page G-4.

Library of Congress Cataloging-in-Publication Data

Bittinger, Marvin L.
    Basic mathematics.—10th ed./Marvin L. Bittinger.
          p. cm.
    Includes index.
    ISBN 0-321-31906-0 (Student's Edition)
    (Hardback Student's Edition)—ISBN 0-321-38823-2
    1. Arithmetic. I. Title.

QA107.2.B57 2005
513'.1—dc22   2005045308

6 7 8 9 10–RRD-JC–09 08

Contents

WHOLE NUMBERS

FRACTION NOTATION: MULTIPLICATION AND DIVISION

RATIO AND PROPORTION

PERCENT NOTATION

DATA, GRAPHS, AND STATISTICS

MEASUREMENT

GEOMETRY

REAL NUMBERS

11 ALGEBRA: SOLVING EQUATIONS AND PROBLEMS

Index of Applications

Agriculture

Acres planted, 277
Acres required to produce tomatoes, 334
Apple farm, 148
Barn silo, 548
Corn production, 146
Farm machinery sales, 387–388
Fertilizer, 184, 298
Garden plots, 164
Grain storage tank, 564
Grass seed coverage, 299, 323
Growth of wheat, 423
Holstein's dairy, 189
Irrigation pipe, 195
Planting corn, 169
Seeded area, 571
Tomato seedlings, 202

Astronomy

Circumference of Earth, 538, 550
Diameter of Mercury, 3
Diameter of Pluto, 671
Distance from the Sun to Pluto, 8
Gravity of Mars, 413
Planet orbits, 157
Planetary moons, 438
Planets, 431
Space shuttle orbits, 197, 340
Surface temperature on Mars, 581, 598
Volume of Earth, 548
Volume of a satellite, 548

Automotive

Acceleration, 206
Accidents by driver age, 456
Automobile violations, 337
Car depreciation, 382, 384
Car value, 374, 379
Corvette accidents, 296
Digital tire gauges, 172
Driving while drowsy, 354
Gallons of gasoline, 140, 216, 319
Gas mileage, 64–65, 70, 112, 114, 192, 198, 258, 261, 275, 277, 279, 286, 289, 299, 320, 322, 333, 336, 338, 339, 340, 419, 424, 455, 459, 462, 507, 612
Most popular car colors, 342
Motor vehicle production, 104, 376–377
Odometer reading, 279
Pickup trucks, 342, 344
Porsche sales, 461
Scrap tires, 104
Students owning a car, 462
Tire tread, 171
Toyota Sienna, 38
Truck bed cover, 62–63
Used cars, 6

Biology

Beehives needed to pollinate apple trees, 86
Bees and honey, 498
Bones in the hands and feet, 74
Deer population, 318, 323
Elephant heartbeat, 304
Fish population, 318, 323
Panda survival, 379
Rhino population, 415, 429
Running speed of animals, 133, 460
Silicon in the Earth's crust, 296
Speed of a black racer snake, 303
Water in the human body, 351

Business

Air conditioner sales, 387
Annual sales of antibiotics, 288
Antique pricing, 389
Battery testing, 423, 457
Book order, 212
Bookmobiles, 413
Business travel, 145
Camera sales, 258
Cartons of soda, 63–64, 86
Clothing sales, 387
Coffee production, 322
Cola testing, 425
Compact disc sales, 278
Daily newspapers, 29
Defective valves, 464
Deli order, 195
Delivering orders, 112
e-books, 84
Excelsior packing material, 489
Exercise equipment sales, 387
Fruit juice sales, 449
Hotel using land for recreation, 118
Kona Jack's Restaurants, 333
Land sale, 378
Light-bulb testing, 425
Machine wrapping candy bars, 674
Magazines, 243, 393
Mailing-list addresses, 133
Manufacturing, 177, 206, 305
Mining company, 86
Motorcycle sales, 388
Newspaper circulation, 198
Oil refinery, 381
Online sales, 240
Owning a business, 464
Packaging, 72, 87, 140
Pencil company, 157
Photo finishing market, 270
Price negotiations, 426
Price of chain link fence, 375–376
Publishing, 321
Quality control, 321, 334
Real estate developer, 117–118
Sales trip, 139, 142
Shipping, 73
Storekeeper getting change, 665

Index of Study Tips

Preface

It is with great pride and excitement that we present to you the tenth edition of *Basic Mathematics*. The text has evolved dramatically over the past 36 years through your comments, responses, and opinions. This feedback, combined with our overall objective of presenting the material in a clear and accurate manner, drives each revision. It is our hope that *Basic Mathematics*, Tenth Edition, and the supporting supplements will help provide an improved teaching and learning experience by matching the needs of instructors and successfully preparing students for their future.

This text is the first in a series that includes the following:

Bittinger: *Basic Mathematics,* Tenth Edition

Bittinger: *Fundamental Mathematics,* Fourth Edition

Bittinger: *Introductory Algebra,* Tenth Edition

Bittinger: *Intermediate Algebra,* Tenth Edition

Bittinger/Beecher: *Introductory and Intermediate Algebra,* Third Edition

Building Understanding through an Interactive Approach

The pedagogy of this text is designed to provide an interactive learning experience between the student and the exposition, annotated examples, art, margin exercises, and exercise sets. This unique approach, which has been developed and refined over ten editions and is illustrated at the right, provides students with a clear set of learning

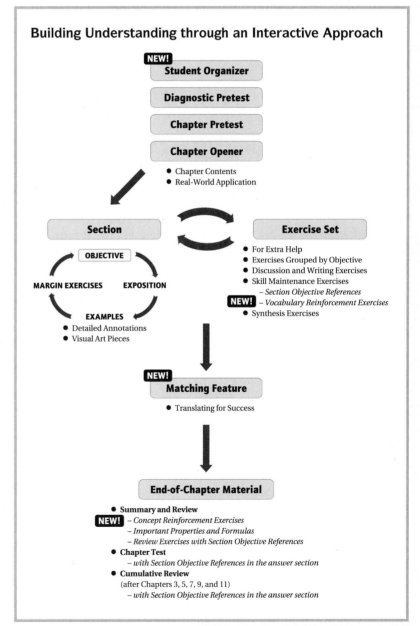

Building Understanding through an Interactive Approach

NEW! **Student Organizer**

Diagnostic Pretest

Chapter Pretest

Chapter Opener
- Chapter Contents
- Real-World Application

Section

OBJECTIVE

MARGIN EXERCISES → EXPOSITION

EXAMPLES
- Detailed Annotations
- Visual Art Pieces

Exercise Set
- For Extra Help
- Exercises Grouped by Objective
- Discussion and Writing Exercises
- Skill Maintenance Exercises
  – *Section Objective References*
  NEW! – *Vocabulary Reinforcement Exercises*
- Synthesis Exercises

NEW! **Matching Feature**
- Translating for Success

End-of-Chapter Material
- **Summary and Review**
  NEW! – *Concept Reinforcement Exercises*
  – *Important Properties and Formulas*
  – *Review Exercises with Section Objective References*
- **Chapter Test**
  – *with Section Objective References in the answer section*
- **Cumulative Review**
  (after Chapters 3, 5, 7, 9, and 11)
  – *with Section Objective References in the answer section*

objectives, involves them with the development of the material, and provides immediate and continual reinforcement and assessment through the margin exercises.

Let's Visit the Tenth Edition

The style, format, and approach of the ninth edition have been strengthened in this new edition in a number of ways. However, the accuracy that the Bittinger books are known for has not changed. This edition, as with all editions, has gone through an exhaustive checking process to ensure accuracy in the problem sets, mathematical art, and accompanying supplements. We know what a critical role the accuracy of a book plays in student learning and we value the reputation for accuracy that we have earned.

NEW! **IN THE TENTH EDITION**

Each revision gives us the opportunity to incorporate new elements and refine existing elements to provide a better experience for students and teachers alike. Below are four new features designed to help students succeed.

- Student Organizer
- Translating for Success matching exercises
- Vocabulary Reinforcement exercises
- Concept Reinforcement exercises

These features, along with the hallmark features of this book, are discussed in the pages that follow.

In addition, the tenth edition has been designed to be open and flexible, helping students focus their attention on details that are critical at this level through prominent headings, boxed definitions and rules, and clearly labeled objectives. *Chapter Pretests*, now located along with the Diagnostic Pretest in the *Printed Test Bank* and in MyMathLab, diagnose at the section and objective level and can be used to place students in a specific section, or objective, of the chapter, allowing them to concentrate on topics with which they have particular difficulty. Answers to these pretests are available in the *Printed Test Bank* and in MyMathLab.

NEW! STUDENT ORGANIZER

Along with the study tips found throughout the text, we have provided a pull-out card that will help students stay organized and increase their ability to be successful in this course. Students can use this card to keep track of important dates and useful contact information and to access information for technology and to plan class time, study time, work time, travel time, family time, and, sometimes most importantly, relaxation time.

CHAPTER OPENERS

To engage students and prepare them for the upcoming chapter material, gateway chapter openers are designed with exceptional artwork that is tied to a motivating real-world application. (See pages 1, 89, and 149.)

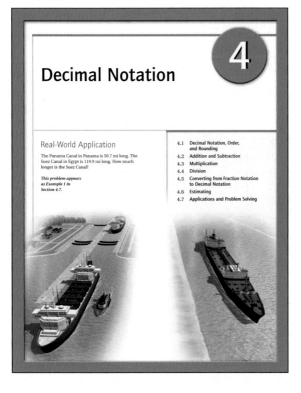

OBJECTIVE BOXES

At the beginning of each section, a boxed list of objectives is keyed by letter not only to section subheadings, but also to the exercises in the Pretest (located in the *Printed Test Bank* and MyMathLab), the section exercise sets, and the Summary and Review exercises, as well as to the answers to the questions in the Chapter Tests and Cumulative Reviews. This correlation enables students to easily find appropriate review material if they need help with a particular exercise or skill at the objective level. (See pages 90, 165, and 218.)

ANNOTATED EXAMPLES

Detailed annotations and color highlights lead the student through the structured steps of the examples. The level of detail in these annotations is a significant reason for students' success with this book. (See pages 168, 192, and 314.)

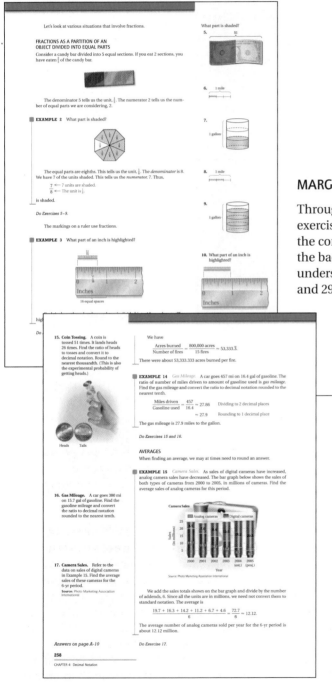

MARGIN EXERCISES

Throughout the text, students are directed to numerous margin exercises that provide immediate practice and reinforcement of the concepts covered in each section. Answers are provided at the back of the book so students can immediately self-assess their understanding of the skill or concept at hand. (See pages 91, 250, and 292.)

REAL-DATA APPLICATIONS

This text encourages students to see and interpret the mathematics that appears every day in the world around them. Throughout the writing process, an extensive and energetic search for real-data applications was conducted, and the result is a variety of examples and exercises that connect the mathematical content with the real world. A large number of the applications are new to this edition, and many are drawn from the fields of business and economics, life and physical sciences, social sciences, and areas of general interest such as sports and daily life. To further encourage students to understand the relevance of mathematics, many applications are enhanced by graphs and drawings similar to those found in today's newspapers and magazines, and feature source lines as well. (See pages 164, 239, and 301.)

Translating for Success The goal of the matching exercises in this new feature is to practice step two, *Translate,* of the five-step problem-solving process. Students translate each of ten problems to an equation and select the correct translation from fifteen given equations. This feature appears once in each chapter and reviews skills and concepts with problems from all preceding chapters. (See pages 195, 277, and 319.)

ART PROGRAM

Today's students are often visually oriented and their approach to a printed page is no exception. The art program is designed to improve the visualization of the mathematical concepts and to enhance the real-data applications. (See pages 107, 270, and 320.)

The use of color is carried out in a methodical and precise manner so that it conveys a consistent meaning, which enhances the readability of the text. For example, when perimeter is considered, figures have a red border to emphasize the perimeter. When area is considered, figures are outlined in black and screened with amber to emphasize the area. Similarly, when volume is considered, figures are three-dimensional and airbrushed blue. When fractional parts are illustrated, those parts are shown in purple.

PHOTOGRAPHS

Often, an application becomes relevant to students when the connection to the real world is illustrated with a photograph. This text has numerous photographs throughout in order to help students see the relevance and visualize the application at hand. (See pages 104, 219, and 303.)

CAUTION BOXES

Found at relevant points throughout the text, boxes with the "Caution!" heading warn students of common misconceptions or errors made in performing a particular mathematics operation or skill. (See pages 125, 190, and 224.)

CALCULATOR CORNERS

Where appropriate throughout the text, students will find optional Calculator Corners. Popular in the ninth edition, these Calculator Corners have been revised to be more accessible to students and to represent current calculators. (See pages 194, 230, and 310.)

STUDY TIPS

A variety of Study Tips throughout the text give students pointers on how to develop good study habits as they progress through the course. At times short snippets and at other times more lengthy discussions, these Study Tips encourage students to get involved in the learning process. (See pages 155, 295, and 308.)

EXERCISE SETS

The exercise sets are a critical part of any math book. To give students ample opportunity to practice what they have learned, each section is followed by an extensive exercise set designed to reinforce the section concepts. In addition, students also have the opportunity to synthesize the objectives from the current section with those from preceding sections.

For Extra Help   Many valuable study aids accompany this text. Located before each exercise set, "For Extra Help" references list appropriate video/CD, tutorial, and Web resources so that students can easily find related support materials.

Exercises Grouped by Objective   Exercises in the section exercise sets are keyed by letter to the section objectives for easy review and remediation. This reinforces the objective-based structure of the book. (See pages 120, 225, and 311.)

Discussion and Writing Exercises Designed to help students develop a deeper comprehension of critical concepts, Discussion and Writing exercises (indicated by D_W) are suitable for individual or group work. These exercises encourage students to both think and write about key mathematical ideas in the chapter. (See pages 121, 189, and 283.)

Skill Maintenance Exercises Found in each exercise set, these exercises review concepts from other sections in the text to prepare students for their final examination. Section and objective codes appear next to each Skill Maintenance exercise for easy reference. (See pages 164, 172, and 284.)

NEW! **Vocabulary Reinforcement Exercises** This new feature checks and reviews students' understanding of the vocabulary introduced throughout the text. It appears once in every chapter, in the Skill Maintenance portion of an exercise set, and is intended to provide a continuing review of the terms that students must know in order to be able to communicate effectively in the language of mathematics. (See pages 143, 200, and 269.)

Synthesis Exercises In most exercise sets, Synthesis exercises help build critical-thinking skills by requiring students to synthesize or combine learning objectives from the current section as well as from preceding text sections. (See pages 114, 208, and 332.)

END-OF-CHAPTER MATERIAL

At the end of each chapter, students can practice all they have learned as well as tie the current chapter content to material covered in earlier chapters.

SUMMARY AND REVIEW

A three-part *Summary and Review* appears at the end of each chapter. The first part includes the Concept Reinforcement Exercises described below. The second part is a list of important properties and formulas, when applicable, and the third part provides an extensive set of review exercises.

NEW! **Concept Reinforcement Exercises** Found in the Summary and Review of every chapter, these true/false exercises are designed to increase understanding of the concepts rather than merely assess students' skill at memorizing procedures. (See pages 209, 285, and 333.)

Important Properties and Formulas A list of the important properties and formulas discussed in the chapter is provided for students in an organized manner to help them prioritize topics learned and prepare for chapter tests. This list is only provided in those chapters in which new properties or formulas are presented. (See pages 410, 510, and 569.)

Review Exercises At the end of each chapter, students are provided with an extensive set of review exercises. Reference codes beside each exercise or direction line allow students to easily refer back to specific, objective-level content for remediation. (See pages 144, 209, and 285.)

CHAPTER TEST

Following the Review Exercises, a sample Chapter Test allows students to review and test comprehension of chapter skills prior to taking an instructor's exam. Answers to all questions in the Chapter Test are given at the back of the book. Section and objective references for each question are included with the answers. (See pages 147, 212, and 288.)

CUMULATIVE REVIEW

Following Chapters 3, 5, 7, 9, and 11, students encounter a Cumulative Review. This exercise set reviews skills and concepts from all preceding chapters to help students recall previously learned material and prepare for a final exam. At the back of the book are answers to all Cumulative Review exercises, together with section and objective references, so that students know exactly what material to study if they miss a review exercise. Additional Cumulative Review Tests for every chapter are available in the *Printed Test Bank*. (See pages 214, 338, and 671.)

Ancillaries

The following ancillaries are available to help both instructors and students use this text more effectively.

Student Supplements

Student's Solutions Manual (ISBN 0-321-29607-9)

- By Judith A. Penna, *Indiana University Purdue University Indianapolis*
- Contains completely worked-out solutions with step-by-step annotations for all the odd-numbered exercises in the text, with the exception of the discussion and writing exercises, as well as completely worked-out solutions to all the exercises in the Chapter Reviews, Chapter Tests, and Cumulative Reviews.

Collaborative Learning Activities Manual
(ISBN 0-321-29604-4)

- Features group activities tied to text sections and includes the focus, time estimate, suggested group size and materials, and background notes for each activity.
- Available as a stand-alone supplement sold in the bookstore, as a textbook bundle component for students, or as a classroom activity resource for instructors.

Videotapes (ISBN 0-321-30569-8)

- Include new chapter openers presented by author Marvin Bittinger.
- Present a series of lectures correlated directly to the content of each section of the text.
- Feature an engaging team of instructors who present material in a format that stresses student interaction, often using examples and exercises from the text.

Digital Video Tutor (ISBN 0-321-29605-2)

- Complete set of digitized videos (as described above) on CD-ROMs for student use at home or on campus.
- Ideal for distance learning or supplemental instruction.
- Are available with captioning on request. Contact your local Addison-Wesley representative for details.

NEW! Work It Out! Chapter Test Video on CD
(ISBN 0-321-41985-5)

- Presented by Judith A. Penna and Barbara Johnson
- Provides step-by-step solutions to every exercise in each Chapter Test from the text.
- Helps students prepare for chapter tests and synthesize content.

Math Study Skills for Students Video on CD
(ISBN 0-321-29745-8)

- Presented by author Marvin Bittinger
- Designed to help students make better use of their math study time and improve their retention of concepts and procedures taught in classes from basic mathematics through intermediate algebra.
- Through carefully crafted graphics and comprehensive on-camera explanation, focuses on study skills that are commonly overlooked.

Instructor Supplements

Annotated Instructor's Edition (ISBN 0-321-30553-1)

- Includes answers to all exercises printed in blue on the same page as those exercises.

Instructor's Solutions Manual (ISBN 0-321-30556-6)

- By Judith A. Penna, *Indiana University Purdue University Indianapolis*
- Contains brief solutions to the even-numbered exercises in the exercise sets, answers to all of the Discussion and Writing exercises, and the completely worked-out solutions to all the exercises in the Chapter Reviews, Chapter Tests, and Cumulative Reviews.

Online Answer Book

- By Judith A. Penna, *Indiana University Purdue University Indianapolis*
- Available in electronic form from the instructor resource center. Contact your local Addison-Wesley representative for details.
- Contains answers to all the section exercises in the text.

Printed Test Bank (ISBN 0-321-30557-4)

- By Laurie Hurley
- Contains one diagnostic test.
- Contains one pretest for each chapter.
- Provides 13 new test forms for every chapter and 8 new test forms for the final exam.
- For the chapter tests, 5 test forms are modeled after the chapter tests in the text, 3 test forms are organized by topic order following the text objectives, 3 test forms are designed for 50-minute class periods and organized so that each objective in the chapter is covered on one of the tests, and 2 test forms are multiple-choice. Chapter tests also include more challenging synthesis questions.
- Contains 2 cumulative tests per chapter beginning with Chapter 2.
- For the final exam, 3 test forms are organized by chapter, 3 test forms are organized by question type, and 2 test forms are multiple-choice.

NEW! Instructor and Adjunct Support Manual
(ISBN 0-321-30554-X)

- Includes Adjunct Support Manual material.
- Features resources and teaching tips designed to help both new and adjunct faculty with course preparation and classroom management.
- Resources include chapter reviews, extra practice sheets, conversion guide, video index, audio index, and transparency masters.
- Also available electronically so course/adjunct coordinators can customize material specific to their schools.

Student Supplements

Audio Recordings

- By Bill Saler
- Lead students through the material in each section of the text, explaining solution steps to examples, pointing out common errors, and focusing on margin exercises and solutions.
- Audio files are available to download in MP3 format. Contact your local Addison-Wesley representative for details.

Addison-Wesley Math Tutor Center

www.aw-bc.com/tutorcenter

- The Addison-Wesley Math Tutor Center is staffed by qualified mathematics instructors who provide students with tutoring on examples and odd-numbered exercises from the textbook. Tutoring is available via toll-free telephone, toll-free fax, e-mail, or the Internet. White Board technology allows tutors and students to actually see problems worked while they "talk" in real time over the Internet during tutoring sessions.

MathXL® Tutorials on CD   (ISBN 0-321-29747-4)

- Provides algorithmically generated practice exercises that correlate at the objective level to the content of the text.
- Includes an example and a guided solution to accompany every exercise and video clips for selected exercises.
- Recognizes student errors and provides feedback; generates printed summaries of students' progress.

Instructor Supplements

TestGen with Quizmaster   (ISBN 0-321-30567-1)

- Enables instructors to build, edit, print, and administer tests.
- Features a computerized bank of questions developed to cover all text objectives.
- Algorithmically based content allows instructors to create multiple but equivalent versions of the same question or test with a click of a button.
- Instructors can also modify test-bank questions or add new questions by using the built-in question editor, which allows users to create graphs, input graphics, and insert math notation, variable numbers, or text.
- Tests can be printed or administered online via the Internet or another network. Quizmaster allows students to take tests on a local area network.
- Available on a dual-platform Windows/Macintosh CD-ROM.

MathXL®   www.mathxl.com

MathXL is a powerful online homework, tutorial, and assessment system that accompanies Addison-Wesley textbooks in mathematics or statistics. With MathXL, instructors can create, edit, and assign online homework and tests using algorithmically generated exercises correlated at the objective level to the textbook. They can also create and assign their own online exercises and import TestGen tests for added flexibility. All student work is tracked in MathXL's online gradebook. Students can take chapter tests in MathXL and receive personalized study plans based on their test results. The study plan diagnoses weaknesses and links students directly to tutorial exercises for the objectives they need to study and retest. Students can also access supplemental animations and video clips directly from selected exercises. MathXL is available to qualified adopters. For more information, visit our Web site at www.mathxl.com or contact your Addison-Wesley sales representative.

MyMathLab   www.mymathlab.com

MyMathLab is a series of text-specific, easily customizable online courses for Addison-Wesley textbooks in mathematics and statistics. Powered by CourseCompass™ (Pearson Education's online teaching and learning environment) and MathXL® (our online homework, tutorial, and assessment system), MyMathLab gives instructors the tools they need to deliver all or a portion of their course online, whether students are in a lab setting or working from home. MyMathLab provides a rich and flexible set of course materials, featuring free-response exercises that are algorithmically generated for unlimited practice and mastery. Students can also use online tools, such as video lectures, animations, and a multimedia textbook, to independently improve their understanding and performance. Instructors can use MyMathLab's homework and test managers to select and assign online exercises correlated directly to the textbook, and they can also create and assign their own online exercises and import TestGen tests for added flexibility. MyMathLab's online gradebook—designed specifically for mathematics and statistics—automatically tracks students' homework and test results and gives the instructor control over how to calculate final grades. Instructors can also add offline (paper-and-pencil) grades to the gradebook. MyMathLab is available to qualified adopters. For more information, visit our Web site at www.mymathlab.com or contact your Addison-Wesley sales representative.

InterAct Math® Tutorial Web site   www.interactmath.com

Get practice and tutorial help online! This interactive tutorial Web site provides algorithmically generated practice exercises that correlate directly to the exercises in the textbook. Students can retry an exercise as many times as they like with new values each time for unlimited practice and mastery. Every exercise is accompanied by an interactive guided solution that provides helpful feedback for incorrect answers, and students can also view a worked-out sample problem that steps them through an exercise similar to the one they're working on.

ADDISON-WESLEY MATH ADJUNCT SUPPORT CENTER

The Addison-Wesley Math Adjunct Support Center is staffed by qualified mathematics instructors with over 50 years of combined experience at both the community college and university level. Assistance is provided for faculty in the following areas:

- Suggested syllabus consultation
- Tips on using materials packaged with your book
- Book-specific content assistance
- Teaching suggestions including advice on classroom strategies

For more information, visit www.aw-bc.com/tutorcenter/math-adjunct.html

Acknowledgments and Reviewers

Many of you helped to shape the tenth edition by reviewing and spending time with us on your campuses. Our deepest appreciation to all of you and in particular to the following:

Sally Clark, *Ozarks Technical Community College*

Penny Deggelman, *Lane Community College*

Molly Misko, *Gadsden State Community College*

Cassonda Thompson, *York Technical College*

Mary Woestman, *Broome Community College*

We wish to express our heartfelt appreciation to a number of people who have contributed in special ways to the development of this textbook. Our editor, Jennifer Crum, and marketing manager, Jay Jenkins, encouraged our vision and provided marketing insight. Kari Heen, the project manager, deserves special recognition for overseeing every phase of the project and keeping it moving. The unwavering support of the Developmental Math group, including Katie Nopper, project editor, Elizabeth Bernardi and Alison Macdonald, editorial assistants, Ron Hampton, managing editor, Dennis Schaefer, cover designer, and Sharon Smith and Ceci Fleming, media producers, and the endless hours of hard work by Kathy Diamond and Geri Davis have led to products of which we are immensely proud.

We also want to thank Judy Penna for writing the *Student's* and *Instructor's Solutions Manuals* and for her strong leadership in the preparation of the printed supplements, videotapes, and MyMathLab. Other strong support has come from Laurie Hurley for the *Printed Test Bank,* Bill Saler for the audio recordings, and Barbara Johnson, Michelle Lanosga, and Jennifer Rosenberg for their accuracy checking of the manuscript. We also wish to recognize those who wrote scripts, presented lessons on camera, and checked the accuracy of the videotapes.

To the Student

As your author, I would like to welcome you to this study of *Basic Mathematics.* Whatever your past experiences, I encourage you to look at this mathematics course as a fresh start. Approach this course with a positive attitude about mathematics. Mathematics is a base for life, for many majors, for personal finances, for most careers, or just for pleasure.

You are the most important factor in the success of your learning. In earlier experiences, you may have allowed yourself to sit back and let the instructor "pour in" the learning, with little or no follow-up on your part. But now you must take a more assertive and proactive stance. This may be the first adjustment you make in college. As soon as possible after class, you should thoroughly read the textbook and the supplements and do all you can to learn on your own. In other words, rid yourself of former habits and take responsibility for your own learning. Then, with all the help you have around you, your hard work will lead to success.

One of the most important suggestions I can make is to allow yourself enough *time* to learn. You can have the best book, the best instructor, and the best supplements, but if you do not give yourself time to learn, how can they be of benefit? Many other helpful suggestions are presented in the Study Tips that you will find throughout the book. You may want to read through all the Study Tips before you begin the text. An Index of Study Tips can be found at the front of the book.

M.L.B.

Bittinger Student Organizer

Study Tips

Throughout this text, you will find a feature called *Study Tips.* We discuss these in the Preface of this text. They are intended to help improve your math study skills. An index of all the *Study Tips* can be found at the front of the book.

For Extra Help

MyMathLab   MathXL

Student's Solutions Manual

Digital Video Tutor CD Videotape

Math Tutor Center   InterAct Math

Additional Resources

Basic Math Review Card (ISBN 0-321-39476-3)

Algebra Review Card (ISBN 0-321-39473-9)

Math for Allied Health Reference Card (ISBN 0-321-39474-7)

Graphing Calculator Reference Card (ISBN 0-321-39475-5)

Math Study Skills for Students Video on CD (ISBN 0-321-29745-8)

Go to www.aw-bc.com/math for more information.

On the first day of class, complete this chart and the weekly planner that follows on the reverse page.

Instructor Information:

Name _____

Office Hours and Location _____

_____

Phone Number _____

Fax Number _____

E-mail Address _____

Find the names of two students whom you could contact for class information or study questions:

1. Name _____
   Phone Number _____
   Fax Number _____
   E-mail Address _____

2. Name _____
   Phone Number _____
   Fax Number _____
   E-mail Address _____

Math Lab on Campus:

Location _____

Hours _____

Phone Number _____

Tutoring:

Campus Location _____

Hours _____

To order the Addison-Wesley Math Tutor Center, call 1-888-777-0463.

(*See the Preface for important information concerning this tutoring.*)

Important Supplements: (*See the Preface for a complete list of available supplements.*)

Supplements recommended by the instructor _____

_____

Online Log-in Information (*include access code, password, Web address, etc.*)

_____

_____

Bittinger Student Organizer

Success is planned. On this page, plan a typical week. Consider time allotments for class, study, work, travel, family, and relaxation.

Important Dates

Midterm Exam

_____

Final Exam

_____

Holidays

_____

_____

_____

_____

Other

_____

_____

_____

TIME	Sun.	Mon.	Tues.	Wed.	Thurs.	Fri.	Sat.
6:00 A.M.							
6:30							
7:00							
7:30							
8:00							
8:30							
9:00							
9:30							
10:00							
10:30							
11:00							
11:30							
12:00 P.M.							
12:30							
1:00							
1:30							
2:00							
2:30							
3:00							
3:30							
4:00							
4:30							
5:00							
5:30							
6:00							
6:30							
7:00							
7:30							
8:00							
8:30							
9:00							
9:30							
10:00							
10:30							
11:00							
11:30							
12:00 A.M.							

For more information about this text or any of its supplements, visit www.aw-bc.com/math

Whole Numbers

Real-World Application

Races in which runners climb the steps inside a building are called "run-up" races. There are 2058 steps in the International Towerthon, Kuala Lumpur, Malaysia. Write expanded notation for the number of steps.

This problem appears as Exercise 13 in Section 1.1.

Objectives

a Give the meaning of digits in standard notation.

b Convert from standard notation to expanded notation.

c Convert between standard notation and word names.

We study mathematics in order to be able to solve problems. In this section, we study how numbers are named. We begin with the concept of place value.

a Place Value

Consider the numbers in the following table.

What does the digit 2 mean in each number?

1. 526,555

2. 265,789

3. 42,789,654

4. 24,789,654

5. 8924

6. 5,643,201

Three Most Populous Countries in the World

COUNTRY	POPULATION 2005
China	1,306,313,812
India	1,080,264,388
United States	295,734,134

Source: *The World Factbook*, July 2005 estimates

A **digit** is a number 0, 1, 2, 3, 4, 5, 6, 7, 8, or 9 that names a place-value location. For large numbers, digits are separated by commas into groups of three, called **periods.** Each period has a name: *ones, thousands, millions, billions, trillions,* and so on. To understand the population of China in the table above, we can use a **place-value chart,** as shown below.

PLACE-VALUE CHART														
Periods →	Trillions			Billions			Millions			Thousands			Ones		
	Hundreds	Tens	Ones	Hundreds	Tens	Ones	Hundreds	Tens	Ones	Hundreds	Tens	Ones	Hundreds	Tens	Ones
						1	3	0	6	3	1	3	8	1	2

1 billion 306 millions 313 thousands 812 ones

EXAMPLES What does the digit 8 mean in each number?

1. 278,342 — 8 thousands

2. 872,342 — 8 hundred thousands

3. 28,343,399,223 — 8 billions

4. 1,023,850 — 8 hundreds

5. 98,413,099 — 8 millions

6. 6328 — 8 ones

Do Margin Exercises 1–6.

Answers on page A-1

EXAMPLE 7 *American Red Cross.* In 2003, private donations to the American Red Cross totaled about $587,492,000. What does each digit name?

Source: *The Chronicle of Philanthropy*

5 8 7, 4 9 2, 0 0 0
— ones
— tens
— hundreds
— thousands
— ten thousands
— hundred thousands
— millions
— ten millions
— hundred millions

Do Exercise 7.

b Converting from Standard Notation to Expanded Notation

To answer questions such as "How many?", "How much?", and "How tall?", we use whole numbers. The set, or collection, of **whole numbers** is

$$0, 1, 2, 3, 4, 5, 6, 7, 8, 9, 10, 11, 12, \ldots.$$

The set goes on indefinitely. There is no largest whole number, and the smallest whole number is 0. Each whole number can be named using various notations. The set $1, 2, 3, 4, 5, \ldots$, without 0, is called the set of **natural numbers.**

Let's look at the data from the bar graph shown here.

Fewer Computer Majors

Year	
1999	20,787
2000	23,416
2001	23,090
2002	23,033
2003	17,706

The number of computer science and computer engineering majors in the fall in the United States and Canada

Source: Computing Research Association Taulbee Survey

The number of computer majors in 2003 was 17,706. **Standard notation** for the number of computer majors is 17,706. We write **expanded notation** for 17,706 as follows:

$$17{,}706 = 1 \text{ ten thousand} + 7 \text{ thousands}$$
$$+ 7 \text{ hundreds} + 0 \text{ tens} + 6 \text{ ones.}$$

7. Presidential Library. In the first year of operation, 280,219 people visited the Ronald Reagan Presidential Library in Simi Valley, California. What does each digit name?

Source: *USA Today* research by Bruce Rosenstein; National Archives & Records Administration; Associated Press

Write expanded notation.

8. 1895

9. 23,416, the number of computer majors in 2000

10. 3031 mi (miles), the diameter of Mercury

Answers on page A-1

11. 4180 mi, the length of the Nile River, the longest river in the world

EXAMPLE 8 Write expanded notation for 4218 mi, the diameter of Mars.

$$4218 = 4 \text{ thousands} + 2 \text{ hundreds} + 1 \text{ ten} + 8 \text{ ones}$$

EXAMPLE 9 Write expanded notation for 3400.

$$3400 = 3 \text{ thousands} + 4 \text{ hundreds} + 0 \text{ tens} + 0 \text{ ones}, \quad \text{or}$$
$$3 \text{ thousands} + 4 \text{ hundreds}$$

EXAMPLE 10 Write expanded notation for 563,384, the population of Washington, D.C.

$$563,384 = 5 \text{ hundred thousands} + 6 \text{ ten thousands}$$
$$+ 3 \text{ thousands} + 3 \text{ hundreds} + 8 \text{ tens} + 4 \text{ ones}$$

12. 154,616, the number of Labrador retrievers registered in 2002

Source: The American Kennel Club

Do Exercises 8–12 (8–10 are on the preceding page).

C Converting Between Standard Notation and Word Names

We often use **word names** for numbers. When we pronounce a number, we are speaking its word name. Russia won 92 medals in the 2004 Summer Olympics in Athens, Greece. A word name for 92 is "ninety-two." Word names for some two-digit numbers like 27, 39, and 92 use hyphens. Others like 17 use only one word, "seventeen."

Write a word name. (Refer to the figure below right.)

13. 49, the total number of medals won by Australia

14. 16, the number of silver medals won by Germany

TOP COUNTRIES IN SUMMER OLYMPICS 2004	MEDAL COUNT			
	GOLD	SILVER	BRONZE	TOTAL
United States of America	35	39	29	103
Russia	27	27	38	92
People's Republic of China	32	17	14	63
Australia	17	16	16	49
Germany	14	16	18	48

Source: 2004 Olympics, Athens, Greece

15. 38, the number of bronze medals won by Russia

Answers on page A-1

EXAMPLES   Write a word name.

11. 35, the total number of gold medals won by the United States

Thirty-five

12. 17, the number of silver medals won by the People's Republic of China

Seventeen

Do Exercises 13–15 on the preceding page.

For word names for larger numbers, we begin at the left with the largest period. The number named in the period is followed by the name of the period; then a comma is written and the next period is named.

EXAMPLE 13   Write a word name for 46,605,314,732.

Forty-six **billion,**

six hundred five **million,**

three hundred fourteen **thousand,**

seven hundred thirty-two

The word "and" *should not* appear in word names for whole numbers. Although we commonly hear such expressions as "two hundred *and* one," the use of "and" is not, strictly speaking, correct in word names for whole numbers. For decimal notation, it is appropriate to use "and" for the decimal point. For example, 317.4 is read as "three hundred seventeen *and* four tenths."

Do Exercises 16–19.

EXAMPLE 14   Write standard notation.

Five hundred six **million,**

three hundred forty-five **thousand,**

two hundred twelve

Standard notation is 506,345,212.

Do Exercise 20.

Write a word name.

16. 204

17. $44,155, the average salary in 2001 for those who have a bachelor's degree or more
Source: U.S. Bureau of the Census

18. 1,879,204

19. 6,449,000,000, the world population in 2005
Source: U.S. Bureau of the Census

20. Write standard notation.

Two hundred thirteen million, one hundred five thousand, three hundred twenty-nine

Answers on page A-1

1.1 EXERCISE SET

For Extra Help

MathXL | MyMathLab | InterAct Math | Math Tutor Center | Digital Video Tutor CD 1 Videotape 1 | Student's Solutions Manual

a   What does the digit 5 mean in each case?

1. 235,888   **2.** 253,777   **3.** 1,488,526   **4.** 500,736

Used Cars.   1,582,370 certified used cars were sold in 2004 in the United States. **Source:** *Motor Trend,* April 2005, p. 26
In the number 1,582,370, what digit names the number of:

5. thousands?   **6.** ones?

7. millions?   **8.** hundred thousands?

b   Write expanded notation.

9. 5702   **10.** 3097   **11.** 93,986   **12.** 38,453

Step-Climbing Races.   Races in which runners climb the steps inside a building are called "run-up" races. The graph below shows the number of steps in four buildings. In Exercises 13–16, write expanded notation for the number of steps in each race.

Step-Climbing Races

International Towerthon, Kuala Lumpur, Malaysia — 2058
CN Tower Run-Up, Toronto — 1776
World Financial Center, New York — 1268
Skytower Run-Up, Auckland, New Zealand — 1081

Source: New York Road Runners Club

13. 2058 steps in the International Towerthon, Kuala Lumpur, Malaysia

14. 1776 steps in the CN Tower Run-Up, Toronto, Ontario, Canada

15. 1268 steps in the World Financial Center, New York City, New York

16. 1081 steps in the Skytower Run-Up, Auckland, New Zealand

Overseas Travelers. The chart below shows the residence and number of overseas travelers to the United States in 2003. In Exercises 17–22, write expanded notation for the number of travelers from each country.

OVERSEAS TRAVELERS TO U.S., 2003

Australia	405,698
Brazil	348,945
Germany	1,180,212
India	272,161
Japan	3,169,682
Spain	284,031

Source: U.S. Department of Commerce, ITA, Office of Travel and Tourism Industries, 2004

17. 405,698 from Australia

18. 3,169,682 from Japan

19. 272,161 from India

20. 284,031 from Spain

21. 1,180,212 from Germany

22. 348,945 from Brazil

C Write a word name.

23. 85

24. 48

25. 88,000

26. 45,987

27. 123,765

28. 111,013

29. 7,754,211,577

30. 43,550,651,808

Write standard notation.

31. Two million, two hundred thirty-three thousand, eight hundred twelve

32. Three hundred fifty-four thousand, seven hundred two

33. Eight billion

34. Seven hundred million

Write a word name for the number in each sentence.

35. *Great Pyramid.* The area of the base of the Great Pyramid in Egypt is 566,280 square feet.

36. *Population of the United States.* The population of the United States in July 2005 was estimated to be 295,734,134.
Source: *The World Factbook*

37. *Busiest Airport.* In 2003, the world's busiest airport, Hartsfield in Atlanta, had 76,086,792 passengers.

Source: Airports Council International World Headquarters, Geneva, Switzerland

Write standard notation for the number in each sentence.

39. Light travels nine trillion, four hundred sixty billion kilometers in one year.

41. *Pacific Ocean.* The area of the Pacific Ocean is sixty-four million, one hundred eighty-six thousand square miles.

38. *Prisoners.* There were 2,078,570 total prisoners, federal, state, and local, in the United States in 2003.

Source: *Prison and Jail Inmates at Midyear 2003,* U.S. Bureau of Justice Statistics

40. The distance from the sun to Pluto is three billion, six hundred sixty-four million miles.

42. *Internet Users.* In a recent year, there were fifty-four million, five hundred thousand Internet users in China.

Source: Computer Industry Almanac, Inc.

To the student and the instructor: The Discussion and Writing exercises are meant to be answered with one or more sentences. They can be discussed and answered collaboratively by the entire class or by small groups. Because of their open-ended nature, the answers to these exercises do not appear at the back of the book. They are denoted by the symbol **D**w.

43. **D**w Explain why we use commas when writing large numbers.

44. **D**w Write an English sentence in which the number 370,000,000 is used.

SYNTHESIS

To the student and the instructor: The Synthesis exercises found at the end of every exercise set challenge students to combine concepts or skills studied in that section or in preceding parts of the text. Exercises marked with a ▦ symbol are meant to be solved using a calculator.

45. How many whole numbers between 100 and 400 contain the digit 2 in their standard notation?

46. ▦ What is the largest number that you can name on your calculator? How many digits does that number have? How many periods?

1.2 ADDITION

a Addition of Whole Numbers

Addition of whole numbers corresponds to combining or putting things together.

The addition that corresponds to the figure above is

$$3 \quad + \quad 4 \quad = \quad 7.$$
$$\downarrow \qquad \downarrow \qquad \downarrow$$
$$\text{Addend} \quad \text{Addend} \quad \text{Sum}$$

We say that the **sum** of 3 and 4 is 7. The numbers added are called **addends.**

Addition also corresponds to moving distances on a number line. The number line below is marked with tick marks at equal distances of 1 *unit*. The sum $3 + 4$ is shown. We first move 3 units from 0, and then 4 more units, and end up at 7. The addition that corresponds to the situation is $3 + 4 = 7$.

How do we do an addition of three numbers, like $2 + 3 + 6$? We do it by adding 3 and 6, and then 2. We can show this with parentheses:

$$2 + (3 + 6) = 2 + 9 = 11. \qquad \text{Parentheses tell what to do first.}$$

We could also add 2 and 3, and then 6:

$$(2 + 3) + 6 = 5 + 6 = 11.$$

Either way we get 11. It does not matter how we group the numbers. This illustrates the **associative law of addition,** $a + (b + c) = (a + b) + c$. We can also add whole numbers in any order. That is, $2 + 3 = 3 + 2$. This illustrates the **commutative law of addition,** $a + b = b + a$. Together, the commutative and associative laws tell us that to add more than two numbers, we can use any order and grouping we wish. Adding 0 to a number does not change the number: $a + 0 = 0 + a = a$. That is, $6 + 0 = 0 + 6 = 6$, or $198 + 0 = 0 + 198 = 198$. We say that 0 is the **additive identity.**

Add.

1.   7 9 6 8
     + 5 4 9 7

To add whole numbers, we add the ones digits first, then the tens, then the hundreds, then the thousands, and so on.

EXAMPLE 1   Add: 6878 + 4995.

Place values are lined up in columns

```
      1
  6 8 7 8
+ 4 9 9 5
        3
```
Add ones. We get 13 ones, or 1 ten + 3 ones. Write 3 in the ones column and 1 above the tens. This is called *carrying*, or *regrouping*.

```
    1 1
  6 8 7 8
+ 4 9 9 5
      7 3
```
Add tens. We get 17 tens, or 1 hundred + 7 tens. Write 7 in the tens column and 1 above the hundreds.

2. 6203 + 3542

```
  1 1 1
  6 8 7 8
+ 4 9 9 5
    8 7 3
```
Add hundreds. We get 18 hundreds, or 1 thousand + 8 hundreds. Write 8 in the hundreds column and 1 above the thousands.

```
  1 1 1
  6 8 7 8
+ 4 9 9 5
1 1 8 7 3
```
Add thousands. We get 11 thousands.

We show you these steps for explanation. You need write only this.

3.   9 8 0 4
     + 6 3 7 8

```
  1 1 1
  6 8 7 8 ← ⎤ Addends
+ 4 9 9 5 ← ⎦
1 1 8 7 3 ← Sum
```

EXAMPLE 2   Add: 391 + 276 + 789 + 498.

```
    2
  3 9 1
  2 7 6
  7 8 9
+ 4 9 8
      4
```
Add ones. We get 24, so we have 2 tens + 4 ones. Write 4 in the ones column and 2 above the tens.

4.   1 9 3 2
       6 7 2 3
       9 8 7 8
     + 8 9 4 1

```
  3 2
  3 9 1
  2 7 6
  7 8 9
+ 4 9 8
    5 4
```
Add tens. We get 35 tens, so we have 30 tens + 5 tens. This is also 3 hundreds + 5 tens. Write 5 in the tens column and 3 above the hundreds.

```
  3 2
  3 9 1
  2 7 6
  7 8 9
+ 4 9 8
1 9 5 4
```
Add hundreds. We get 19 hundreds.

Answers on page A-1

Do Exercises 1–4.

b Finding Perimeter

Addition can be used when finding perimeter.

PERIMETER
The distance around an object is its **perimeter.**

■ **EXAMPLE 3** Find the perimeter of the figure.

19 in.

37 in.

29 in.

44 in.

We add the lengths of the sides:

Perimeter = 29 in. + 19 in. + 37 in. + 44 in.

We carry out the addition as follows.

$$
\begin{array}{r}
\overset{2}{2}\ 9 \\
1\ 9 \\
3\ 7 \\
+\ 4\ 4 \\
\hline
1\ 2\ 9
\end{array}
$$

The perimeter of the figure is 129 in.

Do Exercises 5 and 6.

■ **EXAMPLE 4** Find the perimeter of the octagonal (eight-sided) resort swimming pool.

6 yd

13 yd

6 yd

12 yd

8 yd

16 yd

12 yd

13 yd

Perimeter = 13 yd + 6 yd + 6 yd + 12 yd + 8 yd
             + 12 yd + 13 yd + 16 yd

The perimeter of the pool is 86 yd.

Do Exercise 7.

Find the perimeter of each figure.

5.

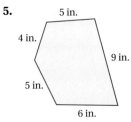

5 in.

4 in.

9 in.

5 in.

6 in.

6.

16 ft

15 ft

15 ft

16 ft

Solve.

7. Index Cards. Two standard sizes for index cards are 3 in. (inches) by 5 in. and 5 in. by 8 in. Find the perimeter of each card.

3 in.

5 in.

5 in.

8 in.

Answers on page A-1

CALCULATOR CORNER

Adding Whole Numbers   *To the student and the instructor:* This is the first of a series of *optional* discussions on using a calculator. A calculator is *not* a requirement for this textbook. There are many kinds of calculators and different instructions for their usage. We have included instructions here for a minimum-cost calculator. Be sure to consult your user's manual as well. Also, check with your instructor about whether you are allowed to use a calculator in the course.

To add whole numbers on a calculator, we use the $+$ and $=$ keys. For example, to add 57 and 34, we press 5 7 $+$ 3 4 $=$. The calculator displays 91, so $57 + 34 = 91$. To find $314 + 259 + 478$, we press 3 1 4 $+$ 2 5 9 $+$ 4 7 8 $=$. The display reads 1051, so $314 + 259 + 478 = 1051$.

Exercises: Use a calculator to find each sum.

1. $19 + 36$

2. $73 + 48$

3. $925 + 677$

4. $276 + 458$

5. $\begin{array}{r} 8\ 2\ 6 \\ 4\ 1\ 5 \\ +\ 6\ 9\ 1 \\ \hline \end{array}$

6. $\begin{array}{r} 2\ 5\ 3 \\ 4\ 9\ 0 \\ +\ 1\ 2\ 1 \\ \hline \end{array}$

Study Tips

USING THIS TEXTBOOK

Throughout this textbook, you will find a feature called "Study Tips." One of the most important ways in which to improve your math study skills is to learn the proper use of the textbook. Here we highlight a few points that we consider most helpful.

- **Be sure to note the symbols** a, b, c, **and so on, that correspond to the objectives you are to master in each section.** The first time you see them is in the margin at the beginning of the section; the second time is in the subheadings of each section; and the third time is in the exercise set for the section. You will also find symbols like [1.1a] or [1.2c] next to the skill maintenance exercises in each exercise set and the review exercises at the end of the chapter, as well as in the answers to the chapter tests and the cumulative reviews. These objective symbols allow you to refer to the appropriate place in the text when you need to review a topic.

- **Read and study each step of each example.** The examples include important side comments that explain each step. These examples and annotations have been carefully chosen so that you will be fully prepared to do the exercises.

- **Stop and do the margin exercises as you study a section.** This gives you immediate reinforcement of each concept as it is introduced and is one of the most effective ways to master the mathematical skills in this text. Don't deprive yourself of this benefit!

- **Note the icons listed at the top of each exercise set.** These refer to the many distinctive multimedia study aids that accompany the book.

a   Add.

1. $\begin{array}{r} 3\ 6\ 4 \\ +\ \ \ 2\ 3 \\ \hline \end{array}$

2. $\begin{array}{r} 1\ 5\ 2\ 1 \\ +\ \ \ 3\ 4\ 8 \\ \hline \end{array}$

3. $\begin{array}{r} 1\ 7\ 1\ 6 \\ +\ 3\ 4\ 8\ 2 \\ \hline \end{array}$

4. $\begin{array}{r} 7\ 5\ 0\ 3 \\ +\ 2\ 6\ 8\ 3 \\ \hline \end{array}$

5. $\begin{array}{r} 8\ 6 \\ +\ 7\ 8 \\ \hline \end{array}$

6. $\begin{array}{r} 7\ 3 \\ +\ 6\ 9 \\ \hline \end{array}$

7. $\begin{array}{r} 9\ 9 \\ +\ \ \ 1 \\ \hline \end{array}$

8. $\begin{array}{r} 9\ 9\ 9 \\ +\ \ \ 1\ 1 \\ \hline \end{array}$

9. 8113 + 390

10. 271 + 3338

11. 356 + 4910

12. 280 + 34,702

13. 3870 + 92 + 7 + 497

14. 10,120 + 12,989 + 5738

15. $\begin{array}{r} 4\ 8\ 2\ 5 \\ +\ 1\ 7\ 8\ 3 \\ \hline \end{array}$

16. $\begin{array}{r} 3\ 6\ 5\ 4 \\ +\ 2\ 7\ 0\ 0 \\ \hline \end{array}$

17. $\begin{array}{r} 2\ 3,4\ 4\ 3 \\ +\ 1\ 0,9\ 8\ 9 \\ \hline \end{array}$

18. $\begin{array}{r} 4\ 5,8\ 7\ 9 \\ +\ 2\ 1,7\ 8\ 6 \\ \hline \end{array}$

19. $\begin{array}{r} 7\ 7,5\ 4\ 3 \\ +\ 2\ 3,7\ 6\ 7 \\ \hline \end{array}$

20. $\begin{array}{r} 9\ 9,9\ 9\ 9 \\ +\ \ \ \ \ \ 1\ 1\ 2 \\ \hline \end{array}$

21. $\begin{array}{r} 4\ 5 \\ 2\ 5 \\ 3\ 6 \\ 4\ 4 \\ +\ 8\ 0 \\ \hline \end{array}$

22. $\begin{array}{r} 3\ 8 \\ 2\ 7 \\ 3\ 2 \\ 1\ 4 \\ +\ 7\ 6 \\ \hline \end{array}$

23. $\begin{array}{r} 1\ 2,0\ 7\ 0 \\ 2,9\ 5\ 4 \\ +\ \ \ 3,4\ 0\ 0 \\ \hline \end{array}$

24. $\begin{array}{r} 4\ 2,4\ 8\ 7 \\ 8\ 3,1\ 4\ 1 \\ +\ 3\ 6,7\ 1\ 2 \\ \hline \end{array}$

25. $\begin{array}{r} 4\ 8\ 3\ 5 \\ 7\ 2\ 9 \\ 9\ 2\ 0\ 4 \\ 8\ 9\ 8\ 6 \\ +\ 7\ 9\ 3\ 1 \\ \hline \end{array}$

26. $\begin{array}{r} 9\ 8\ 9 \\ 5\ 6\ 6 \\ 8\ 3\ 4 \\ 9\ 2\ 0 \\ +\ 7\ 0\ 3 \\ \hline \end{array}$

b Find the perimeter of each figure.

27.

22 mi   14 mi   13 mi   8 mi
47 mi   10 mi

28.

62 yd
28 yd
39 yd
46 yd
54 yd

29. Find the perimeter of a standard hockey rink.

200 ft

85 ft

30. In Major League Baseball, how far does a batter travel in circling the bases when a home run has been hit?

90 ft

90 ft

31. **D_W** Explain in your own words what the associative law of addition means.

32. **D_W** Describe a situation that corresponds to this mathematical expression:

$$80 \text{ mi} + 245 \text{ mi} + 336 \text{ mi.}$$

The exercises that follow begin an important feature called *Skill Maintenance exercises*. These exercises provide an ongoing review of topics previously covered in the book. You will see them in virtually every exercise set. It has been found that this kind of continuing review can significantly improve your performance on a final examination.

33. What does the digit 8 mean in 486,205?   [1.1a]

34. Write a word name for the number in the following sentence: [1.1c]

In fiscal year 2004, Starbucks Corporation had total net revenues of $5,294,247,000.

Source: Starbucks Corporation

35. A fast way to add all the numbers from 1 to 10 inclusive is to pair 1 with 9, 2 with 8, and so on. Use a similar approach to add all numbers from 1 to 100 inclusive.

1.3 SUBTRACTION

Objectives

a Convert between addition sentences and subtraction sentences.

b Subtract whole numbers.

a Subtraction and Related Sentences

TAKE AWAY

Subtraction of whole numbers applies to two kinds of situations. The first is called "take away." Consider the following example.

A bowler starts with 10 pins and knocks down 8 of them.

From 10 pins, the bowler "takes away" 8 pins. There are 2 pins left. The subtraction is $10 - 8 = 2$.

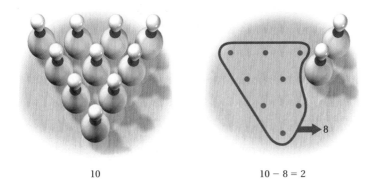

$$10 \qquad\qquad\qquad\qquad 10 - 8 = 2$$

We use the following terminology with subtraction:

$$10 \quad - \quad 8 \quad = \quad 2 \; .$$

Minuend   Subtrahend   Difference

The **minuend** is the number from which another number is being subtracted. The **subtrahend** is the number being subtracted. The **difference** is the result of subtracting the subtrahend from the minuend.

Subtraction also corresponds to moving distances on a number line. The number line below is marked with tick marks at equal distances of 1 unit. The difference $10 - 8$ is shown. We first move from 0 right 10 units, and then left 8 units, and end up at 2. The subtraction that corresponds to the situation is $10 - 8 = 2$.

This leads us to the following definition of subtraction.

SUBTRACTION

The difference $a - b$ is that unique whole number c for which $a = c + b$.

For example, $13 - 4$ is the number 9 since $13 = 9 + 4$.

Write a related addition sentence.

1. $7 - 5 = 2$

2. $17 - 8 = 9$

Write two related subtraction
sentences.

3. $5 + 8 = 13$

4. $11 + 3 = 14$

RELATED SENTENCES

Subtraction is defined in terms of addition. For example, $5 - 2$ is that number which, when added to 2, gives 5. Thus for the subtraction sentence

$$5 - 2 = 3, \qquad \text{Taking away 2 from 5 gives 3.}$$

there is a *related addition sentence*

$$5 = 3 + 2. \qquad \text{Putting back the 2 gives 5 again.}$$

In fact, we know that answers we find to subtractions are correct only because of the related addition, which provides a handy way to *check* a subtraction.

EXAMPLE 1 Write a related addition sentence: $8 - 5 = 3$.

$$8 - 5 = 3$$

This number
gets added.

$$8 = 3 + 5$$

By the commutative law of addition, there is also another addition sentence:

$$8 = 5 + 3.$$

The related addition sentence is $8 = 3 + 5$.

Do Exercises 1 and 2.

EXAMPLE 2 Write two related subtraction sentences: $4 + 3 = 7$.

$$4 + 3 = 7$$

This addend gets
subtracted from
the sum.

$$4 = 7 - 3$$

(7 take away 3 is 4.)

$$4 + 3 = 7$$

This addend gets
subtracted from
the sum.

$$3 = 7 - 4$$

(7 take away 4 is 3.)

The related subtraction sentences are $4 = 7 - 3$ and $3 = 7 - 4$.

Do Exercises 3 and 4.

HOW MANY MORE

The second kind of situation to which subtraction can apply is called "how many more." You have 2 notebooks, but you need 7. You can think of this as "how many do I need to add to 2 to get 7?" Finding the answer can be thought of as finding a missing addend, and can be found by subtracting 2 from 7.

Need 7 notebooks

Have 2
notebooks

5 notebooks

What must be added to 2 to get 7? The answer is 5.

Missing addend

Difference

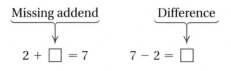

$$2 + \square = 7 \qquad 7 - 2 = \square$$

Let's look at the following example in which a missing addend occurs: Jillian wants to buy the roll-on luggage shown in this ad. She has $30. She needs $79. How much more does she need in order to buy the luggage?

Thinking of this situation in terms of a missing addend, we have:

Plus

$30   30 + ☐ = 79 $79

To find the answer, we think of the related subtraction sentence:

$$30 + \square = 79$$
$$\square = 79 - 30.$$

b Subtraction of Whole Numbers

To subtract numbers, we subtract the ones digits first, then the tens digits, then the hundreds, then the thousands, and so on.

 EXAMPLE 3 Subtract: $9768 - 4320$.

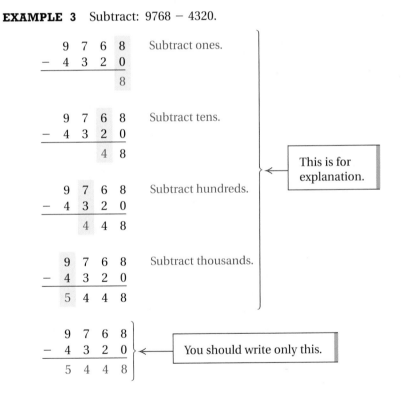

```
    9  7  6  8     Subtract ones.
 -  4  3  2  0
             8
```

```
    9  7  6  8     Subtract tens.
 -  4  3  2  0
          4  8
```

```
    9  7  6  8     Subtract hundreds.
 -  4  3  2  0
       4  4  8
```

```
    9  7  6  8     Subtract thousands.
 -  4  3  2  0
    5  4  4  8
```

This is for explanation.

```
    9  7  6  8
 -  4  3  2  0
    5  4  4  8
```

You should write only this.

5. Subtract.

```
    7  8  9  3
 -  4  0  9  2
```

Answer on page A-1

We have considered the subtraction $9768 - 4320 = \square$. That is, we have found the missing addend in the sentence $9768 = 4320 + \square$. If 5448 is indeed the missing addend, then if we add it to 4320, the answer should be 9768. The related addition sentence is the basis for adding as a *check*.

Subtraction:

$$
\begin{array}{r}
9\ 7\ 6\ 8 \\
-\ 4\ 3\ 2\ 0 \\
\hline
5\ 4\ 4\ 8
\end{array}
$$

Check by Addition:

$$
\begin{array}{r}
5\ 4\ 4\ 8 \\
+\ 4\ 3\ 2\ 0 \\
\hline
9\ 7\ 6\ 8
\end{array}
$$

Do Exercise 5 on the preceding page.

EXAMPLE 4   Subtract: $348 - 165$.

We have

$$
\begin{array}{rll}
3 \text{ hundreds} + 4 \text{ tens} + 8 \text{ ones} = & 2 \text{ hundreds} + 14 \text{ tens} + 8 \text{ ones} \\
-\ 1 \text{ hundred} \ -\ 6 \text{ tens} - 5 \text{ ones} = & -\ 1 \text{ hundred} \ -\quad 6 \text{ tens} - 5 \text{ ones} \\
\hline
= & 1 \text{ hundred} \ +\quad 8 \text{ tens} + 3 \text{ ones} \\
= & 183.
\end{array}
$$

Note that in this case, although we can subtract the ones ($8 - 5 = 3$), we cannot do so with the tens, because $4 - 6$ is *not* a whole number. To see why, consider

$$4 - 6 = \square \quad \text{and the related addition sentence} \quad 4 = \square + 6.$$

There is no whole number that when added to 6 gives 4. To complete the subtraction, we must *borrow* 1 hundred from 3 hundreds and regroup it with 4 tens. Then we can do the subtraction $14 \text{ tens} - 6 \text{ tens} = 8 \text{ tens}$. Below we consider a shortened form.

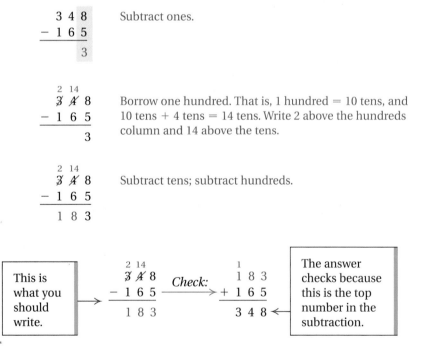

$$
\begin{array}{r}
3\ 4\ 8 \\
-\ 1\ 6\ 5 \\
\hline
3
\end{array}
$$
Subtract ones.

$$
\begin{array}{r}
{}^{2}\ {}^{14}\ \\
\cancel{3}\ \cancel{4}\ 8 \\
-\ 1\ 6\ 5 \\
\hline
3
\end{array}
$$
Borrow one hundred. That is, 1 hundred = 10 tens, and 10 tens + 4 tens = 14 tens. Write 2 above the hundreds column and 14 above the tens.

$$
\begin{array}{r}
{}^{2}\ {}^{14}\ \\
\cancel{3}\ \cancel{4}\ 8 \\
-\ 1\ 6\ 5 \\
\hline
1\ 8\ 3
\end{array}
$$
Subtract tens; subtract hundreds.

This is what you should write.

$$
\begin{array}{r}
{}^{2}\ {}^{14}\ \\
\cancel{3}\ \cancel{4}\ 8 \\
-\ 1\ 6\ 5 \\
\hline
1\ 8\ 3
\end{array}
$$

Check:

$$
\begin{array}{r}
{}^{1}\ \\
1\ 8\ 3 \\
+\ 1\ 6\ 5 \\
\hline
3\ 4\ 8
\end{array}
$$

The answer checks because this is the top number in the subtraction.

Answers on page A-1

EXAMPLE 5   Subtract: 6246 − 1879.

$$\begin{array}{r} {}^{3}\ 16 \\ 6\ 2\ \not{4}\ \not{6} \\ -\ 1\ 8\ 7\ 9 \\ \hline 7 \end{array}$$

We cannot subtract 9 ones from 6 ones, but we can subtract 9 ones from 16 ones. We borrow 1 ten to get 16 ones.

$$\begin{array}{r} {}^{13} \\ {}^{1}\ \not{3}\ 16 \\ 6\ \not{2}\ \not{4}\ \not{6} \\ -\ 1\ 8\ 7\ 9 \\ \hline 6\ 7 \end{array}$$

We cannot subtract 7 tens from 3 tens, but we can subtract 7 tens from 13 tens. We borrow 1 hundred to get 13 tens.

$$\begin{array}{r} {}^{11}\ {}^{13} \\ {}^{5}\ \not{1}\ \not{3}\ 16 \\ \not{6}\ \not{2}\ \not{4}\ \not{6} \\ -\ 1\ 8\ 7\ 9 \\ \hline 4\ 3\ 6\ 7 \end{array}$$

We cannot subtract 8 hundreds from 1 hundred, but we can subtract 8 hundreds from 11 hundreds. We borrow 1 thousand to get 11 hundreds.

| This is what you should write. | $\begin{array}{r} {}^{11}\ {}^{13} \\ {}^{5}\ \not{1}\ \not{3}\ 16 \\ \not{6}\ \not{2}\ \not{4}\ \not{6} \\ -\ 1\ 8\ 7\ 9 \\ \hline 4\ 3\ 6\ 7 \end{array}$ | *Check:* | $\begin{array}{r} {}^{1}\ {}^{1}\ {}^{1} \\ 4\ 3\ 6\ 7 \\ +\ 1\ 8\ 7\ 9 \\ \hline 6\ 2\ 4\ 6 \end{array}$ | The answer checks because this is the top number in the subtraction. |

Do Exercises 6 and 7.

EXAMPLE 6   Subtract: 902 − 477.

$$\begin{array}{r} {}^{8}\ {}^{9}\ 12 \\ \not{9}\ \not{0}\ \not{2} \\ -\ 4\ 7\ 7 \\ \hline 4\ 2\ 5 \end{array}$$

We cannot subtract 7 ones from 2 ones. We have 9 hundreds, or 90 tens. We borrow 1 ten to get 12 ones. We then have 89 tens.

Do Exercises 8 and 9.

EXAMPLE 7   Subtract: 8003 − 3667.

$$\begin{array}{r} {}^{7}\ {}^{9}\ {}^{9}\ 13 \\ \not{8}\ \not{0}\ \not{0}\ \not{3} \\ -\ 3\ 6\ 6\ 7 \\ \hline 4\ 3\ 3\ 6 \end{array}$$

We have 8 thousands, or 800 tens. We borrow 1 ten to get 13 ones. We then have 799 tens.

EXAMPLES

8. Subtract: 6000 − 3762.

$$\begin{array}{r} {}^{5}\ {}^{9}\ {}^{9}\ 10 \\ \not{6}\ \not{0}\ \not{0}\ \not{0} \\ -\ 3\ 7\ 6\ 2 \\ \hline 2\ 2\ 3\ 8 \end{array}$$

9. Subtract: 6024 − 2968.

$$\begin{array}{r} {}^{11} \\ {}^{5}\ {}^{9}\ \not{1}\ 14 \\ \not{6}\ \not{0}\ \not{2}\ \not{4} \\ -\ 2\ 9\ 6\ 8 \\ \hline 3\ 0\ 5\ 6 \end{array}$$

Do Exercises 10–12.

Subtract. Check by adding.

6.
$$\begin{array}{r} 8\ 6\ 8\ 6 \\ -\ 2\ 3\ 5\ 8 \\ \hline \end{array}$$

7.
$$\begin{array}{r} 7\ 1\ 4\ 5 \\ -\ 2\ 3\ 9\ 8 \\ \hline \end{array}$$

Subtract.

8.
$$\begin{array}{r} 7\ 0 \\ -\ 1\ 4 \\ \hline \end{array}$$

9.
$$\begin{array}{r} 5\ 0\ 3 \\ -\ 2\ 9\ 8 \\ \hline \end{array}$$

Subtract.

10.
$$\begin{array}{r} 7\ 0\ 0\ 7 \\ -\ 6\ 3\ 4\ 9 \\ \hline \end{array}$$

11.
$$\begin{array}{r} 6\ 0\ 0\ 0 \\ -\ 3\ 1\ 4\ 9 \\ \hline \end{array}$$

12.
$$\begin{array}{r} 9\ 0\ 3\ 5 \\ -\ 7\ 4\ 8\ 9 \\ \hline \end{array}$$

Answers on page A-1

1.3

EXERCISE SET

MathXL   MyMathLab   InterAct Math   Math Tutor Center   Digital Video Tutor CD 1 Videotape 1   Student's Solutions Manual

a Write a related addition sentence.

1. $7 - 4 = 3$ **2.** $12 - 5 = 7$ **3.** $13 - 8 = 5$ **4.** $9 - 9 = 0$

5. $23 - 9 = 14$ **6.** $20 - 8 = 12$ **7.** $43 - 16 = 27$ **8.** $51 - 18 = 33$

Write two related subtraction sentences.

9. $6 + 9 = 15$ **10.** $7 + 9 = 16$ **11.** $8 + 7 = 15$ **12.** $8 + 0 = 8$

13. $17 + 6 = 23$ **14.** $11 + 8 = 19$ **15.** $23 + 9 = 32$ **16.** $42 + 10 = 52$

b Subtract.

17. $\begin{array}{r} 65 \\ -21 \\ \hline \end{array}$ **18.** $\begin{array}{r} 87 \\ -34 \\ \hline \end{array}$ **19.** $\begin{array}{r} 866 \\ -333 \\ \hline \end{array}$ **20.** $\begin{array}{r} 526 \\ -323 \\ \hline \end{array}$

21. $86 - 47$ **22.** $73 - 28$ **23.** $981 - 747$ **24.** $887 - 698$

25. $\begin{array}{r} 7769 \\ -2387 \\ \hline \end{array}$ **26.** $\begin{array}{r} 6431 \\ -2896 \\ \hline \end{array}$ **27.** $\begin{array}{r} 7640 \\ -3809 \\ \hline \end{array}$ **28.** $\begin{array}{r} 8003 \\ -599 \\ \hline \end{array}$

29. $\begin{array}{r} 12,647 \\ -4,899 \\ \hline \end{array}$ **30.** $\begin{array}{r} 16,222 \\ -5,888 \\ \hline \end{array}$ **31.** $90,237 - 47,209$ **32.** $84,703 - 298$

20

CHAPTER 1: Whole Numbers

Copyright © 2007 Pearson Education, Inc.

33.   8 0
      − 2 4

34.   9 0
      − 7 8

35.   6 9 0
      − 2 3 6

36.   8 0 3
      − 4 1 8

37.   6 8 0 8
      − 3 0 5 9

38.   6 4 0 8
      −   2 5 8

39.   2 3 0 0
      −   1 0 9

40.   6 0 0 7
      − 1 5 8 9

41. 101,734 − 5760

42. 15,017 − 7809

43. 10,008 − 19

44. 21,043 − 8909

45.   7 0 0 0
      − 2 7 9 4

46.   8 0 0 1
      − 6 5 4 3

47.   4 8,0 0 0
      − 3 7,6 9 5

48.   1 7,0 4 3
      − 1 1,5 9 8

49. D_W Describe two situations that correspond to the subtraction $20 − $17, one "take away" and one "how many more."

50. D_W Is subtraction commutative (is there a commutative law of subtraction)? Why or why not?

SKILL MAINTENANCE

Add. [1.2a]

51.   9 4 6
      +   7 8

52.   9 0 7 8
      + 3 6 5 4

53.   5 7,8 7 7
      + 3 2,4 0 6

54.   8 0 0 4
         6 7 8 9
         7 7 2 0
      + 6 8 5 1

55. 567 + 778

56. 901 + 23

57. 12,885 + 9807

58. 9909 + 1011

59. Write a word name for 6,375,602. [1.1c]

60. What does the digit 7 mean in 6,375,602? [1.1a]

SYNTHESIS

61. Fill in the missing digits to make the subtraction true:
   9,☐48,621 − 2,097,☐81 = 7,251,140.

62. ▦ Subtract: 3,928,124 − 1,098,947.

ROUNDING AND ESTIMATING; ORDER

Objectives

a Round to the nearest ten, hundred, or thousand.

b Estimate sums and differences by rounding.

c Use < or > for ▢ to write a true sentence in a situation like 6 ▢ 10.

a Rounding

We round numbers in various situations when we do not need an exact answer. For example, we might round to see if we are being charged the correct amount in a store. We might also round to check if an answer to a problem is reasonable or to check a calculation done by hand or on a calculator.

To understand how to round, we first look at some examples using number lines, even though this is not the way we generally do rounding.

Round to the nearest ten.

1. 37

2. 52

3. 73

4. 98

EXAMPLE 1 Round 47 to the nearest ten.

Here is a part of a number line; 47 is between 40 and 50. Since 47 is closer to 50, we round up to 50.

EXAMPLE 2 Round 42 to the nearest ten.

42 is between 40 and 50. Since 42 is closer to 40, we round down to 40.

Do Exercises 1–4.

EXAMPLE 3 Round 45 to the nearest ten.

45 is halfway between 40 and 50. We could round 45 down to 40 or up to 50. We agree to round up to 50.

Round to the nearest ten.

5. 35

6. 75

7. 85

When a number is halfway between rounding numbers, round up.

Do Exercises 5–7.

Here is a rule for rounding.

ROUNDING WHOLE NUMBERS

To round to a certain place:

a) Locate the digit in that place.

b) Consider the next digit to the right.

c) If the digit to the right is 5 or higher, round up. If the digit to the right is 4 or lower, round down.

d) Change all digits to the right of the rounding location to zeros.

Answers on page A-2

EXAMPLE 4 Round 6485 to the nearest ten.

a) Locate the digit in the tens place, 8.

    6 4 **8** 5
         ↑

b) Consider the next digit to the right, 5.

    6 4 8 **5**
         ↑

c) Since that digit, 5, is 5 or higher, round 8 tens up to 9 tens.

d) Change all digits to the right of the tens digit to zeros.

    6 4 9 0 ← This is the answer.

EXAMPLE 5 Round 6485 to the nearest hundred.

a) Locate the digit in the hundreds place, 4.

    6 **4** 8 5
     ↑

b) Consider the next digit to the right, 8.

    6 4 **8** 5
     ↑

c) Since that digit, 8, is 5 or higher, round 4 hundreds up to 5 hundreds.

d) Change all digits to the right of hundreds to zeros.

    6 5 0 0 ← This is the answer.

EXAMPLE 6 Round 6485 to the nearest thousand.

a) Locate the digit in the thousands place, 6.

    6 4 8 5
  ↑

b) Consider the next digit to the right, 4.

    6 **4** 8 5
  ↑

c) Since that digit, 4, is 4 or lower, round down, meaning that 6 thousands stays as 6 thousands.

d) Change all digits to the right of thousands to zeros.

    6 0 0 0 ← This is the answer.

Do Exercises 8–19.

CAUTION!

7000 is not a correct answer to Example 6. It is incorrect to round from the ones digit over, as follows:

    6485, → 6490, → 6500, → 7000.

Note that 6485 is closer to 6000 than it is to 7000.

Round to the nearest ten.

8. 137

9. 473

10. 235

11. 285

Round to the nearest hundred.

12. 641

13. 759

14. 750

15. 9325

Round to the nearest thousand.

16. 7896

17. 8459

18. 19,343

19. 68,500

Answers on page A-2

20. Round 48,968 to the nearest ten, hundred, and thousand.

21. Round 269,582 to the nearest ten, hundred, and thousand.

Refer to the chart on the next page to answer Margin Exercises 22 and 23.

22. By eliminating options, find a way that Ethan and Olivia can buy the ION·2 and stay within their $16,500 budget.

23. Tara and Alex are shopping for a new car. They are considering a Saturn ION·3 and have allowed a budget of $19,000.

   a) Estimate by first rounding to the nearest hundred the cost of an ION·3 with all the options.

   b) Can they afford this car with a budget of $19,000?

Sometimes rounding involves changing more than one digit in a number.

EXAMPLE 7   Round 78,595 to the nearest ten.

a) Locate the digit in the tens place, 9.

$$7\ 8,5\ \underset{\uparrow}{9}\ 5$$

b) Consider the next digit to the right, 5.

$$7\ 8,5\ 9\ \underset{\uparrow}{5}$$

c) Since that digit, 5, is 5 or higher, round 9 tens to 10 tens. To carry this out, we think of 10 tens as 1 hundred + 0 tens, and increase the hundreds digit by 1, to get 6 hundreds + 0 tens. We then write 6 in the hundreds place and 0 in the tens place.

d) Change the digit to the right of the tens digit to zero.

$$7\ 8,6\ 0\ 0 \leftarrow \text{This is the answer.}$$

Note that if we round this number to the nearest hundred, we get the same answer.

Do Exercises 20 and 21.

b   Estimating

Estimating can be done in many ways. In general, an estimate made by rounding to the nearest ten is more accurate than one rounded to the nearest hundred, and an estimate rounded to the nearest hundred is more accurate than one rounded to the nearest thousand, and so on.

In the following example, we see how estimation can be used in making a purchase.

EXAMPLE 8   *Estimating the Cost of an Automobile Purchase.*   Ethan and Olivia Benson are shopping for a new car. They are considering a Saturn ION. There are three basic models of this car, and each has options beyond the basic price, as shown in the chart on the following page. Ethan and Olivia have allowed themselves a budget of $16,500. They look at the list of options and want to make a quick estimate of the cost of model ION·2 with all the options.

Estimate by rounding to the nearest hundred the cost of the ION·2 with all the options and decide whether it will fit into their budget.

MODEL ION·1 SEDAN (4 DOOR) 2.2-LITER ENGINE, 4-SPEED AUTOMATIC TRANSMISSION	MODEL ION·2 SEDAN (4 DOOR) 2.2-LITER ENGINE, 5-SPEED MANUAL TRANSMISSION	MODEL ION·3 SEDAN (4 DOOR) 2.2-LITER ENGINE, 5-SPEED MANUAL TRANSMISSION
Base Price: $12,975	Base Price: $14,945	Base Price: $16,470

Each of these vehicles comes with several options. Note that some of the options are standard on certain models. Others are not available for all models.

Antilock Braking System with Traction Control:	$400
Head Curtain Side Air Bags:	$395
Power Sunroof (Not available for ION·1):	$725
Rear Spoiler (Not available for ION·1):	$250
Air Conditioning with Dust and Pollen Filtration (Standard on ION·2 and ION·3):	$960
CD/MP3 Player with AM/FM Stereo and 4 Coaxial Speakers (Standard on ION·3):	ION·1—$510 ION·2—$220
Power Package: Power Windows, Power Exterior Mirrors, Remote Keyless Entry, and Cruise Control (Not available for ION·1 and Standard for ION·3):	$825

Source: Saturn

First, we list the base price of the ION·2 and then the cost of each of the options. We then round each number to the nearest hundred and add.

```
  1 4,9 4 5        1 4,9 0 0
      4 0 0            4 0 0
      3 9 5            4 0 0
      7 2 5            7 0 0
      2 5 0            3 0 0
      2 2 0            2 0 0
  +   8 2 5      +     8 0 0
                   1 7,7 0 0  ← Estimated answer
```

Air conditioning is standard on the ION·2, so we do not include that cost. The estimated cost is $17,700. Since Ethan and Olivia have allowed themselves a budget of $16,500 for the car, they will need to forgo some options.

Do Exercises 22 and 23 on the preceding page.

EXAMPLE 9 Estimate this sum by first rounding to the nearest ten:

78 + 49 + 31 + 85.

We round each number to the nearest ten. Then we add.

```
  7 8          8 0
  4 9          5 0
  3 1          3 0
+ 8 5        + 9 0
            2 5 0  ← Estimated answer
```

Do Exercises 24 and 25.

24. Estimate the sum by first rounding to the nearest ten. Show your work.

```
    7 4
    2 3
    3 5
  + 6 6
```

25. Estimate the sum by first rounding to the nearest hundred. Show your work.

```
    6 5 0
    6 8 5
    2 3 8
  + 1 6 8
```

Answers on page A-2

26. Estimate the difference by first rounding to the nearest hundred. Show your work.

$$\begin{array}{r} 9\ 2\ 8\ 5 \\ -\ 6\ 7\ 3\ 9 \\ \hline \end{array}$$

27. Estimate the difference by first rounding to the nearest thousand. Show your work.

$$\begin{array}{r} 2\ 3{,}2\ 7\ 8 \\ -\ 1\ 1{,}6\ 9\ 8 \\ \hline \end{array}$$

Use < or > for ☐ to write a true sentence. Draw a number line if necessary.

28. 8 ☐ 12

29. 12 ☐ 8

30. 76 ☐ 64

31. 64 ☐ 76

32. 217 ☐ 345

33. 345 ☐ 217

Answers on page A-2

▌**EXAMPLE 10** Estimate the difference by first rounding to the nearest thousand: 9324 − 2849.

We have

$$\begin{array}{r} 9\ 3\ 2\ 4 \\ -\ 2\ 8\ 4\ 9 \\ \hline \end{array} \qquad \begin{array}{r} 9\ 0\ 0\ 0 \\ -\ 3\ 0\ 0\ 0 \\ \hline 6\ 0\ 0\ 0 \end{array}$$

Do Exercises 26 and 27.

The sentence $7 - 5 = 2$ says that $7 - 5$ is the same as 2. When we round, the result is rarely the same as the number we started with. Thus we use the symbol $\approx$ when rounding. This symbol means "**is approximately equal to.**" For example, when 687 is rounded to the nearest ten, we can write $687 \approx 690$.

C Order

We know that 2 is not the same as 5. We express this by the sentence $2 \neq 5$. We also know that 2 is less than 5. We symbolize this by the expression $2 < 5$. We can see this order on the number line: 2 is to the left of 5. The number 0 is the smallest whole number.

ORDER OF WHOLE NUMBERS

For any whole numbers a and b:

1. $a < b$ (read "a is less than b") is true when a is to the left of b on the number line.
2. $a > b$ (read "a is greater than b") is true when a is to the right of b on the number line.

We call < and > **inequality symbols.**

▌**EXAMPLE 11** Use < or > for ☐ to write a true sentence: 7 ☐ 11.

$$\begin{array}{ccccccc} + & + & + & + & + & + & + \\ 6 & 7 & 8 & 9 & 10 & 11 & 12 & 13 \end{array}$$

Since 7 is to the left of 11 on the number line, $7 < 11$.

▌**EXAMPLE 12** Use < or > for ☐ to write a true sentence: 92 ☐ 87.

$$\begin{array}{ccccccc} + & + & + & + & + & + & + \\ 86 & 87 & 88 & 89 & 90 & 91 & 92 & 93 \end{array}$$

Since 92 is to the right of 87 on the number line, $92 > 87$.

A sentence like $8 + 5 = 13$ is called an **equation.** It is a *true* equation. The equation $4 + 8 = 11$ is a *false* equation. A sentence like $7 < 11$ is called an **inequality.** The sentence $7 < 11$ is a *true* inequality. The sentence $23 > 69$ is a *false* inequality.

Do Exercises 28–33.

EXERCISE SET

For Extra Help

a Round to the nearest ten.

1. 48

2. 532

3. 467

4. 8945

5. 731

6. 17

7. 895

8. 798

Round to the nearest hundred.

9. 146

10. 874

11. 957

12. 650

13. 9079

14. 4645

15. 32,850

16. 198,402

Round to the nearest thousand.

17. 5876

18. 4500

19. 7500

20. 2001

21. 45,340

22. 735,562

23. 373,405

24. 6,713,855

b Estimate the sum or difference by first rounding to the nearest ten. Show your work.

25.
```
   7 8
 + 9 7
```

26.
```
   6 2
   9 7
   4 6
 + 8 8
```

27.
```
  8 0 7 4
- 2 3 4 7
```

28.
```
  6 7 3
-   2 8
```

Estimate the sum by first rounding to the nearest ten. Do any of the given sums seem to be incorrect when compared to the estimate? Which ones?

29.
```
   4 5
   7 7
   2 5
 + 5 6
   3 4 3
```

30.
```
   4 1
   2 1
   5 5
 + 6 0
   1 7 7
```

31.
```
   6 2 2
     7 8
     8 1
 + 1 1 1
   9 3 2
```

32.
```
   8 3 6
   3 7 4
   7 9 4
 + 9 3 8
   3 9 4 7
```

Estimate the sum or difference by first rounding to the nearest hundred. Show your work.

33.
```
   7 3 4 8
 + 9 2 4 7
```

34.
```
   5 6 8
   4 7 2
   9 3 8
 + 4 0 2
```

35.
```
  6 8 5 2
- 1 7 4 8
```

36.
```
  9 4 3 8
- 2 7 8 7
```

Planning a Kitchen. Perfect Kitchens offers custom kitchen packages with three choices for each of four items: cabinets, countertops, appliances, and flooring. The chart below lists the price for each choice.

Perfect Kitchens lets you customize your kitchen with one choice from each of the following four features:

CABINETS		TYPE	PRICE
	(a)	Oak	$7450
	(b)	Cherry	8820
	(c)	Painted	9630

COUNTERTOPS		TYPE	PRICE
	(d)	Laminate	$1595
	(e)	Solid surface	2870
	(f)	Granite	3528

APPLIANCES		PRICE RANGE	PRICE
	(g)	Low	$1540
	(h)	Medium	3575
	(i)	High	6245

FLOORING		TYPE	PRICE
	(j)	Vinyl	$ 625
	(k)	Ceramic tile	985
	(l)	Hardwood	1160

37. Estimate the cost of remodeling a kitchen with choices (a), (d), (g), and (j) by rounding to the nearest hundred dollars.

38. Estimate the cost of a kitchen with choices (c), (f), (i), and (l) by rounding to the nearest hundred dollars.

39. Sara and Ben are planning to remodel their kitchen and have a budget of $17,700. Estimate by rounding to the nearest hundred dollars the cost of their kitchen remodeling project if they choose options (b), (e), (i), and (k). Can they afford their choices?

40. The Davidsons must make a final decision on the kitchen choices for their new home. The allotted kitchen budget is $16,000. Estimate by rounding to the nearest hundred dollars the kitchen cost if they choose options (a), (f), (h), and (l). Does their budget allotment cover the cost?

41. Suppose you are planning a new kitchen and must stay within a budget of $14,500. Decide on the options you would like and estimate the cost by rounding to the nearest hundred dollars. Does your budget support your choices?

42. Suppose you are planning a new kitchen and must stay within a budget of $18,500. Decide on the options you would like and estimate the cost by rounding to the nearest hundred dollars. Does your budget support your choices?

Estimate the sum by first rounding to the nearest hundred. Do any of the given sums seem to be incorrect when compared to the estimate? Which ones?

43.
```
    2 1 6
      8 4
    7 4 5
  + 5 9 5
  -------
  1 6 4 0
```

44.
```
    4 8 1
    7 0 2
    6 2 3
  + 1 0 4 3
  ---------
  1 8 4 9
```

45.
```
    7 5 0
    4 2 8
      6 3
  + 2 0 5
  -------
  1 4 4 6
```

46.
```
    3 2 6
    2 7 5
    7 5 8
  + 9 4 3
  -------
  2 3 0 2
```

Estimate the sum or difference by first rounding to the nearest thousand. Show your work.

47.
```
    9 6 4 3
    4 8 2 1
    8 9 4 3
  + 7 0 0 4
```

48.
```
    7 6 4 8
    9 3 4 8
    7 8 4 2
  + 2 2 2 2
```

49.
```
    9 2,1 4 9
  - 2 2,5 5 5
```

50.
```
    8 4,8 9 0
  - 1 1,1 1 0
```

C Use < or > for ☐ to write a true sentence. Draw a number line if necessary.

51. 0 ☐ 17

52. 32 ☐ 0

53. 34 ☐ 12

54. 28 ☐ 18

55. 1000 ☐ 1001

56. 77 ☐ 117

57. 133 ☐ 132

58. 999 ☐ 997

59. 460 ☐ 17

60. 345 ☐ 456

61. 37 ☐ 11

62. 12 ☐ 32

Daily Newspapers.   The top three daily newspapers in the United States are *USA Today,* the *Wall Street Journal,* and the *New York Times.* The daily circulation of each as of September 30, 2002, is listed in the table below. Use this table when answering Exercises 63 and 64.

NEWSPAPER	DAILY CIRCULATION
USA Today	2,136,068
Wall Street Journal	1,800,607
New York Times	1,113,000

Source: Editor and Publisher International Year Book 2003

63. Use an inequality to compare the daily circulation of the *Wall Street Journal* and *USA Today.*

64. Use an inequality to compare the daily circulation of *USA Today* and the *New York Times.*

65. *Pedestrian Fatalities.* The annual number of pedestrians killed when hit by a motor vehicle has declined from 6482 in 1990 to 4827 in 2003. Use an inequality to compare these annual totals of pedestrian fatalities.

Pedestrian Fatalities on the Decline

Source: National Center for Statistics and Analysis

66. *Life Expectancy.* The life expectancy of a female in 2010 is predicted to be about 82 yr and of a male about 76 yr. Use an inequality to compare these life expectancies.

Life Expectancy

Source: U.S. Bureau of the Census

67. **D**w Explain how estimating and rounding can be useful when shopping for groceries.

68. **D**w When rounding 748 to the nearest hundred, a student rounds to 750 and then to 800. What mistake is he making?

Write expanded notation.

69. 7992   [1.1b]

70. 23,000,000   [1.1b]

71. Write a word name for 246,605,004,032.   [1.1c]

72. Write a word name for 1,005,100.   [1.1c]

Add.   [1.2a]

73.
```
  6 7,7 8 9
+ 1 8,9 6 5
```

74.
```
  9 0 0 2
+ 4 5 8 7
```

Subtract.   [1.3b]

75.
```
  6 7,7 8 9
− 1 8,9 6 5
```

76.
```
  9 0 0 2
− 4 5 8 7
```

77.–80. 🖩 Use a calculator to find the sums and differences in each of Exercises 47–50. Then compare your answers with those found using estimation. Even when using a calculator it is possible to make an error if you press the wrong buttons, so it is a good idea to check by estimating.

CHAPTER 1: Whole Numbers

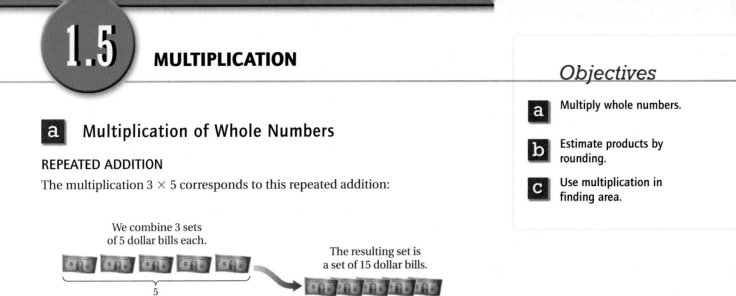

1.5 MULTIPLICATION

a Multiplication of Whole Numbers

REPEATED ADDITION

The multiplication 3×5 corresponds to this repeated addition:

We combine 3 sets of 5 dollar bills each.

The resulting set is a set of 15 dollar bills.

$$3 \times 5 = 5 + 5 + 5 = 15$$

3 addends; each is 5

The numbers that we multiply are called **factors.** The result of the multiplication is called a **product.**

$$3 \quad \times \quad 5 \quad = \quad 15$$

Factor   Factor   Product

RECTANGULAR ARRAYS

Multiplications can also be thought of as rectangular arrays. Each of the following corresponds to the multiplication 3×5.

3 rows with 5 bills in each row;
$3 \times 5 = 15$

5 columns with 3 bills in each column;
$3 \times 5 = 15$

When you write a multiplication sentence corresponding to a real-world situation, you should think of either a rectangular array or repeated addition. In some cases, it may help to think both ways.

We have used an "$\times$" to denote multiplication. A dot "$\cdot$" is also commonly used. (Use of the dot is attributed to the German mathematician Gottfried Wilhelm von Leibniz in 1698.) Parentheses are also used to denote multiplication. For example,

$$3 \times 5 = 3 \cdot 5 = (3)(5) = 3(5) = 15.$$

31

Multiply.

1. $\begin{array}{r} 5\,8 \\ \times\ \ 2 \\ \hline \end{array}$

The product of 0 and any whole number is 0: $0 \cdot a = a \cdot 0 = 0$. Multiplying a number by 1 does not change the number: $1 \cdot a = a \cdot 1 = a$. We say that 1 is the **multiplicative identity.** For example, $0 \cdot 3 = 3 \cdot 0 = 0$ and $1 \cdot 3 = 3 \cdot 1 = 3$.

EXAMPLE 1 Multiply: 5×734.

We have

$$\begin{array}{r} 7\ 3\ 4 \\ \times\qquad\ 5 \\ \hline 2\ 0 \\ 1\ 5\ 0 \\ 3\ 5\ 0\ 0 \\ \hline 3\ 6\ 7\ 0 \end{array}$$

$\leftarrow$ Multiply the 4 ones by 5: $5 \times 4 = 20$.
$\leftarrow$ Multiply the 3 tens by 5: $5 \times 30 = 150$.
$\leftarrow$ Multiply the 7 hundreds by 5: $5 \times 700 = 3500$.
$\leftarrow$ Add.

Instead of writing each product on a separate line, we can use a shorter form.

2. $\begin{array}{r} 3\,7 \\ \times\ \ 4 \\ \hline \end{array}$

$$\begin{array}{r} \overset{2}{}\ \\ 7\ 3\ 4 \\ \times\qquad\ 5 \\ \hline 0 \end{array}$$

Multiply the ones by 5: $5 \cdot (4 \text{ ones}) = 20 \text{ ones} = 2 \text{ tens} + 0 \text{ ones}$. Write 0 in the ones column and 2 above the tens.

$$\begin{array}{r} \overset{1}{}\overset{2}{}\ \\ 7\ 3\ 4 \\ \times\qquad\ 5 \\ \hline 7\ 0 \end{array}$$

Multiply the 3 tens by 5 and add 2 tens: $5 \cdot (3 \text{ tens}) = 15 \text{ tens}$, $15 \text{ tens} + 2 \text{ tens} = 17 \text{ tens} = 1 \text{ hundred} + 7 \text{ tens}$. Write 7 in the tens column and 1 above the hundreds.

3. $\begin{array}{r} 8\,2\,3 \\ \times\qquad 6 \\ \hline \end{array}$

$$\begin{array}{r} \overset{1}{}\overset{2}{}\ \\ 7\ 3\ 4 \\ \times\qquad\ 5 \\ \hline 3\ 6\ 7\ 0 \end{array}$$

Multiply the 7 hundreds by 5 and add 1 hundred: $5 \cdot (7 \text{ hundreds}) = 35 \text{ hundreds}$, $35 \text{ hundreds} + 1 \text{ hundred} = 36 \text{ hundreds}$.

$$\left.\begin{array}{r} \overset{1}{}\overset{2}{}\ \\ 7\ 3\ 4 \\ \times\qquad\ 5 \\ \hline 3\ 6\ 7\ 0 \end{array}\right\}$$ You should write only this.

Do Exercises 1–4.

Let's find the product

$$\begin{array}{r} 5\ 4 \\ \times\ 3\ 2 \\ \hline \end{array}$$

To do this, we multiply 54 by 2, then 54 by 30, and then add.

4. $\begin{array}{r} 1\,3\,4\,8 \\ \times\qquad\ 5 \\ \hline \end{array}$

$$\begin{array}{r} 5\ 4 \\ \times\ \ 2 \\ \hline 1\ 0\ 8 \end{array} \qquad \begin{array}{r} \overset{1}{}\ \\ 5\ 4 \\ \times\ 3\ 0 \\ \hline 1\ 6\ 2\ 0 \end{array}$$

Since we are going to add the results, let's write the work this way.

$$\begin{array}{r} 5\ 4 \\ \times\ 3\ 2 \\ \hline 1\ 0\ 8 \\ 1\ 6\ 2\ 0 \\ \hline 1\ 7\ 2\ 8 \end{array}$$

Multiplying by 2
Multiplying by 30
Adding to obtain the product

The fact that we can do this is based on a property called the **distributive law.** It says that to multiply a number by a sum, $a \cdot (b + c)$, we can multiply each addend by a and then add like this: $(a \cdot b) + (a \cdot c)$. Thus, $a \cdot (b + c) = (a \cdot b) + (a \cdot c)$. Applied to the example above, the distributive law gives us

$$54 \cdot 32 = 54 \cdot (30 + 2) = (54 \cdot 30) + (54 \cdot 2).$$

EXAMPLE 2 Multiply: 43×57.

```
        2
      5 7
  ×   4 3
  ─────────
    1 7 1      Multiplying by 3


        2
        2
      5 7
  ×   4 3
  ─────────
    1 7 1
  2 2 8 0      Multiplying by 40. (We write a 0 and then multiply 57
               by 4).
```

You may have learned that such a 0 does not have to be written. You may omit it if you wish. If you do omit it, remember, when multiplying by tens, to start writing the answer in the tens place.

```
        2
        2
      5 7
  ×   4 3
  ─────────
    1 7 1
  2 2 8 0
  ─────────
  2 4 5 1      Adding to obtain the product
```

Do Exercises 5 and 6.

EXAMPLE 3 Multiply: 457×683.

```
      5   2
      6 8 3
  ×   4 5 7
  ─────────
  4 7 8 1      Multiplying 683 by 7


        4 1
        5 2
      6 8 3
  ×   4 5 7
  ─────────
    4 7 8 1
  3 4 1 5 0    Multiplying 683 by 50


      3 1
      4 1 2
      5
      6 8 3
  ×   4 5 7
  ─────────
      4 7 8 1
    3 4 1 5 0
  2 7 3 2 0 0  Multiplying 683 by 400
  ───────────
  3 1 2 , 1 3 1  Adding
```

Do Exercises 7 and 8.

Multiply.

5.
```
    4 5
  × 2 3
```

6. 48×63

Multiply.

7.
```
    7 4 6
  ×   6 2
```

8. 245×837

Multiply.

9.
```
    4 7 2
  × 3 0 6
```

10. 408×704

11.
```
    2 3 4 4
  × 6 0 0 5
```

Answers on page A-2

Multiply.

12.
$$\begin{array}{r} 4\ 7\ 2 \\ \times\ 8\ 3\ 0 \\ \hline \end{array}$$

13.
$$\begin{array}{r} 2\ 3\ 4\ 4 \\ \times\ 7\ 4\ 0\ 0 \\ \hline \end{array}$$

14. 100×562

15. 1000×562

16. a) Find $23 \cdot 47$.

b) Find $47 \cdot 23$.

c) Compare your answers to parts (a) and (b).

Multiply.

17. $5 \cdot 2 \cdot 4$

18. $5 \cdot 1 \cdot 3$

EXAMPLE 4 Multiply: 306×274.

Note that $306 = 3$ hundreds $+ 6$ ones.

$$\begin{array}{r} 2\ 7\ 4 \\ \times\ 3\ 0\ 6 \\ \hline 1\ 6\ 4\ 4 \\ 8\ 2\ 2\ 0\ 0 \\ \hline 8\ 3\,,8\ 4\ 4 \end{array}$$

Multiplying by 6

Multiplying by 3 hundreds. (We write 00 and then multiply 274 by 3.)

Adding

Do Exercises 9–11 on the preceding page.

EXAMPLE 5 Multiply: 360×274.

Note that $360 = 3$ hundreds $+ 6$ tens.

$$\begin{array}{r} 2\ 7\ 4 \\ \times\ 3\ 6\ 0 \\ \hline 1\ 6\ 4\ 4\ 0 \\ 8\ 2\ 2\ 0\ 0 \\ \hline 9\ 8\,,6\ 4\ 0 \end{array}$$

Multiplying by 6 tens. (We write 0 and then multiply 274 by 6.)

Multiplying by 3 hundreds. (We write 00 and then multiply 274 by 3.)

Adding

Do Exercises 12–15.

Check on your own that $17 \cdot 37 = 629$ and that $37 \cdot 17 = 629$. This illustrates the **commutative law of multiplication.** It says that we can multiply two numbers in any order, $a \cdot b = b \cdot a$, and get the same answer.

Do Exercise 16.

To multiply three or more numbers, we generally group them so that we multiply two at a time. Consider $2 \cdot (3 \cdot 4)$ and $(2 \cdot 3) \cdot 4$. The parentheses tell what to do first:

$$2 \cdot (3 \cdot 4) = 2 \cdot (12) = 24. \qquad \text{We multiply 3 and 4, then 2.}$$

We can also multiply 2 and 3, then 4:

$$(2 \cdot 3) \cdot 4 = (6) \cdot 4 = 24.$$

Either way we get 24. It does not matter how we group the numbers. This illustrates the **associative law of multiplication:** $a \cdot (b \cdot c) = (a \cdot b) \cdot c$.

Do Exercises 17 and 18.

Answers on page A-2

b Estimating Products by Rounding

EXAMPLE 6 *Lawn Tractors.* Leisure Lawn Care is buying new lawn tractors that cost $1534 each. By rounding to the nearest ten, estimate the cost of purchasing 18 tractors.

Exact	*Nearest ten*
1 5 3 4	1 5 3 0
× 1 8	× 2 0
2 7,6 1 2	3 0,6 0 0

The lawn tractors will cost about $30,600.

Do Exercise 19.

EXAMPLE 7 Estimate the following product by first rounding to the nearest ten and to the nearest hundred: 683 × 457.

Nearest ten	*Nearest hundred*	*Exact*
6 8 0	7 0 0	6 8 3
× 4 6 0	× 5 0 0	× 4 5 7
4 0 8 0 0	3 5 0,0 0 0	4 7 8 1
2 7 2 0 0 0		3 4 1 5 0
3 1 2,8 0 0		2 7 3 2 0 0
		3 1 2,1 3 1

Do Exercise 20.

c Finding Area

The area of a rectangular region can be considered to be the number of square units needed to fill it. Here is a rectangle 4 cm (centimeters) long and 3 cm wide. It takes 12 square centimeters (sq cm) to fill it.

This is a square centimeter (a square unit).

In this case, we have a rectangular array of 3 rows, each of which contains 4 squares. The number of square units is given by 3 · 4, or 12. That is, $A = l \cdot w = 3 \text{ cm} \cdot 4 \text{ cm} = 12 \text{ sq cm}$.

Multiplying Whole Numbers To multiply whole numbers on a calculator, we use the ⊠ and ⊜ keys. For example, to find 13 × 47, we press ⊡ ⊡ ⊠ ⊡ ⊡ ⊜. The calculator displays 611, so 13 × 47 = 611.

Exercises: Use a calculator to find each product.

1. 56 × 8
2. 845 × 26
3. 5 · 1276
4. 126(314)
5. 3 7 6 0
   × 4 8
6. 5 2 1 8
   × 4 5 3

19. Lawn Tractors. By rounding to the nearest ten, estimate the cost to Leisure Lawn Care of 12 lawn tractors.

20. Estimate the product by first rounding to the nearest ten and to the nearest hundred. Show your work.

  8 3 7
× 2 4 5

Answers on page A-2

21. Table Tennis. Find the area of a standard table tennis table that has dimensions of 9 ft by 5 ft.

EXAMPLE 8 *Professional Pool Table.* The playing area of a standard pool table has dimensions of 50 in. by 100 in. (There are rails 6 in. wide on the outside that are not included in the playing area.) Find the playing area.

If we think of filling the rectangle with square inches, we have a rectangular array. The length $l = 100$ in. and the width $w = 50$ in. Thus the area A is given by the formula

$$A = l \cdot w = 100 \cdot 50 = 5000 \text{ sq in.}$$

Do Exercise 21.

Study Tips — TIME MANAGEMENT (PART 1)

Time is the most critical factor in your success in learning mathematics. Have reasonable expectations about the time you need to study math.

- **Juggling time.** Working 40 hours per week and taking 12 credit hours is equivalent to working two full-time jobs. Can you handle such a load? Your ratio of number of work hours to number of credit hours should be about 40/3, 30/6, 20/9, 10/12, or 5/14.

- **A rule of thumb on study time.** Budget about 2–3 hours for homework and study for every hour you spend in class each week.

- **Scheduling your time.** Use the Student Organizer card to make an hour-by-hour schedule of your typical week. Include work, school, home, sleep, study, and leisure times. Try to schedule time for study when you are most alert. Choose a setting that will enable you to focus and concentrate. Plan for success and it will happen!

"You cannot increase the quality or quantity of your achievement or performance except to the degree in which you increase your ability to use time effectively."

Brian Tracy, motivational/inspirational speaker

a   Multiply.

1. 87
× 10

2. 100
× 96

3. 2340
× 1000

4. 800
× 70

5. 65
× 8

6. 87
× 4

7. 94
× 6

8. 76
× 9

9. 3 · 509

10. 7 · 806

11. 7(9229)

12. 4(7867)

13. 90(53)

14. 60(78)

15. (47)(85)

16. (34)(87)

17. 640
× 72

18. 777
× 77

19. 444
× 33

20. 509
× 88

21. 509
× 408

22. 432
× 375

23. 853
× 936

24. 346
× 650

25. 6428
× 3224

26. 8928
× 3172

27. 3482
× 104

28. 6408
× 6064

29.
```
    5 0 0 6
×   4 0 0 8
```

30.
```
    6 7 8 9
×   2 3 3 0
```

31.
```
    5 6 0 8
×   4 5 0 0
```

32.
```
    4 5 6 0
×   7 8 9 0
```

33.
```
      8 7 6
×     3 4 5
```

34.
```
      3 5 5
×     2 9 9
```

35.
```
    7 8 8 9
×   6 2 2 4
```

36.
```
    6 5 0 1
×   3 4 4 9
```

b Estimate the product by first rounding to the nearest ten. Show your work.

37.
```
    4 5
×   6 7
```

38.
```
    5 1
×   7 8
```

39.
```
    3 4
×   2 9
```

40.
```
    6 3
×   5 4
```

Estimate the product by first rounding to the nearest hundred. Show your work.

41.
```
    8 7 6
×   3 4 5
```

42.
```
    3 5 5
×   2 9 9
```

43.
```
    4 3 2
×   1 9 9
```

44.
```
    7 8 9
×   4 3 4
```

45. *Toyota Sienna.* A pharmaceutical company buys a Toyota Sienna for each of its 112 sales representatives. Each car costs $27,896 plus an additional $540 per car in destination charges.

   a) Estimate the total cost of the purchase by rounding the cost of each car, the destination charge, and the number of sales representatives to the nearest hundred.

   b) Estimate the total cost of the purchase by rounding the cost of each car to the nearest thousand and the destination charge and the number of reps to the nearest hundred.

   Source: Toyota

46. A travel club of 176 people decides to fly from Los Angeles to Tokyo. The cost of a round-trip ticket is $643.

   a) Estimate the total cost of the trip by rounding the cost of the airfare and the number of travelers to the nearest ten.

   b) Estimate the total cost of the trip by rounding the cost of the airfare and the number of travelers to the nearest hundred.

C Find the area of the region.

47.

728 mi

728 mi

48.

129 yd

65 yd

49. Find the area of the region formed by the base lines on a Major League Baseball diamond.

90 ft

90 ft

50. Find the area of a standard-sized hockey rink.

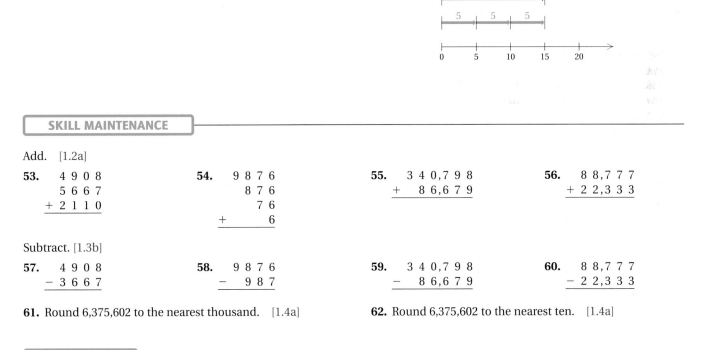

200 ft

85 ft

51. **D**_W Describe a situation that corresponds to each multiplication: 4 · $150; $4 · 150.

52. **D**_W Explain the multiplication illustrated in the diagram below.

3 · 5

5   5   5

0   5   10   15   20

Add.   [1.2a]

53.   4 9 0 8
    5 6 6 7
  + 2 1 1 0

54.   9 8 7 6
      8 7 6
       7 6
  +   6

55.   3 4 0,7 9 8
  +   8 6,6 7 9

56.   8 8,7 7 7
  + 2 2,3 3 3

Subtract. [1.3b]

57.   4 9 0 8
  − 3 6 6 7

58.   9 8 7 6
  −   9 8 7

59.   3 4 0,7 9 8
  −   8 6,6 7 9

60.   8 8,7 7 7
  − 2 2,3 3 3

61. Round 6,375,602 to the nearest thousand.   [1.4a]

62. Round 6,375,602 to the nearest ten.   [1.4a]

63. ▦ An 18-story office building is box-shaped. Each floor measures 172 ft by 84 ft with a 20-ft by 35-ft rectangular area lost to an elevator and a stairwell. How much area is available as office space?

Objectives

a Convert between division sentences and multiplication sentences.

b Divide whole numbers.

1. Consider $54 \div 6 = 9$. Express this division in two other ways.

Answer on page A-2

a Division and Related Sentences

REPEATED SUBTRACTION

Division of whole numbers applies to two kinds of situations. The first is repeated subtraction. Suppose we have 20 notebooks in a pile, and we want to find out how many sets of 5 there are. One way to do this is to repeatedly subtract sets of 5 as follows.

How many sets of 5 notebooks each?

20 notebooks

Since there are 4 sets of 5 notebooks each, we have

$$
\underset{\displaystyle \text{Dividend}}{20} \quad \underset{\displaystyle \text{Divisor}}{\div \quad 5} \quad \underset{\displaystyle \text{Quotient}}{= \quad 4.}
$$

The division $20 \div 5$, read "20 divided by 5," corresponds to the figure above. We say that the **dividend** is 20, the **divisor** is 5, and the **quotient** is 4. We divide the *dividend* by the *divisor* to get the *quotient*.

We can also express the division $20 \div 5 = 4$ as

$$\frac{20}{5} = 4 \quad \text{or} \quad 5\overline{)20}^{\,4}$$

Do Exercise 1.

RECTANGULAR ARRAYS

We can also think of division in terms of rectangular arrays. Consider again the pile of 20 notebooks and division by 5. We can arrange the notebooks in a rectangular array with 5 rows and ask, "How many are in each row?"

We can also consider a rectangular array with 5 notebooks in each column and ask, "How many columns are there?" The answer is still 4.

In each case, we are asking, "What do we multiply 5 by in order to get 20?"

$$\underset{\displaystyle \text{Missing factor}}{5 \cdot \square = 20} \qquad \underset{\displaystyle \text{Quotient}}{20 \div 5 = \square}$$

This leads us to the following definition of division.

DIVISION

The quotient $a \div b$, where $b \neq 0$, is that unique number c for which $a = b \cdot c$.

RELATED SENTENCES

By looking at rectangular arrays, we can see how multiplication and division are related. The array of notebooks on the preceding page shows that $5 \cdot 4 = 20$. The array also shows the following:

$$20 \div 5 = 4 \quad \text{and} \quad 20 \div 4 = 5.$$

The division $20 \div 5$ is defined to be the number that when multiplied by 5 gives 20. Thus, for every division sentence, there is a related multiplication sentence.

$20 \div 5 = 4$ Division sentence

$20 = 4 \cdot 5$ Related multiplication sentence

To get the related multiplication sentence, we use
Dividend = Quotient · Divisor.

EXAMPLE 1 Write a related multiplication sentence: $12 \div 6 = 2$.

We have

$12 \div 6 = 2$ Division sentence

$12 = 2 \cdot 6.$ Related multiplication sentence

The related multiplication sentence is $12 = 2 \cdot 6$.

By the commutative law of multiplication, there is also another multiplication sentence: $12 = 6 \cdot 2$.

Do Exercises 2 and 3.

For every multiplication sentence, we can write related division sentences.

EXAMPLE 2 Write two related division sentences: $7 \cdot 8 = 56$.

We have

$7 \cdot 8 = 56$ $7 \cdot 8 = 56$

This factor becomes a divisor.   This factor becomes a divisor.

$7 = 56 \div 8.$ $8 = 56 \div 7.$

The related division sentences are $7 = 56 \div 8$ and $8 = 56 \div 7$.

Do Exercises 4 and 5.

Write a related multiplication sentence.

2. $15 \div 3 = 5$

3. $72 \div 8 = 9$

Write two related division sentences.

4. $6 \cdot 2 = 12$

5. $7 \cdot 6 = 42$

Answers on page A-2

Division of Whole Numbers

Before we consider division with remainders, let's recall four basic facts about division.

DIVIDING BY 1

Any number divided by 1 is that same number:

$$a \div 1 = \frac{a}{1} = a.$$

DIVIDING A NUMBER BY ITSELF

Any nonzero number divided by itself is 1:

$$\frac{a}{a} = 1, \quad a \neq 0.$$

DIVIDENDS OF 0

Zero divided by any nonzero number is 0:

$$\frac{0}{a} = 0, \quad a \neq 0.$$

EXCLUDING DIVISION BY 0

Division by 0 is not defined. (We agree not to divide by 0.)

$$\frac{a}{0} \text{ is } \textbf{not defined.}$$

Why can't we divide by 0? Suppose the number 4 could be divided by 0. Then if $\square$ were the answer,

$$4 \div 0 = \square$$

and since 0 times any number is 0, we would have

$$4 = \square \cdot 0 = 0. \quad \text{False!}$$

Thus, $a \div 0$ would be some number $\square$ such that $a = \square \cdot 0 = 0$. So the only possible number that could be divided by 0 would be 0 itself.

But such a division would give us any number we wish, for

$$0 \div 0 = 8 \quad \text{because} \quad 0 = 8 \cdot 0;$$
$$0 \div 0 = 3 \quad \text{because} \quad 0 = 3 \cdot 0; \quad \left.\right\} \quad \text{All true!}$$
$$0 \div 0 = 7 \quad \text{because} \quad 0 = 7 \cdot 0.$$

We avoid the preceding difficulties by agreeing to exclude division by 0.

Suppose we have 18 cans of soda and want to pack them in cartons of 6 cans each. How many cartons will we fill? We can determine this by repeated subtraction. We keep track of the number of times we subtract. We stop when the number of objects remaining, the **remainder,** is smaller than the divisor.

Divide by repeated subtraction. Then check.

6. $54 \div 9$

EXAMPLE 3 Divide by repeated subtraction: $18 \div 6$.

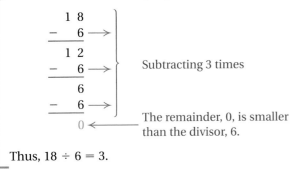

$$
\left.\begin{array}{r}
1\ 8 \\
-\ \ \ 6 \longrightarrow \\
\hline
1\ 2 \\
-\ \ \ 6 \longrightarrow \\
\hline
6 \\
-\ \ \ 6 \longrightarrow
\end{array}\right\} \text{Subtracting 3 times}
$$

$0 \longleftarrow$ The remainder, 0, is smaller than the divisor, 6.

Thus, $18 \div 6 = 3$.

7. $61 \div 9$

Suppose we have 22 cans of soda and want to pack them in cartons of 6 cans each. We end up with 3 cartons with 4 cans left over.

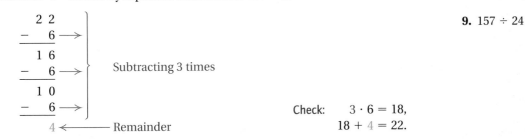

1

2

3

4 left over

8. $53 \div 12$

EXAMPLE 4 Divide by repeated subtraction: $22 \div 6$.

9. $157 \div 24$

$$
\left.\begin{array}{r}
2\ 2 \\
-\ \ \ 6 \longrightarrow \\
\hline
1\ 6 \\
-\ \ \ 6 \longrightarrow \\
\hline
1\ 0 \\
-\ \ \ 6 \longrightarrow
\end{array}\right\} \text{Subtracting 3 times}
$$

$4 \longleftarrow$ Remainder

Check: $3 \cdot 6 = 18$,
$18 + 4 = 22$.

We can write this division as follows.

$$
\begin{array}{r}
3 \\
6\overline{)2\ 2} \\
1\ 8 \\
\hline
4
\end{array}
$$

Answers on page A-2

Divide and check.

10. $4 \overline{)\ 2\ 3\ 9}$

Note that

$$\text{Quotient} \cdot \text{Divisor} + \text{Remainder} = \text{Dividend}.$$

We write answers to a division sentence as follows.

$$22 \div 6 = 3 \text{ R } 4$$

Dividend   Divisor   Quotient   Remainder

Do Exercises 6–9 on the preceding page.

EXAMPLE 5 Divide and check: $3642 \div 5$.

$$5 \overline{)\ 3\ 6\ 4\ 2}$$?

1. Find the number of thousands in the quotient. Consider 3 thousands ÷ 5 and think 3 ÷ 5. Since 3 ÷ 5 is not a whole number, move to hundreds.

11. $6 \overline{)\ 8\ 8\ 5\ 5}$

$$\begin{array}{r} 7 \\ 5 \overline{)\ 3\ 6\ 4\ 2} \\ 3\ 5\ 0\ 0 \\ \hline 1\ 4\ 2 \end{array}$$

2. Find the number of hundreds in the quotient. Consider 36 hundreds ÷ 5 and think 36 ÷ 5. The estimate is about 7 hundreds. Multiply 700 by 5 and subtract.
← The remainder is larger than the divisor.

$$\begin{array}{r} 7\ 2 \\ 5 \overline{)\ 3\ 6\ 4\ 2} \\ 3\ 5\ 0\ 0 \\ \hline 1\ 4\ 2 \\ 1\ 0\ 0 \\ \hline 4\ 2 \end{array}$$

3. Find the number of tens in the quotient using 142, the first remainder. Consider 14 tens ÷ 5 and think 14 ÷ 5. The estimate is about 2 tens. Multiply 20 by 5 and subtract. (If our estimate had been 3 tens, we could not have subtracted 150 from 142.)
← The remainder is larger than the divisor.

12. $5 \overline{)\ 5\ 0\ 7\ 5}$

$$\begin{array}{r} 7\ 2\ 8 \\ 5 \overline{)\ 3\ 6\ 4\ 2} \\ 3\ 5\ 0\ 0 \\ \hline 1\ 4\ 2 \\ 1\ 0\ 0 \\ \hline 4\ 2 \\ 4\ 0 \\ \hline 2 \end{array}$$

4. Find the number of ones in the quotient using 42, the second remainder. Consider 42 ones ÷ 5 and think 42 ÷ 5. The estimate is about 8 ones. Multiply 8 by 5 and subtract. The remainder, 2, is less than the divisor, 5, so we are finished.

2 ← The remainder is less than the divisor.

> You may have learned to divide like this, not writing the extra zeros. You may omit them if desired.

$$\begin{array}{r} 7\ 2\ 8 \\ 5 \overline{)\ 3\ 6\ 4\ 2} \\ 3\ 5 \downarrow \\ \hline 1\ 4 \\ 1\ 0 \downarrow \\ \hline 4\ 2 \\ 4\ 0 \\ \hline 2 \end{array}$$

Check: $728 \cdot 5 = 3640$,
$3640 + 2 = 3642$.

The answer is 728 R 2.

Answers on page A-2

Do Exercises 10–12.

We can summarize our division procedure as follows.

To do division of whole numbers:

a) Estimate.
b) Multiply.
c) Subtract.

Sometimes rounding the divisor helps us find estimates.

EXAMPLE 6 Divide: $8904 \div 42$.

We mentally round 42 to 40.

```
           2
    4 2 ) 8 9 0 4    ← Think: 89 hundreds ÷ 40.
          8 4 0 0         Estimate 2 hundreds. Multiply 200 · 42
          ───────         and subtract.
            5 0 4
```

```
           2 1
    4 2 ) 8 9 0 4
          8 4 0 0
          ───────
            5 0 4    ← Think: 50 tens ÷ 40.
            4 2 0         Estimate 1 ten. Multiply 10 · 42
            ─────         and subtract.
              8 4
```

```
           2 1 2
    4 2 ) 8 9 0 4
          8 4 0 0
          ───────
            5 0 4
            4 2 0
            ─────
              8 4    ← Think: 84 ones ÷ 40.
              8 4         Estimate 2 ones. Multiply 2 · 42
              ───         and subtract.
                0
```

Caution!

Be careful to keep the digits lined up correctly.

The answer is 212. *Remember*: If after estimating and multiplying you get a number that is larger than the number from which it is being subtracted, lower your estimate.

Do Exercises 13 and 14.

Divide.

13. $4 5 \overline{) 6 0 3 0}$

14. $5 2 \overline{) 3 2 8 8}$

Divide.

15. $6 \overline{) 4 8 4 6}$

16. $7 \overline{) 7 6 1 6}$

Answers on page A-2

Divide.

17. $2\ 7\ \overline{)\ 9\ 7\ 2\ 4}$

18. $5\ 6\ \overline{)\ 4\ 4{,}8\ 4\ 7}$

Answers on page A-2

Study Tips

EXERCISES

■ **Odd-numbered exercises.** Usually an instructor assigns some odd-numbered exercises as homework. When you complete these, you can check your answers at the back of the book. If you miss any, check your work in the *Student's Solutions Manual* or ask your instructor for guidance.

■ **Even-numbered exercises.** Whether or not your instructor assigns the even-numbered exercises, always do some on your own. Remember, there are no answers given for the chapter tests, so you need to practice doing exercises without answers. Check your answers later with a friend or your instructor.

ZEROS IN QUOTIENTS

EXAMPLE 7 Divide: $6341 \div 7$.

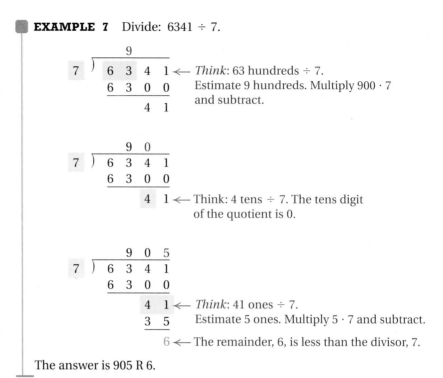

Think: 63 hundreds ÷ 7. Estimate 9 hundreds. Multiply $900 \cdot 7$ and subtract.

Think: 4 tens ÷ 7. The tens digit of the quotient is 0.

Think: 41 ones ÷ 7. Estimate 5 ones. Multiply $5 \cdot 7$ and subtract.

6 ← The remainder, 6, is less than the divisor, 7.

The answer is 905 R 6.

Do Exercises 15 and 16 on the preceding page.

EXAMPLE 8 Divide: $8889 \div 37$.

We round 37 to 40.

Think: $37 \approx 40$; 88 hundreds ÷ 40. Estimate 2 hundreds. Multiply $200 \cdot 37$ and subtract.

Think: 148 tens ÷ 40. Estimate 4 tens. Multiply $40 \cdot 37$ and subtract.

9 ← The remainder, 9, is less than the divisor, 37.

The answer is 240 R 9.

Do Exercises 17 and 18.

CHAPTER 1: Whole Numbers

CALCULATOR CORNER

Dividing Whole Numbers: Finding Remainders   To divide whole numbers on a calculator, we use the $\div$ and $=$ keys. For example, to divide 711 by 9, we press $\boxed{7}\,\boxed{1}\,\boxed{1}\,\boxed{\div}\,\boxed{9}\,\boxed{=}$. The display reads $\boxed{79}$, so $711 \div 9 = 79$.

When we enter $453 \div 15$, the display reads $\boxed{30.2}$. Note that the result is not a whole number. This tells us that there is a remainder. The number 30.2 is expressed in decimal notation. The symbol "." is called a decimal point. (Decimal notation will be studied in Chapter 4.) The number to the left of the decimal point, 30, is the quotient. We can use the remaining part of the result to find the remainder. To do this, first subtract 30 from 30.2. Then multiply the difference by the divisor, 15. We get 3. This is the remainder. Thus, $453 \div 15 = 30$ R 3. The steps that we performed to find this result can be summarized as follows:

$$453 \div 15 = 30.2,$$
$$30.2 - 30 = .2,$$
$$0.2 \times 15 = 3.$$

To follow these steps on a calculator, we press $\boxed{4}\,\boxed{5}\,\boxed{3}\,\boxed{\div}$ $\boxed{1}\,\boxed{5}\,\boxed{=}$ and write the number that appears to the left of the decimal point. This is the quotient. Then we continue by pressing $\boxed{-}\,\boxed{3}\,\boxed{0}\,\boxed{=}$ $\boxed{\times}\,\boxed{1}\,\boxed{5}\,\boxed{=}$. The last number that appears is the remainder. In some cases, it will be necessary to round the remainder to the nearest one.

To check this result, we multiply the quotient by the divisor and then add the remainder.

$$30 \times 15 = 450,$$
$$450 + 3 = 453$$

Exercises: Use a calculator to perform each division. Check the results with a calculator also.

1. $92 \div 27$
2. $9\overline{)532}$
3. $6\overline{)746}$
4. $3817 \div 29$
5. $126\overline{)35{,}715}$
6. $308\overline{)259{,}831}$

a Write a related multiplication sentence.

1. $18 \div 3 = 6$ **2.** $72 \div 9 = 8$ **3.** $22 \div 22 = 1$ **4.** $32 \div 1 = 32$

5. $54 \div 6 = 9$ **6.** $90 \div 10 = 9$ **7.** $37 \div 1 = 37$ **8.** $28 \div 28 = 1$

Write two related division sentences.

9. $9 \times 5 = 45$ **10.** $2 \cdot 7 = 14$ **11.** $37 \cdot 1 = 37$ **12.** $4 \cdot 12 = 48$

13. $8 \times 8 = 64$ **14.** $9 \cdot 7 = 63$ **15.** $11 \cdot 6 = 66$ **16.** $1 \cdot 43 = 43$

b Divide, if possible. If not possible, write "not defined."

17. $72 \div 6$ **18.** $54 \div 9$ **19.** $\dfrac{23}{23}$ **20.** $\dfrac{37}{37}$

21. $22 \div 1$ **22.** $\dfrac{56}{1}$ **23.** $\dfrac{16}{0}$ **24.** $74 \div 0$

Divide.

25. $277 \div 5$ **26.** $699 \div 3$ **27.** $864 \div 8$ **28.** $869 \div 8$

29. $4 \overline{)1228}$ **30.** $3 \overline{)2124}$ **31.** $6 \overline{)4521}$ **32.** $9 \overline{)9110}$

33. $297 \div 4$

34. $389 \div 2$

35. $738 \div 8$

36. $881 \div 6$

37. $5 \overline{)8\ 5\ 1\ 5}$

38. $3 \overline{)6\ 0\ 2\ 7}$

39. $9 \overline{)8\ 8\ 8\ 8}$

40. $8 \overline{)4\ 1\ 3\ 9}$

41. $127,000 \div 10$

42. $127,000 \div 100$

43. $127,000 \div 1000$

44. $4260 \div 10$

45. $7\ 0 \overline{)3\ 6\ 9\ 2}$

46. $2\ 0 \overline{)5\ 7\ 9\ 8}$

47. $3\ 0 \overline{)8\ 7\ 5}$

48. $4\ 0 \overline{)9\ 8\ 7}$

49. $852 \div 21$

50. $942 \div 23$

51. $8\ 5 \overline{)7\ 6\ 7\ 2}$

52. $5\ 4 \overline{)2\ 7\ 2\ 9}$

53. $111 \overline{)3219}$ **54.** $102 \overline{)5612}$ **55.** $8 \overline{)843}$ **56.** $7 \overline{)749}$

57. $5 \overline{)8047}$ **58.** $9 \overline{)7273}$ **59.** $5 \overline{)5036}$ **60.** $7 \overline{)7074}$

61. $1058 \div 46$ **62.** $7242 \div 24$ **63.** $3425 \div 32$ **64.** $48 \overline{)4899}$

65. $24 \overline{)8880}$ **66.** $36 \overline{)7563}$ **67.** $28 \overline{)17,067}$ **68.** $36 \overline{)28,929}$

69. $80 \overline{)24,320}$ **70.** $90 \overline{)88,560}$ **71.** $285 \overline{)999,999}$

72. $306 \overline{)888,888}$ **73.** $456 \overline{)3,679,920}$ **74.** $803 \overline{)5,622,606}$

75. D_W Is division associative? Why or why not? Give an example.

76. D_W Suppose a student asserts that "$0 \div 0 = 0$ because nothing divided by nothing is nothing." Devise an explanation to persuade the student that the assertion is false.

VOCABULARY REINFORCEMENT

In each of Exercises 77–84, fill in the blank with the correct term from the given list. Some of the choices may not be used and some may be used more than once.

77. The distance around an object is its _____ .
[1.2b]

78. A sentence like $10 - 3 = 7$ is called a(n) _____ ;
a sentence like $31 < 33$ is called a(n) _____ .
[1.4c]

79. For large numbers, _____ are separated by commas into groups of three, called _____ .
[1.1a]

80. The number 0 is called the _____ identity.
[1.2a]

81. In the sentence $28 \div 7 = 4$, the _____ is 28.
[1.6a]

82. In the sentence $10 \times 1000 = 10{,}000$, 10 and 1000 are called _____ and 10,000 is called the _____ .
[1.5a]

83. The _____ is the number from which another number is being subtracted.   [1.3a]

84. The sentence $3 \times (6 \times 2) = (3 \times 6) \times 2$ illustrates the _____ law of multiplication.   [1.5a]

associative	minuend
commutative	subtrahend
addends	digits
factors	periods
perimeter	additive
dividend	multiplicative
quotient	equation
product	inequality

85. Complete the following table.

a	b	$a \cdot b$	$a + b$
	68	3672	
84			117
		32	12

86. Find a pair of factors whose product is 36 and:
a) whose sum is 13.
b) whose difference is 0.
c) whose sum is 20.
d) whose difference is 9.

87. A group of 1231 college students is going to take buses for a field trip. Each bus can hold only 42 students. How many buses are needed?

88. ▦ Fill in the missing digits to make the equation true:
$34{,}584{,}132 \div 76\square = 4\square{,}386.$

51

Objectives

a Solve simple equations by trial.

b Solve equations like
$x + 28 = 54$,
$28 \cdot x = 168$, and
$98 \cdot 2 = y$.

Find a number that makes the sentence true.

1. $8 = 1 + \square$

2. $\square + 2 = 7$

3. Determine whether 7 is a solution of $\square + 5 = 9$.

4. Determine whether 4 is a solution of $\square + 5 = 9$.

Answers on page A-2

1.7 SOLVING EQUATIONS

a Solutions by Trial

Let's find a number that we can put in the blank to make this sentence true:

$$9 = 3 + \square.$$

We are asking "9 is 3 plus what number?" The answer is 6.

$$9 = 3 + \;6$$

Do Exercises 1 and 2.

A sentence with = is called an **equation.** A **solution** of an equation is a number that makes the sentence true. Thus, 6 is a solution of

$$9 = 3 + \square \quad \text{because} \quad 9 = 3 + \;6 \;\text{ is true.}$$

However, 7 is not a solution of

$$9 = 3 + \square \quad \text{because} \quad 9 = 3 + \;7 \;\text{ is false.}$$

Do Exercises 3 and 4.

We can use a letter instead of a blank. For example,

$$9 = 3 + n.$$

We call n a **variable** because it can represent any number. If a replacement for a variable makes an equation true, it is a **solution** of the equation.

SOLUTIONS OF AN EQUATION

A **solution** is a replacement for the variable that makes the equation true. When we find all the solutions, we say that we have **solved** the equation.

EXAMPLE 1 Solve $y + 12 = 27$ by trial.

We replace y with several numbers.

If we replace y with 13, we get a false equation: $13 + 12 = 27$.
If we replace y with 14, we get a false equation: $14 + 12 = 27$.
If we replace y with 15, we get a true equation: $15 + 12 = 27$.

No other replacement makes the equation true, so the solution is 15.

EXAMPLES Solve.

2. $7 + n = 22$
(7 plus what number is 22?)
The solution is 15.

3. $63 = 3 \cdot x$
(63 is 3 times what number?)
The solution is 21.

Do Exercises 5–8 on the following page.

52

CHAPTER 1: Whole Numbers

b Solving Equations

We now begin to develop more efficient ways to solve certain equations. When an equation has a variable alone on one side, it is easy to see the solution or to compute it. When a calculation is on one side and the variable is alone on the other, we can find the solution by carrying out the calculation.

EXAMPLE 4 Solve: $x = 245 \times 34$.

To solve the equation, we carry out the calculation.

$$
\begin{array}{r}
2\ 4\ 5 \\
\times\ \ \ 3\ 4 \\
\hline
9\ 8\ 0 \\
7\ 3\ 5\ 0 \\
\hline
8\ 3\ 3\ 0
\end{array}
\qquad
\begin{aligned}
x &= 245 \times 34 \\
x &= 8330
\end{aligned}
$$

The solution is 8330.

Do Exercises 9–12.

Look at the equation

$$x + 12 = 27.$$

We can get x alone by writing a related subtraction sentence:

$x = 27 - 12$ 12 gets subtracted to find the related subtraction sentence.

$x = 15.$ Doing the subtraction

It is useful in our later study of algebra to think of this as "subtracting 12 *on both sides.*" Thus

$x + 12 - 12 = 27 - 12$ Subtracting 12 on both sides

$x + 0 = 15$ Carrying out the subtraction

$x = 15.$

SOLVING $x + a = b$

To solve $x + a = b$, subtract a on both sides.

If we can get an equation in a form with the variable alone on one side, we can "see" the solution.

EXAMPLE 5 Solve: $t + 28 = 54$.

We have

$$
\begin{aligned}
t + 28 &= 54 \\
t + 28 - 28 &= 54 - 28 \qquad \text{Subtracting 28 on both sides} \\
t + 0 &= 26 \\
t &= 26.
\end{aligned}
$$

Solve by trial.

5. $n + 3 = 8$

6. $x - 2 = 8$

7. $45 \div 9 = y$

8. $10 + t = 32$

Solve.

9. $346 \times 65 = y$

10. $x = 2347 + 6675$

11. $4560 \div 8 = t$

12. $x = 6007 - 2346$

Answers on page A-2

Solve. Be sure to check.

13. $x + 9 = 17$

To check the answer, we substitute 26 for t in the original equation.

Check:
$$t + 28 = 54$$
$$\overline{26 + 28 \overset{?}{=} 54}$$
$$54 \mid \quad \text{TRUE}$$

The solution is 26.

Do Exercises 13 and 14.

14. $77 = m + 32$

EXAMPLE 6   Solve: $182 = 65 + n$.

We have

$$182 = 65 + n$$
$$182 - 65 = 65 + n - 65 \qquad \text{Subtracting 65 on both sides}$$
$$117 = 0 + n \qquad\qquad 65 \text{ plus } n \text{ minus 65 is } 0 + n.$$
$$117 = n.$$

Check:
$$182 = 65 + n$$
$$\overline{182 \overset{?}{=} 65 + 117}$$
$$\mid 182 \qquad \text{TRUE}$$

15. Solve: $155 = t + 78$. Be sure to check.

The solution is 117.

Do Exercise 15.

EXAMPLE 7   Solve: $7381 + x = 8067$.

We have

$$7381 + x = 8067$$
$$7381 + x - 7381 = 8067 - 7381 \qquad \text{Subtracting 7381 on both sides}$$
$$x = 686.$$

Solve. Be sure to check.

16. $4566 + x = 7877$

The check is left to the student. The solution is 686.

Do Exercises 16 and 17.

We now learn to solve equations like $8 \cdot n = 96$. Look at

$$8 \cdot n = 96.$$

17. $8172 = h + 2058$

We can get n alone by writing a related division sentence:

$$n = 96 \div 8 = \frac{96}{8} \qquad 96 \text{ is divided by 8.}$$
$$n = 12. \qquad\qquad \text{Doing the division}$$

It is useful in our later study of algebra to think of the preceding as "dividing by 8 *on both sides.*" Thus,

$$\frac{8 \cdot n}{8} = \frac{96}{8} \qquad \text{Dividing by 8 on both sides}$$
$$n = 12. \qquad 8 \text{ times } n \text{ divided by 8 is } n.$$

Answers on page A-2

SOLVING $a \cdot x = b$
To solve $a \cdot x = b$, divide by a on both sides.

EXAMPLE 8   Solve: $10 \cdot x = 240$.

We have

$$10 \cdot x = 240$$

$$\frac{10 \cdot x}{10} = \frac{240}{10} \qquad \text{Dividing by 10 on both sides}$$

$$x = 24.$$

Check: $$\frac{10 \cdot x = 240}{10 \cdot 24 \ ? \ 240}$$
$$\qquad\qquad 240 \ \Big| \qquad \text{TRUE}$$

The solution is 24.

EXAMPLE 9   Solve: $5202 = 9 \cdot t$.

We have

$$5202 = 9 \cdot t$$

$$\frac{5202}{9} = \frac{9 \cdot t}{9} \qquad \text{Dividing by 9 on both sides}$$

$$578 = t.$$

The check is left to the student. The solution is 578.

Do Exercises 18–20.

EXAMPLE 10   Solve: $14 \cdot y = 1092$.

We have

$$14 \cdot y = 1092$$

$$\frac{14 \cdot y}{14} = \frac{1092}{14} \qquad \text{Dividing by 14 on both sides}$$

$$y = 78.$$

The check is left to the student. The solution is 78.

EXAMPLE 11   Solve: $n \cdot 56 = 4648$.

We have

$$n \cdot 56 = 4648$$

$$\frac{n \cdot 56}{56} = \frac{4648}{56} \qquad \text{Dividing by 56 on both sides}$$

$$n = 83.$$

The check is left to the student. The solution is 83.

Do Exercises 21 and 22.

Solve. Be sure to check.

18. $8 \cdot x = 64$

19. $144 = 9 \cdot n$

20. $5152 = 8 \cdot t$

Solve. Be sure to check.

21. $18 \cdot y = 1728$

22. $n \cdot 48 = 4512$

Answers on page A-2

a   Solve by trial.

1. $x + 0 = 14$

2. $x - 7 = 18$

3. $y \cdot 17 = 0$

4. $56 \div m = 7$

b   Solve. Be sure to check.

5. $13 + x = 42$

6. $15 + t = 22$

7. $12 = 12 + m$

8. $16 = t + 16$

9. $3 \cdot x = 24$

10. $6 \cdot x = 42$

11. $112 = n \cdot 8$

12. $162 = 9 \cdot m$

13. $45 \cdot 23 = x$

14. $23 \cdot 78 = y$

15. $t = 125 \div 5$

16. $w = 256 \div 16$

17. $p = 908 - 458$

18. $9007 - 5667 = m$

19. $x = 12{,}345 + 78{,}555$

20. $5678 + 9034 = t$

21. $3 \cdot m = 96$

22. $4 \cdot y = 96$

23. $715 = 5 \cdot z$

24. $741 = 3 \cdot t$

25. $10 + x = 89$

26. $20 + x = 57$

27. $61 = 16 + y$

28. $53 = 17 + w$

29. $6 \cdot p = 1944$

30. $4 \cdot w = 3404$

31. $5 \cdot x = 3715$

32. $9 \cdot x = 1269$

33. $47 + n = 84$

34. $56 + p = 92$

35. $x + 78 = 144$

36. $z + 67 = 133$

37. $165 = 11 \cdot n$

38. $660 = 12 \cdot n$

39. $624 = t \cdot 13$

40. $784 = y \cdot 16$

41. $x + 214 = 389$

42. $x + 221 = 333$

43. $567 + x = 902$

44. $438 + x = 807$

45. $18 \cdot x = 1872$

46. $19 \cdot x = 6080$

47. $40 \cdot x = 1800$

48. $20 \cdot x = 1500$

49. $2344 + y = 6400$

50. $9281 = 8322 + t$

51. $8322 + 9281 = x$

52. $9281 - 8322 = y$

53. $234 \cdot 78 = y$

54. $10{,}534 \div 458 = q$

55. $58 \cdot m = 11{,}890$

56. $233 \cdot x = 22{,}135$

57. $\mathbf{D_W}$ Describe a procedure that can be used to convert any equation of the form $a \cdot b = c$ to a related division equation.

58. $\mathbf{D_W}$ Describe a procedure that can be used to convert any equation of the form $a + b = c$ to a related subtraction equation.

SKILL MAINTENANCE

59. Write two related subtraction sentences: $7 + 8 = 15$. [1.3a]

60. Write two related division sentences: $6 \cdot 8 = 48$. [1.6a]

Use $>$ or $<$ for $\square$ to write a true sentence. [1.4c]

61. $123 \; \square \; 789$

62. $342 \; \square \; 339$

63. $688 \; \square \; 0$

64. $0 \; \square \; 11$

Divide. [1.6b]

65. $1283 \div 9$

66. $1278 \div 9$

67. $1\,7 \;) \overline{5\,6\,7\,8}$

68. $1\,7 \;) \overline{5\,6\,8\,9}$

SYNTHESIS

Solve.

69. ▦ $23{,}465 \cdot x = 8{,}142{,}355$

70. ▦ $48{,}916 \cdot x = 14{,}332{,}388$

Objective

a Solve applied problems involving addition, subtraction, multiplication, or division of whole numbers.

a A Problem-Solving Strategy

Applications and problem solving are the most important uses of mathematics. To solve a problem using the operations on the whole numbers, we first look at the situation. We try to translate the problem to an equation. Then we solve the equation. We check to see if the solution of the equation is a solution of the original problem. We are using the following five-step strategy.

FIVE STEPS FOR PROBLEM SOLVING

1. *Familiarize* yourself with the situation.

   a) Carefully read and reread until you understand *what* you are being asked to find.
   b) Draw a diagram or see if there is a formula that applies to the situation.
   c) Assign a letter, or *variable,* to the unknown.

2. *Translate* the problem to an equation using the letter or variable.
3. *Solve* the equation.
4. *Check* the answer in the original wording of the problem.
5. *State* the answer to the problem clearly with appropriate units.

EXAMPLE 1 *International Adoptions.* The top ten countries of origin for United States international adoptions in 2003 and 2002 are listed in the table below. Find the total number of adoptions from China, South Korea, India, and Vietnam in 2003.

INTERNATIONAL ADOPTIONS

RANK	2003 COUNTRY OF ORIGIN	NUMBER	2002 COUNTRY OF ORIGIN	NUMBER
1	China (mainland)	6859	China (mainland)	5053
2	Russia	5209	Russia	4939
3	Guatemala	2328	Guatemala	2219
4	South Korea	1790	South Korea	1779
5	Kazakhstan	825	Ukraine	1106
6	Ukraine	702	Kazakhstan	819
7	India	472	Vietnam	766
8	Vietnam	382	India	466
9	Colombia	272	Colombia	334
10	Haiti	250	Bulgaria	260

Source: U.S. Department of State

1. Familiarize. We can make a drawing or at least visualize the situation.

6859	+	1790	+	472	+	382
from		from		from		from
China		South Korea		India		Vietnam

Since we are combining numbers of adoptions, addition can be used. First, we define the unknown. We let $n =$ the total number of adoptions from China, South Korea, India, and Vietnam.

2. Translate. We translate to an equation:

$$6859 + 1790 + 472 + 382 = n.$$

3. Solve. We solve the equation by carrying out the addition.

$$
\begin{array}{r}
{\scriptstyle 2\;3\;1} \\
6\,8\,5\,9 \\
1\,7\,9\,0 \\
4\,7\,2 \\
+\quad 3\,8\,2 \\
\hline
9\,5\,0\,3
\end{array}
$$

$$6859 + 1790 + 472 + 382 = n$$
$$9503 = n$$

4. Check. We check 9503 in the original problem. There are many ways in which this can be done. For example, we can repeat the calculation. (We leave this to the student.) Another way is to check whether the answer is reasonable. In this case, we would expect the total to be greater than the number of adoptions from any of the individual countries, which it is. We can also estimate by rounding. Here we round to the nearest hundred.

$$6859 + 1790 + 472 + 382 \approx 6900 + 1800 + 500 + 400$$
$$= 9600$$

Since $9600 \approx 9503$, we have a partial check. If we had an estimate like 4800 or 7500, we might be suspicious that our calculated answer is incorrect. Since our estimated answer is close to our calculation, we are further convinced that our answer checks.

5. State. The total number of adoptions from China, South Korea, India, and Vietnam in 2003 is 9503.

Do Exercises 1–3.

Refer to the table on the preceding page to answer Margin Exercises 1–3.

1. Find the total number of adoptions from China, Russia, Kazakhstan, and India in 2002.

2. Find the total number of adoptions from Russia, Guatemala, Vietnam, Colombia, and Haiti in 2003.

3. Find the total number of adoptions from the top ten countries in 2003.

Answers on page A-3

1.8 Applications and Problem Solving

4. Checking Account Balance.
The balance in Heidi's checking account is $2003. She uses her debit card to buy the same Roto Zip Spiral Saw Combo, featured in Example 2, that Tyler did. Find the new balance in her checking account.

EXAMPLE 2 *Checking Account Balance.* The balance in Tyler's checking account is $528. He uses his debit card to buy the Roto Zip Spiral Saw Combo shown in this ad. Find the new balance in his checking account.

NOW $129 00

Source: Reproduced with permission from the copyright owner (©2004 by Robert Bosch Tool Corporation). Further reproductions strictly prohibited.

1. **Familiarize.** We first make a drawing or at least visualize the situation. We let M = the new balance in his account. This gives us the following:

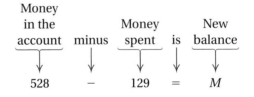

Take away
$129
$528   New balance

2. **Translate.** We can think of this as a "take-away" situation. We translate to an equation.

$$528 - 129 = M$$

3. **Solve.** This sentence tells us what to do. We subtract.

$$
\begin{array}{r}
{\scriptstyle 11} \\
{\scriptstyle 4\ \not{1}\ 18} \\
\not{5}\ \not{2}\ \not{8} \\
-\ 1\ 2\ 9 \\
\hline
3\ 9\ 9
\end{array}
$$

$$528 - 129 = M$$
$$399 = M$$

4. **Check.** To check our answer of $399, we can repeat the calculation. We note that the answer should be less than the original amount, $528, which it is. We can add the difference, 399, to the subtrahend, 129: $129 + 399 = 528$. We can also estimate:

$$528 - 129 \approx 530 - 130 = 400 \approx 399.$$

5. **State.** Tyler has a new balance of $399 in his checking account.

Answer on page A-3

Do Exercise 4.

In the real world, problems may not be stated in written words. You must still become familiar with the situation before you can solve the problem.

EXAMPLE 3   *Travel Distance.*   Abigail is driving from Indianapolis to Salt Lake City to interview for a news anchor position. The distance from Indianapolis to Salt Lake City is 1634 mi. She travels 1154 mi to Denver. How much farther must she travel?

1. **Familiarize.**   We first make a drawing or at least visualize the situation. We let x = the remaining distance to Salt Lake City.

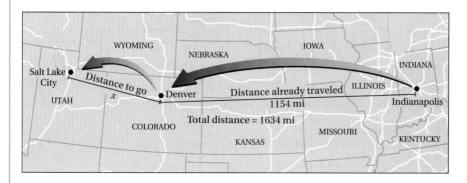

2. **Translate.**   We see that this is a "how much more" situation. We translate to an equation.

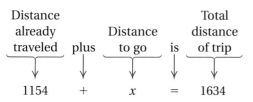

Distance already traveled plus Distance to go is Total distance of trip

$$1154 + x = 1634$$

3. **Solve.**   To solve the equation, we subtract 1154 on both sides:

$$1154 + x = 1634$$
$$1154 + x - 1154 = 1634 - 1154$$
$$x = 480.$$

$$\begin{array}{r} {\scriptstyle 5\ 13} \\ 1\,\cancel{6}\,\cancel{3}\,4 \\ -\ 1\ 1\ 5\ 4 \\ \hline 4\ 8\ 0 \end{array}$$

4. **Check.**   We check our answer of 480 mi in the original problem. This number should be less than the total distance, 1634 mi, which it is. We can add the difference, 480, to the subtrahend, 1154: $1154 + 480 = 1634$. We can also estimate:

$$1634 - 1154 \approx 1600 - 1200$$
$$= 400 \approx 480.$$

The answer, 480 mi, checks.

5. **State.**   Abigail must travel 480 mi farther to Salt Lake City.

Do Exercise 5.

5. **Home Vacuum.**   William has $228. He wants to purchase the home vacuum shown in the ad below. How much more does he need?

Answer on page A-3

6. Total Cost of Gas Grills. What is the total cost of 14 gas grills, each with 520 sq in. of total cooking surface and porcelain cast-iron cooking grates, if each one costs $398?

EXAMPLE 4 *Total Cost of Chairs.* What is the total cost of 6 Logan side chairs from Restoration Hardware if each one costs $210?

1. **Familiarize.** We first make a drawing or at least visualize the situation. We let $T =$ the cost of 6 chairs. Repeated addition works well in this case.

$210 $210 $210 $210 $210 $210

2. **Translate.** We translate to an equation.

Number of chairs	times	Cost of each chair	is	Total cost
↓	↓	↓	↓	↓
6	×	$210	=	T

3. **Solve.** This sentence tells us what to do. We multiply.

$$\begin{array}{r} 2\ 1\ 0 \\ \times\quad 6 \\ \hline 1\ 2\ 6\ 0 \end{array}$$

$$6 \times 210 = T$$
$$1260 = T$$

4. **Check.** We have an answer, 1260, that is much greater than the cost of any individual chair, which is reasonable. We can repeat our calculation. We can also check by estimating:

$$6 \times 210 \approx 6 \times 200 = 1200 \approx 1260.$$

The answer checks.

5. **State.** The total cost of 6 chairs is $1260.

Do Exercise 6.

EXAMPLE 5 *Truck Bed Cover.* The dimensions of a fiberglass truck bed cover for a pickup truck are 79 in. by 68 in. What is the area of the cover?

79 in.

68 in.

Answer on page A-3

1. **Familiarize.** The truck bed cover is a rectangle that measures 79 in. by 68 in. We let A = the area and use the area formula $A = l \cdot w$.

2. **Translate.** Using this formula, we have

$$A = \text{length} \cdot \text{width} = l \cdot w = 79 \cdot 68.$$

3. **Solve.** We carry out the multiplication.

$$
\begin{array}{r}
7\ 9 \\
\times\ 6\ 8 \\
\hline
6\ 3\ 2 \\
4\ 7\ 4\ 0 \\
\hline
5\ 3\ 7\ 2
\end{array}
\qquad
\begin{aligned}
A &= 79 \cdot 68 \\
A &= 5372
\end{aligned}
$$

4. **Check.** We repeat our calculation. We also note that the answer is greater than either the length or the width, which it should be. (This might not be the case if we were using fractions or decimals.) The answer checks.

5. **State.** The area of the truck bed cover is 5372 sq in.

Do Exercise 7.

EXAMPLE 6 *Cartons of Soda.* A bottling company produces 3304 cans of soda. How many 12-can cartons can be filled? How many cans will be left over?

1. **Familiarize.** We first make a drawing. We let n = the number of 12-can cartons that can be filled. The problem can be considered as repeated subtraction, taking successive sets of 12 cans and putting them into n cartons.

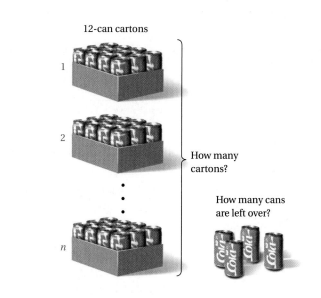

12-can cartons

1

2

n

How many cartons?

How many cans are left over?

2. **Translate.** We translate to an equation.

Number of cans	divided by	Number in each carton	is	Number of cartons
↓	↓	↓	↓	↓
3304	÷	12	=	n

Answer on page A-3

8. Cartons of Soda. The bottling company in Example 6 also uses 6-can cartons. How many 6-can cartons can be filled with 2269 cans of cola? How many will be left over?

3. Solve. We solve the equation by carrying out the division.

$$
\begin{array}{r}
2\ 7\ 5 \\
1\ 2\ )\overline{3\ 3\ 0\ 4} \\
2\ 4\ 0\ 0 \\
\hline
9\ 0\ 4 \\
8\ 4\ 0 \\
\hline
6\ 4 \\
6\ 0 \\
\hline
4
\end{array}
$$

$3304 \div 12 = n$

$275\ \text{R}\ 4 = n$

4. Check. We can check by multiplying the number of cartons by 12 and adding the remainder, 4:

$12 \cdot 275 = 3300,$

$3300 + 4 = 3304.$

5. State. Thus, 275 twelve-can cartons can be filled. There will be 4 cans left over.

Do Exercise 8.

EXAMPLE 7 *Automobile Mileage.* The Pontiac G6 GT gets 21 miles to the gallon (mpg) in city driving. How many gallons will it use in 3843 mi of city driving?

Source: General Motors

1. Familiarize. We first make a drawing. It is often helpful to be descriptive about how we define a variable. In this case, we let g = the number of gallons ("g" comes from "gallons").

3843 mi to drive

2. Translate. Repeated addition applies here. Thus the following multiplication applies to the situation.

Number of miles per gallon	times	Number of gallons needed	is	Number of miles to drive
21	·	g	=	3843

3. Solve. To solve the equation, we divide by 21 on both sides.

$21 \cdot g = 3843$

$\dfrac{21 \cdot g}{21} = \dfrac{3843}{21}$

$g = 183$

$$
\begin{array}{r}
1\ 8\ 3 \\
2\ 1\ )\overline{3\ 8\ 4\ 3} \\
2\ 1\ 0\ 0 \\
\hline
1\ 7\ 4\ 3 \\
1\ 6\ 8\ 0 \\
\hline
6\ 3 \\
6\ 3 \\
\hline
0
\end{array}
$$

Answer on page A-3

4. Check. To check, we multiply 183 by 21: $21 \times 183 = 3843$.

5. State. The Pontiac G6 GT will use 183 gal.

Do Exercise 9.

Multistep Problems

Sometimes we must use more than one operation to solve a problem, as in the following example.

EXAMPLE 8 *Aircraft Seating.* Boeing Corporation builds commercial aircraft. A Boeing 767 has a seating configuration with 4 rows of 6 seats across in first class and 35 rows of 7 seats across in economy class. Find the total seating capacity of the plane.

Sources: The Boeing Company; Delta Airlines

1. Familiarize. We first make a drawing.

Economy class: 35 rows of 7 seats

First class: 4 rows of 6 seats

2. Translate. There are three parts to the problem. We first find the number of seats in each class. Then we add.

First-class: Repeated addition applies here. Thus the following multiplication corresponds to the situation. We let F = the number of seats in first class.

Number of rows	times	Seats in each row	is	Total number
4	$\cdot$	6	=	F

Economy class: Repeated addition applies here. Thus the following multiplication corresponds to the situation. We let E = the number of seats in economy class.

Number of rows	times	Seats in each row	is	Total number
35	$\cdot$	7	=	E

We let T = the total number of seats in both classes.

Number of seats in first class	plus	Number of seats in economy class	is	Total number of seats on plane
F	+	E	=	T

9. Automobile Mileage. The Pontiac G6 GT gets 29 miles to the gallon (mpg) in highway driving. How many gallons will it take to drive 2291 mi of highway driving?

Source: General Motors

Answer on page A-3

10. Aircraft Seating. A Boeing 767 used for foreign travel has three classes of seats. First class has 3 rows of 5 seats across; business class has 6 rows with 6 seats across and 1 row with 2 seats on each of the outside aisles. Economy class has 18 rows with 7 seats across. Find the total seating capacity of the plane.

Sources: The Boeing Company; Delta Airlines

Economy class: 18 rows of 7 seats   First class: 3 rows of 5 seats

Business class: 6 rows of 6 seats...   ...with 2 seats on each outside aisle

3. Solve. We solve each equation and add the solutions.

$$4 \cdot 6 = F \qquad 35 \cdot 7 = E \qquad F + E = T$$
$$24 = F \qquad 245 = E \qquad 24 + 245 = T$$
$$269 = T$$

4. Check. To check, we repeat our calculations. (We leave this to the student.) We could also check by rounding, multiplying, and adding.

5. State. There are 269 seats in a Boeing 767.

Do Exercise 10.

As you consider the following exercises, here are some words and phrases that may be helpful to look for when you are translating problems to equations.

KEY WORDS, PHRASES, AND CONCEPTS

ADDITION (+)	SUBTRACTION (−)
add	subtract
added to	subtracted from
sum	difference
total	minus
plus	less than
more than	decreased by
increased by	take away
	how much more
	missing addend

MULTIPLICATION (·)	DIVISION (÷)
multiply	divide
multiplied by	divided by
product	quotient
times	repeated subtraction
of	missing factor
repeated addition	finding equal quantities
rectangular arrays	

Answer on page A-3

Translating for Success

1. *Brick-mason Expense.* A commercial contractor is building 30 two-unit condominiums in a retirement community. The brick-mason expense for each building is $10,860. What is the total cost of bricking the buildings?

2. *Heights.* Dean's sons are on the high school basketball team. Their heights are 73 in., 69 in., and 76 in. How much taller is the tallest son than the shortest son?

3. *Account Balance.* You have $423 in your checking account. Then you deposit $73 and use your debit card for purchases of $76 and $69. How much is left in your account?

4. *Purchasing Camcorder.* A camcorder is on sale for $423. Jenny has only $69. How much more does she need to buy the camcorder?

5. *Purchasing Coffee Makers.* Sara purchases 8 coffee makers for the newly remodeled bed-and-breakfast hotel that she manages. If she pays $52 for each coffee maker, what is the total cost of her purchase?

The goal of these matching questions is to practice step (2), *Translate*, of the five-step problem-solving process. Translate each word problem to an equation and select a correct translation from equations A–O.

A. $8 \cdot 52 = n$

B. $69 \cdot n = 76$

C. $73 - 76 - 69 = n$

D. $423 + 73 - 76 - 69 = n$

E. $30 \cdot 10,860 = n$

F. $15 \cdot n = 195$

G. $69 + n = 423$

H. $n = 10,860 - 300$

I. $n = 423 \div 69$

J. $30 \cdot n = 10,860$

K. $15 \cdot 195 = n$

L. $n = 52 - 8$

M. $69 + n = 76$

N. $15 \div 195 = n$

O. $52 + n = 60$

Answers on page A-3

6. *Hourly Rate.* Miller Auto Repair charges $52 an hour for labor. Jackson Auto Care charges $60 per hour. How much more does Jackson charge than Miller?

7. *College Band.* A college band with 195 members marches in a 15-row formation in the homecoming halftime performance. How many members are in each row?

8. *Shoe Purchase.* A professional football team purchases 15 pairs of shoes at $195 a pair. What is the total cost of this purchase?

9. *Loan Payment.* Kendra borrows $10,860 for a new boat. The loan is to be paid off in 30 payments. How much is each payment?

10. *College Enrollment.* At the beginning of the fall term, the total enrollment in Lakeview Community College was 10,860. By the end of the first two weeks, 300 students withdrew. How many students were then enrolled?

a Solve.

Longest Broadway Run. The bar graph below lists the five broadway shows with the greatest number of performances as of May 3, 2004.

Source: League of American Theatres and Producers, Inc.

1. What was the total number of performances of all five shows?

2. What was the total number of performances of the top three shows?

3. How many more performances of *Cats* were there than performances of *The Phantom of the Opera*?

4. How many more performances of *Les Misérables* were there than performances of *Oh! Calcutta!*?

5. *Boundaries Between Countries.* The boundary between mainland United States and Canada including the Great Lakes is 3987 miles long. The length of the boundary between the United States and Mexico is 1933 miles. How much longer is the Canadian border?
Source: U.S. Geological Survey

6. *Caffeine.* Hershey's 6-oz milk chocolate almond bar contains 25 milligrams of caffeine. A 20-oz bottle of Coca-Cola has 32 more milligrams of caffeine than the Hershey bar. How many milligrams of caffeine does the 20-oz bottle of Coca-Cola have?
Source: *National Geographic,* "Caffeine," by T. R. Reid, January 2005

7. A carpenter drills 216 holes in a rectangular array in a pegboard. There are 12 holes in each row. How many rows are there?

8. Lou works as a CPA. He arranges 504 entries on a spreadsheet in a rectangular array that has 36 rows. How many entries are in each row?

Bachelor's Degree. The line graph below illustrates data about bachelor's degrees awarded to men and women from 1970 to 2002. Use this graph when answering Exercises 9–12.

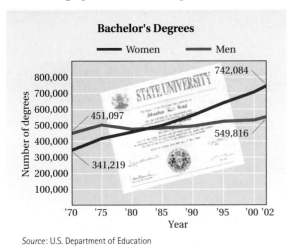

Bachelor's Degrees

— Women — Men

742,084
451,097
549,816
341,219

Number of degrees: 800,000 700,000 600,000 500,000 400,000 300,000 200,000 100,000

Year: '70 '75 '80 '85 '90 '95 '00 '02

Source: U.S. Department of Education

9. Find the total number of bachelor's degrees awarded in 1970 and the total number awarded in 2002.

10. Determine how many more bachelor's degrees were awarded in 2002 than in 1970.

11. How many more bachelor's degrees were awarded to women than to men in 2002?

12. How many more bachelor's degrees were awarded to men than to women in 1970?

13. *Median Mortgage Debt.* The median mortgage debt in 2001 was $29,475 more than the median mortgage debt in 1989. The debt in 1989 was $39,802. What was the median mortgage debt in 2001?

Source: Federal Reserve Board Survey of Consumer Finances; Consumer Bank Association 2003 Home Equity Study; Freddie Mac

14. *Olympics in Athens.* In the first modern Olympics in Athens, Greece, in 1896, there were 43 events. In 2004, there were 258 more events in the Summer Olympics in Athens. How many events were there in 2004?

Source: *USA Today* research; The Olympic Games: Athens 1896–Athens 2004

15. *Longest Rivers.* The longest river in the world is the Nile in Egypt at 4100 mi. The longest river in the United States is the Missouri–Mississippi at 3860 mi. How much longer is the Nile?

16. *Speeds on Interstates.* Recently, speed limits on interstate highways in many Western states were raised from 65 mph to 75 mph. By how many miles per hour were they raised?

17. There are 24 hours (hr) in a day and 7 days in a week. How many hours are there in a week?

18. There are 60 min in an hour and 24 hr in a day. How many minutes are there in a day?

19. *Crossword.* The *USA Today* crossword puzzle is a rectangle containing 15 rows with 15 squares in each row. How many squares does the puzzle have altogether?

20. *Pixels.* A computer screen consists of small rectangular dots called *pixels.* How many pixels are there on a screen that has 600 rows with 800 pixels in each row?

Pixel

21. *Refrigerator Purchase.* Gourmet Deli has a chain of 24 restaurants. It buys a commercial refrigerator for each store at a cost of $1019 each. Find the total cost of the purchase.

22. *Microwave Purchase.* Bridgeway College is constructing new dorms, in which each room has a small kitchen. It buys 96 microwave ovens at $88 each. Find the total cost of the purchase.

23. *"Seinfeld" Episodes.* "Seinfeld" was a long-running television comedy with 177 episodes. A local station picks up the syndicated reruns. If the station runs 5 episodes per week, how many full weeks will pass before it must start over with past episodes? How many episodes will be left for the last week?

24. A lab technician separates a vial containing 70 cubic centimeters (cc) of blood into test tubes, each of which contains 3 cc of blood. How many test tubes can be filled? How much blood is left over?

25. *Automobile Mileage.* The 2005 Hyundai Tucson GLS gets 26 miles to the gallon (mpg) in highway driving. How many gallons will it use in 6136 mi of highway driving?

Source: Hyundai

26. *Automobile Mileage.* The 2005 Volkswagen Jetta (5 cylinder) gets 24 miles to the gallon (mpg) in city driving. How many gallons will it use in 3960 mi of city driving?

Source: Volkswagen of America, Inc.

Boeing Jets. Use the chart below in Exercises 27–32 to compare the Boeing 747 jet with its main competitor, the Boeing 777 jet.

		232 ft

BOEING 747–400

Passenger capacity	416
Nonstop flight distance	8826 miles
Cruising speed	567 mph
Gallons of fuel used per hour	3201

Costs (to fly one hour):	
Crew	$1948
Fuel	$2867

BOEING 777–200

Passenger capacity	368
Nonstop flight distance	5210 miles
Cruising speed	615 mph
Gallons of fuel used per hour	2021

Costs (to fly one hour):	
Crew	$1131
Fuel	$1816

Source: Éclat Consulting; Boeing

27. The nonstop flight distance of the Boeing 747 jet is how much greater than the nonstop flight distance of the Boeing 777 jet?

28. How much larger is the passenger capacity of the Boeing 747 jet than the passenger capacity of the Boeing 777 jet?

29. How many gallons of fuel are needed for a 4-hour flight of the Boeing 747?

30. How much longer is the Boeing 747 than the Boeing 777?

31. What is the total cost for the crew and fuel for a 3-hour flight of the Boeing 747 jet?

32. What is the total cost for the crew and fuel for a 2-hour flight of the Boeing 777 jet?

33. Dana borrows $5928 for a used car. The loan is to be paid off in 24 equal monthly payments. How much is each payment (excluding interest)?

34. A family borrows $7824 to build a sunroom on the back of their home. The loan is to be paid off in equal monthly payments of $163 (excluding interest). How many months will it take to pay off the loan?

35. *High School Court.* The standard basketball court used by high school players has dimensions of 50 ft by 84 ft.
  a) What is its area?
  b) What is its perimeter?

36. *NBA Court.* The standard basketball court used by college and NBA players has dimensions of 50 ft by 94 ft.
  a) What is its area?
  b) What is its perimeter?
  c) How much greater is the area of an NBA court than a high school court? (See Exercise 35.)

37. *Clothing Imports and Exports.* In the United States, the exports of clothing in 2003 totaled $2,596,000,000 while the imports totaled $31,701,000,000. How much more were the imports than the exports?

Source: U.S. Census Bureau, Foreign Trade Division

38. *Corn Imports and Exports.* In the United States, the exports of corn in 2003 totaled $2,264,000,000 while the imports totaled $130,000,000. How much more were the exports than the imports?

Source: U.S. Census Bureau, Foreign Trade Division

39. *Colonial Population.* Before the establishment of the U.S. Census in 1790, it was estimated that the Colonial population in 1780 was 2,780,400. This was an increase of 2,628,900 from the population in 1680. What was the Colonial population in 1680?

Source: *Time Almanac,* 2005

40. *Deaths by Firearms.* Deaths by firearms totaled 29,573 in 2001. This was a decrease of 10,022 from the deaths by firearms in 1993. How many deaths by firearms were there in 1993?

Source: Center for Disease Control and Prevention, *National Vital Statistics Reports,* vol 52, no. 3, September 18, 2003

41. *Hershey Bars.* Hershey Chocolate USA makes small, fun-size chocolate bars. How many 20-bar packages can be filled with 11,267 bars? How many bars will be left over?

42. *Reese's Peanut Butter Cups.* H. B. Reese Candy Co. makes small, fun-size peanut butter cups. The company manufactures 23,579 cups and fills 1025 packages. How many cups are in a package? How many cups will be left over?

43. *Map Drawing.* A map has a scale of 64 mi to the inch. How far apart *in reality* are two cities that are 6 in. apart on the map? How far apart *on the map* are two cities that, in reality, are 1728 mi apart?

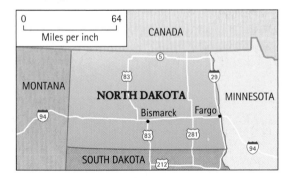

44. *Map Drawing.* A map has a scale of 150 mi to the inch. How far apart *on the map* are two cities that, in reality, are 2400 mi apart? How far apart *in reality* are two cities that are 13 in. apart on the map?

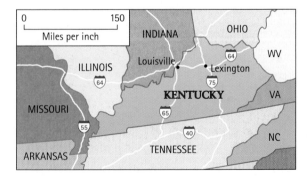

45. *Crossword.* The *Los Angeles Times* crossword puzzle is a rectangle containing 441 squares arranged in 21 rows. How many columns does the puzzle have?

46. *Sheet of Stamps.* A sheet of 100 stamps typically has 10 rows of stamps. How many stamps are in each row?

47. Copies of this book are generally shipped from the warehouse in cartons containing 24 books each. How many cartons are needed to ship 1355 books?

48. According to the H. J. Heinz Company, 16-oz bottles of catsup are generally shipped in cartons containing 12 bottles each. How many cartons are needed to ship 528 bottles of catsup?

49. Elena buys 5 video games at $64 each and pays for them with $10 bills. How many $10 bills does it take?

50. Pedro buys 5 video games at $64 each and pays for them with $20 bills. How many $20 bills does it take?

51. You have $568 in your bank account. You use your debit card for $46, $87, and $129. Then you deposit $94 back in the account after the return of some books. How much is left in your account?

52. The balance in your bank account is $749. You use your debit card for $34 and $65. Then you make a deposit of $123 from your paycheck. What is your new balance?

Weight Loss. Many Americans exercise for weight control. It is known that one must burn off about 3500 calories in order to lose one pound. The chart shown here details how much of certain types of exercise is required to burn 100 calories. Use this chart for Exercises 53–56.

To burn off 100 calories, you must:

- Run for 8 minutes at a brisk pace, or
- Swim for 2 minutes at a brisk pace, or
- Bicycle for 15 minutes at 9 mph, or
- Do aerobic exercises for 15 minutes.

53. How long must you run at a brisk pace in order to lose one pound?

54. How long must you swim in order to lose one pound?

55. How long must you do aerobic exercises in order to lose one pound?

56. How long must you bicycle at 9 mph in order to lose one pound?

57. *Bones in the Hands and Feet.* There are 27 bones in each human hand and 26 bones in each human foot. How many bones are there in all in the hands and feet?

58. *Subway Travel.* The distance to Mars is about 303,000,000 miles. The number of miles that people in the United States traveled on subways in 2003 approximately equaled 22 round trips to Mars. What was the total distance traveled on subways?

Source: American Public Transportation Association, NASA

59. *Index Cards.* Index cards of dimension 3 in. by 5 in. are normally shipped in packages containing 100 cards each. How much writing area is available if one uses the front and back sides of a package of these cards?

60. An office for adjunct instructors at a community college has 6 bookshelves, each of which is 3 ft wide. The office is moved to a new location that has dimensions of 16 ft by 21 ft. Is it possible for the bookshelves to be put side by side on the 16-ft wall?

61. **D**_W In the newspaper article "When Girls Play, Knees Fail," the author discusses the fact that female athletes have six times the number of knee injuries that male athletes have. What information would be needed if you were to write a math problem based on the article? What might the problem be?

Source: *The Arizona Republic,* 2/9/00, p. C1

62. **D**_W Write a problem for a classmate to solve. Design the problem so that the solution is "The driver still has 329 mi to travel."

SKILL MAINTENANCE

Round 234,562 to the nearest: [1.4a]

63. Hundred.

64. Ten.

65. Thousand.

Estimate the computation by rounding to the nearest thousand. [1.4b]

66. $2783 + 4602 + 5797 + 8111$

67. $28,430 - 11,977$

68. $2100 + 5800$

69. $5800 - 2100$

Estimate the product by rounding to the nearest hundred. [1.5b]

70. $787 \cdot 363$

71. $887 \cdot 799$

72. $10,362 \cdot 4531$

SYNTHESIS

73. 🖩 *Speed of Light.* Light travels about 186,000 miles per second (mi/sec) in a vacuum as in outer space. In ice it travels about 142,000 mi/sec, and in glass it travels about 109,000 mi/sec. In 18 sec, how many more miles will light travel in a vacuum than in ice? than in glass?

74. Carney Community College has 1200 students. Each professor teaches 4 classes and each student takes 5 classes. There are 30 students and 1 teacher in each classroom. How many professors are there at Carney Community College?

CHAPTER 1: Whole Numbers

1.9 EXPONENTIAL NOTATION AND ORDER OF OPERATIONS

Objectives

a Write exponential notation for products such as $4 \cdot 4 \cdot 4$.

b Evaluate exponential notation.

c Simplify expressions using the rules for order of operations.

d Remove parentheses within parentheses.

a Writing Exponential Notation

Consider the product $3 \cdot 3 \cdot 3 \cdot 3$. Such products occur often enough that mathematicians have found it convenient to create a shorter notation, called **exponential notation,** for them. For example,

$\underbrace{3 \cdot 3 \cdot 3 \cdot 3}_{\text{4 factors}}$ is shortened to $3^4 \leftarrow$ exponent

$\uparrow$ base

We read exponential notation as follows.

NOTATION	WORD DESCRIPTION
3^4	"three to the fourth power," or "the fourth power of three"
5^3	"five-cubed," or "the cube of five," or "five to the third power," or "the third power of five"
7^2	"seven squared," or "the square of seven," or "seven to the second power," or "the second power of seven"

The wording "seven squared" for 7^2 comes from the fact that a square with side s has area A given by $A = s^2$.

s

$A = s^2$ s

An expression like $3 \cdot 5^2$ is read "three times five squared," or "three times the square of five."

EXAMPLE 1 Write exponential notation for $10 \cdot 10 \cdot 10 \cdot 10 \cdot 10$.

Exponential notation is 10^5.   5 is the *exponent.*
                                    10 is the *base.*

EXAMPLE 2 Write exponential notation for $2 \cdot 2 \cdot 2$.

Exponential notation is 2^3.

Do Exercises 1–4.

Write exponential notation.

1. $5 \cdot 5 \cdot 5 \cdot 5$

2. $5 \cdot 5 \cdot 5 \cdot 5 \cdot 5$

3. $10 \cdot 10$

4. $10 \cdot 10 \cdot 10 \cdot 10$

Evaluate.

5. 10^4 **6.** 10^2

7. 8^3 **8.** 2^5

Answers on page A-3

Simplify.

9. $93 - 14 \cdot 3$

10. $104 \div 4 + 4$

11. $25 \cdot 26 - (56 + 10)$

12. $75 \div 5 + (83 - 14)$

b Evaluating Exponential Notation

We evaluate exponential notation by rewriting it as a product and computing the product.

EXAMPLE 3 Evaluate: 10^3.

$$10^3 = 10 \cdot 10 \cdot 10 = 1000$$

Caution!

10^3 does not mean $10 \cdot 3$.

EXAMPLE 4 Evaluate: 5^4.

$$5^4 = 5 \cdot 5 \cdot 5 \cdot 5 = 625$$

Do Exercises 5–8 on the preceding page.

c Simplifying Expressions

Suppose we have a calculation like the following:

$$3 + 4 \cdot 8.$$

How do we find the answer? Do we add 3 to 4 and then multiply by 8, or do we multiply 4 by 8 and then add 3? In the first case, the answer is 56. In the second, the answer is 35. We agree to compute as in the second case.

Consider the calculation

$$7 \cdot 14 - (12 + 18).$$

What do the parentheses mean? To deal with these questions, we must make some agreement regarding the order in which we perform operations. The rules are as follows.

RULES FOR ORDER OF OPERATIONS

1. Do all calculations within parentheses ( ), brackets [ ], or braces { } before operations outside.
2. Evaluate all exponential expressions.
3. Do all multiplications and divisions in order from left to right.
4. Do all additions and subtractions in order from left to right.

It is worth noting that these are the rules that computers and most scientific calculators use to do computations.

EXAMPLE 5 Simplify: $16 \div 8 \cdot 2$.

There are no parentheses or exponents, so we start with the third step.

$$\left.\begin{array}{l} 16 \div 8 \cdot 2 = 2 \cdot 2 \\ = 4 \end{array}\right\} \quad \text{Doing all multiplications and divisions in order from left to right}$$

Answers on page A-3

EXAMPLE 6 Simplify: $7 \cdot 14 - (12 + 18)$.

$$7 \cdot 14 - (12 + 18) = 7 \cdot 14 - 30 \qquad \text{Carrying out operations inside parentheses}$$
$$= 98 - 30 \qquad \text{Doing all multiplications and divisions}$$
$$= 68 \qquad \text{Doing all additions and subtractions}$$

Do Exercises 9–12 on the preceding page.

EXAMPLE 7 Simplify and compare: $23 - (10 - 9)$ and $(23 - 10) - 9$.

We have

$$23 - (10 - 9) = 23 - 1 = 22;$$
$$(23 - 10) - 9 = 13 - 9 = 4.$$

We can see that $23 - (10 - 9)$ and $(23 - 10) - 9$ represent different numbers. Thus subtraction is not associative.

Do Exercises 13 and 14.

EXAMPLE 8 Simplify: $7 \cdot 2 - (12 + 0) \div 3 - (5 - 2)$.

$$7 \cdot 2 - (12 + 0) \div 3 - (5 - 2) = 7 \cdot 2 - 12 \div 3 - 3$$
$$\text{Carrying out operations inside parentheses}$$
$$= 14 - 4 - 3$$
$$\text{Doing all multiplications and divisions in order from left to right}$$
$$= 7 \qquad \text{Doing all additions and subtractions in order from left to right}$$

Do Exercise 15.

EXAMPLE 9 Simplify: $15 \div 3 \cdot 2 \div (10 - 8)$.

$$15 \div 3 \cdot 2 \div (10 - 8) = 15 \div 3 \cdot 2 \div 2 \qquad \text{Carrying out operations inside parentheses}$$
$$\left. \begin{array}{l} = 5 \cdot 2 \div 2 \\ = 10 \div 2 \\ = 5 \end{array} \right\} \begin{array}{l} \text{Doing all multiplications and divisions in order from left to right} \end{array}$$

Do Exercises 16–18.

Simplify and compare.

13. $64 \div (32 \div 2)$ and $(64 \div 32) \div 2$

14. $(28 + 13) + 11$ and $28 + (13 + 11)$

15. Simplify:

$9 \times 4 - (20 + 4) \div 8 - (6 - 2)$.

Simplify.

16. $5 \cdot 5 \cdot 5 + 26 \cdot 71 - (16 + 25 \cdot 3)$

17. $30 \div 5 \cdot 2 + 10 \cdot 20 + 8 \cdot 8 - 23$

18. $95 - 2 \cdot 2 \cdot 2 \cdot 5 \div (24 - 4)$

Answers on page A-3

Simplify.

19. $5^3 + 26 \cdot 71 - (16 + 25 \cdot 3)$

EXAMPLE 10 Simplify: $4^2 \div (10 - 9 + 1)^3 \cdot 3 - 5$.

$$4^2 \div (10 - 9 + 1)^3 \cdot 3 - 5$$

$\qquad = 4^2 \div (1 + 1)^3 \cdot 3 - 5$ Subtracting inside parentheses

$\qquad = 4^2 \div 2^3 \cdot 3 - 5$ Adding inside parentheses

$\qquad = 16 \div 8 \cdot 3 - 5$ Evaluating exponential expressions

$\qquad = 2 \cdot 3 - 5$ ⎫
$\qquad \qquad \qquad \qquad$ ⎬ Doing all multiplications and divisions
$\qquad = 6 - 5$ ⎭ in order from left to right

$\qquad = 1$ Subtracting

Do Exercises 19–21.

EXAMPLE 11 Simplify: $2^9 \div 2^6 \cdot 2^3$.

20. $(1 + 3)^3 + 10 \cdot 20 + 8^2 - 23$

$2^9 \div 2^6 \cdot 2^3 = 512 \div 64 \cdot 8$ There are no parentheses. Evaluating
$\qquad \qquad \qquad \qquad \qquad \qquad \qquad$ exponential expressions

$\qquad = 8 \cdot 8$ ⎫
$\qquad \qquad \qquad \qquad$ ⎬ Doing all multiplications and divisions
$\qquad = 64$ ⎭ in order from left to right

Do Exercise 22.

21. $81 - 3^2 \cdot 2 \div (12 - 9)$

CALCULATOR CORNER

Order of Operations To determine whether a calculator is programmed to follow the rules for order of operations, we can enter a simple calculation that requires using those rules. For example, we enter ③ ⊞ ④ ⊠ ② ⊜. If the result is 11, we know that the rules for order of operations have been followed. That is, the multiplication $4 \times 2 = 8$ was performed first and then 3 was added to produce a result of 11. If the result is 14, we know that the calculator performs operations as they are entered rather than following the rules for order of operations. That means, in this case, that 3 and 4 were added first to get 7 and then that sum was multiplied by 2 to produce the result of 14. For such calculators, we would have to enter the operations in the order in which we want them performed. In this case, we would press ④ ⊠ ② ⊞ ③ ⊜.

Many calculators have parenthesis keys that can be used to enter an expression containing parentheses. To enter $5(4 + 3)$, for example, we press ⑤ ⦅ ④ ⊞ ③ ⦆ ⊜. The result is 35.

22. Simplify: $2^3 \cdot 2^8 \div 2^9$.

Exercises: Simplify.

1. $84 - 5 \cdot 7$

2. $80 + 50 \div 10$

3. $3^2 + 9^2 \div 3$

4. $4^4 \div 64 - 4$

5. $15 \cdot 7 - (23 + 9)$

6. $(4 + 3)^2$

Answers on page A-3

AVERAGES

In order to find the average of a set of numbers, we use addition and then division. For example, the average of 2, 3, 6, and 9 is found as follows.

$$\text{Average} = \frac{2 + 3 + 6 + 9}{4} = \frac{20}{4} = 5$$

The number of addends is 4.

Divide by 4.

The fraction bar acts as a pair of grouping symbols so

$\frac{2 + 3 + 6 + 9}{4}$ is equivalent to $(2 + 3 + 6 + 9) \div 4$.

Thus we are using order of operations when we compute an average.

AVERAGE

The **average** of a set of numbers is the sum of the numbers divided by the number of addends.

EXAMPLE 12 *Average Number of Career Hits.*   The number of career hits of five Hall of Fame baseball players are given in the bar graph below. Find the average number of career hits of all five.

Career Hits

Mel Ott	2876
Jake Beckley	2930
Dave Winfield	3110
Eddie Murray	3255
Carl Yastrzemski	3419

Sources: Associated Press; Major League Baseball

The average is given by

$$\frac{3419 + 3255 + 3110 + 2930 + 2876}{5} = \frac{15,590}{5} = 3118.$$

Thus the average number of career hits of these five Hall of Fame players is 3118.

Do Exercise 23.

 Removing Parentheses within Parentheses

When parentheses occur within parentheses, we can make them different shapes, such as [ ] (also called "brackets") and { } (also called "braces"). All of these have the same meaning. When parentheses occur within parentheses, computations in the innermost ones are to be done first.

---

23. World's Tallest Buildings. The heights, in feet, of the four tallest buildings in the world are given in the bar graph below. Find the average height of these buildings.

World's Tallest Buildings

Source: Council on Tall Buildings and Urban Habitat, Lehigh University, 2004

Answer on page A-3

Simplify.

24. $9 \times 5 + \{6 \div [14 - (5 + 3)]\}$

25. $[18 - (2 + 7) \div 3]$
$- (31 - 10 \times 2)$

Answers on page A-3

■ **EXAMPLE 13** Simplify: $[25 - (4 + 3) \cdot 3] \div (11 - 7)$.

$$[25 - (4 + 3) \cdot 3] \div (11 - 7)$$

$= [25 - 7 \cdot 3] \div (11 - 7)$	Doing the calculations in the innermost parentheses first
$= [25 - 21] \div (11 - 7)$	Doing the multiplication in the brackets
$= 4 \div 4$	Subtracting
$= 1$	Dividing

■ **EXAMPLE 14** Simplify: $16 \div 2 + \{40 - [13 - (4 + 2)]\}$.

$$16 \div 2 + \{40 - [13 - (4 + 2)]\}$$

$= 16 \div 2 + \{40 - [13 - 6]\}$	Doing the calculations in the innermost parentheses first
$= 16 \div 2 + \{40 - 7\}$	Again, doing the calculations in the innermost parentheses
$= 16 \div 2 + 33$	Subtracting inside the braces
$= 8 + 33$	Doing all multiplications and divisions in order from left to right
$= 41$	Adding

Do Exercises 24 and 25.

Study Tips

TEST PREPARATION

You are probably ready to begin preparing for your first test. Here are some test-taking study tips.

- **Make up your own test questions as you study.** After you have done your homework over a particular objective, write one or two questions on your own that you think might be on a test. You will be amazed at the insight this will provide.

- **Do an overall review of the chapter, focusing on the objectives and the examples.** This should be accompanied by a study of any class notes you may have taken.

- **Do the review exercises at the end of the chapter.** Check your answers at the back of the book. If you have trouble with an exercise, use the objective symbol as a guide to go back and do further study of that objective.

- **Call the AW Math Tutor Center at 1-888-777-0463 if you need extra help.**

- **Take the chapter test at the end of the chapter.** Check the answers and use the objective symbols at the back of the book as a reference for where to review.

- **Ask former students for old exams.** Working such exams can be very helpful and allows you to see what various professors think is important.

- **When taking a test, read each question carefully and try to do all the questions the first time through, but pace yourself.** Answer all the questions, and mark those to recheck if you have time at the end. Very often, your first hunch will be correct.

- **Try to write your test in a neat and orderly manner.** Very often, your instructor tries to give you partial credit when grading an exam. If your test paper is sloppy and disorderly, it is difficult to verify the partial credit. Doing your work neatly can ease such a task for the instructor.

a   Write exponential notation.

1. $3 \cdot 3 \cdot 3 \cdot 3$

2. $2 \cdot 2 \cdot 2 \cdot 2 \cdot 2$

3. $5 \cdot 5$

4. $13 \cdot 13 \cdot 13$

5. $7 \cdot 7 \cdot 7 \cdot 7 \cdot 7$

6. $10 \cdot 10$

7. $10 \cdot 10 \cdot 10$

8. $1 \cdot 1 \cdot 1 \cdot 1$

b   Evaluate.

9. 7^2

10. 5^3

11. 9^3

12. 10^2

13. 12^4

14. 10^5

15. 11^2

16. 6^3

c   Simplify.

17. $12 + (6 + 4)$

18. $(12 + 6) + 18$

19. $52 - (40 - 8)$

20. $(52 - 40) - 8$

21. $1000 \div (100 \div 10)$

22. $(1000 \div 100) \div 10$

23. $(256 \div 64) \div 4$

24. $256 \div (64 \div 4)$

25. $(2 + 5)^2$

26. $2^2 + 5^2$

27. $(11 - 8)^2 - (18 - 16)^2$

28. $(32 - 27)^3 + (19 + 1)^3$

29. $16 \cdot 24 + 50$

30. $23 + 18 \cdot 20$

31. $83 - 7 \cdot 6$

32. $10 \cdot 7 - 4$

33. $10 \cdot 10 - 3 \cdot 4$

34. $90 - 5 \cdot 5 \cdot 2$

35. $4^3 \div 8 - 4$

36. $8^2 - 8 \cdot 2$

37. $17 \cdot 20 - (17 + 20)$

38. $1000 \div 25 - (15 + 5)$

39. $6 \cdot 10 - 4 \cdot 10$

40. $3 \cdot 8 + 5 \cdot 8$

41. $300 \div 5 + 10$

42. $144 \div 4 - 2$

43. $3 \cdot (2 + 8)^2 - 5 \cdot (4 - 3)^2$

44. $7 \cdot (10 - 3)^2 - 2 \cdot (3 + 1)^2$

45. $4^2 + 8^2 \div 2^2$

46. $6^2 - 3^4 \div 3^3$

47. $10^3 - 10 \cdot 6 - (4 + 5 \cdot 6)$

48. $7^2 + 20 \cdot 4 - (28 + 9 \cdot 2)$

49. $6 \cdot 11 - (7 + 3) \div 5 - (6 - 4)$

50. $8 \times 9 - (12 - 8) \div 4 - (10 - 7)$

51. $120 - 3^3 \cdot 4 \div (5 \cdot 6 - 6 \cdot 4)$

52. $80 - 2^4 \cdot 15 \div (7 \cdot 5 - 45 \div 3)$

53. $2^3 \cdot 2^8 \div 2^6$

54. $2^7 \div 2^5 \cdot 2^4 \div 2^2$

55. Find the average of $64, $97, and $121.

56. Find the average of four test grades of 86, 92, 80, and 78.

57. Find the average of 320, 128, 276, and 880.

58. Find the average of $1025, $775, $2062, $942, and $3721.

d Simplify.

59. $8 \times 13 + \{42 \div [18 - (6 + 5)]\}$

60. $72 \div 6 - \{2 \times [9 - (4 \times 2)]\}$

61. $[14 - (3 + 5) \div 2] - [18 \div (8 - 2)]$

62. $[92 \times (6 - 4) \div 8] + [7 \times (8 - 3)]$

63. $(82 - 14) \times [(10 + 45 \div 5) - (6 \cdot 6 - 5 \cdot 5)]$

64. $(18 \div 2) \cdot \{[(9 \cdot 9 - 1) \div 2] - [5 \cdot 20 - (7 \cdot 9 - 2)]\}$

65. $4 \times \{(200 - 50 \div 5) - [(35 \div 7) \cdot (35 \div 7) - 4 \times 3]\}$

66. $15(23 - 4 \cdot 2)^3 \div (3 \cdot 25)$

67. $\{[18 - 2 \cdot 6] - [40 \div (17 - 9)]\} + \{48 - 13 \times 3 + [(50 - 7 \cdot 5) + 2]\}$

68. $(19 - 2^4)^5 - (141 \div 47)^2$

69. $\mathbf{D_W}$ Consider the problem in Example 8 of Section 1.8. How can you translate the problem to a single equation involving what you have learned about order of operations? How does the single equation relate to how we solved the problem?

70. $\mathbf{D_W}$ Consider the expressions $9 - (4 \cdot 2)$ and $(3 \cdot 4)^2$. Are the parentheses necessary in each case? Explain.

SKILL MAINTENANCE

Solve. [1.7b]

71. $x + 341 = 793$

72. $4197 + x = 5032$

73. $7 \cdot x = 91$

74. $1554 = 42 \cdot y$

75. $3240 = y + 898$

76. $6000 = 1102 + t$

77. $25 \cdot t = 625$

78. $10{,}000 = 100 \cdot t$

Solve. [1.8a]

79. *Colorado.*   The state of Colorado is roughly the shape of a rectangle that is 273 mi by 382 mi. What is its area?

80. On a long four-day trip, a family bought the following amounts of gasoline for their motor home:

> 23 gallons,   24 gallons,
> 26 gallons,   25 gallons.

How much gasoline did they buy in all?

SYNTHESIS

Each of the answers in Exercises 81–83 is incorrect. First find the correct answer. Then place as many parentheses as needed in the expression in order to make the incorrect answer correct.

81. $1 + 5 \cdot 4 + 3 = 36$

82. $12 \div 4 + 2 \cdot 3 - 2 = 2$

83. $12 \div 4 + 2 \cdot 3 - 2 = 4$

84. Use one occurrence each of 1, 2, 3, 4, 5, 6, 7, 8, and 9 and any of the symbols $+$, $-$, $\times$, $\div$, and ( ) to represent 100.

The review that follows is meant to prepare you for a chapter exam. It consists of two parts. The first part, Concept Reinforcement, is designed to increase understanding of the concepts through true/false exercises. The second part is the Review Exercises. These provide practice exercises for the exam, together with references to section objectives so you can go back and review. Before beginning, stop and look back over the skills you have obtained. What skills in mathematics do you have now that you did not have before studying this chapter?

CONCEPT REINFORCEMENT

Determine whether the statement is true or false. Answers are given at the back of the book.

_____ **1.** The product of two natural numbers is always greater than either of the factors.

_____ **2.** Zero divided by any nonzero number is 0.

_____ **3.** Each member of the set of natural numbers is a member of the set of whole numbers

_____ **4.** The sum of two natural numbers is always greater than either of the addends.

_____ **5.** Any number divided by 1 is the number 1.

_____ **6.** The number 0 is the smallest natural number.

Review Exercises

The review exercises that follow are for practice. Answers are given at the back of the book. If you miss an exercise, restudy the objective indicated in red next to the exercise or direction line that precedes it.

1. What does the digit 8 mean in 4,678,952? [1.1a]

2. In 13,768,940, what digit tells the number of millions? [1.1a]

Write expanded notation. [1.1b]

3. 2793

4. 56,078

5. 4,007,101

Write a word name. [1.1c]

6. 67,819

7. 2,781,427

8. 1,065,070,607, the population of India in 2004.
Source: U.S. Census Bureau, International Database

Write standard notation. [1.1c]

9. Four hundred seventy-six thousand, five hundred eighty-eight

10. *e-books.* The publishing industry predicts that sales of digital books will reach two billion, four hundred thousand by 2005.
Source: Andersen Consulting

Add. [1.2a]

11. 7304 + 6968

12. 27,609 + 38,415

13. 2703 + 4125 + 6004 + 8956

14. $\begin{array}{r} 9\,1{,}4\,2\,6 \\ +\quad 7{,}4\,9\,5 \\ \hline \end{array}$

15. Write a related addition sentence: [1.3a]
$$10 - 6 = 4.$$

16. Write two related subtraction sentences: [1.3a]
$$8 + 3 = 11.$$

Subtract. [1.3b]

17. 8045 − 2897

18. 9001 − 7312

19. 6003 − 3729

20. $\begin{array}{r} 3\,7{,}4\,0\,5 \\ -\;1\,9{,}6\,4\,8 \\ \hline \end{array}$

Round 345,759 to the nearest: [1.4a]

21. Hundred.   **22.** Ten.

23. Thousand.   **24.** Hundred thousand.

Estimate the sum, difference, or product by first rounding to the nearest hundred. Show your work. [1.4b], [1.5b]

25. 41,348 + 19,749   **26.** 38,652 − 24,549

27. 396 · 748

Use < or > for ☐ to write a true sentence. [1.4c]

28. 67 ☐ 56   **29.** 1 ☐ 23

Multiply. [1.5a]

30. 17,000 · 300   **31.** 7846 · 800

32. 726 · 698   **33.** 587 · 47

34. $\begin{array}{r} 8\,3\,0\,5 \\ \times\quad 6\,4\,2 \\ \hline \end{array}$

35. Write a related multiplication sentence: [1.6a]
$$56 \div 8 = 7.$$

36. Write two related division sentences: [1.6a]
$$13 \cdot 4 = 52.$$

Divide. [1.6b]

37. 63 ÷ 5   **38.** 80 ÷ 16

39. $7\,\overline{)\,6\,3\,9\,4}$ **40.** 3073 ÷ 8

41. $6\,0\,\overline{)\,2\,8\,6}$ **42.** 4266 ÷ 79

43. $3\,8\,\overline{)\,1\,7{,}1\,7\,6}$ **44.** $1\,4\,\overline{)\,7\,0{,}1\,1\,2}$

45. 52,668 ÷ 12

Solve. [1.7b]

46. $46 \cdot n = 368$ **47.** $47 + x = 92$

48. $1 \cdot y = 58$ **49.** $24 = x + 24$

50. Write exponential notation: 4 · 4 · 4. [1.9a]

Evaluate. [1.9b]

51. 10^4 **52.** 6^2

Simplify. [1.9c, d]

53. 8 · 6 + 17

54. $10 \cdot 24 - (18 + 2) \div 4 - (9 - 7)$

55. $7 + (4 + 3)^2$

56. $7 + 4^2 + 3^2$

57. $(80 \div 16) \times [(20 - 56 \div 8) + (8 \cdot 8 - 5 \cdot 5)]$

58. Find the average of 157, 170, and 168.

Solve.   [1.8a]

59. *Computer Workstation.*   Natasha has $196 and wants to buy a computer workstation for $698. How much more does she need?

Computer Workstation

Raised monitor platform, sliding keyboard shelf, and mobile CPU shelf. Locking 3-drawer file cabinet. Maple and honey finish with durable melamine work surface.

Workstation just...
$698

60. Toni has $406 in her checking account. She is paid $78 for a part-time job and deposits that in her checking account. How much is then in her account?

61. *Lincoln-Head Pennies.*   In 1909, the first Lincoln-head pennies were minted. Seventy-three years later, these pennies were first minted with a decreased copper content. In what year was the copper content reduced?

62. A beverage company packed 228 cans of soda into 12-can cartons. How many cartons did they fill?

63. An apple farmer keeps bees in her orchard to help pollinate the apple blossoms so more apples will be produced. The bees from an average beehive can pollinate 30 surrounding trees during one growing season. A farmer has 420 trees. How many beehives does she need to pollinate them all?

Source: Jordan Orchards, Westminster, PA

64. An apartment builder bought 13 gas stoves at $425 each and 13 refrigerators at $620 each. What was the total cost?

65. A family budgets $7825 for food and clothing and $2860 for entertainment. The yearly income of the family was $38,283. How much of this income remained after these two allotments?

66. A chemist has 2753 mL of alcohol. How many 20-mL beakers can be filled? How much will be left over?

67. *Olympic Trampoline.*   Shown below is an Olympic trampoline. Find the area and the perimeter of the trampoline.   [1.2b], [1.5c]

Source: International Trampoline Industry Association, Inc.

14 ft

7 ft

68. **D**W Write a problem for a classmate to solve. Design the problem so that the solution is "Each of the 144 bottles will contain 8 oz of hot sauce."   [1.8a]

69. **D**W Is subtraction associative? Why or why not?   [1.3b]

SYNTHESIS

70. ▦ Determine the missing digit *d*.   [1.5a]

$$\begin{array}{r} 9\,d \\ \times\ \ d\,2 \\ \hline 8\,0\,3\,6 \end{array}$$

71. ▦ Determine the missing digits *a* and *b*.   [1.6b]

$$2\,b\,1\,\overline{)\,2\,3\,6{,}4\,2\,1\ }^{9\,a\,1}$$

72. A mining company estimates that a crew must tunnel 2000 ft into a mountain to reach a deposit of copper ore. Each day the crew tunnels about 500 ft. Each night about 200 ft of loose rocks roll back into the tunnel. How many days will it take the mining company to reach the copper deposit?   [1.8a]

1. In the number 546,789, which digit tells the number of hundred thousands?

2. Write expanded notation: 8843.

3. Write a word name: 38,403,277.

Add.

4.
```
    6 8 1 1
  + 3 1 7 8
```

5.
```
    4 5,8 8 9
  + 1 7,9 0 2
```

6.
```
    1 2
     8
     3
     7
  +  4
```

7.
```
    6 2 0 3
  + 4 3 1 2
```

Subtract.

8.
```
    7 9 8 3
  - 4 3 5 3
```

9.
```
    2 9 7 4
  - 1 9 3 5
```

10.
```
    8 9 0 7
  - 2 0 5 9
```

11.
```
    2 3,0 6 7
  - 1 7,8 9 2
```

Multiply.

12.
```
    4 5 6 8
  ×       9
```

13.
```
    8 8 7 6
  ×     6 0 0
```

14.
```
      6 5
  ×   3 7
```

15.
```
      6 7 8
  ×   7 8 8
```

Divide.

16. 15 ÷ 4

17. 420 ÷ 6

18. 8 9)‾8̅ ̅6̅ ̅3̅ ̅3̅

19. 4 4)‾3̅ ̅5̅,̅4̅ ̅2̅ ̅8̅

Solve.

20. *Hostess Ding Dongs®.* Hostess packages its Ding Dong® snack products in 12-packs. It manufactures 22,231 cakes. How many 12-packs can it fill? How many will be left over?

21. *Largest States.* The following table lists the five largest states in terms of their land area. Find the total land area of these states.

STATE	AREA (In Square Miles)
Alaska	571,951
Texas	261,797
California	155,959
Montana	145,552
New Mexico	121,356

Source: Department of Commerce, Bureau of the Census

22. *Pool Tables.* The Hartford™ pool table made by Brunswick Billiards comes in three sizes of playing area, 50 in. by 100 in., 44 in. by 88 in., and 38 in. by 76 in.

a) Find the perimeter and the area of the playing area of each table.

b) By how much area does the large table exceed the small table?

Source: Brunswick Billiards

23. *Voting Early.* In the 2004 Presidential Election, 345,689 Nevada voters voted early. This was 139,359 more than in the 2000 election. How many voted early in 2000?

Source: National Association of Secretaries of State

24. A sack of oranges weighs 27 lb. A sack of apples weighs 32 lb. Find the total weight of 16 bags of oranges and 43 bags of apples.

25. A box contains 5000 staples. How many staplers can be filled from the box if each stapler holds 250 staples?

Solve.

26. $28 + x = 74$

27. $169 \div 13 = n$

28. $38 \cdot y = 532$

29. $381 = 0 + a$

Round 34,578 to the nearest:

30. Thousand.

31. Ten.

32. Hundred.

Estimate the sum, difference, or product by first rounding to the nearest hundred. Show your work.

33.
$$\begin{array}{r} 2\,3{,}6\,4\,9 \\ +\ 5\,4{,}7\,4\,6 \\ \hline \end{array}$$

34.
$$\begin{array}{r} 5\,4{,}7\,5\,1 \\ -\ 2\,3{,}6\,4\,9 \\ \hline \end{array}$$

35.
$$\begin{array}{r} 8\,2\,4 \\ \times\ 4\,8\,9 \\ \hline \end{array}$$

Use < or > for ☐ to write a true sentence.

36. 34 ☐ 17

37. 117 ☐ 157

38. Write exponential notation: $12 \cdot 12 \cdot 12 \cdot 12$.

Evaluate.

39. 7^3

40. 10^5

41. 25^2

Simplify.

42. $35 - 1 \cdot 28 \div 4 + 3$

43. $10^2 - 2^2 \div 2$

44. $(25 - 15) \div 5$

45. $8 \times \{(20 - 11) \cdot [(12 + 48) \div 6 - (9 - 2)]\}$

46. $2^4 + 24 \div 12$

47. Find the average of 97, 98, 87, and 86.

| SYNTHESIS |

48. An open cardboard container is 8 in. wide, 12 in. long, and 6 in. high. How many square inches of cardboard are used?

49. Use trials to find the single-digit number a for which
$$359 - 46 + a \div 3 \times 25 - 7^2 = 339.$$

50. Cara spends $229 a month to repay her student loan. If she has already paid $9160 on the 10-yr loan, how many payments remain?

Fraction Notation: Multiplication and Division

2

Real-World Application

The length of a rectangular ice-skating rink in the atrium of a shopping mall is $\frac{7}{100}$ mi. The width is $\frac{3}{100}$ mi. What is the area of the skating rink?

This problem appears as Example 10 in Section 2.4

FACTORIZATIONS

Objectives

a Determine whether one number is a factor of another, and find the factors of a number.

b Find some multiples of a number, and determine whether a number is divisible by another.

c Given a number from 1 to 100, tell whether it is prime, composite, or neither.

d Find the prime factorization of a composite number.

To the student:

In the preface, at the front of the text, you will find a Student Organizer card. This pullout card will help you keep track of important dates and useful contact information. You can also use it to plan time for class, study, work, and relaxation. By managing your time wisely, you will provide yourself the best possible opportunity to be successful in this course.

In this chapter, we begin our work with fractions and fraction notation. Certain skills make such work easier. For example, in order to simplify $\frac{12}{32}$, it is important that we be able to *factor* 12 and 32, as follows:

$$\frac{12}{32} = \frac{4 \cdot 3}{4 \cdot 8}.$$

Then we "remove" a factor of 1:

$$\frac{4 \cdot 3}{4 \cdot 8} = \frac{4}{4} \cdot \frac{3}{8} = 1 \cdot \frac{3}{8} = \frac{3}{8}.$$

Thus factoring is an important skill in working with fractions.

a Factors and Factorization

In Sections 2.1 and 2.2, we consider only the **natural numbers** 1, 2, 3, and so on.

Let's look at the product $3 \cdot 4 = 12$. We say that 3 and 4 are **factors** of 12. When we divide 12 by 3, we get a remainder of 0. We say that the divisor 3 is a **factor** of the dividend 12.

FACTOR

- In the product $a \cdot b$, a and b are called **factors.**
- If we divide Q by d and get a remainder of 0, then the divisor d is a **factor** of the dividend Q.

EXAMPLE 1 Determine by long division whether 6 is a factor of 72.

$$\begin{array}{r} 12 \\ 6\overline{)72} \\ \underline{60} \\ 12 \\ \underline{12} \\ 0 \end{array}$$

The remainder is 0, so 6 is a factor of 72. We sometimes say that 6 divides 72 "evenly" because there is a remainder of 0.

EXAMPLE 2 Determine by long division whether 15 is a factor of 7894.

$$\begin{array}{r} 526 \\ 15\overline{)7894} \\ \underline{7500} \\ 394 \\ \underline{300} \\ 94 \\ \underline{90} \\ 4 \end{array} \leftarrow \text{Not } 0$$

The remainder is *not* 0, so 15 is not a factor of 7894.

Do Exercises 1 and 2 on the following page.

Consider $12 = 3 \cdot 4$. We say that $3 \cdot 4$ is a **factorization** of 12. Similarly, $6 \cdot 2$, $12 \cdot 1$, $2 \cdot 2 \cdot 3$, and $1 \cdot 3 \cdot 4$ are also factorizations of 12. Since $a = a \cdot 1$, every number has a factorization, and every number has factors. For some numbers, the factors consist of only the number itself and 1. For example, the only factorization of 17 is $17 \cdot 1$, so the only factors of 17 are 17 and 1.

EXAMPLE 3 Find all the factors of 70.

We find as many "two-factor" factorizations as we can. We check sequentially the numbers 1, 2, 3, and so on, to see if we can form any factorizations:

70

$1 \cdot 70$
$2 \cdot 35$
$5 \cdot 14$
$7 \cdot 10.$

> Note that all but one of the factors of a natural number are *less* than the number.

Note that 3, 4, and 6 are not factors. If there are additional factors, they must be between 7 and 10. Since 8 and 9 are not factors, we are finished. The factors of 70 are 1, 2, 5, 7, 10, 14, 35, and 70.

Do Exercises 3–6.

b Multiples and Divisibility

A **multiple** of a natural number is a product of that number and some natural number. For example, some multiples of 2 are:

2 (because $2 = 1 \cdot 2$);
4 (because $4 = 2 \cdot 2$);
6 (because $6 = 3 \cdot 2$);
8 (because $8 = 4 \cdot 2$);
10 (because $10 = 5 \cdot 2$).

> Note that all but one of the multiples of a number are *larger* than the number.

We find multiples of 2 by counting by twos: 2, 4, 6, 8, and so on. We can find multiples of 3 by counting by threes: 3, 6, 9, 12, and so on.

EXAMPLE 4 Show that each of the numbers 8, 12, 20, and 36 is a multiple of 4.

$8 = 2 \cdot 4 \qquad 12 = 3 \cdot 4 \qquad 20 = 5 \cdot 4 \qquad 36 = 9 \cdot 4$

Do Exercises 7 and 8.

EXAMPLE 5 Multiply by 1, 2, 3, and so on, to find ten multiples of 7.

$1 \cdot 7 = 7 \qquad\quad 6 \cdot 7 = 42$
$2 \cdot 7 = 14 \qquad\quad 7 \cdot 7 = 49$
$3 \cdot 7 = 21 \qquad\quad 8 \cdot 7 = 56$
$4 \cdot 7 = 28 \qquad\quad 9 \cdot 7 = 63$
$5 \cdot 7 = 35 \qquad\quad 10 \cdot 7 = 70$

Do Exercise 9.

Determine whether the second number is a factor of the first.

1. 72; 8

2. 2384; 28

Find all the factors of the number.

3. 10   1, 2, 5, 10

4. 45   1, 5, 10, 9, 25, 45

5. 62   1, 2, 31, 62

6. 24   1, 2, 12, 3, 6, 8

7. Show that each of the numbers 5, 45, and 100 is a multiple of 5.

5·1
5·8
5·10

8. Show that each of the numbers 10, 60, and 110 is a multiple of 10.

10·1   10·6   10·11

9. Multiply by 1, 2, 3, and so on, to find ten multiples of 5.

Answers on page A-4

10. Determine whether 16 is divisible by 2.

DIVISIBILITY

The number a is **divisible** by another number b if there exists a number c such that $a = b \cdot c$. The statements "a is **divisible** by b," "a is a **multiple** of b," and "b is a **factor** of a" all have the same meaning.

Thus 27 is *divisible* by 3 because $27 = 3 \cdot 9$. We can also say that 27 is a *multiple* of 3, and 3 is a *factor* of 27.

EXAMPLE 6 Determine whether 45 is divisible by 9.

We divide 45 by 9 to determine if 9 is a factor of 45. If so, then 45 is divisible by 9.

$$
\begin{array}{r}
5 \\
9\overline{)45} \\
\underline{45} \\
0
\end{array}
$$

11. Determine whether 125 is divisible by 5.

Because the remainder is 0, 45 is divisible by 9.

EXAMPLE 7 Determine whether 98 is divisible by 4.

We divide 98 by 4:

$$
\begin{array}{r}
24 \\
4\overline{)98} \\
\underline{80} \\
18 \\
\underline{16} \\
2 \leftarrow \text{Not } 0
\end{array}
$$

12. Determine whether 125 is divisible by 6.

Since the remainder is not 0, 98 is *not* divisible by 4.

Answers on page A-4

Do Exercises 10–12.

CALCULATOR CORNER

Divisibility and Factors We can use a calculator to determine whether one number is divisible by another number or whether one number is a factor of another number. For example, to determine whether 387 is divisible by 18, we first press [3] [8] [7] [÷] [1] [8] [=]. The display reads [21.5]. Note that the result is not a natural number. (For a brief discussion of decimal notation, see the Calculator Corner on page 47. Decimal notation will be studied in detail in Chapter 4.) Thus we know that 387 is not a multiple of 18; that is, 387 is not divisible by 18 and 18 is not a factor of 387.

When we divide 387 by 9, the result is [43]. Since 43 is a natural number, we know that 387 is a multiple of 9; that is, $387 = 43 \cdot 9$. Thus, 387 is divisible by 9, and 9 is a factor of 387.

Exercises: For each pair of numbers, determine whether the first number is divisible by the second number.

1. 722; 19

2. 845; 7

3. 1047; 14

4. 5283; 9

For each pair of numbers, determine whether the second number is a factor of the first number.

5. 502; 8

6. 651; 21

7. 3875; 25

8. 8464; 12

9. 32,768; 256

10. 32,768; 864

CHAPTER 2: Fraction Notation:
Multiplication and Division

c | Prime and Composite Numbers

> **PRIME AND COMPOSITE NUMBERS**
>
> - A natural number that has exactly two *different* factors, only itself and 1, is called a **prime number.**
> - The number 1 is *not* prime.
> - A natural number, other than 1, that is not prime is **composite.**

EXAMPLE 8 Determine whether the numbers 1, 2, 3, 4, 5, 6, 7, 9, 10, 11, and 63 are prime, composite, or neither.

The number 1 is not prime. It does not have *two* different factors.

The number 2 is prime. It has only the factors 2 and 1.

The numbers 3, 5, 7, and 11 are prime. Each has only two factors, itself and 1.

The number 4 is not prime. It has the factors 1, 2, and 4 and is composite.

The numbers 6, 9, 10, and 63 are composite. Each has more than two factors.

Thus we have:

Prime: 2, 3, 5, 7, 11;

Composite: 4, 6, 9, 10, 63;

Neither: 1.

The number 2 is the *only* even prime number. It is also the smallest prime number. The number 0 is also neither prime nor composite, but 0 is *not* a natural number and thus is not considered here. We are considering only natural numbers.

Do Exercise 13.

The following is a table of the prime numbers from 2 to 157. There are more extensive tables, but these prime numbers will be the most helpful to you in this text.

> **A TABLE OF PRIMES FROM 2 TO 157**
>
> 2, 3, 5, 7, 11, 13, 17, 19, 23, 29, 31, 37, 41, 43, 47, 53, 59, 61, 67, 71, 73, 79, 83, 89, 97, 101, 103, 107, 109, 113, 127, 131, 137, 139, 149, 151, 157

d | Prime Factorizations

When we factor a composite number into a product of primes, we find the **prime factorization** of the number. To do this, we consider the primes

   2, 3, 5, 7, 11, 13, 17, 19, 23, and so on,

and determine whether a given number is divisible by the primes.

---

13. Tell whether each number is prime, composite, or neither.

1, 2, 6, 12, 13, 19, 41, 65, 73, 99

Answer on page A-4

Study Tips

STUDY TIPS REVIEW

The following Study Tips were presented in Chapter 1. If you have not studied them, go back and do so now.

- Using This Textbook (Section 1.2)
- Learning Resources (Section 1.3)
- Time Management (Part 1) (Section 1.5)
- Highlighting (Section 1.6)
- Exercises (Section 1.6)
- Test Preparation (Section 1.9)

EXAMPLE 9 Find the prime factorization of 39.

a) We divide by the first prime, 2.

$$\begin{array}{r} 19 \\ 2\overline{)39} \\ 38 \\ \hline 1 \end{array} \quad R = 1$$

Because the remainder is not 0, 2 is not a factor of 39, and 39 is not divisible by 2.

b) We divide by the next prime, 3.

$$\begin{array}{r} 13 \\ 3\overline{)39} \end{array} \quad R = 0$$

The remainder is 0, so we know that $39 = 3 \cdot 13$. Because 13 is a prime, we are finished. The prime factorization is

$$39 = 3 \cdot 13.$$

EXAMPLE 10 Find the prime factorization of 220.

a) We divide by the first prime, 2.

$$\begin{array}{r} 110 \\ 2\overline{)220} \end{array} \quad R = 0 \qquad 220 = 2 \cdot 110$$

b) Because 110 is composite, we start with 2 again.

$$\begin{array}{r} 55 \\ 2\overline{)110} \end{array} \quad R = 0 \qquad 220 = 2 \cdot 2 \cdot 55$$

c) Since 55 is composite and is not divisible by 2 or 3, we divide by the next prime, 5.

$$\begin{array}{r} 11 \\ 5\overline{)55} \end{array} \quad R = 0 \qquad 220 = 2 \cdot 2 \cdot 5 \cdot 11$$

Because 11 is prime, we are finished. The prime factorization is

$$220 = 2 \cdot 2 \cdot 5 \cdot 11.$$

We abbreviate our procedure as follows.

$$\begin{array}{r} 11 \\ 5\overline{)55} \\ 2\overline{)110} \\ 2\overline{)220} \end{array}$$
$$220 = 2 \cdot 2 \cdot 5 \cdot 11$$

Multiplication is commutative so a factorization such as $2 \cdot 2 \cdot 5 \cdot 11$ could also be expressed as $5 \cdot 2 \cdot 2 \cdot 11$ or $2 \cdot 5 \cdot 11 \cdot 2$ (or, in exponential notation, as $2^2 \cdot 5 \cdot 11$ or $11 \cdot 2^2 \cdot 5$), but the prime factors are the same. For this reason, we agree that any of these is "the" prime factorization of 220.

Every number has just one (unique) prime factorization.

EXAMPLE 11 Find the prime factorization of 72.

We can do divisions "up" as follows:

$$3 \leftarrow \text{Prime quotient}$$
$$3\overline{)\,9}$$
$$2\overline{)18}$$
$$2\overline{)36}$$
$$2\overline{)72} \leftarrow \text{Begin here.}$$

Or, we can do divisions "down":

$$2\overline{)72} \leftarrow \text{Begin here.}$$
$$2\overline{)36}$$
$$2\overline{)18}$$
$$3\overline{)\,9}$$
$$\quad 3 \leftarrow \text{Prime quotient}$$

$$72 = 2 \cdot 2 \cdot 2 \cdot 3 \cdot 3$$

Another way to find the prime factorization of 72 uses a **factor tree** as follows. Begin by determining any factorization you can, and then continue factoring.

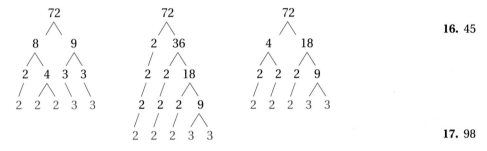

EXAMPLE 12 Find the prime factorization of 189.

We can use a string of successive divisions or a factor tree. Since 189 is not divisible by 2, we begin with 3.

$$7$$
$$3\overline{)21}$$
$$3\overline{)63}$$
$$3\overline{)189}$$

$$189 = 3 \cdot 3 \cdot 3 \cdot 7$$

189
3   63
3   7   9
3   7   3   3

Caution!

Keep in mind the difference between finding all the factors of a number and finding the prime factorization. In Example 12, the prime factorization is $3 \cdot 3 \cdot 3 \cdot 7$. The factors of 189 are 1, 3, 7, 9, 21, 27, 63, and 189.

EXAMPLE 13 Find the prime factorization of 1638.

We can use a string of successive divisions.

$$13$$
$$7\overline{)91}$$
$$3\overline{)273}$$
$$3\overline{)819}$$
$$2\overline{)1638}$$

$$1638 = 2 \cdot 3 \cdot 3 \cdot 7 \cdot 13$$

Do Exercises 14–21.

Find the prime factorization of the number.

14. 6

15. 12

16. 45

17. 98

18. 126

19. 144

20. 1960

21. 1925

Answers on page A-4

a Determine whether the second number is a factor of the first.

1. 52; 14   **2.** 52; 13   **3.** 625; 25   **4.** 680; 16

Find all the factors of the number.

5. 18   **6.** 16   **7.** 54   **8.** 48

9. 4   **10.** 9   **11.** 1   **12.** 13

13. 98   **14.** 100   **15.** 255   **16.** 120

b Multiply by 1, 2, 3, and so on, to find ten multiples of the number.

17. 4   **18.** 11   **19.** 20   **20.** 50

21. 3   **22.** 5   **23.** 12   **24.** 13

25. 10   **26.** 6   **27.** 9   **28.** 14

29. Determine whether 26 is divisible by 6.   **30.** Determine whether 48 is divisible by 8.

31. Determine whether 1880 is divisible by 8.   **32.** Determine whether 4227 is divisible by 3.

33. Determine whether 256 is divisible by 16.   **34.** Determine whether 102 is divisible by 4.

35. Determine whether 4227 is divisible by 9.   **36.** Determine whether 200 is divisible by 25.

37. Determine whether 8650 is divisible by 16.   **38.** Determine whether 4143 is divisible by 7.

c Determine whether the number is prime, composite, or neither.

39. 1   **40.** 2   **41.** 9   **42.** 19

43. 11   **44.** 27   **45.** 29   **46.** 49

d   Find the prime factorization of the number.

47. 8   **48.** 16   **49.** 14   **50.** 15   **51.** 42

52. 32   **53.** 25   **54.** 40   **55.** 50   **56.** 62

57. 169   **58.** 140   **59.** 100   **60.** 110   **61.** 35

62. 70   **63.** 72   **64.** 86   **65.** 77   **66.** 99

67. 2884   **68.** 484   **69.** 51   **70.** 91   **71.** 1200

72. 1800   **73.** 273   **74.** 675   **75.** 1122   **76.** 6435

77. D_W Is every natural number a multiple of 1? Explain.

78. D_W Explain a method for finding a composite number that contains exactly two factors other than itself and 1.

SKILL MAINTENANCE

Multiply.   [1.5a]

79. $2 \cdot 13$ **80.** $8 \cdot 32$ **81.** $17 \cdot 25$ **82.** $25 \cdot 168$

Divide.   [1.6b]

83. $0 \div 22$ **84.** $22 \div 1$ **85.** $22 \div 22$ **86.** $66 \div 22$

Solve.   [1.8a]

87. Find the total cost of 7 shirts at $48 each and 4 pairs of pants at $69 each.

88. Sandy can type 62 words per minute. How long will it take her to type 12,462 words?

SYNTHESIS

89. *Factors and Sums.* The top number in each column of the table below can be factored as a product of two numbers whose sum is the bottom number in the column. For example, in the first column, 56 has been factored as $7 \cdot 8$, and $7 + 8 = 15$. Fill in the blank spaces in the table.

Product	56	63	36	72	140	96		168	110			
Factor	7									9	24	3
Factor	8						8	8		10	18	
Sum	15	16	20	38	24	20	14		21			24

Objective

a   Determine whether a number is divisible by 2, 3, 4, 5, 6, 8, 9, or 10.

Determine whether the number is divisible by 2.

1. 84

2. 59

3. 998

4. 2225

Suppose you are asked to find the simplest fraction notation for

$$\frac{117}{225}.$$

Since the numbers are quite large, you might feel that the task is difficult. However, both the numerator and the denominator are divisible by 9. If you knew this, you could factor and simplify quickly as follows:

$$\frac{117}{225} = \frac{9 \cdot 13}{9 \cdot 25} = \frac{9}{9} \cdot \frac{13}{25} = 1 \cdot \frac{13}{25} = \frac{13}{25}.$$

How did we know that both numbers have 9 as a factor? There is a simple test for determining this.

In this section, we learn quick ways to determine whether a number is divisible by 2, 3, 4, 5, 6, 8, 9, or 10. This will make simplifying fraction notation much easier.

a   Rules for Divisibility

DIVISIBILITY BY 2

You may already know the test for divisibility by 2.

BY 2

A number is **divisible by 2** (is *even*) if it has a ones digit of 0, 2, 4, 6, or 8 (that is, it has an even ones digit).

Let's see why. Consider 354, which is

3 hundreds + 5 tens + 4.

Hundreds and tens are both multiples of 2. If the last digit is a multiple of 2, then the entire number is a multiple of 2.

EXAMPLES   Determine whether the number is divisible by 2.

1.   355   is *not* a multiple of 2; 5 is *not* even.

2. 4786   is a multiple of 2; 6 is even.

3. 8990   is a multiple of 2; 0 is even.

4. 4261   is *not* a multiple of 2; 1 is *not* even.

Do Exercises 1–4.

DIVISIBILITY BY 3

Determine whether the number is divisible by 3.

BY 3

> A number is **divisible by 3** if the sum of its digits is divisible by 3.

5. 111

Let's illustrate why the test for divisibility by 3 works. Consider 852; since $852 = 3 \cdot 284$, 852 is divisible by 3.

$$852 = 8 \cdot 100 + 5 \cdot 10 + 2 \cdot 1$$
$$= 8(99 + 1) + 5(9 + 1) + 2(1)$$
$$= 8 \cdot 99 + 8 \cdot 1 + 5 \cdot 9 + 5 \cdot 1 + 2 \cdot 1 \qquad \text{Using the Distributive Law,}$$
$$a(b + c) = a \cdot b + a \cdot c$$

6. 1111

Since 99 and 9 are each a multiple of 3, we see that $8 \cdot 99$ and $5 \cdot 9$ are multiples of 3. This leaves $8 \cdot 1 + 5 \cdot 1 + 2 \cdot 1$, or $8 + 5 + 2$. If $8 + 5 + 2$, the sum of the digits, is divisible by 3, then 852 is divisible by 3.

7. 309

EXAMPLES Determine whether the number is divisible by 3.

5. 18 $1 + 8 = 9$

6. 93 $9 + 3 = 12$ } Each is divisible by 3 because the sum of its digits is divisible by 3.

7. 201 $2 + 0 + 1 = 3$

8. 256 $2 + 5 + 6 = 13$ The sum of the digits, 13, is *not* divisible by 3, so 256 is *not* divisible by 3.

8. 17,216

Do Exercises 5–8.

DIVISIBILITY BY 6

A number divisible by 6 is a multiple of 6. But $6 = 2 \cdot 3$, so the number is also a multiple of 2 and 3. Thus we have the following.

Determine whether the number is divisible by 6.

9. 420

BY 6

> A number is **divisible by 6** if its ones digit is 0, 2, 4, 6, or 8 (is even) and the sum of its digits is divisible by 3.

10. 106

EXAMPLES Determine whether the number is divisible by 6.

9. 720

Because 720 is even, it is divisible by 2. Also, $7 + 2 + 0 = 9$, so 720 is divisible by 3. Thus, 720 is divisible by 6.

720    $7 + 2 + 0 = 9$
↑   ↑
Even   Divisible by 3

11. 321

10. 73

73 is *not* divisible by 6 because it is *not* even.

11. 256

Although 256 is even, it is *not* divisible by 6 because the sum of its digits, $2 + 5 + 6$, or 13, is *not* divisible by 3.

12. 444

Answers on page A-4

Do Exercises 9–12.

Determine whether the number is divisible by 9.

13. 16

14. 117

15. 930

16. 29,223

Determine whether the number is divisible by 10.

17. 305

18. 300

19. 847

20. 8760

Determine whether the number is divisible by 5.

21. 5780

22. 3427

23. 34,678

24. 7775

Answers on page A-4

DIVISIBILITY BY 9

The test for divisibility by 9 is similar to the test for divisibility by 3.

> **BY 9**
>
> A number is **divisible by 9** if the sum of its digits is divisible by 9.

EXAMPLES Determine whether the number is divisible by 9.

12. 6984

Because $6 + 9 + 8 + 4 = 27$ and 27 is divisible by 9, 6984 is divisible by 9.

13. 322

Because $3 + 2 + 2 = 7$ and 7 is *not* divisible by 9, 322 is *not* divisible by 9.

Do Exercises 13–16.

DIVISIBILITY BY 10

> **BY 10**
>
> A number is **divisible by 10** if its ones digit is 0.

We know that this test works because the product of 10 and *any* number has a ones digit of 0.

EXAMPLES Determine whether the number is divisible by 10.

14. 3440   is divisible by 10 because the ones digit is 0.

15. 3447   is *not* divisible by 10 because the ones digit is not 0.

Do Exercises 17–20.

DIVISIBILITY BY 5

> **BY 5**
>
> A number is **divisible by 5** if its ones digit is 0 or 5.

EXAMPLES Determine whether the number is divisible by 5.

16.   220   is divisible by 5 because the ones digit is 0.

17.   475   is divisible by 5 because the ones digit is 5.

18. 6514   is *not* divisible by 5 because the ones digit is neither 0 nor 5.

Do Exercises 21–24.

Let's see why the test for 5 works. Consider 7830:

$$7830 = 10 \cdot 783 = 5 \cdot 2 \cdot 783.$$

Since 7830 is divisible by 10 and 5 is a factor of 10, 7830 is divisible by 5.

Consider 6734:

$$6734 = 673 \text{ tens} + 4.$$

Tens are multiples of 5, so the only number that must be checked is the ones digit. If the last digit is a multiple of 5, the entire number is. In this case, 4 is *not* a multiple of 5, so 6734 is *not* divisible by 5.

DIVISIBILITY BY 4

The test for divisibility by 4 is similar to the test for divisibility by 2.

> **BY 4**
>
> A number is **divisible by 4** if the number named by its last *two* digits is divisible by 4.

EXAMPLES Determine whether the number is divisible by 4.

19. 8212   is divisible by 4 because 12 is divisible by 4.
20. 5216   is divisible by 4 because 16 is divisible by 4.
21. 8211   is *not* divisible by 4 because 11 is *not* divisible by 4.
22. 7538   is *not* divisible by 4 because 38 is *not* divisible by 4.

Do Exercises 25–28.

To see why the test for divisibility by 4 works, consider 516:

$$516 = 5 \text{ hundreds} + 16.$$

Hundreds are multiples of 4. If the number named by the last two digits is a multiple of 4, then the entire number is a multiple of 4.

DIVISIBILITY BY 8

The test for divisibility by 8 is an extension of the tests for divisibility by 2 and 4.

> **BY 8**
>
> A number is **divisible by 8** if the number named by its last *three* digits is divisible by 8.

EXAMPLES Determine whether the number is divisible by 8.

23.   5648   is divisible by 8 because 648 is divisible by 8.
24. 96,088   is divisible by 8 because 88 is divisible by 8.
25.   7324   is *not* divisible by 8 because 324 is *not* divisible by 8.
26. 13,420   is *not* divisible by 8 because 420 is *not* divisible by 8.

Do Exercises 29–32.

A NOTE ABOUT DIVISIBILITY BY 7

There are several tests for divisibility by 7, but all of them are more complicated than simply dividing by 7. So if you want to test for divisibility by 7, simply divide by 7, either by hand or using a calculator.

Determine whether the number is divisible by 4.

25. 216

26. 217

27. 5862

28. 23,524

Determine whether the number is divisible by 8.

29. 7564

30. 7864

31. 17,560

32. 25,716

Answers on page A-4

EXERCISE SET

For Extra Help

a   To answer Exercises 1–8, consider the following numbers.

46	300	85	256
224	36	711	8064
19	45,270	13,251	1867
555	4444	254,765	21,568

1. Which of the above are divisible by 2?

2. Which of the above are divisible by 3?

3. Which of the above are divisible by 4?

4. Which of the above are divisible by 5?

5. Which of the above are divisible by 6?

6. Which of the above are divisible by 8?

7. Which of the above are divisible by 9?

8. Which of the above are divisible by 10?

To answer Exercises 9–16, consider the following numbers.

56	200	75	35
324	42	812	402
784	501	2345	111,111
55,555	3009	2001	1005

9. Which of the above are divisible by 3?

10. Which of the above are divisible by 2?

11. Which of the above are divisible by 5?

12. Which of the above are divisible by 4?

13. Which of the above are divisible by 9?

14. Which of the above are divisible by 6?

15. Which of the above are divisible by 10?

16. Which of the above are divisible by 8?

To answer Exercises 17–24, consider the following numbers.

305	313,332	876	64,000
1101	7624	1110	9990
13,025	111,126	5128	126,111

17. Which of the above are divisible by 2?

18. Which of the above are divisible by 3?

19. Which of the above are divisible by 6?

20. Which of the above are divisible by 5?

21. Which of the above are divisible by 9?

22. Which of the above are divisible by 8?

23. Which of the above are divisible by 10?

24. Which of the above are divisible by 4?

25. D_W How can the divisibility tests be used to find prime factorizations?

26. D_W Which of the years from 2000 to 2020, if any, also happen to be prime numbers? Explain at least two ways in which you might go about solving this problem.

SKILL MAINTENANCE

Solve. [1.7b]

27. $56 + x = 194$

28. $y + 124 = 263$

29. $3008 = x + 2134$

30. $18 \cdot t = 1008$

31. $24 \cdot m = 624$

32. $338 = a \cdot 26$

Divide. [1.6b]

33. $2106 \div 9$

34. $4\,5\,\overline{)\,1\,8\,0,1\,3\,5}$

Solve. [1.8a]

35. An automobile with a 5-speed transmission gets 33 mpg in city driving. How many gallons of gas will it use to travel 1485 mi?

36. There are 60 min in 1 hr. How many minutes are there in 72 hr?

SYNTHESIS

Find the prime factorization of the number. Use divisibility tests where applicable.

37. 7800

38. 2520

39. 2772

40. 1998

41. ▦ Fill in the missing digits of the number

$$95, \square\ \square\ 8$$

so that it is divisible by 99.

42. A passenger in a taxicab asks for the driver's company number. The driver says abruptly, "Sure—you can have my number. Work it out: If you divide it by 2, 3, 4, 5, or 6, you will get a remainder of 1. If you divide it by 11, the remainder will be 0 and no driver has a company number that meets these requirements and is smaller than this one." Determine the number.

2.3 FRACTIONS AND FRACTION NOTATION

Objectives

a Identify the numerator and the denominator of a fraction and write fraction notation for part of an object or part of a set of objects and as a ratio.

b Simplify fraction notation like n/n to 1, $0/n$ to 0, and $n/1$ to n.

For each fraction, identify the numerator and the denominator.

1. $\dfrac{83}{100}$ of all scrap tires are reused or recycled.
   Source: Rubber Manufacturers Association

2. $\dfrac{27}{50}$ of all kids prefer white bread.
   Source: Bruce Horovitz, *USA Today*, 8/9/04

3. $\dfrac{11}{25}$ of all adults eat in restaurants on a given day.
   Source: *AARP Bulletin*, November 2004

4. $\dfrac{21}{1000}$ of stay-at-home parents are men.
   Source: U.S. Census Bureau

The study of arithmetic begins with the set of whole numbers

0, 1, 2, 3, 4, 5, 6, 7, 8, 9, 10, 11, and so on.

But we also need to be able to use fractional parts of numbers such as halves, thirds, fourths, and so on. Here are some examples:

In 1950, about $\frac{3}{4}$ of the motor vehicles produced in the world were produced in the United States. In 2003, only $\frac{1}{5}$ were produced in the United States.

Source: American Automobile Manufacturers Association; Automotive News Data Center and Marketing Systems GmbH

More than $\frac{2}{5}$ of Americans are taking at least one prescription daily, and $\frac{1}{6}$ are taking three or more.

Source: Randolph Schmid, Associated Press, *Indianapolis Star*, 12/3/04

a Fractions and the Real World

Numbers like those above and the ones below are written in **fraction notation.** The top number is called the **numerator** and the bottom number is called the **denominator.**

$$\frac{1}{2}, \quad \frac{3}{4}, \quad \frac{8}{5}, \quad \frac{11}{23}$$

EXAMPLE 1 Identify the numerator and the denominator.

$$\frac{7}{8} \begin{array}{l} \leftarrow \text{Numerator} \\ \leftarrow \text{Denominator} \end{array}$$

Do Exercises 1–4.

Answers on page A-4

Let's look at various situations that involve fractions.

FRACTIONS AS A PARTITION OF AN OBJECT DIVIDED INTO EQUAL PARTS

Consider a candy bar divided into 5 equal sections. If you eat 2 sections, you have eaten $\frac{2}{5}$ of the candy bar.

The denominator 5 tells us the unit, $\frac{1}{5}$. The numerator 2 tells us the number of equal parts we are considering, 2.

EXAMPLE 2 What part is shaded?

The equal parts are eighths. This tells us the unit, $\frac{1}{8}$. The *denominator* is 8. We have 7 of the units shaded. This tells us the *numerator*, 7. Thus,

$$\frac{7}{8} \begin{array}{l} \leftarrow \text{ 7 units are shaded.} \\ \leftarrow \text{ The unit is } \frac{1}{8}. \end{array}$$

is shaded.

Do Exercises 5–9.

The markings on a ruler use fractions.

EXAMPLE 3 What part of an inch is highlighted?

Each inch on the ruler shown above is divided into 16 equal parts. The highlighting extends to the 11th mark. Thus, $\frac{11}{16}$ of an inch is highlighted.

Do Exercise 10.

What part is shaded?

5. $1

6. 1 mile

7.

1 gallon

8. 1 mile

9.

1 gallon

10. What part of an inch is highlighted?

Answers on page A-4

11. What part is shaded?

What part is shaded?

12.   1 mile

13.

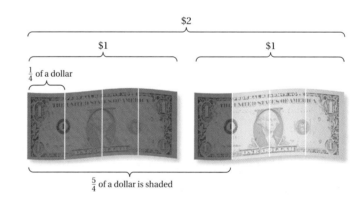

Fractions greater than or equal to 1, such as $\frac{24}{24}$, $\frac{10}{3}$, and $\frac{5}{4}$, correspond to situations like the following.

EXAMPLE 4 What part is shaded?

The rectangle is divided into 24 equal parts. The unit is $\frac{1}{24}$. The denominator is 24. All 24 equal parts are shaded. This tells us that the numerator is 24. Thus, $\frac{24}{24}$ are shaded.

Do Exercise 11.

EXAMPLE 5 What part is shaded?

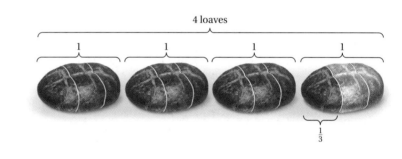

Each loaf of bread is divided into 3 equal parts. The unit is $\frac{1}{3}$. The *denominator* is 3. We have 10 of the units shaded. This tells us the *numerator* is 10. Thus, $\frac{10}{3}$ are shaded.

EXAMPLE 6 What part is shaded?

We can regard this as two objects of 4 parts each and take 5 of those parts. The unit is $\frac{1}{4}$. The denominator is 4, and the numerator is 5. Thus $\frac{5}{4}$ are shaded.

Do Exercises 12 and 13.

Fractions larger than or equal to 1, such as $\frac{13}{6}$ or $\frac{9}{9}$, are sometimes referred to as "improper" fractions. We will not use this terminology because notation such as $\frac{27}{8}$, $\frac{11}{3}$, and $\frac{4}{4}$ is quite "proper" and very common in algebra.

FRACTIONS AS RATIOS

A **ratio** is a quotient of two quantities. We can express a ratio with fraction notation. (We will consider ratios in more detail in Chapter 5.)

EXAMPLE 7 What part of this set, or collection, of people are actresses who have won an Oscar? U.S. Senators?

| Nicole Kidman | Halle Berry | Elizabeth Dole | Hilary Swank | Gwyneth Paltrow | Julia Roberts | Diane Feinstein |

There are 7 people in the set. We know that 5 of them, Nicole Kidman, Halle Berry, Julia Roberts, Hilary Swank, and Gwyneth Paltrow, are actresses who have won an Oscar. Thus, 5 of 7, or $\frac{5}{7}$, are Oscar-winning actresses. The 2 remaining are U.S. Senators. Thus, $\frac{2}{7}$ are U.S. Senators.

Do Exercise 14.

EXAMPLE 8 *Baseball Standings.* The following are the final standings in the American League East for 2004, when the division was won by the New York Yankees. Find the ratio of Yankee wins to losses, wins to total games, and losses to total games.

EAST	W	L	Pct.	GB	STRK	LAST 10	VS. DIV.	HOME	AWAY	VS. NL
New York*	101	61	.623	—	W1	6–4	49–27	57–24	44–37	10–8
Boston	98	64	.605	3.0	L1	7–3	48–28	55–26	43–38	9–9
Baltimore	78	84	.481	23.0	W1	6–4	37–39	38–43	40–41	5–13
Tampa Bay	70	91	.435	30.5	W1	6–4	26–49	41–39	29–52	15–3
Toronto	67	94	.416	33.5	L1	4–6	29–46	40–41	27–53	8–10

*clinched division

Source: Major League Baseball

The Yankees won 101 games and lost 61 games. They played a total of 101 + 61, or 162 games. Thus we have the following.

The ratio of wins to losses is $\frac{101}{61}$.
The ratio of wins to total games is $\frac{101}{162}$.
The ratio of losses to total games is $\frac{61}{162}$.

Do Exercise 15.

14. What part of this set, or collection, are clocks? thermometers?

15. Baseball Standings. Refer to the table in Example 8. The Boston Red Sox finished second in the American League East in 2004. Find the ratio of Red Sox wins to losses, wins to total games, and losses to total games.

Source: Major League Baseball

Answers on page A-4

Simplify.

16. $\dfrac{1}{1}$ **17.** $\dfrac{4}{4}$

18. $\dfrac{34}{34}$ **19.** $\dfrac{100}{100}$

20. $\dfrac{2347}{2347}$ **21.** $\dfrac{103}{103}$

b   ## Some Fraction Notation for Whole Numbers

FRACTION NOTATION FOR 1

The number 1 corresponds to situations like those shown here.

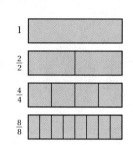

If we divide an object into n parts and take n of them, we get all of the object (1 whole object).

THE NUMBER 1 IN FRACTION NOTATION
$\dfrac{n}{n} = 1,$ for any whole number n that is not 0.

EXAMPLES Simplify.

9. $\dfrac{5}{5} = 1$ **10.** $\dfrac{9}{9} = 1$ **11.** $\dfrac{23}{23} = 1$

Do Exercises 16–21.

FRACTION NOTATION FOR 0

Consider the fraction $\frac{0}{4}$. This corresponds to dividing an object into 4 parts and taking none of them. We get 0.

THE NUMBER 0 IN FRACTION NOTATION
$\dfrac{0}{n} = 0,$ for any whole number n that is not 0.

EXAMPLES Simplify.

12. $\dfrac{0}{1} = 0$ **13.** $\dfrac{0}{9} = 0$ **14.** $\dfrac{0}{23} = 0$

Answers on page A-4

Fraction notation with a denominator of 0, such as $n/0$, is meaningless because we cannot speak of an object being divided into *zero* parts. See also the discussion of excluding division by 0 in Section 1.6.

A DENOMINATOR OF 0

$\dfrac{n}{0}$ is not defined for any whole number n.

Do Exercises 22–27.

OTHER WHOLE NUMBERS

Consider the fraction $\frac{4}{1}$. This corresponds to taking 4 objects and dividing each into 1 part. (We do not divide them.) We have 4 objects.

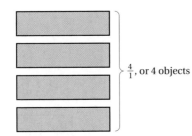

$\frac{4}{1}$, or 4 objects

ANY WHOLE NUMBER IN FRACTION NOTATION

Any whole number divided by 1 is the whole number. That is,

$$\frac{n}{1} = n, \quad \text{for any whole number } n.$$

EXAMPLES Simplify.

15. $\dfrac{2}{1} = 2$ **16.** $\dfrac{9}{1} = 9$ **17.** $\dfrac{34}{1} = 34$

Do Exercises 28–31.

Simplify, if possible.

22. $\dfrac{0}{1}$ **23.** $\dfrac{0}{8}$

24. $\dfrac{0}{107}$ **25.** $\dfrac{4-4}{567}$

26. $\dfrac{15}{0}$ **27.** $\dfrac{0}{3-3}$

Simplify.

28. $\dfrac{8}{1}$ **29.** $\dfrac{10}{1}$

30. $\dfrac{346}{1}$ **31.** $\dfrac{24-1}{23-22}$

Answers on page A-4

a Identify the numerator and the denominator.

1. $\dfrac{3}{4}$ **2.** $\dfrac{9}{10}$ **3.** $\dfrac{11}{2}$ **4.** $\dfrac{18}{5}$ **5.** $\dfrac{0}{7}$ **6.** $\dfrac{1}{13}$

What part of the object or set of objects is shaded?

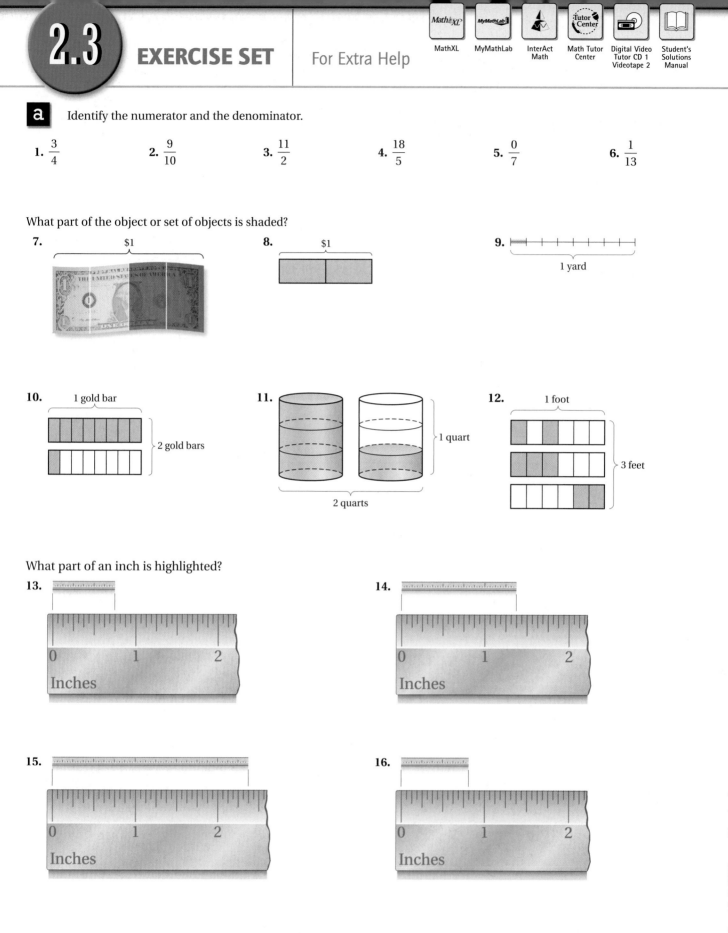

7. $1

8. $1

9. 1 yard

10. 1 gold bar / 2 gold bars

11. 1 quart / 2 quarts

12. 1 foot / 3 feet

What part of an inch is highlighted?

13. Inches

14. Inches

15. Inches

16. Inches

17.

18. } 1 year

19.

20.

21. } 1 pie

22.

23.

1 acre

24. } 1 square inch

25.

26.

27.

28.

For each of Exercises 29–32, give fraction notation for the amount of gas (a) in the tank and (b) used from a full tank.

29.

30.

31.

32.

33. For the following set of people, what is the ratio of:

    a) women to the total number of people?
    b) women to men?
    c) men to the total number of people?
    d) men to women?

34. For the following set of nuts and bolts, what is the ratio of:

    a) nuts to bolts?
    b) bolts to nuts?
    c) nuts to the total number of elements?
    d) total number of elements to nuts?

For Exercises 35 and 36, use the bar graph below that lists the police agency and the number of full-time police officers per 10,000 residents.

Source: Law Enforcement and Administrative Statistics, 2000

35. What is the ratio of police officers per 10,000 residents in the given city?

    a) Orlando   **b)** New York
    c) Detroit   **d)** Washington
    e) St. Louis   **f)** Santa Fe

36. What is the ratio of police officers per 10,000 residents in the given city?

    a) Chicago   **b)** Boston
    c) Newark   **d)** Louisville
    e) Cincinnati   **f)** Fort Worth

37. *Gas Mileage.* A 2005 Mini Cooper S will go 390 mi on 13 gal of gasoline. What is the ratio of:

    a) miles driven to gasoline burned?
    b) gasoline burned to miles driven?

38. Jake delivers car parts to auto service centers. On Thursday he had 15 deliveries. By noon he had delivered only 4 orders. What is the ratio of:

    a) orders delivered to total number of orders?
    b) orders delivered to orders not delivered?
    c) orders not delivered to total number of orders?

39. *Veterinary Care.* Of every 1000 households that own dogs, 850 obtain veterinary care. What is the ratio of households who own dogs that seek veterinary care to all households that own dogs?

Source: *U.S. Pet Ownership and Demographics Sourcebook,* 2002

40. *Moviegoers.* Of every 1000 people who attend movies, 340 are in the 18–24 age group. What is the ratio of moviegoers in the 18–24 age group to all moviegoers?

Source: American Demographics

b   Simplify.

41. $\dfrac{0}{8}$

42. $\dfrac{8}{8}$

43. $\dfrac{8-1}{9-8}$

44. $\dfrac{16}{1}$

45. $\dfrac{20}{20}$

46. $\dfrac{20}{1}$

47. $\dfrac{45}{45}$

48. $\dfrac{11-1}{10-9}$

49. $\dfrac{0}{238}$

50. $\dfrac{238}{1}$

51. $\dfrac{238}{238}$

52. $\dfrac{0}{16}$

53. $\dfrac{3}{3}$

54. $\dfrac{56}{56}$

55. $\dfrac{87}{87}$

56. $\dfrac{98}{98}$

57. $\dfrac{18}{18}$

58. $\dfrac{0}{18}$

59. $\dfrac{18}{1}$

60. $\dfrac{8-8}{1247}$

61. $\dfrac{729}{0}$

62. $\dfrac{1317}{0}$

63. $\dfrac{5}{6-6}$

64. $\dfrac{13}{10-10}$

65. D_W Write a sentence to describe $\frac{5}{8}$ in each of the following situations.

a) As a part of a whole
b) As a part of a set
c) As a ratio

66. D_W Explain in your own words why $n/n = 1$, for any natural number n.

SKILL MAINTENANCE

Round 34,562 to the nearest: [1.4a]

67. Ten.

68. Hundred.

69. Thousand.

70. Ten thousand.

Solve. [1.8a]

71. *Salaries and Education.* In 2001, the average annual salary for a person who completed only high school was $25,303. The average for someone with a bachelor's degree was $40,994. How much more did the person who had a bachelor's degree earn than the person who was only a high school graduate?

Source: U.S. Bureau of the Census

72. *Gas Mileage.* The Pontiac G6 GT gets 29 miles per gal (mpg) in highway driving. How many gallons will it use in 2784 miles of highway driving?

Subtract. [1.3b]

73. $9001 - 6798$

74. $2037 - 1189$

75. $67,113 - 29,874$

76. $12,327 - 476$

SYNTHESIS

What part of the object is shaded?

77.

78.

79.

80.

Shade or mark the figure to show $\frac{3}{5}$.

81.

82.

83.

84.

2.4 MULTIPLICATION AND APPLICATIONS

Objectives

a Multiply a whole number and a fraction.

b Multiply a fraction by a fraction.

c Solve applied problems involving multiplication of fractions.

a Multiplication by a Whole Number

We can find $3 \cdot \frac{1}{4}$ by thinking of repeated addition. We add three $\frac{1}{4}$'s. We see that $3 \cdot \frac{1}{4}$ is $\frac{3}{4}$.

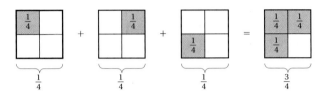

Do Exercises 1 and 2.

We find a product such as $6 \cdot \frac{4}{5}$ as follows.

> To multiply a fraction by a whole number,
>
> **a)** multiply the numerator by the whole number, and
>
> $$6 \cdot \frac{4}{5} = \frac{6 \cdot 4}{5} = \frac{24}{5}$$
>
> **b)** keep the same denominator.

EXAMPLES Multiply.

1. $5 \times \frac{3}{8} = \frac{5 \times 3}{8} = \frac{15}{8}$

2. $\frac{2}{5} \cdot 13 = \frac{2 \cdot 13}{5} = \frac{26}{5}$

3. $10 \cdot \frac{1}{3} = \frac{10}{3}$

Do Exercises 3–5.

b Multiplication Using Fraction Notation

When neither factor is a whole number, multiplication using fraction notation does not correspond to repeated addition. Let's visualize the product of two fractions. We consider the multiplication

$$\frac{3}{5} \cdot \frac{3}{4}.$$

This is equivalent to finding $\frac{3}{5}$ of $\frac{3}{4}$. We first consider an object and take $\frac{3}{4}$ of it. We divide it into 4 equal parts using vertical lines and take 3 of them. That is shown by the shading at right.

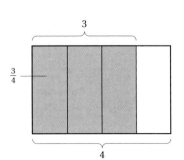

1. Find $2 \cdot \frac{1}{3}$.

1 day   1 day

2. Find $5 \cdot \frac{1}{8}$.

Multiply.

3. $5 \times \frac{2}{3}$

4. $11 \times \frac{3}{8}$

5. $23 \cdot \frac{2}{5}$

Answers on page A-5

6. Draw a diagram like the one at right to show the multiplication $\frac{2}{3} \cdot \frac{4}{5}$.

Multiply.

7. $\frac{3}{8} \cdot \frac{5}{7}$

8. $\frac{4}{3} \times \frac{8}{5}$

9. $\frac{3}{10} \cdot \frac{1}{10}$

10. $7 \cdot \frac{2}{3}$

Answers on page A-5

Next, we take $\frac{3}{5}$ of the shaded area. We divide it into 5 equal parts using horizontal lines and take 3 of them. That is shown by the darker shading below.

The entire object has been divided into 20 parts, and we have shaded 9 of them for a second time. Thus we see that $\frac{3}{5}$ of $\frac{3}{4}$ is $\frac{9}{20}$, or

$$\frac{3}{5} \cdot \frac{3}{4} = \frac{9}{20}.$$

The figure above shows a rectangular array inside a rectangular array. The number of pieces in the entire array is $5 \cdot 4$ (the product of the denominators). The number of pieces shaded a second time is $3 \cdot 3$ (the product of the numerators). And the product is represented by 9 pieces out of a set of 20, or $\frac{9}{20}$, or the product of the numerators over the product of the denominators. This leads us to a statement of the procedure for multiplying a fraction by a fraction.

Do Exercise 6.

We find a product such as $\frac{9}{7} \cdot \frac{3}{4}$ as follows.

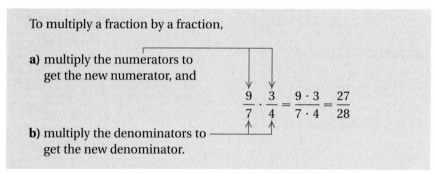

To multiply a fraction by a fraction,

a) multiply the numerators to get the new numerator, and

b) multiply the denominators to get the new denominator.

$$\frac{9}{7} \cdot \frac{3}{4} = \frac{9 \cdot 3}{7 \cdot 4} = \frac{27}{28}$$

EXAMPLES Multiply.

4. $\frac{5}{6} \times \frac{7}{4} = \frac{5 \times 7}{6 \times 4} = \frac{35}{24}$

Skip writing this step whenever you can.

5. $\frac{3}{5} \cdot \frac{7}{8} = \frac{3 \cdot 7}{5 \cdot 8} = \frac{21}{40}$

6. $\frac{3}{5} \cdot \frac{3}{4} = \frac{9}{20}$

7. $\frac{1}{4} \cdot \frac{1}{3} = \frac{1}{12}$

8. $6 \cdot \frac{4}{5} = \frac{6}{1} \cdot \frac{4}{5} = \frac{24}{5}$

Do Exercises 7–10.

C | Applications and Problem Solving

Many problems that can be solved by multiplying fractions can be thought of in terms of rectangular arrays.

EXAMPLE 9 A real estate developer owns a plot of land that measures 1 square mile. He plans to use $\frac{4}{5}$ of the plot for a small strip mall and parking lot. Of this, $\frac{2}{3}$ will be needed for the parking lot. What part of the plot will be used for parking?

1. **Familiarize.** We first make a drawing to help familiarize ourselves with the problem. The land may not be rectangular. It could be in a shape like A or B below. But to think out the problem, we can think of it as a rectangle, as shown in shape C.

1 square mile   1 square mile   1 square mile

The strip mall including the parking lot uses $\frac{4}{5}$ of the plot. We shade $\frac{4}{5}$.

$\frac{4}{5}$

The parking lot alone takes $\frac{2}{3}$ of the preceding part. We shade that.

$\frac{2}{3} \cdot \frac{4}{5}$

2. **Translate.** We let n = the part of the plot that is used for parking. We are taking "two-thirds of four-fifths." Recall from Section 1.8 that the word "of" corresponds to multiplication. Thus the following multiplication sentence corresponds to the situation:

$$\frac{2}{3} \cdot \frac{4}{5} = n.$$

11. A resort hotel uses $\frac{3}{4}$ of its extra land for recreational purposes. Of that, $\frac{1}{2}$ is used for swimming pools. What part of the land is used for swimming pools?

3. Solve. The number sentence tells us what to do. We multiply:

$$\frac{2}{3} \cdot \frac{4}{5} = \frac{2 \cdot 4}{3 \cdot 5} = \frac{8}{15}.$$

4. Check. We can check partially by noting that the answer is smaller than the original area, 1, which we expect since the developer is using only part of the original plot of land. Thus, $\frac{8}{15}$ is a reasonable answer. We can also check this in the figure above, where we see that 8 of 15 parts have been shaded a second time.

5. State. The parking lot takes $\frac{8}{15}$ of the square mile of land.

Do Exercise 11.

Example 9 and the preceding discussion indicate that the area of a rectangular region can be found by multiplying length by width. That is true whether length and width are whole numbers or not. Remember, the area of a rectangular region is given by the formula

$$A = l \cdot w.$$

12. Area of a Ceramic Tile. The length of a rectangular ceramic tile on an inlaid ceramic counter is $\frac{4}{9}$ ft. The width is $\frac{2}{9}$ ft. What is the area of one tile?

EXAMPLE 10 *Area of an Ice-Skating Rink.* The length of a rectangular ice-skating rink in the atrium of a shopping mall is $\frac{7}{100}$ mi. The width is $\frac{3}{100}$ mi. What is the area of the rink?

1. Familiarize. Recall that area is length times width. We make a drawing and let A = the area of the ice-skating rink.

2. Translate. Then we translate.

$$\underbrace{\text{Area}}_{\downarrow} \quad \underbrace{\text{is}}_{\downarrow} \quad \underbrace{\text{Length}}_{\downarrow} \quad \underbrace{\text{times}}_{\downarrow} \quad \underbrace{\text{Width}}_{\downarrow}$$
$$A \quad = \quad \frac{7}{100} \quad \times \quad \frac{3}{100}$$

3. Solve. The sentence tells us what to do. We multiply:

$$\frac{7}{100} \cdot \frac{3}{100} = \frac{7 \cdot 3}{100 \cdot 100} = \frac{21}{10,000}.$$

4. Check. We check by repeating the calculation. This is left to the student.

5. State. The area is $\frac{21}{10,000}$ mi^2.

Do Exercise 12.

Answers on page A-5

EXAMPLE 11 A recipe for Banana Oat Pancakes calls for $\frac{3}{4}$ cup of old fashioned oats. A chef is making $\frac{1}{2}$ of the recipe. How much oats should the chef use?

Source: Reprinted with permission from *Taste of Home* magazine, www.tasteofhome.com

1. **Familiarize.** We first make a drawing or at least visualize the situation. We let $n =$ the amount of oats the chef should use.

$\frac{3}{4}$ cup in recipe

$\frac{1}{2}$ of $\frac{3}{4}$ cup

2. **Translate.** We are finding $\frac{1}{2}$ of $\frac{3}{4}$, so the multiplication sentence $\frac{1}{2} \cdot \frac{3}{4} = n$ corresponds to the situation.

3. **Solve.** We carry out the multiplication:

$$\frac{1}{2} \cdot \frac{3}{4} = \frac{1 \cdot 3}{2 \cdot 4} = \frac{3}{8}.$$

4. **Check.** We check by repeating the calculation. This is left to the student.

5. **State.** The chef should use $\frac{3}{8}$ cup of oats.

Do Exercise 13.

13. Of the students at Overton Junior College, $\frac{1}{8}$ participate in sports and $\frac{3}{5}$ of these play football. What fractional part of the students play football?

Answer on page A-5

Study Tips   BETTER TEST TAKING

How often do you make the following statement after taking a test: "I was able to do the homework, but I froze during the test"? Here are two tips to help you with this difficulty. Both are intended to make test taking less stressful by getting you to practice good test-taking habits on a daily basis.

■ **Treat every homework exercise as if it were a test question.**   If you had to work a problem at your job with no answer provided, what would you do? You would probably work it very deliberately, checking and rechecking every step. You might work it more than one time, or you might try to work it another way to check the result. Try to use this approach when doing your homework. Treat every exercise as though it were a test question with no answer at the back of the book.

■ **Be sure that you do questions without answers as part of every homework assignment whether or not the instructor has assigned them!**   One reason a test may seem such a different task is that questions on a test lack answers. That is the reason for taking a test: to see if you can do the questions without assistance. As part of your test preparation, be sure you do some exercises for which you do not have the answers. Thus when you take a test, you are doing a more familiar task.

The purpose of doing your homework using these approaches is to give you more test-taking practice beforehand. Let's make a sports analogy here. At a basketball game, the players take lots of practice shots before the game. They play the first half, go to the locker room, and come out for the second half. What do they do before the second half, even though they have just played 20 minutes of basketball? They shoot baskets again! We suggest the same approach here. Create more and more situations in which you practice taking test questions by treating each homework exercise like a test question and by doing exercises for which you have no answers. Good luck!

119

a   Multiply.

1. $3 \cdot \dfrac{1}{5}$

2. $2 \cdot \dfrac{1}{3}$

3. $5 \times \dfrac{1}{8}$

4. $4 \times \dfrac{1}{5}$

5. $\dfrac{2}{11} \cdot 4$

6. $\dfrac{2}{5} \cdot 3$

7. $10 \cdot \dfrac{7}{9}$

8. $9 \cdot \dfrac{5}{8}$

9. $\dfrac{2}{5} \cdot 1$

10. $\dfrac{3}{8} \cdot 1$

11. $\dfrac{2}{5} \cdot 3$

12. $\dfrac{3}{5} \cdot 4$

13. $7 \cdot \dfrac{3}{4}$

14. $7 \cdot \dfrac{2}{5}$

15. $17 \times \dfrac{5}{6}$

16. $\dfrac{3}{7} \cdot 40$

b   Multiply.

17. $\dfrac{1}{2} \cdot \dfrac{1}{3}$

18. $\dfrac{1}{6} \cdot \dfrac{1}{4}$

19. $\dfrac{1}{4} \times \dfrac{1}{10}$

20. $\dfrac{1}{3} \times \dfrac{1}{10}$

21. $\dfrac{2}{3} \times \dfrac{1}{5}$

22. $\dfrac{3}{5} \times \dfrac{1}{5}$

23. $\dfrac{2}{5} \cdot \dfrac{2}{3}$

24. $\dfrac{3}{4} \cdot \dfrac{3}{5}$

25. $\dfrac{3}{4} \cdot \dfrac{3}{4}$

26. $\dfrac{3}{7} \cdot \dfrac{4}{5}$

27. $\dfrac{2}{3} \cdot \dfrac{7}{13}$

28. $\dfrac{3}{11} \cdot \dfrac{4}{5}$

29. $\dfrac{1}{10} \cdot \dfrac{7}{10}$

30. $\dfrac{3}{10} \cdot \dfrac{3}{10}$

31. $\dfrac{7}{8} \cdot \dfrac{7}{8}$

32. $\dfrac{4}{5} \cdot \dfrac{4}{5}$

33. $\dfrac{1}{10} \cdot \dfrac{1}{100}$

34. $\dfrac{3}{10} \cdot \dfrac{7}{100}$

35. $\dfrac{14}{15} \cdot \dfrac{13}{19}$

36. $\dfrac{12}{13} \cdot \dfrac{12}{13}$

c   Solve.

37. *Tossed Salad.* The recipe for Cherry Brie Tossed Salad calls for $\frac{3}{4}$ cup of sliced almonds. How much is needed to make $\frac{1}{2}$ of the recipe?

Source: Reprinted with permission from *Taste of Home* magazine, www.tasteofhome.com

38. It takes $\frac{2}{3}$ yd of ribbon to make a bow. How much ribbon is needed to make 5 bows?

$\leftarrow\!\!-\!\!-\ \frac{2}{3}\text{ yd}\ -\!\!-\!\!\rightarrow$

CHAPTER 2: Fraction Notation: Multiplication and Division

39. *Basement Carpet.* A basement floor is being covered with carpet. An area $\frac{7}{8}$ of the length and $\frac{3}{4}$ of the width is covered by lunch time. What fraction of the floor has been completed?

40. *Basketball: High School to Pro.* One of 35 high school basketball players plays college basketball. One of 75 college players plays professional basketball. What fractional part of high school players play professional basketball?

Source: National Basketball Association

41. A rectangular table top measures $\frac{4}{5}$ m long by $\frac{3}{5}$ m wide. What is its area?

42. If each piece of pie is $\frac{1}{6}$ of a pie, how much of the pie is $\frac{1}{2}$ of a piece?

43. A gasoline can holds $\frac{7}{8}$ liter (L). How much will the can hold when it is $\frac{1}{2}$ full?

44. *Floor Tiling.* The floor of a room is being covered with tile. An area $\frac{3}{5}$ of the length and $\frac{3}{4}$ of the width is covered. What fraction of the floor has been tiled?

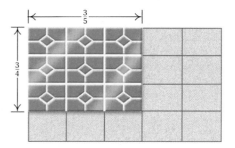

45. $\mathbf{D_W}$ Write a problem for a classmate to solve. Design the problem so that the solution is "About $\frac{1}{30}$ of the students are left-handed women."

46. $\mathbf{D_W}$ On pp. 115–116, we explained, using words and pictures, why $\frac{3}{5} \cdot \frac{3}{4}$ equals $\frac{9}{20}$. Present a similar explanation of why $\frac{2}{3} \cdot \frac{4}{7}$ equals $\frac{8}{21}$.

SKILL MAINTENANCE

Divide. [1.6b]

47. $7140 \div 35$

48. $32,200 \div 46$

49. $9 \overline{)27,009}$

50. $35 \overline{)7148}$

What does the digit 8 mean in each number? [1.1a]

51. 4,678,952

52. 8,473,901

53. 7148

54. 23,803

Simplify. [1.9c]

55. $12 - 3^2$

56. $(12 - 3)^2$

57. $8 \cdot 12 - (63 \div 9 + 13 \cdot 3)$

58. $(10 - 3)^4 + 10^3 \cdot 4 - 10 \div 5$

SYNTHESIS

Multiply. Write the answer using fraction notation.

59. ▦ $\dfrac{341}{517} \cdot \dfrac{209}{349}$

60. ▦ $\left(\dfrac{57}{61}\right)^3$

61. ▦ $\left(\dfrac{2}{5}\right)^3 \left(\dfrac{7}{9}\right)$

62. ▦ $\left(\dfrac{1}{2}\right)^5 \left(\dfrac{3}{5}\right)$

Objectives

 a Use multiplying by 1 to find different fraction notation for a number.

b Simplify fraction notation.

c Use the test for equality to determine whether two fractions name the same number.

Multiply.

1. $\dfrac{1}{2} \cdot \dfrac{8}{8}$ **2.** $\dfrac{3}{5} \cdot \dfrac{10}{10}$

3. $\dfrac{13}{25} \cdot \dfrac{4}{4}$ **4.** $\dfrac{8}{3} \cdot \dfrac{25}{25}$

Answers on page A-5

122

2.5 SIMPLIFYING

a Multiplying by 1

Recall the following:

$$1 = \frac{1}{1} = \frac{2}{2} = \frac{3}{3} = \frac{4}{4} = \frac{10}{10} = \frac{45}{45} = \frac{100}{100} = \frac{n}{n}.$$

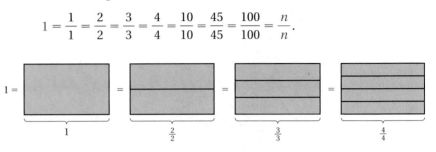

Any nonzero number divided by itself is 1. (See Section 1.6.)

Now recall the multiplicative identity from Section 1.5. For any whole number a, $1 \cdot a = a \cdot 1 = a$. This holds for numbers of arithmetic as well.

MULTIPLICATIVE IDENTITY FOR FRACTIONS

When we multiply a number by 1, we get the same number:

$$\frac{3}{5} = \frac{3}{5} \cdot 1 = \frac{3}{5} \cdot \frac{4}{4} = \frac{12}{20}.$$

Since $\frac{3}{5} = \frac{12}{20}$, we know that $\frac{3}{5}$ and $\frac{12}{20}$ are two names for the same number. We also say that $\frac{3}{5}$ and $\frac{12}{20}$ are **equivalent.**

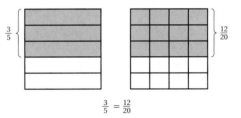

$$\frac{3}{5} = \frac{12}{20}$$

Do Exercises 1–4.

Suppose we want to find a name for $\frac{2}{3}$, but one that has a denominator of 9. We can multiply by 1 to find equivalent fractions:

$$\frac{2}{3} = \frac{2}{3} \cdot \frac{3}{3} = \frac{2 \cdot 3}{3 \cdot 3} = \frac{6}{9}.$$

Since $9 = 3 \cdot 3$, we chose $\frac{3}{3}$ for 1 in order to get a denominator of 9.

$$\frac{2}{3} = \frac{6}{9}$$

EXAMPLE 1 Find a name for $\frac{1}{4}$ with a denominator of 24.

Since $4 \cdot 6 = 24$, we multiply by $\frac{6}{6}$:

$$\frac{1}{4} = \frac{1}{4} \cdot \frac{6}{6} = \frac{1 \cdot 6}{4 \cdot 6} = \frac{6}{24}.$$

EXAMPLE 2 Find a name for $\frac{2}{5}$ with a denominator of 35.

Since $5 \cdot 7 = 35$, we multiply by $\frac{7}{7}$:

$$\frac{2}{5} = \frac{2}{5} \cdot \frac{7}{7} = \frac{2 \cdot 7}{5 \cdot 7} = \frac{14}{35}.$$

Do Exercises 5–9.

b Simplifying Fraction Notation

All of the following are names for three-fourths:

$$\frac{3}{4}, \frac{6}{8}, \frac{9}{12}, \frac{12}{16}, \frac{15}{20}.$$

We say that $\frac{3}{4}$ is **simplest** because it has the smallest numerator and the smallest denominator. That is, the numerator and the denominator have no common factor other than 1.

To simplify, we reverse the process of multiplying by 1:

$$\frac{12}{18} = \frac{2 \cdot 6}{3 \cdot 6} \quad \leftarrow \text{Factoring the numerator}$$
$$\leftarrow \text{Factoring the denominator}$$

$$= \frac{2}{3} \cdot \frac{6}{6} \quad \text{Factoring the fraction}$$

$$= \frac{2}{3} \cdot 1 \qquad \frac{6}{6} = 1$$

$$= \frac{2}{3}. \qquad \text{Removing a factor of 1: } \frac{2}{3} \cdot 1 = \frac{2}{3}$$

EXAMPLES Simplify.

3. $\dfrac{8}{20} = \dfrac{2 \cdot 4}{5 \cdot 4} = \dfrac{2}{5} \cdot \dfrac{4}{4} = \dfrac{2}{5}$

4. $\dfrac{2}{6} = \dfrac{1 \cdot 2}{3 \cdot 2} = \dfrac{1}{3} \cdot \dfrac{2}{2} = \dfrac{1}{3}$

> The number 1 allows for pairing of factors in the numerator and the denominator.

5. $\dfrac{30}{6} = \dfrac{5 \cdot 6}{1 \cdot 6} = \dfrac{5}{1} \cdot \dfrac{6}{6} = \dfrac{5}{1} = 5$ $\leftarrow$

> We could also simplify $\frac{30}{6}$ by doing the division $30 \div 6$. That is, $\frac{30}{6} = 30 \div 6 = 5$.

Do Exercises 10–13.

Find another name for the number, but with the denominator indicated. Use multiplying by 1.

5. $\dfrac{4}{3} = \dfrac{?}{9}$

6. $\dfrac{3}{4} = \dfrac{?}{24}$

7. $\dfrac{9}{10} = \dfrac{?}{100}$

8. $\dfrac{3}{15} = \dfrac{?}{45}$

9. $\dfrac{8}{7} = \dfrac{?}{49}$

Simplify.

10. $\dfrac{2}{8}$

11. $\dfrac{10}{12}$

12. $\dfrac{40}{8}$

13. $\dfrac{24}{18}$

Answers on page A-5

Simplify.

14. $\dfrac{35}{40}$ **15.** $\dfrac{801}{702}$

16. $\dfrac{24}{21}$ **17.** $\dfrac{75}{300}$

18. $\dfrac{280}{960}$ **19.** $\dfrac{1332}{2880}$

20. Simplify each fraction in this circle graph.

Days Spent Shopping for the Holidays

8 to 14 days $\dfrac{14}{100}$

1 to 2 days $\dfrac{20}{100}$

15 or more days $\dfrac{16}{100}$

Less than a day $\dfrac{8}{100}$

3 to 7 days $\dfrac{42}{100}$

The use of prime factorizations can be helpful for simplifying when numerators and/or denominators are larger numbers.

EXAMPLE 6   Simplify: $\dfrac{90}{84}$.

$\dfrac{90}{84} = \dfrac{2 \cdot 3 \cdot 3 \cdot 5}{2 \cdot 2 \cdot 3 \cdot 7}$ Factoring the numerator and the denominator into primes

$= \dfrac{2 \cdot 3 \cdot 3 \cdot 5}{2 \cdot 3 \cdot 2 \cdot 7}$ Changing the order so that like primes are above and below each other

$= \dfrac{2}{2} \cdot \dfrac{3}{3} \cdot \dfrac{3 \cdot 5}{2 \cdot 7}$ Factoring the fraction

$= 1 \cdot 1 \cdot \dfrac{3 \cdot 5}{2 \cdot 7}$

$= \dfrac{3 \cdot 5}{2 \cdot 7}$ Removing factors of 1

$= \dfrac{15}{14}$

The tests for divisibility (Section 2.2) are very helpful in simplifying fraction notation. We could have shortened the preceding example had we noted that 6 is a factor of both the numerator and the denominator. Then we have

$$\dfrac{90}{84} = \dfrac{6 \cdot 15}{6 \cdot 14} = \dfrac{6}{6} \cdot \dfrac{15}{14} = \dfrac{15}{14}.$$

EXAMPLE 7   Simplify: $\dfrac{603}{207}$.

At first glance this looks difficult. But note, using the test for divisibility by 9 (sum of digits divisible by 9), that both the numerator and the denominator are divisible by 9. Thus we can factor 9 from both numbers:

$$\dfrac{603}{207} = \dfrac{9 \cdot 67}{9 \cdot 23} = \dfrac{9}{9} \cdot \dfrac{67}{23} = \dfrac{67}{23}.$$

EXAMPLE 8   Simplify: $\dfrac{660}{1140}$.

Using the tests for divisibility, we have

$\dfrac{660}{1140} = \dfrac{10 \cdot 66}{10 \cdot 114} = \dfrac{10}{10} \cdot \dfrac{66}{114} = \dfrac{66}{114}$ Both 660 and 1140 are divisible by 10.

$= \dfrac{6 \cdot 11}{6 \cdot 19} = \dfrac{6}{6} \cdot \dfrac{11}{19} = \dfrac{11}{19}.$ Both 66 and 114 are divisible by 6.

Do Exercises 14–20.

Answers on page A-5

CANCELING

Canceling is a shortcut that you may have used for removing a factor of 1 when working with fraction notation. With *great* concern, we mention it as a possibility for speeding up your work. Canceling may be done only when removing common factors in numerators and denominators. Each such pair allows us to remove a factor of 1 in a product.

Our concern is that canceling be done with care and understanding. In effect, slashes are used to indicate factors of 1 that have been removed. For instance, Example 6 might have been done faster as follows:

$$\frac{90}{84} = \frac{2 \cdot 3 \cdot 3 \cdot 5}{2 \cdot 2 \cdot 3 \cdot 7} \qquad \text{Factoring the numerator and the denominator}$$

$$= \frac{\cancel{2} \cdot \cancel{3} \cdot 3 \cdot 5}{2 \cdot \cancel{2} \cdot \cancel{3} \cdot 7} \qquad \text{When a factor of 1 is noted,}$$

$$\qquad\qquad\qquad\text{it is "canceled" as shown: } \frac{2 \cdot 3}{2 \cdot 3} = 1.$$

$$= \frac{3 \cdot 5}{2 \cdot 7} = \frac{15}{14}.$$

Caution!

The difficulty with canceling is that it is often applied incorrectly in situations like the following:

$$\frac{\cancel{2} + 3}{\cancel{2}} = 3; \qquad \frac{\cancel{4} + 1}{\cancel{4} + 2} = \frac{1}{2}; \qquad \frac{1\cancel{5}}{\cancel{5}4} = \frac{1}{4}.$$

Wrong!   Wrong!   Wrong!

The correct answers are

$$\frac{2 + 3}{2} = \frac{5}{2}; \qquad \frac{4 + 1}{4 + 2} = \frac{5}{6}; \qquad \frac{15}{54} = \frac{3 \cdot 5}{3 \cdot 18} = \frac{3}{3} \cdot \frac{5}{18} = \frac{5}{18}.$$

In each situation, the number canceled was not a factor of 1. Factors are parts of products. For example, in $2 \cdot 3$, 2 and 3 are factors, but in $2 + 3$, 2 and 3 are *not* factors. Canceling may not be done when sums or differences are in numerators or denominators, as shown here. **If you cannot factor, you cannot cancel! If in doubt, do not cancel!**

C   A Test for Equality

When denominators are the same, we say that fractions have a **common denominator**. When fractions have a common denominator, we can compare them by comparing numerators. Suppose we want to compare $\frac{3}{6}$ and $\frac{2}{4}$. First we find a common denominator. To do this, we multiply each fraction by 1, using the denominator of the other fraction to form the symbol for 1.

The "unit" is $\frac{1}{6}$.

$$\frac{3}{6} = \frac{3}{6} \cdot \frac{4}{4} = \frac{3 \cdot 4}{6 \cdot 4} = \frac{12}{24}$$

$$\frac{2}{4} = \frac{2}{4} \cdot \frac{6}{6} = \frac{2 \cdot 6}{4 \cdot 6} = \frac{12}{24}$$

The "unit" is $\frac{1}{4}$.

Both "units" are $\frac{1}{24}$.

We multiply $\frac{3}{6}$ by $\frac{4}{4}$ and $\frac{2}{4}$ by $\frac{6}{6}$.

Once we have a common denominator, 24, we compare the numerators. And since these numerators are both 12, the fractions are equal:

$$\frac{3}{6} = \frac{2}{4}.$$

125

Use = or ≠ for ☐ to write a true sentence.

21. $\dfrac{2}{6}$ ☐ $\dfrac{3}{9}$

Note in the preceding that if

$$\frac{3}{6} = \frac{2}{4}, \quad \text{then} \quad 3 \cdot 4 = 6 \cdot 2.$$

This tells us that we need to check only the products $3 \cdot 4$ and $6 \cdot 2$ to compare the fractions.

A TEST FOR EQUALITY

We multiply these two numbers: $3 \cdot 4$.　　　　We multiply these two numbers: $6 \cdot 2$.

We call $3 \cdot 4$ and $6 \cdot 2$ **cross products.** Since the cross products are the same, that is, $3 \cdot 4 = 6 \cdot 2$, we know that

$$\frac{3}{6} = \frac{2}{4}.$$

If a sentence $a = b$ is true, it means that a and b name the same number. If a sentence $a \neq b$ (read "a is not equal to b") is true, it means that a and b do *not* name the same number.

22. $\dfrac{2}{3}$ ☐ $\dfrac{14}{20}$

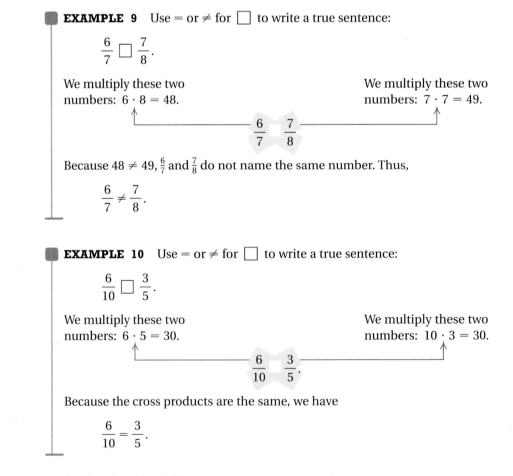

EXAMPLE 9 Use = or ≠ for ☐ to write a true sentence:

$$\frac{6}{7} \ \square \ \frac{7}{8}.$$

We multiply these two numbers: $6 \cdot 8 = 48$.　　　　We multiply these two numbers: $7 \cdot 7 = 49$.

Because $48 \neq 49$, $\frac{6}{7}$ and $\frac{7}{8}$ do not name the same number. Thus,

$$\frac{6}{7} \neq \frac{7}{8}.$$

EXAMPLE 10 Use = or ≠ for ☐ to write a true sentence:

$$\frac{6}{10} \ \square \ \frac{3}{5}.$$

We multiply these two numbers: $6 \cdot 5 = 30$.　　　　We multiply these two numbers: $10 \cdot 3 = 30$.

Because the cross products are the same, we have

$$\frac{6}{10} = \frac{3}{5}.$$

Answers on page A-5

Do Exercises 21 and 22.

CHAPTER 2: Fraction Notation: Multiplication and Division

MathXL MyMathLab InterAct Math Math Tutor Center Digital Video Tutor CD 1 Videotape 2 Student's Solutions Manual

a Find another name for the given number, but with the denominator indicated. Use multiplying by 1.

1. $\dfrac{1}{2} = \dfrac{?}{10}$

2. $\dfrac{1}{6} = \dfrac{?}{18}$

3. $\dfrac{5}{8} = \dfrac{?}{32}$

4. $\dfrac{2}{9} = \dfrac{?}{18}$

5. $\dfrac{9}{10} = \dfrac{?}{30}$

6. $\dfrac{5}{6} = \dfrac{?}{48}$

7. $\dfrac{7}{8} = \dfrac{?}{32}$

8. $\dfrac{2}{5} = \dfrac{?}{25}$

9. $\dfrac{5}{12} = \dfrac{?}{48}$

10. $\dfrac{3}{8} = \dfrac{?}{56}$

11. $\dfrac{17}{18} = \dfrac{?}{54}$

12. $\dfrac{11}{16} = \dfrac{?}{256}$

13. $\dfrac{5}{3} = \dfrac{?}{45}$

14. $\dfrac{11}{5} = \dfrac{?}{30}$

15. $\dfrac{7}{22} = \dfrac{?}{132}$

16. $\dfrac{10}{21} = \dfrac{?}{126}$

b Simplify.

17. $\dfrac{2}{4}$

18. $\dfrac{4}{8}$

19. $\dfrac{6}{8}$

20. $\dfrac{8}{12}$

21. $\dfrac{3}{15}$

22. $\dfrac{8}{10}$

23. $\dfrac{24}{8}$

24. $\dfrac{36}{9}$

25. $\dfrac{18}{24}$

26. $\dfrac{42}{48}$

27. $\dfrac{14}{16}$

28. $\dfrac{15}{25}$

29. $\dfrac{12}{10}$

30. $\dfrac{16}{14}$

31. $\dfrac{16}{48}$

32. $\dfrac{100}{20}$

33. $\dfrac{150}{25}$

34. $\dfrac{19}{76}$

35. $\dfrac{17}{51}$

36. $\dfrac{425}{525}$

37. $\dfrac{220}{4125}$

38. $\dfrac{540}{810}$

39. $\dfrac{1575}{3920}$

40. $\dfrac{1000}{1080}$

c Use = or ≠ for ☐ to write a true sentence.

41. $\dfrac{3}{4} \;\square\; \dfrac{9}{12}$

42. $\dfrac{4}{8} \;\square\; \dfrac{3}{6}$

43. $\dfrac{1}{5} \;\square\; \dfrac{2}{9}$

44. $\dfrac{1}{4} \;\square\; \dfrac{2}{9}$

45. $\dfrac{3}{8} \;\square\; \dfrac{6}{16}$

46. $\dfrac{2}{6} \;\square\; \dfrac{6}{18}$

47. $\dfrac{2}{5} \;\square\; \dfrac{3}{7}$

48. $\dfrac{1}{3} \;\square\; \dfrac{1}{4}$

49. $\dfrac{12}{9}$ □ $\dfrac{8}{6}$

50. $\dfrac{16}{14}$ □ $\dfrac{8}{7}$

51. $\dfrac{5}{2}$ □ $\dfrac{17}{7}$

52. $\dfrac{3}{10}$ □ $\dfrac{7}{24}$

53. $\dfrac{3}{10}$ □ $\dfrac{30}{100}$

54. $\dfrac{700}{1000}$ □ $\dfrac{70}{100}$

55. $\dfrac{5}{10}$ □ $\dfrac{520}{1000}$

56. $\dfrac{49}{100}$ □ $\dfrac{50}{1000}$

57. $\mathbf{D_W}$ Explain in your own words when it *is* possible to "cancel" and when it *is not* possible to "cancel."

58. $\mathbf{D_W}$ Can fraction notation be simplified if its numerator and its denominator are two different prime numbers? Why or why not?

SKILL MAINTENANCE

Solve. [1.8a]

59. A playing field is 78 ft long and 64 ft wide. What is its area? its perimeter?

60. A landscaper buys 13 small maple trees and 17 small oak trees for a project. A maple costs $23 and an oak costs $37. How much is spent altogether for the trees?

Subtract. [1.3b]

61. $34 - 23$

62. $50 - 18$

63. $803 - 617$

64. $8344 - 5607$

Solve. [1.7b]

65. $30 \cdot x = 150$

66. $10{,}947 = 123 \cdot y$

67. $5280 = 1760 + t$

68. $x + 2368 = 11{,}369$

SYNTHESIS

Simplify. Use the list of prime numbers on p. 93.

69. ▦ $\dfrac{2603}{2831}$

70. ▦ $\dfrac{3197}{3473}$

71. *Shy People.* Sociologists have found that 4 of 10 people are shy. Write fraction notation for the part of the population that is shy; the part that is not shy. Simplify.

72. *Left-Handed People.* Sociologists estimate that 3 of 20 people are left-handed. In a crowd of 460 people, how many would you expect to be left-handed?

73. *Batting Averages.* For the 2004 season, Ichiro Suzuki, of the Seattle Mariners, won the American League batting title with 262 hits in 704 times at bat. Barry Bonds, of the San Francisco Giants, won the National League title with 135 hits in 373 times at bat. Did they have the same fraction of hits in times at bat (batting average)? Why or why not?
Source: Major League Baseball

74. ▦ On a test of 82 questions, a student got 63 correct. On another test of 100 questions, she got 77 correct. Did she get the same portion of each test correct? Why or why not?

a Multiplying and Simplifying Using Fraction Notation

It is often possible to simplify after we multiply. To make such simplifying easier, it is generally best not to carry out the products in the numerator and the denominator, but to factor and simplify first. Consider the product

$$\frac{3}{8} \cdot \frac{4}{9}.$$

We proceed as follows:

$$\frac{3}{8} \cdot \frac{4}{9} = \frac{3 \cdot 4}{8 \cdot 9}$$ We write the products in the numerator and the denominator, but we do not carry them out.

$$= \frac{3 \cdot 2 \cdot 2}{2 \cdot 2 \cdot 2 \cdot 3 \cdot 3}$$ Factoring the numerator and the denominator

$$= \frac{3 \cdot 2 \cdot 2 \cdot 1}{2 \cdot 2 \cdot 2 \cdot 3 \cdot 3}$$ Using the identity property of 1 to insert the number 1 as a factor

$$= \frac{3 \cdot 2 \cdot 2}{3 \cdot 2 \cdot 2} \cdot \frac{1}{2 \cdot 3}$$ Factoring the fraction

$$= 1 \cdot \frac{1}{2 \cdot 3}$$

$$= \frac{1}{2 \cdot 3}$$ Removing a factor of 1

$$= \frac{1}{6}.$$

The procedure could have been shortened had we noticed that 4 is a factor of the 8 in the denominator:

$$\frac{3}{8} \cdot \frac{4}{9} = \frac{3 \cdot 4}{8 \cdot 9} = \frac{3 \cdot 4}{4 \cdot 2 \cdot 3 \cdot 3} = \frac{3 \cdot 4}{3 \cdot 4} \cdot \frac{1}{2 \cdot 3} = 1 \cdot \frac{1}{2 \cdot 3} = \frac{1}{2 \cdot 3} = \frac{1}{6}.$$

> **To multiply and simplify:**
>
> **a)** Write the products in the numerator and the denominator, but do not carry out the products.
> **b)** Factor the numerator and the denominator.
> **c)** Factor the fraction to remove a factor of 1.
> **d)** Carry out the remaining products.

EXAMPLES Multiply and simplify.

1. $\dfrac{2}{3} \cdot \dfrac{9}{4} = \dfrac{2 \cdot 9}{3 \cdot 4} = \dfrac{2 \cdot 3 \cdot 3}{3 \cdot 2 \cdot 2} = \dfrac{2 \cdot 3}{2 \cdot 3} \cdot \dfrac{3}{2} = 1 \cdot \dfrac{3}{2} = \dfrac{3}{2}$

2. $\dfrac{6}{7} \cdot \dfrac{5}{3} = \dfrac{6 \cdot 5}{7 \cdot 3} = \dfrac{3 \cdot 2 \cdot 5}{7 \cdot 3} = \dfrac{3}{3} \cdot \dfrac{2 \cdot 5}{7} = 1 \cdot \dfrac{2 \cdot 5}{7} = \dfrac{2 \cdot 5}{7} = \dfrac{10}{7}$

3. $40 \cdot \dfrac{7}{8} = \dfrac{40 \cdot 7}{8} = \dfrac{8 \cdot 5 \cdot 7}{8 \cdot 1} = \dfrac{8}{8} \cdot \dfrac{5 \cdot 7}{1} = 1 \cdot \dfrac{5 \cdot 7}{1} = \dfrac{5 \cdot 7}{1} = 35$

Multiply and simplify.

1. $\dfrac{2}{3} \cdot \dfrac{7}{8}$

2. $\dfrac{4}{5} \cdot \dfrac{5}{12}$

3. $16 \cdot \dfrac{3}{8}$

4. $\dfrac{5}{8} \cdot 4$

5. A landscaper uses $\frac{2}{3}$ lb of peat moss for a rosebush. How much will be needed for 21 rosebushes?

Canceling can be used as follows for these examples.

1. $\dfrac{2}{3} \cdot \dfrac{9}{4} = \dfrac{2 \cdot 9}{3 \cdot 4} = \dfrac{\cancel{2} \cdot \cancel{3} \cdot 3}{\cancel{3} \cdot \cancel{2} \cdot 2} = \dfrac{3}{2}$ Removing a factor of 1: $\dfrac{2 \cdot 3}{2 \cdot 3} = 1$

2. $\dfrac{6}{7} \cdot \dfrac{5}{3} = \dfrac{6 \cdot 5}{7 \cdot 3} = \dfrac{\cancel{3} \cdot 2 \cdot 5}{7 \cdot \cancel{3}} = \dfrac{2 \cdot 5}{7} = \dfrac{10}{7}$ Removing a factor of 1: $\dfrac{3}{3} = 1$

3. $40 \cdot \dfrac{7}{8} = \dfrac{40 \cdot 7}{8} = \dfrac{\cancel{8} \cdot 5 \cdot 7}{\cancel{8} \cdot 1} = \dfrac{5 \cdot 7}{1} = 35$ Removing a factor of 1: $\dfrac{8}{8} = 1$

Remember, if you can't factor, you can't cancel!

Do Exercises 1–4.

b   Applications and Problem Solving

EXAMPLE 4   Elite Elegance is preparing souvenir favors for a charity fund-raising dinner. How many pounds of caramels will be needed to fill 235 boxes if each box contains $\frac{2}{5}$ lb?

1. Familiarize. We first make a drawing or at least visualize the situation. Repeated addition will work here.

235 boxes

$\frac{2}{5}$ of a pound in each box

We let $n =$ the number of pounds of caramels.

2. Translate. The problem translates to the following equation:

$$n = 235 \cdot \dfrac{2}{5}.$$

3. Solve. To solve the equation, we carry out the multiplication:

$$n = 235 \cdot \dfrac{2}{5} = \dfrac{235 \cdot 2}{5} \qquad \text{Multiplying}$$

$$= \dfrac{5 \cdot 47 \cdot 2}{5 \cdot 1}$$

$$= \dfrac{5}{5} \cdot \dfrac{47 \cdot 2}{1}$$

$$= 94. \qquad \text{Simplifying}$$

4. Check. We check by repeating the calculation. (The check is left to the student.) We can also think about the reasonableness of the answer. We are putting less than a pound of caramels in each box, so the answer should be less than 235. Since 94 is less than 235, we have a partial check of the reasonableness of the answer. The number 94 checks.

5. State. Thus, 94 lb of caramels will be needed.

Do Exercise 5.

a   Multiply and simplify.   Don't forget to simplify!

1. $\dfrac{2}{3} \cdot \dfrac{1}{2}$

2. $\dfrac{3}{8} \cdot \dfrac{1}{3}$

3. $\dfrac{7}{8} \cdot \dfrac{1}{7}$

4. $\dfrac{4}{9} \cdot \dfrac{1}{4}$

5. $\dfrac{1}{8} \cdot \dfrac{4}{5}$

6. $\dfrac{2}{5} \cdot \dfrac{1}{6}$

7. $\dfrac{1}{4} \cdot \dfrac{2}{3}$

8. $\dfrac{4}{6} \cdot \dfrac{1}{6}$

9. $\dfrac{12}{5} \cdot \dfrac{9}{8}$

10. $\dfrac{16}{15} \cdot \dfrac{5}{4}$

11. $\dfrac{10}{9} \cdot \dfrac{7}{5}$

12. $\dfrac{25}{12} \cdot \dfrac{4}{3}$

13. $9 \cdot \dfrac{1}{9}$

14. $4 \cdot \dfrac{1}{4}$

15. $\dfrac{1}{3} \cdot 3$

16. $\dfrac{1}{6} \cdot 6$

17. $\dfrac{7}{10} \cdot \dfrac{10}{7}$

18. $\dfrac{8}{9} \cdot \dfrac{9}{8}$

19. $\dfrac{7}{5} \cdot \dfrac{5}{7}$

20. $\dfrac{2}{11} \cdot \dfrac{11}{2}$

21. $\dfrac{1}{4} \cdot 8$

22. $\dfrac{1}{3} \cdot 18$

23. $24 \cdot \dfrac{1}{6}$

24. $16 \cdot \dfrac{1}{2}$

25. $12 \cdot \dfrac{3}{4}$

26. $18 \cdot \dfrac{5}{6}$

27. $\dfrac{3}{8} \cdot 24$

28. $\dfrac{2}{9} \cdot 36$

29. $13 \cdot \dfrac{2}{5}$

30. $15 \cdot \dfrac{1}{6}$

31. $\dfrac{7}{10} \cdot 28$

32. $\dfrac{5}{8} \cdot 34$

33. $\dfrac{1}{6} \cdot 360$

34. $\dfrac{1}{3} \cdot 120$

35. $240 \cdot \dfrac{1}{8}$

36. $150 \cdot \dfrac{1}{5}$

37. $\dfrac{4}{10} \cdot \dfrac{5}{10}$

38. $\dfrac{7}{10} \cdot \dfrac{34}{150}$

39. $\dfrac{8}{10} \cdot \dfrac{45}{100}$

40. $\dfrac{3}{10} \cdot \dfrac{8}{10}$

41. $\dfrac{11}{24} \cdot \dfrac{3}{5}$

42. $\dfrac{15}{22} \cdot \dfrac{4}{7}$

43. $\dfrac{10}{21} \cdot \dfrac{3}{4}$

44. $\dfrac{17}{18} \cdot \dfrac{3}{5}$

b   Solve.

The *pitch* of a screw is the distance between its threads. With each complete rotation, the screw goes in or out a distance equal to its pitch. Use this information to answer Exercises 45 and 46.

}Pitch = $\frac{1}{16}$ in. ➡

Each rotation moves the screw in or out $\frac{1}{16}$ in.

45. The pitch of a screw is $\frac{1}{16}$ in. How far will it go into a piece of oak when it is turned 10 complete rotations clockwise?

46. The pitch of a screw is $\frac{3}{32}$ in. How far will it go out of a piece of plywood when it is turned 10 complete rotations counterclockwise?

47. *Running Speeds.* The maximum running speed of an elk over approximately quarter-mile distances is about 45 mph. The maximum speed of a grizzly bear is about $\frac{2}{3}$ that of an elk. Find the running speed of a grizzly bear.

Source: *Natural History Magazine,* ©American Museum of Natural History

$\frac{2}{3}x$

$x = 45$ mph

48. After Jack completes 60 hr of teacher training at college, he can earn $75 for working a full day as a substitute teacher. How much will he receive for working $\frac{1}{5}$ of a day?

49. *Mailing-List Addresses.* Business people have determined that $\frac{1}{4}$ of the addresses on a mailing list will change in one year. A business has a mailing list of 2500 people. After one year, how many addresses on that list will be incorrect?

50. *Shy People.* Sociologists have determined that $\frac{2}{5}$ of the people in the world are shy. A sales manager is interviewing 650 people for an aggressive sales position. How many of these people might be shy?

51. A recipe for piecrust calls for $\frac{2}{3}$ cup of flour. A chef is making $\frac{1}{2}$ of the recipe. How much flour should the chef use?

52. Of the students in the freshman class, $\frac{2}{5}$ have digital cameras; $\frac{1}{4}$ of these students also join the college photography club. What fraction of the students in the freshman class join the photography club?

53. A house worth $154,000 is assessed for $\frac{3}{4}$ of its value. What is the assessed value of the house?

54. Roxanne's tuition was $4600. A loan was obtained for $\frac{3}{4}$ of the tuition. How much was the loan?

55. *Map Scaling.* On a map, 1 in. represents 240 mi. How much does $\frac{2}{3}$ in. represent?

56. *Map Scaling.* On a map, 1 in. represents 120 mi. How much does $\frac{3}{4}$ in. represent?

57. *Household Budgets.* A family has an annual income of $36,000. Of this, $\frac{1}{4}$ is spent for food, $\frac{1}{5}$ for housing, $\frac{1}{10}$ for clothing, $\frac{1}{9}$ for savings, $\frac{1}{4}$ for taxes, and the rest for other expenses. How much is spent for each?

58. *Household Budgets.* A family has an annual income of $29,700. Of this, $\frac{1}{4}$ is spent for food, $\frac{1}{5}$ for housing, $\frac{1}{10}$ for clothing, $\frac{1}{9}$ for savings, $\frac{1}{4}$ for taxes, and the rest for other expenses. How much is spent for each?

Family Income

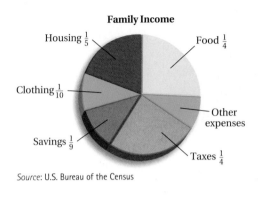

Source: U.S. Bureau of the Census

59. DW When multiplying using fraction notation, we form products in the numerator and the denominator, but do not immediately calculate the products. Why?

60. DW If a fraction's numerator and denominator have no factors (other than 1) in common, can the fraction be simplified? Why or why not?

Solve. [1.7b]

61. $48 \cdot t = 1680$

62. $74 \cdot x = 6290$

63. $3125 = 25 \cdot t$

64. $2880 = 24 \cdot y$

65. $t + 28 = 5017$

66. $456 + x = 9002$

67. $8797 = y + 2299$

68. $10,000 = 3593 + m$

Subtract. [1.3b]

69. $\begin{array}{r} 9\ 0\ 6\ 0 \\ -\ 4\ 3\ 8\ 7 \\ \hline \end{array}$

70. $\begin{array}{r} 7\ 8\ 0\ 0 \\ -\ 2\ 4\ 6\ 2 \\ \hline \end{array}$

Multiply and simplify. Use the list of prime numbers on p. 93 or a fraction calculator.

71. ▦ $\dfrac{201}{535} \cdot \dfrac{4601}{6499}$

72. ▦ $\dfrac{5767}{3763} \cdot \dfrac{159}{395}$

73. *College Profile.* Of students entering a college, $\frac{7}{8}$ have completed high school and $\frac{2}{3}$ are older than 20. If $\frac{1}{7}$ of all students are left-handed, what fraction of students entering the college are left-handed high school graduates over the age of 20?

74. *College Profile.* Refer to the information in Exercise 73. If 480 students are entering the college, how many of them are left-handed high school graduates 20 yr old or younger?

75. *College Profile.* Refer to Exercise 73. What fraction of students entering the college did not graduate high school, are 20 yr old or younger, and are left-handed?

2.7 DIVISION AND APPLICATIONS

Objectives

a Find the reciprocal of a number.

b Divide and simplify using fraction notation.

c Solve equations of the type $a \cdot x = b$ and $x \cdot a = b$, where a and b may be fractions.

d Solve applied problems involving division of fractions.

a Reciprocals

Look at these products:

$$8 \cdot \frac{1}{8} = \frac{8 \cdot 1}{8} = \frac{8}{8} = 1; \qquad \frac{2}{3} \cdot \frac{3}{2} = \frac{2 \cdot 3}{3 \cdot 2} = \frac{6}{6} = 1.$$

RECIPROCALS

If the product of two numbers is 1, we say that they are **reciprocals** of each other. To find a reciprocal of a fraction, interchange the numerator and the denominator.

$$\text{Number} \longrightarrow \frac{3}{4} \rightarrow \frac{4}{3} \longrightarrow \text{Reciprocal}$$

EXAMPLES Find the reciprocal.

1. The reciprocal of $\frac{4}{5}$ is $\frac{5}{4}$. $\quad \frac{4}{5} \cdot \frac{5}{4} = \frac{20}{20} = 1$

2. The reciprocal of $\frac{8}{7}$ is $\frac{7}{8}$. $\quad \frac{8}{7} \cdot \frac{7}{8} = \frac{56}{56} = 1$

3. The reciprocal of 24 is $\frac{1}{24}$. $\quad$ Think of 24 as $\frac{24}{1}$: $\frac{24}{1} \cdot \frac{1}{24} = \frac{24}{24} = 1$.

4. The reciprocal of $\frac{1}{3}$ is 3. $\quad \frac{1}{3} \cdot 3 = \frac{3}{3} = 1$

Do Exercises 1–4.

Does 0 have a reciprocal? If it did, it would have to be a number x such that

$$0 \cdot x = 1.$$

But 0 times any number is 0. Thus we have the following.

0 HAS NO RECIPROCAL

The number 0, or $\frac{0}{n}$, has no reciprocal. $\left(\text{Recall that } \frac{n}{0} \text{ is not defined.} \right)$

Find the reciprocal.

1. $\frac{2}{5}$

2. $\frac{10}{7}$

3. 9

4. $\frac{1}{5}$

Answers on page A-5

b Division

Consider the division $\frac{3}{4} \div \frac{1}{8}$. We are asking how many $\frac{1}{8}$'s are in $\frac{3}{4}$. We can answer this by looking at the figure below.

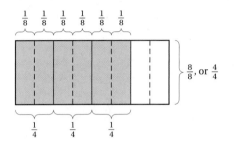

We see that there are six $\frac{1}{8}$'s in $\frac{3}{4}$. Thus,

$$\frac{3}{4} \div \frac{1}{8} = 6.$$

We can check this by multiplying:

$$6 \cdot \frac{1}{8} = \frac{6}{8} = \frac{3}{4}.$$

Here is a faster way to do this division:

$$\frac{3}{4} \div \frac{1}{8} = \frac{3}{4} \cdot \frac{8}{1} = \frac{24}{4} = 6.$$ Multiplying by the reciprocal of the divisor

---

To divide fractions, multiply the dividend by the reciprocal of the divisor:

$$\frac{2}{5} \div \frac{3}{4} = \frac{2}{5} \cdot \frac{4}{3} = \frac{2 \cdot 4}{5 \cdot 3} = \frac{8}{15}.$$

Multiply by the reciprocal of the divisor.

EXAMPLES   Divide and simplify.

5. $\dfrac{5}{6} \div \dfrac{2}{3} = \dfrac{5}{6} \cdot \dfrac{3}{2} = \dfrac{5 \cdot 3}{6 \cdot 2} = \dfrac{5 \cdot 3}{3 \cdot 2 \cdot 2} = \dfrac{3}{3} \cdot \dfrac{5}{2 \cdot 2} = \dfrac{5}{2 \cdot 2} = \dfrac{5}{4}$

6. $\dfrac{7}{8} \div \dfrac{1}{16} = \dfrac{7}{8} \cdot 16 = \dfrac{7 \cdot 16}{8} = \dfrac{7 \cdot 2 \cdot 8}{8 \cdot 1} = \dfrac{8}{8} \cdot \dfrac{7 \cdot 2}{1} = \dfrac{7 \cdot 2}{1} = 14$

7. $\dfrac{2}{5} \div 6 = \dfrac{2}{5} \cdot \dfrac{1}{6} = \dfrac{2 \cdot 1}{5 \cdot 6} = \dfrac{2 \cdot 1}{5 \cdot 2 \cdot 3} = \dfrac{2}{2} \cdot \dfrac{1}{5 \cdot 3} = \dfrac{1}{5 \cdot 3} = \dfrac{1}{15}$

8. $\dfrac{3}{5} \div \dfrac{1}{2} = \dfrac{3}{5} \cdot 2 = \dfrac{3 \cdot 2}{5} = \dfrac{6}{5}$

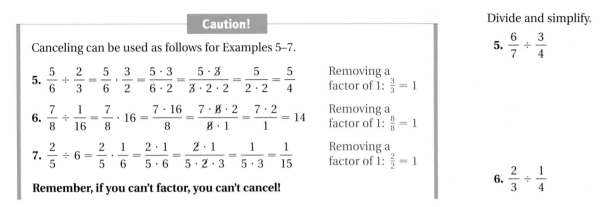

Caution!

Canceling can be used as follows for Examples 5–7.

5. $\dfrac{5}{6} \div \dfrac{2}{3} = \dfrac{5}{6} \cdot \dfrac{3}{2} = \dfrac{5 \cdot 3}{6 \cdot 2} = \dfrac{5 \cdot \cancel{3}}{\cancel{3} \cdot 2 \cdot 2} = \dfrac{5}{2 \cdot 2} = \dfrac{5}{4}$ Removing a factor of 1: $\frac{3}{3} = 1$

6. $\dfrac{7}{8} \div \dfrac{1}{16} = \dfrac{7}{8} \cdot 16 = \dfrac{7 \cdot 16}{8} = \dfrac{7 \cdot \cancel{8} \cdot 2}{\cancel{8} \cdot 1} = \dfrac{7 \cdot 2}{1} = 14$ Removing a factor of 1: $\frac{8}{8} = 1$

7. $\dfrac{2}{5} \div 6 = \dfrac{2}{5} \cdot \dfrac{1}{6} = \dfrac{2 \cdot 1}{5 \cdot 6} = \dfrac{\cancel{2} \cdot 1}{5 \cdot \cancel{2} \cdot 3} = \dfrac{1}{5 \cdot 3} = \dfrac{1}{15}$ Removing a factor of 1: $\frac{2}{2} = 1$

Remember, if you can't factor, you can't cancel!

Do Exercises 5–9.

What is the explanation for multiplying by a reciprocal when dividing? Let's consider $\frac{2}{3} \div \frac{7}{5}$. We multiply by 1. The name for 1 that we will use is $(5/7)/(5/7)$; it comes from the reciprocal of $\frac{7}{5}$.

$$\frac{2}{3} \div \frac{7}{5} = \frac{\dfrac{2}{3}}{\dfrac{7}{5}}$$ Writing fraction notation for the division

$$= \frac{\dfrac{2}{3}}{\dfrac{7}{5}} \cdot 1$$ Multiplying by 1

$$= \frac{\dfrac{2}{3}}{\dfrac{7}{5}} \cdot \frac{\dfrac{5}{7}}{\dfrac{5}{7}}$$ Multiplying by 1; $\frac{5}{7}$ is the reciprocal of $\frac{7}{5}$ and $\dfrac{\frac{5}{7}}{\frac{5}{7}} = 1$

$$= \frac{\dfrac{2}{3} \cdot \dfrac{5}{7}}{\dfrac{7}{5} \cdot \dfrac{5}{7}}$$ Multiplying the numerators and the denominators

$$= \frac{\dfrac{2}{3} \cdot \dfrac{5}{7}}{1}$$ After we multiplied, we got 1 for the denominator. The numerator shows the multiplication by the reciprocal.

$$= \frac{2}{3} \cdot \frac{5}{7} = \frac{10}{21}$$

Thus,

$$\frac{2}{3} \div \frac{7}{5} = \frac{2}{3} \cdot \frac{5}{7} = \frac{10}{21}.$$

Do Exercise 10.

Divide and simplify.

5. $\dfrac{6}{7} \div \dfrac{3}{4}$

6. $\dfrac{2}{3} \div \dfrac{1}{4}$

7. $\dfrac{4}{5} \div 8$

8. $60 \div \dfrac{3}{5}$

9. $\dfrac{3}{5} \div \dfrac{3}{5}$

10. Divide by multiplying by 1:

$$\frac{\dfrac{4}{5}}{\dfrac{6}{7}}.$$

Answers on page A-5

137

2.7   Division and Applications

Solve.

11. $\dfrac{5}{6} \cdot y = \dfrac{2}{3}$

12. $\dfrac{3}{4} \cdot n = 24$

c Solving Equations

Now let's solve equations $a \cdot x = b$ and $x \cdot a = b$, where a and b may be fractions. We proceed as we did with equations involving whole numbers. We divide by a on both sides.

EXAMPLE 9 Solve: $\frac{4}{3} \cdot x = \frac{6}{7}$.

We have

$$\frac{4}{3} \cdot x = \frac{6}{7}$$

$$\frac{\frac{4}{3} \cdot x}{\frac{4}{3}} = \frac{\frac{6}{7}}{\frac{4}{3}} \qquad \text{Dividing by } \tfrac{4}{3} \text{ on both sides}$$

$$x = \frac{6}{7} \cdot \frac{3}{4} \qquad \text{Multiplying by the reciprocal}$$

$$= \frac{6 \cdot 3}{7 \cdot 4} = \frac{2 \cdot 3 \cdot 3}{7 \cdot 2 \cdot 2} = \frac{2}{2} \cdot \frac{3 \cdot 3}{7 \cdot 2} = \frac{3 \cdot 3}{7 \cdot 2} = \frac{9}{14}.$$

The solution is $\frac{9}{14}$.

EXAMPLE 10 Solve: $t \cdot \frac{4}{5} = 80$.

Dividing by $\frac{4}{5}$ on both sides, we get

$$t = 80 \div \frac{4}{5} = 80 \cdot \frac{5}{4} = \frac{80 \cdot 5}{4} = \frac{4 \cdot 20 \cdot 5}{4 \cdot 1} = \frac{4}{4} \cdot \frac{20 \cdot 5}{1} = \frac{20 \cdot 5}{1} = 100.$$

The solution is 100.

Do Exercises 11 and 12.

d Applications and Problem Solving

EXAMPLE 11 *Syringes.* How many syringes, each containing $\frac{3}{5}$ mL, can a clinical pharmacist fill from a 30 mL vial?

1. **Familiarize.** We are asking the question, "How many $\frac{3}{5}$'s are in 30?" Repeated addition will apply here. We make a drawing. We let $n =$ the number of syringes in all.

$\frac{3}{5}$ of a milliliter in each syringe

n syringes in all

2. **Translate.** The equation that corresponds to the situation is

$$n = 30 \div \frac{3}{5}.$$

3. Solve. We solve the equation by carrying out the division:

$$n = 30 \div \frac{3}{5} = 30 \cdot \frac{5}{3} = \frac{30 \cdot 5}{3} = \frac{3 \cdot 10 \cdot 5}{3 \cdot 1}$$

$$= \frac{3}{3} \cdot \frac{10 \cdot 5}{1} = 50.$$

4. Check. We check by repeating the calculation.

5. State. The pharmacist can fill 50 syringes.

Do Exercise 13.

EXAMPLE 12 *Hand-Knit Scarves.* Nayah knits winter scarves for the Quick-Knit Boutique. After she knits 63 inches of a scarf, $\frac{7}{8}$ of a scarf is complete. What is the total length of this scarf?

1. Familiarize. We ask: "63 in. is $\frac{7}{8}$ of what length?" We make a drawing or at least visualize the problem. We let $s =$ the length of the scarf.

2. Translate. We translate to an equation.

Fraction completed	of	Total length of scarf	is	Amount already knitted
$\frac{7}{8}$	$\cdot$	s	$=$	63

3. Solve. The equation that corresponds to the situation is $\frac{7}{8} \cdot s = 63$. We divide by $\frac{7}{8}$ on both sides and carry out the division:

$$s = 63 \div \frac{7}{8} = 63 \cdot \frac{8}{7} = \frac{63 \cdot 8}{7} = \frac{7 \cdot 9 \cdot 8}{7} = \frac{7}{7} \cdot \frac{9 \cdot 8}{1} = 72.$$

4. Check. Since $\frac{7}{8} \cdot 72 = \frac{7 \cdot 8 \cdot 9}{8} = \frac{8}{8} \cdot \frac{7 \cdot 9}{1} = 63$, the answer, 72 checks.

5. State. The completed scarf is 72 in. long.

Do Exercise 14.

---

13. Each loop in a spring uses $\frac{3}{8}$ in. of wire. How many loops can be made from 120 in. of wire?

14. Sales Trip. Miles Lanosga sells soybean seeds to seed companies. After he had driven 210 mi, $\frac{5}{6}$ of his sales trip was completed. How long was the total trip?

$\frac{5}{6}$ of the trip
210 mi

Answers on page A-5

Translating for Success

1. *Valentine Boxes.* Jane's Fudge Shop is preparing Valentine boxes. How many pounds of fudge will be needed to fill 80 boxes if each box contains $\frac{5}{16}$ lb?

2. *Gallons of Gasoline.* On the third day of a business trip, a sales representative used $\frac{4}{5}$ of a full tank of gasoline. If the tank is a 20-gallon tank, how many gallons were used on the third day?

3. *Purchasing a Shirt.* Tom received $36 for his birthday. If he spends $\frac{3}{4}$ of the gift on a new shirt, what is the cost of the shirt?

4. *Checkbook Balance.* The balance in Sam's checking account is $1456. He writes a check for $28 and makes a deposit of $52. What is the new balance?

5. *Valentine Boxes.* Jane's Fudge Shop prepared 80 lb of fudge for Valentine boxes. If each box contains $\frac{5}{16}$ lb, how many boxes can be filled?

The goal of these matching questions is to practice step (2), *Translate*, of the five-step problem-solving process. Translate each word problem to an equation and select a correct translation from equations A–O.

A. $x = \frac{3}{4} \cdot 36$

B. $28 \cdot x = 52$

C. $x = 80 \cdot \frac{5}{16}$

D. $x = 1456 \div 28$

E. $x = \frac{5}{4} \cdot 20$

F. $20 = \frac{4}{5} \cdot x$

G. $x = 12 \cdot 28$

H. $x = \frac{4}{5} \cdot 20$

I. $\frac{3}{4} \cdot x = 36$

J. $x = 1456 - 52 - 28$

K. $x \div 28 = 1456$

L. $x = 52 - 28$

M. $x = 52 \cdot 28$

N. $x = 1456 - 28 + 52$

O. $\frac{5}{16} \cdot x = 80$

Answers on page A-5

6. *Gasoline Tank.* A gasoline tank contains 20 gal when it is $\frac{4}{5}$ full. How many gallons can it hold when full?

7. *Knitting a Scarf.* It takes Rachel 36 hr to knit a scarf. She can knit only $\frac{3}{4}$ hr per day because she is taking 16 hr of college classes. How many days will it take her to knit the scarf?

8. *Bicycle Trip.* On a recent 52-mile bicycle trip, David stopped to make a cellphone call after completing 28 miles. How many more miles does he bicycle after the call?

9. *Crème de Menthe Thins.* Andes Candies L.P. makes Crème de Menthe Thins. How many 28-piece packages can be filled with 1456 pieces?

10. *Cereal Donations.* The Williams family donates 28 boxes of cereal weekly to the local Family in Crisis Center. How many boxes does this family donate in one year?

2.7

EXERCISE SET

For Extra Help

Math⫶XL   MyMathLab   InterAct   Math Tutor   Digital Video   Student's
                         Math   Center   Tutor CD 1   Solutions
MathXL   MyMathLab   Videotape 2   Manual

a Find the reciprocal.

1. $\dfrac{5}{6}$

2. $\dfrac{7}{8}$

3. 6

4. 4

5. $\dfrac{1}{6}$

6. $\dfrac{1}{4}$

7. $\dfrac{10}{3}$

8. $\dfrac{17}{4}$

b Divide and simplify. | Don't forget to simplify!

9. $\dfrac{3}{5} \div \dfrac{3}{4}$

10. $\dfrac{2}{3} \div \dfrac{3}{4}$

11. $\dfrac{3}{5} \div \dfrac{9}{4}$

12. $\dfrac{6}{7} \div \dfrac{3}{5}$

13. $\dfrac{4}{3} \div \dfrac{1}{3}$

14. $\dfrac{10}{9} \div \dfrac{1}{3}$

15. $\dfrac{1}{3} \div \dfrac{1}{6}$

16. $\dfrac{1}{4} \div \dfrac{1}{5}$

17. $\dfrac{3}{8} \div 3$

18. $\dfrac{5}{6} \div 5$

19. $\dfrac{12}{7} \div 4$

20. $\dfrac{18}{5} \div 2$

21. $12 \div \dfrac{3}{2}$

22. $24 \div \dfrac{3}{8}$

23. $28 \div \dfrac{4}{5}$

24. $40 \div \dfrac{2}{3}$

25. $\dfrac{5}{8} \div \dfrac{5}{8}$

26. $\dfrac{2}{5} \div \dfrac{2}{5}$

27. $\dfrac{8}{15} \div \dfrac{4}{5}$

28. $\dfrac{6}{13} \div \dfrac{3}{26}$

29. $\dfrac{9}{5} \div \dfrac{4}{5}$

30. $\dfrac{5}{12} \div \dfrac{25}{36}$

31. $120 \div \dfrac{5}{6}$

32. $360 \div \dfrac{8}{7}$

c Solve.

33. $\dfrac{4}{5} \cdot x = 60$

34. $\dfrac{3}{2} \cdot t = 90$

35. $\dfrac{5}{3} \cdot y = \dfrac{10}{3}$

36. $\dfrac{4}{9} \cdot m = \dfrac{8}{3}$

37. $x \cdot \dfrac{25}{36} = \dfrac{5}{12}$

38. $p \cdot \dfrac{4}{5} = \dfrac{8}{15}$

39. $n \cdot \dfrac{8}{7} = 360$

40. $y \cdot \dfrac{5}{6} = 120$

d Solve.

41. Benny uses $\frac{2}{5}$ gram (g) of toothpaste each time he brushes his teeth. If Benny buys a 30-gram tube, how many times will he be able to brush his teeth?

42. *Gasoline Tanker.* A tanker that delivers gasoline to gas stations had 1400 gal of gasoline when it was $\frac{7}{9}$ full. How much could the tanker hold when it is full?

43. A pair of basketball shorts requires $\frac{3}{4}$ yd of nylon. How many pairs of shorts can be made from 24 yd of nylon?

44. A child's baseball shirt requires $\frac{5}{6}$ yd of fabric. How many shirts can be made from 25 yd of the fabric?

45. How many $\frac{2}{3}$-cup sugar bowls can be filled from 16 cups of sugar?

46. How many $\frac{2}{3}$-cup cereal bowls can be filled from 10 cups of cornflakes?

47. A bucket had 12 L of water in it when it was $\frac{3}{4}$ full. How much could it hold altogether?

48. A tank had 20 L of gasoline in it when it was $\frac{4}{5}$ full. How much could it hold altogether?

49. Yoshi Teramoto sells hardware tools. After driving 180 kilometers (km), he completes $\frac{5}{8}$ of a sales trip. How long is the total trip? How many kilometers are left to drive?

50. A piece of coaxial cable $\frac{4}{5}$ meter (m) long is to be cut into 8 pieces of the same length. What is the length of each piece?

Pitch of a screw. Refer to Exercises 45 and 46 in Exercise Set 2.6.

51. After a screw has been turned 8 complete rotations, it is extended $\frac{1}{2}$ in. into a piece of wallboard. What is the pitch of the screw?

52. The pitch of a screw is $\frac{3}{32}$ in. How many complete rotations are necessary to drive the screw $\frac{3}{4}$ in. into a piece of pine wood?

53. **D_W** Without performing the division, explain why $5 \div \frac{1}{7}$ is a greater number than $5 \div \frac{2}{3}$.

54. **D_W** A student incorrectly insists that $\frac{2}{5} \div \frac{3}{4}$ is $\frac{15}{8}$. What mistake is he probably making?

VOCABULARY REINFORCEMENT

In each of Exercises 55–62, fill in the blank with the correct term from the given list. Some of the choices may not be used.

55. The equation $14 + (2 + 30) = (14 + 2) + 30$ illustrates the _____ law of addition.   [1.2a]

56. In the product $10 \cdot \frac{3}{4}$, 10 and $\frac{3}{4}$ are called _____.   [2.1a]

57. A natural number that has exactly two different factors, only itself and 1, is called a _____ number.   [2.1c]

58. In the fraction $\frac{4}{17}$, we call 17 the _____.   [2.3a]

59. For any number a, $a + 0 = a$. The number 0 is the _____ identity.   [1.2a]

60. The product of 6 and $\frac{1}{6}$ is 1; we say that 6 and $\frac{1}{6}$ are _____ of each other.   [2.7a]

61. The set of _____ numbers is 0, 1, 2, 3, 4, ...   [1.1b]

62. A sentence with = is called a(n) _____.   [1.7a]

associative

commutative

additive

multiplicative

reciprocals

factors

prime

composite

numerator

denominator

equation

variables

whole

natural

Simplify. Use the list of prime numbers on p. 93.

63. ▦ $\dfrac{711}{1957} \div \dfrac{10{,}033}{13{,}081}$

64. ▦ $\dfrac{8633}{7387} \div \dfrac{485}{581}$

65. $\left(\dfrac{9}{10} \div \dfrac{2}{5} \div \dfrac{3}{8} \right)^2$

66. $\dfrac{\left(\dfrac{3}{7}\right)^2 \div \dfrac{12}{5}}{\left(\dfrac{2}{9}\right)\left(\dfrac{9}{2}\right)}$

67. If $\frac{1}{3}$ of a number is $\frac{1}{4}$, what is $\frac{1}{2}$ of the number?

143

The review that follows is meant to prepare you for a chapter exam. It consists of two parts. The first part, Concept Reinforcement, is designed to increase understanding of the concepts through true/false exercises. The second part is the Review Exercises. These provide practice exercises for the exam, together with references to section objectives so you can go back and review. Before beginning, stop and look back over the skills you have obtained. What skills in mathematics do you have now that you did not have before studying this chapter?

✎ CONCEPT REINFORCEMENT

Determine whether the statement is true or false. Answers are given at the back of the book.

_____ **1.** A number a is divisible by another number b if b is a factor of a.

_____ **2.** If a number is not divisible by 6, then it is not divisible by 3.

_____ **3.** The number 1 is not prime.

_____ **4.** For any natural number n, $\dfrac{n}{n} > \dfrac{0}{n}$.

_____ **5.** A number is divisible by 10 if its ones digit is 0 or 5.

_____ **6.** If a number is divisible by 9, then it is also divisible by 3.

_____ **7.** The fraction $\dfrac{13}{6}$ is larger than the fraction $\dfrac{11}{6}$.

_____ **8.** The fraction $\dfrac{13}{7}$ is larger than the fraction $\dfrac{13}{6}$.

Review Exercises

Find all the factors of the number.   [2.1a]

1. 60   **2.** 176

3. Multiply by 1, 2, 3, and so on, to find ten multiples of 8.
[2.1b]

4. Determine whether 924 is divisible by 11.   [2.1b]

5. Determine whether 1800 is divisible by 16.   [2.1b]

Determine whether the number is prime, composite, or neither.   [2.1c]

6. 37   **7.** 1   **8.** 91

Find the prime factorization of the number.   [2.1d]

9. 70   **10.** 30

11. 45   **12.** 150

13. 648   **14.** 5250

To answer Exercises 15–22, consider the following numbers:
[2.2a]

140	716	93	2802
95	2432	330	711
182	4344	255,555	
475	600	780	

Which of the above are divisible by the given number?

15. 3   **16.** 2

17. 4   **18.** 8

19. 5   **20.** 6

21. 9   **22.** 10

23. Identify the numerator and the denominator of $\frac{2}{7}$. [2.3a]

What part of the object or set of objects is shaded? [2.3a]

24.

25. } 2 bars

26.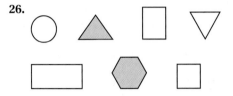

27. For a committee in the United States Senate that consists of 3 Democrats and 5 Republicans, what is the ratio of: [2.3a]

   a) Democrats to Republicans?
   b) Republicans to Democrats?
   c) Democrats to the total number of members of the committee?

Simplify. [2.3b], [2.5b]

28. $\frac{0}{4}$ **29.** $\frac{23}{23}$ **30.** $\frac{48}{1}$

31. $\frac{48}{8}$ **32.** $\frac{10}{15}$ **33.** $\frac{7}{28}$

34. $\frac{21}{21}$ **35.** $\frac{0}{25}$ **36.** $\frac{12}{30}$

37. $\frac{18}{1}$ **38.** $\frac{32}{8}$ **39.** $\frac{9}{27}$

40. $\frac{18}{0}$ **41.** $\frac{5}{8-8}$ **42.** $\frac{88}{184}$

43. $\frac{140}{490}$ **44.** $\frac{1170}{1200}$ **45.** $\frac{288}{2025}$

46. Simplify, if possible, the fractions on this circle graph. [2.5b]

How the Business Travel Dollar Is Spent

Airfare $\frac{43}{100}$

Other $\frac{3}{100}$

Car rental $\frac{8}{100}$

Lodging $\frac{21}{100}$

Entertainment $\frac{10}{100}$

Meals $\frac{15}{100}$

Use = or ≠ for ☐ to write a true sentence. [2.5c]

47. $\frac{3}{5}$ ☐ $\frac{4}{6}$ **48.** $\frac{4}{7}$ ☐ $\frac{8}{14}$

49. $\frac{4}{5}$ ☐ $\frac{5}{6}$ **50.** $\frac{4}{3}$ ☐ $\frac{28}{21}$

Multiply and simplify. [2.6a]

51. $4 \cdot \frac{3}{8}$ **52.** $\frac{7}{3} \cdot 24$

53. $9 \cdot \frac{5}{18}$ **54.** $\frac{6}{5} \cdot 20$

55. $\frac{3}{4} \cdot \frac{8}{9}$ **56.** $\frac{5}{7} \cdot \frac{1}{10}$

57. $\frac{3}{7} \cdot \frac{14}{9}$ **58.** $\frac{1}{4} \cdot \frac{2}{11}$

59. $\frac{4}{25} \cdot \frac{15}{16}$ **60.** $\frac{11}{3} \cdot \frac{30}{77}$

Find the reciprocal. [2.7a]

61. $\dfrac{4}{5}$

62. 3

63. $\dfrac{1}{9}$

64. $\dfrac{47}{36}$

Divide and simplify. [2.7b]

65. $6 \div \dfrac{4}{3}$

66. $\dfrac{5}{9} \div \dfrac{5}{18}$

67. $\dfrac{1}{6} \div \dfrac{1}{11}$

68. $\dfrac{3}{14} \div \dfrac{6}{7}$

69. $\dfrac{1}{4} \div \dfrac{1}{9}$

70. $180 \div \dfrac{3}{5}$

71. $\dfrac{23}{25} \div \dfrac{23}{25}$

72. $\dfrac{2}{3} \div \dfrac{3}{2}$

Solve. [2.7c]

73. $\dfrac{5}{4} \cdot t = \dfrac{3}{8}$

74. $x \cdot \dfrac{2}{3} = 160$

Solve. [2.6b], [2.7d]

75. A road crew repaves $\frac{1}{12}$ mi of road each day. How long will it take the crew to repave a $\frac{3}{4}$-mile stretch of road?

76. After driving 600 km, the Buxton family has completed $\frac{3}{5}$ of their vacation. How long is the total trip?

77. Molly is making a pepper steak recipe that calls for $\frac{2}{3}$ cup of green bell peppers. How much would be needed to make $\frac{1}{2}$ recipe? 3 recipes?

78. Bernardo usually earns $105 for working a full day. How much does he receive for working $\frac{1}{7}$ of a day?

79. A book bag requires $\frac{4}{5}$ yd of fabric. How many bags can be made from 48 yd?

80. *Corn Production.* In 2003, the United States produced approximately $\frac{2}{5}$ of the world production of corn. The total world corn production was about 640,000,000 metric tons. How much corn did the United States produce?

Source: UN Food and Agriculture Organization

81. **Dw** Write, in your own words, a series of steps that can be used when simplifying fraction notation. [2.5b]

82. **Dw** A student claims that "taking $\frac{1}{2}$ of a number is the same as dividing by $\frac{1}{2}$." Explain the error in this reasoning. [2.7b]

83. **Dw** Use the number 9432 to explain why the test for divisibility by 9 works. [2.2a]

SYNTHESIS

84. In the division below, find a and b. [2.7b]
$$\dfrac{19}{24} \div \dfrac{a}{b} = \dfrac{187{,}853}{268{,}224}$$

85. A prime number that becomes a prime number when its digits are reversed is called a **palindrome prime.** For example, 17 is a palindrome prime because both 17 and 71 are primes. Which of the following numbers are palindrome primes? [2.1c]

13, 91, 16, 11, 15, 24, 29, 101, 201, 37

1. Find all the factors of 300.

Determine whether the number is prime, composite, or neither.

2. 41

3. 14

Find the prime factorization of the number.

4. 18

5. 60

6. Determine whether 1784 is divisible by 8.

7. Determine whether 784 is divisible by 9.

8. Determine whether 5552 is divisible by 5.

9. Determine whether 2322 is divisible by 6.

10. Identify the numerator and the denominator of $\frac{4}{5}$.

11. What part is shaded?

12. What part of the set is shaded?

13. *Pass Completion Ratio.* In the 2004 regular season, Peyton Manning, of the Indianapolis Colts, completed 336 of 497 passes to become the No. 1 rated quarterback for the second year in a row as Associated Press NFL most valuable player.

a) What was the ratio of pass completions to attempts?
b) What was the ratio of incomplete passes to attempts?

Simplify.

14. $\frac{26}{1}$

15. $\frac{12}{12}$

16. $\frac{0}{16}$

17. $\frac{12}{24}$

18. $\frac{42}{7}$

19. $\frac{2}{28}$

20. $\frac{9}{0}$

21. $\frac{7}{2-2}$

22. $\frac{35}{140}$

23. $\frac{72}{108}$

Use = or ≠ for ☐ to write a true sentence.

24. $\frac{3}{4}$ ☐ $\frac{6}{8}$

25. $\frac{5}{4}$ ☐ $\frac{9}{7}$

Multiply and simplify.

26. $\frac{4}{3} \cdot 24$

27. $5 \cdot \frac{3}{10}$

28. $\frac{2}{3} \cdot \frac{15}{4}$

29. $\frac{3}{5} \cdot \frac{1}{6}$

30. $\frac{22}{15} \cdot \frac{5}{33}$

Find the reciprocal.

31. $\dfrac{5}{8}$

32. $\dfrac{1}{4}$

33. 18

Divide and simplify.

34. $\dfrac{3}{8} \div \dfrac{5}{4}$

35. $\dfrac{1}{5} \div \dfrac{1}{8}$

36. $12 \div \dfrac{2}{3}$

37. $\dfrac{24}{5} \div \dfrac{28}{15}$

Solve.

38. $\dfrac{7}{8} \cdot x = 56$

39. $t \cdot \dfrac{2}{5} = \dfrac{7}{10}$

40. There are 7000 students at La Poloma College, and $\frac{5}{8}$ of them live in dorms. How many live in dorms?

41. A strip of taffy $\frac{9}{10}$ m long is cut into 12 equal pieces. What is the length of each piece?

42. A thermos of iced tea had 3 qt of tea when it was $\frac{3}{5}$ full. How much tea could it hold when full?

43. The pitch of a screw is $\frac{1}{8}$ in. How far will it go into a piece of walnut when it is turned 6 complete rotations?

┌─ **SYNTHESIS** ───

44. A recipe for a batch of buttermilk pancakes calls for $\frac{3}{4}$ teaspoon (tsp) of salt. Jacqueline plans to cut the amount of salt in half for each of 5 batches of pancakes. How much salt will she need?

45. Grandma Hammons left $\frac{2}{3}$ of her $\frac{7}{8}$-acre apple farm to Karl. Karl gave $\frac{1}{4}$ of his share to his oldest daughter, Eileen. How much land did Eileen receive?

46. Simplify: $\left(\dfrac{3}{8}\right)^2 \div \dfrac{6}{7} \cdot \dfrac{2}{9} \div 5$.

47. Solve: $\dfrac{33}{38} \cdot \dfrac{34}{55} = \dfrac{17}{35} \cdot \dfrac{15}{19}x$.

148

Fraction Notation and Mixed Numerals

3

Real-World Application

Arie Luyendyk won the Indianapolis 500 in 1990 with the highest average speed of about 186 mph. This record high through 2004 is about $2\frac{12}{25}$ times the average speed of the first winner, Ray Harroun, in 1911. What was the average speed in the first Indianapolis 500?

Source: Indianapolis Motor Speedway

This problem appears as Example 9 in Section 3.6.

Objective

a Find the least common multiple, LCM, of two or more numbers.

1. By examining lists of multiples, find the LCM of 9 and 15.

In this chapter, we study addition and subtraction using fraction notation. Suppose we want to add $\frac{2}{3}$ and $\frac{1}{2}$. To do so, we rewrite the fractions with a common denominator. The number we choose for the common denominator is the least common multiple of the denominators, 6; $\frac{2}{3} + \frac{1}{2} = \frac{4}{6} + \frac{3}{6}$. Then we add the numerators and keep the common denominator. In order to do this, we must be able to find the **least common denominator (LCD),** or **least common multiple (LCM)** of the denominators. (A review of Section 2.1b might be helpful.)

a Finding Least Common Multiples

LEAST COMMON MULTIPLE, LCM

The **least common multiple,** or LCM, of two natural numbers is the smallest number that is a multiple of both numbers.

EXAMPLE 1 Find the LCM of 20 and 30.

First list some multiples of 20 by multiplying 20 by 1, 2, 3, and so on:

20, 40, 60, 80, 100, 120, 140, 160, 180, 200, 220, 240, ...

Then list some multiples of 30 by multiplying 30 by 1, 2, 3, and so on:

30, 60, 90, 120, 150, 180, 210, 240, ...

Now we determine the smallest number common to both lists. The LCM of 20 and 30 is 60.

Do Exercise 1.

Next we develop three methods that are more efficient for finding LCMs. You may choose to learn only one method (consult with your instructor), but if you are going to study algebra, you should definitely learn method 2.

METHOD 1: FINDING LCMS USING ONE LIST OF MULTIPLES

One method for finding LCMs uses *one* list of multiples. Let's consider finding the LCM of 9 and 12. The largest number, 12, is not a multiple of 9; so we check multiples of 12 until we find a number that is also a multiple of 9:

$1 \cdot 12 = 12$, not a multiple of 9;

$2 \cdot 12 = 24$, not a multiple of 9;

$3 \cdot 12 = 36$, a multiple of 9: $4 \cdot 9 = 36$.

The LCM of 9 and 12 is 36.

Answer on page A-6

CHAPTER 3: Fraction Notation and Mixed Numerals

Method 1. To find the LCM of a set of numbers using a list of multiples:

a) Determine whether the largest number is a multiple of the others. If it is, it is the LCM. That is, if the largest number has the others as factors, the LCM is that number.

b) If not, check multiples of the largest number until you get one that is a multiple of each of the others.

2. By examining lists of multiples, find the LCM of 8 and 10.

EXAMPLE 2   Find the LCM of 12 and 15.

a) 15 is not a multiple of 12.

b) Check multiples of 15:

$1 \cdot 15 = 15,$ Not a multiple of 12. When we divide 15 by 12, we get a nonzero remainder.

$2 \cdot 15 = 30,$ Not a multiple of 12

$3 \cdot 15 = 45,$ Not a multiple of 12

$4 \cdot 15 = 60.$ A multiple of 12: $5 \cdot 12 = 60$

The LCM $= 60$.

Do Exercise 2.

Find the LCM.

3. 10, 15

EXAMPLE 3   Find the LCM of 4 and 14.

a) 14 is not a multiple of 4.

b) Check multiples of 14:

$1 \cdot 14 = 14,$

$2 \cdot 14 = 28.$ A multiple of 4: $7 \cdot 4 = 28$

The LCM $= 28$.

4. 6, 8

EXAMPLE 4   Find the LCM of 8 and 32.

a) 32 is a multiple of 8 $(4 \cdot 8 = 32)$, so the LCM $= 32$.

5. 5, 10

EXAMPLE 5   Find the LCM of 10, 100, and 1000.

a) 1000 is a multiple of 10 $(100 \cdot 10 = 1000)$ and of 100 $(10 \cdot 100 = 1000)$, so the LCM $= 1000$.

Do Exercises 3–6.

6. 20, 40, 80

METHOD 2: FINDING LCMS USING PRIME FACTORIZATIONS

A second method for finding LCMs uses prime factorizations. Consider again 20 and 30. Their prime factorizations are $20 = 2 \cdot 2 \cdot 5$ and $30 = 2 \cdot 3 \cdot 5$. Let's look at these prime factorizations in order to find the LCM. Any multiple of 20 will have to have *two* 2's as factors and *one* 5 as a factor. Any multiple of 30 will have to have *one* 2, *one* 3, and *one* 5 as factors. The smallest number satisfying these conditions is

Two 2's, one 5; makes 20 a factor

$2 \cdot 2 \cdot 3 \cdot 5.$

One 2, one 3, one 5; makes 30 a factor

Answers on page A-6

Use prime factorizations to find the LCM.

7. 8, 10

Thus the LCM of 20 and 30 is 2 · 2 · 3 · 5, or 60. It has all the factors of 20 and all the factors of 30, but the factors are not repeated when they are common to both numbers.

Observe that the greatest number of times 2 occurs as a factor of either 20 or 30 is two, and the LCM has 2 as a factor twice. The greatest number of times 3 occurs as a factor of either 20 or 30 is one, and the LCM has 3 as a factor once. The greatest number of times 5 occurs as a factor of either 20 or 30 is one, and the LCM has 5 as a factor once.

> *Method 2.* To find the LCM of a set of numbers using prime factorizations:
>
> **a)** Find the prime factorization of each number.
> **b)** Create a product of factors, using each factor the greatest number of times that it occurs in any one factorization.

8. 18, 40

EXAMPLE 6 Find the LCM of 6 and 8.

a) Find the prime factorization of each number.

$$6 = 2 \cdot 3, \qquad 8 = 2 \cdot 2 \cdot 2$$

b) Create a product by writing factors that appear in the factorizations of 6 and 8, using each the greatest number of times that it occurs in any one factorization.

Consider the factor 2. The greatest number of times that 2 occurs in any one factorization is three. We write 2 as a factor three times.

$$2 \cdot 2 \cdot 2 \cdot ?$$

Consider the factor 3. The greatest number of times that 3 occurs in any one factorization is one. We write 3 as a factor one time.

$$2 \cdot 2 \cdot 2 \cdot 3 \cdot ?$$

Since there are no other prime factors in either factorization, the

LCM is 2 · 2 · 2 · 3, or 24.

9. 32, 54

EXAMPLE 7 Find the LCM of 24 and 36.

a) Find the prime factorization of each number.

$$24 = 2 \cdot 2 \cdot 2 \cdot 3, \qquad 36 = 2 \cdot 2 \cdot 3 \cdot 3$$

b) Create a product by writing factors, using each the greatest number of times that it occurs in any one factorization.

Consider the factor 2. The greatest number of times that 2 occurs in any one factorization is three. We write 2 as a factor three times:

$$2 \cdot 2 \cdot 2 \cdot ?$$

Consider the factor 3. The greatest number of times that 3 occurs in any one factorization is two. We write 3 as a factor two times:

$$2 \cdot 2 \cdot 2 \cdot 3 \cdot 3 \cdot ?$$

Since there are no other prime factors in either factorization, the

LCM is 2 · 2 · 2 · 3 · 3, or 72.

Answers on page A-6

CHAPTER 3: Fraction Notation
and Mixed Numerals

Do Exercises 7–9.

EXAMPLE 8   Find the LCM of 27, 90, and 84.

a) Find the prime factorization of each number.

$$27 = 3 \cdot 3 \cdot 3, \qquad 90 = 2 \cdot 3 \cdot 3 \cdot 5, \qquad 84 = 2 \cdot 2 \cdot 3 \cdot 7$$

b) Create a product by writing factors, using each the greatest number of times that it occurs in any one factorization.

Consider the factor 2. The greatest number of times that 2 occurs in any one factorization is two. We write 2 as a factor two times:

$$2 \cdot 2 \cdot ?$$

Consider the factor 3. The greatest number of times that 3 occurs in any one factorization is three. We write 3 as a factor three times:

$$2 \cdot 2 \cdot 3 \cdot 3 \cdot 3 \cdot ?$$

Consider the factor 5. The greatest number of times that 5 occurs in any one factorization is one. We write 5 as a factor one time:

$$2 \cdot 2 \cdot 3 \cdot 3 \cdot 3 \cdot 5 \cdot ?$$

Consider the factor 7. The greatest number of times that 7 occurs in any one factorization is one. We write 7 as a factor one time:

$$2 \cdot 2 \cdot 3 \cdot 3 \cdot 3 \cdot 5 \cdot 7 \cdot ?$$

Since there are no other prime factors in any of the factorizations, the

LCM is $2 \cdot 2 \cdot 3 \cdot 3 \cdot 3 \cdot 5 \cdot 7$, or 3780.

Do Exercise 10.

The use of exponents might be helpful to you as an extension of the factorization method. Let's reconsider Example 8. We want to find the LCM of 27, 90, and 84. We factor and then convert to exponential notation:

$$27 = 3 \cdot 3 \cdot 3 = 3^3,$$
$$90 = 2 \cdot 3 \cdot 3 \cdot 5 = 2^1 \cdot 3^2 \cdot 5^1, \quad \text{and}$$
$$84 = 2 \cdot 2 \cdot 3 \cdot 7 = 2^2 \cdot 3^1 \cdot 7^1.$$

Note that in 84, the 2 in 2^2 is the largest exponent of 2 in any of the factorizations. It is also the exponent of 2 in the LCM. It indicates the greatest number of times that 2 occurs as a factor of any of the numbers. Similarly, in 27, the 3 in 3^3 is the largest exponent of 3 in any of the factorizations. It is also the exponent of 3 in the LCM. Likewise, the 1's in 5^1 and 7^1 tell us the exponents of 5 and 7 in the LCM. They indicate the greatest number of times that 5 and 7 occur as factors. Thus the

LCM is $2^2 \cdot 3^3 \cdot 5^1 \cdot 7^1$, or 3780.

Similarly, the LCMs in Examples 6 and 7 can be expressed with exponents as $2^3 \cdot 3$ and $2^3 \cdot 3^2$, respectively.

Do Exercises 11 and 12.

10. Find the LCM of 24, 35, and 45.

11. Use exponents to find the LCM of 24, 35, and 45.

12. Redo Margin Exercises 7–9 using exponents.

Answers on page A-6

Find the LCM.

13. 3, 18

14. 12, 24

Find the LCM.

15. 4, 9

16. 5, 6, 7

EXAMPLE 9   Find the LCM of 7 and 21.

We find the prime factorization of each number. Because 7 is prime, it has no prime factorization.

$$7 = 7, \qquad 21 = 3 \cdot 7$$

Note that 7 is a factor of 21. We stated earlier that if one number is a factor of another, the LCM is the larger of the numbers. Thus the LCM is $7 \cdot 3$, or 21.

Do Exercises 13 and 14.

EXAMPLE 10   Find the LCM of 8 and 9.

We find the prime factorization of each number.

$$8 = 2 \cdot 2 \cdot 2, \qquad 9 = 3 \cdot 3$$

Note that the two numbers, 8 and 9, have no common prime factor. When this is the case, the LCM is just the product of the two numbers. Thus the LCM is $2 \cdot 2 \cdot 2 \cdot 3 \cdot 3$, or $8 \cdot 9$, or 72.

Do Exercises 15 and 16.

Let's compare the two methods considered so far for finding LCMs: the multiples method and the factorization method.

Method 1, the **multiples method,** can be longer than the factorization method when the LCM is large or when there are more than two numbers. But this method can be faster and easier to use mentally for two numbers.

Method 2, the **factorization method,** works well for several numbers. It is just like a method used in algebra. If you are going to study algebra, you should definitely learn the factorization method.

METHOD 3: FINDING LCMS USING DIVISION BY PRIMES

Here is another method for finding LCMs that is especially useful for three or more numbers. For example, to find the LCM of 48, 72, and 80, we first look for any prime that divides any two of the numbers with no remainder. Then we divide as follows.

$$2 \; | \; \underline{48 \;\; 72 \;\; 80}$$
$$24 \;\; 36 \;\; 40$$

We repeat the process, bringing down any numbers not divisible by the prime, until we can divide no more, that is, until there are no two numbers divisible by the same prime:

$$
\begin{array}{r|lll}
2 & 48 & 72 & 80 \\
3 & 24 & 36 & 40 \\
2 & 8 & 12 & 40 \\
2 & 4 & 6 & 20 \\
2 & 2 & 3 & 10 \\
\hline
 & 1 & 3 & 5
\end{array}
$$

40 is not divisible by 3.

3 is not divisible by 2.

No two of 1, 3, and 5 are divisible by the same prime. We stop here. The LCM is

$$2 \cdot 3 \cdot 2 \cdot 2 \cdot 2 \cdot 1 \cdot 3 \cdot 5, \text{ or } 720.$$

Method 3: To find the LCM using division by primes:

a) First look for any prime that divides at least two of the numbers with no remainder. Then divide, bringing down any numbers not divisible by the prime.

b) Repeat the process until you can divide no more, that is, until there are no two numbers divisible by the same prime.

EXAMPLE 11   Find the LCM of 24, 35, and 45.

$$
\begin{array}{c|ccc}
5 & 24 & 35 & 45 \\
3 & 24 & 7 & 9 \\
  & 8 & 7 & 3
\end{array}
\qquad
\begin{array}{l}
\\
\text{24 is not divisible by 5.} \\
\text{7 is not divisible by 3.}
\end{array}
$$

The LCM is $5 \cdot 3 \cdot 8 \cdot 7 \cdot 3$, or 2520.

EXAMPLE 12   Find the LCM of 12, 18, 20, and 21.

$$
\begin{array}{c|cccc}
3 & 12 & 18 & 20 & 21 \\
2 & 4 & 6 & 20 & 7 \\
2 & 2 & 3 & 10 & 7 \\
  & 1 & 3 & 5 & 7
\end{array}
$$

The LCM is $3 \cdot 2 \cdot 2 \cdot 1 \cdot 3 \cdot 5 \cdot 7$, or 1260.

Do Exercises 17–19.

Find the LCM using division by primes.

17. 12, 75, 120

18. 27, 90, 84

19. 12, 24, 75, 120

Answers on page A-6

Study Tips

TIPS FROM A FORMER STUDENT

A former student of Professor Bittinger, Mike Rosenborg, earned a master's degree in mathematics and now teaches mathematics. Here are some of his study tips.

- Because working problems is the best way to learn math, instructors generally assign lots of problems. Never let yourself get behind in your math homework.

- If you are struggling with a math concept, do not give up. Ask for help from your friends and your instructor. Since each concept is built on previous concepts, any gaps in your understanding will follow you through the entire course, so make sure you understand each concept as you go along.

- Read your textbook! It will often contain the help and tips you need to solve any problem with which you are struggling. It may also bring out points that you missed in class or that your instructor may not have covered.

- Learn to use scratch paper to jot down your thoughts and to draw pictures. Don't try to figure everything out "in your head." You will think more clearly and accurately this way.

- When preparing for a test, it is often helpful to work at least two problems per section as practice: one easy and one difficult. Write out all the new rules and procedures your test will cover, and then read through them twice. Doing so will enable you to both learn and retain them better.

- Most schools have classrooms set up where you can get free help from math tutors. Take advantage of this, but be sure you do the work first. Don't let your tutor do all the work for you—otherwise you'll never learn the material.

- In math, as in many other areas of life, patience and persistence are virtues—cultivate them. "Cramming" for an exam will not help you learn and retain the material.

a   Find the LCM of the set of numbers.

1. 2, 4

2. 3, 15

3. 10, 25

4. 10, 15

5. 20, 40

6. 8, 12

7. 18, 27

8. 9, 11

9. 30, 50

10. 24, 36

11. 30, 40

12. 21, 27

13. 18, 24

14. 12, 18

15. 60, 70

16. 35, 45

17. 16, 36

18. 18, 20

19. 32, 36

20. 36, 48

21. 2, 3, 5

22. 5, 18, 3

23. 3, 5, 7

24. 6, 12, 18

25. 24, 36, 12

26. 8, 16, 22

27. 5, 12, 15

28. 12, 18, 40

29. 9, 12, 6

30. 8, 16, 12

31. 180, 100, 450, 60

32. 18, 30, 50, 48

33. 8, 48

34. 16, 32

35. 5, 50

36. 12, 72

37. 11, 13

38. 13, 14

39. 12, 35

40. 23, 25

41. 54, 63

42. 56, 72

43. 81, 90

44. 75, 100

45. 36, 54, 80

46. 22, 42, 51

47. 39, 91, 108, 26

48. 625, 75, 500, 25

Applications of LCMs: Planet Orbits.   The earth, Jupiter, Saturn, and Uranus all revolve around the sun. The earth takes 1 yr, Jupiter 12 yr, Saturn 30 yr, and Uranus 84 yr to make a complete revolution. On a certain night, you look at those three distant planets and wonder how many years it will take before they have the same position again. (*Hint*: To find out, you find the LCM of 12, 30, and 84. It will be that number of years.)

Source: *The Handy Science Answer Book*

49. How often will Jupiter and Saturn appear in the same direction in the night sky as seen from the earth?

50. How often will Jupiter and Uranus appear in the same direction in the night sky as seen from the earth?

51. How often will Saturn and Uranus appear in the same direction in the night sky as seen from the earth?

52. How often will Jupiter, Saturn, and Uranus appear in the same direction in the night sky as seen from the earth?

53. $\mathbf{D_W}$ Use both methods 1 and 2 to find the LCM of each of the following sets of numbers.

a) 6, 8   **b)** 6, 7   **c)** 6, 21   **d)** 24, 36

Which method do you consider more efficient? Explain why.

54. $\mathbf{D_W}$ Is the LCM of two numbers always larger than either number? Why or why not?

SKILL MAINTENANCE

Solve.

55. Joy uses $\frac{1}{2}$ yd of dental floss each day. How long will a 45-yd container of dental floss last for Joy?   [2.7d]

56. *Broadway Runs.*   As of May 3, 2004, *Cats* had the longest Broadway run with 7485 performances from October 1982 to September 2000. From June 1988 to the present, *The Phantom of the Opera* has had 6781 performances. How many more performances of *The Phantom of the Opera* are needed to equal the record set by the *Cats* show?   [1.8a]

Source: League of American Theatres and Producers, Inc.

57. Add: 23,456 + 5677 + 4002.   [1.2a]

58. Subtract: 10,007 − 3068.   [1.3b]

59. Multiply and simplify: $\dfrac{4}{5} \cdot \dfrac{10}{12}$.   [2.6a]

60. Divide and simplify: $\dfrac{4}{5} \div \dfrac{7}{10}$.   [2.7b]

SYNTHESIS

61. Find the LCM of 27, 90, 84, 210, 108, and 50.

62. Find the LCM of 18, 21, 24, 36, 63, 56, and 20.

63. A pencil company uses two sizes of boxes, 5 in. by 6 in. and 5 in. by 8 in. These boxes are packed in bigger cartons for shipping. Find the width and the length of the smallest carton that will accommodate boxes of either size without any room left over. (Each carton can contain only one type of box and all boxes must point in the same direction.)

64. Consider 8 and 12. Determine whether each of the following is the LCM of 8 and 12. Tell why or why not.

a) $2 \cdot 2 \cdot 3 \cdot 3$
b) $2 \cdot 2 \cdot 3$
c) $2 \cdot 3 \cdot 3$
d) $2 \cdot 2 \cdot 2 \cdot 3$

Objectives

a Add using fraction notation.

b Solve applied problems involving addition with fraction notation.

1. Find $\dfrac{1}{5} + \dfrac{3}{5}$.

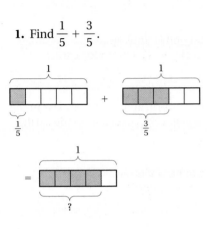

Add and simplify.

2. $\dfrac{1}{3} + \dfrac{2}{3}$

3. $\dfrac{5}{12} + \dfrac{1}{12}$

4. $\dfrac{9}{16} + \dfrac{3}{16}$

Answers on page A-6

158

CHAPTER 3: Fraction Notation and Mixed Numerals

 3.2 **ADDITION AND APPLICATIONS**

a **Addition Using Fraction Notation**

LIKE DENOMINATORS

Addition using fraction notation corresponds to combining or putting like things together, just as addition with whole numbers does. For example,

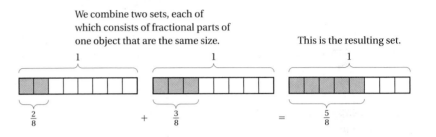

We combine two sets, each of which consists of fractional parts of one object that are the same size.

This is the resulting set.

$$\tfrac{2}{8} \qquad + \qquad \tfrac{3}{8} \qquad = \qquad \tfrac{5}{8}$$

2 eighths + 3 eighths = 5 eighths,

or $\quad 2 \cdot \dfrac{1}{8} + 3 \cdot \dfrac{1}{8} = 5 \cdot \dfrac{1}{8}, \quad$ or $\quad \dfrac{2}{8} + \dfrac{3}{8} = \dfrac{5}{8}.$

We see that to add when denominators are the same, we add the numerators and keep the denominator.

Do Exercise 1.

To add when denominators are the same,

a) add the numerators,

b) keep the denominator, and

$$\dfrac{2}{6} + \dfrac{5}{6} = \dfrac{2+5}{6} = \dfrac{7}{6}$$

c) simplify, if possible.

EXAMPLES Add and simplify.

1. $\dfrac{2}{4} + \dfrac{1}{4} = \dfrac{2+1}{4} = \dfrac{3}{4} \qquad$ No simplifying is possible.

2. $\dfrac{11}{6} + \dfrac{3}{6} = \dfrac{11+3}{6} = \dfrac{14}{6} = \dfrac{2 \cdot 7}{2 \cdot 3} = \dfrac{2}{2} \cdot \dfrac{7}{3} = 1 \cdot \dfrac{7}{3} = \dfrac{7}{3} \qquad$ Here we simplified.

3. $\dfrac{3}{12} + \dfrac{5}{12} = \dfrac{3+5}{12} = \dfrac{8}{12} = \dfrac{4 \cdot 2}{4 \cdot 3} = \dfrac{4}{4} \cdot \dfrac{2}{3} = 1 \cdot \dfrac{2}{3} = \dfrac{2}{3}$

Do Exercises 2–4.

DIFFERENT DENOMINATORS

What do we do when denominators are different? We can find a common denominator by multiplying by 1. Consider adding $\frac{1}{6}$ and $\frac{3}{4}$. There are many common denominators that can be obtained. Let's look at two possibilities.

A.
$$\frac{1}{6} + \frac{3}{4} = \frac{1}{6} \cdot 1 + \frac{3}{4} \cdot 1$$

$$= \frac{1}{6} \cdot \frac{4}{4} + \frac{3}{4} \cdot \frac{6}{6}$$

$$= \frac{4}{24} + \frac{18}{24}$$

$$= \frac{22}{24}$$

$$= \frac{11}{12}$$

B.
$$\frac{1}{6} + \frac{3}{4} = \frac{1}{6} \cdot 1 + \frac{3}{4} \cdot 1$$

$$= \frac{1}{6} \cdot \frac{2}{2} + \frac{3}{4} \cdot \frac{3}{3}$$

$$= \frac{2}{12} + \frac{9}{12}$$

$$= \frac{11}{12}$$

5. Add. (Find the least common denominator.)

$$\frac{2}{3} + \frac{1}{6}$$

We had to simplify in (A). We didn't have to simplify in (B). In (B), we used the least common multiple of the denominators, 12. That number is called the **least common denominator,** or **LCD.** As we will see in Example 6, we may still need to simplify when using the LCD, but it is usually easier.

> To add when denominators are different:
>
> **a)** Find the least common multiple of the denominators. That number is the least common denominator, LCD.
> **b)** Multiply by 1, using an appropriate notation, n/n, to express each number in terms of the LCD.
> **c)** Add the numerators, keeping the same denominator.
> **d)** Simplify, if possible.

EXAMPLE 4 Add: $\dfrac{3}{4} + \dfrac{1}{8}$.

The LCD is 8.   4 is a factor of 8, so the LCM of 4 and 8 is 8.

$$\frac{3}{4} + \frac{1}{8} = \frac{3}{4} \cdot 1 + \frac{1}{8} \longleftarrow \text{This fraction already has the LCD as its denominator.}$$

$$= \frac{3}{4} \cdot \frac{2}{2} + \frac{1}{8} \qquad \begin{array}{l} \textit{Think: } 4 \times \square = 8. \text{ The answer is 2, so we} \\ \text{multiply by 1, using } \frac{2}{2}. \end{array}$$

$$= \frac{6}{8} + \frac{1}{8} = \frac{7}{8}$$

6. Add: $\dfrac{3}{8} + \dfrac{5}{6}$.

Do Exercise 5.

EXAMPLE 5 Add: $\dfrac{1}{9} + \dfrac{5}{6}$.

The LCD is 18. $9 = 3 \cdot 3$ and $6 = 2 \cdot 3$, so the LCM of 9 and 6 is $2 \cdot 3 \cdot 3$, or 18.

$$\frac{1}{9} + \frac{5}{6} = \frac{1}{9} \cdot 1 + \frac{5}{6} \cdot 1 = \frac{1}{9} \cdot \frac{2}{2} + \frac{5}{6} \cdot \frac{3}{3} \qquad \begin{array}{l} \textit{Think: } 6 \times \square = 18. \\ \text{The answer is 3, so we} \\ \text{multiply by 1 using } \frac{3}{3}. \end{array}$$

$$\begin{array}{l} \textit{Think: } 9 \times \square = 18. \\ \text{The answer is 2, so we} \\ \text{multiply by 1 using } \frac{2}{2}. \end{array}$$

$$= \frac{2}{18} + \frac{15}{18} = \frac{17}{18}$$

Answers on page A-6

Do Exercise 6.

7. Add: $\dfrac{1}{6} + \dfrac{7}{18}$.

EXAMPLE 6 Add: $\dfrac{5}{9} + \dfrac{11}{18}$.

The LCD is 18.

$$\dfrac{5}{9} + \dfrac{11}{18} = \dfrac{5}{9} \cdot \dfrac{2}{2} + \dfrac{11}{18} = \dfrac{10}{18} + \dfrac{11}{18}$$

$$= \dfrac{21}{18} = \dfrac{3 \cdot 7}{3 \cdot 6} = \dfrac{3}{3} \cdot \dfrac{7}{6} \left.\right\} \quad \text{Simplifying}$$

$$= \dfrac{7}{6}$$

Do Exercise 7.

8. Add: $\dfrac{4}{10} + \dfrac{1}{100} + \dfrac{3}{1000}$.

EXAMPLE 7 Add: $\dfrac{1}{10} + \dfrac{3}{100} + \dfrac{7}{1000}$.

Since 10 and 100 are factors of 1000, the LCD is 1000. Then

$$\dfrac{1}{10} + \dfrac{3}{100} + \dfrac{7}{1000} = \dfrac{1}{10} \cdot \dfrac{100}{100} + \dfrac{3}{100} \cdot \dfrac{10}{10} + \dfrac{7}{1000}$$

$$= \dfrac{100}{1000} + \dfrac{30}{1000} + \dfrac{7}{1000} = \dfrac{137}{1000}.$$

Do Exercise 8.

When denominators are large, we most often use the prime factorization of each denominator to find the LCD. This is shown in Example 8. Using the prime factorization in this manner is similar to what is done in algebra.

Add.

9. $\dfrac{7}{10} + \dfrac{2}{21} + \dfrac{1}{7}$

EXAMPLE 8 Add: $\dfrac{13}{70} + \dfrac{11}{21} + \dfrac{6}{15}$.

We have

$$\dfrac{13}{70} + \dfrac{11}{21} + \dfrac{6}{15} = \dfrac{13}{2 \cdot 5 \cdot 7} + \dfrac{11}{3 \cdot 7} + \dfrac{6}{3 \cdot 5}. \quad \text{Factoring denominators}$$

The LCD is $2 \cdot 3 \cdot 5 \cdot 7$, or 210. Then

$$\dfrac{13}{70} + \dfrac{11}{21} + \dfrac{6}{15} = \dfrac{13}{2 \cdot 5 \cdot 7} \cdot \dfrac{3}{3} + \dfrac{11}{3 \cdot 7} \cdot \dfrac{2 \cdot 5}{2 \cdot 5} + \dfrac{6}{3 \cdot 5} \cdot \dfrac{7 \cdot 2}{7 \cdot 2}$$

The LCM of 70, 21, and 15 is $2 \cdot 3 \cdot 5 \cdot 7$. In each case, think of which factors are needed to get the LCD. Then multiply by 1 to obtain the LCD in each denominator.

10. $\dfrac{7}{18} + \dfrac{5}{24} + \dfrac{11}{36}$

$$= \dfrac{13 \cdot 3}{2 \cdot 5 \cdot 7 \cdot 3} + \dfrac{11 \cdot 2 \cdot 5}{3 \cdot 7 \cdot 2 \cdot 5} + \dfrac{6 \cdot 7 \cdot 2}{3 \cdot 5 \cdot 7 \cdot 2}$$

$$= \dfrac{39}{3 \cdot 5 \cdot 7 \cdot 2} + \dfrac{110}{3 \cdot 5 \cdot 7 \cdot 2} + \dfrac{84}{3 \cdot 5 \cdot 7 \cdot 2}$$

$$= \dfrac{233}{3 \cdot 5 \cdot 7 \cdot 2}$$

$$= \dfrac{233}{210}. \quad \text{We left 210 factored until we knew we could not simplify.}$$

Answers on page A-6

Do Exercises 9 and 10.

b ▸ Applications and Problem Solving

EXAMPLE 9 *Subflooring.* Matt Beecher Builders and Developers requires their subcontractors to use two layers of subflooring under a ceramic tile floor. First the contractors install a $\frac{3}{4}$-in. $\left(\frac{3''}{4}\right)$ layer of oriented strand board (OSB). Then a $\frac{1}{2}$-in. $\left(\frac{1''}{2}\right)$ sheet of cement board is mortared to the OSB. The mortar is $\frac{1}{8}$-in. thick. What is the total thickness of the two installed subfloors?

1. **Familiarize.** We first make a drawing. We let T = the total thickness of the subfloors.

2. **Translate.** The problem can be translated to an equation as follows.

$$\underbrace{\frac{3}{4}}_{\text{OSB}} \; + \; \underbrace{\frac{1}{8}}_{\text{Mortar}} \; + \; \underbrace{\frac{1}{2}}_{\substack{\text{Cement} \\ \text{board}}} \; = \; \underbrace{T}_{\substack{\text{Total} \\ \text{thickness}}}$$

3. **Solve.** To solve the equation, we carry out the addition. The LCM of the denominators is 8 because 2 and 4 are factors of 8. We multiply by 1 in order to obtain the LCD:

$$\frac{3}{4} + \frac{1}{8} + \frac{1}{2} = T$$

$$\frac{3}{4} \cdot \frac{2}{2} + \frac{1}{8} + \frac{1}{2} \cdot \frac{4}{4} = T$$

$$\frac{6}{8} + \frac{1}{8} + \frac{4}{8} = T$$

$$\frac{11}{8} = T.$$

4. **Check.** We check by repeating the calculation. We also note that the sum should be larger than any of the individual measurements, which it is. This tells us that the answer is reasonable.

5. **State.** The total thickness of the installed subfloors is $\frac{11}{8}$ in.

Do Exercise 11.

11. Sally jogs for $\frac{4}{5}$ mi, rests, and then jogs for another $\frac{1}{10}$ mi. How far does she jog in all?

Answer on page A-6

MathXL MyMathLab InterAct Math Math Tutor Center Digital Video Tutor CD 2 Videotape 3 Student's Solutions Manual

a Add and simplify.

1. $\dfrac{7}{8} + \dfrac{1}{8}$

2. $\dfrac{2}{5} + \dfrac{3}{5}$

3. $\dfrac{1}{8} + \dfrac{5}{8}$

4. $\dfrac{3}{10} + \dfrac{3}{10}$

5. $\dfrac{2}{3} + \dfrac{5}{6}$

6. $\dfrac{5}{6} + \dfrac{1}{9}$

7. $\dfrac{1}{8} + \dfrac{1}{6}$

8. $\dfrac{1}{6} + \dfrac{3}{4}$

9. $\dfrac{4}{5} + \dfrac{7}{10}$

10. $\dfrac{3}{4} + \dfrac{1}{12}$

11. $\dfrac{5}{12} + \dfrac{3}{8}$

12. $\dfrac{7}{8} + \dfrac{1}{16}$

13. $\dfrac{3}{20} + \dfrac{3}{4}$

14. $\dfrac{2}{15} + \dfrac{2}{5}$

15. $\dfrac{5}{6} + \dfrac{7}{9}$

16. $\dfrac{5}{8} + \dfrac{5}{6}$

17. $\dfrac{3}{10} + \dfrac{1}{100}$

18. $\dfrac{9}{10} + \dfrac{3}{100}$

19. $\dfrac{5}{12} + \dfrac{4}{15}$

20. $\dfrac{3}{16} + \dfrac{1}{12}$

21. $\dfrac{9}{10} + \dfrac{99}{100}$

22. $\dfrac{3}{10} + \dfrac{27}{100}$

23. $\dfrac{7}{8} + \dfrac{0}{1}$

24. $\dfrac{0}{1} + \dfrac{5}{6}$

25. $\dfrac{3}{8} + \dfrac{1}{6}$

26. $\dfrac{5}{8} + \dfrac{1}{6}$

27. $\dfrac{5}{12} + \dfrac{7}{24}$

28. $\dfrac{1}{18} + \dfrac{7}{12}$

29. $\dfrac{3}{16} + \dfrac{5}{16} + \dfrac{4}{16}$

30. $\dfrac{3}{8} + \dfrac{1}{8} + \dfrac{2}{8}$

31. $\dfrac{8}{10} + \dfrac{7}{100} + \dfrac{4}{1000}$

32. $\dfrac{1}{10} + \dfrac{2}{100} + \dfrac{3}{1000}$

33. $\dfrac{3}{8} + \dfrac{5}{12} + \dfrac{8}{15}$

34. $\dfrac{1}{2} + \dfrac{3}{8} + \dfrac{1}{4}$

35. $\dfrac{15}{24} + \dfrac{7}{36} + \dfrac{91}{48}$

36. $\dfrac{5}{7} + \dfrac{25}{52} + \dfrac{7}{4}$

CHAPTER 3: Fraction Notation and Mixed Numerals

Solve.

37. Rene bought $\frac{1}{3}$ lb of orange pekoe tea and $\frac{1}{2}$ lb of English cinnamon tea. How many pounds of tea did he buy?

38. Stan bought $\frac{1}{4}$ lb of gumdrops and $\frac{1}{2}$ lb of caramels. How many pounds of candy did he buy?

39. Russ walked $\frac{7}{6}$ mi to a friend's dormitory, and then $\frac{3}{4}$ mi to class. How far did he walk?

40. Elaine walked $\frac{7}{8}$ mi to the student union, and then $\frac{2}{5}$ mi to class. How far did she walk?

41. *Concrete Mix.* A cubic meter of concrete mix contains 420 kilograms (kg) of cement, 150 kg of stone, and 120 kg of sand. What is the total weight of a cubic meter of the mix? What part is cement? stone? sand? Add these fractional amounts. What is the result?

42. *Punch Recipe.* A recipe for strawberry punch calls for $\frac{1}{5}$ quart (qt) of ginger ale and $\frac{3}{5}$ qt of strawberry soda. How much liquid is needed? If the recipe is doubled, how much liquid is needed? If the recipe is halved, how much liquid is needed?

43. *Carpentry.* To cut expenses, a carpenter sometimes glues two kinds of plywood together. He glues a $\frac{1}{4}''$ piece of walnut plywood to a $\frac{3}{8}''$ piece of less expensive plywood. What is the total thickness of these pieces?

44. *Iced Cookies.* A chef prepared cookies for the freshmen orientation reception. He frosted the $\frac{5}{16}''$ cookies with a $\frac{5}{32}''$ layer of icing. What is the thickness of the iced cookie?

45. A baker used $\frac{1}{2}$ lb of flour for rolls, $\frac{1}{4}$ lb for donuts, and $\frac{1}{3}$ lb for cookies. How much flour was used?

46. A tile $\frac{5}{8}$ in. thick is glued to a board $\frac{7}{8}$ in. thick. The glue is $\frac{3}{32}$ in. thick. How thick is the result?

47. **D**w Explain the role of multiplication when adding using fraction notation with different denominators.

48. **D**w To add numbers with different denominators, a student consistently uses the product of the denominators as a common denominator. Is this correct? Why or why not?

Multiply. [1.5a]

49. 408 · 516

50. 1125 · 3728

51. 423 · 8009

52. 2025 · 174

Presidential Elections. The results of the five closest presidential elections are listed in the following table. Use these data for Exercises 53–58. [1.3b]

DATE	PRESIDENT	ELECTORAL VOTES	POPULAR VOTES
1916	Woodrow Wilson (D)	277	9,129,606
	Charles E. Hughes (R)	254	8,538,221
1960	John F. Kennedy (D)	303	34,226,731
	Richard M. Nixon (R)	219	34,108,157
1968	Richard M. Nixon (R)	301	31,785,480
	Hubert H. Humphrey (D)	191	31,275,166
	George C. Wallace (American Independent)	46	9,906,473
1976	Jimmy Carter (D)	297	40,830,763
	Gerald R. Ford (R)	240	39,147,973
2000	George W. Bush (R)	271	50,455,156
	Albert A. Gore (D)	266	50,992,335

Source: *Time Almanac 2005*, p. 55

53. How many more popular votes did Albert A. Gore have than George W. Bush in the 2000 presidential election?

54. How many more electoral votes did George W. Bush have than Albert A. Gore in the 2000 presidential election?

55. In the 1960 presidential election, how many more electoral votes did John F. Kennedy have than Richard M. Nixon?

56. In the 1968 presidential election, how many more popular votes did Richard M. Nixon have than Hubert H. Humphrey?

57. How much greater was the total of the popular vote counts in the 2000 presidential election than in the 1976 election?

58. How much greater was the total of the popular vote counts in the 2000 presidential election than in the 1960 election?

59. Elsa has $9 to spend on ride tickets at the school carnival. If the tickets cost 75¢, or $\frac{3}{4}$, each, how many tickets can she purchase? [2.7d]

60. The Bingham community garden is to be split into 16 equally sized plots. If the garden occupies $\frac{3}{4}$ acre of land, how large will each plot be? [2.7d]

61. A guitarist's band is booked for Friday and Saturday nights at a local club. The guitarist is part of a trio on Friday and part of a quintet on Saturday. Thus the guitarist is paid one-third of one-half the weekend's pay for Friday and one-fifth of one-half the weekend's pay for Saturday. What fractional part of the band's pay did the guitarist receive for the weekend's work? If the band was paid $1200, how much did the guitarist receive?

3.3 SUBTRACTION, ORDER, AND APPLICATIONS

a Subtraction Using Fraction Notation

Objectives

 a Subtract using fraction notation.

b Use $<$ or $>$ with fraction notation to write a true sentence.

c Solve equations of the type $x + a = b$ and $a + x = b$, where a and b may be fractions.

d Solve applied problems involving subtraction with fraction notation.

LIKE DENOMINATORS

We can consider the difference $\frac{4}{8} - \frac{3}{8}$ as we did before, as either "take away" or "how many more." Let's consider "take away."

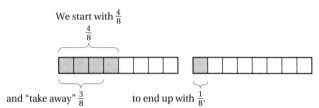

We start with $\frac{4}{8}$

$\frac{4}{8}$

and "take away" $\frac{3}{8}$ to end up with $\frac{1}{8}$.

We start with 4 eighths and take away 3 eighths:

4 eighths $-$ 3 eighths $=$ 1 eighth,

or $4 \cdot \frac{1}{8} - 3 \cdot \frac{1}{8} = \frac{1}{8}$,   or $\frac{4}{8} - \frac{3}{8} = \frac{1}{8}$.

> To subtract when denominators are the same,
>
> **a)** subtract the numerators,
>
> **b)** keep the denominator, and
>
> **c)** simplify, if possible.
>
> $$\frac{7}{10} - \frac{4}{10} = \frac{7 - 4}{10} = \frac{3}{10}$$

Subtract and simplify.

1. $\frac{7}{8} - \frac{3}{8}$

EXAMPLES Subtract and simplify.

1. $\frac{7}{10} - \frac{3}{10} = \frac{7 - 3}{10} = \frac{4}{10} = \frac{2 \cdot 2}{5 \cdot 2} = \frac{2}{5} \cdot \frac{2}{2} = \frac{2}{5} \cdot 1 = \frac{2}{5}$

2. $\frac{8}{9} - \frac{2}{9} = \frac{8 - 2}{9} = \frac{6}{9} = \frac{2 \cdot 3}{3 \cdot 3} = \frac{2}{3} \cdot \frac{3}{3} = \frac{2}{3} \cdot 1 = \frac{2}{3}$

3. $\frac{32}{12} - \frac{25}{12} = \frac{32 - 25}{12} = \frac{7}{12}$

Do Exercises 1–3.

DIFFERENT DENOMINATORS

The procedure for subtraction with different denominators is similar to the procedure for addition.

2. $\frac{10}{16} - \frac{4}{16}$

3. $\frac{8}{10} - \frac{3}{10}$

> To subtract when denominators are different:
>
> **a)** Find the least common multiple of the denominators. That number is the least common denominator, LCD.
> **b)** Multiply by 1, using an appropriate notation, n/n, to express each number in terms of the LCD.
> **c)** Subtract the numerators, keeping the same denominator.
> **d)** Simplify, if possible.

Answers on page A-6

165

3.3 Subtraction, Order, and Applications

4. Subtract: $\dfrac{3}{4} - \dfrac{2}{3}$.

Subtract.

5. $\dfrac{5}{6} - \dfrac{1}{9}$

6. $\dfrac{4}{5} - \dfrac{3}{10}$

7. Subtract: $\dfrac{11}{28} - \dfrac{5}{16}$.

8. Use $<$ or $>$ for $\square$ to write a true sentence:

$$\dfrac{3}{8} \ \square \ \dfrac{5}{8}.$$

CHAPTER 3: Fraction Notation
and Mixed Numerals

■ **EXAMPLE 4** Subtract: $\dfrac{2}{5} - \dfrac{3}{8}$.

The LCM of 5 and 8 is 40, so the LCD is 40.

$$\dfrac{2}{5} - \dfrac{3}{8} = \dfrac{2}{5} \cdot \dfrac{8}{8} - \dfrac{3}{8} \cdot \dfrac{5}{5} \longleftarrow$$

Think: $8 \times \square = 40$. The answer is 5, so we multiply by 1, using $\dfrac{5}{5}$.

Think: $5 \times \square = 40$. The answer is 8, so we multiply by 1, using $\dfrac{8}{8}$.

$$= \dfrac{16}{40} - \dfrac{15}{40} = \dfrac{16 - 15}{40} = \dfrac{1}{40}$$

Do Exercise 4.

■ **EXAMPLE 5** Subtract: $\dfrac{5}{6} - \dfrac{7}{12}$.

Since 12 is a multiple of 6, the LCM of 6 and 12 is 12. The LCD is 12.

$$\dfrac{5}{6} - \dfrac{7}{12} = \dfrac{5}{6} \cdot \dfrac{2}{2} - \dfrac{7}{12}$$

$$= \dfrac{10}{12} - \dfrac{7}{12} = \dfrac{10 - 7}{12} = \dfrac{3}{12}$$

$$= \dfrac{3 \cdot 1}{3 \cdot 4} = \dfrac{3}{3} \cdot \dfrac{1}{4} = \dfrac{1}{4}$$

Do Exercises 5 and 6.

■ **EXAMPLE 6** Subtract: $\dfrac{17}{24} - \dfrac{4}{15}$.

We have

$$\dfrac{17}{24} - \dfrac{4}{15} = \dfrac{17}{3 \cdot 2 \cdot 2 \cdot 2} - \dfrac{4}{5 \cdot 3}.$$

The LCD is $3 \cdot 2 \cdot 2 \cdot 2 \cdot 5$, or 120. Then

$$\dfrac{17}{24} - \dfrac{4}{15} = \dfrac{17}{3 \cdot 2 \cdot 2 \cdot 2} \cdot \dfrac{5}{5} - \dfrac{4}{5 \cdot 3} \cdot \dfrac{2 \cdot 2 \cdot 2}{2 \cdot 2 \cdot 2}$$

The LCM of 24 and 15 is $2 \cdot 2 \cdot 2 \cdot 3 \cdot 5$. In each case, we multiply by 1 to obtain the LCD.

$$= \dfrac{17 \cdot 5}{3 \cdot 2 \cdot 2 \cdot 2 \cdot 5} - \dfrac{4 \cdot 2 \cdot 2 \cdot 2}{5 \cdot 3 \cdot 2 \cdot 2 \cdot 2}$$

$$= \dfrac{85}{120} - \dfrac{32}{120} = \dfrac{53}{120}.$$

Do Exercise 7.

b Order

We see from this figure that $\frac{4}{5}$ is greater than $\frac{3}{5}$, and $\frac{3}{5}$ is less than $\frac{4}{5}$. That is, $\frac{4}{5} > \frac{3}{5}$, and $\frac{3}{5} < \frac{4}{5}$.

To determine which of two numbers is greater when there is a common denominator, compare the numerators:

$$\frac{4}{5}, \frac{3}{5}; \quad 4 > 3; \quad \frac{4}{5} > \frac{3}{5}.$$

Do Exercises 8 and 9. (Exercise 8 is on the preceding page.)

When denominators are different, we cannot compare numerators. We multiply by 1 to make the denominators the same.

EXAMPLE 7 Use $<$ or $>$ for $\square$ to write a true sentence:

$$\frac{2}{5} \, \square \, \frac{3}{4}.$$

The LCD is 20. We have

$$\frac{2}{5} \cdot \frac{4}{4} = \frac{8}{20}; \quad \text{We multiply by 1 using } \frac{4}{4} \text{ to get the LCD.}$$

$$\frac{3}{4} \cdot \frac{5}{5} = \frac{15}{20}. \quad \text{We multiply by 1 using } \frac{5}{5} \text{ to get the LCD.}$$

Now that the denominators are the same, 20, we can compare the numerators. Since $8 < 15$, it follows that $\frac{8}{20} < \frac{15}{20}$, so

$$\frac{2}{5} < \frac{3}{4}.$$

EXAMPLE 8 Use $<$ or $>$ for $\square$ to write a true sentence:

$$\frac{9}{10} \, \square \, \frac{89}{100}.$$

The LCD is 100. We write $\frac{9}{10}$ with a denominator of 100 to make the denominators the same.

$$\frac{9}{10} \cdot \frac{10}{10} = \frac{90}{100} \quad \text{We multiply by } \frac{10}{10} \text{ to get the LCD.}$$

Since $90 > 89$, it follows that $\frac{90}{100} > \frac{89}{100}$, so

$$\frac{9}{10} > \frac{89}{100}.$$

Do Exercises 10–12.

c Solving Equations

Now let's solve equations of the form $x + a = b$ or $a + x = b$, where a and b may be fractions. Proceeding as we have before, we subtract a on both sides of the equation.

---

9. Use $<$ or $>$ for $\square$ to write a true sentence:

$$\frac{7}{10} \, \square \, \frac{6}{10}.$$

Use $<$ or $>$ for $\square$ to write a true sentence.

10. $\frac{2}{3} \, \square \, \frac{5}{8}$

11. $\frac{3}{4} \, \square \, \frac{8}{12}$

12. $\frac{5}{6} \, \square \, \frac{7}{8}$

Solve.

13. $x + \frac{2}{3} = \frac{5}{6}$

14. $\frac{3}{5} + t = \frac{7}{8}$

Answers on page A-6

15. Natasha has run for $\frac{2}{3}$ mi and will stop when she has run for $\frac{7}{8}$ mi. How much farther does she have to go?

$\frac{2}{3}$ mi

$\frac{7}{8}$ mi d mi

EXAMPLE 9 Solve: $x + \frac{1}{4} = \frac{3}{5}$.

$$x + \frac{1}{4} - \frac{1}{4} = \frac{3}{5} - \frac{1}{4}$$ Subtracting $\frac{1}{4}$ on both sides

$$x + 0 = \frac{3}{5} \cdot \frac{4}{4} - \frac{1}{4} \cdot \frac{5}{5}$$ The LCD is 20. We multiply by 1 to get the LCD.

$$x = \frac{12}{20} - \frac{5}{20}$$

$$x = \frac{7}{20}$$

Do Exercises 13 and 14 on the preceding page.

d Applications and Problem Solving

EXAMPLE 10 *Pendant Necklace.* At one time Coldwater Creek offered the pendant necklace illustrated at left. The sterling silver capping at the top measures $\frac{11}{32}$ in. and the total length of the pendant is $\frac{7}{8}$ in. Find the length, or diameter, w of the pearl ball on the pendant.

1. **Familiarize.** We let $w = $ the length of the pearl on the pendant.

2. **Translate.** We see that this is a "how many more" situation. We can translate to an equation.

$$\underbrace{\text{Length of silver capping}}_{\frac{11}{32}} \underbrace{\text{plus}}_{+} \underbrace{\text{Length of pearl ball}}_{w} \underbrace{\text{is}}_{=} \underbrace{\text{Total length of pendant}}_{\frac{7}{8}}$$

3. **Solve.** To solve the equation, we subtract $\frac{11}{32}$ on both sides:

$$\frac{11}{32} + w = \frac{7}{8}$$

$$\frac{11}{32} + w - \frac{11}{32} = \frac{7}{8} - \frac{11}{32}$$ Subtracting $\frac{11}{32}$ on both sides

$$w + 0 = \frac{7}{8} \cdot \frac{4}{4} - \frac{11}{32}$$ The LCD is 32. We multiply by 1 to obtain the LCD.

$$w = \frac{28}{32} - \frac{11}{32}$$

$$= \frac{17}{32}.$$

4. **Check.** To check, we return to the original problem and add:

$$\frac{11}{32} + \frac{17}{32} = \frac{28}{32} = \frac{7}{8} \cdot \frac{4}{4} = \frac{7}{8}.$$

5. **State.** The length of the pearl ball on the pendant is $\frac{17}{32}$ in.

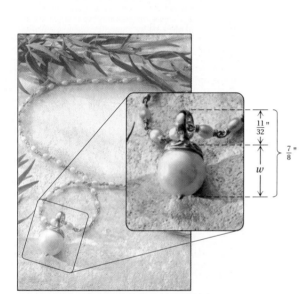

Source: ©Coldwater Creek Inc. www.coldwatercreek.com

$\frac{11}{32}$"

$\frac{7}{8}$"

w

Answer on page A-6   *Do Exercise 15.*

Translating for Success

1. *Bubble Wrap.* One-Stop Postal Center orders bubble wrap in 64-yd rolls. On the average, $\frac{3}{4}$ yd is used per small package. How many small packages can be prepared with 2 rolls of bubble wrap?

2. *Distance from College.* The post office is $\frac{7}{9}$ mi from the community college. The medical clinic is $\frac{2}{5}$ as far from the college as the post office is. How far is the clinic from the college?

3. *Swimming.* Andrew swims $\frac{7}{9}$ mi every day. One day he swims $\frac{2}{5}$ mi by 11:00 A.M. How much farther must Andrew swim to reach his daily goal?

4. *Tuition.* The average tuition at Waterside University is $12,000. If a loan is obtained for $\frac{1}{3}$ of the tuition, how much is the loan?

5. *Thermos Bottle Capacity.* A thermos bottle holds $\frac{11}{12}$ gal. How much is in the bottle when it is $\frac{4}{7}$ full?

The goal of these matching questions is to practice step (2), *Translate*, of the five-step problem-solving process. Translate each word problem to an equation and select a correct translation from equations A–O.

A. $\frac{3}{4} \cdot 64 = x$

B. $\frac{1}{3} \cdot 12{,}000 = x$

C. $\frac{1}{3} + \frac{2}{5} = x$

D. $\frac{2}{5} + x = \frac{7}{9}$

E. $\frac{2}{5} \cdot \frac{7}{9} = x$

F. $\frac{3}{4} \cdot x = 64$

G. $\frac{4}{7} = x + \frac{11}{12}$

H. $\frac{2}{5} = x + \frac{7}{9}$

I. $\frac{4}{7} \cdot \frac{11}{12} = x$

J. $\frac{3}{4} \cdot x = 128$

K. $\frac{1}{3} \cdot x = 12{,}000$

L. $\frac{1}{3} + \frac{2}{5} + x = 1$

M. $\frac{4}{3} \cdot x = 64$

N. $\frac{4}{7} + x = \frac{11}{12}$

O. $\frac{1}{3} + x = \frac{2}{5}$

Answers on page A-6

6. *Cutting Rope.* A piece of rope $\frac{11}{12}$ yd long is cut into two pieces. One piece is $\frac{4}{7}$ yd long. How long is the other piece?

7. *Planting Corn.* Each year, Prairie State Farm plants 64 acres of corn. With good weather, $\frac{3}{4}$ of the planting can be completed by April 20. How many acres can be planted with good weather?

8. *Painting Trim.* A painter used $\frac{1}{3}$ gal of white paint for the trim in the library and $\frac{2}{5}$ gal in the family room. How much paint was used for the trim in the two rooms?

9. *Lottery Winnings.* Sally won $12,000 in a state lottery and decides to give the net amount after taxes to three charities. One received $\frac{1}{3}$ of the money, and a second received $\frac{2}{5}$. What fractional part did the third charity receive?

10. *Reading Assignment.* Lowell read 64 pages of his political science assignment. This completed $\frac{3}{4}$ of his required reading. How many total pages were assigned?

a Subtract and simplify.

1. $\dfrac{5}{6} - \dfrac{1}{6}$

2. $\dfrac{5}{8} - \dfrac{3}{8}$

3. $\dfrac{11}{12} - \dfrac{2}{12}$

4. $\dfrac{17}{18} - \dfrac{11}{18}$

5. $\dfrac{3}{4} - \dfrac{1}{8}$

6. $\dfrac{2}{3} - \dfrac{1}{9}$

7. $\dfrac{1}{8} - \dfrac{1}{12}$

8. $\dfrac{1}{6} - \dfrac{1}{8}$

9. $\dfrac{4}{3} - \dfrac{5}{6}$

10. $\dfrac{7}{8} - \dfrac{1}{16}$

11. $\dfrac{3}{4} - \dfrac{3}{28}$

12. $\dfrac{2}{5} - \dfrac{2}{15}$

13. $\dfrac{3}{4} - \dfrac{3}{20}$

14. $\dfrac{5}{6} - \dfrac{1}{2}$

15. $\dfrac{3}{4} - \dfrac{1}{20}$

16. $\dfrac{3}{4} - \dfrac{4}{16}$

17. $\dfrac{5}{12} - \dfrac{2}{15}$

18. $\dfrac{9}{10} - \dfrac{11}{16}$

19. $\dfrac{6}{10} - \dfrac{7}{100}$

20. $\dfrac{9}{10} - \dfrac{3}{100}$

21. $\dfrac{7}{15} - \dfrac{3}{25}$

22. $\dfrac{18}{25} - \dfrac{4}{35}$

23. $\dfrac{99}{100} - \dfrac{9}{10}$

24. $\dfrac{78}{100} - \dfrac{11}{20}$

25. $\dfrac{2}{3} - \dfrac{1}{8}$

26. $\dfrac{3}{4} - \dfrac{1}{2}$

27. $\dfrac{3}{5} - \dfrac{1}{2}$

28. $\dfrac{5}{6} - \dfrac{2}{3}$

29. $\dfrac{5}{12} - \dfrac{3}{8}$

30. $\dfrac{7}{12} - \dfrac{2}{9}$

31. $\dfrac{7}{8} - \dfrac{1}{16}$

32. $\dfrac{5}{12} - \dfrac{5}{16}$

33. $\dfrac{17}{25} - \dfrac{4}{15}$

34. $\dfrac{11}{18} - \dfrac{7}{24}$

35. $\dfrac{23}{25} - \dfrac{112}{150}$

36. $\dfrac{89}{90} - \dfrac{53}{120}$

b Use < or > for ☐ to write a true sentence.

37. $\dfrac{5}{8}$ ☐ $\dfrac{6}{8}$

38. $\dfrac{7}{9}$ ☐ $\dfrac{5}{9}$

39. $\dfrac{1}{3}$ ☐ $\dfrac{1}{4}$

40. $\dfrac{1}{8}$ ☐ $\dfrac{1}{6}$

41. $\dfrac{2}{3}$ ☐ $\dfrac{5}{7}$

42. $\dfrac{3}{5}$ ☐ $\dfrac{4}{7}$

43. $\dfrac{4}{5}$ ☐ $\dfrac{5}{6}$

44. $\dfrac{3}{2}$ ☐ $\dfrac{7}{5}$

45. $\dfrac{19}{20}$ ☐ $\dfrac{4}{5}$

46. $\dfrac{5}{6}$ ☐ $\dfrac{13}{16}$

47. $\dfrac{19}{20}$ ☐ $\dfrac{9}{10}$

48. $\dfrac{3}{4}$ ☐ $\dfrac{11}{15}$

49. $\dfrac{31}{21}$ ☐ $\dfrac{41}{13}$

50. $\dfrac{12}{7}$ ☐ $\dfrac{132}{49}$

CHAPTER 3: Fraction Notation
and Mixed Numerals

c Solve.

51. $x + \frac{1}{30} = \frac{1}{10}$

52. $y + \frac{9}{12} = \frac{11}{12}$

53. $\frac{2}{3} + t = \frac{4}{5}$

54. $\frac{2}{3} + p = \frac{7}{8}$

55. $x + \frac{1}{3} = \frac{5}{6}$

56. $m + \frac{5}{6} = \frac{9}{10}$

d Solve.

57. Jaci spent $\frac{3}{4}$ hr listening to DVDs of Maroon 5 and U2. She spent $\frac{1}{3}$ hr listening to Maroon 5. How many hours were spent listening to U2?

58. As part of a fitness program, Deb swims $\frac{1}{2}$ mi every day. One day she had already swum $\frac{1}{5}$ mi. How much farther should Deb swim?

59. A server has a bowl containing $\frac{11}{12}$ cup of grated Parmesan cheese and puts $\frac{1}{8}$ cup on a customer's spaghetti and meatballs. How much remains in the bowl?

60. *Tire Tread.* A long-life tire has a tread depth of $\frac{3}{8}$ in. instead of a more typical $\frac{11}{32}$ in. depth. How much deeper is the long-life tread depth?

Source: *Popular Science*

61. A Subway franchise is owned by Alicia, Erica, and Hannah. Alicia owns $\frac{7}{12}$ of the business and Erica owns $\frac{1}{6}$. What part of the business does Hannah own?

62. An estate was left to four children. One received $\frac{1}{4}$ of the estate, the second $\frac{1}{16}$, and the third $\frac{3}{8}$. How much did the fourth receive?

63. From a $\frac{4}{5}$-lb wheel of cheese, a $\frac{1}{4}$-lb piece was served. How much cheese remained on the wheel?

64. Jovan has an $\frac{11}{10}$-lb mixture of cashews and peanuts that includes $\frac{3}{5}$ lb of cashews. How many pounds of peanuts are in the mixture?

65. **D**_W A fellow student made the following error:

$$\frac{8}{5} - \frac{8}{2} = \frac{8}{3}.$$

Find at least two ways to convince him of the mistake.

66. **D**_W Explain how one could use pictures to convince someone that $\frac{7}{29}$ is larger than $\frac{13}{57}$.

Divide, if possible. If not possible, write "not defined."   [1.6b], [2.3b]

67. $\dfrac{38}{38}$ **68.** $\dfrac{38}{0}$ **69.** $\dfrac{124}{0}$ **70.** $\dfrac{124}{31}$

Divide and simplify.   [2.7b]

71. $\dfrac{3}{7} \div \dfrac{9}{4}$ **72.** $\dfrac{9}{10} \div \dfrac{3}{5}$ **73.** $7 \div \dfrac{1}{3}$ **74.** $\dfrac{1}{4} \div 8$

Solve.

75. *Digital Tire Gauges.*   A factory produces 3885 digital tire gauges per day. How long will it take to fill an order for 66,045 tire gauges?   [1.8a]

76. A batch of fudge requires $\dfrac{3}{4}$ cup of sugar. How much sugar is needed to make 12 batches?   [2.6b]

Solve.

77. $x + \dfrac{16}{323} = \dfrac{10}{187}$

78. $x + \dfrac{7}{253} = \dfrac{12}{299}$

79. As part of a rehabilitation program, an athlete must swim and then walk a total of $\dfrac{9}{10}$ km each day. If one lap in the swimming pool is $\dfrac{3}{80}$ km, how far must the athlete walk after swimming 10 laps?

80. *Mountain Climbing.*   A mountain climber, beginning at sea level, climbs $\dfrac{3}{5}$ km, descends $\dfrac{1}{4}$ km, climbs $\dfrac{1}{3}$ km, and then descends $\dfrac{1}{7}$ km. At what elevation does the climber finish?

Simplify. Use the rules for order of operations given in Section 1.9.

81. $\dfrac{7}{8} - \dfrac{1}{10} \times \dfrac{5}{6}$ **82.** $\dfrac{2}{5} + \dfrac{1}{6} \div 3$ **83.** $\left(\dfrac{2}{3}\right)^2 + \left(\dfrac{3}{4}\right)^2$ **84.** $5 \times \dfrac{3}{7} - \dfrac{1}{7} \times \dfrac{4}{5}$

Use <, >, or = for ☐ to write a true sentence.

85. $\dfrac{37}{157} + \dfrac{19}{107}$ ☐ $\dfrac{6941}{16,799}$

86. $\dfrac{12}{97} + \dfrac{67}{139}$ ☐ $\dfrac{8167}{13,289}$

87. *Microsoft Interview.*   The following is a question taken from an employment interview with Microsoft. Try to answer it.

"Given a gold bar that can be cut exactly twice and a contractor who must be paid one-seventh of a gold bar every day for seven days, how should the bar be cut?"

Source: *Fortune Magazine,* January 22, 2001

3.4 MIXED NUMERALS

Objectives

a Convert between mixed numerals and fraction notation.

b Divide whole numbers, writing the quotient as a mixed numeral.

a Mixed Numerals

The following figure illustrates the use of a **mixed numeral.** The bolt shown is $2\frac{3}{8}$ in. long. The length is given as a whole-number part, 2, and a fractional part less than 1, $\frac{3}{8}$. We can represent the measurement of the bolt with fraction notation as $\frac{19}{8}$, but the meaning or interpretation of such a symbol is less understandable or visual.

$\frac{8}{8}$, or 1 $\frac{8}{8}$, or 1 $\frac{3}{8}$

$2 + \frac{3}{8}$

A mixed numeral $2\frac{3}{8}$ represents a sum:

$$2\frac{3}{8} \quad \text{means} \quad 2 + \frac{3}{8}$$

This is a whole number.   This is a fraction less than 1.

EXAMPLES Convert to a mixed numeral.

1. $7 + \frac{2}{5} = 7\frac{2}{5}$ **2.** $4 + \frac{3}{10} = 4\frac{3}{10}$

Do Exercises 1–4.

The notation $2\frac{3}{4}$ has a plus sign left out. To aid in understanding, we sometimes write the missing plus sign.

EXAMPLES Convert to fraction notation.

3. $2\frac{3}{4} = 2 + \frac{3}{4}$ Inserting the missing plus sign

$= \frac{2}{1} + \frac{3}{4}$ $2 = \frac{2}{1}$

$= \frac{2}{1} \cdot \frac{4}{4} + \frac{3}{4}$ Finding a common denominator

$= \frac{8}{4} + \frac{3}{4} = \frac{11}{4}$

4. $4\frac{3}{10} = 4 + \frac{3}{10} = \frac{4}{1} + \frac{3}{10} = \frac{4}{1} \cdot \frac{10}{10} + \frac{3}{10} = \frac{40}{10} + \frac{3}{10} = \frac{43}{10}$

Convert to a mixed numeral.

1. $1 + \frac{2}{3} = \square\frac{\square}{\square}$

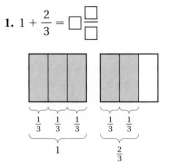

$\frac{1}{3}$ $\frac{1}{3}$ $\frac{1}{3}$ $\frac{1}{3}$ $\frac{1}{3}$

1      $\frac{2}{3}$

2. $2 + \frac{3}{4} = \square\frac{\square}{\square}$

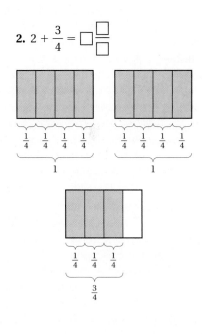

$\frac{1}{4}$ $\frac{1}{4}$ $\frac{1}{4}$ $\frac{1}{4}$ $\frac{1}{4}$ $\frac{1}{4}$ $\frac{1}{4}$ $\frac{1}{4}$

1   1

$\frac{1}{4}$ $\frac{1}{4}$ $\frac{1}{4}$

$\frac{3}{4}$

3. $8 + \frac{3}{4}$ **4.** $12 + \frac{2}{3}$

Answers on page A-7

Convert to fraction notation.

5. $4\frac{2}{5}$ **6.** $6\frac{1}{10}$

Do Exercises 5 and 6.

Let's now consider a faster method for converting a mixed numeral to fraction notation.

To convert from a mixed numeral to fraction notation:

ⓐ Multiply the whole number by the denominator: $4 \cdot 10 = 40$.
ⓑ Add the result to the numerator: $40 + 3 = 43$.
ⓒ Keep the denominator.

Convert to fraction notation. Use the faster method.

7. $4\frac{5}{6}$

EXAMPLES Convert to fraction notation.

5. $6\frac{2}{3} = \frac{20}{3}$ $6 \cdot 3 = 18, \ 18 + 2 = 20$

6. $8\frac{2}{9} = \frac{74}{9}$ $8 \cdot 9 = 72, \ 72 + 2 = 74$

7. $10\frac{7}{8} = \frac{87}{8}$ $10 \cdot 8 = 80, \ 80 + 7 = 87$

Do Exercises 7–9.

WRITING MIXED NUMERALS

We can find a mixed numeral for $\frac{5}{3}$ as follows:

$$\frac{5}{3} = \frac{3}{3} + \frac{2}{3} = 1 + \frac{2}{3} = 1\frac{2}{3}.$$

8. $9\frac{1}{4}$

In terms of objects, we can think of $\frac{5}{3}$ as $\frac{3}{3}$, or 1, plus $\frac{2}{3}$, as shown below.

$$\frac{5}{3} = \qquad \frac{3}{3}, \text{or } 1 \qquad + \qquad \frac{2}{3}$$

Fraction symbols like $\frac{5}{3}$ also indicate division; $\frac{5}{3}$ means $5 \div 3$. Let's divide the numerator by the denominator.

$$\begin{array}{r} 1 \\ 3\overline{)5} \\ \underline{3} \\ 2 \end{array} \leftarrow 2 \div 3 = \frac{2}{3}$$

9. $20\frac{2}{3}$

Thus, $\frac{5}{3} = 1\frac{2}{3}$.

To convert from fraction notation to a mixed numeral, divide.

Answers on page A-7

EXAMPLES Convert to a mixed numeral.

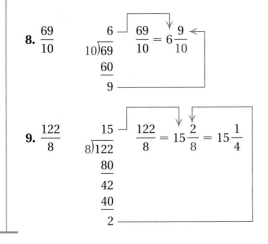

8. $\dfrac{69}{10}$

$$10\overline{)69} \qquad \dfrac{69}{10} = 6\dfrac{9}{10}$$
$$\underline{60}$$
$$9$$

9. $\dfrac{122}{8}$

$$8\overline{)122} \qquad \dfrac{122}{8} = 15\dfrac{2}{8} = 15\dfrac{1}{4}$$
$$\underline{80}$$
$$42$$
$$\underline{40}$$
$$2$$

Do Exercises 10–12.

b Writing Mixed Numerals for Quotients

It is quite common when dividing whole numbers to write the quotient using a mixed numeral. The remainder is the numerator of the fractional part of the mixed numeral.

EXAMPLE 10 Divide. Write a mixed numeral for the quotient.

$$7\overline{)6\ 3\ 4\ 1}$$

We first divide as usual.

$$\begin{array}{r} 9\ 0\ 5 \\ 7\overline{)6\ 3\ 4\ 1} \\ \underline{6\ 3\ 0\ 0} \\ 4\ 1 \\ \underline{3\ 5} \\ 6 \end{array}$$

The answer is 905 R 6. We write a mixed numeral for the answer as follows:

$$905\dfrac{6}{7}.$$

The division $6341 \div 7$ can be expressed using fraction notation or a mixed numeral as follows:

$$\dfrac{6341}{7} = 905\dfrac{6}{7}.$$

Convert to a mixed numeral.

10. $\dfrac{7}{3}$

11. $\dfrac{11}{10}$

12. $\dfrac{110}{6}$

Answers on page A-7

Study Tips

FORMING A STUDY GROUP

Consider forming a study group with some of your fellow students. Exchange e-mail addresses, telephone numbers, and schedules so that you can coordinate study time for homework and tests.

Divide. Write a mixed numeral for the answer.

13. $6\overline{)4846}$

EXAMPLE 11 Divide. Write a mixed numeral for the answer.

$42\overline{)8915}$

We first divide as usual.

$$
\begin{array}{r}
212 \\
42\overline{)8915} \\
8400 \\
\hline
515 \\
420 \\
\hline
95 \\
84 \\
\hline
11
\end{array}
$$

$$\frac{8915}{42} = 212\frac{11}{42}$$

The answer is $212\frac{11}{42}$.

Do Exercises 13 and 14.

A fraction larger than 1, such as $\frac{27}{8}$, is sometimes referred to as an "improper" fraction. We will not use this terminology because notation such as $\frac{27}{8}$, $\frac{11}{9}$, and $\frac{89}{10}$ is quite "proper" and very common in algebra.

As we will see in subsequent sections, mixed numerals have many real-world applications, especially in the areas of garment manufacturing and carpentry, as well as many other fields.

14. $45\overline{)6053}$

CALCULATOR CORNER

Writing Quotients as Mixed Numerals When using a calculator to divide whole numbers, we can express the result using a mixed numeral. To do so, we first find the quotient and the remainder as shown in the Calculator Corner on p. 47. The quotient is the whole-number part of the mixed numeral, the remainder is the numerator of the fractional part, and the divisor is the denominator of the fractional part. For example, on p. 47, we saw that $453 \div 15 = 30$ R 3. Then we can also write

$$453 \div 15 = 30\frac{3}{15} = 30\frac{1}{5}.$$

Exercises: Use a calculator to divide. Write the result as a mixed numeral.

1. $6\overline{)8857}$ 6. $32\overline{)234{,}567}$

2. $9\overline{)6088}$ 7. $45\overline{)6033}$

3. $56\overline{)44{,}851}$ 8. $213\overline{)567{,}988}$

4. $18\overline{)234{,}567}$ 9. $112\overline{)400{,}003}$

5. $11\overline{)567{,}895}$ 10. $908\overline{)11{,}234}$

Answers on page A-7

CHAPTER 3: Fraction Notation
and Mixed Numerals

a

1. *Alarm Clock Dimensions.* A world time clock/radio with NOAA (National Oceanic and Atmospheric Administration) Weather Band measures $14\frac{1}{4}$ in. $\times$ $6\frac{3}{4}$ in. $\times$ $2\frac{1}{4}$ in. Convert $14\frac{1}{4}$, $6\frac{3}{4}$, and $2\frac{1}{4}$ to fraction notation.

2. *Carpentry.* Dick Bonewitz, master carpenter, is making a jewelry box according to the design below. Convert each mixed numeral to fraction notation.

3. *Garment Manufacturing.* A tailoring shop determines that for a certain size dress, it must use $\frac{29}{8}$ yd of fabric that is 45 in. wide. To make the same dress with fabric that is 60 in. wide, it needs $\frac{11}{4}$ yd. Convert $\frac{29}{8}$ and $\frac{11}{4}$ to mixed numerals.

4. *Bake Sale* A fire department organized a bake sale for a charitable fund-raiser. Each pie was cut into 8 pieces and each cake into 12 pieces. They sold 73 pieces of pie, or $\frac{73}{8}$ pies, and 55 pieces of cake, or $\frac{55}{12}$ cakes. Convert $\frac{73}{8}$ and $\frac{55}{12}$ to mixed numerals.

Convert to fraction notation.

5. $5\frac{2}{3}$

6. $3\frac{4}{5}$

7. $3\frac{1}{4}$

8. $6\frac{1}{2}$

9. $10\frac{1}{8}$

10. $20\frac{1}{5}$

11. $5\frac{1}{10}$

12. $9\frac{1}{10}$

13. $20\frac{3}{5}$

14. $30\frac{4}{5}$

15. $9\frac{5}{6}$

16. $8\frac{7}{8}$

17. $7\frac{3}{10}$

18. $6\frac{9}{10}$

19. $1\frac{5}{8}$

20. $1\frac{3}{5}$

21. $12\frac{3}{4}$

22. $15\frac{2}{3}$

23. $4\frac{3}{10}$

24. $5\frac{7}{10}$

25. $2\frac{3}{100}$

26. $5\frac{7}{100}$

27. $66\frac{2}{3}$

28. $33\frac{1}{3}$

29. $5\frac{29}{50}$

30. $84\frac{3}{8}$

Convert to a mixed numeral.

31. $\frac{18}{5}$

32. $\frac{17}{4}$

33. $\frac{14}{3}$

34. $\frac{39}{8}$

35. $\frac{27}{6}$

36. $\frac{30}{9}$

37. $\frac{57}{10}$

38. $\frac{89}{10}$

39. $\frac{53}{7}$

40. $\frac{59}{8}$

41. $\frac{45}{6}$

42. $\frac{50}{8}$

43. $\frac{46}{4}$

44. $\frac{39}{9}$

45. $\frac{12}{8}$

46. $\frac{28}{6}$

47. $\frac{757}{100}$

48. $\frac{467}{100}$

49. $\frac{345}{8}$

50. $\frac{223}{4}$

b Divide. Write a mixed numeral for the answer.

51. $8 \overline{)\ 8\ 6\ 9}$

52. $3 \overline{)\ 2\ 1\ 2\ 6}$

53. $5 \overline{)\ 3\ 0\ 9\ 1}$

54. $9 \overline{)\ 9\ 1\ 1\ 0}$

55. $2\ 1 \overline{)\ 8\ 5\ 2}$

56. $8\ 5 \overline{)\ 7\ 6\ 7\ 2}$

57. $1\ 0\ 2 \overline{)\ 5\ 6\ 1\ 2}$

58. $4\ 6 \overline{)\ 1\ 0\ 8\ 1}$

59. **D$_W$** Describe in your own words a method for rewriting a fraction as a mixed numeral.

60. **D$_W$** Are the numbers $2\frac{1}{3}$ and $2 \cdot \frac{1}{3}$ equal? Why or why not?

SKILL MAINTENANCE

61. Round to the nearest hundred: 45,765. [1.4a]

62. Round to the nearest ten: 45,765. [1.4a]

Simplify. [2.5b]

63. $\dfrac{200}{375}$

64. $\dfrac{63}{75}$

65. $\dfrac{160}{270}$

66. $\dfrac{6996}{8028}$

Multiply and simplify. [2.4a], [2.6a]

67. $\dfrac{6}{5} \cdot 15$

68. $\dfrac{5}{12} \cdot 6$

69. $\dfrac{7}{10} \cdot \dfrac{5}{14}$

70. $\dfrac{1}{10} \cdot \dfrac{20}{5}$

Divide and simplify. [2.7b]

71. $\dfrac{2}{3} \div \dfrac{1}{36}$

72. $28 \div \dfrac{4}{7}$

73. $200 \div \dfrac{15}{64}$

74. $\dfrac{3}{4} \div \dfrac{9}{16}$

SYNTHESIS

Write a mixed numeral.

75. ▦ $\dfrac{128,236}{541}$

76. ▦ $\dfrac{103,676}{349}$

77. $\dfrac{56}{7} + \dfrac{2}{3}$

78. $\dfrac{72}{12} + \dfrac{5}{6}$

79. There are $\frac{366}{7}$ weeks in a leap year.

80. There are $\frac{365}{7}$ weeks in a year.

Objectives

 a Add using mixed numerals.

 b Subtract using mixed numerals.

 c Solve applied problems involving addition and subtraction with mixed numerals.

1. Add.

$$2\frac{3}{10}$$
$$+\,5\frac{1}{10}$$

2. Add.

$$8\frac{2}{5}$$
$$+\,3\frac{7}{10}$$

3. Add.

$$9\frac{3}{4}$$
$$+\,3\frac{5}{6}$$

CHAPTER 3: Fraction Notation and Mixed Numerals

a Addition Using Mixed Numerals

To find the sum $1\frac{5}{8} + 3\frac{1}{8}$, we first add the fractions. Then we add the whole numbers.

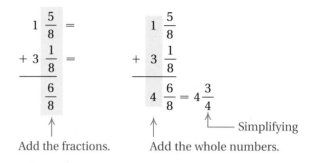

$$1\frac{5}{8} \quad\quad\quad 1\frac{5}{8}$$
$$+\,3\frac{1}{8} \quad\quad +\,3\frac{1}{8}$$
$$\overline{\quad\frac{6}{8}\quad} \quad\quad \overline{4\frac{6}{8} = 4\frac{3}{4}}$$

↑ — Simplifying

Add the fractions.   Add the whole numbers.

Do Exercise 1.

Sometimes we must write the fractional parts with a common denominator before we can add.

EXAMPLE 1 Add: $5\frac{2}{3} + 3\frac{5}{6}$. Write a mixed numeral for the answer.

The LCD is 6.

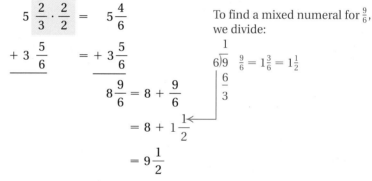

$$5\,\frac{2}{3}\cdot\frac{2}{2} = \quad 5\frac{4}{6}$$
$$+\,3\,\frac{5}{6} \;= +\,3\frac{5}{6}$$
$$\overline{\qquad\qquad\qquad 8\frac{9}{6}} = 8 + \frac{9}{6}$$
$$= 8 + 1\frac{1}{2}$$
$$= 9\frac{1}{2}$$

To find a mixed numeral for $\frac{9}{6}$, we divide:

$$\begin{array}{r} 1 \\ 6\overline{)9} \\ \underline{6} \\ 3 \end{array} \qquad \frac{9}{6} = 1\frac{3}{6} = 1\frac{1}{2}$$

Do Exercise 2.

EXAMPLE 2 Add: $10\frac{5}{6} + 7\frac{3}{8}$.

The LCD is 24.

$$10\,\frac{5}{6}\cdot\frac{4}{4} = \quad 10\frac{20}{24}$$
$$+\,7\,\frac{3}{8}\cdot\frac{3}{3} = +\,7\frac{9}{24}$$
$$\overline{\qquad\qquad\qquad 17\frac{29}{24}} = 17 + \frac{29}{24}$$
$$= 17 + 1\frac{5}{24} \qquad \text{Writing } \frac{29}{24} \text{ as a mixed numeral, } 1\frac{5}{24}$$
$$= 18\frac{5}{24}$$

Do Exercise 3.

b Subtraction Using Mixed Numerals

Subtraction is a lot like addition; we subtract the fractions and then the whole numbers.

EXAMPLE 3 Subtract: $7\frac{3}{4} - 2\frac{1}{4}$.

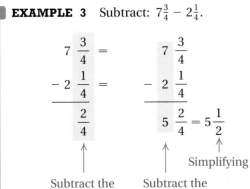

$$7\frac{3}{4} = 7\frac{3}{4}$$
$$-2\frac{1}{4} = -2\frac{1}{4}$$
$$\frac{2}{4} \qquad 5\frac{2}{4} = 5\frac{1}{2}$$

Subtract the fractions.

Subtract the whole numbers.

Simplifying

EXAMPLE 4 Subtract: $9\frac{4}{5} - 3\frac{1}{2}$.

The LCD is 10.

$$9\frac{4}{5} \cdot \frac{2}{2} = 9\frac{8}{10}$$
$$-3\frac{1}{2} \cdot \frac{5}{5} = -3\frac{5}{10}$$
$$6\frac{3}{10}$$

Do Exercises 4 and 5.

EXAMPLE 5 Subtract: $7\frac{1}{6} - 2\frac{1}{4}$.

The LCD is 12.

$$7\frac{1}{6} \cdot \frac{2}{2} = 7\frac{2}{12}$$
$$-2\frac{1}{4} \cdot \frac{3}{3} = -2\frac{3}{12}$$

We cannot subtract $\frac{3}{12}$ from $\frac{2}{12}$.
We borrow 1, or $\frac{12}{12}$, from 7:
$7\frac{2}{12} = 6 + 1 + \frac{2}{12} = 6 + \frac{12}{12} + \frac{2}{12} = 6\frac{14}{12}$.

We can write this as

$$7\frac{2}{12} = 6\frac{14}{12}$$
$$-2\frac{3}{12} = -2\frac{3}{12}$$
$$4\frac{11}{12}$$

Do Exercise 6.

Subtract.

4. $10\frac{7}{8}$
$\quad -9\frac{3}{8}$

5. $8\frac{2}{3}$
$\quad -5\frac{1}{2}$

6. Subtract.

$8\frac{1}{9}$
$-4\frac{5}{6}$

Answers on page A-7

7. Subtract.

$$\begin{array}{r} 5 \\ - 1\dfrac{1}{3} \\ \hline \end{array}$$

EXAMPLE 6 Subtract: $12 - 9\frac{3}{8}$.

$$\begin{array}{r} 12 \quad = \quad 11\dfrac{8}{8} \\ - \; 9\dfrac{3}{8} = - \; 9\dfrac{3}{8} \\ \hline 2\dfrac{5}{8} \end{array} \qquad 12 = 11 + 1 = 11 + \dfrac{8}{8} = 11\dfrac{8}{8}$$

Do Exercise 7.

8. Travel Distance. Chrissy Jenkins is a college textbook sales representative for Addison-Wesley. On two business days, Chrissy drove $144\frac{9}{10}$ mi and $87\frac{1}{4}$ mi. What was the total distance Chrissy drove?

ⓒ Applications and Problem Solving

EXAMPLE 7 *Widening a Driveway.* Sherry and Woody are widening their existing $17\frac{1}{4}$-ft driveway by adding $5\frac{9}{10}$ ft on one side. What is the width of the new driveway?

1. **Familiarize.** We let $w =$ the width of the new driveway.
2. **Translate.** We translate as follows.

$$\underbrace{\text{Width of existing driveway}} + \underbrace{\text{Width of additional driveway}} = \underbrace{\text{Width of new driveway}}$$

$$17\dfrac{1}{4} \quad + \quad 5\dfrac{9}{10} \quad = \quad w$$

3. **Solve.** The translation tells us what to do. We add. The LCD is 20.

$$17\dfrac{1}{4} = \quad 17\; \dfrac{1}{4} \cdot \dfrac{5}{5} = \quad 17\dfrac{5}{20}$$

$$+ \; 5\dfrac{9}{10} = + \; 5 \; \dfrac{9}{10} \cdot \dfrac{2}{2} = + \; 5\dfrac{18}{20}$$

$$\overline{\qquad\qquad\qquad\qquad\qquad 22\dfrac{23}{20} = 23\dfrac{3}{20}}$$

Thus, $w = 23\frac{3}{20}$.

4. **Check.** We check by repeating the calculation. We also note that the answer is larger than either of the widths, which means that the answer is reasonable.
5. **State.** The width of the new driveway is $23\frac{3}{20}$ ft.

Answers on page A-7

Do Exercise 8.

EXAMPLE 8 *Communication Cable.* Celebrity Cable has two crews who install Internet communication cable. Crew A can install $38\frac{1}{8}$ ft per hour. Crew B can install $31\frac{2}{3}$ ft per hour. How many fewer feet can Crew B install per hour than Crew A?

1. **Familiarize.** The phrase "how many fewer" indicates subtraction. We let c = the difference in the numbers of feet per hour.

2. **Translate.** We translate as follows.

$$\underbrace{\text{Crew A:}}_{\text{ft/hr}} \quad - \quad \underbrace{\text{Crew B:}}_{\text{ft/hr}} \quad = \quad \underbrace{\text{Difference in}}_{\text{ft/hr}}$$

$$38\frac{1}{8} \quad - \quad 31\frac{2}{3} \quad = \quad c$$

3. **Solve.** To solve the equation, we carry out the subtraction. The LCD = 24.

$$38\frac{1}{8} = 38\,\frac{1}{8}\cdot\frac{3}{3} = 38\frac{3}{24} = 37\frac{27}{24}$$

$$-31\frac{2}{3} = -31\,\frac{2}{3}\cdot\frac{8}{8} = -31\frac{16}{24} = -31\frac{16}{24}$$

$$\overline{\phantom{-31\frac{2}{3}}}\qquad\overline{\phantom{=-31\,\frac{2}{3}\cdot\frac{8}{8}}}\qquad\overline{\phantom{=-31\frac{16}{24}}}\qquad\overline{6\frac{11}{24}}$$

Thus, $c = 6\frac{11}{24}$.

4. **Check.** To check, we add the difference, $6\frac{11}{24}$, to Crew B's rate:

$$31\frac{2}{3} + 6\frac{11}{24} = 31\frac{16}{24} + 6\frac{11}{24}$$

$$= 37\frac{27}{24}$$

$$= 38\frac{3}{24}$$

$$= 38\frac{1}{8}.$$

This checks.

5. **State.** Crew B installs $6\frac{11}{24}$ fewer feet per hour than Crew A.

Do Exercise 9.

9. NCAA Football Goal Posts. In college football, the distance between goal posts was reduced from $23\frac{1}{3}$ ft to $18\frac{1}{2}$ ft in 1991. By how much was it reduced?

Source: NCAA

Answer on page A-7

10. Liquid Fertilizer. There are $283\frac{5}{8}$ gal of liquid fertilizer in a fertilizer application tank. After applying $178\frac{2}{3}$ gal to a soybean field, the farmer requests that Braden's Farm Supply deliver an additional 250 gal. How many gallons of fertilizer are in the tank after the delivery?

MULTISTEP PROBLEMS

EXAMPLE 9 *Carpentry.* The following diagram shows the layout for the construction of a desk drawer. Find the missing length a.

Middle Drawer / Back Layout

1. **Familiarize.** The length a is shown in the drawing.

2. **Translate.** From the drawing, we see that the length a is the full length of the drawer minus the sum of the lengths $10\frac{15}{32}$" and $8\frac{19}{32}$". Thus we have

$$a = 19\frac{7}{8} - \left(10\frac{15}{32} + 8\frac{19}{32}\right).$$

3. **Solve.** This is a two-step problem.

   a) We first add the two lengths, $10\frac{15}{32}$ and $8\frac{19}{32}$.

$$
\begin{array}{r}
10\dfrac{15}{32} \\[2mm]
+\ 8\dfrac{19}{32} \\[1mm]
\hline
18\dfrac{34}{32} = 18\dfrac{17}{16} = 19\dfrac{1}{16}
\end{array}
$$

   b) Next, we subtract $19\frac{1}{16}$ from $19\frac{7}{8}$.

$$
\begin{array}{r}
19\dfrac{7}{8} = \quad 19\dfrac{14}{16} \\[2mm]
-19\dfrac{1}{16} = -19\dfrac{1}{16} \\[1mm]
\hline
\dfrac{13}{16}
\end{array}
$$

   Thus, $a = \frac{13}{16}$.

4. **Check.** We check by repeating the calculation, or adding the three measurements:

$$10\frac{15}{32} + 8\frac{19}{32} + \frac{13}{16} = 10\frac{15}{32} + 8\frac{19}{32} + \frac{26}{32}$$

$$= 18\frac{60}{32} = 19\frac{28}{32} = 19\frac{7}{8}.$$

5. **State.** The length a in the diagram is $\frac{13}{16}$".

Do Exercise 10.

Answer on page A-7

a Add. Write a mixed numeral for the answer.

1. 20
$\quad + \ 8\frac{3}{4}$

2. 37
$\quad + \ 18\frac{2}{3}$

3. $129\frac{7}{8}$
$\quad + \ \ 56$

4. $2003\frac{4}{11}$
$\quad + \ \ \ \ 59$

5. $2\frac{7}{8}$
$\quad + \ 3\frac{5}{8}$

6. $4\frac{5}{6}$
$\quad + \ 3\frac{5}{6}$

7. $1\frac{1}{4} + 1\frac{2}{3}$

8. $4\frac{1}{3} + 5\frac{2}{9}$

9. $8\frac{3}{4}$
$\quad + \ 5\frac{5}{6}$

10. $4\frac{3}{8}$
$\quad + \ 6\frac{5}{12}$

11. $3\frac{2}{5}$
$\quad + \ 8\frac{7}{10}$

12. $5\frac{1}{2}$
$\quad + \ 3\frac{7}{10}$

13. $5\frac{3}{8}$
$\quad + \ 10\frac{5}{6}$

14. $\frac{5}{8}$
$\quad + \ 1\frac{5}{6}$

15. $12\frac{4}{5}$
$\quad + \ \ 8\frac{7}{10}$

16. $15\frac{5}{8}$
$\quad + \ 11\frac{3}{4}$

17. $14\frac{5}{8}$
$\quad + \ 13\frac{1}{4}$

18. $16\frac{1}{4}$
$\quad + \ 15\frac{7}{8}$

19. $7\frac{1}{8}$
$\quad \ \ 9\frac{2}{3}$
$\quad + \ 10\frac{3}{4}$

20. $45\frac{2}{3}$
$\quad \ \ 31\frac{3}{5}$
$\quad + \ 12\frac{1}{4}$

b Subtract. Write a mixed numeral for the answer.

21. $4\frac{1}{5}$
$\quad - \ 2\frac{3}{5}$

22. $5\frac{1}{8}$
$\quad - \ 2\frac{3}{8}$

23. $6\frac{3}{5} - 2\frac{1}{2}$

24. $7\frac{2}{3} - 6\frac{1}{2}$

25. $34\frac{1}{3}$
$\quad - \ 12\frac{5}{8}$

26. $23\frac{5}{16}$
$\quad - \ 16\frac{3}{4}$

27. 21
$\quad - \ 8\frac{3}{4}$

28. 42
$\quad - \ 3\frac{7}{8}$

29. $\quad 34$
$\quad\quad -\, 18\frac{5}{8}$
$\quad\quad \overline{\phantom{-18\frac{5}{8}}}$

30. $\quad 23$
$\quad\quad -\, 19\frac{3}{4}$
$\quad\quad \overline{\phantom{-19\frac{3}{4}}}$

31. $\quad 21\frac{1}{6}$
$\quad\quad -\, 13\frac{3}{4}$
$\quad\quad \overline{\phantom{-13\frac{3}{4}}}$

32. $\quad 42\frac{1}{10}$
$\quad\quad -\, 23\frac{7}{12}$
$\quad\quad \overline{\phantom{-23\frac{7}{12}}}$

33. $\quad 14\frac{1}{8}$
$\quad\quad -\, \frac{3}{4}$
$\quad\quad \overline{\phantom{-1\frac{3}{4}}}$

34. $\quad 28\frac{1}{6}$
$\quad\quad -\, 5$
$\quad\quad \overline{}$

35. $\quad 25\frac{1}{9}$
$\quad\quad -\, 13\frac{5}{6}$
$\quad\quad \overline{\phantom{-13\frac{5}{6}}}$

36. $\quad 23\frac{5}{16}$
$\quad\quad -\, 14\frac{7}{12}$
$\quad\quad \overline{\phantom{-14\frac{7}{12}}}$

C   Solve.

37. *Sewing from a Pattern.*   Suppose you want to make an outfit in size 8. Using 45-in. fabric, you need $1\frac{3}{8}$ yd for the dress, $\frac{5}{8}$ yd of contrasting fabric for the band at the bottom, and $3\frac{3}{8}$ yd for the jacket. How many yards of 45-in. fabric are needed to make the outfit?

38. *Sewing from a Pattern.*   Suppose you want to make an outfit in size 12. Using 45-in. fabric, you need $2\frac{3}{4}$ yd for the dress and $3\frac{1}{2}$ yd for the jacket. How many yards of 45-in. fabric are needed to make the outfit?

39. For a family barbecue, Cayla bought packages of hamburger weighing $1\frac{2}{3}$ lb and $5\frac{3}{4}$ lb. What was the total weight of the meat?

40. Marsha's Butcher Shop sold packages of sliced turkey breast weighing $1\frac{1}{3}$ lb and $4\frac{3}{5}$ lb. What was the total weight of the meat?

41. Tara is 66 in. tall and her son, Tom, is $59\frac{7}{12}$ in. tall. How much taller is Tara?

42. Nicholas is $73\frac{2}{3}$ in. tall and his daughter, Kendra, is $71\frac{5}{16}$ in. tall. How much shorter is Kendra?

186

43. *Cutco Knives.* The Kitchen Classics knife set sold by Cutco contains three basic knives: $7\frac{5}{8}$" Petite Chef, $4\frac{3}{4}$" Trimmer, and $2\frac{3}{4}$" Paring Knife. How much larger is the blade in the Petite Chef than in the Trimmer? than in the Paring Knife?

Source: Cutco Cutlery Corporation

44. *Upholstery Fabric.* Executive Car Care sells 45-in. upholstery fabric for car restoration. Art buys $9\frac{1}{4}$ yd and $10\frac{5}{6}$ yd for two car projects. How many total yards did Art buy?

45. A plumber uses pipes of lengths $10\frac{5}{16}$ in. and $8\frac{3}{4}$ in. in the installation of a sink. How much pipe was used?

46. *Writing Supplies.* The standard pencil is $6\frac{7}{8}$ in. wood and $\frac{1}{2}$ in. eraser. What is the total length of the standard pencil?

Source: Eberhard Faber American

47. Kim Park is a computer technician. One day, she drove $180\frac{7}{10}$ mi away from Los Angeles for a service call. The next day, she drove $85\frac{1}{2}$ mi back toward Los Angeles for another service call. How far was she from Los Angeles?

48. Pilar is $4\frac{1}{2}$ in. taller than her daughter, Teresa. Teresa is $66\frac{2}{3}$ in. tall. How tall is Pilar?

49. *Book Size.* One standard book size is $8\frac{1}{2}$ in. by $9\frac{3}{4}$ in. What is the total distance around (perimeter of) the front cover of such a book?

50. *Copier Paper.* A standard sheet of copier paper is $8\frac{1}{2}$ in. by 11 in. What is the total distance around (perimeter of) the paper?

51. *Carpentry.* When cutting wood with a saw, a carpenter must take into account the thickness of the saw blade. Suppose that from a piece of wood 36 in. long, a carpenter cuts a $15\frac{3}{4}$-in. length with a saw blade that is $\frac{1}{8}$ in. thick. How long is the piece that remains?

52. *Painting.* A painter used $1\frac{3}{4}$ gal of paint for the Garcias' living room and $1\frac{1}{3}$ gal for their family room. How much paint was used in all?

53. Rene is $5\frac{1}{4}$ in. taller than his son, who is $72\frac{5}{6}$ in. tall. How tall is Rene?

54. A Boeing 767 flew 640 mi on a nonstop flight. On the return flight, it landed after having flown $320\frac{3}{10}$ mi. How far was the plane from its original point of departure?

55. *Interior Design.* Sue worked $10\frac{1}{2}$ hr over a three-day period on an interior design project. If she worked $2\frac{1}{2}$ hr on the first day and $4\frac{1}{5}$ hr on the second, how many hours did Sue work on the third day?

56. *Painting.* Geri had $3\frac{1}{2}$ gal of paint. It took $2\frac{3}{4}$ gal to paint the family room. She estimated that it would take $2\frac{1}{4}$ gal to paint the living room. How much more paint did Geri need?

Find the perimeter of (distance around) the figure.

57.
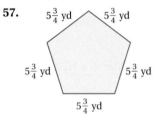
$5\frac{3}{4}$ yd $5\frac{3}{4}$ yd
$5\frac{3}{4}$ yd $5\frac{3}{4}$ yd
$5\frac{3}{4}$ yd

58.

$3\frac{7}{16}$ ft
$3\frac{7}{16}$ ft
$6\frac{7}{8}$ ft
$6\frac{7}{8}$ ft

Find the length d in the figure.

59.

$2\frac{3}{4}$ ft d $2\frac{3}{4}$ ft
$12\frac{7}{8}$ ft

60.

$2\frac{1}{5}$ in. d $2\frac{1}{5}$ in.
$10\frac{1}{2}$ in.

61. Find the smallest length of a bolt that will pass through a piece of tubing with an outside diameter of $\frac{1}{2}$ in., a washer $\frac{1}{16}$ in. thick, a piece of tubing with a $\frac{3}{4}$-in. outside diameter, another washer, and a nut $\frac{3}{16}$ in. thick.

62. The front of the stage at the Lagrange Town Hall is $6\frac{1}{2}$ yd wide. If renovation work succeeds in adding $2\frac{3}{4}$ yd in width, how wide is the renovated stage?

63. $\mathbf{D_W}$ Write a problem for a classmate to solve. Design the problem so the solution is "The larger package holds $4\frac{1}{2}$ oz more than the smaller package."

64. $\mathbf{D_W}$ Is the sum of two mixed numerals always a mixed numeral? Why or why not?

Solve.

65. Rick's Market sells Swiss cheese in $\frac{3}{4}$-lb packages. How many packages can be made from a 12-lb slab of cheese? [2.7d]

66. Holstein's Dairy produced 4578 oz of milk one morning. How many 16-oz cartons could be filled? How much milk would be left over? [1.8a]

Determine whether the first number is divisible by the second. [2.2a]

67. 9993 by 3

68. 9993 by 9

69. 2345 by 9

70. 2345 by 5

71. 2335 by 10

72. 7764 by 6

73. 18,888 by 8

74. 18,888 by 4

75. Multiply and simplify: $\dfrac{15}{9} \cdot \dfrac{18}{39}$. [2.6a]

76. Divide and simplify: $\dfrac{12}{25} \div \dfrac{24}{5}$. [2.7b]

Calculate each of the following. Write the result as a mixed numeral.

77. ▦ $3289\frac{1047}{1189} + 5278\frac{32}{41}$

78. ▦ $5798\frac{17}{53} - 3909\frac{1957}{2279}$

79. A post for a pier is 29 ft long. Half of the post extends above the water's surface and $8\frac{3}{4}$ ft of the post is buried in mud. How deep is the water at that point?

80. Solve: $47\frac{2}{3} + n = 56\frac{1}{4}$.

MULTIPLICATION AND DIVISION USING MIXED NUMERALS; APPLICATIONS

Objectives

a Multiply using mixed numerals.

b Divide using mixed numerals.

c Solve applied problems involving multiplication and division with mixed numerals.

1. Multiply: $6 \cdot 3\frac{1}{3}$.

2. Multiply: $2\frac{1}{2} \cdot \frac{3}{4}$.

3. Multiply: $2 \cdot 6\frac{2}{5}$.

4. Multiply: $3\frac{1}{3} \cdot 2\frac{1}{2}$.

Answers on page A-7

a Multiplication Using Mixed Numerals

Carrying out addition and subtraction with mixed numerals is usually easier if the numbers are left as mixed numerals. With multiplication and division, however, it is easier to convert the numbers to fraction notation first.

> **MULTIPLICATION USING MIXED NUMERALS**
>
> To multiply using mixed numerals, first convert to fraction notation and multiply. Then convert the answer to a mixed numeral, if appropriate.

EXAMPLE 1 Multiply: $6 \cdot 2\frac{1}{2}$.

$$6 \cdot 2\frac{1}{2} = \frac{6}{1} \cdot \frac{5}{2} = \frac{6 \cdot 5}{1 \cdot 2} = \frac{2 \cdot 3 \cdot 5}{2 \cdot 1} = \frac{2}{2} \cdot \frac{3 \cdot 5}{1} = 15$$

Note that fraction notation is needed to carry out the multiplication.

Do Exercise 1.

EXAMPLE 2 Multiply: $3\frac{1}{2} \cdot \frac{3}{4}$.

$$3\frac{1}{2} \cdot \frac{3}{4} = \frac{7}{2} \cdot \frac{3}{4} = \frac{21}{8} = 2\frac{5}{8}$$

Here we write fraction notation.

Do Exercise 2.

EXAMPLE 3 Multiply: $8 \cdot 4\frac{2}{3}$.

$$8 \cdot 4\frac{2}{3} = \frac{8}{1} \cdot \frac{14}{3} = \frac{112}{3} = 37\frac{1}{3}$$

Do Exercise 3.

EXAMPLE 4 Multiply: $2\frac{1}{4} \cdot 3\frac{2}{5}$.

$$2\frac{1}{4} \cdot 3\frac{2}{5} = \frac{9}{4} \cdot \frac{17}{5} = \frac{153}{20} = 7\frac{13}{20}$$

Caution!

$2\frac{1}{4} \cdot 3\frac{2}{5} \neq 6\frac{2}{20}$. A common error is to multiply the whole numbers and then the fractions. This does not give the correct answer, $7\frac{13}{20}$, which is found by converting to fraction notation first.

Do Exercise 4.

b Division Using Mixed Numerals

The division $1\frac{1}{2} \div \frac{1}{6}$ is shown here. *Think*: "How many $\frac{1}{6}$'s are in $1\frac{1}{2}$?" We see that the answer is 9.

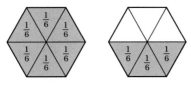

When we divide using mixed numerals, we convert to fraction notation first.

$$1\frac{1}{2} \div \frac{1}{6} = \frac{3}{2} \div \frac{1}{6} = \frac{3}{2} \cdot 6$$

$$= \frac{3 \cdot 6}{2} = \frac{3 \cdot 3 \cdot 2}{2 \cdot 1} = \frac{3 \cdot 3}{1} \cdot \frac{2}{2} = \frac{3 \cdot 3}{1} \cdot 1 = 9$$

> **DIVISION USING MIXED NUMERALS**
>
> To divide using mixed numerals, first write fraction notation and divide. Then convert the answer to a mixed numeral, if appropriate.

EXAMPLE 5 Divide: $32 \div 3\frac{1}{5}$.

$$32 \div 3\frac{1}{5} = \frac{32}{1} \div \frac{16}{5} \qquad \text{Writing the mixed numeral in fraction notation}$$

$$= \frac{32}{1} \cdot \frac{5}{16} = \frac{32 \cdot 5}{1 \cdot 16} = \frac{2 \cdot 16 \cdot 5}{1 \cdot 16} = \frac{16}{16} \cdot \frac{2 \cdot 5}{1} = 1 \cdot \frac{2 \cdot 5}{1} = 10$$

$$\underset{\text{Remember to multiply by the reciprocal.}}{\uparrow}$$

Do Exercise 5.

EXAMPLE 6 Divide: $35 \div 4\frac{1}{3}$.

$$35 \div 4\frac{1}{3} = \frac{35}{1} \div \frac{13}{3} = \frac{35}{1} \cdot \frac{3}{13} = \frac{105}{13} = 8\frac{1}{13}$$

Do Exercise 6.

EXAMPLE 7 Divide: $2\frac{1}{3} \div 1\frac{3}{4}$.

$$2\frac{1}{3} \div 1\frac{3}{4} = \frac{7}{3} \div \frac{7}{4} = \frac{7}{3} \cdot \frac{4}{7} = \frac{7 \cdot 4}{7 \cdot 3} = \frac{7}{7} \cdot \frac{4}{3} = 1 \cdot \frac{4}{3} = \frac{4}{3} = 1\frac{1}{3}$$

Caution!

The reciprocal of $1\frac{3}{4}$ is not $1\frac{4}{3}$!

EXAMPLE 8 Divide: $1\frac{3}{5} \div 3\frac{1}{3}$.

$$1\frac{3}{5} \div 3\frac{1}{3} = \frac{8}{5} \div \frac{10}{3} = \frac{8}{5} \cdot \frac{3}{10} = \frac{8 \cdot 3}{5 \cdot 10} = \frac{2 \cdot 4 \cdot 3}{5 \cdot 2 \cdot 5} = \frac{2}{2} \cdot \frac{4 \cdot 3}{5 \cdot 5} = \frac{12}{25}$$

Do Exercises 7 and 8.

5. Divide: $84 \div 5\frac{1}{4}$.

6. Divide: $26 \div 3\frac{1}{2}$.

Divide.

7. $2\frac{1}{4} \div 1\frac{1}{5}$

8. $1\frac{3}{4} \div 2\frac{1}{2}$

Answers on page A-7

3.6 Multiplication and Division Using
Mixed Numerals; Applications

9. Kyle's pickup truck travels on an interstate highway at 65 mph for $3\frac{1}{2}$ hr. How far does it travel?

  Applications and Problem Solving

EXAMPLE 9 *Average Speed in Indianapolis 500.* Arie Luyendyk won the Indianapolis 500 in 1990 with the highest average speed of about 186 mph. This record high through 2004 is about $2\frac{12}{25}$ times the average speed of the first winner, Ray Harroun, in 1911. What was the average speed in the first Indianapolis 500?

Source: Indianapolis Motor Speedway

1. **Familiarize.** We ask the question, "186 is $2\frac{12}{25}$ times what number? We let s = the average speed in 1911. Then the average speed in 1990 was $2\frac{12}{25} \cdot s$.

2. **Translate.** The problem can be translated to an equation as follows.

$$186 = 2\frac{12}{25} \cdot s$$

10. Holly's minivan travels 302 mi on $15\frac{1}{10}$ gal of gas. How many miles per gallon did it get?

3. **Solve.** To solve the equation, we divide on both sides.

$$186 = \frac{62}{25} \cdot s \qquad \text{Converting } 2\frac{12}{25} \text{ to fraction notation}$$

$$\frac{186}{\frac{62}{25}} = \frac{\frac{62}{25} \cdot s}{\frac{62}{25}} \qquad \text{Dividing by } \frac{62}{25} \text{ on both sides}$$

$$\frac{186}{\frac{62}{25}} = s \qquad \begin{array}{l}\text{Factoring and removing a factor of 1:}\\ ((62/25)/(62/25)) = 1\end{array}$$

$$186 \cdot \frac{25}{62} = s \qquad \text{Multiplying by the reciprocal}$$

$$75 = s \qquad \text{Simplifying: } 186 \cdot \frac{25}{62} = 3 \cdot 25 = 75$$

4. **Check.** If the average speed in 1911 was about 75 mph, we find the average speed in 1990 by multiplying 75 by $2\frac{12}{25}$:

$$2\frac{12}{25} \cdot 75 = \frac{62}{25} \cdot 75 = \frac{62 \cdot 75}{25} = \frac{62 \cdot 25 \cdot 3}{25} = 62 \cdot 3 = 186.$$

The answer checks.

5. **State.** The average speed in the first Indianapolis 500 was about 75 mph.

Answers on page A-7

Do Exercises 9 and 10.

EXAMPLE 10 *Mirror Area.* The mirror-backed candle shelf, shown below with a carpenter's diagram, was designed and built by Harry Cooper. Such shelves were popular in Colonial times because the mirror provided extra lighting from the candle. A rectangular walnut board is used to make the back of the shelf. Find the area of the original board and the amount left over after the opening for the mirror has been cut out.

Source: Popular Science Woodworking Projects

Front View

11. A room measures $22\frac{1}{2}$ ft by $15\frac{1}{2}$ ft. A 9-ft by 12-ft Oriental rug is placed in the center of the room. How much area is not covered by the rug?

1. **Familiarize.** Refer to the figure above. We let h = the height of the back of the shelf and B = the area of the original board. We know the width of the original board, $8\frac{1}{2}$". (Remember, $8\frac{1}{2}$" means $8\frac{1}{2}$ in.) We let A = the area left over after the mirror has been cut out.

2. **Translate.** This is a multistep problem. To find B, which equals $8\frac{1}{2} \cdot h$, we first need to calculate h. We read the dimensions $5\frac{3}{8}$", $11\frac{1}{2}$", and $6\frac{3}{8}$" from the diagram and add them to find h:

$$h = 5\frac{3}{8} + 11\frac{1}{2} + 6\frac{3}{8}.$$

The dimensions of the mirror are $11\frac{1}{2}$" and $5\frac{1}{2}$". Then A is the area of the original board minus the area of the mirror. That is,

$$A = B - 11\frac{1}{2} \cdot 5\frac{1}{2}.$$

Answer on page A-7

3. Solve. We carry out each calculation as follows:

$$h = 5\frac{3}{8} + 11\frac{1}{2} + 6\frac{3}{8} \qquad\qquad B = 8\frac{1}{2} \cdot h$$

$$= 5\frac{3}{8} + 11\frac{4}{8} + 6\frac{3}{8} \qquad\qquad = 8\frac{1}{2} \cdot 23\frac{1}{4} = \frac{17}{2} \cdot \frac{93}{4}$$

$$= 22\frac{10}{8} = 22\frac{5}{4} = 23\frac{1}{4}; \qquad\qquad = \frac{1581}{8} = 197\frac{5}{8};$$

$$A = B - 11\frac{1}{2} \cdot 5\frac{1}{2}$$

$$= 197\frac{5}{8} - 11\frac{1}{2} \cdot 5\frac{1}{2}$$

$$= 197\frac{5}{8} - \frac{23}{2} \cdot \frac{11}{2} = 197\frac{5}{8} - \frac{253}{4}$$

$$= 197\frac{5}{8} - 63\frac{1}{4} = 197\frac{5}{8} - 63\frac{2}{8} = 134\frac{3}{8}.$$

4. Check. We perform a check by repeating the calculations.

5. State. The area of the original board is $197\frac{5}{8}$ in^2. The area left over is $134\frac{3}{8}$ in^2.

Do Exercise 11 on the preceding page.

CALCULATOR CORNER

Operations on Fractions and Mixed Numerals   Fraction calculators can add, subtract, multiply, and divide fractions and mixed numerals. The $\boxed{a^{b}/_{c}}$ key is used to enter fractions and mixed numerals. To find $\frac{3}{4} + \frac{1}{2}$, for example, we press $\boxed{3}$ $\boxed{a^{b}/_{c}}$ $\boxed{4}$ $\boxed{+}$ $\boxed{1}$ $\boxed{a^{b}/_{c}}$ $\boxed{2}$ $\boxed{=}$. Note that 3/4 and 1/2 appear on the display as $\boxed{ 3 \lrcorner 4}$ and $\boxed{ 1 \lrcorner 2}$, respectively. The result is given as the mixed numeral $1\frac{1}{4}$ and is displayed as $\boxed{ 1 \lrcorner 1 \lrcorner 4}$. Fraction results that are greater than 1 are always displayed as mixed numerals. To express this result as a fraction, we press $\boxed{\text{SHIFT}}$ $\boxed{\text{d/c}}$. We get $\boxed{ 5 \lrcorner 4}$, or 5/4.

   To find $3\frac{2}{3} \cdot 4\frac{1}{5}$, we press $\boxed{3}$ $\boxed{a^{b}/_{c}}$ $\boxed{2}$ $\boxed{a^{b}/_{c}}$ $\boxed{3}$ $\boxed{\times}$ $\boxed{4}$ $\boxed{a^{b}/_{c}}$ $\boxed{1}$ $\boxed{a^{b}/_{c}}$ $\boxed{5}$ $\boxed{=}$. The calculator displays $\boxed{ 15 \lrcorner 2 \lrcorner 5}$, so the product is $15\frac{2}{5}$.

   Some calculators are capable of displaying mixed numerals in the way in which we write them, as shown below.

Exercises: Perform each calculation. Give the answer in fraction notation.

1. $\dfrac{1}{3} + \dfrac{1}{4}$ 2. $\dfrac{7}{5} - \dfrac{3}{10}$

3. $\dfrac{15}{4} \cdot \dfrac{7}{12}$ 4. $\dfrac{4}{5} \div \dfrac{8}{3}$

Perform each calculation. Give the answer as a mixed numeral.

5. $4\dfrac{1}{3} + 5\dfrac{4}{5}$ 6. $9\dfrac{2}{7} - 8\dfrac{1}{4}$

7. $2\dfrac{1}{3} \cdot 4\dfrac{3}{5}$ 8. $10\dfrac{7}{10} \div 3\dfrac{5}{6}$

Translating for Success

1. *Raffle Tickets.* At the Happy Hollow Camp Fall Festival, Rico and Becca, together, spent $270 on raffle tickets that sell for $$\frac{9}{20}$$ each. How many tickets did they buy?

2. *Irrigation Pipe.* Jed uses two pipes, one of which measures $5\frac{1}{3}$ ft, to repair the irrigation system in the Buxtons' lawn. The total length of the two pipes is $8\frac{7}{12}$ ft. How long is the other pipe?

3. *Vacation Days.* Together, Helmut and Claire have 36 vacation days a year. Helmut has 22 vacation days per year. How many does Claire have?

4. *Enrollment in Japanese Classes.* Last year at the Lakeside Community College, 225 students enrolled in basic mathematics. This number is $4\frac{1}{2}$ times as many as the number who enrolled in Japanese. How many enrolled in Japanese?

5. *Bicycling.* Cole rode his bicycle $5\frac{1}{3}$ mi on Saturday and $8\frac{7}{12}$ mi on Sunday. How far did he ride on the weekend?

The goal of these matching questions is to practice step (2), *Translate,* of the five-step problem-solving process. Translate each word problem to an equation and select a correct translation from equations A–O.

A. $13\frac{11}{12} = x + 5\frac{1}{3}$

B. $\frac{3}{4} \cdot x = 1\frac{2}{3}$

C. $\frac{20}{9} \cdot x = 270$

D. $225 = 4\frac{1}{2} \cdot x$

E. $98 \div 2\frac{1}{3} = x$

F. $22 + x = 36$

G. $x = 4\frac{1}{2} \cdot 225$

H. $x = 5\frac{1}{3} + 8\frac{7}{12}$

I. $22 \cdot x = 36$

J. $x = \frac{3}{4} \cdot 1\frac{2}{3}$

K. $5\frac{1}{3} + x = 8\frac{7}{12}$

L. $\frac{9}{20} \cdot 270 = x$

M. $1\frac{2}{3} + \frac{3}{4} = x$

N. $98 - 2\frac{1}{3} = x$

O. $\frac{9}{20} \cdot x = 270$

Answers on page A-7

6. *Deli Order.* For a promotional open house for contractors last year, the Bayside Builders Association ordered 225 turkey sandwiches. Due to increased registrations this year, $4\frac{1}{2}$ times as many sandwiches will be needed. How many sandwiches are ordered?

7. *Dog Ownership.* In Sam's community, $\frac{9}{20}$ of the households own at least one dog. There are 270 households. How many own dogs?

8. *Magic Tricks.* A magic trick requires a piece of rope $2\frac{1}{3}$ ft long. Gerry, a magician, has 98 ft of rope and needs to divide it into $2\frac{1}{3}$-ft pieces. How many pieces can be cut from the rope?

9. *Painting.* Laura needs $1\frac{2}{3}$ gal of paint to paint the ceiling of the exercise room and $\frac{3}{4}$ gal of the same paint for the bathroom. How much paint does Laura need?

10. *Chocolate Fudge Bars.* A recipe for chocolate fudge bars that serves 16 includes $1\frac{2}{3}$ cups of sugar. How much sugar is needed for $\frac{3}{4}$ of this recipe?

a   Multiply. Write a mixed numeral for the answer.

1. $8 \cdot 2\frac{5}{6}$

2. $5 \cdot 3\frac{3}{4}$

3. $3\frac{5}{8} \cdot \frac{2}{3}$

4. $6\frac{2}{3} \cdot \frac{1}{4}$

5. $3\frac{1}{2} \cdot 2\frac{1}{3}$

6. $4\frac{1}{5} \cdot 5\frac{1}{4}$

7. $3\frac{2}{5} \cdot 2\frac{7}{8}$

8. $2\frac{3}{10} \cdot 4\frac{2}{5}$

9. $4\frac{7}{10} \cdot 5\frac{3}{10}$

10. $6\frac{3}{10} \cdot 5\frac{7}{10}$

11. $20\frac{1}{2} \cdot 10\frac{1}{5} \cdot 4\frac{2}{3}$

12. $21\frac{1}{3} \cdot 11\frac{1}{3} \cdot 3\frac{5}{8}$

b   Divide. Write a mixed numeral for the answer.

13. $20 \div 3\frac{1}{5}$

14. $18 \div 2\frac{1}{4}$

15. $8\frac{2}{5} \div 7$

16. $3\frac{3}{8} \div 3$

17. $4\frac{3}{4} \div 1\frac{1}{3}$

18. $5\frac{4}{5} \div 2\frac{1}{2}$

19. $1\frac{7}{8} \div 1\frac{2}{3}$

20. $4\frac{3}{8} \div 2\frac{5}{6}$

21. $5\frac{1}{10} \div 4\frac{3}{10}$

22. $4\frac{1}{10} \div 2\frac{1}{10}$

23. $20\frac{1}{4} \div 90$

24. $12\frac{1}{2} \div 50$

C Solve.

25. *Beagles.* There are about 155,000 Labrador retrievers registered with The American Kennel Club. This is $3\frac{4}{9}$ times the number of beagles registered. How many beagles are registered?

Source: The American Kennel Club

26. *Exercise.* At one point during an aerobics class, Kea's bicycle wheel was completing $76\frac{2}{3}$ revolutions per minute. How many revolutions did the wheel complete in 6 min?

27. *Population.* The population of Louisiana is $2\frac{1}{2}$ times the population of West Virginia. The population of West Virginia is approximately 1,800,000. What is the population of Louisiana?

Source: U.S. Bureau of the Census

28. *Population.* The population of New York is $3\frac{1}{3}$ times that of Missouri. The population of New York is approximately 19,000,000. What is the population of Missouri?

Source: U.S. Bureau of the Census

29. *Mural.* Cecilia hired an artist to paint a mural on the wall in her twin sons' bedroom. Dimensions of the mural must be $6\frac{2}{3}$ ft by $9\frac{3}{8}$ ft. What is the area of the mural?

30. *Sidewalk.* A sidewalk alongside a garden at the conservatory is to be $14\frac{2}{5}$ yd long. Rectangular stone tiles that are each $1\frac{1}{8}$ yd long are used to form the sidewalk. How many tiles are used?

31. *Sodium Consumption.* The average American woman consumes $1\frac{1}{3}$ tsp of sodium each day. How much sodium do 10 average American women consume in one day?

Source: *Nutrition Action Health Letter,* March 1994, p. 6. 1875 Connecticut Ave., N.W., Washington, DC 20009-5728

32. *Aeronautics.* Most space shuttles orbit the earth once every $1\frac{1}{2}$ hr. How many orbits are made every 24 hr?

33. *Weight of Water.* The weight of water is $62\frac{1}{2}$ lb per cubic foot. What is the weight of $5\frac{1}{2}$ cubic feet of water?

34. *Weight of Water.* The weight of water is $62\frac{1}{2}$ lb per cubic foot. What is the weight of $2\frac{1}{4}$ cubic feet of water?

35. *Temperatures.* Fahrenheit temperature can be obtained from Celsius (centigrade) temperature by multiplying by $1\frac{4}{5}$ and adding 32°. What Fahrenheit temperature corresponds to a Celsius temperature of 20°?

36. *Temperature.* Fahrenheit temperature can be obtained from Celsius (centigrade) temperature by multiplying by $1\frac{4}{5}$ and adding 32°. What Fahrenheit temperature corresponds to the Celsius temperature of boiling water, which is 100°?

37. *Newspaper Circulation.* In 2002, daily circulation of the *Wall Street Journal* (New York, N.Y.) was about $5\frac{1}{4}$ times the daily circulation of the *Star Tribune* (Minneapolis). The circulation of the *Star Tribune* was about 343,000. What was the daily circulation of the *Wall Street Journal*?

Source: *Editor and Publisher International Year Book* 2003

38. *Language Enrollments.* In institutions of higher education in Fall 2002, the number of students who enrolled in Spanish was $14\frac{3}{10}$ times the number who enrolled in Japanese. About 746,270 students enrolled in Spanish. About how many enrolled in Japanese?

Source: Association of Departments of Foreign Languages at the Modern Language Association

39. *Creamy Peach Coffee Cake.* Listed below is the recipe for creamy peach coffee cake. What are the ingredients for $\frac{1}{2}$ cake? for 4 cakes?

Source: Reprinted with permission from *Taste of Home* magazine, www.tasteofhome.com

Creamy Peach Coffee Cake

$2\frac{1}{4}$ cups all-purpose flour
$\frac{3}{4}$ cup sugar
$\frac{3}{4}$ cup cold butter
$\frac{3}{4}$ cup sour cream
$\frac{1}{2}$ teaspoon baking powder
$\frac{1}{2}$ teaspoon baking soda
1 egg
1 teaspoon almond extract
FILLING:
1 package (8 ounces) cream cheese, softened
$\frac{1}{4}$ cup sugar
1 egg
$\frac{3}{4}$ cup peach preserves
$\frac{1}{2}$ cup sliced almonds

40. *Butterscotch Hard Candy.* Listed below is the recipe for butterscotch hard candy. What are the ingredients for $\frac{1}{2}$ batch? for 3 batches?

Source: Reprinted with permission from *Taste of Home* magazine, www.tasteofhome.com

Butterscotch Hard Candy

$2\frac{1}{2}$ cups sugar
$\frac{3}{4}$ cup water
$\frac{1}{2}$ cup light corn syrup
1 cup butter, cubed
$\frac{1}{4}$ cup honey
$\frac{1}{2}$ teaspoon salt
$\frac{1}{2}$ teaspoon rum extract

41. A car traveled 213 mi on $14\frac{2}{10}$ gal of gas. How many miles per gallon did it get?

42. A car traveled 385 mi on $15\frac{4}{10}$ gal of gas. How many miles per gallon did it get?

43. *Weight of Water.* The weight of water is $62\frac{1}{2}$ lb per cubic foot. How many cubic feet would be occupied by 250 lb of water?

44. *Weight of Water.* The weight of water is $62\frac{1}{2}$ lb per cubic foot. How many cubic feet would be occupied by 375 lb of water?

45. *Servings of Salmon.* A serving of filleted fish is generally considered to be about $\frac{1}{3}$ lb. How many servings can be prepared from $5\frac{1}{2}$ lb of salmon fillet?

46. *Servings of Tuna.* A serving of fish steak (cross section) is generally $\frac{1}{2}$ lb. How many servings can be prepared from a cleaned $18\frac{3}{4}$-lb tuna?

Find the area of the shaded region.

47.

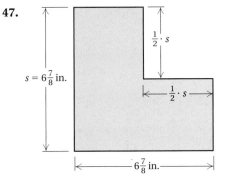

$s = 6\frac{7}{8}$ in.

$\frac{1}{2} \cdot s$

$\frac{1}{2} \cdot s$

$6\frac{7}{8}$ in.

48.

$10\frac{1}{2}$ ft

$8\frac{1}{2}$ ft

4 ft

$10\frac{1}{2}$ ft

49. *Construction.* A rectangular lot has dimensions of $302\frac{1}{2}$ ft by $205\frac{1}{4}$ ft. A building with dimensions of 100 ft by $25\frac{1}{2}$ ft is built on the lot. How much area is left over?

50. *Word Processing.* Kelly wants to create a table using Microsoft® Word software for word processing. She needs to have two columns, each $1\frac{1}{2}$ in. wide, and five columns, each $\frac{3}{4}$ in. wide. Will this table fit on a piece of standard paper that is $8\frac{1}{2}$ in. wide? If so, how wide will each margin be if her margins on each side are to be of equal width?

51. **D**_{**W**} Write a problem for a classmate to solve. Design the problem so that its solution is found by performing the multiplication $4\frac{1}{2} \cdot 33\frac{1}{3}$.

52. **D**_{**W**} Under what circumstances is a pair of mixed numerals more easily added than multiplied?

✎ VOCABULARY REINFORCEMENT

In each of Exercises 53–60, fill in the blank with the correct term from the given list. Some of the choices may not be used and some may be used more than once.

53. In the equation $420 \div 60 = 7$, 60 is called the
_____ , 7 the _____ , and 420 the
_____ .   [1.6a]

54. When denominators are the same, we say that fractions have a _____ denominator.   [2.5c]

55. The numbers 91, 95, and 111 are examples of _____ numbers.   [2.1c]

56. The number $22,223,133$ is _____ by 9 because the sum of its digits is _____ by 9.   [2.2a]

57. When simplifying $24 \div 4 + 4 \times 12 - 6 \div 2$, do all _____ and _____ in order from left to right before doing all _____ and _____ in order from left to right.   [1.9c]

58. In the equation $2 + 3 = 5$, 2 and 3 are called _____ .   [1.2a]

59. In the expression $\dfrac{c}{d}$, we call c the _____ .   [2.3a]

60. The number 0 has no _____ .   [2.7a]

identity
reciprocal
divisor
dividend
quotient
additions
subtractions
multiplications
divisions
numerator
denominator
equal
common
prime
composite
product
divisible
digits
factors
addends

Multiply. Write the answer as a mixed numeral whenever possible.

61. ▦ $15\dfrac{2}{11} \cdot 23\dfrac{31}{43}$

62. ▦ $17\dfrac{23}{31} \cdot 19\dfrac{13}{15}$

Simplify.

63. $8 \div \dfrac{1}{2} + \dfrac{3}{4} + \left(5 - \dfrac{5}{8}\right)^2$

64. $\left(\dfrac{5}{9} - \dfrac{1}{4}\right) \times 12 + \left(4 - \dfrac{3}{4}\right)^2$

65. $\dfrac{1}{3} \div \left(\dfrac{1}{2} - \dfrac{1}{5}\right) \times \dfrac{1}{4} + \dfrac{1}{6}$

66. $\dfrac{7}{8} - 1\dfrac{1}{8} \times \dfrac{2}{3} + \dfrac{9}{10} \div \dfrac{3}{5}$

67. $4\dfrac{1}{2} \div 2\dfrac{1}{2} + 8 - 4 \div \dfrac{1}{2}$

68. $6 - 2\dfrac{1}{3} \times \dfrac{3}{4} + \dfrac{5}{8} \div \dfrac{2}{3}$

CHAPTER 3: Fraction Notation
and Mixed Numerals

Objectives

a Simplify expressions using the rules for order of operations.

b Estimate with fraction and mixed-numeral notation.

a Order of Operations; Fraction Notation and Mixed Numerals

The rules for order of operations that we use with whole numbers (see Section 1.9) apply when we are simplifying expressions involving fraction notation and mixed numerals. For review, these rules are listed below.

RULES FOR ORDER OF OPERATIONS

1. Do all calculations within parentheses before operations outside.
2. Evaluate all exponential expressions.
3. Do all multiplications and divisions in order from left to right.
4. Do all additions and subtractions in order from left to right.

Simplify.

1. $\dfrac{2}{5} \cdot \dfrac{5}{8} + \dfrac{1}{4}$

EXAMPLE 1 Simplify: $\dfrac{1}{6} + \dfrac{2}{3} \div \dfrac{1}{2} \cdot \dfrac{5}{8}$.

$$\frac{1}{6} + \frac{2}{3} \div \frac{1}{2} \cdot \frac{5}{8} = \frac{1}{6} + \frac{2}{3} \cdot \frac{2}{1} \cdot \frac{5}{8}$$

Doing the division first by multiplying by the reciprocal of $\frac{1}{2}$

$$= \frac{1}{6} + \frac{2 \cdot 2 \cdot 5}{3 \cdot 1 \cdot 8}$$

Doing the multiplications in order from left to right

$$= \frac{1}{6} + \frac{\cancel{2} \cdot \cancel{2} \cdot 5}{3 \cdot 1 \cdot \cancel{2} \cdot \cancel{2} \cdot 2}$$

Factoring in order to simplify

$$= \frac{1}{6} + \frac{5}{6}$$

Removing a factor of 1: $\dfrac{2 \cdot 2}{2 \cdot 2} = 1$; simplifying

$$= \frac{6}{6}, \quad \text{or } 1$$

Doing the addition

2. $\dfrac{1}{3} \cdot \dfrac{3}{4} \div \dfrac{5}{8} - \dfrac{1}{10}$

Do Exercises 1 and 2.

EXAMPLE 2 Simplify: $\dfrac{2}{3} \cdot 24 - 11\dfrac{1}{2}$.

$$\frac{2}{3} \cdot 24 - 11\frac{1}{2} = \frac{2 \cdot 24}{3} - 11\frac{1}{2}$$

Doing the multiplication first

$$= \frac{2 \cdot \cancel{3} \cdot 8}{\cancel{3}} - 11\frac{1}{2}$$

Factoring the numerator

$$= 2 \cdot 8 - 11\frac{1}{2}$$

Removing a factor of 1: $\dfrac{3}{3} = 1$

$$= 16 - 11\frac{1}{2}$$

Completing the multiplication

$$= 4\frac{1}{2}, \quad \text{or } \frac{9}{2}$$

Doing the subtraction

3. Simplify: $\dfrac{3}{4} \cdot 16 + 8\dfrac{2}{3}$.

Answers on page A-7

4. After two weeks, Kurt's tomato seedlings measure $9\frac{1}{2}$ in., $10\frac{3}{4}$ in., $10\frac{1}{4}$ in., and 9 in. tall. Find their average height.

5. Find the average of

$$\frac{1}{2}, \ \frac{1}{3}, \ \text{ and } \ \frac{5}{6}.$$

6. Find the average of $\dfrac{3}{4}$ and $\dfrac{4}{5}$.

7. Simplify:

$$\left(\frac{2}{3} + \frac{3}{4}\right) \div 2\frac{1}{3} - \left(\frac{1}{2}\right)^3.$$

Answers on page A-7

CHAPTER 3: Fraction Notation and Mixed Numerals

Do Exercise 3 on the preceding page.

EXAMPLE 3 Melody has triplets. Their birth weights were $3\frac{1}{2}$ lb, $2\frac{3}{4}$ lb, and $3\frac{1}{8}$ lb. What was the average weight of her babies?

Recall that to compute an **average,** we add the numbers and then divide the sum by the number of addends (see Section 1.9). We have

$$\frac{3\frac{1}{2} + 2\frac{3}{4} + 3\frac{1}{8}}{3}.$$

We first add in the numerator:

$$\frac{3\frac{1}{2} + 2\frac{3}{4} + 3\frac{1}{8}}{3} = \frac{3\frac{4}{8} + 2\frac{6}{8} + 3\frac{1}{8}}{3}$$

$$= \frac{8\frac{11}{8}}{3} = \frac{\frac{75}{8}}{3} \quad \text{Doing the additions; converting } 8\frac{11}{8} \text{ to fraction notation, } \frac{75}{8}$$

$$= \frac{75}{8} \cdot \frac{1}{3} = \frac{75}{24} = \frac{25}{8} \quad \text{Multiplying by the reciprocal and simplifying}$$

$$= 3\frac{1}{8}. \quad \text{Converting to a mixed numeral}$$

The average weight of the three babies is $3\frac{1}{8}$ lb.

Do Exercises 4–6.

EXAMPLE 4 Simplify: $\left(\dfrac{7}{8} - \dfrac{1}{3}\right) \times 48 + \left(13 + \dfrac{4}{5}\right)^2.$

$$\left(\frac{7}{8} - \frac{1}{3}\right) \times 48 + \left(13 + \frac{4}{5}\right)^2$$

$$= \left(\frac{7}{8} \cdot \frac{3}{3} - \frac{1}{3} \cdot \frac{8}{8}\right) \times 48 + \left(13 \cdot \frac{5}{5} + \frac{4}{5}\right)^2 \quad \begin{array}{l}\text{Carrying out operations inside parentheses first.} \\ \text{To do so, we first multiply by 1 to obtain the LCD.}\end{array}$$

$$= \left(\frac{21}{24} - \frac{8}{24}\right) \times 48 + \left(\frac{65}{5} + \frac{4}{5}\right)^2$$

$$= \frac{13}{24} \times 48 + \left(\frac{69}{5}\right)^2 \quad \text{Completing the operations within parentheses}$$

$$= \frac{13}{24} \times 48 + \frac{4761}{25} \quad \text{Evaluating the exponential expression next}$$

$$= 26 + \frac{4761}{25} \quad \text{Doing the multiplication}$$

$$= 26 + 190\frac{11}{25} \quad \text{Converting to a mixed numeral}$$

$$= 216\frac{11}{25}, \quad \text{or} \quad \frac{5411}{25} \quad \text{Adding}$$

Answers can be given using either fraction notation or mixed numerals.

Do Exercise 7.

b | Estimation with Fraction Notation and Mixed Numerals

We now estimate with fraction notation and mixed numerals.

EXAMPLES Estimate each of the following as 0, $\frac{1}{2}$, or 1.

5. $\frac{2}{17}$

A fraction is close to 0 when the numerator is small in comparison to the denominator. Thus, 0 is an estimate for $\frac{2}{17}$ because 2 is small in comparison to 17. Thus, $\frac{2}{17} \approx 0$.

6. $\frac{11}{23}$

A fraction is close to $\frac{1}{2}$ when the denominator is about twice the numerator. Thus, $\frac{1}{2}$ is an estimate for $\frac{11}{23}$ because $2 \cdot 11 = 22$ and 22 is close to 23. Thus, $\frac{11}{23} \approx \frac{1}{2}$.

7. $\frac{37}{38}$

A fraction is close to 1 when the numerator is nearly equal to the denominator. Thus, 1 is an estimate for $\frac{37}{38}$ because 37 is nearly equal to 38. Thus, $\frac{37}{38} \approx 1$.

8. $\frac{43}{41}$

As in the preceding example, the numerator 43 is nearly equal to the denominator 41. Thus, $\frac{43}{41} \approx 1$.

Do Exercises 8–11.

EXAMPLE 9 Find a number for the blank so that $\frac{9}{\square}$ is close to but less than 1. Answers may vary.

If the number in the blank were 9, we would have 1, so we increase 9 to 10. An answer is 10; $\frac{9}{10}$ is close to 1. The number 11 would also be a correct answer; $\frac{9}{11}$ is close to 1.

Do Exercises 12 and 13.

Estimate each of the following as 0, $\frac{1}{2}$, or 1.

8. $\frac{3}{59}$

9. $\frac{61}{59}$

10. $\frac{29}{59}$

11. $\frac{57}{59}$

Find a number for the blank so that the fraction is close to but less than 1.

12. $\frac{11}{\square}$

13. $\frac{\square}{33}$

Answers on page A-7

Study Tips

SKILL MAINTENANCE EXERCISES

It is never too soon to begin reviewing for the final examination. The Skill Maintenance exercises found in each exercise set review and reinforce skills taught in earlier sections. Include all of these exercises in your weekly preparation. Answers to both odd-numbered and even-numbered exercises along with section references appear at the back of the book.

Find a number for the blank so that the fraction is close to but less than $\frac{1}{2}$.

14. $\dfrac{13}{\square}$

15. $\dfrac{\square}{31}$

Find a number for the blank so that the fraction is close to but greater than 0.

16. $\dfrac{\square}{37}$

17. $\dfrac{13}{\square}$

Estimate each part of the following as a whole number or as a mixed numeral where the fractional part is $\frac{1}{2}$.

18. $5\dfrac{9}{10} + 26\dfrac{1}{2} - 10\dfrac{3}{29}$

19. $10\dfrac{7}{8} \cdot \left(25\dfrac{11}{13} - 14\dfrac{1}{9} \right)$

20. $\left(10\dfrac{4}{5} + 7\dfrac{5}{9} \right) \div \dfrac{17}{30}$

EXAMPLE 10 Find a number for the blank so that $\dfrac{9}{\square}$ is close to but less than $\frac{1}{2}$. Answers may vary.

If we double 9 to get 18 and use it for the blank, we have $\frac{1}{2}$. If we increase that denominator by 1, to get 19, and use it for the blank, we get a number less than $\frac{1}{2}$ but close to $\frac{1}{2}$. Thus, $\frac{9}{19} \approx \frac{1}{2}$.

Do Exercises 14 and 15.

EXAMPLE 11 Find a number for the blank so that $\dfrac{\square}{50}$ is close to but greater than 0.

Since 50 is rather large, any small number such as 1, 2, or 3 will make the fraction close to 0. For example, $\frac{1}{50} \approx 0$.

Do Exercises 16 and 17.

EXAMPLE 12 Estimate $16\frac{8}{9} + 11\frac{2}{13} - 4\frac{22}{43}$ as a whole number or as a mixed numeral where the fractional part is $\frac{1}{2}$.

We estimate each fraction as 0, $\frac{1}{2}$, or 1. Then we calculate:

$$16\dfrac{8}{9} + 11\dfrac{2}{13} - 4\dfrac{22}{43} \approx 17 + 11 - 4\dfrac{1}{2}$$

$$= 28 - 4\dfrac{1}{2}$$

$$= 23\dfrac{1}{2}.$$

Do Exercises 18–20.

Answers on page A-7

CHAPTER 3: Fraction Notation and Mixed Numerals

a Simplify.

1. $\dfrac{1}{2} \cdot \dfrac{1}{3} \cdot \dfrac{1}{4}$

2. $\dfrac{1}{3} \cdot \dfrac{1}{4} \cdot \dfrac{1}{5}$

3. $6 \div 3 \div 5$

4. $12 \div 4 \div 8$

5. $\dfrac{2}{3} \div \dfrac{4}{3} \div \dfrac{7}{8}$

6. $\dfrac{5}{6} \div \dfrac{3}{4} \div \dfrac{2}{5}$

7. $\dfrac{5}{8} \div \dfrac{1}{4} - \dfrac{2}{3} \cdot \dfrac{4}{5}$

8. $\dfrac{4}{7} \cdot \dfrac{7}{15} + \dfrac{2}{3} \div 8$

9. $\dfrac{3}{4} - \dfrac{2}{3} \cdot \left(\dfrac{1}{2} + \dfrac{2}{5} \right)$

10. $\dfrac{3}{4} \div \dfrac{1}{2} \cdot \left(\dfrac{8}{9} - \dfrac{2}{3} \right)$

11. $28\dfrac{1}{8} - 5\dfrac{1}{4} + 3\dfrac{1}{2}$

12. $10\dfrac{3}{5} - 4\dfrac{1}{10} - 1\dfrac{1}{2}$

13. $\dfrac{7}{8} \div \dfrac{1}{2} \cdot \dfrac{1}{4}$

14. $\dfrac{7}{10} \cdot \dfrac{4}{5} \div \dfrac{2}{3}$

15. $\left(\dfrac{2}{3} \right)^2 - \dfrac{1}{3} \cdot 1\dfrac{1}{4}$

16. $\left(\dfrac{3}{4} \right)^2 + 3\dfrac{1}{2} \div 1\dfrac{1}{4}$

17. $\dfrac{1}{2} - \left(\dfrac{1}{2} \right)^2 + \left(\dfrac{1}{2} \right)^3$

18. $1 + \dfrac{1}{4} + \left(\dfrac{1}{4} \right)^2 - \left(\dfrac{1}{4} \right)^3$

19. Find the average of $\dfrac{2}{3}$ and $\dfrac{7}{8}$.

20. Find the average of $\dfrac{1}{4}$ and $\dfrac{1}{5}$.

21. Find the average of $\dfrac{1}{6}, \dfrac{1}{8}$, and $\dfrac{3}{4}$.

22. Find the average of $\dfrac{4}{5}, \dfrac{1}{2}$, and $\dfrac{1}{10}$.

23. Find the average of $3\dfrac{1}{2}$ and $9\dfrac{3}{8}$.

24. Find the average of $10\dfrac{2}{3}$ and $24\dfrac{5}{6}$.

25. *Birth Weights.* The Piper quadruplets of Great Britain weighed $2\frac{9}{16}$ lb, $2\frac{9}{32}$ lb, $2\frac{1}{8}$ lb, and $2\frac{5}{16}$ lb at birth. Find their average birth weight.

Source: *The Guinness Book of Records,* 1998

26. *Vertical Leaps.* Eight-year-old Zachary registered vertical leaps of $12\frac{3}{4}$ in., $13\frac{3}{4}$ in., $13\frac{1}{2}$ in., and 14 in. Find his average vertical leap.

27. *Acceleration* The results of a *Motor Trend* road acceleration test for five cars are given in the graph below. The test measures the time in seconds required to go from 0 mph to 60 mph. What was the average time?

Source: *Motor Trend,* March 2005, pp. 134–142

28. *Manufacturing.* A test of five light bulbs showed that they burned for the lengths of time given on the graph below. For how many days, on average, did the bulbs burn?

Simplify.

29. $\left(\dfrac{2}{3} + \dfrac{3}{4}\right) \div \left(\dfrac{5}{6} - \dfrac{1}{3}\right)$

30. $\left(\dfrac{3}{5} - \dfrac{1}{2}\right) \div \left(\dfrac{3}{4} - \dfrac{3}{10}\right)$

31. $\left(\dfrac{1}{2} + \dfrac{1}{3}\right)^2 \cdot 144 - \dfrac{5}{8} \div 10\dfrac{1}{2}$

32. $\left(3\dfrac{1}{2} - 2\dfrac{1}{3}\right)^2 + 6 \cdot 2\dfrac{1}{2} \div 32$

b Estimate each of the following as $0, \frac{1}{2},$ or 1.

33. $\dfrac{2}{47}$

34. $\dfrac{4}{5}$

35. $\dfrac{1}{13}$

36. $\dfrac{7}{8}$

37. $\dfrac{6}{11}$

38. $\dfrac{10}{13}$

39. $\dfrac{7}{15}$

40. $\dfrac{1}{16}$

41. $\dfrac{7}{100}$

42. $\dfrac{5}{9}$

43. $\dfrac{19}{20}$

44. $\dfrac{5}{12}$

Find a number for the blank so that the fraction is close to but greater than $\frac{1}{2}$. Answers may vary.

45. $\dfrac{\square}{11}$ **46.** $\dfrac{\square}{8}$ **47.** $\dfrac{\square}{23}$ **48.** $\dfrac{\square}{35}$

49. $\dfrac{10}{\square}$ **50.** $\dfrac{51}{\square}$

Find a number for the blank so that the fraction is close to but greater than 1. Answers may vary.

51. $\dfrac{7}{\square}$ **52.** $\dfrac{11}{\square}$ **53.** $\dfrac{13}{\square}$ **54.** $\dfrac{27}{\square}$ **55.** $\dfrac{\square}{15}$ **56.** $\dfrac{\square}{100}$

Estimate each part of the following as a whole number, $\frac{1}{2}$, or as a mixed numeral where the fractional part is $\frac{1}{2}$.

57. $2\dfrac{7}{8}$ **58.** $1\dfrac{1}{3}$ **59.** $12\dfrac{5}{6}$

60. $26\dfrac{6}{13}$ **61.** $\dfrac{4}{5} + \dfrac{7}{8}$ **62.** $\dfrac{1}{12} \cdot \dfrac{7}{15}$

63. $\dfrac{2}{3} + \dfrac{7}{13} + \dfrac{5}{9}$ **64.** $\dfrac{8}{9} + \dfrac{4}{5} + \dfrac{11}{12}$ **65.** $\dfrac{43}{100} + \dfrac{1}{10} - \dfrac{11}{1000}$

66. $\dfrac{23}{24} + \dfrac{37}{39} + \dfrac{51}{50}$ **67.** $7\dfrac{29}{60} + 10\dfrac{12}{13} \cdot 24\dfrac{2}{17}$ **68.** $5\dfrac{13}{14} - 1\dfrac{5}{8} + 1\dfrac{23}{28} \cdot 6\dfrac{35}{74}$

69. $24 \div 7\dfrac{8}{9}$ **70.** $43\dfrac{16}{17} \div 11\dfrac{2}{13}$ **71.** $76\dfrac{3}{14} + 23\dfrac{19}{20}$

72. $76\dfrac{13}{14} \cdot 23\dfrac{17}{20}$ **73.** $16\dfrac{1}{5} \div 2\dfrac{1}{11} + 25\dfrac{9}{10} - 4\dfrac{11}{23}$ **74.** $96\dfrac{2}{13} \div 5\dfrac{19}{20} + 3\dfrac{1}{7} \cdot 5\dfrac{18}{21}$

75. **D_W** A student insists that $3\frac{2}{5} \cdot 1\frac{3}{7} = 3\frac{6}{35}$. What mistake is he making and how should he have proceeded?

76. **D_W** A student insists that $5 \cdot 3\frac{2}{7} = (5 \cdot 3) \cdot \left(5 \cdot \frac{2}{7}\right)$. What mistake is she making and how should she have proceeded?

77. Multiply: $27 \cdot 126$.   [1.5a]

78. Multiply: $132 \cdot 7865$.   [1.5a]

79. Divide: $7865 \div 132$.   [1.6b]

Multiply and simplify.   [2.4a], [2.6a]

80. $\dfrac{2}{3} \cdot 522$

81. $\dfrac{3}{2} \cdot 522$

Divide and simplify.   [2.7b]

82. $\dfrac{4}{5} \div \dfrac{3}{10}$

83. $\dfrac{3}{10} \div \dfrac{4}{5}$

84. Classify the given numbers as prime, composite, or neither.   [2.1c]

$1, 5, 7, 9, 14, 23, 43$

Solve.

85. *Luncheon Servings.*   Ian purchased 6 lb of cold cuts for a luncheon. If Ian has allowed $\frac{3}{8}$ lb per person, how many people did he invite to the luncheon?   [2.7d]

86. *Cholesterol.*   A 3-oz serving of crabmeat contains 85 milligrams (mg) of cholesterol. A 3-oz serving of shrimp contains 128 mg of cholesterol. How much more cholesterol is in the shrimp?   [1.8a]

87. a) Find an expression for the sum of the areas of the two rectangles shown here.
   b) Simplify the expression.
   c) How is the computation in part (b) related to the rules for order of operations?

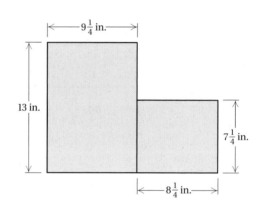

88. Find r if

$$\frac{1}{r} = \frac{1}{100} + \frac{1}{150} + \frac{1}{200}.$$

89. In the sum below, a and b are digits. Find a and b.

$$\frac{a}{17} + \frac{1b}{23} = \frac{35a}{391}$$

90. Consider only the numbers 3, 4, 5, and 6. Assume each can be placed in a blank in the following.

$$\square + \frac{\square}{\square} \cdot \square = ?$$

What placement of the numbers in the blanks yields the largest number?

91. Consider only the numbers 2, 3, 4, and 5. Assume each is placed in a blank in the following.

$$\frac{\square}{\square} + \frac{\square}{\square} = ?$$

What placement of the numbers in the blanks yields the largest sum?

92. Use a standard calculator. Arrange the following in order from smallest to largest.

$$\frac{3}{4}, \frac{17}{21}, \frac{13}{15}, \frac{7}{9}, \frac{15}{17}, \frac{13}{12}, \frac{19}{22}$$

Summary and Review

The review that follows is meant to prepare you for a chapter exam. It consists of two parts. The first part, Concept Reinforcement, is designed to increase understanding of the concepts through true/false exercises. The second part is the Review Exercises. These provide practice exercises for the exam, together with references to section objectives so you can go back and review. Before beginning, stop and look back over the skills you have obtained. What skills in mathematics do you have now that you did not have before studying this chapter?

✎ CONCEPT REINFORCEMENT

Determine whether the statement is true or false. Answers are given at the back of the book.

_____ **1.** If $\dfrac{a}{b} > \dfrac{c}{b}$, $b \neq 0$, then $a > c$.

_____ **2.** All mixed numerals represent numbers larger than 1.

_____ **3.** The least common multiple of two natural numbers is the smallest number that is a factor of both.

_____ **4.** The mixed numeral $5\dfrac{2}{3}$ can be represented by the sum $5 \cdot \dfrac{3}{3} + \dfrac{2}{3}$.

_____ **5.** The least common multiple of two numbers is always larger than or equal to the larger number.

_____ **6.** To add fractions when denominators are the same, we keep the numerator and add the denominators.

Review Exercises

Find the LCM. [3.1a]

1. 12 and 18

2. 18 and 45

3. 3, 6, and 30

4. 26, 36, and 54

Add and simplify. [3.2a]

5. $\dfrac{6}{5} + \dfrac{3}{8}$

6. $\dfrac{5}{16} + \dfrac{1}{12}$

7. $\dfrac{6}{5} + \dfrac{11}{15} + \dfrac{3}{20}$

8. $\dfrac{1}{1000} + \dfrac{19}{100} + \dfrac{7}{10}$

Subtract and simplify. [3.3a]

9. $\dfrac{5}{9} - \dfrac{2}{9}$

10. $\dfrac{7}{8} - \dfrac{3}{4}$

11. $\dfrac{11}{27} - \dfrac{2}{9}$

12. $\dfrac{5}{6} - \dfrac{2}{9}$

Use $<$ or $>$ for $\square$ to write a true sentence. [3.3b]

13. $\dfrac{4}{7} \,\square\, \dfrac{5}{9}$

14. $\dfrac{8}{9} \,\square\, \dfrac{11}{13}$

Solve. [3.3c]

15. $x + \dfrac{2}{5} = \dfrac{7}{8}$

16. $\dfrac{1}{2} + y = \dfrac{9}{10}$

Convert to fraction notation. [3.4a]

17. $7\dfrac{1}{2}$

18. $8\dfrac{3}{8}$

19. $4\dfrac{1}{3}$

20. $10\dfrac{5}{7}$

Convert to a mixed numeral. [3.4a]

21. $\dfrac{7}{3}$

22. $\dfrac{27}{4}$

23. $\dfrac{63}{5}$

24. $\dfrac{7}{2}$

Divide. Write a mixed numeral for the answer. [3.4b]

25. $9 \overline{)7896}$ **26.** $23 \overline{)10493}$

Add. Write a mixed numeral for the answer. [3.5a]

27. $5\frac{3}{5}$ **28.** $8\frac{1}{3}$
$+ 4\frac{4}{5}$ $+ 3\frac{2}{5}$
_____ _____

29. $5\frac{5}{6}$ **30.** $2\frac{3}{4}$
$+ 4\frac{5}{6}$ $+ 5\frac{1}{2}$
_____ _____

Subtract. Write a mixed numeral for the answer where appropriate. [3.5b]

31. 12 **32.** $9\frac{3}{5}$
$- 4\frac{2}{9}$ $- 4\frac{13}{15}$
_____ _____

33. $10\frac{1}{4}$ **34.** 24
$- 6\frac{1}{10}$ $- 10\frac{5}{8}$
_____ _____

Multiply. Write a mixed numeral for the answer where appropriate. [3.6a]

35. $6 \cdot 2\frac{2}{3}$ **36.** $5\frac{1}{4} \cdot \frac{2}{3}$

37. $2\frac{1}{5} \cdot 1\frac{1}{10}$ **38.** $2\frac{2}{5} \cdot 2\frac{1}{2}$

Divide. Write a mixed numeral for the answer where appropriate. [3.6b]

39. $27 \div 2\frac{1}{4}$ **40.** $2\frac{2}{5} \div 1\frac{7}{10}$

41. $3\frac{1}{4} \div 26$ **42.** $4\frac{1}{5} \div 4\frac{2}{3}$

Solve. [3.2b], [3.5c], [3.6c]

43. *Sewing.* Gloria wants to make a dress and jacket. She needs $1\frac{5}{8}$ yd of 60-in. fabric for the dress and $2\frac{5}{8}$ yd for the jacket. How many yards in all does Gloria need to make the outfit?

44. What is the sum of the areas in the figure below?

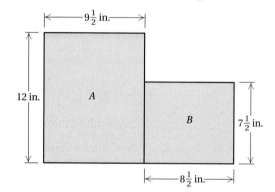

45. In the figure above, how much larger is the area of rectangle *A* than the area of rectangle *B*?

46. *Snapper Provençal.* Listed below is the recipe for Snapper Provençal. It serves 4. What are the quantities needed to prepare this dish when serving only 2? serving 12?

Source: Reprinted from: *The South Beach Diet: The Delicious, Doctor-Designed, Foolproof Plan for Fast and Healthy Weight Loss* by Arthur Agatston, M.D. ©2003 by Arthur Agatston, M.D. Permission granted by Rodale, Inc., Emmaus, PA 18098. Available wherever books are sold or directly from the publisher by calling (800) 848-4735 or visit their website at www.rodalestore.com. More information on the diet available at www.sbdiet.com.

Snapper Provençal

$\frac{1}{4}$ cup extra-virgin olive oil

$1\frac{1}{2}$ pounds fresh red snapper fillets

$\frac{1}{3}$ cup kalamata olives

$2\frac{1}{2}$ tablespoons capers

1 cup canned tomatoes

3 tablespoons chopped shallots

$\frac{1}{2}$ tablespoon fresh rosemary leaves

$\frac{1}{2}$ tablespoon minced garlic

$\frac{1}{3}$ cup white wine

CHAPTER 3: Fraction Notation
and Mixed Numerals

47. *Carpentry.* A board $\frac{9}{10}$ in. thick is glued to a board $\frac{8}{10}$ in. thick. The glue is $\frac{3}{100}$ in. thick. How thick is the result?

48. *Turkey Servings.* Turkey contains $1\frac{1}{3}$ servings per pound. How many pounds are needed for 32 servings?

49. *Weightlifting.* In 1998, Sun Tianni of China snatched 111 kg. This amount was about $1\frac{3}{5}$ times her body weight. How much did Tianni weigh?

Source: *The Guinness Book of Records,* 2000

50. *Cake Recipe.* A wedding-cake recipe requires 12 cups of shortening. Being calorie-conscious, the wedding couple decides to reduce the shortening by $3\frac{5}{8}$ cups and replace it with prune purée. How many cups of shortening are used in their new recipe?

51. *Firefighters' Pie Sale.* Green River's Volunteer Fire Department recently hosted its annual ice cream social. Each of the 83 pies donated was cut into 6 pieces. At the end of the evening, they had sold 382 pieces of pie. How many pies did they sell? How many were left over? Express your answers in mixed numerals.

Simplify the expression using the rules for order of operations. [3.7a]

52. $\frac{1}{8} \div \frac{1}{4} + \frac{1}{2}$

53. $\frac{4}{5} - \frac{1}{2} \cdot \left(1 + \frac{1}{4}\right)$

54. $20\frac{3}{4} - 1\frac{1}{2} \times 12 + \left(\frac{1}{2}\right)^2$

55. Find the average of $\frac{1}{2}, \frac{1}{4}, \frac{1}{3}$, and $\frac{1}{5}$. [3.7a]

Estimate each of the following as 0, $\frac{1}{2}$, or 1. [3.7b]

56. $\frac{29}{59}$ **57.** $\frac{2}{59}$ **58.** $\frac{61}{59}$

Estimate each of the following as a whole number or as a mixed numeral where the fractional part is $\frac{1}{2}$. [3.7b]

59. $6\frac{7}{8}$ **60.** $10\frac{2}{17}$

61. $\frac{3}{10} + \frac{5}{6} + \frac{31}{29}$

62. $32\frac{14}{15} + 27\frac{3}{4} - 4\frac{25}{28} \cdot 6\frac{37}{76}$

63. **Dw** Discuss the role of least common multiples in adding and subtracting with fraction notation. [3.2a], [3.3a]

64. **Dw** Find a real-world situation that fits this equation: [3.5c], [3.6c]

$$2 \cdot 15\frac{3}{4} + 2 \cdot 28\frac{5}{8} = 88\frac{3}{4}.$$

SYNTHESIS

65. *Orangutan Circus Act.* Yuri and Olga are orangutans who perform in a circus by riding bicycles around a circular track. It takes Yuri 6 min and Olga 4 min to make one trip around the track. Suppose they start at the same point and then complete their act when they again reach the same point. How long is their act? [3.1a]

66. Place the numbers 3, 4, 5, and 6 in the boxes in order to make a true equation: [3.5a]

$$\frac{\square}{\square} + \frac{\square}{\square} = 3\frac{1}{4}.$$

Find the LCM.

1. 16 and 12

2. 15, 40, and 50

Add and simplify.

3. $\dfrac{1}{2} + \dfrac{5}{2}$

4. $\dfrac{7}{8} + \dfrac{2}{3}$

5. $\dfrac{7}{10} + \dfrac{19}{100} + \dfrac{31}{1000}$

Subtract and simplify.

6. $\dfrac{5}{6} - \dfrac{3}{6}$

7. $\dfrac{5}{6} - \dfrac{3}{4}$

8. $\dfrac{17}{24} - \dfrac{1}{15}$

Solve.

9. $\dfrac{1}{4} + y = 4$

10. $x + \dfrac{2}{3} = \dfrac{11}{12}$

11. Use $<$ or $>$ for $\square$ to write a true sentence:
$$\dfrac{6}{7} \,\square\, \dfrac{21}{25}.$$

Convert to fraction notation.

12. $3\dfrac{1}{2}$

13. $9\dfrac{7}{8}$

Convert to a mixed numeral.

14. $\dfrac{9}{2}$

15. $\dfrac{74}{9}$

Divide. Write a mixed numeral for the answer.

16. $1\,1 \,\overline{)\,1\,7\,8\,9}$

Add. Write a mixed numeral for the answer.

17. $\begin{aligned} 6\dfrac{2}{5} \\ + 7\dfrac{4}{5} \\ \hline \end{aligned}$

18. $\begin{aligned} 9\dfrac{1}{4} \\ + 5\dfrac{1}{6} \\ \hline \end{aligned}$

Subtract. Write a mixed numeral for the answer.

19. $\begin{aligned} 10\dfrac{1}{6} \\ - 5\dfrac{7}{8} \\ \hline \end{aligned}$

20. $\begin{aligned} 14 \\ - 7\dfrac{5}{6} \\ \hline \end{aligned}$

Multiply. Write a mixed numeral for the answer.

21. $9 \cdot 4\dfrac{1}{3}$

22. $6\dfrac{3}{4} \cdot \dfrac{2}{3}$

Divide. Write a mixed numeral for the answer.

23. $2\dfrac{1}{3} \div 1\dfrac{1}{6}$

24. $2\dfrac{1}{12} \div 75$

25. *Weightlifting.* In 2002, Hossein Rezazadeh of Iran did a clean and jerk of 263 kg. This amount was about $2\dfrac{1}{2}$ times his body weight. How much did Rezazadeh weigh?

Source: *The Guinness Book of Records,* 2005

26. *Book Order.* An order of books for a math course weighs 220 lb. Each book weighs $2\dfrac{3}{4}$ lb. How many books are in the order?

27. *Carpentry.* The following diagram shows a middle drawer support guide for a cabinet drawer. Find each of the following.

a) The short length a across the top
b) The length b across the bottom

28. *Carpentry.* In carpentry, some pieces of plywood that are called "$\frac{3}{4}$-inch" plywood are actually $\frac{11}{16}$ in. thick. How much thinner is such a piece than its name indicates?

29. *Women's Dunks.* The first three women in the history of college basketball able to dunk a basketball are listed below. Their names, heights, and universities are:

Michelle Snow, $6\frac{5}{12}$ ft, Tennessee;

Charlotte Smith, $5\frac{11}{12}$ ft, North Carolina;

Georgeann Wells, $6\frac{7}{12}$ ft, West Virginia.

Find the average height of these women.

Source: *USA Today,* 11/30/00, p. 3C

Simplify.

30. $\dfrac{2}{3} + 1\dfrac{1}{3} \cdot 2\dfrac{1}{8}$

31. $1\dfrac{1}{2} - \dfrac{1}{2}\left(\dfrac{1}{2} \div \dfrac{1}{4}\right) + \left(\dfrac{1}{2}\right)^2$

Estimate each of the following as 0, $\frac{1}{2}$, or 1.

32. $\dfrac{3}{82}$

33. $\dfrac{93}{91}$

Estimate each of the following as a whole number or as a mixed numeral where the fractional part is $\frac{1}{2}$.

34. $3\dfrac{8}{9}$

35. $18\dfrac{9}{17}$

36. $256 \div 15\dfrac{19}{21}$

37. $43\dfrac{15}{31} \cdot 27\dfrac{5}{6} - 9\dfrac{15}{28} + 6\dfrac{5}{76}$

SYNTHESIS

38. The students in a math class can be organized into study groups of 8 each so that no students are left out. The same class of students can also be organized into groups of 6 so that no students are left out.

a) Find some class sizes for which this will work.
b) Find the smallest such class size.

39. Rebecca walks 17 laps at her health club. Trent walks 17 laps at his health club. If the track at Rebecca's health club is $\frac{1}{7}$ mi long, and the track at Trent's is $\frac{1}{8}$ mi long, who walks farther? How much farther?

Solve.

1. *Excelsior Made from Aspen.* Excelsior consists of slender, curved wood shavings and is often used for packing. Shown at right are examples of excelsior and the saw blades used to cut it, as made by Western Excelsior Corporation of Mancos, CO. The width of strips for craft decoration is either $\frac{1}{16}$ in. or $\frac{1}{8}$ in. The width for erosion control mats used for stabilizing soil and nourishing young crops is $\frac{1}{24}$ in.

   a) How much wider is the $\frac{1}{16}$-in. craft decoration excelsior than the erosion control excelsior?
   b) How much wider is the $\frac{1}{8}$-in. craft decoration excelsior than the erosion control excelsior?

   Source: Western Excelsior Corporation

2. *DVD Storage.* Gregory is making a home entertainment center. He is planning a 27-in. shelf that holds DVDs that are each $\frac{7}{16}$ in. thick. How many DVDs will the shelf hold?

3. *Cross-Country Skiing.* During a three-day holiday weekend trip, David and Sally Jean cross-country skied $3\frac{2}{3}$ mi on Friday, $6\frac{1}{8}$ mi on Saturday, and $4\frac{3}{4}$ mi on Sunday.

   a) Find the total miles they skied.
   b) Find the average miles they skied per day. Express your answer as a mixed numeral.

4. *Room Carpeting.* The Chandlers are carpeting an L-shaped family room consisting of a rectangle that is $8\frac{1}{2}$ ft by 11 ft and one that is $6\frac{1}{2}$ ft by $7\frac{1}{2}$ ft.

   a) Find the area of the carpet.
   b) Find the perimeter of the carpet.

5. How many people can get equal $16 shares from a total of $496?

6. An emergency food pantry fund contains $423. From this fund, $148 and $167 are withdrawn for expenses. How much is left in the fund?

7. A recipe calls for $\frac{4}{5}$ tsp of salt. How much salt should be used for $\frac{1}{2}$ recipe? for 5 recipes?

8. A book weighs $2\frac{3}{5}$ lb. How much do 15 books weigh?

9. How many pieces, each $2\frac{3}{8}$ ft long, can be cut from a piece of wire 38 ft long?

10. In a walkathon, Jermaine walked $\frac{9}{10}$ mi and Oleta walked $\frac{75}{100}$ mi. What was the total distance they walked?

11. In the number 2753, what digit names tens?

12. Write expanded notation for 6075.

13. Write a word name for the number in the following sentence: The diameter of Uranus is 29,500 miles.

14. What part is shaded?

Calculate and simplify.

15.
$$\begin{array}{r} 6\ 2\ 8 \\ +\ 2\ 7\ 1 \\ \hline \end{array}$$

16.
$$\begin{array}{r} 3\ 7\ 0\ 4 \\ +\ 5\ 2\ 7\ 8 \\ \hline \end{array}$$

17. $\dfrac{3}{8} + \dfrac{1}{24}$

18.
$$\begin{array}{r} 2\dfrac{3}{4} \\ +\ 5\dfrac{1}{2} \\ \hline \end{array}$$

19.
$$\begin{array}{r} 7\ 4\ 6\ 9 \\ -\ 2\ 3\ 4\ 5 \\ \hline \end{array}$$

20.
$$\begin{array}{r} 7\ 6\ 0\ 5 \\ -\ 3\ 0\ 8\ 7 \\ \hline \end{array}$$

21. $\dfrac{3}{4} - \dfrac{1}{3}$

22.
$$\begin{array}{r} 2\dfrac{1}{3} \\ -\ 1\dfrac{1}{6} \\ \hline \end{array}$$

23.
$$\begin{array}{r} 2\ 7\ 8 \\ \times\ \ \ 1\ 8 \\ \hline \end{array}$$

24.
$$\begin{array}{r} 8\ 9\ 4 \\ \times\ 3\ 2\ 8 \\ \hline \end{array}$$

25. $\dfrac{9}{10} \cdot \dfrac{5}{3}$

26. $18 \cdot \dfrac{5}{6}$

27. $2\dfrac{1}{3} \cdot 3\dfrac{1}{7}$

Divide. Write the answer with the remainder in the form 34 R 7.

28. $6\overline{)\ 4\ 2\ 9\ 0}$

29. $4\ 5\overline{)\ 2\ 5\ 3\ 1}$

30. In Question 29, write a mixed numeral for the answer.

31. Simplify:
$$\left(\dfrac{1}{2} + \dfrac{2}{5}\right)^2 \div 3 + 6 \times \left(2 + \dfrac{1}{4}\right).$$

Divide and simplify, where appropriate.

32. $\dfrac{2}{5} \div \dfrac{7}{10}$

33. $2\dfrac{1}{5} \div \dfrac{3}{10}$

34. Round 38,478 to the nearest hundred.

35. Find the LCM of 18 and 24.

36. Determine whether 3718 is divisible by 8.

37. Find all factors of 16.

Use $<$, $>$, or $=$ for $\square$ to write a true sentence.

38. $\dfrac{4}{5}\ \square\ \dfrac{4}{6}$

39. $\dfrac{3}{13}\ \square\ \dfrac{9}{39}$

40. $\dfrac{5}{12}\ \square\ \dfrac{3}{7}$

Simplify.

41. $\dfrac{36}{45}$

42. $\dfrac{0}{27}$

43. $\dfrac{320}{10}$

44. Convert to fraction notation: $4\dfrac{5}{8}$.

45. Convert to a mixed numeral: $\dfrac{17}{3}$.

Solve.

46. $x + 24 = 117$

47. $x + \dfrac{7}{9} = \dfrac{4}{3}$

48. $\dfrac{7}{9} \cdot t = \dfrac{4}{3}$

49. $y = 32{,}580 \div 36$

Estimate each of the following as $0, \frac{1}{2},$ or 1.

50. $\dfrac{29}{30}$

51. $\dfrac{15}{29}$

52. $\dfrac{2}{43}$

Estimate each of the following as a whole number or as a mixed numeral where the fractional part is $\frac{1}{2}$.

53. $30\dfrac{4}{53}$

54. $\dfrac{9}{10} - \dfrac{7}{8} + \dfrac{41}{39}$

55. $78\dfrac{14}{15} - 28\dfrac{7}{8} - 7\dfrac{25}{28} \div \dfrac{65}{66}$

56. *Matching.* Match each item in the first column with the appropriate item in the second column by drawing connecting lines.

Factors of 68	12, 54, 72, 300
Factorization of 68	2, 3, 17, 19, 23, 31, 47, 101
Prime factorization of 68	$2 \cdot 2 \cdot 17$
Numbers divisible by 6	$2 \cdot 34$
Numbers divisible by 8	8, 16, 24, 32, 40, 48, 64, 864
Numbers divisible by 5	1, 2, 4, 17, 34, 68
Prime numbers	70, 95, 215

For each of Exercises 57–60, choose the correct answer from the selections given.

57. In Arizona, people often install desert landscaping to conserve water. In a development, each home lot requires $\frac{4}{5}$ ton of gravel. How many tons of gravel are required for 40 lots?

a) $\dfrac{80}{100}$ **b)** 50   **c)** $\dfrac{80}{5}$
d) 32   **e)** None

58. A gasoline tank contains 20 gal when it is $\frac{3}{4}$ full. How many gallons can it hold when full?

a) $\dfrac{80}{3}$ **b)** $\dfrac{23}{4}$ **c)** 15
d) $\dfrac{83}{4}$ **e)** None

59. A gallon of ice cream provides $28\frac{1}{2}$ servings. How many gallons of ice cream are needed to serve 228 guests at a wedding reception?

a) 8   **b)** $8\dfrac{1}{2}$ **c)** 10
d) $10\dfrac{1}{2}$ **e)** None

60. For a certain type of load of dishes in a Kitchen-Aid® dishwasher, $1\frac{3}{4}$ oz of Cascade® detergent are required. How many ounces of detergent are needed for 8 loads?

a) 14   **b)** $9\dfrac{3}{4}$ **c)** $4\dfrac{4}{7}$
d) $7\dfrac{3}{32}$ **e)** None

SYNTHESIS

61. a) Simplify each of the following, using fraction notation for your answers.

$$\dfrac{1}{1 \cdot 2}$$

$$\dfrac{1}{1 \cdot 2} + \dfrac{1}{2 \cdot 3}$$

$$\dfrac{1}{1 \cdot 2} + \dfrac{1}{2 \cdot 3} + \dfrac{1}{3 \cdot 4}$$

$$\dfrac{1}{1 \cdot 2} + \dfrac{1}{2 \cdot 3} + \dfrac{1}{3 \cdot 4} + \dfrac{1}{4 \cdot 5}$$

b) Look for a pattern in your answers to part (a). Then find the following without carrying out the computations.

$$\dfrac{1}{1 \cdot 2} + \dfrac{1}{2 \cdot 3} + \dfrac{1}{3 \cdot 4} + \dfrac{1}{4 \cdot 5} + \dfrac{1}{5 \cdot 6}$$
$$+ \dfrac{1}{6 \cdot 7} + \dfrac{1}{7 \cdot 8} + \dfrac{1}{8 \cdot 9} + \dfrac{1}{9 \cdot 10}$$

62. Find the smallest prime number that is larger than 2000.

Decimal Notation

4

Real-World Application

The Panama Canal in Panama is 50.7 mi long. The Suez Canal in Egypt is 119.9 mi long. How much longer is the Suez Canal?

This problem appears as Example 1 in Section 4.7.

Objectives

a   Given decimal notation, write a word name.

b   Convert between fraction notation and decimal notation.

c   Given a pair of numbers in decimal notation, tell which is larger.

d   Round decimal notation to the nearest thousandth, hundredth, tenth, one, ten, hundred, or thousand.

The set of **arithmetic numbers,** or **nonnegative rational numbers,** consists of the whole numbers 0, 1, 2, 3, 4, 5, 6, 7, 8, 9, 10, and so on, and fractions like $\frac{1}{2}, \frac{2}{3}, \frac{7}{8}, \frac{17}{10}$, and so on. Note that we can write the whole numbers using fraction notation. For example, 3 can be written as $\frac{3}{1}$. We studied the use of fraction notation for arithmetic numbers in Chapters 2 and 3.

In Chapter 4, we will study the use of *decimal notation.* The word *decimal* comes from the Latin word *decima,* meaning a tenth part. Although we are using different notation, we are still considering the nonnegative rational numbers. Using decimal notation, we can write 0.875 for $\frac{7}{8}$, for example, or 48.97 for $48\frac{97}{100}$.

a   Decimal Notation and Word Names

A 5-quart stand mixer sells for $249.98. The dot in $249.98 is called a **decimal point.** Since $0.98, or 98¢, is $\frac{98}{100}$ of a dollar, it follows that

$$\$249.98 = 249 + \frac{98}{100} \text{ dollars.}$$

Also, since $0.98, or 98¢, has the same value as

    9 dimes + 8 cents

and 1 dime is $\frac{1}{10}$ of a dollar and 1 cent is $\frac{1}{100}$ of a dollar, we can write

$$249.98 = 2 \cdot 100 + 4 \cdot 10 + 9 \cdot 1 + 9 \cdot \frac{1}{10} + 8 \cdot \frac{1}{100}.$$

This is an extension of the expanded notation for whole numbers that we used in Chapter 1. The place values are 100, 10, 1, $\frac{1}{10}$, $\frac{1}{100}$, and so on. We can see this on a **place-value chart.** The value of each place is $\frac{1}{10}$ as large as the one to its left.

Let's consider decimal notation using a place-value chart to represent 26.3385 min, the men's 10,000-meter run record held by Kenenisa Bekele from Ethiopia.

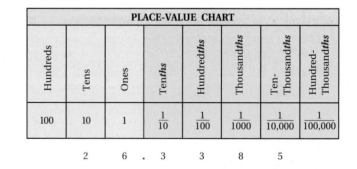

PLACE-VALUE CHART							
Hundreds	Tens	Ones	Ten*ths*	Hundred*ths*	Thousand*ths*	Ten-Thousand*ths*	Hundred-Thousand*ths*
100	10	1	$\frac{1}{10}$	$\frac{1}{100}$	$\frac{1}{1000}$	$\frac{1}{10,000}$	$\frac{1}{100,000}$
	2	6 .	3	3	8	5	

Study Tips

QUIZ-TEST FOLLOW-UP

You may have just completed a chapter quiz or test. Immediately after each chapter quiz or test, write out a step-by-step solution of each question you missed. Visit your instructor or tutor for help with problems that are still giving you trouble. When the week of the final examination arrives, you will be glad to have the excellent study guide these corrected tests provide.

The decimal notation 26.3385 means

$$20 + 6 + \frac{3}{10} + \frac{3}{100} + \frac{8}{1000} + \frac{5}{10,000}, \text{ or } 26\frac{3385}{10,000}.$$

We read both 26.3385 and $26\frac{3385}{10,000}$ as

"Twenty-six and three thousand three hundred eighty-five ten-thousandths."

We can also read 26.3385 as

"Two six *point* three three eight five, or twenty-six *point* three three eight five."

To write a word name from decimal notation,

a) write a word name for the whole number (the number named to the left of the decimal point),

397.685 ⟶ Three hundred ninety-seven

b) write the word "and" for the decimal point, and

397.685 Three hundred ninety-seven and

c) write a word name for the number named to the right of the decimal point, followed by the place value of the last digit.

397.685 Three hundred ninety-seven and six hundred eighty-five *thousandths*

EXAMPLE 1 *Ice Cream.* Write a word name for the number in this sentence: The average person eats 26.3 servings of ice cream a year.
Source: NPD Group; J. M. Hirsch, Associated Press

Twenty-six and three tenths

EXAMPLE 2 Write a word name for 410.87.

Four hundred ten and eighty-seven hundredths

Write a word name for the numbers.

1. **Life Expectancy.** In 2004, the life expectancy at birth in Switzerland was 80.31. In the United States, it was 77.43.
Source: *CIA World Factbook*, 2004

2. **Kentucky Derby.** In 2004, racehorse Smarty Jones won the Kentucky Derby in a time of 2.06767 min.
Source: *Time Almanac*, 2005

Answers on page A-8

3. 245.89

EXAMPLE 3 *5000-m Run.* Write a word name for the number in this sentence: The world outdoor record in the women's 5000-m run is 14.4113 min, held by Elvan Abeylegesse of Turkey.

> Fourteen and four thousand one hundred thirteen ten-thousandths

EXAMPLE 4 Write a word name for 1788.405.

> One thousand, seven hundred eighty-eight and four hundred five thousandths

Do Exercises 1–5. (Exercises 1 and 2 are on the preceding page.)

b Converting Between Decimal Notation and Fraction Notation

Given decimal notation, we can convert to fraction notation as follows:

$$9.875 = 9 + \frac{8}{10} + \frac{7}{100} + \frac{5}{1000}$$

$$= 9 \cdot \frac{1000}{1000} + \frac{8}{10} \cdot \frac{100}{100} + \frac{7}{100} \cdot \frac{10}{10} + \frac{5}{1000}$$

$$= \frac{9000}{1000} + \frac{800}{1000} + \frac{70}{1000} + \frac{5}{1000} = \frac{9875}{1000}.$$

4. 34.0064

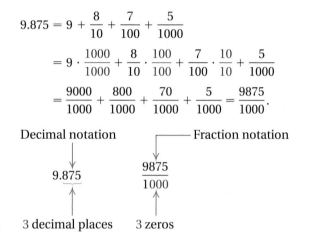

Decimal notation   Fraction notation

$$9.875 \qquad \frac{9875}{1000}$$

3 decimal places   3 zeros

5. 31,079.764

> To convert from decimal to fraction notation,
>
> **a)** count the number of decimal places,   4.98
>   2 places
> **b)** move the decimal point that many   4.98.   Move
>   places to the right, and   2 places.
> **c)** write the answer over a denominator $\dfrac{498}{100}$
>   with a 1 followed by that number of zeros.   2 zeros

Answers on page A-8

EXAMPLE 5 Write fraction notation for 0.876. Do not simplify.

$$0.876 \qquad 0.876, \qquad 0.876 = \frac{876}{1000}$$

3 places   3 zeros

For a number like 0.876, we generally write a 0 before the decimal point to draw attention to the presence of the decimal point.

EXAMPLE 6 Write fraction notation for 56.23. Do not simplify.

$$56.23 \qquad 56.23, \qquad 56.23 = \frac{5623}{100}$$

2 places   2 zeros

EXAMPLE 7 Write fraction notation for 1.5018. Do not simplify.

$$1.5018 \qquad 1.5018, \qquad 1.5018 = \frac{15,018}{10,000}$$

4 places   4 zeros

Do Exercises 6–9.

If fraction notation has a denominator that is a power of ten, such as 10, 100, 1000, and so on, we reverse the procedure we used before.

To convert from fraction notation to decimal notation when the denominator is 10, 100, 1000, and so on,

a) count the number of zeros, and
$$\frac{8679}{1000}$$
3 zeros

b) move the decimal point that number of places to the left. Leave off the denominator.
$$8.679.$$ Move 3 places.
$$\frac{8679}{1000} = 8.679$$

EXAMPLE 8 Write decimal notation for $\frac{47}{10}$.

$$\frac{47}{10} \qquad 4.7. \qquad \frac{47}{10} = 4.7$$

1 zero   1 place

Write fraction notation.

6. 0.896

7. 23.78

8. 5.6789

9. 1.9

Answers on page A-8

Write decimal notation.

10. $\dfrac{743}{100}$

11. $\dfrac{406}{1000}$

12. $\dfrac{67,089}{10,000}$

13. $\dfrac{9}{10}$

14. $\dfrac{57}{1000}$

15. $\dfrac{830}{10,000}$

Write decimal notation.

16. $4\dfrac{3}{10}$

17. $283\dfrac{71}{100}$

18. $456\dfrac{13}{1000}$

EXAMPLE 9 Write decimal notation for $\dfrac{123,067}{10,000}$.

$$\dfrac{123,067}{10,000} \qquad 12.3067. \qquad \dfrac{123,067}{10,000} = 12.3067$$

$\quad$ 4 zeros $\qquad$ 4 places

EXAMPLE 10 Write decimal notation for $\dfrac{13}{1000}$.

$$\dfrac{13}{1000} \qquad 0.013. \qquad \dfrac{13}{1000} = 0.013$$

$\quad$ 3 zeros $\qquad$ 3 places

EXAMPLE 11 Write decimal notation for $\dfrac{570}{100,000}$.

$$\dfrac{570}{100,000} \qquad 0.00570. \qquad \dfrac{570}{100,000} = 0.0057$$

$\quad$ 5 zeros $\qquad$ 5 places

Do Exercises 10–15.

When denominators are numbers other than 10, 100, and so on, we will use another method for conversion. It will be considered in Section 4.5.

If a mixed numeral has a fractional part with a denominator that is a power of ten, such as 10, 100, or 1000, and so on, we first write the mixed numeral as a sum of a whole number and a fraction. Then we convert to decimal notation.

EXAMPLE 12 Write decimal notation for $23\dfrac{59}{100}$.

$$23\dfrac{59}{100} = 23 + \dfrac{59}{100} = 23 \text{ and } \dfrac{59}{100} = 23.59$$

EXAMPLE 13 Write decimal notation for $772\dfrac{129}{10,000}$.

$$772\dfrac{129}{10,000} = 772 + \dfrac{129}{10,000} = 772 \text{ and } \dfrac{129}{10,000} = 772.0129$$

Do Exercises 16–18.

Answers on page A-8

C Order

To understand how to compare numbers in decimal notation, consider 0.85 and 0.9. First note that $0.9 = 0.90$ because $\frac{9}{10} = \frac{90}{100}$. Then $0.85 = \frac{85}{100}$ and $0.90 = \frac{90}{100}$. Since $\frac{85}{100} < \frac{90}{100}$, it follows that $0.85 < 0.90$. This leads us to a quick way to compare two numbers in decimal notation.

> **COMPARING NUMBERS IN DECIMAL NOTATION**
>
> To compare two numbers in decimal notation, start at the left and compare corresponding digits moving from left to right. If two digits differ, the number with the larger digit is the larger of the two numbers. To ease the comparison, extra zeros can be written to the right of the last decimal place.

EXAMPLE 14 Which of 2.109 and 2.1 is larger?

```
                    Think.
     2.109          2.109
     2.1    ─────→  2.100
                    Same  └ Different; 9 > 0
```

Thus, 2.109 is larger than 2.1. That is, $2.109 > 2.1$.

EXAMPLE 15 Which of 0.09 and 0.108 is larger?

```
                     Think.
     0.09            0.090
     0.108  ─────→   0.108
                     Same └── Different; 1 > 0
```

Thus, 0.108 is larger than 0.09. That is, $0.108 > 0.09$.

Do Exercises 19–24.

d Rounding

Rounding is done as for whole numbers. To understand, we first consider an example using a number line. It might help to review Section 1.4.

EXAMPLE 16 Round 0.37 to the nearest tenth.

Here is part of a number line.

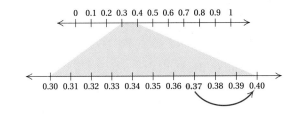

We see that 0.37 is closer to 0.40 than to 0.30. Thus, 0.37 rounded to the nearest tenth is 0.4.

Which number is larger?

19. 2.04, 2.039

20. 0.06, 0.008

21. 0.5, 0.58

22. 1, 0.9999

23. 0.8989, 0.09898

24. 21.006, 21.05

Answers on page A-8

Round to the nearest tenth.

25. 2.76   **26.** 13.85

27. 234.448   **28.** 7.009

Round to the nearest hundredth.

29. 0.636   **30.** 7.834

31. 34.675   **32.** 0.025

Round to the nearest thousandth.

33. 0.9434   **34.** 8.0038

35. 43.1119   **36.** 37.4005

Round 7459.3548 to the nearest:

37. Thousandth.

38. Hundredth.

39. Tenth.

40. One.

41. Ten. (*Caution:* "Tens" are not "tenths.")

42. Hundred.

43. Thousand.

Answers on page A-9

ROUNDING DECIMAL NOTATION

To round to a certain place:

a) Locate the digit in that place.
b) Consider the next digit to the right.
c) If the digit to the right is 5 or higher, round up; if the digit to the right is 4 or lower, round down.

EXAMPLE 17   Round 3872.2459 to the nearest tenth.

a) Locate the digit in the tenths place, 2.

3 8 7 2.2 4 5 9
            ↑

b) Consider the next digit to the right, 4.

3 8 7 2.2 4 5 9
              ↑

Caution!

3872.3 is not a correct answer to Example 17. It is *incorrect* to round from the ten-thousandths digit over to the tenths digit, as follows:

3872.246 → 3872.25 → 3872.3.

c) Since that digit, 4, is less than 5, round down.

3 8 7 2.2 ← This is the answer.

EXAMPLE 18   Round 3872.2459 to the nearest thousandth, hundredth, tenth, one, ten, hundred, and thousand.

Thousandth:	3872.246	Ten:	3870
Hundredth:	3872.25	Hundred:	3900
Tenth:	3872.2	Thousand:	4000
One:	3872		

EXAMPLE 19   Round 14.8973 to the nearest hundredth.

a) Locate the digit in the hundredths place, 9.   1 4.8 9 7 3
                                                                    ↑

b) Consider the next digit to the right, 7.   1 4.8 9 7 3
                                                                ↑

c) Since that digit, 7, is 5 or higher, round up. When we make the hundredths digit a 10, we carry 1 to the tenths place.

The answer is 14.90. Note that the 0 in 14.90 indicates that the answer is correct to the nearest hundredth.

EXAMPLE 20   Round 0.008 to the nearest tenth.

a) Locate the digit in the tenths place, 0.   0.0 0 8
                                                              ↑

b) Consider the next digit to the right, 0.   0.0 0 8
                                                            ↑

c) Since that digit, 0, is less than 5, round down.

The answer is 0.0.

Do Exercises 25–43.

a Write a word name for the number in the sentence.

NYSE INDEX
MOST ACTIVE: SHARE VOLUME

	VOL. (000s)	LAST	CHANGE
Elan	97,202	6.65	−1.29
Lucent	46,527	3.08	−.08
NewsCpA n	32,126	17.28	+.48
BostonSci	24,330	30.90	−.66
Pfizer	21,024	26.59	−.02
ExxonMbl	20,149	63.05	+.37
WalMart	19,358	52.86	+.91
GenElec	17,824	35.55	−.05
Motorola	17,355	15.20	−.38
Texinst	17,234	26.58	−.39

1. *ExxonMbl.* Recently, the stock of Exxon Mobil sold for $63.05 per share.

2. *Motorola.* Recently, the stock of Motorola sold for $15.20 per share.

3. *Pfizer.* Recently, the stock of Pfizer sold for $26.59 per share.

4. *Wal-Mart.* Recently, the stock of Wal-Mart sold for $52.86 per share.

5. *Water Weight.* One gallon of water weighs 8.35 lb.

6. One gallon of paint is equal to 3.785 liters of paint.

7. *Tires.* A tire was on sale for $86.89. This was the lowest price of the year.

8. *Digital Audio Jukebox.* The Apple iPod mini 4GB digital audio jukebox for Mac & Windows stores over 65 hours of music. Recently, it sold for $249.99.

Write a word name.

9. 34.891

10. 27.1245

b Write fraction notation. Do not simplify.

11. 8.3

12. 0.17

13. 3.56

14. 203.6

15. 46.03   **16.** 1.509   **17.** 0.00013   **18.** 0.0109

19. 1.0008   **20.** 2.0114   **21.** 20.003   **22.** 4567.2

Write decimal notation.

23. $\dfrac{8}{10}$ **24.** $\dfrac{51}{10}$ **25.** $\dfrac{889}{100}$ **26.** $\dfrac{92}{100}$ **27.** $\dfrac{3798}{1000}$

28. $\dfrac{780}{1000}$ **29.** $\dfrac{78}{10,000}$ **30.** $\dfrac{56,788}{100,000}$ **31.** $\dfrac{19}{100,000}$ **32.** $\dfrac{2173}{100}$

33. $\dfrac{376,193}{1,000,000}$ **34.** $\dfrac{8,953,074}{1,000,000}$ **35.** $99\dfrac{44}{100}$ **36.** $4\dfrac{909}{1000}$ **37.** $3\dfrac{798}{1000}$

38. $67\dfrac{83}{100}$ **39.** $2\dfrac{1739}{10,000}$ **40.** $9243\dfrac{1}{10}$ **41.** $8\dfrac{953,073}{1,000,000}$ **42.** $2256\dfrac{3059}{10,000}$

c   Which number is larger?

43. 0.06,   0.58   **44.** 0.008,   0.8   **45.** 0.905,   0.91   **46.** 42.06,   42.1

47. 0.0009,   0.001   **48.** 7.067,   7.054   **49.** 234.07,   235.07   **50.** 0.99999,   1

51. 0.004, $\dfrac{4}{100}$ **52.** $\dfrac{73}{10}$,   0.73   **53.** 0.432,   0.4325   **54.** 0.8437,   0.84384

d   Round to the nearest tenth.

55. 0.11   **56.** 0.85   **57.** 0.49   **58.** 0.5794

59. 2.7449   **60.** 4.78   **61.** 123.65   **62.** 36.049

Round to the nearest hundredth.

63. 0.893   **64.** 0.675   **65.** 0.6666   **66.** 6.529

67. 0.995   **68.** 207.9976   **69.** 0.094   **70.** 11.4246

Round to the nearest thousandth.

71. 0.3246

72. 0.6666

73. 17.0015

74. 123.4562

75. 10.1011

76. 0.1161

77. 9.9989

78. 67.100602

Round 809.4732 to the nearest:

79. Hundred.

80. Tenth.

81. Thousandth.

82. Hundredth.

83. One.

84. Ten.

Round 34.54389 to the nearest:

85. Ten-thousandth.

86. Thousandth.

87. Hundredth.

88. Tenth.

89. One.

90. Ten.

91. D_W Describe in your own words a procedure for converting from decimal notation to fraction notation.

92. D_W A fellow student rounds 236.448 to the nearest one and gets 237. Explain the possible error.

SKILL MAINTENANCE

Round 6172 to the nearest: [1.4a]

93. Ten.

94. Hundred.

95. Thousand.

Find the prime factorization.   [2.1d]

96. 2000

97. 1530

98. 2002

99. 4312

SYNTHESIS

100. Arrange the following numbers in order from smallest to largest.

0.99, 0.099, 1, 0.9999, 0.89999, 1.00009, 0.909, 0.9889

101. Arrange the following numbers in order from smallest to largest.

2.1, 2.109, 2.108, 2.018, 2.0119, 2.0302, 2.000001

Truncating.   There are other methods of rounding decimal notation. To round using **truncating,** we drop off all decimal places past the rounding place, which is the same as changing all digits to the right to zeros. For example, rounding 6.78093456285102 to the ninth decimal place, using truncating, gives us 6.780934562. Use truncating to round each of the following to the fifth decimal place, that is, the hundred-thousandth.

102. 6.78346623

103. 6.783465902

104. 99.999999999

105. 0.030303030303

Objectives

a Add using decimal notation.

b Subtract using decimal notation.

c Solve equations of the type $x + a = b$ and $a + x = b$, where a and b may be in decimal notation.

d Balance a checkbook.

Add.

1.
```
    0.8 4 7
+  1 0.0 7
```

2.
```
     2.1
     0.7 3 9
+  3 1.3 6 8 9
```

Add.

3. $0.02 + 4.3 + 0.649$

4. $0.12 + 3.006 + 0.4357$

5. $0.4591 + 0.2374 + 8.70894$

Answers on page A-9

a Addition

Adding with decimal notation is similar to adding whole numbers. First we line up the decimal points so that we can add corresponding place-value digits. Then we add digits from the right. For example, we add the thousandths, then the hundredths, and so on, carrying if necessary. If desired, we can write extra zeros to the right of the decimal point so that the number of places is the same in all of the addends.

EXAMPLE 1   Add: $56.314 + 17.78$.

```
   5 6 . 3 1 4      Lining up the decimal points in order to add
+  1 7 . 7 8 0      Writing an extra zero to the right
                    of the decimal point
```

```
   5 6 . 3 1 4      Adding thousandths
+  1 7 . 7 8 0
             4
```

```
   5 6 . 3 1 4      Adding hundredths
+  1 7 . 7 8 0
           9 4
```

```
         1
   5 6 . 3 1 4      Adding tenths
+  1 7 . 7 8 0      We get 10 tenths = 1 one + 0 tenths,
       . 0 9 4      so we carry the 1 to the ones column.
                    Write a decimal point in the answer.
```

```
   1   1
   5 6 . 3 1 4      Adding ones
+  1 7 . 7 8 0      We get 14 ones = 1 ten + 4 ones,
     4 . 0 9 4      so we carry the 1 to the tens column.
```

```
   1   1
   5 6 . 3 1 4      Adding tens
+  1 7 . 7 8 0
   7 4 . 0 9 4
```

Do Exercises 1 and 2.

EXAMPLE 2   Add: $3.42 + 0.237 + 14.1$.

```
     3.4 2 0      Lining up the decimal points
     0.2 3 7      and writing extra zeros
+  1 4.1 0 0
   1 7.7 5 7      Adding
```

Do Exercises 3–5.

Consider the addition 3456 + 19.347. Keep in mind that any whole number has an "unwritten" decimal point at the right that can be followed by zeros. For example, 3456 can also be written 3456.000. When adding, we can always write in the decimal point and extra zeros if desired.

EXAMPLE 3 Add: 3456 + 19.347.

$$
\begin{array}{r}
\overset{1}{3\,4\,5\,6}.0\,0\,0 \\
+\quad 1\,9.3\,4\,7 \\
\hline
3\,4\,7\,5.3\,4\,7
\end{array}
$$

Writing in the decimal point and extra zeros
Lining up the decimal points
Adding

Do Exercises 6 and 7.

b   Subtraction

Subtracting with decimal notation is similar to subtracting whole numbers. First we line up the decimal points so that we can subtract corresponding place-value digits. Then we subtract digits from the right. For example, we subtract the thousandths, then the hundredths, the tenths, and so on, borrowing if necessary.

EXAMPLE 4 Subtract: 56.314 − 17.78.

$$
\begin{array}{r}
5\,6.3\,1\,4 \\
-\,1\,7.7\,8\,0 \\
\end{array}
$$
Lining up the decimal points in order to subtract
Writing an extra 0

$$
\begin{array}{r}
5\,6.3\,1\,4 \\
-\,1\,7.7\,8\,0 \\
\hline
4
\end{array}
$$
Subtracting thousandths

$$
\begin{array}{r}
5\,6.\overset{2}{\cancel{3}}\,\overset{11}{\cancel{1}}\,4 \\
-\,1\,7.7\,8\,0 \\
\hline
3\,4
\end{array}
$$
Borrowing tenths to subtract hundredths

$$
\begin{array}{r}
\overset{12}{5\,\overset{\cancel{6}}{\cancel{7}}}.\overset{5\,\,\overset{11}{}}{\cancel{3}\,\cancel{1}}\,4 \\
-\,1\,7.7\,8\,0 \\
\hline
.5\,3\,4
\end{array}
$$
Borrowing ones to subtract tenths
Writing a decimal point

$$
\begin{array}{r}
\overset{15\ \ 12}{\overset{4\ \ \cancel{5}\ \ \cancel{7}}{\cancel{5}\,\cancel{6}}}.\overset{11}{\cancel{3}}\,\cancel{1}\,4 \\
-\,1\,7.7\,8\,0 \\
\hline
8.5\,3\,4
\end{array}
$$
Borrowing tens to subtract ones

$$
\begin{array}{r}
\overset{15\ \ 12}{\overset{4\ \ \cancel{5}\ \ \cancel{7}}{\cancel{5}\,\cancel{6}}}.\overset{11}{\cancel{3}}\,\cancel{1}\,4 \\
-\,1\,7.7\,8\,0 \\
\hline
3\,8.5\,3\,4
\end{array}
$$
Subtracting tens

Check:
$$
\begin{array}{r}
\overset{1\ \ 1\ \ 1}{3\,8.5\,3\,4} \\
+\,1\,7.7\,8\,0 \\
\hline
5\,6.3\,1\,4
\end{array}
$$

Do Exercises 8 and 9.

Add.

6. 789 + 123.67

7. 45.78 + 2467 + 1.993

Subtract.

8. 37.428 − 26.674

9.
$$
\begin{array}{r}
0.3\,4\,7 \\
-\,0.0\,0\,8
\end{array}
$$

Answers on page A-9

Subtract.

10. $1.2345 - 0.7$

11. $0.9564 - 0.4392$

12. $7.37 - 0.00008$

Subtract.

13. $1277 - 82.78$

14. $5 - 0.0089$

EXAMPLE 5 Subtract: $13.07 - 9.205$.

$$
\begin{array}{r}
\overset{12}{\cancel{1}}\ \overset{10}{\cancel{3}}.\overset{}{\cancel{0}}\ \overset{6}{7}\ \overset{10}{\cancel{0}} \\
-\quad 9.2\ 0\ 5 \\
\hline
3.8\ 6\ 5
\end{array}
$$

Writing an extra zero

Subtracting

EXAMPLE 6 Subtract: $23.08 - 5.0053$.

$$
\begin{array}{r}
\overset{1}{\cancel{2}}\ \overset{13}{\cancel{3}}.0\ \overset{7}{\cancel{8}}\ \overset{9}{\cancel{0}}\ \overset{10}{\cancel{0}} \\
-\quad 5.0\ 0\ 5\ 3 \\
\hline
1\ 8.0\ 7\ 4\ 7
\end{array}
$$

Writing two extra zeros

Subtracting

Check by addition:

$$
\begin{array}{r}
\overset{1}{1}\ 8.0\ \overset{1}{7}\ \overset{1}{4}\ 7 \\
+\quad 5.0\ 0\ 5\ 3 \\
\hline
2\ 3.0\ 8\ 0\ 0
\end{array}
$$

The answer checks because this is the top number in the subtraction.

Do Exercises 10–12.

When subtraction involves a whole number, again keep in mind that there is an "unwritten" decimal point that can be written in if desired. Extra zeros can also be written in to the right of the decimal point.

EXAMPLE 7 Subtract: $456 - 2.467$.

$$
\begin{array}{r}
4\ 5\ \overset{5}{\cancel{6}}.\overset{9}{\cancel{0}}\ \overset{9}{\cancel{0}}\ \overset{10}{\cancel{0}} \\
-\quad 2.4\ 6\ 7 \\
\hline
4\ 5\ 3.5\ 3\ 3
\end{array}
$$

Writing in the decimal point and extra zeros

Subtracting

Do Exercises 13 and 14.

CALCULATOR CORNER

Addition and Subtraction with Decimal Notation To use a calculator to add and subtract with decimal notation, we use the $\boxed{\cdot}$, $\boxed{+}$, $\boxed{-}$, and $\boxed{=}$ keys. To find $47.046 - 28.193$, for example, we press $\boxed{4}\boxed{7}$ $\boxed{\cdot}\boxed{0}\boxed{4}\boxed{6}\boxed{-}\boxed{2}\boxed{8}\boxed{\cdot}\boxed{1}\boxed{9}\boxed{3}\boxed{=}$. The display reads $\boxed{18.853}$, so $47.046 - 28.193 = 18.853$.

Exercises:

Use a calculator to add.

1. 274.159
    $+\ \ \ 43.486$

2. $\ \ \ \ 19.805$
    $+\ 486.748$

3. $1.7 + 14.56 + 0.89$

4. $3.4 + 45 + 0.68$

Use a calculator to subtract.

5. 9.2
    $-\ 4.8$

6. 52.34
    $-\ 18.51$

7. $489 - 34.26$

8. $6.09 - 5.1$

c Solving Equations

Now let's solve equations $x + a = b$ and $a + x = b$, where a and b may be in decimal notation. Proceeding as we have before, we subtract a on both sides.

EXAMPLE 8   Solve: $x + 28.89 = 74.567$.

We have

$$x + 28.89 - 28.89 = 74.567 - 28.89 \qquad \text{Subtracting 28.89 on both sides}$$
$$x = 45.677.$$

$$\begin{array}{r} {\scriptstyle 6\ \ 13\ 1416} \\ 7\,\cancel{4}.\cancel{5}\,\cancel{6}\,7 \\ -\ 2\,8.8\,9\,0 \\ \hline 4\,5.6\,7\,7 \end{array}$$

The solution is 45.677.

EXAMPLE 9   Solve: $0.8879 + y = 9.0026$.

We have

$$0.8879 + y - 0.8879 = 9.0026 - 0.8879 \qquad \text{Subtracting 0.8879 on both sides}$$
$$y = 8.1147.$$

$$\begin{array}{r} {\scriptstyle 8\ \ 9\ \ 9\ \ 11\ 16} \\ 9.\cancel{0}\,\cancel{0}\,\cancel{2}\,\cancel{6} \\ -\ 0.8\,8\,7\,9 \\ \hline 8.1\,1\,4\,7 \end{array}$$

The solution is 8.1147.

Do Exercises 15 and 16.

EXAMPLE 10   Solve: $120 + x = 4380.6$.

We have

$$120 + x - 120 = 4380.6 - 120 \qquad \text{Subtracting 120 on both sides}$$
$$x = 4260.6$$

$$\begin{array}{r} 4\,3\,8\,0.6 \\ -\ \ \ 1\,2\,0.0 \\ \hline 4\,2\,6\,0.6 \end{array}$$

The solution is 4260.6.

Do Exercise 17.

d Balancing a Checkbook

Let's use addition and subtraction with decimals to balance a checkbook.

EXAMPLE 11   Find the errors, if any, in the balances in this checkbook.

20___		RECORD ALL CHARGES OR CREDITS THAT AFFECT YOUR ACCOUNT							
DATE	CHECK NUMBER	TRANSACTION DESCRIPTION	√ T	(−) PAYMENT/ DEBIT		(+ OR −) OTHER	(+) DEPOSIT/ CREDIT	BALANCE FORWARD	
								8767	73
8/16	432	Burch Laundry		23	56			8744	16
8/19	433	Rogers TV		20	49			8764	65
8/20		Deposit					85 00	8848	65
8/21	434	Galaxy Records		48	60			8801	05
8/22	435	Electric Works		267	95			8533	09

Solve.

15. $x + 17.78 = 56.314$

16. $8.906 + t = 23.07$

17. Solve: $241 + y = 2374.5$.

Answers on page A-9

18. Find the errors, if any, in this checkbook.

20___	RECORD ALL CHARGES OR CREDITS THAT AFFECT YOUR ACCOUNT						
DATE	CHECK NUMBER	TRANSACTION DESCRIPTION	√ T	(−) PAYMENT/ DEBIT	(+ OR −) OTHER	(+) DEPOSIT/ CREDIT	BALANCE FORWARD
							3078 92
12/1	888	H.H. Gregg Appliances		340 69			2738 23
12/3	889	Marie Callendar's Pies		78 56			2659 66
12/5		Deposit <Paycheck>				230 80	2890 46
12/6	890	Chili's Restaurant		13 14			2877 32
12/8	891	Stonecreek Golf Course		48 00			2829 32
12/8		Deposit <Molly>				39 58	2868 90
12/10	892	Galyan's Trading Post		102 87			2766 83
12/14	893	Goody's Music		68 59			2697 45
12/15	894	Salvation Army		100 00			2497 45

There are two ways to determine whether there are errors. We assume that the amount $8767.73 in the "Balance forward" column is correct. If we can determine that the ending balance is correct, we have some assurance that the checkbook is correct. But two errors could offset each other to give us that balance.

METHOD 1

a) We add the debits: $23.56 + 20.49 + 48.60 + 267.95 = 360.60$.

b) We add the deposits/credits. In this case, there is only one deposit, 85.00.

c) We add the total of the deposits to the balance brought forward:

$$8767.73 + 85.00 = 8852.73.$$

d) We subtract the total of the debits: $8852.73 - 360.60 = 8492.13$.

The result should be the ending balance, 8533.09. We see that $8492.13 \neq 8533.09$. Since the numbers are not equal, we proceed to method 2.

METHOD 2   We successively add or subtract deposit/credits and debits, and check the result in the "Balance forward" column.

$$8767.73 - 23.56 = 8744.17.$$

We have found our first error. The subtraction was incorrect. We correct it and continue, using 8744.17 as the corrected balance forward:

$$8744.17 - 20.49 = 8723.68.$$

It looks as though 20.49 was added instead of subtracted. Actually, we would have to correct this line even if it had been subtracted, because the error of 1¢ in the first step has been carried through successive calculations. We correct that balance line and continue, using 8723.68 as the balance and adding the deposit 85.00:

$$8723.68 + 85.00 = 8808.68.$$

We make the correction and continue subtracting the last two debits:

$$8808.68 - 48.60 = 8760.08.$$

Then

$$8760.08 - 267.95 = 8492.13.$$

The corrected checkbook is below.

20___	RECORD ALL CHARGES OR CREDITS THAT AFFECT YOUR ACCOUNT							
DATE	CHECK NUMBER	TRANSACTION DESCRIPTION	√ T	(−) PAYMENT/ DEBIT	(+ OR −) OTHER	(+) DEPOSIT/ CREDIT	BALANCE FORWARD 8767 73	
8/16	432	Burch Laundry		23 56			8744 16	→ 8744.17
8/19	433	Rogers TV		20 49			8764 65	→ 8723.68
8/20		Deposit				85 00	8848 65	→ 8808.68
8/21	434	Galaxy Records		48 60			8801 05	→ 8760.08
8/22	435	Electric Works		267 95			8533 09	→ 8492.13

Do Exercise 18.

There are other ways in which errors can be made in checkbooks, such as forgetting to record a transaction or writing the amounts incorrectly, but we will not consider those here.

Answer on page A-9

EXERCISE SET

For Extra Help

MathXL MyMathLab InterAct Math Math Tutor Center Digital Video Tutor CD 2 Videotape 4 Student's Solutions Manual

a Add.

1.
```
   3 1 6.2 5
 +    1 8.1 2
```

2.
```
   6 4 1.8 0 3
 +    1 4.9 3 5
```

3.
```
   6 5 9.4 0 3
 + 9 1 6.8 1 2
```

4.
```
   4 2 0 3.2 8
 +        3.3 9
```

5.
```
        9.1 0 4
 + 1 2 3.4 5 6
```

6.
```
   8 1.0 0 8
 +    3.4 0 9
```

7. 20.0124 + 30.0124

8. 0.687 + 0.9

9. 39 + 1.007

10. 2.3 + 0.729 + 23

11.
```
       4 7.8
     2 1 9.8 5 2
       4 3.5 9
 +   6 6 6.7 1 3
```

12.
```
       1 3.7 2
        9.1 1 2
   6 5 4 2.7 9 0 8
 +     2 3.9 0 1
```

13. 0.34 + 3.5 + 0.127 + 768

14. 17 + 3.24 + 0.256 + 0.3689

15. 99.6001 + 7285.18 + 500.042 + 870

16. 65.987 + 9.4703 + 6744.02 + 1.0003 + 200.895

b Subtract.

17.
```
   5 1.3 1
 -    2.2 9
```

18.
```
   4 4.3 4 5
 -    3.1 0 5
```

19.
```
   9 2.3 4 1
 -    6.4 2
```

20.
```
   9 7.0 1
 -    3.1 5
```

21.
```
   2.5
 - 0.0 0 2 5
```

22.
```
   3 9.0
 -    0.2 8
```

23.
```
   3.4
 - 0.0 0 3
```

24.
```
   2.8
 - 2.0 8
```

25. 28.2 − 19.35

26. 100.16 − 0.118

27. 34.07 − 30.7

28. 36.2 − 16.28

29. $8.45 - 7.405$ **30.** $3.801 - 2.81$ **31.** $6.003 - 2.3$ **32.** $9.087 - 8.807$

33. $1 - 0.0098$ **34.** $2 - 1.0908$ **35.** $100 - 0.34$ **36.** $624 - 18.79$

37. $7.48 - 2.6$ **38.** $18.4 - 5.92$ **39.** $3 - 2.006$ **40.** $263.7 - 102.08$

41. $19 - 1.198$ **42.** $2548.98 - 2.007$ **43.** $65 - 13.87$ **44.** $45 - 0.999$

45.
$$\begin{array}{r} 3\,2.7\,9\,7\,8 \\ -\ \ 0.0\,5\,9\,2 \\ \hline \end{array}$$

46.
$$\begin{array}{r} 0.4\,9\,6\,3\,4 \\ -\ 0.1\,2\,6\,7\,8 \\ \hline \end{array}$$

47.
$$\begin{array}{r} 6.0\,7 \\ -\ 2.0\,0\,7\,8 \\ \hline \end{array}$$

48.
$$\begin{array}{r} 1.0 \\ -\ 0.9\,9\,9\,9 \\ \hline \end{array}$$

C   Solve.

49. $x + 17.5 = 29.15$ **50.** $t + 50.7 = 54.07$ **51.** $17.95 + p = 402.63$

52. $w + 1.3004 = 47.8$ **53.** $13{,}083.3 = x + 12{,}500.33$ **54.** $100.23 = 67.8 + z$

55. $x + 2349 = 17{,}684.3$ **56.** $1830.4 + t = 23{,}067$

d Find the errors, if any, in each checkbook.

57.

20 ___		RECORD ALL CHARGES OR CREDITS THAT AFFECT YOUR ACCOUNT					
DATE	CHECK NUMBER	TRANSACTION DESCRIPTION	√ T	(−) PAYMENT/ DEBIT	(+ OR −) OTHER	(+) DEPOSIT/ CREDIT	BALANCE FORWARD 9704 56
8/8	342	Bill Rydman		27 44			9677 12
8/9		Deposit <Beauty Contest>				1000 00	10,677 12
8/12	343	Chuck Taylor		123 95			10,553 17
8/14	344	Jennifer Crum		124 02			10,677 19
8/22	345	Neon Johnny's Pizza		12 43			10,664 76
8/24		Deposit <Bowling Tournament>				2500 00	13,164 76
8/29	346	Border's Bookstore		137 78			13,302 54
9/2		Deposit <Bodybuilder Contest>				18 88	13,283 66
9/3	347	Fireman's Fund		2800 00			10,483 66

58.

20 ___		RECORD ALL CHARGES OR CREDITS THAT AFFECT YOUR ACCOUNT					
DATE	CHECK NUMBER	TRANSACTION DESCRIPTION	√ T	(−) PAYMENT/ DEBIT	(+ OR −) OTHER	(+) DEPOSIT/ CREDIT	BALANCE FORWARD 1876 43
4/1	500	Bart Kaufman		500 12			1376 31
4/3	501	Jim Lawler		28 56			1347 75
4/3		Deposit <State Lottery>				10,000 00	11,347 75
4/3	502	Victoria Montoya		464 00			10,883 75
4/3		Deposit <Jewelry Sale>				2500 00	8383 75
4/4	503	Baskin & Robbins		1600 00			6783 75
4/8	504	Golf Galaxy		1349 98			5433 77
4/12	505	Don Mitchell Pro Shops		658 97			4774 80
4/13		Deposit <Publisher's Clearing House>				100000 00	104,774 80
4/15	506	American Airlines		6885 58			98,889 22

59. **D_W** Explain the error in the following:
Add.

$$13.07 + 9.205 = 10.512$$

60. **D_W** Explain the error in the following:
Subtract.

$$73.089 - 5.0061 = 2.3028$$

SKILL MAINTENANCE

61. Round 34,567 to the nearest thousand.   [1.4a]

62. Round 34,496 to the nearest thousand.   [1.4a]

Subtract.

63. $\dfrac{13}{24} - \dfrac{3}{8}$ [3.3a]

64. $\dfrac{8}{9} - \dfrac{2}{15}$ [3.3a]

65. $8805 - 2639$ [1.3b]

66. $8005 - 2639$ [1.3b]

Solve.

67. A serving of filleted fish is generally considered to be about $\frac{1}{3}$ lb. How many servings can be prepared from $5\frac{1}{2}$ lb of flounder fillet?   [3.6c]

68. A photocopier technician drove $125\frac{7}{10}$ mi away from Scottsdale for a repair call. The next day he drove $65\frac{1}{2}$ mi back toward Scottsdale for another service call. How far was the technician from Scottsdale?   [3.5c]

SYNTHESIS

69. A student presses the wrong button when using a calculator and adds 235.7 instead of subtracting it. The incorrect answer is 817.2. What is the correct answer?

4.3 MULTIPLICATION

Objectives

a Multiply using decimal notation.

b Convert from notation like 45.7 million to standard notation, and between dollars and cents.

a Multiplication

Let's find the product

2.3 × 1.12.

To understand how we find such a product, we first convert each factor to fraction notation. Next, we multiply the whole numbers 23 and 112, and then divide by 1000.

$$2.3 \times 1.12 = \frac{23}{10} \times \frac{112}{100} = \frac{23 \times 112}{10 \times 100} = \frac{2576}{1000} = 2.576$$

Note the number of decimal places.

$$
\begin{array}{rl}
1.1\,2 & \text{(2 decimal places)} \\
\times \quad 2.3 & \text{(1 decimal place)} \\
\hline
2.5\,7\,6 & \text{(3 decimal places)}
\end{array}
$$

Now consider

$$0.011 \times 15.0002 = \frac{11}{1000} \times \frac{150{,}002}{10{,}000} = \frac{1{,}650{,}022}{10{,}000{,}000} = 0.1650022.$$

Note the number of decimal places.

$$
\begin{array}{rl}
1\,5.0\,0\,0\,2 & \text{(4 decimal places)} \\
\times \qquad 0.0\,1\,1 & \text{(3 decimal places)} \\
\hline
0.1\,6\,5\,0\,0\,2\,2 & \text{(7 decimal places)}
\end{array}
$$

To multiply using decimals:

a) Ignore the decimal points and multiply as though both factors were whole numbers.

b) Then place the decimal point in the result. The number of decimal places in the product is the sum of the numbers of places in the factors (count places from the right).

0.8 × 0.43

$$
\begin{array}{r}
{\scriptstyle 2} \\
0.4\,3 \\
\times \quad 0.8 \\
\hline
3\,4\,4
\end{array}
\quad \text{Ignore the decimal points for now.}
$$

$$
\begin{array}{rl}
0.4\,3 & \text{(2 decimal places)} \\
\times \quad 0.8 & \text{(1 decimal place)} \\
\hline
0.3\,4\,4 & \text{(3 decimal places)}
\end{array}
$$

EXAMPLE 1 Multiply: 8.3 × 74.6.

a) Ignore the decimal points and multiply as though factors were whole numbers.

$$
\begin{array}{r}
{\scriptstyle 3\;\;4} \\
{\scriptstyle 1\;\;1} \\
7\,4.6 \\
\times \quad 8.3 \\
\hline
2\,2\,3\,8 \\
5\,9\,6\,8\,0 \\
\hline
6\,1\,9\,1\,8
\end{array}
$$

b) Place the decimal point in the result. The number of decimal places in the product is the sum of the number of places in the factors, 1 + 1, or 2.

```
      7 4.6        (1 decimal place)
   ×     8.3       (1 decimal place)
   ---------
      2 2 3 8
    5 9 6 8 0
   ---------
    6 1 9.1 8      (2 decimal places)
```

Do Exercise 1.

EXAMPLE 2   Multiply: 0.0032×2148.

```
         2 1 4 8      (0 decimal places)
   × 0.0 0 3 2        (4 decimal places)
   -----------
         4 2 9 6
       6 4 4 4 0
   -----------
       6.8 7 3 6      (4 decimal places)
```

EXAMPLE 3   Multiply: 0.14×0.867.

```
      0.8 6 7         (3 decimal places)
   ×      0.1 4       (2 decimal places)
   ---------
      3 4 6 8
      8 6 7 0
   ---------
   0.1 2 1 3 8        (5 decimal places)
```

Do Exercises 2 and 3.

MULTIPLYING BY 0.1, 0.01, 0.001, AND SO ON

Now let's consider some special kinds of products. The first involves multiplying by a tenth, hundredth, thousandth, or ten-thousandth. Let's look at those products.

$$0.1 \times 38 = \frac{1}{10} \times 38 = \frac{38}{10} = 3.8$$

$$0.01 \times 38 = \frac{1}{100} \times 38 = \frac{38}{100} = 0.38$$

$$0.001 \times 38 = \frac{1}{1000} \times 38 = \frac{38}{1000} = 0.038$$

$$0.0001 \times 38 = \frac{1}{10,000} \times 38 = \frac{38}{10,000} = 0.0038$$

Note in each case that the product is *smaller* than 38. That is, the decimal point in each product is farther to the left than the unwritten decimal point in 38.

1. Multiply.

```
   8 5.4
 ×   6.2
```

Multiply.

2.
```
       1 2 3 4
   × 0.0 0 4 1
```

3.
```
    4 2.6 5
  × 0.8 0 4
```

Answers on page A-9

Multiply.

4. 0.1×3.48

5. 0.01×3.48

6. 0.001×3.48

7. 0.0001×3.48

To multiply any number by 0.1, 0.01, 0.001, and so on,

a) count the number of decimal places in the tenth, hundredth, or thousandth, and so on, and

b) move the decimal point that many places to the left.

0.001×34.45678
$\qquad \to 3$ places

$0.001 \times 34.45678 = 0.034.45678$

Move 3 places to the left.

$0.001 \times 34.45678 = 0.03445678$

EXAMPLES Multiply.

4. $0.1 \times 14.605 = 1.4605$ $\qquad 1.4.605$

5. $0.01 \times 14.605 = 0.14605$

6. $0.001 \times 14.605 = 0.014605$
$\qquad \qquad \uparrow$ We write an extra zero.

7. $0.0001 \times 14.605 = 0.0014605$
$\qquad \qquad \uparrow$ We write two extra zeros.

Do Exercises 4–7.

MULTIPLYING BY 10, 100, 1000, AND SO ON

Next, let's consider multiplying by 10, 100, 1000, and so on. Let's look at those products.

Multiply.

8. 10×3.48

9. 100×3.48

10. 1000×3.48

11. $10,000 \times 3.48$

$$10 \times 97.34 = 973.4$$
$$100 \times 97.34 = 9734$$
$$1000 \times 97.34 = 97,340$$
$$10,000 \times 97.34 = 973,400$$

Note in each case that the product is *larger* than 97.34. That is, the decimal point in each product is farther to the right than the decimal point in 97.34.

To multiply any number by 10, 100, 1000, and so on,

a) count the number of zeros, and

b) move the decimal point that many places to the right.

1000×34.45678
$\qquad \to 3$ zeros

$1000 \times 34.45678 = 34.456.78$

Move 3 places to the right.

$1000 \times 34.45678 = 34,456.78$

EXAMPLES Multiply.

8. $10 \times 14.605 = 146.05$ $\qquad 14.6.05$

9. $100 \times 14.605 = 1460.5$

10. $1000 \times 14.605 = 14,605$

11. $10,000 \times 14.605 = 146,050$ $\qquad 14.6050.$ We write an extra zero.

Answers on page A-9

Do Exercises 8–11 on the preceding page.

b Applications Using Multiplication with Decimal Notation

NAMING LARGE NUMBERS

We often see notation like the following in newspapers and magazines and on television.

- The largest building in the world is the Pentagon, which has 3.7 million square feet of floor space.
- In 2003, consumers spent $51.7 billion on online retail products. Total spending in specific categories is listed in the following table.

ONLINE CONSUMER SPENDING
(in billions of dollars)

CATEGORY	2003	CATEGORY	2003
1. PCs	$8.8	7. Event Tickets	$2.7
2. Peripherals	$2.5	8. Apparel	$6.1
3. Software	$2.9	9. Jewelry	$1.3
4. Consumer Electronics	$2.5	10. Toys	$1.3
5. Books	$3.1	11. Housewares/ Small Appliances	$1.9
6. Music	$1.3	12. Office Products	$1.7

Source: Jupiter Media Metrix, Inc.

To understand such notation, consider the information in the following table.

NAMING LARGE NUMBERS

1 hundred = 100 = $10 \cdot 10 = 10^2$
↳ 2 zeros

1 thousand = 1000 = $10 \cdot 10 \cdot 10 = 10^3$
↳ 3 zeros

1 million = 1,000,000 = $10 \cdot 10 \cdot 10 \cdot 10 \cdot 10 \cdot 10 = 10^6$
↳ 6 zeros

1 billion = 1,000,000,000 = 10^9
↳ 9 zeros

1 trillion = 1,000,000,000,000 = 10^{12}
↳ 12 zeros

CALCULATOR CORNER

Multiplication with Decimal Notation To use a calculator to multiply with decimal notation, we use the $\boxed{\cdot}$, $\boxed{\times}$, and $\boxed{=}$ keys. To find 4.78 × 0.34, for example, we press $\boxed{4}$ $\boxed{\cdot}$ $\boxed{7}$ $\boxed{8}$ $\boxed{\times}$ $\boxed{\cdot}$ $\boxed{3}$ $\boxed{4}$ $\boxed{=}$. The display reads $\boxed{1.6252}$, so 4.78 × 0.34 = 1.6252.

Exercises: Use a calculator to multiply.

1. $\begin{array}{r} 5.4 \\ \times \quad 9 \\ \hline \end{array}$

2. $\begin{array}{r} 4\ 1\ 5 \\ \times\ 1\ 6.7 \\ \hline \end{array}$

3. $\begin{array}{r} 1\ 7.6\ 3 \\ \times \quad\ \ 8.1 \\ \hline \end{array}$

4. 0.04 × 12.69

5. 586.4 × 13.5

6. 4.003 × 5.1

Convert the number in the sentence to standard notation.

12. The largest building in the world is the Pentagon, which has 3.7 million square feet of floor space.

13. Online Apparel Sales. In 2003, consumers spent $6.1 billion for apparel.
Source: Jupiter Media Metrix, Inc.

14. Online Ticket Sales. In 2003, consumers spent $2.7 billion for event tickets.
Source: Jupiter Media Metrix, Inc.

Convert from dollars to cents.

15. $15.69

16. $0.17

Convert from cents to dollars.

17. 35¢

18. 577¢

Answers on page A-9

To convert a large number to standard notation, we proceed as follows.

EXAMPLE 12 Convert the number in this sentence to standard notation: In 1999, the U.S. Mint produced 11.6 billion pennies.
Source: U.S. Mint

$$11.6 \text{ billion} = 11.6 \times 1 \text{ billion}$$
$$= 11.6 \times 1{,}000{,}000{,}000$$

9 zeros

$$= 11{,}600{,}000{,}000 \qquad \text{Moving the decimal point 9 places to the right}$$

Do Exercises 12–14.

MONEY CONVERSION

Converting from dollars to cents is like multiplying by 100. To see why, consider $19.43.

$19.43 = 19.43 \times \$1$	We think of $19.43 as 19.43×1 dollar, or $19.43 \times \$1$.
$= 19.43 \times 100¢$	Substituting 100¢ for $1: \$1 = 100¢$
$= 1943¢$	Multiplying

DOLLARS TO CENTS

To convert from dollars to cents, move the decimal point two places to the right and change the $ sign in front to a ¢ sign at the end.

EXAMPLES Convert from dollars to cents.

13. $189.64 = 18,964¢

14. $0.75 = 75¢

Do Exercises 15 and 16.

Converting from cents to dollars is like multiplying by 0.01. To see why, consider 65¢.

$65¢ = 65 \times 1¢$	We think of 65¢ as 65×1 cent, or $65 \times 1¢$.
$= 65 \times \$0.01$	Substituting $0.01 for 1¢: $1¢ = \$0.01$
$= \$0.65$	Multiplying

CENTS TO DOLLARS

To convert from cents to dollars, move the decimal point two places to the left and change the ¢ sign at the end to a $ sign in front.

EXAMPLES Convert from cents to dollars.

15. 395¢ = $3.95

16. 8503¢ = $85.03

Do Exercises 17 and 18.

4.3

EXERCISE SET

For Extra Help

MathXL   MyMathLab   InterAct   Math Tutor   Digital Video   Student's
                       Math   Center   Tutor CD 2   Solutions
                                                 Videotape 4   Manual

a   Multiply.

1.   8.6
      $\times$ 7

2.   5.7
      $\times$ 0.8

3.   0.8 4
      $\times$ 8

4.   9.4
      $\times$ 0.6

5.   6.3
      $\times$ 0.0 4

6.   9.8
      $\times$ 0.0 8

7.   8 7
      $\times$ 0.0 0 6

8.   1 8.4
      $\times$ 0.0 7

9. 10×23.76

10. 100×3.8798

11. 1000×583.686852

12. 0.34×1000

13. 7.8×100

14. 0.00238×10

15. 0.1×89.23

16. 0.01×789.235

17. 0.001×97.68

18. 8976.23×0.001

19. 78.2×0.01

20. 0.0235×0.1

21.   3 2.6
      $\times$ 1 6

22.   9.2 8
      $\times$ 8.6

23.   0.9 8 4
      $\times$ 3.3

24.   8.4 8 9
      $\times$ 7.4

25.   3 7 4
      $\times$ 2.4

26.   8 6 5
      $\times$ 1.0 8

27.   7 4 9
      $\times$ 0.4 3

28.   9 7 8
      $\times$ 2 0.5

29.   0.8 7
      $\times$ 6 4

30.   7.2 5
      $\times$ 6 0

31.   4 6.5 0
      $\times$ 7 5

32.   8.2 4
      $\times$ 7 0 3

33.
$$\begin{array}{r} 8\,1.7 \\ \times\,0.6\,1\,2 \\ \hline \end{array}$$

34.
$$\begin{array}{r} 3\,1.8\,2 \\ \times\;\;\;7.1\,5 \\ \hline \end{array}$$

35.
$$\begin{array}{r} 1\,0.1\,0\,5 \\ \times\,1\,1.3\,2\,4 \\ \hline \end{array}$$

36.
$$\begin{array}{r} 1\,5\,1.2 \\ \times\,4.5\,5\,5 \\ \hline \end{array}$$

37.
$$\begin{array}{r} 1\,2.3 \\ \times\,1.0\,8 \\ \hline \end{array}$$

38.
$$\begin{array}{r} 7.8\,2 \\ \times\,0.0\,2\,4 \\ \hline \end{array}$$

39.
$$\begin{array}{r} 3\,2.4 \\ \times\;\;\;2.8 \\ \hline \end{array}$$

40.
$$\begin{array}{r} 8.0\,9 \\ \times\,0.0\,0\,7\,5 \\ \hline \end{array}$$

41.
$$\begin{array}{r} 0.0\,0\,3\,4\,2 \\ \times\;\;\;\;\;0.8\,4 \\ \hline \end{array}$$

42.
$$\begin{array}{r} 2.0\,0\,5\,6 \\ \times\;\;\;\;\;3.8 \\ \hline \end{array}$$

43.
$$\begin{array}{r} 0.3\,4\,7 \\ \times\;\;\;2.0\,9 \\ \hline \end{array}$$

44.
$$\begin{array}{r} 2.5\,3\,2 \\ \times\,1.0\,6\,7 \\ \hline \end{array}$$

45.
$$\begin{array}{r} 3.0\,0\,5 \\ \times\,0.6\,2\,3 \\ \hline \end{array}$$

46.
$$\begin{array}{r} 1\,6.3\,4 \\ \times\,0.0\,0\,0\,5\,1\,2 \\ \hline \end{array}$$

47. 1000×45.678

48. 0.001×45.678

[b] Convert from dollars to cents.

49. $28.88

50. $67.43

51. $0.66

52. $1.78

Convert from cents to dollars.

53. 34¢

54. 95¢

55. 3445¢

56. 933¢

Convert the number in the sentence to standard notation.

57. *Wal-Mart.* In 2003, Wal-Mart stores had revenues of about $258.7 billion.
Source: Fortune 500

58. *Magazine Circulation.* The total paid circulation in 2003 of the magazine *Better Homes and Gardens* was about 7.6 million copies.
Source: Audit Bureau of Circulations, tabulated by Magazine Publishers of America

59. *Broadway.* Broadway productions grossed $748.9 million in 2004.
Source: League of American Theatres and Producers

60. *Rail Transportation.* People in the United States traveled 13.6 billion miles on city rail transportation in 2003.
Source: American Public Transportation Association, NASA

61. **D**_W If two rectangles have the same perimeter, will they also have the same area? Experiment with different dimensions. Be sure to use decimals. Explain your answer.

62. **D**_W A student insists that $346.708 \times 0.1 = 3467.08$. How could you convince him that a mistake had been made without checking on a calculator?

SKILL MAINTENANCE

Calculate.

63. $2\frac{1}{3} \cdot 4\frac{4}{5}$ [3.6a]

64. $2\frac{1}{3} \div 4\frac{4}{5}$ [3.6b]

65. $4\frac{4}{5} - 2\frac{1}{3}$ [3.5b]

66. $4\frac{4}{5} + 2\frac{1}{3}$ [3.5a]

Divide. [1.6b]

67. $2\,4\,\overline{)\,8\,2\,0\,8}$

68. $4\,\overline{)\,3\,4\,8}$

69. $7\,\overline{)\,3\,1{,}9\,6\,2}$

70. $1\,8\,\overline{)\,2\,2{,}6\,2\,6}$

71. $4\,0\,\overline{)\,3\,4\,8\,0}$

72. $1\,7\,\overline{)\,2\,0{,}0\,0\,6}$

SYNTHESIS

Consider the following names for large numbers in addition to those already discussed in this section:

$$1 \text{ quadrillion} = 1{,}000{,}000{,}000{,}000{,}000 = 10^{15};$$
$$1 \text{ quintillion} = 1{,}000{,}000{,}000{,}000{,}000{,}000 = 10^{18};$$
$$1 \text{ sextillion} = 1{,}000{,}000{,}000{,}000{,}000{,}000{,}000 = 10^{21};$$
$$1 \text{ septillion} = 1{,}000{,}000{,}000{,}000{,}000{,}000{,}000{,}000 = 10^{24}.$$

Find each of the following. Express the answer with a name that is a power of 10.

73. (1 trillion) · (1 billion)

74. (1 million) · (1 billion)

75. (1 trillion) · (1 trillion)

76. Is a billion millions the same as a million billions? Explain.

Objectives

a Divide using decimal notation.

b Solve equations of the type $a \cdot x = b$, where a and b may be in decimal notation.

c Simplify expressions using the rules for order of operations.

Divide.

1. $9 \overline{)\, 5.4}$

2. $15 \overline{)\, 2\,2.5}$

3. $82 \overline{)\, 3\,8.5\,4}$

4.4 DIVISION

a Division

WHOLE-NUMBER DIVISORS

We use the following method when we divide a decimal quantity by a whole number.

To divide by a whole number,

a) place the decimal point directly above the decimal point in the dividend, and

b) divide as though dividing whole numbers.

$$
\begin{array}{r}
0.8\,4 \leftarrow \text{Quotient} \\
\text{Divisor} \rightarrow 7 \overline{)\, 5.8\,8} \leftarrow \text{Dividend} \\
5\,6\,0 \\
\hline
2\,8 \\
2\,8 \\
\hline
0 \leftarrow \text{Remainder}
\end{array}
$$

EXAMPLE 1 Divide: $379.2 \div 8$.

Place the decimal point.

$$
\begin{array}{r}
4\,7.4 \\
8 \overline{)\, 3\,7\,9.2} \\
3\,2\,0\,0 \\
\hline
5\,9\,2 \\
5\,6\,0 \\
\hline
3\,2 \\
3\,2 \\
\hline
0
\end{array}
$$

Divide as though dividing whole numbers.

EXAMPLE 2 Divide: $82.08 \div 24$.

Place the decimal point.

$$
\begin{array}{r}
3.4\,2 \\
2\,4 \overline{)\, 8\,2.0\,8} \\
7\,2\,0\,0 \\
\hline
1\,0\,0\,8 \\
9\,6\,0 \\
\hline
4\,8 \\
4\,8 \\
\hline
0
\end{array}
$$

Divide as though dividing whole numbers.

Do Exercises 1–3.

Suppose the dividend is a whole number. We can think of it as having a decimal point at the end with as many 0's as we wish after the decimal point. For example,

$$12 = 12. = 12.0 = 12.00 = 12.000, \text{ and so on.}$$

Divide.

4. 2 5 $\overline{)8}$

EXAMPLE 3 Divide: $30 \div 8$.

$$\begin{array}{r} 3. \\ 8 \overline{)30.} \\ \underline{24} \\ 6 \end{array}$$

Place the decimal point and divide to find how many ones.

$$\begin{array}{r} 3. \\ 8 \overline{)30.0} \\ \underline{24} \downarrow \\ 6\ 0 \end{array}$$

Write an extra zero.

$$\begin{array}{r} 3.7 \\ 8 \overline{)30.0} \\ \underline{24} \\ 6\ 0 \\ \underline{5\ 6} \\ 4 \end{array}$$

Divide to find how many tenths.

5. 4 $\overline{)15}$

$$\begin{array}{r} 3.7 \\ 8 \overline{)30.00} \\ \underline{24} \\ 6\ 0 \\ \underline{5\ 6} \downarrow \\ 4\ 0 \end{array}$$

Write another zero.

$$\begin{array}{r} 3.75 \\ 8 \overline{)30.00} \\ \underline{24} \\ 6\ 0 \\ \underline{5\ 6} \\ 4\ 0 \\ \underline{4\ 0} \\ 0 \end{array}$$

Divide to find how many hundredths.

6. 8 6 $\overline{)21.5}$

Check: $\begin{array}{r} \overset{6\ 4}{3.75} \\ \times\qquad 8 \\ \hline 30.00 \end{array}$

EXAMPLE 4 Divide: $4 \div 25$.

$$\begin{array}{r} 0.16 \\ 25 \overline{)4.00} \\ \underline{2\ 5} \\ 1\ 5\ 0 \\ \underline{1\ 5\ 0} \\ 0 \end{array}$$

Check: $\begin{array}{r} \overset{1}{\overset{3}{}} \\ 0.16 \\ \times\quad 25 \\ \hline 80 \\ 320 \\ \hline 4.00 \end{array}$

Do Exercises 4–6.

Answers on page A-9

7. a) Complete.

$$\frac{3.75}{0.25} = \frac{3.75}{0.25} \times \frac{100}{100}$$

$$= \frac{(\quad)}{25}$$

DIVISORS THAT ARE NOT WHOLE NUMBERS

Consider the division

$$0.2\,4 \overline{\smash{\big)}\,8.2\,0\,8}$$

We write the division as $\dfrac{8.208}{0.24}$. Then we multiply by 1 to change to a whole-number divisor:

$$\frac{8.208}{0.24} = \frac{8.208}{0.24} \times \frac{100}{100} = \frac{820.8}{24}.$$

The division $0.24\overline{\smash{\big)}8.208}$ is the same as $24\overline{\smash{\big)}820.8}$.

The divisor is now a whole number.

b) Divide.

$$0.2\,5 \overline{\smash{\big)}\,3.7\,5}$$

> To divide when the divisor is not a whole number,
>
> **a)** move the decimal point (multiply by 10, 100, and so on) to make the divisor a whole number;
>
> **b)** move the decimal point (multiply the same way) in the dividend the same number of places; and
>
> **c)** place the decimal point directly above the new decimal point in the dividend and divide as though dividing whole numbers.
>
> $$0.2\,4 \overline{\smash{\big)}\,8.2\,0\,8}$$
> Move 2 places to the right.
>
> $$0.2\,4 \overline{\smash{\big)}\,8.2\,0\,8}$$
> Move 2 places to the right.
>
> $$\begin{array}{r} 3\,4.2 \\ 0.2\,4 \overline{\smash{\big)}\,8.2\,0{\scriptstyle\wedge}8} \\ 7\,2\,0\,0 \\ \hline 1\,0\,0\,8 \\ 9\,6\,0 \\ \hline 4\,8 \\ 4\,8 \\ \hline 0 \end{array}$$
>
> (The new decimal point in the dividend is indicated by a caret.)

Divide.

8. $0.8\,3 \overline{\smash{\big)}\,4.0\,6\,7}$

EXAMPLE 5 Divide: $5.848 \div 8.6$.

$$8.6 \overline{\smash{\big)}\,5.8\,4\,8}$$

Multiply the divisor by 10 (move the decimal point 1 place). Multiply the same way in the dividend (move 1 place).

9. $3.5 \overline{\smash{\big)}\,4\,4.8}$

$$\begin{array}{r} 0.6\,8 \\ 8.6 \overline{\smash{\big)}\,5.8{\scriptstyle\wedge}4\,8} \\ 5\,1\,6\,0 \\ \hline 6\,8\,8 \\ 6\,8\,8 \\ \hline 0 \end{array}$$

Place a decimal point above the new decimal point and then divide.
Note: $\frac{5.848}{8.6} = \frac{5.848}{8.6} \cdot \frac{10}{10} = \frac{58.48}{86}$.

Do Exercises 7–9.

Answers on page A-9

Sometimes it helps to write extra zeros to the right of the decimal point in the dividend.

EXAMPLE 6 Divide: $12 \div 0.64$.

$$0.6\,4\,\overline{)\,1\,2\,.}$$

Place a decimal point at the end of the whole number.

$$0.6\,4\,\overline{)\,1\,2\,.0\,0}$$

Multiply the divisor by 100 (move the decimal point 2 places). Multiply the same way in the dividend (move 2 places).

```
              1 8.7 5
0.6 4 ) 1 2.0 0,0 0
              6 4 0
              5 6 0
              5 1 2
                4 8 0
                4 4 8
                  3 2 0
                  3 2 0
                      0
```

Place a decimal point above and then divide.

Do Exercise 10.

DIVIDING BY 10, 100, 1000, AND SO ON

It is often helpful to be able to divide quickly by a ten, hundred, or thousand, or by a tenth, hundredth, or thousandth. Each procedure we use is based on multiplying by 1. Consider the following example:

$$\frac{23.789}{1000} = \frac{23.789}{1000} \cdot \frac{1000}{1000} = \frac{23{,}789}{1{,}000{,}000} = 0.023789.$$

We are dividing by a number greater than 1: The result is *smaller* than 23.789.

To divide by 10, 100, 1000, and so on,

a) count the number of zeros in the divisor, and

$$\frac{713.49}{100}$$

↳ 2 zeros

b) write the quotient by moving the decimal point in the dividend that number of places to the left.

$$\frac{713.49}{100}, \qquad 7.13.49 \qquad \frac{713.49}{100} = 7.1349$$

2 places to the left

EXAMPLE 7 Divide: $\dfrac{0.0104}{10}$.

$$\frac{0.0104}{10}, \qquad 0.0.0104, \qquad \frac{0.0104}{10} = 0.00104$$

1 zero   1 place to the left

---

10. Divide.

$$1.6\,\overline{)\,2\,5}$$

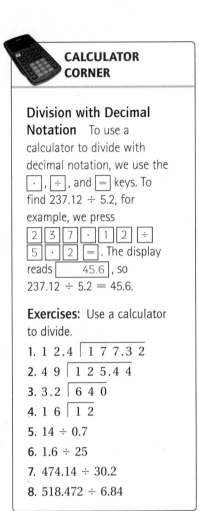
Answer on page A-9

Study Tips

HOMEWORK TIPS

Prepare for your homework assignment by reading the explanations of concepts and following the step-by-step solutions of examples in the text. The time you spend preparing will save valuable time when you do your assignment.

Divide.

11. $\dfrac{0.1278}{0.01}$

12. $\dfrac{0.1278}{100}$

13. $\dfrac{98.47}{1000}$

14. $\dfrac{6.7832}{0.1}$

DIVIDING BY 0.1, 0.01, 0.001, AND SO ON

Now consider the following example:

$$\frac{23.789}{0.01} = \frac{23.789}{0.01} \cdot \frac{100}{100} = \frac{2378.9}{1} = 2378.9.$$

We are dividing by a number less than 1: The result is *larger* than 23.789. We use the following procedure.

> To divide by 0.1, 0.01, 0.001, and so on,
>
> a) count the number of decimal places in the divisor, and
>
> $$\frac{713.49}{0.001}$$
>
> $\searrow$ 3 places
>
> b) write the quotient by moving the decimal point in the dividend that number of places to the right.
>
> $$\frac{713.49}{0.001},\quad 713.490.\quad \frac{713.49}{0.001} = 713{,}490$$
>
> 3 places to the right

EXAMPLE 8   Divide: $\dfrac{23.738}{0.001}$.

$$\frac{23.738}{0.001},\quad 23.738.\quad \frac{23.738}{0.001} = 23{,}738$$

3 places   3 places to the right to change 0.001 to 1

Do Exercises 11–14.

 Solving Equations

Now let's solve equations of the type $a \cdot x = b$, where a and b may be in decimal notation. Proceeding as before, we divide by a on both sides.

EXAMPLE 9   Solve: $8 \cdot x = 27.2$.

We have

$$\frac{8 \cdot x}{8} = \frac{27.2}{8}\qquad \text{Dividing by 8 on both sides}$$

$$x = 3.4.$$

$$
\begin{array}{r}
3.4 \\
8\,\overline{)\,2\,7.2} \\
2\,4\,0 \\
\hline
3\,2 \\
3\,2 \\
\hline
0
\end{array}
$$

The solution is 3.4.

Answers on page A-9

EXAMPLE 10 Solve: $2.9 \cdot t = 0.14616$.

We have

$$\frac{2.9 \cdot t}{2.9} = \frac{0.14616}{2.9} \qquad \text{Dividing by 2.9 on both sides}$$

$$t = 0.0504.$$

```
         0.0 5 0 4
   2.9 ) 0.1ʌ4 6 1 6
         1 4 5 0 0
             1 1 6
             1 1 6
                 0
```

The solution is 0.0504.

Do Exercises 15 and 16.

Solve.

15. $100 \cdot x = 78.314$

C Order of Operations: Decimal Notation

The same rules for order of operations used with whole numbers and fraction notation apply when simplifying expressions with decimal notation.

> **RULES FOR ORDER OF OPERATIONS**
>
> 1. Do all calculations within grouping symbols before operations outside.
> 2. Evaluate all exponential expressions.
> 3. Do all multiplications and divisions in order from left to right.
> 4. Do all additions and subtractions in order from left to right.

16. $0.25 \cdot y = 276.4$

EXAMPLE 11 Simplify: $2.56 \times 25.6 \div 25{,}600 \times 256$.

There are no exponents or parentheses, so we multiply and divide from left to right:

$$2.56 \times 25.6 \div 25{,}600 \times 256 = 65.536 \div 25{,}600 \times 256 \qquad \text{Doing all}$$

multiplications and
divisions in order
from left to right

$$= 0.00256 \times 256$$

$$= 0.65536.$$

EXAMPLE 12 Simplify: $(5 - 0.06) \div 2 + 3.42 \times 0.1$.

$$(5 - 0.06) \div 2 + 3.42 \times 0.1 = 4.94 \div 2 + 3.42 \times 0.1 \qquad \text{Carrying out the}$$

operation inside
parentheses

$$= 2.47 + 0.342 \qquad \text{Doing all multiplications and}$$

divisions in order from left
to right

$$= 2.812$$

Simplify.

17. $625 \div 62.5 \times 25 \div 6250$

18. $0.25 \cdot (1 + 0.08) - 0.0274$

19. $20^2 - 3.4^2 +$
$\{2.5[20(9.2 - 5.6)] + 5(10 - 5)\}$

20. Mountains in Peru. Refer to the figure in Example 14. Find the average height of the mountains, in meters.

EXAMPLE 13 Simplify: $10^2 \times \{[(3 - 0.24) \div 2.4] - (0.21 - 0.092)\}$.

$10^2 \times \{[(3 - 0.24) \div 2.4] - (0.21 - 0.092)\}$

$= 10^2 \times \{[2.76 \div 2.4] - 0.118\}$ Doing the calculations in the innermost parentheses first

$= 10^2 \times \{1.15 - 0.118\}$ Again, doing the calculations in the innermost parentheses

$= 10^2 \times 1.032$ Subtracting inside the parentheses

$= 100 \times 1.032$ Evaluating the exponential expression

$= 103.2$

Do Exercises 17–19.

EXAMPLE 14 *Mountains in Peru.* The following figure shows a range of very high mountains in Peru, together with their altitudes, given both in feet and in meters. Find the average height of these mountains, in feet.
Source: *National Geographic*, July 1968, p. 130

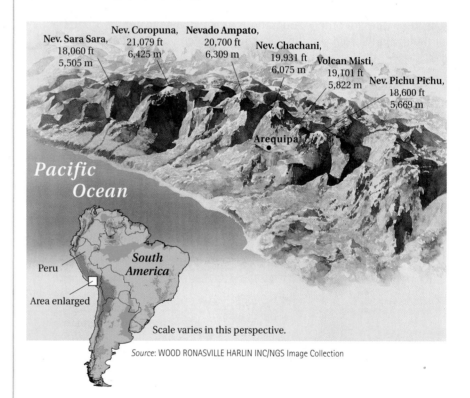

Nev. Sara Sara, 18,060 ft 5,505 m
Nev. Coropuna, 21,079 ft 6,425 m
Nevado Ampato, 20,700 ft 6,309 m
Nev. Chachani, 19,931 ft 6,075 m
Volcan Misti, 19,101 ft 5,822 m
Nev. Pichu Pichu, 18,600 ft 5,669 m
Arequipa
Pacific Ocean
Peru
South America
Area enlarged
Scale varies in this perspective.
Source: WOOD RONASVILLE HARLIN INC/NGS Image Collection

The **average** of a set of numbers is the sum of the numbers divided by the number of addends. (See Section 1.9.) We find the sum of the heights divided by the number of addends, 6:

$$\frac{18{,}060 + 21{,}079 + 20{,}700 + 19{,}931 + 19{,}101 + 18{,}600}{6} = \frac{117{,}471}{6} = 19{,}578.5.$$

Thus the average height of these mountains is 19,578.5 ft.

Do Exercise 20.

Answers on page A-9

a Divide.

1. $2 \overline{) 5.9\ 8}$

2. $5 \overline{) 1\ 8}$

3. $4 \overline{) 9\ 5.1\ 2}$

4. $8 \overline{) 2\ 5.9\ 2}$

5. $1\ 2 \overline{) 8\ 9.7\ 6}$

6. $2\ 3 \overline{) 2\ 5.0\ 7}$

7. $3\ 3 \overline{) 2\ 3\ 7.6}$

8. $12.4 \div 4$

9. $9.144 \div 8$

10. $4.5 \div 9$

11. $12.123 \div 3$

12. $7 \overline{) 5.6}$

13. $5 \overline{) 0.3\ 5}$

14. $0.0\ 4 \overline{) 1.6\ 8}$

15. $0.1\ 2 \overline{) 8.4}$

16. $0.3\ 6 \overline{) 2.8\ 8}$

17. $3.4 \overline{) 6\ 8}$

18. $0.2\ 5 \overline{) 5}$

19. $1\ 5 \overline{) 6}$

20. $1\ 2 \overline{) 1.8}$

21. $3\ 6 \overline{) 1\ 4.7\ 6}$

22. $5\ 2 \overline{) 1\ 1\ 9.6}$

23. $3.2 \overline{) 2\ 7.2}$

24. $8.5 \overline{) 2\ 7.2}$

25. $4.2 \overline{) 3\ 9.0\ 6}$

26. $4.8 \overline{) 0.1\ 1\ 0\ 4}$

27. $8 \overline{) 5}$

28. $8 \overline{) 3}$

29. $0.4\ 7 \overline{) 0.1\ 2\ 2\ 2}$

30. $1.0\ 8 \overline{) 0.5\ 4}$

31. $4.8 \overline{) 7\ 5}$

32. $0.28 \overline{)63}$

33. $0.032 \overline{)0.07488}$

34. $0.017 \overline{)1.581}$

35. $82 \overline{)38.54}$

36. $34 \overline{)0.1462}$

37. $\dfrac{213.4567}{1000}$

38. $\dfrac{213.4567}{100}$

39. $\dfrac{213.4567}{10}$

40. $\dfrac{100.7604}{0.1}$

41. $\dfrac{1.0237}{0.001}$

42. $\dfrac{1.0237}{0.01}$

b Solve.

43. $4.2 \cdot x = 39.06$

44. $36 \cdot y = 14.76$

45. $1000 \cdot y = 9.0678$

46. $789.23 = 0.25 \cdot q$

47. $1048.8 = 23 \cdot t$

48. $28.2 \cdot x = 423$

c Simplify.

49. $14 \times (82.6 + 67.9)$

50. $(26.2 - 14.8) \times 12$

51. $0.003 + 3.03 \div 0.01$

52. $9.94 + 4.26 \div (6.02 - 4.6) - 0.9$

53. $42 \times (10.6 + 0.024)$

54. $(18.6 - 4.9) \times 13$

55. $4.2 \times 5.7 + 0.7 \div 3.5$

56. $123.3 - 4.24 \times 1.01$

57. $9.0072 + 0.04 \div 0.1^2$

58. $12 \div 0.03 - 12 \times 0.03^2$

59. $(8 - 0.04)^2 \div 4 + 8.7 \times 0.4$

60. $(5 - 2.5)^2 \div 100 + 0.1 \times 6.5$

61. $86.7 + 4.22 \times (9.6 - 0.03)^2$

62. $2.48 \div (1 - 0.504) + 24.3 - 11 \times 2$

63. $4 \div 0.4 + 0.1 \times 5 - 0.1^2$

64. $6 \times 0.9 + 0.1 \div 4 - 0.2^3$

65. $5.5^2 \times [(6 - 4.2) \div 0.06 + 0.12]$

66. $12^2 \div (12 + 2.4) - [(2 - 1.6) \div 0.8]$

67. $200 \times \{[(4 - 0.25) \div 2.5] - (4.5 - 4.025)\}$

68. $0.03 \times \{1 \times 50.2 - [(8 - 7.5) \div 0.05]\}$

69. Find the average of $1276.59, $1350.49, $1123.78, and $1402.58.

70. Find the average weight of two wrestlers who weigh 308 lb and 296.4 lb.

71. *Income Taxes.* The amount paid in United States individual income taxes in a 5-year period is shown in the bar graph below. Find the average amount paid per year in individual income tax over the 5-year period.

United States Individual Income Taxes

*Figures are estimates.

Sources: Department of the Treasury; Office of Management and Budget

72. *Herbal Supplement.* Biologically based therapies in complementary and alternative medicine use substances found in nature such as herbs, food, and vitamins. The graph below shows spending on the herbal supplement echinacea, also known as purple coneflower, which is thought to strengthen the immune system. Find the average amount spent per year from 1999 to 2003.

Spending on Herbal Supplement Echinacea

Source: Nutrition Business Journal

73. **D**$_W$ How is division with decimal notation similar to division of whole numbers? How is it different?

74. **D**$_W$ Kayla made these two computational mistakes:

$$0.247 \div 0.1 = 0.0247; \qquad 0.247 \div 10 = 2.47.$$

In each case, how could you convince her that a mistake has been made?

SKILL MAINTENANCE

Simplify.   [2.5b]

75. $\dfrac{36}{42}$

76. $\dfrac{56}{64}$

77. $\dfrac{38}{146}$

78. $\dfrac{114}{438}$

Find the prime factorization.   [2.1d]

79. 684

80. 162

81. 2007

82. 2005

83. Add: $10\frac{1}{2} + 4\frac{5}{8}$.   [3.5a]

84. Subtract: $10\frac{1}{2} - 4\frac{5}{8}$.   [3.5b]

SYNTHESIS

Simplify.

85. ▦ $9.0534 - 2.041^2 \times 0.731 \div 1.043^2$

86. ▦ $23.042(7 - 4.037 \times 1.46 - 0.932^2)$

In Exercises 87–90, find the missing value.

87. $439.57 \times 0.01 \div 1000 \times \square = 4.3957$

88. $5.2738 \div 0.01 \times 1000 \div \square = 52.738$

89. $0.0329 \div 0.001 \times 10^4 \div \square = 3290$

90. $0.0047 \times 0.01 \div 10^4 \times \square = 4.7$

254

4.5 CONVERTING FROM FRACTION NOTATION TO DECIMAL NOTATION

Objectives

a Convert from fraction notation to decimal notation.

b Round numbers named by repeating decimals in problem solving.

c Calculate using fraction and decimal notation together.

a Fraction Notation to Decimal Notation

When a denominator has no prime factors other than 2's and 5's, we can find decimal notation by multiplying by 1. We multiply to get a denominator that is a power of ten, like 10, 100, or 1000.

EXAMPLE 1 Find decimal notation for $\frac{3}{5}$.

$$\frac{3}{5} = \frac{3}{5} \cdot \frac{2}{2} = \frac{6}{10} = 0.6$$ We use $\frac{2}{2}$ for 1 to get a denominator of 10.

EXAMPLE 2 Find decimal notation for $\frac{7}{20}$.

$$\frac{7}{20} = \frac{7}{20} \cdot \frac{5}{5} = \frac{35}{100} = 0.35$$ We use $\frac{5}{5}$ for 1 to get a denominator of 100.

EXAMPLE 3 Find decimal notation for $\frac{87}{25}$.

$$\frac{87}{25} = \frac{87}{25} \cdot \frac{4}{4} = \frac{348}{100} = 3.48$$ We use $\frac{4}{4}$ for 1 to get a denominator of 100.

EXAMPLE 4 Find decimal notation for $\frac{9}{40}$.

$$\frac{9}{40} = \frac{9}{40} \cdot \frac{25}{25} = \frac{225}{1000} = 0.225$$ We use $\frac{25}{25}$ for 1 to get a denominator of 1000.

Do Exercises 1–4.

We can always divide to find decimal notation.

EXAMPLE 5 Find decimal notation for $\frac{3}{5}$.

$$\frac{3}{5} = 3 \div 5 \qquad 5 \overline{)\begin{array}{l} 0.6 \\ 3.0 \\ \underline{3\ 0} \\ 0 \end{array}} \qquad \frac{3}{5} = 0.6$$

EXAMPLE 6 Find decimal notation for $\frac{7}{8}$.

$$\frac{7}{8} = 7 \div 8 \qquad 8 \overline{)\begin{array}{l} 0.8\ 7\ 5 \\ 7.0\ 0\ 0 \\ \underline{6\ 4} \\ 6\ 0 \\ \underline{5\ 6} \\ 4\ 0 \\ \underline{4\ 0} \\ 0 \end{array}} \qquad \frac{7}{8} = 0.875$$

Do Exercises 5 and 6.

Find decimal notation. Use multiplying by 1.

1. $\dfrac{4}{5}$

2. $\dfrac{9}{20}$

3. $\dfrac{11}{40}$

4. $\dfrac{33}{25}$

Find decimal notation.

5. $\dfrac{2}{5}$

6. $\dfrac{3}{8}$

Answers on page A-10

Find decimal notation.

7. $\dfrac{1}{6}$

8. $\dfrac{2}{3}$

Find decimal notation.

9. $\dfrac{5}{11}$

10. $\dfrac{12}{11}$

In Examples 5 and 6, the division *terminated,* meaning that eventually we got a remainder of 0. A **terminating decimal** occurs when the denominator has only 2's or 5's, or both, as factors, as in $\frac{17}{25}$, $\frac{5}{8}$, or $\frac{83}{100}$. This assumes that the fraction notation has been simplified.

Consider a different situation:

$$\frac{5}{6}, \quad \text{or} \quad \frac{5}{2 \cdot 3}.$$

Since 6 has a 3 as a factor, the division will not terminate. Although we can still use division to get decimal notation, the answer will be a **repeating decimal,** as follows.

EXAMPLE 7 Find decimal notation for $\frac{5}{6}$.

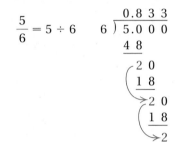

$$\frac{5}{6} = 5 \div 6$$

Since 2 keeps reappearing as a remainder, the digits repeat and will continue to do so; therefore,

$$\frac{5}{6} = 0.83333\ldots.$$

The red dots indicate an endless sequence of digits in the quotient. When there is a repeating pattern, the dots are often replaced by a bar to indicate the repeating part—in this case, only the 3:

$$\frac{5}{6} = 0.8\overline{3}.$$

Do Exercises 7 and 8.

EXAMPLE 8 Find decimal notation for $\frac{4}{11}$.

$$\frac{4}{11} = 4 \div 11 \qquad
\begin{array}{r}
0.3\ 6\ 3\ 6 \\
11\overline{)4.0\ 0\ 0\ 0} \\
3\ 3 \\
\hline
7\ 0 \\
6\ 6 \\
\hline
4\ 0 \\
3\ 3 \\
\hline
7\ 0 \\
6\ 6 \\
\hline
4
\end{array}$$

Since 7 and 4 keep repeating as remainders, the sequence of digits "36" repeats in the quotient, and

$$\frac{4}{11} = 0.363636\ldots, \quad \text{or} \quad 0.\overline{36}.$$

Do Exercises 9 and 10.

EXAMPLE 9 Find decimal notation for $\frac{5}{7}$.

```
        0.7 1 4 2 8 5
    7 ) 5.0 0 0 0 0 0
        4 9
          1 0
            7
          3 0
          2 8
            2 0
            1 4
              6 0
              5 6
                4 0
                3 5
                  5
```

Since 5 appears as a remainder, the sequence of digits "714285" repeats in the quotient, and

$$\frac{5}{7} = 0.714285714285\ldots, \quad \text{or} \quad 0.\overline{714285}.$$

The length of a repeating part can be very long—too long to find on a calculator. An example is $\frac{5}{97}$, which has a repeating part of 96 digits.

Do Exercise 11.

b Rounding in Problem Solving

In applied problems, repeating decimals are rounded to get approximate answers. To round a repeating decimal, we can extend the decimal notation at least one place past the rounding digit, and then round as before.

EXAMPLES Round each of the following to the nearest tenth, hundredth, and thousandth.

	Nearest tenth	Nearest hundredth	Nearest thousandth
10. $0.8\overline{3} = 0.83333\ldots$	0.8	0.83	0.833
11. $0.\overline{09} = 0.090909\ldots$	0.1	0.09	0.091
12. $0.\overline{714285} = 0.714285714285\ldots$	0.7	0.71	0.714

Do Exercises 12–14.

CONVERTING RATIOS TO DECIMAL NOTATION

When solving applied problems, we often convert ratios to decimal notation.

EXAMPLE 13 *Forest Fires.* In October 2003, 15 devastating forest fires burned in Southern California for two weeks, burning 800,000 acres. Find the ratio of number of acres burned to number of fires and convert it to decimal notation. Round to the nearest thousandth.
Source: National Forest Service

11. Find decimal notation for $\frac{3}{7}$.

Round each to the nearest tenth, hundredth, and thousandth.

12. $0.\overline{6}$

13. $0.\overline{80}$

14. $6.\overline{245}$

Answers on page A-10

15. Coin Tossing. A coin is tossed 51 times. It lands heads 26 times. Find the ratio of heads to tosses and convert it to decimal notation. Round to the nearest thousandth. (This is also the experimental probability of getting heads.)

Heads   Tails

16. Gas Mileage. A car goes 380 mi on 15.7 gal of gasoline. Find the gasoline mileage and convert the ratio to decimal notation rounded to the nearest tenth.

17. Camera Sales. Refer to the data on sales of digital cameras in Example 15. Find the average sales of these cameras for the 6-yr period.

Source: Photo Marketing Association International

We have

$$\frac{\text{Acres burned}}{\text{Number of fires}} = \frac{800{,}000 \text{ acres}}{15 \text{ fires}} \approx 53{,}333.\overline{3}.$$

There were about 53,333.333 acres burned per fire.

EXAMPLE 14 *Gas Mileage.* A car goes 457 mi on 16.4 gal of gasoline. The ratio of number of miles driven to amount of gasoline used is *gas mileage.* Find the gas mileage and convert the ratio to decimal notation rounded to the nearest tenth.

$$\frac{\text{Miles driven}}{\text{Gasoline used}} = \frac{457}{16.4} \approx 27.86 \qquad \text{Dividing to 2 decimal places}$$

$$\approx 27.9 \qquad \text{Rounding to 1 decimal place}$$

The gas mileage is 27.9 miles to the gallon.

Do Exercises 15 and 16.

AVERAGES

When finding an average, we may at times need to round an answer.

EXAMPLE 15 *Camera Sales.* As sales of digital cameras have increased, analog camera sales have decreased. The bar graph below shows the sales of both types of cameras from 2000 to 2005, in millions of cameras. Find the average sales of analog cameras for this period.

Camera Sales

■ Analog cameras   ■ Digital cameras

Source: Photo Marketing Association International

We add the sales totals shown on the bar graph and divide by the number of addends, 6. Since all the units are in millions, we need not convert them to standard notation. The average is

$$\frac{19.7 + 16.3 + 14.2 + 11.2 + 6.7 + 4.6}{6} = \frac{72.7}{6} \approx 12.12.$$

The average number of analog cameras sold per year for the 6-yr period is about 12.12 million.

Do Exercise 17.

C Calculations with Fraction and Decimal Notation Together

In certain kinds of calculations, fraction and decimal notation might occur together. In such cases, there are at least three ways in which we might proceed.

EXAMPLE 16 Calculate: $\frac{2}{3} \times 0.576$.

METHOD 1 One way to do this calculation is to convert the fraction notation to decimal notation so that both numbers are in decimal notation. Since $\frac{2}{3}$ converts to repeating decimal notation, it is first rounded to some chosen decimal place. We choose three decimal places. Then, using decimal notation, we multiply.

$$\frac{2}{3} \times 0.576 = 0.\overline{6} \times 0.576 \approx 0.667 \times 0.576 = 0.384192$$

METHOD 2 A second way to do this calculation is to convert the decimal notation to fraction notation so that both numbers are in fraction notation. The answer can be left in fraction notation and simplified, or we can convert to decimal notation and round, if appropriate.

$$\frac{2}{3} \times 0.576 = \frac{2}{3} \cdot \frac{576}{1000} = \frac{2 \cdot 576}{3 \cdot 1000}$$

$$= \frac{2 \cdot 2 \cdot 2 \cdot 2 \cdot 2 \cdot 2 \cdot 2 \cdot 3 \cdot 3}{2 \cdot 2 \cdot 2 \cdot 3 \cdot 5 \cdot 5 \cdot 5}$$

$$= \frac{2 \cdot 2 \cdot 2 \cdot 3}{2 \cdot 2 \cdot 2 \cdot 3} \cdot \frac{2 \cdot 2 \cdot 2 \cdot 3}{5 \cdot 5 \cdot 5}$$

$$= 1 \cdot \frac{2 \cdot 2 \cdot 2 \cdot 2 \cdot 3}{5 \cdot 5 \cdot 5}$$

$$= \frac{2 \cdot 2 \cdot 2 \cdot 2 \cdot 3}{5 \cdot 5 \cdot 5} = \frac{48}{125}, \text{ or } 0.384$$

METHOD 3 A third way to do this calculation is to treat 0.576 as $\frac{0.576}{1}$. Then we multiply 0.576 by 2, and divide the result by 3.

$$\frac{2}{3} \times 0.576 = \frac{2}{3} \times \frac{0.576}{1} = \frac{2 \times 0.576}{3} = \frac{1.152}{3} = 0.384$$

Note that we get an exact answer with methods 2 and 3, but method 1 gives an approximation since we rounded decimal notation for $\frac{2}{3}$.

Do Exercise 18.

EXAMPLE 17 Calculate: $\frac{2}{3} \times 0.576 + 3.287 \div \frac{4}{5}$.

We use the rules for order of operations, doing first the multiplication and then the division. Then we add.

$$\frac{2}{3} \times 0.576 + 3.287 \div \frac{4}{5} = 0.384 + 3.287 \cdot \frac{5}{4}$$

$$= 0.384 + 4.10875$$

$$= 4.49275$$

Method 3:
$\frac{2}{3} \times \frac{0.576}{1} = 0.384;$
$\frac{3.287}{1} \times \frac{5}{4} = 4.10875$

Do Exercises 19 and 20.

18. Calculate: $\frac{5}{6} \times 0.864$.

Calculate.

19. $\frac{1}{3} \times 0.384 + \frac{5}{8} \times 0.6784$

20. $\frac{5}{6} \times 0.864 + 14.3 \div \frac{8}{5}$

Answers on page A-10

a   Find decimal notation.

1. $\dfrac{23}{100}$

2. $\dfrac{9}{100}$

3. $\dfrac{3}{5}$

4. $\dfrac{19}{20}$

5. $\dfrac{13}{40}$

6. $\dfrac{3}{16}$

7. $\dfrac{1}{5}$

8. $\dfrac{4}{5}$

9. $\dfrac{17}{20}$

10. $\dfrac{11}{20}$

11. $\dfrac{3}{8}$

12. $\dfrac{7}{8}$

13. $\dfrac{39}{40}$

14. $\dfrac{31}{40}$

15. $\dfrac{13}{25}$

16. $\dfrac{61}{125}$

17. $\dfrac{2502}{125}$

18. $\dfrac{181}{200}$

19. $\dfrac{1}{4}$

20. $\dfrac{1}{2}$

21. $\dfrac{29}{25}$

22. $\dfrac{37}{25}$

23. $\dfrac{19}{16}$

24. $\dfrac{5}{8}$

25. $\dfrac{4}{15}$

26. $\dfrac{7}{9}$

27. $\dfrac{1}{3}$

28. $\dfrac{1}{9}$

29. $\dfrac{4}{3}$

30. $\dfrac{8}{9}$

31. $\dfrac{7}{6}$

32. $\dfrac{7}{11}$

33. $\dfrac{4}{7}$

34. $\dfrac{14}{11}$

35. $\dfrac{11}{12}$

36. $\dfrac{5}{12}$

b

37.–47. Odds. Round each answer of the odd-numbered Exercises 25–35 to the nearest tenth, hundredth, and thousandth.

38.–48. Evens. Round each answer of the even-numbered Exercises 26–36 to the nearest tenth, hundredth, and thousandth.

Round each to the nearest tenth, hundredth, and thousandth.

49. $0.\overline{18}$

50. $0.\overline{83}$

51. $0.2\overline{7}$

52. $3.5\overline{4}$

53. For this set of people, what is the ratio, in decimal notation rounded to the nearest thousandth, where appropriate, of:

a) women to the total number of people?
b) women to men?
c) men to the total number of people?
d) men to women?

54. For this set of nuts and bolts, what is the ratio, in decimal notation rounded to the nearest thousandth, where appropriate, of:

a) nuts to bolts?
b) bolts to nuts?
c) nuts to the total?
d) total number to nuts?

Gas Mileage.   In each of Exercises 55–58, find the gas mileage rounded to the nearest tenth.

55. 285 mi; 18 gal

56. 396 mi; 17 gal

57. 324.8 mi; 18.2 gal

58. 264.8 mi; 12.7 gal

59. *Windy Cities.*   Although nicknamed the Windy City, Chicago is not the windiest city in the United States. Listed in the table below are the six windiest cities and their average wind speeds. Find the average of these wind speeds and round your answer to the nearest tenth.

Source: *The Handy Geography Answer Book*

CITY	AVERAGE WIND SPEED (in miles per hour)
Mt. Washington, NH	35.3
Boston, MA	12.5
Honolulu, HI	11.3
Dallas TX	10.7
Kansas City, MO	10.7
Chicago, IL	10.4

60. *Areas of the New England States.*   The table below lists the areas of the New England states. Find the average area and round your answer to the nearest tenth.

Source: *The New York Times Almanac*

STATE	TOTAL AREA (in square miles)
Maine	33,265
New Hampshire	9,279
Vermont	9,614
Massachusetts	8,284
Connecticut	5,018
Rhode Island	1,211

Stock Prices. At one time stock prices were given using mixed numerals involving halfs, fourths, eighths, and, more recently, sixteenths. The Securities and Exchange Commission has mandated the use of decimal notation. Thus a price of $23\frac{13}{16}$ is now converted to decimal notation rounded to the nearest hundredth, that is, $23.81. Complete the following table.

Sources: *The Indianapolis Star,* 1/30/01; www.yahoo.com

	STOCK	PRICE PER SHARE	DECIMAL NOTATION	ROUNDED TO NEAREST HUNDREDTH
61.	Pfizer	$24\frac{9}{16}$		
62.	Lucent	$3\frac{31}{64}$		
63.	Qwest	$3\frac{47}{64}$		
64.	Merck	$32\frac{5}{8}$		
65.	Exxon Mobil	$59\frac{7}{8}$		
66.	Hewlett Packard	$21\frac{7}{16}$		

Source: New York Stock Exchange

C Calculate.

67. $\frac{7}{8} \times 12.64$

68. $\frac{4}{5} \times 384.8$

69. $2\frac{3}{4} + 5.65$

70. $4\frac{4}{5} + 3.25$

71. $\frac{47}{9} \times 79.95$

72. $\frac{7}{11} \times 2.7873$

73. $\frac{1}{2} - 0.5$

74. $3\frac{1}{8} - 2.75$

75. $4.875 - 2\frac{1}{16}$

76. $55\frac{3}{5} - 12.22$

77. $\frac{5}{6} \times 0.0765 + \frac{5}{4} \times 0.1124$

78. $\frac{3}{5} \times 6384.1 - \frac{3}{8} \times 156.56$

79. $\frac{4}{5} \times 384.8 + 24.8 \div \frac{8}{3}$

80. $102.4 \div \frac{2}{5} - 12 \times \frac{5}{6}$

81. $\dfrac{7}{8} \times 0.86 - 0.76 \times \dfrac{3}{4}$

82. $17.95 \div \dfrac{5}{8} + \dfrac{3}{4} \times 16.2$

83. $3.375 \times 5\dfrac{1}{3}$

84. $2.5 \times 3\dfrac{5}{8}$

85. $6.84 \div 2\dfrac{1}{2}$

86. $8\dfrac{1}{2} \div 2.125$

87. D_W When is long division *not* the fastest way to convert from fraction notation to decimal notation?

88. D_W Examine Example 16 of this section. How could the problem be changed so that method 1 would give a result that is completely accurate?

SKILL MAINTENANCE

Multiply. [3.6a]

89. $9 \cdot 2\dfrac{1}{3}$

90. $10\dfrac{1}{2} \cdot 22\dfrac{3}{4}$

Divide. [3.6b]

91. $84 \div 8\dfrac{2}{5}$

92. $8\dfrac{3}{5} \div 10\dfrac{2}{5}$

Add. [3.5a]

93. $17\dfrac{5}{6} + 32\dfrac{3}{8}$

94. $14\dfrac{3}{5} + 16\dfrac{1}{10}$

Subtract. [3.5b]

95. $16\dfrac{1}{10} - 14\dfrac{3}{5}$

96. $32\dfrac{3}{8} - 17\dfrac{5}{6}$

Solve. [3.5c]

97. A recipe for bread calls for $\dfrac{2}{3}$ cup of water, $\dfrac{1}{4}$ cup of milk, and $\dfrac{1}{8}$ cup of oil. How many cups of liquid ingredients does the recipe call for?

98. A board $\dfrac{7}{10}$ in. thick is glued to a board $\dfrac{3}{5}$ in. thick. The glue is $\dfrac{3}{100}$ in. thick. How thick is the result?

SYNTHESIS

▦ Find decimal notation.

99. $\dfrac{1}{7}$

100. $\dfrac{2}{7}$

101. $\dfrac{3}{7}$

102. $\dfrac{4}{7}$

103. $\dfrac{5}{7}$

104. ▦ From the pattern of Exercises 99–103, guess the decimal notation for $\dfrac{6}{7}$. Check on your calculator.

▦ Find decimal notation.

105. $\dfrac{1}{9}$

106. $\dfrac{1}{99}$

107. $\dfrac{1}{999}$

108. ▦ From the pattern of Exercises 105–107, guess the decimal notation for $\dfrac{1}{9999}$. Check on your calculator.

263

Objective

a Estimate sums, differences, products, and quotients.

1. Estimate by rounding to the nearest ten the total cost of one refrigerator and one printer/copier. Which of the following is an appropriate estimate?

   a) $500   b) $510

   c) $5100   d) $51

2. About how much more does the printer/copier cost than the TV/DVD combo? Estimate by rounding to the nearest ten. Which of the following is an appropriate estimate?

   a) $60   b) $600

   c) $340   d) $6000

Answers on page A-10

a Estimating Sums, Differences, Products, and Quotients

Estimating has many uses. It can be done before a problem is even attempted in order to get an idea of the answer. It can be done afterward as a check, even when we are using a calculator. In many situations, an estimate is all we need. We usually estimate by rounding the numbers so that there are one or two nonzero digits, depending on how accurate we want our estimate. Consider the following advertisements for Examples 1–3.

EXAMPLE 1   Estimate by rounding to the nearest ten the total cost of one all-in-one printer/copier and one TV/DVD combo.

We are estimating the sum

$199.98 + $141.99 = Total cost.

The estimate found by rounding the addends to the nearest ten is

$200 + $140 = $340.   (Estimated total cost)

Do Exercise 1.

EXAMPLE 2   About how much more does the refrigerator cost than the TV/DVD combo? Estimate by rounding to the nearest ten.

We are estimating the difference

$309.95 − $141.99 = Price difference.

The estimate to the nearest ten is

$310 − $140 = $170.   (Estimated price difference)

Do Exercise 2.

EXAMPLE 3 Estimate the total cost of 4 TV/DVD combos.

We are estimating the product

$4 \times \$141.99 =$ Total cost.

The estimate is found by rounding 141.99 to the nearest ten:

$4 \times \$140 = \$560.$

Do Exercise 3.

EXAMPLE 4 About how many Xbox™ Online Ready Video Game Systems at $149.99 each can be purchased for $1480?

We estimate the quotient

$\$1480 \div \$149.99.$

Since we want a whole-number estimate, we choose our rounding appropriately. Rounding $149.99 to the nearest one, we get $150. Since $1480 is close to $1500, which is a multiple of 150, we estimate

$\$1500 \div \$150,$

so the answer is 10.

Do Exercise 4.

EXAMPLE 5 Estimate: 4.8×52. Do not find the actual product. Which of the following is an appropriate estimate?

a) 25   **b)** 250   **c)** 2500   **d)** 360

We round 4.8 to the nearest one and 52 to the nearest ten.

$5 \times 50 = 250.$ (Estimated product)

Thus an approximate estimate is (b).

Other estimates we might have used in Example 5 are

$5 \times 52 = 260$ or $4.8 \times 50 = 240.$

The estimate in Example 5, $5 \times 50 = 250$, is the easiest to do because it has the fewest nonzero digits. You could probably do it mentally. In general, we try to round so that a computation has as few nonzero digits as possible while still keeping the estimated value close to the original value.

Do Exercises 5–10.

3. Estimate the total cost of 6 refrigerators. Which of the following is an appropriate estimate?

a) $1200   b) $186

c) $18,600   d) $1860

4. About how many Xbox™ Game Systems can be purchased for $4530? Which of the following is an appropriate estimate?

a) 90   b) 30

c) 300   d) 45

Estimate the product. Do not find the actual product. Which of the following is an appropriate estimate?

5. 2.4×8

a) 16   b) 34

c) 125   d) 5

6. 24×0.6

a) 200   b) 5

c) 110   d) 20

7. 0.86×0.432

a) 0.04   b) 0.4

c) 1.1   d) 4

8. 0.82×0.1

a) 800   b) 8

c) 0.08   d) 80

9. 0.12×18.248

a) 180   b) 1.8

c) 0.018   d) 18

10. 24.234×5.2

a) 200   b) 120

c) 12.5   d) 234

Answers on page A-10

Estimate the quotient. Which of the following is an appropriate estimate?

11. $59.78 \div 29.1$

a) 200   b) 20

c) 2   d) 0.2

EXAMPLE 6 Estimate: $82.08 \div 24$. Which of the following is an appropriate estimate?

a) 400   b) 16   c) 40   d) 4

This is about $80 \div 20$, so the answer is about 4. Thus an appropriate estimate is (d).

EXAMPLE 7 Estimate: $94.18 \div 3.2$. Which of the following is an appropriate estimate?

a) 30   b) 300   c) 3   d) 60

This is about $90 \div 3$, so the answer is about 30. Thus an appropriate estimate is (a).

12. $82.08 \div 2.4$

a) 40   b) 4.0

c) 400   d) 0.4

EXAMPLE 8 Estimate: $0.0156 \div 1.3$. Which of the following is an appropriate estimate?

a) 0.2   b) 0.002   c) 0.02   d) 20

This is about $0.02 \div 1$, so the answer is about 0.02. Thus an appropriate estimate is (c).

Do Exercises 11–13.

In some cases, it is easier to estimate a quotient directly rather than by rounding the divisor and the dividend.

13. $0.1768 \div 0.08$

a) 8   b) 10

c) 2   d) 20

EXAMPLE 9 Estimate: $0.0074 \div 0.23$. Which of the following is an appropriate estimate?

a) 0.3   b) 0.03   c) 300   d) 3

We estimate 3 for a quotient. We check by multiplying.

$$0.23 \times 3 = 0.69$$

We make the estimate smaller. We estimate 0.3 and check by multiplying.

$$0.23 \times 0.3 = 0.069$$

We make the estimate smaller. We estimate 0.03 and check by multiplying.

$$0.23 \times 0.03 = 0.0069$$

This is about 0.0074, so the quotient is about 0.03. Thus an appropriate estimate is (b).

14. Estimate: $0.0069 \div 0.15$. Which of the following is an appropriate estimate?

a) 0.5   b) 50

c) 0.05   d) 0.004

Do Exercise 14.

Answers on page A-10

4.6

EXERCISE SET

For Extra Help

MathXL   MyMathLab   InterAct   Math Tutor   Digital Video   Student's
                        Math   Center   Tutor CD 2   Solutions
                                                  Videotape 4   Manual

a Consider the following advertisements for Exercises 1–8. Estimate the sums, differences, products, or quotients involved in these problems. Indicate which of the choices is an appropriate estimate.

Weatherband Radio X4402
Rechargeable Two-Way Radios with weather band and up to a five-mile range.
$79⁹⁵
For Two

Satellite Radio
Receiver and car dock bundle. Over 120 digital channels of music, sports, entertainment and news including 65 channels of 100% commercial-free music.
$149⁹⁹

Upright Vacuum Cleaner
Features a lifetime HEPA filter and 17 ft quick draw hose.
$279

1. Estimate the total cost of one vacuum cleaner and one satellite radio.

   a) $43   **b)** $4300   **c)** $360   **d)** $430

2. Estimate the total cost of one satellite radio and one set of two-way radios.

   a) $230   **b)** $23   **c)** $2300   **d)** $400

3. About how much more does the vacuum cleaner cost than the satellite radio?

   a) $1300   **b)** $200   **c)** $130   **d)** $13

4. About how much more does the satellite radio cost than the two-way radios?

   a) $7000   **b)** $70   **c)** $130   **d)** $700

5. Estimate the total cost of 6 sets of two-way radios.

   a) $480   **b)** $48   **c)** $240   **d)** $4800

6. Estimate the total cost of 4 vacuum cleaners.

   a) $1200   **b)** $1120   **c)** $11,200   **d)** $600

7. About how many sets of two-way radios can be purchased for $830?

   a) 120   **b)** 100   **c)** 10   **d)** 1000

8. About how many vacuum cleaners can be purchased for $5627?

   a) 200   **b)** 20   **c)** 1800   **d)** 2000

Estimate by rounding as directed.

9. $0.02 + 1.31 + 0.34$; nearest tenth

10. $0.88 + 2.07 + 1.54$; nearest one

11. $6.03 + 0.007 + 0.214$; nearest one

12. $1.11 + 8.888 + 99.94$; nearest one

13. $52.367 + 1.307 + 7.324$; nearest one

14. $12.9882 + 1.0115$; nearest tenth

15. 2.678 − 0.445; nearest tenth

16. 12.9882 − 1.0115; nearest one

17. 198.67432 − 24.5007; nearest ten

Estimate. Choose a rounding digit that gives one or two nonzero digits. Indicate which of the choices is an appropriate estimate.

18. 234.12321 − 200.3223
  a) 600   **b)** 60
  c) 300   **d)** 30

19. 49 × 7.89
  a) 400   **b)** 40
  c) 4   **d)** 0.4

20. 7.4 × 8.9
  a) 95   **b)** 63
  c) 124   **d)** 6

21. 98.4 × 0.083
  a) 80   **b)** 12
  c) 8   **d)** 0.8

22. 78 × 5.3
  a) 400   **b)** 800
  c) 40   **d)** 8

23. 3.6 ÷ 4
  a) 10   **b)** 1
  c) 0.1   **d)** 0.01

24. 0.0713 ÷ 1.94
  a) 3.5   **b)** 0.35
  c) 0.035   **d)** 35

25. 74.68 ÷ 24.7
  a) 9   **b)** 3
  c) 12   **d)** 120

26. 914 ÷ 0.921
  a) 10   **b)** 100
  c) 1000   **d)** 1

27. *Fence Posts*   A zoo plans to construct a fence around its proposed African wildlife exhibit. The perimeter of the area to be fenced is 1760 ft. Estimate the number of wooden fence posts needed if the posts are placed 8.625 ft apart?

28. *Ticketmaster.*   Recently, Ticketmaster stock sold for $8.63 per share. Estimate how many shares can be purchased for $27,000.

29. D_W Describe a situation in which an estimation is made by rounding to the nearest 10,000 and then multiplying.

30. D_W A roll of fiberglass insulation costs $21.95. Describe two situations involving estimating and the cost of fiberglass insulation. Devise one situation so that $21.95 is rounded to $22. Devise the other situation so that $21.95 is rounded to $20.

SKILL MAINTENANCE

VOCABULARY REINFORCEMENT

In each of Exercises 31–38, fill in the blank with the correct term from the given list. Some of the choices may not be used and some may be used more than once.

31. The decimal $0.57\overline{3}$ is an example of a _____ decimal.   [4.5a]

32. The least common _____ of two natural numbers is the smallest number that is a multiple of both. [3.1a]

33. The sentence $5(3 + 8) = 5 \cdot 3 + 5 \cdot 8$ illustrates the _____ law.   [1.5a]

34. A _____ of an equation is a replacement for the variable that makes the equation true.   [1.7a]

35. The number 1 is the _____ identity.   [1.5a]

36. The sentence $13 + 7 = 7 + 13$ illustrates the _____ law of addition.   [1.2a]

37. The least common _____ of two or more fractions is the least common _____ of their denominators.   [3.2a]

38. The number 3728 is _____ by 4 if the number named by the last two digits is _____ by 4. [2.2a]

additive

multiplicative

numerator

denominator

commutative

associative

distributive

solution

divisible

terminating

repeating

multiple

factor

SYNTHESIS

The following were done on a calculator. Estimate to determine whether the decimal point was placed correctly.

39. $178.9462 \times 61.78 = 11,055.29624$

40. $14,973.35 \div 298.75 = 501.2$

41. $19.7236 - 1.4738 \times 4.1097 = 1.366672414$

42. $28.46901 \div 4.9187 - 2.5081 = 3.279813473$

43. Use one of $+$, $-$, $\times$, and $\div$ in each blank to make a true sentence.

  a) $(0.37 \ \square \ 18.78) \ \square \ 2^{13} = 156,876.8$

  b) $2.56 \ \square \ 6.4 \ \square \ 51.2 \ \square \ 17.4 = 312.84$

Objective

a Solve applied problems involving decimals.

1. Photofinishing Market.

Because an increasing number of people are printing their own digital photographs at home, the commercial photofinishing industry's revenue is declining. Between 1994 and 2005, the total revenue from film processing ranged in value from a low of $3.7 billion to a high of $6.2 billion. By how much did the high value differ from the low value?

Photofinishing Market

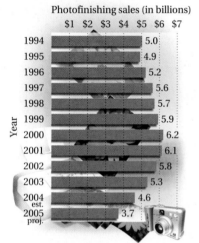

Photofinishing sales (in billions)

Year	Sales
1994	5.0
1995	4.9
1996	5.2
1997	5.6
1998	5.7
1999	5.9
2000	6.2
2001	6.1
2002	5.8
2003	5.3
2004 est.	4.6
2005 proj.	3.7

Source: PMA Marketing Research

a Solving Applied Problems

Solving applied problems with decimals is like solving applied problems with whole numbers. We translate first to an equation that corresponds to the situation. Then we solve the equation.

EXAMPLE 1 *Canals.* The Panama Canal in Panama is 50.7 mi long. The Suez Canal in Egypt is 119.9 mi long. How much longer is the Suez Canal?

Panama Canal   Suez Canal

1. **Familiarize.** This is a "how much more" situation. We let l = the distance in miles that the length of the longer canal differs from the length of the shorter canal.

2. **Translate.** We translate as follows, using the given information.

Length of Panama Canal, the shorter canal	plus	Additional length	is	Length of Suez Canal, the longer canal
↓	↓	↓	↓	↓
50.7 mi	+	l	=	119.9 mi

3. **Solve.** We solve the equation by subtracting 50.7 mi on both sides:

$$50.7 + l = 119.9$$
$$50.7 + l - 50.7 = 119.9 - 50.7$$
$$l = 69.2.$$

4. **Check.** We can check by adding 69.2 to 50.7 to get 119.9.

5. **State.** The Suez Canal is 69.2 mi longer than the Panama Canal.

Do Exercise 1.

Answer on page A-10

EXAMPLE 2 *Injections of Medication.* A patient was given injections of 2.8 mL, 1.35 mL, 2.0 mL, and 1.88 mL over a 24-hr period. What was the total amount of the injections?

1. **Familiarize.** We make a drawing or at least visualize the situation. We let t = the amount of the injections.

2.8 mL   1.35 mL   2.0 mL   1.88 mL

2. **Translate.** Amounts are being combined. We translate to an equation.

First plus Second plus Third plus Fourth is Total

2.8 + 1.35 + 2.0 + 1.88 = t

3. **Solve.** To solve, we carry out the addition.

```
  2 1
  2.8 0
  1.3 5
  2.0 0
+ 1.8 8
  8.0 3
```

Thus, $t = 8.03$.

4. **Check.** We can check by repeating our addition. We can also see whether our answer is reasonable by first noting that it is indeed larger than any of the numbers being added. We can also partially check by rounding:

$$2.8 + 1.35 + 2.0 + 1.88 \approx 3 + 1 + 2 + 2$$
$$= 8 \approx 8.03.$$

If we had gotten an answer like 80.3 or 0.803, then our estimate, 8, would have told us that we did something wrong, like not lining up the decimal points.

5. **State.** The total amount of the injections was 8.03 mL.

Do Exercise 2.

2. **Meat and Seafood Consumption.** In 2002, the average American consumed about 64.5 lb of beef, 48.2 lb of pork, 56.8 lb of chicken, 14.0 lb of turkey, and 15.6 lb of seafood. What is the total amount of meat and seafood that the average American consumed?

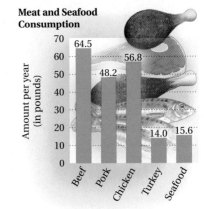

Meat and Seafood Consumption

Source: *Statistical Abstract of the United States*, 2004

Answer on page A-10

3. Printing Costs. At a printing company, the cost of copying is 11 cents per page. How much, in dollars, would it cost to make 466 copies?

EXAMPLE 3 *IRS Driving Allowance.*   In 2004, the Internal Revenue Service allowed a tax deduction of 37.5¢ per mile for mileage driven for business purposes. What deduction, in dollars, would be allowed for driving 9143 mi?

Source: Internal Revenue Service

1. **Familiarize.**   We first make a drawing or at least visualize the situation. Repeated addition fits this situation. We let d = the deduction, in dollars, allowed for driving 9143 mi.

9143 mi

2. **Translate.**   We translate as follows.

Deduction for each mile	times	Number of miles driven	is	Total deduction
$0.375	×	9143	=	d

Converting 37.5 cents to dollars gives us $0.375.

3. **Solve.**   To solve the equation, we carry out the multiplication.

```
        9 1 4 3
   ×    0.3 7 5
        4 5 7 1 5
      6 4 0 0 1 0
    2 7 4 2 9 0 0
    3 4 2 8.6 2 5
```
Thus, d = 3428.625 ≈ $3428.63.

4. **Check.**   We can obtain a partial check by rounding and estimating:

$$9143 \times 0.375 \approx 9000 \times 0.4 = 3600 \approx 3428.63.$$

5. **State.**   The total allowable deduction would be $3428.63.

Do Exercise 3.

EXAMPLE 4 *Loan Payments.*   A car loan of $7382.52 is to be paid off in 36 monthly payments. How much is each payment?

1. **Familiarize.**   We first make a drawing. We let n = the amount of each payment.

There may be some fractional part of $1.

$7382.52

36 rows. How many in each row?

n = the amount of payment in each row

Answer on page A-10

2. **Translate.** The problem can be translated to the following equation, thinking that

(Total loan) ÷ (Number of payments) = Amount of each payment
$$\$7382.52 \div 36 = n.$$

3. **Solve.** To solve the equation, we carry out the division.

```
          2 0 5.0 7
   3 6 ) 7 3 8 2.5 2
         7 2 0 0 0 0
         1 8 2 5 2
         1 8 0 0 0
               2 5 2
               2 5 2
                   0
```

Thus, $n = 205.07$.

4. **Check.** A partial check can be obtained by estimating the quotient: $\$7382.56 \div 36 \approx 8000 \div 40 = 200 \approx 205.07$. The estimate checks.

5. **State.** Each payment is $205.07.

Do Exercise 4.

EXAMPLE 5 *Jackie Robinson Poster.* A limited-edition poster by sports artist Leroy Neiman commemorates Jackie Robinson's entrance into Major League Baseball in 1947. Robinson was the first African-American player in the major leagues. The dimensions of the poster are 19.3 in. by 27.4 in. Find the area.

Source: Barton L. Kaufman, private collection

1. **Familiarize.** We first make a drawing. We let A = the area.

4. **Loan Payments.** A loan of $4425 is to be paid off in 12 monthly payments. How much is each payment?

Answer on page A-10

Study Tips

FIVE STEPS FOR PROBLEM SOLVING

Are you using the five steps for problem solving that were developed in Section 1.8?

1. **Familiarize** yourself with the situation.

   a) Carefully read and reread until you understand what you are being asked to find.

   b) Draw a diagram or see if there is a formula that applies.

   c) Assign a letter, or *variable*, to the unknown.

2. **Translate** the problem to an equation using the letter or variable.

3. **Solve** the equation.

4. **Check** the answer in the original wording of the problem.

5. **State** the answer to the problem clearly with appropriate units.

5. Index Cards. A standard-size index card measures 12.7 cm by 7.6 cm. Find its area.

7.6 cm

12.7 cm

2. Translate. We use the formula $A = l \cdot w$ and substitute.

$$A = l \cdot w$$
$$A = 27.4 \times 19.3.$$

3. Solve. We solve by carrying out the multiplication.

```
        2 7.4
   ×    1 9.3
      ─────────
        8 2 2
    2 4 6 6 0
    2 7 4 0 0
  ─────────────
    5 2 8.8 2
```

4. Check. We obtain a partial check by estimating the product:

$$A = 27.4 \times 19.3 \approx 25 \times 20 = 500.$$

This estimate is close to 528.82 so it's a good check.

5. State. The area of the Jackie Robinson poster is 528.82 in².

Do Exercise 5.

EXAMPLE 6 *Digital Camera Purchase.* Dawson Real Estate spent $10,399.74 on a set of 26 Nikon COOLPIX 5200 cameras, so that its realtors can place photos of properties on the firm's Web site. How much did each camera cost?

1. Familiarize. We let c = the cost of each camera.

2. Translate. We translate as follows.

Cost of each camera	is	Total cost of purchase	divided by	Number of cameras purchased
c	=	$10,399.74	÷	26

6. One pound of lean boneless ham contains 4.5 servings. It costs $5.99 per pound. What is the cost per serving? Round to the nearest cent.

3. Solve. To solve, we carry out the division.

```
              3 9 9.9 9
    2 6 ) 1 0,3 9 9.7 4
          7 8
          ─────
          2 5 9
          2 3 4
          ─────
            2 5 9
            2 3 4
            ─────
              2 5 7
              2 3 4
              ─────
                2 3 4
                2 3 4
                ─────
                    0
```

4. Check. We check by estimating

$$10,399.74 \div 26 \approx 10,400 \div 25 = 416.$$

Since 416 is close to 399.99, the answer is probably correct.

5. State. The cost of each camera was $399.99.

Do Exercise 6.

Answers on page A-10

MULTISTEP PROBLEMS

EXAMPLE 7 *Gas Mileage.* A driver filled the gasoline tank and noted that the odometer read 67,507.8. After the next filling, the odometer read 68,006.1. It took 16.5 gal to fill the tank. How many miles per gallon did the driver get?

1. Familiarize. We first make a drawing.

n miles, 16.5 gallons

This is a two-step problem. First, we find the number of miles that have been driven between fillups. We let n = the number of miles driven.

2., 3. Translate and **Solve.** This is a "how many more" situation. We translate and solve as follows.

First odometer reading	plus	Number of miles driven	is	Second odometer reading
67,507.8	+	n	=	68,006.1

To solve the equation, we subtract 67,507.8 on both sides:

$n = 68,006.1 - 67,507.8$

$\quad = 498.3.$

$$\begin{array}{r} 6\,8,0\,0\,6.1 \\ -\ 6\,7,5\,0\,7.8 \\ \hline 4\,9\,8.3 \end{array}$$

Second, we divide the total number of miles driven by the number of gallons. This gives us m = the number of miles per gallon—that is, the mileage. The division that corresponds to the situation is

$\quad 498.3 \div 16.5 = m.$

To find the number m, we divide.

$$\begin{array}{r} 3\,0.2 \\ 1\,6.5\,)\overline{\,4\,9\,8.3_{\wedge}0} \\ 4\,9\,5\,0 \\ \hline 3\,3\,0 \\ 3\,3\,0 \\ \hline 0 \end{array}$$

Thus, $m = 30.2$.

4. Check. To check, we first multiply the number of miles per gallon times the number of gallons:

$\quad 16.5 \times 30.2 = 498.3.$

Then we add 498.3 to 67,507.8:

$\quad 67,507.8 + 498.3 = 68,006.1.$

The mileage 30.2 checks.

5. State. The driver got 30.2 miles per gallon.

Do Exercise 7.

7. Gas Mileage. A driver filled the gasoline tank and noted that the odometer read 38,320.8. After the next filling, the odometer read 38,735.5. It took 14.5 gal to fill the tank. How many miles per gallon did the driver get?

Answer on page A-10

EXAMPLE 8 *Home-Cost Comparison.* Suppose you own a home like the one shown here and it is valued at $225,000 in Tulsa, Oklahoma. What would it cost to buy a similar (replacement) home in Boston, Massachusetts? To find out, we can use an index table prepared by Coldwell Banker Real Estate Corporation. (For a complete index table, contact your local representative.) We use the following formula:

$$\left(\begin{array}{c}\text{Cost of your}\\ \text{home in new city}\end{array}\right) = \frac{\left(\begin{array}{c}\text{Value of}\\ \text{your home}\end{array}\right) \times \left(\begin{array}{c}\text{Index of}\\ \text{new city}\end{array}\right)}{\left(\begin{array}{c}\text{Index of}\\ \text{your city}\end{array}\right)}.$$

Find the cost of your Tulsa home in Boston. Round to the nearest one.

STATE	CITY	INDEX	STATE	CITY	INDEX
Massachusetts	Boston	297	Texas	Dallas	67
Illinois	Chicago	215	Oklahoma	Tulsa	39
California	Palo Alto	342	Florida	Miami/Coral Gables	143
	Sacramento	98		Orlando	69
	San Diego	171		Key West	214
Alaska	Juneau	119	Colorado	Boulder	125
Kentucky	Louisville	63	North Carolina	Charlotte	54

Source: Coldwell Banker Real Estate Corporation

Refer to the table in Example 8 to answer Margin Exercises 8 and 9.

8. Home-Cost Comparison. Find the cost of a $180,000 home in Charlotte if you were to try to replace it when moving to Chicago. Round to the nearest one.

9. Find the cost of a $315,000 home in Boulder if you were to try to replace it when moving to Louisville. Round to the nearest one.

1. Familiarize. We let C = the cost of the home in Boston. We use the table and look up the indexes of the city in which you now live and the city to which you are moving.

2. Translate. Using the formula, we translate to the following equation:

$$C = \frac{\$225{,}000 \times 297}{39}.$$

3. Solve. To solve, we carry out the computations using the rules for order of operations (see Section 4.4).

$$C = \frac{\$225{,}000 \times 297}{39}$$

$$= \frac{66{,}825{,}000}{39} \qquad \text{Carrying out the multiplication first}$$

$$\approx \$1{,}713{,}462. \qquad \text{Carrying out the division and rounding to the nearest one}$$

On a calculator, the computation could be done in one step.

4. Check. We can repeat our computations.

5. State. A home that sells for $225,000 in Tulsa would cost about $1,713,462 in Boston.

Do Exercises 8 and 9.

Answers on page A-10

Translating
for Success

1. *Gas Mileage.* Art filled his SUV's gas tank and noted that the odometer read 38,271.8. At the next filling, the odometer read 38,677.92. It took 28.4 gal to fill the tank. How many miles per gallon did the SUV get?

2. *Dimensions of a Parking Lot.* Seals' parking lot is a rectangle that measures 85.2 ft by 52.3 ft. What is the area of the parking lot?

3. *Game Snacks.* Three students pay $18.40 for snacks at a football game. What is each person's share?

4. *Electrical Wiring.* An electrician needs 1314 ft of wiring cut into $2\frac{1}{2}$-ft pieces. How many pieces will she have?

5. *College Tuition.* Wayne needs $4638 for the fall semester's tuition. On the day of registration, he had only $3092. How much does he need to borrow?

The goal of these matching questions is to practice step (2), *Translate*, of the five-step problem-solving process. Translate each word problem to an equation and select a correct translation from equations A–O.

A. $2\frac{1}{2} \cdot n = 1314$

B. $18.4 \times 1.87 = n$

C. $n = 85.2 \times 52.3$

D. $1314.28 - 437 = n$

E. $3 \times 18.40 = n$

F. $2\frac{1}{2} \cdot 1314 = n$

G. $3092 + n = 4638$

H. $18.4 \cdot n = 1.87$

I. $\dfrac{406.12}{28.4} = n$

J. $52.3 \cdot n = 85.2$

K. $n = 1314.28 + 437$

L. $52.3 + n = 85.2$

M. $3092 + 4638 = n$

N. $3 \cdot n = 18.40$

O. $85.2 + 52.3 = n$

Answers on page A-10

6. *Cost of Gasoline.* What is the cost, in dollars, of 18.4 gal of gasoline at $1.87 per gallon?

7. *Savings Account Balance.* Margaret has $1314.28 in her savings account. Before using her debit card to buy an office chair, she transferred $437 to her checking account. How much was left in her savings account?

8. *Acres Planted.* This season Sam planted 85.2 acres of corn and 52.3 acres of soybeans. Find the total number of acres that he planted.

9. *Amount Inherited.* Tara inherited $2\frac{1}{2}$ times as much as her cousin. Her cousin received $1314. How much did Tara receive?

10. *Travel Funds.* The athletic department needs travel funds of $4638 for the tennis team and $3092 for the golf team. What is the total amount needed for travel?

a Solve.

1. *Hurricanes.* The five most costly hurricanes in the United States are listed in the following chart. The costs, adjusted to 2003 dollars by the Insurance Information Institute, ranged from a low of $2.5 billion to a high of $20.3 billion. By how much did the high value differ from the low value?

COSTLY HURRICANES

RANK	HURRICANE	DATE	IN 2003 DOLLARS (in billions)
1	Andrew	Aug. 23–26, 1992	$20.3
2	Charley	Aug. 13–15, 2004	$6.8
3	Hugo	Sept. 17–22, 1989	$6.2
4	Georges	Sept. 21–28, 1998	$3.3
5	Opal	Oct. 4, 1995	$2.5

Sources: Insurance Services Office; Insurance Information Institute

2. *Compact Discs* Because of increased sales of DVDs, the sales of recorded music in compact disc (CD) format are declining. Between 1996 and 2003, the total sales of recorded music on CDs ranged from a high of 942.5 million to a low of 745.9 million. By how much did the high value differ from the low value?

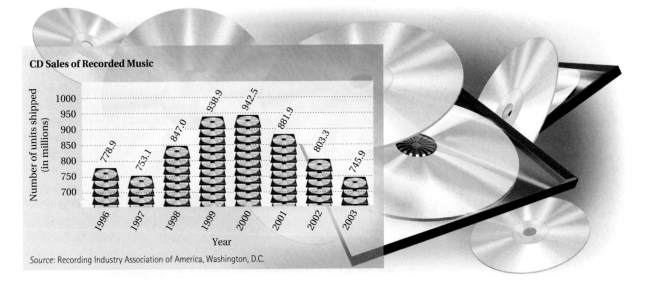

CD Sales of Recorded Music

Source: Recording Industry Association of America, Washington, D.C.

3. Andrew bought a DVD of the complete ninth season of the TV show "Friends" for $29.24 plus $1.61 sales tax. He paid for it with a $50 bill. How much change did he receive?

4. Jaden bought the CD "Feels Like Home" by Norah Jones for $13.49 plus $0.81 sales tax. She paid for it with a $20 bill. How much change did she receive?

5. *Body Temperature.* Normal body temperature is 98.6°F. During an illness, a patient's temperature rose 4.2°. What was the new temperature?

6. *Gasoline Cost.* What is the cost, in dollars, of 20.4 gal of gasoline at $2.13 per gallon? Round the answer to the nearest cent.

7. *Lottery Winnings.* The largest lotto jackpot ever won in California totaled $193,000,000 and was shared equally by 3 winners. How much was each winner's share? Round to the cent.

Source: California State Lottery

8. *Lunch Costs.* A group of 4 students pays $47.84 for lunch. What is each person's share?

9. *Stamp.* Find the area and the perimeter of the stamp shown here.

2.5 cm

3.25 cm

10. *Pole Vault Pit.* Find the area and the perimeter of the landing area of the pole vault pit shown here.

16.4 ft

16.4 ft

Landing Area

11. *Odometer Reading.* A family checked the odometer before starting a trip. It read 22,456.8 and they know that they will be driving 234.7 mi. What will the odometer read at the end of the trip?

12. *Miles Driven.* Petra bought gasoline when the odometer read 14,296.3. At the next gasoline purchase, the odometer read 14,515.8. How many miles had been driven?

13. *Gas Mileage.* Peggy filled her van's gas tank and noted that the odometer read 26,342.8. After the next filling, the odometer read 26,736.7. It took 19.5 gal to fill the tank. How many miles per gallon did the van get?

14. *Gas Mileage.* Henry filled his Honda's gas tank and noted that the odometer read 18,943.2. After the next filling, the odometer read 19,306.2. It took 13.2 gal to fill the tank. How many miles per gallon did the car get?

15. *Cost of Video Game.* A certain video game costs 75 cents and runs for 1.5 min. Assuming a player does not win any free games and plays continuously, how much money, in dollars, does it cost to play the video game for 1 hr?

16. *Property Taxes.* The Colavitos own a house with an assessed value of $184,500. For every $1000 of assessed value, they pay $7.68 in taxes. How much do they pay in taxes?

17. *Chemistry.* The water in a filled tank weighs 748.45 lb. One cubic foot of water weighs 62.5 lb. How many cubic feet of water does the tank hold?

18. *Highway Routes.* You can drive from home to work using either of two routes:

> *Route A*: Via interstate highway, 7.6 mi, with a speed limit of 65 mph.

> *Route B*: Via a country road, 5.6 mi, with a speed limit of 50 mph.

Assuming you drive at the posted speed limit, which route takes less time? (Use the formula *Distance = Speed × Time.*)

Find the perimeter of the figure.

19. 8.9 cm   23.8 cm   4.7 cm   18.6 cm   22.1 cm

20. 104.8 yd   111.9 yd   68.9 yd   56.7 yd   49.2 yd

21. 2.5 cm   2.25 cm

22. 2.5 cm   4.0 cm

Find the length *d* in the figure.

23. 0.8 cm   0.8 cm   *d*   3.91 cm

24. 0.9 cm   0.9 cm   *d*   4.52 cm

25. *Calories Burned Mowing.* A person weighing 150 lb burns 7.3 calories per minute while mowing a lawn with a power lawnmower. How many calories would be burned in 2 hr of mowing?
Source: *The Handy Science Answer Book*

26. Lot A measures 250.1 ft by 302.7 ft. Lot B measures 389.4 ft by 566.2 ft. What is the total area of the two lots?

27. Holly had $1123.56 in her checking account. She used her debit card to pay bills of $23.82, $507.88, and $98.32. She then deposited a bonus check of $678.20. How much is in her account after these changes?

28. Natalie had $185.00 to spend for fall clothes: $44.95 was spent on shoes, $71.95 for a jacket, and $55.35 for pants. How much was left?

29. A rectangular yard is 20 ft by 15 ft. The yard is covered with grass except for an 8.5-ft square flower garden. How much grass is in the yard?

30. Rita earns a gross paycheck (before deductions) of $495.72. Her deductions are $59.60 for federal income tax, $29.00 for FICA, and $29.00 for medical insurance. What is her take-home paycheck?

31. *Batting Averages.* For the 2004 season, Barry Bonds of the San Francisco Giants won the National League batting title with 135 hits in 373 times at bat. What part of his at-bats were hits? Give decimal notation to the nearest thousandth. (This is a player's *batting average.*)
Source: Major League Baseball

32. *Batting Averages.* For the 2004 season, Ichiro Suzuki of the Seattle Mariners won the American League batting title with 262 hits in 704 times at bat. What part of his at-bats were hits? Give decimal notation to the nearest thousandth.
Source: Major League Baseball

33. *Verizon Wireless.* In 2005, Verizon Wireless offered its America's Choice cellular phone plan with 450 anytime minutes per month for a monthly access fee of $39.99. Minutes in excess of 450 were charged at the rate of $0.45 per minute. In June, Leila used her cellphone for 479 minutes. What was she charged?
Source: www.vzwshop.com

34. *Verizon Wireless.* In 2005, Verizon Wireless offered its America's Choice cellular phone plan with 900 anytime minutes per month for a monthly access fee of $59.99. Minutes in excess of 900 were charged at the rate of $0.40 per minute. One month Jeff used his cellphone for 946 minutes. What was the charge?
Source: www.vzwshop.com

35. *Field Dimensions.* The dimensions of a World Cup soccer field are 114.9 yd by 74.4 yd. The dimensions of a standard football field are 120 yd by 53.3 yd. How much greater is the area of a World Cup soccer field?

36. *Loan Payment.* In order to make money on loans, financial institutions are paid back more money than they loan. You borrow $120,000 to buy a house and agree to make monthly payments of $880.52 for 30 yr. How much do you pay back altogether? How much more do you pay back than the amount of the loan?

World Cup Soccer Field

Football Field

114.9 yd

74.4 yd

120 yd

53.3 yd

37. Egg Costs. A restaurant owner bought 20 dozen eggs for $25.80. Find the cost of each egg to the nearest tenth of a cent (thousandth of a dollar).

38. Weight Loss. A person weighing 170 lb burns 8.6 calories per minute while mowing a lawn. One must burn about 3500 calories in order to lose 1 lb. How many pounds would be lost by mowing for 2 hr? Round to the nearest tenth.

39. Construction Pay. A construction worker is paid $18.50 per hour for the first 40 hr of work, and time and a half, or $27.75 per hour, for any overtime exceeding 40 hr per week. One week she works 46 hr. How much is her pay?

40. Summer Work. Zachary worked 53 hr during a week one summer. He earned $6.50 per hour for the first 40 hr and $9.75 per hour for overtime (hours exceeding 40). How much did Zachary earn during the week?

41. Projected World Population. Using the information in the following bar graph, determine the average population of the world for the years 1950 through 2040.

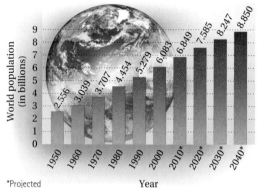

World Population (historical and projected)

World population (in billions): 2.556, 3.039, 3.707, 4.454, 5.279, 6.083, 6.849, 7.585, 8.247, 8.850

Years: 1950, 1960, 1970, 1980, 1990, 2000, 2010*, 2020*, 2030*, 2040*

*Projected

Source: U.S. Bureau of the Census

42. Volunteer Hours. The chart at right lists the average number of hours per month that volunteers of specific age groups spend doing volunteer work. What is the average number of volunteer hours of all age groups?

Source: *Statistical Abstract of the United States*

HOURS SPENT VOLUNTEERING

AGE	AVERAGE HOURS PER MONTH
21–24 yrs	12.1
25–34 yrs	15.9
35–44 yrs	16.1
45–54 yrs	14.7
55–64 yrs	12.1
65–74 yrs	14.1
75 yrs and older	19.5

43. *Body Temperature.* Normal body temperature is 98.6°F. A baby's bath water should be 100°F. How many degrees above normal body temperature is this?

44. *Body Temperature.* Normal body temperature is 98.6°F. The lowest temperature at which a patient has survived is 69°F. How many degrees below normal is this?

Home-Cost Comparison. Use the table and formula from Example 8. In each of the following cases, find the value of the house in the new location.

Source: Coldwell Banker Real Estate Corporation

	VALUE	PRESENT LOCATION	NEW LOCATION	NEW VALUE
45.	$125,000	Dallas	Miami/Coral Gables	
46.	$180,000	San Diego	Key West	
47.	$96,000	Orlando	Juneau	
48.	$300,000	Palo Alto	Dallas	
49.	$240,000	Louisville	Boston	
50.	$160,000	Charlotte	Sacramento	

Powells.com logo used with permission.

Comparison Shopping. The Internet now provides many sites to shop for a product rather than shopping in a local store. The following lists various Web sites for purchasing the recent best-selling novel *The Da Vinci Code* by Dan Brown.

51. D_W Complete the total costs in the table below for each merchant. (Assume you live in a state in which sales taxes are not charged on Internet purchases.) Then decide which site you would use to make a purchase. Discuss possible ways in which answers might vary.

MERCHANT	PRICE	SHIPPING AND HANDLING	SALES TAX	TOTAL COST
Amazon.com www.amazon.com	$14.97	$3.00 per order, plus $0.99 per item	$0	?
Barnes & Noble www.bn.com	$14.97	$3.00 per order, plus $0.99 per item	$0	?
Powell's Books www.powells.com	$24.95	$2.50 per order, plus $1.00 per item	$0	?
Costco Wholesale www.costco.com	$21.99	$5.00	$0	?
Overstock.com www.overstock.com	$13.47	$1.40	$0	?
Borders Bookstore, local store	$17.46	$0	$1.05	?

52. **D**_W *Internet Project.* Consider using the Internet to buy a copy of the book *The Broker* by John Grisham. Then compare your costs with buying it at a local bookstore. Decide which way you would make a purchase. Discuss possible ways in which answers might vary.

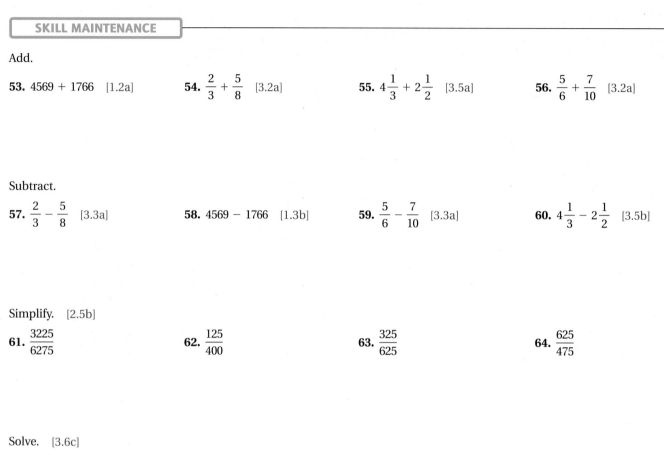

Add.

53. $4569 + 1766$ [1.2a]

54. $\dfrac{2}{3} + \dfrac{5}{8}$ [3.2a]

55. $4\dfrac{1}{3} + 2\dfrac{1}{2}$ [3.5a]

56. $\dfrac{5}{6} + \dfrac{7}{10}$ [3.2a]

Subtract.

57. $\dfrac{2}{3} - \dfrac{5}{8}$ [3.3a]

58. $4569 - 1766$ [1.3b]

59. $\dfrac{5}{6} - \dfrac{7}{10}$ [3.3a]

60. $4\dfrac{1}{3} - 2\dfrac{1}{2}$ [3.5b]

Simplify.   [2.5b]

61. $\dfrac{3225}{6275}$

62. $\dfrac{125}{400}$

63. $\dfrac{325}{625}$

64. $\dfrac{625}{475}$

Solve.   [3.6c]

65. If a water wheel made 469 revolutions at a rate of $16\dfrac{3}{4}$ revolutions per minute, how long did it rotate?

66. If a bicycle wheel made 480 revolutions at a rate of $66\dfrac{2}{3}$ revolutions per minute, how long did it rotate?

67. *Calories in Pie.* A piece of pecan pie $\left(\dfrac{1}{8} \text{ of a 9-in. pie}\right)$ has 502 calories. A piece of pumpkin pie has 316 calories. How many more calories does a piece of pecan pie have than a piece of pumpkin pie?   [1.8a]

68. *Calories in Turkey.* Dark meat turkey contains 187 calories per 3.5-oz serving. White meat turkey contains 157 calories per 3.5-oz serving. How many more calories per 3.5-oz serving does the dark turkey contain than the white turkey?   [1.8a]

69. You buy a half-dozen packs of basketball cards with a dozen cards in each pack. The cost is twelve dozen cents for each half-dozen cards. How much do you pay for the cards?

284

The review that follows is meant to prepare you for a chapter exam. It consists of two parts. The first part, Concept Reinforcement, is designed to increase understanding of the concepts through true/false exercises. The second part is the Review Exercises. These provide practice exercises for the exam, together with references to section objectives so you can go back and review. Before beginning, stop and look back over the skills you have obtained. What skills in mathematics do you have now that you did not have before studying this chapter?

CONCEPT REINFORCEMENT

Determine whether the statement is true or false. Answers are given at the back of the book.

_____ **1.** In the number 308.00567, the digit 6 names the hundreds place.

_____ **2.** To multiply any number by a multiple of 10, count the number of zeros and move the decimal point that many places to the right.

_____ **3.** One thousand billions is one trillion.

_____ **4.** The number of decimal places in the product of two numbers is the product of the number of places in the factors.

_____ **5.** When writing a word name for decimal notation, we write the word "and" for the decimal point.

Review Exercises

Convert the number in the sentence to standard notation.

1. Russia has the largest total area of any country in the world, at 6.59 million square miles. [4.3b]

RUSSIA

2. The total weight of the turkeys consumed by Americans during the Thanksgiving holidays is about 6.9 million pounds. [4.3b]

Write a word name. [4.1a]

3. 3.47   **4.** 0.031

5. 27.00011   **6.** 0.000007

Write fraction notation. [4.1b]

7. 0.09   **8.** 4.561

9. 0.089   **10.** 3.0227

Write decimal notation. [4.1b]

11. $\dfrac{34}{1000}$ **12.** $\dfrac{42,603}{10,000}$

13. $27\dfrac{91}{100}$ **14.** $867\dfrac{6}{1000}$

Which number is larger? [4.1c]

15. 0.034, 0.0185   **16.** 0.91, 0.19

17. 0.741, 0.6943   **18.** 1.038, 1.041

Round 17.4287 to the nearest: [4.1d]

19. Tenth.   **20.** Hundredth.

21. Thousandth.   **22.** One.

Add. [4.2a]

23.
```
    2.0 4 8
   6 5.3 7 1
 + 5 0 7.1
```

24.
```
   0.6
   0.0 0 4
   0.0 7
 +0.0 0 9 8
```

25. 219.3 + 2.8 + 7

26. 0.41 + 4.1 + 41 + 0.041

Subtract. [4.2b]

27.
```
   3 0.0
 −  0.7 9 0 8
```

28.
```
   8 4 5.0 8
 −   5 4.7 9
```

29. 37.645 − 8.497

30. 70.8 − 0.0109

Multiply. [4.3a]

31.
```
      4 8
  × 0.2 7
```

32.
```
   0.1 7 4
 ×   0.8 3
```

33. 100 × 0.043

34. 0.001 × 24.68

Divide. [4.4a]

35. $8\,\overline{)\,6\,0}$

36. $5\,2\,\overline{)\,2\,3.4}$

37. $2.6\,\overline{)\,1\,1\,7.5\,2}$

38. $2.1\,4\,\overline{)\,2.1\,8\,7\,0\,8}$

39. $\dfrac{276.3}{1000}$

40. $\dfrac{13.892}{0.01}$

Solve. [4.2c], [4.4b]

41. $x + 51.748 = 548.0275$

42. $3 \cdot x = 20.85$

43. $10 \cdot y = 425.4$

44. $0.0089 + y = 5$

Solve. [4.7a]

45. Stacia, a coronary intensive care nurse, earned $620.74 during a recent 40-hr week. What was her hourly wage? Round to the nearest cent.

46. *Nutrition.* The average person eats 683.6 lb of fruits and vegetables in a year. What is the average consumption in one day (Use 1 year = 365 days.) Round to the nearest tenth of a pound.

47. Derek had $1283.67 in his checking account. He used $370.99 to buy a Sony home theater system with his debit card. How much was left in his account?

48. *Verizon Wireless.* In 2005, Verizon Wireless offered its America's Choice cellular phone plan with 1350 anytime minutes per month for a monthly access fee of $79.99. Minutes in excess of 1350 were charged $0.35 per minute. One month Maria used her cellphone for 2000 min. What was the charge?
Source: www.vzwshop.com

49. *Gas Mileage.* A driver wants to estimate gas mileage per gallon. At 36,057.1 mi, the tank is filled with 10.7 gal. At 36,217.6 mi, the tank is filled with 11.1 gal. Find the mileage per gallon. Round to the nearest tenth.

50. *Seafood Consumption.* The following graph shows the annual consumption, in pounds, of seafood per person in the United States in recent years. [4.4c]

a) Find the total per capita consumption for the seven years.

b) Find the average per capita consumption.

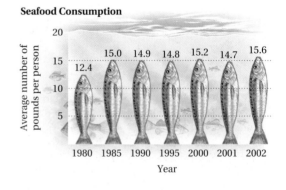

Seafood Consumption

Estimate each of the following. [4.6a]

51. The product 7.82×34.487 by rounding to the nearest one

52. The difference $219.875 - 4.478$ by rounding to the nearest one

53. The quotient $82.304 \div 17.287$ by rounding to the nearest ten

54. The sum $\$45.78 + \$78.99$ by rounding to the nearest one

Find decimal notation. Use multiplying by 1. [4.5a]

55. $\dfrac{13}{5}$ **56.** $\dfrac{32}{25}$ **57.** $\dfrac{11}{4}$

Find decimal notation. Use division. [4.5a]

58. $\dfrac{13}{4}$ **59.** $\dfrac{7}{6}$ **60.** $\dfrac{17}{11}$

Round the answer to Exercise 60 to the nearest: [4.5b]

61. Tenth.   **62.** Hundredth.   **63.** Thousandth.

Convert from cents to dollars. [4.3b]

64. 8273 cents   **65.** 487 cents

Convert from dollars to cents. [4.3b]

66. $24.93   **67.** $9.86

Calculate. [4.4c], [4.5c]

68. $(8 - 1.23) \div 4 + 5.6 \times 0.02$

69. $(1 + 0.07)^2 + 10^3 \div 10^2$
$+ [4(10.1 - 5.6) + 8(11.3 - 7.8)]$

70. $\dfrac{3}{4} \times 20.85$

71. $\dfrac{1}{3} \times 123.7 + \dfrac{4}{9} \times 0.684$

72. D_W Consider finding decimal notation for $\frac{44}{125}$. Discuss as many ways as you can for finding such notation and give the answer. [4.5a]

73. D_W Explain how we can use fraction notation to understand why we count decimal places when multiplying with decimal notation. [4.3a]

74. ▦ In each of the following, use one of $+$, $-$, $\times$, and $\div$ in each blank to make a true sentence. [4.4c]
a) 2.56 ☐ 6.4 ☐ 51.2 ☐ 17.4 ☐ $89.7 = 72.62$
b) $(11.12$ ☐ $0.29)$ ☐ $3^4 = 877.23$

75. Find repeating decimal notation for 1 and explain. Use the following hints. [4.5a]

$$\frac{1}{3} = 0.33333333\ldots,$$

$$\frac{2}{3} = 0.66666666\ldots$$

76. Find repeating decimal notation for 2. [4.5a]

77. ▦ In the subtraction below, a and b are digits. Find a and b. [4.2b]

$$\begin{array}{r} b876.a4321 \\ -\,1234.a678b \\ \hline 8641.b7a32 \end{array}$$

Convert the number in the sentence to standard notation.

1. The annual sales of antibiotics in the United States is $8.9 billion.
Source: IMS Health

2. There are 3.756 million people enrolled in bowling organizations in the United States.
Source: *Bowler's Journal International*, December 2000

Write a word name.

3. 2.34

4. 105.0005

Write fraction notation.

5. 0.91

6. 2.769

Write decimal notation.

7. $\dfrac{74}{1000}$

8. $\dfrac{37,047}{10,000}$

9. $756\dfrac{9}{100}$

10. $91\dfrac{703}{1000}$

Which number is larger?

11. 0.07, 0.162

12. 0.078, 0.06

13. 0.09, 0.9

Round 5.6783 to the nearest:

14. One.

15. Hundredth.

16. Thousandth.

17. Tenth.

Calculate.

18.
```
   0.7
   0.0 8
   0.0 0 9
 + 0.0 0 1 2
```

19. $102.4 + 6.1 + 78$

20. $0.93 + 9.3 + 93 + 930$

21.
```
  5 2.6 7 8
 −   4.3 2 1
```

22.
```
  2 0.0
 −   0.9 0 9 9
```

23. $234.6788 − 81.7854$

24.
```
   0.1 2 5
 ×   0.2 4
```

25. 0.001×213.45

26. 1000×73.962

27. $4\overline{)1\ 9}$

28. $3.3 \overline{) 1\ 0\ 0.3\ 2}$

29. $8\ 2 \overline{) 1\ 5.5\ 8}$

30. $\dfrac{346.89}{1000}$

31. $\dfrac{346.89}{0.01}$

Solve.

32. $4.8 \cdot y = 404.448$

33. $x + 0.018 = 9$

34. *Verizon Wireless.* In 2005, Verizon Wireless offered its America's Choice cellular phone plan with 2000 anytime minutes per month for a monthly access fee of $99.99. Minutes in excess of 2000 were charged $0.25 per minute. One month Trey used his cell phone for 2860 min. What was the charge?

Source: www.vzwshop.com

35. *Gas Mileage.* Tina wants to estimate the gas mileage per gallon in her economy car. At 76,843 mi, the tank is filled with 14.3 gal of gasoline. At 77,310 mi, the tank is filled with 16.5 gal of gasoline. Find the mileage per gallon. Round to the nearest tenth.

36. *Checking Account Balance.* Nicholas has a balance of $10,200 in his checking account before making purchases of $123.89, $56.68, and $3446.98 with his debit card. What was the balance after making the purchases?

37. The Drake, Smith, and Nicholas law firm buys 7 cases of copy paper at $25.99 per case. What is the total cost?

38. *Airport Passengers.* The following graph shows the number of passengers in 2003 who traveled through the country's busiest airports. Find the average number of passengers through these airports.

Busiest Airports in the United States

Atlanta (Hartsfield) 76.1
Chicago (O'Hare) 69.4
Los Angeles 55.0
Dallas/Fort Worth 53.2
Denver 37.5

20 30 40 50 60 70 80
Passengers (in millions)

Source: Airports Council International, www.airports.org

Estimate each of the following.

39. The product 8.91×22.457 by rounding to the nearest one

40. The quotient $78.2209 \div 16.09$ by rounding to the nearest ten

Find decimal notation. Use multiplying by 1.

41. $\dfrac{8}{5}$

42. $\dfrac{22}{25}$

43. $\dfrac{21}{4}$

Find decimal notation. Use division.

44. $\dfrac{3}{4}$

45. $\dfrac{11}{9}$

46. $\dfrac{15}{7}$

Round the answer to Question 46 to the nearest:

47. Tenth.

48. Hundredth.

49. Thousandth.

50. Convert from cents to dollars: 949 cents.

Calculate.

51. $256 \div 3.2 \div 2 - 1.56 + 78.325 \times 0.02$

52. $(1 - 0.08)^2 + 6[5(12.1 - 8.7) + 10(14.3 - 9.6)]$

53. $\dfrac{7}{8} \times 345.6$

54. $\dfrac{2}{3} \times 79.95 - \dfrac{7}{9} \times 1.235$

55. The Silver's Health Club generally charges a $79 membership fee and $42.50 a month. Allise has a coupon that will allow her to join the club for $299 for six months. How much will Allise save if she uses the coupon?

56. ▦ Arrange from smallest to largest.

$$\dfrac{2}{3}, \ \dfrac{15}{19}, \ \dfrac{11}{13}, \ \dfrac{5}{7}, \ \dfrac{13}{15}, \ \dfrac{17}{20}$$

CHAPTER 4: Decimal Notation

Ratio and Proportion

Real-World Application

The number of women attending Purdue University's veterinary school of medicine has grown to surpass the number of men in the past three decades. In 1971, 53 men and 12 women were enrolled. In 1979, 36 men and 36 women were enrolled, and in 2004, there were 58 women and 12 men. What was the ratio of women to men in 1971, in 1979, and in 2004? What was the ratio of men to total enrollment in 2004?

Source: Purdue University School of Veterinary Medicine, *Indianapolis Star*

This problem appears as Example 7 in Section 5.1.

5.1 INTRODUCTION TO RATIOS

Objectives

a Find fraction notation for ratios.

b Simplify ratios.

1. Find the ratio of 5 to 11.

2. Find the ratio of 57.3 to 86.1.

3. Find the ratio of $6\frac{3}{4}$ to $7\frac{2}{5}$.

4. **Rainfall.** The greatest amount of rainfall ever recorded for a 12-month period was 739 in. in Kukui, Maui, Hawaii, from December 1981 to December 1982. Find the ratio of rainfall to time in months.

Source: *The Handy Science Answer Book*

a Ratios

> **RATIO**
>
> A **ratio** is the quotient of two quantities.

In the 2004–2005 season, the Detroit Pistons basketball team averaged 93.3 points per game and allowed their opponents an average of 89.5 points per game. The *ratio* of points earned to points allowed is given by the fraction notation

Points earned $\longrightarrow \dfrac{93.3}{89.5}$ Points allowed $\longrightarrow$ or by the colon notation $93.3 : 89.5$.

We read both forms of notation as "the ratio of 93.3 to 89.5," listing the numerator first and the denominator second.

> **RATIO NOTATION**
>
> The **ratio** of a to b is given by the fraction notation $\dfrac{a}{b}$, where a is the numerator and b is the denominator, or by the colon notation $a:b$.

EXAMPLE 1 Find the ratio of 7 to 8.

The ratio is $\dfrac{7}{8}$,   or $7:8$.

EXAMPLE 2 Find the ratio of 31.4 to 100.

The ratio is $\dfrac{31.4}{100}$,   or $31.4:100$.

EXAMPLE 3 Find the ratio of $4\frac{2}{3}$ to $5\frac{7}{8}$. You need not simplify.

The ratio is $\dfrac{4\frac{2}{3}}{5\frac{7}{8}}$,   or $4\frac{2}{3} : 5\frac{7}{8}$.

Do Exercises 1–3.

In most of our work, we will use fraction notation for ratios.

EXAMPLE 4 *Wind Speeds.* The average wind speed in Chicago is 10.4 mph. The average wind speed in Boston is 12.5 mph. Find the ratio of the wind speed in Chicago to the wind speed in Boston.

Source: *The Handy Geography Answer Book*

The ratio is $\dfrac{10.4}{12.5}$.

EXAMPLE 5 *Batting.* In the 2004 season, Vladimir Guerrero of the Anaheim Angels got 206 hits in 612 at-bats. What was the ratio of hits to at-bats? of at-bats to hits?

Source: Major League Baseball

The ratio of hits to at-bats is

$$\frac{206}{612}.$$

The ratio of at-bats to hits is

$$\frac{612}{206}.$$

5. Fat Grams. In one serving ($\frac{1}{2}$-cup) of fried scallops, there are 12 g of fat. In one serving ($\frac{1}{2}$-cup) of fried oysters, there are 14 g of fat. What is the ratio of grams of fat in one serving of scallops to grams of fat in one serving of oysters?

Source: *Better Homes and Gardens: A New Cook Book*

Do Exercises 4–6. (Exercise 4 is on the preceding page.)

EXAMPLE 6 Refer to the triangle below.

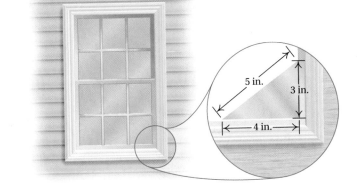

6. Earned Runs. In the 2004 season, Roger Clemens of the Houston Astros gave up 71 earned runs in 214.1 innings pitched. What was the ratio of earned runs to innings pitched? of innings pitched to earned runs?

Source: Major League Baseball

a) What is the ratio of the length of the longest side to the length of the shortest side?

$$\frac{5}{3}$$

b) What is the ratio of the length of the shortest side to the length of the longest side?

$$\frac{3}{5}$$

7. In the triangle below, what is the ratio of the length of the shortest side to the length of the longest side?

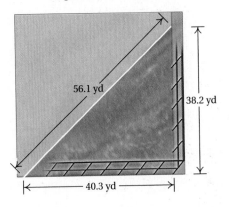

Do Exercise 7.

Answers on page A-11

8. Soap Box Derby. Participation in the All-American Soap Box Derby World Championship has increased by more than 300 competitors since 1985. In 2004, there were 483 participants, 278 boys and 205 girls. What was the ratio of girls to boys? of boys to girls? of boys to total number of participants?

Source: All-American Soap Box Derby

EXAMPLE 7 *Veterinary Medicine.* The number of women attending Purdue University's veterinary school of medicine has grown to surpass the number of men in the past three decades.

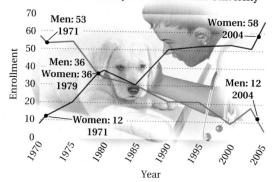

Enrollment in Veterinary Medicine: Purdue University

Sources: Purdue University School of Veterinary Medicine; *Indianapolis Star*

a) What was the ratio of women to men in 1971? in 1979? in 2004?

b) What was the ratio of men to women in 1971? in 1979? in 2004?

c) What was the ratio of women to total enrollment in 2004?

d) What was the ratio of men to total enrollment in 2004?

a) The ratio of women to men

in 1971: $\dfrac{12}{53}$; in 1979: $\dfrac{36}{36}$; in 2004: $\dfrac{58}{12}$.

b) The ratio of men to women

in 1971: $\dfrac{53}{12}$; in 1979: $\dfrac{36}{36}$; in 2004: $\dfrac{12}{58}$.

c) The ratio of women to total enrollment

in 2004: $\dfrac{58}{70}$.

d) The ratio of men to total enrollment

in 2004: $\dfrac{12}{70}$.

Do Exercise 8.

b Simplifying Notation for Ratios

Sometimes a ratio can be simplified. This provides a means of finding other numbers with the same ratio.

EXAMPLE 8 Find the ratio of 6 to 8. Then simplify and find two other numbers in the same ratio.

Answer on page A-11

We write the ratio in fraction notation and then simplify:

$$\frac{6}{8} = \frac{2 \cdot 3}{2 \cdot 4} = \frac{2}{2} \cdot \frac{3}{4} = 1 \cdot \frac{3}{4} = \frac{3}{4}.$$

Thus, 3 and 4 have the same ratio as 6 and 8. We can express this by saying "6 is to 8" as "3 is to 4."

Do Exercise 9.

EXAMPLE 9   Find the ratio of 2.4 to 10. Then simplify and find two other numbers in the same ratio.

We first write the ratio in fraction notation. Next, we multiply by 1 to clear the decimal from the numerator. Then we simplify.

$$\frac{2.4}{10} = \frac{2.4}{10} \cdot \frac{10}{10} = \frac{24}{100} = \frac{4 \cdot 6}{4 \cdot 25} = \frac{4}{4} \cdot \frac{6}{25} = \frac{6}{25}$$

Thus, 2.4 is to 10 as 6 is to 25.

Do Exercises 10 and 11.

EXAMPLE 10   A standard HDTV screen with a width of 40 in. has a height of $22\frac{1}{2}$ in. Find the ratio of width to height and simplify.

$$\text{The ratio is } \frac{40}{22\frac{1}{2}} = \frac{40}{22.5} = \frac{400}{225}$$

$$= \frac{25 \cdot 16}{25 \cdot 9} = \frac{25}{25} \cdot \frac{16}{9}$$

$$= \frac{16}{9}.$$

Thus we can say the ratio of width to height is 16 to 9, which can also be expressed as 16:9.

Do Exercise 12.

9. Find the ratio of 18 to 27. Then simplify and find two other numbers in the same ratio.

10. Find the ratio of 3.6 to 12. Then simplify and find two other numbers in the same ratio.

11. Find the ratio of 1.2 to 1.5. Then simplify and find two other numbers in the same ratio.

12. In Example 10, find the ratio of the height of the HDTV screen to the width and simplify.

Answers on page A-11

Study Tips

TIME MANAGEMENT (PART 2)

Here are some additional tips to help you with time management. (See also the Study Tips on time management in Sections 1.5 and 5.5.)

- **Avoid "time killers."**   We live in a media age, and the Internet, e-mail, television, and movies all are time killers. Allow yourself a break to enjoy some college and outside activities. But keep track of the time you spend on such activities and compare it to the time you spend studying.

- **Prioritize your tasks.**   Be careful about taking on too many college activities that fall outside of academics. Examples of such activities are decorating a homecoming float, joining a fraternity or sorority, and participating on a student council committee. Any of these is important but keep your involvement to a minimum to be sure that you have enough time for your studies.

- **Be aggressive about your study tasks.**   Instead of worrying over your math homework or test preparation, do something to get yourself started. Work a problem here and a problem there, and before long you will accomplish the task at hand. If the task is large, break it down into smaller parts, and do one at a time. You will be surprised at how quickly the large task can then be completed.

EXERCISE SET

For Extra Help

MathXL   MyMathLab   InterAct Math   Math Tutor Center   Digital Video Tutor CD 3 Videotape 5   Student's Solutions Manual

a   Find fraction notation for the ratio. You need not simplify.

1. 4 to 5

2. 3 to 2

3. 178 to 572

4. 329 to 967

5. 0.4 to 12

6. 2.3 to 22

7. 3.8 to 7.4

8. 0.6 to 0.7

9. 56.78 to 98.35

10. 456.2 to 333.1

11. $8\frac{3}{4}$ to $9\frac{5}{6}$

12. $10\frac{1}{2}$ to $43\frac{1}{4}$

13. *Corvette Accidents.*   Of every 5 fatal accidents involving a Corvette, 4 do not involve another vehicle. Find the ratio of fatal accidents involving just a Corvette to those involving a Corvette and at least one other vehicle.
Source: *Harper's Magazine*

14. *Price of a Book.*   The recent paperback book *A Short History of Nearly Everything* by Bill Bryson had a list price of $15.95 but was sold by Amazon.com for $10.85. What was the ratio of the sale price to the list price? of the list price to the sale price?
Source: www.amazon.com

15. *Physicians.*   In 2001, there were 356 physicians in Connecticut per 100,000 residents. In Wyoming, there were 173 physicians per 100,000 residents. Find the ratio of the number of physicians to residents in Connecticut and in Wyoming.
Source: U.S. Census Bureau, American Medical Association

16. *Silicon in the Earth's Crust.*   Of every 100 tons of the earth's crust, there will be about 28 tons of silicon in its content. What is the ratio of silicon to the weight of crust? of the weight of crust to the weight of silicon?
Source: *The Handy Science Answer Book*

17. *Heart Disease.*   In the state of Minnesota, of every 1000 people, 93.2 will die of heart disease. Find the ratio of those who die of heart disease to all people.
Source: "Reforming the Health Care System; State Profiles 1999," AARP

18. *Cancer Deaths.*   In the state of Texas, of every 1000 people, 122.8 will die of cancer. Find the ratio of those who die of cancer to all people.
Source: "Reforming the Health Care System; State Profiles 1999," AARP

19. _Batting._ In the 2004 season, Todd Helton of the Colorado Rockies got 190 hits in 547 at-bats. What was the ratio of hits to at-bats? of at-bats to hits?

Source: Major League Baseball

20. _Batting._ In the 2004 season, Manny Ramirez of the Boston Red Sox got 175 hits in 568 at-bats. What was the ratio of hits to at-bats? of at-bats to hits?

Source: Major League Baseball

21. _Field Hockey._ A diagram of the playing area for field hockey is shown below. What is the ratio of width to length? of length to width?

Source: _Sports: The Complete Visual Reference_

22. _The Leaning Tower of Pisa._ At the time of this writing, the Leaning Tower of Pisa is still standing. It is 184.5 ft tall but leans about 17 ft out from its base. What is the ratio of the distance it leans to its height? its height to the distance it leans?

Source: _The Handy Science Answer Book_

b Find the ratio of the first number to the second and simplify.

23. 4 to 6

24. 6 to 10

25. 18 to 24

26. 28 to 36

27. 4.8 to 10

28. 5.6 to 10

29. 2.8 to 3.6

30. 4.8 to 6.4

31. 20 to 30   **32.** 40 to 60   **33.** 56 to 100   **34.** 42 to 100

35. 128 to 256   **36.** 232 to 116   **37.** 0.48 to 0.64   **38.** 0.32 to 0.96

39. In this rectangle, find the ratios of length to width and of width to length.

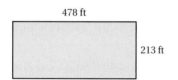

478 ft

213 ft

40. In this right triangle, find the ratios of shortest length to longest length and of longest length to shortest length.

47.5 m   107.3 m

96.2 m

41. D_W Can every ratio be written as the ratio of some number to 1? Why or why not?

42. D_W What can be concluded about a rectangle's width if the ratio of length to perimeter is 1 to 3? Make some sketches and explain your reasoning.

SKILL MAINTENANCE

Use = or ≠ for ☐ to write a true sentence.   [2.5c]

43. $\dfrac{12}{8}$ ☐ $\dfrac{6}{4}$ **44.** $\dfrac{4}{7}$ ☐ $\dfrac{5}{9}$ **45.** $\dfrac{7}{2}$ ☐ $\dfrac{31}{9}$ **46.** $\dfrac{17}{25}$ ☐ $\dfrac{68}{100}$

Divide. Write decimal notation for the answer.   [4.4a]

47. 200 ÷ 4   **48.** 95 ÷ 10   **49.** 232 ÷ 16   **50.** 342 ÷ 2.25

Solve.   [3.5c]

51. Rocky is $187\frac{1}{10}$ cm tall and his daughter is $180\frac{3}{4}$ cm tall. How much taller is Rocky?

52. Aunt Louise is $168\frac{1}{4}$ cm tall and her son is $150\frac{7}{10}$ cm tall. How much taller is Aunt Louise?

SYNTHESIS

53. Find the ratio of $3\dfrac{3}{4}$ to $5\dfrac{7}{8}$ and simplify.

Fertilizer. Exercises 54 and 55 refer to a common lawn fertilizer known as "5, 10, 15." This mixture contains 5 parts of potassium for every 10 parts of phosphorus and 15 parts of nitrogen (this is often denoted 5:10:15).

54. Find the ratio of potassium to nitrogen and of nitrogen to phosphorus.

55. Simplify the ratio 5:10:15.

5.2 RATES AND UNIT PRICES

Objectives

 a Give the ratio of two different measures as a rate.

b Find unit prices and use them to compare purchases.

a Rates

A 2005 Kia Sportage EX 4WD can go 414 miles on 18 gallons of gasoline. Let's consider the ratio of miles to gallons:

Source: Kia Motors America, Inc.

$$\frac{414 \text{ mi}}{18 \text{ gal}} = \frac{414}{18} \frac{\text{miles}}{\text{gallon}} = \frac{23}{1} \frac{\text{miles}}{\text{gallon}}$$

$$= 23 \text{ miles per gallon} = 23 \text{ mpg.}$$

"per" means "division," or "for each."

The ratio

$$\frac{414 \text{ mi}}{18 \text{ gal}}, \quad \text{or} \quad \frac{414}{18} \frac{\text{mi}}{\text{gal}}, \quad \text{or } 23 \text{ mpg}$$

is called a **rate**.

RATE

When a ratio is used to compare two different kinds of measure, we call it a **rate**.

Suppose David says his car goes 392.4 mi on 16.8 gal of gasoline. Is the mpg (mileage) of his car better than that of the Kia Sportage above? To determine this, it helps to convert the ratio to decimal notation and perhaps round. Then we have

$$\frac{392.4 \text{ miles}}{16.8 \text{ gallons}} = \frac{392.4}{16.8} \text{ mpg} \approx 23.357 \text{ mpg.}$$

Since 23.357 > 23, David's car gets better mileage than the Kia Sportage does.

EXAMPLE 1 It takes 60 oz of grass seed to seed 3000 sq ft of lawn. What is the rate in ounces per square foot?

$$\frac{60 \text{ oz}}{3000 \text{ sq ft}} = \frac{1}{50} \frac{\text{oz}}{\text{sq ft}}, \quad \text{or} \quad 0.02 \frac{\text{oz}}{\text{sq ft}}$$

EXAMPLE 2 A cook buys 10 lb of potatoes for $3.69. What is the rate in cents per pound?

$$\frac{\$3.69}{10 \text{ lb}} = \frac{369 \text{ cents}}{10 \text{ lb}}, \quad \text{or} \quad 36.9 \frac{\text{cents}}{\text{lb}}$$

Study Tips

RECORDING YOUR LECTURES

Consider recording your lectures and playing them back when convenient, say, while commuting to campus. (Be sure to get permission from your instructor before doing so, however.) Important points can be emphasized verbally. We consider this idea so worthwhile that we provide a series of audio recordings that accompany the book. (See the Preface for more information.)

A ratio of distance traveled to time is called *speed.* What is the rate, or speed, in miles per hour?

1. 45 mi, 9 hr

2. 120 mi, 10 hr

3. 89 km, 13 hr (Round to the nearest hundredth.)

What is the rate, or speed, in feet per second?

4. 2200 ft, 2 sec

5. 52 ft, 13 sec

6. 242 ft, 16 sec

7. Ratio of Home Runs to Strikeouts. Referring to Example 4, determine Rolen's home-run to strikeout rate.

Source: Major League Baseball

EXAMPLE 3 A pharmacy student working as a pharmacist's assistant earned $3690 for working 3 months one summer. What was the rate of pay per month?

The rate of pay is the ratio of money earned per length of time worked, or

$$\frac{\$3690}{3 \text{ mo}} = 1230 \frac{\text{dollars}}{\text{month}}, \quad \text{or}$$
$1230 per month.

EXAMPLE 4 *Ratio of Strikeouts to Home Runs.* In the 2004 season, Scott Rolen of the St. Louis Cardinals had 92 strikeouts and 34 home runs. What was his strikeout to home-run rate?

Source: Major League Baseball

$$\frac{92 \text{ strikeouts}}{34 \text{ home runs}} = \frac{92}{34} \frac{\text{strikeouts}}{\text{home runs}} = \frac{92}{34} \text{ strikeouts per home run}$$
$$\approx 2.71 \text{ strikeouts per home run}$$

Do Exercises 1–8. (Exercise 8 is on the following page.)

b Unit Pricing

> **UNIT PRICE**
>
> A **unit price,** or **unit rate,** is the ratio of price to the number of units.

EXAMPLE 5 *Unit Price of Pears.* A consumer bought a $15\frac{1}{4}$-oz can of pears for $1.07. What is the unit price in cents per ounce?

Often it is helpful to change the cost to cents so we can compare unit prices more easily:

$$\text{Unit price} = \frac{\text{Price}}{\text{Number of units}}$$
$$= \frac{\$1.07}{15\frac{1}{4} \text{ oz}} = \frac{107 \text{ cents}}{15.25 \text{ oz}} = \frac{107}{15.25} \frac{\text{cents}}{\text{oz}}$$
$$\approx 7.016 \text{ cents per ounce.}$$

Answers on page A-11

Do Exercise 9 on the following page.

To do comparison shopping, it helps to compare unit prices.

EXAMPLE 6 *Unit price of Heinz Ketchup.* Many factors can contribute to determining unit pricing in food, such as variations in store pricing and special discounts. Heinz produces ketchup in containers of various sizes. The table below lists several examples of pricing for these packages from a Meijer store. Starting with the price given for each package, compute the unit prices and decide which is the best purchase on the basis of unit price per ounce alone.

Source: Meijer Stores

PACKAGE	PRICE	UNIT PRICE
14 oz	$1.29	9.214¢/oz
24 oz	$1.47	6.125¢/oz
36 oz	$2.49	6.917¢/oz
64 oz	$4.45	6.953¢/oz
101-oz twin pack (two 50½-oz packages)	$5.69	5.634¢/oz Lowest unit price

We compute the unit price for the 24-oz package and leave the remaining prices to the student to check. The unit price for the 24-oz, $1.47 package is given by

$$\frac{\$1.47}{24 \text{ oz}} = \frac{147 \text{ cents}}{24 \text{ oz}} = \frac{147}{24}\frac{\text{cents}}{\text{oz}} = 6.125 \text{ cents per ounce} = 6.125 \text{ ¢/oz}.$$

On the basis of unit price alone, we see that the 101-oz twin pack is the best buy.

Sometimes, as you will see in Margin Exercise 10, a larger size may not have the lower unit price. It is also worth noting that "bigger" is not always "cheaper." (For example, you may not have room for larger packages or the food may go to waste before it is used.)

Do Exercise 10.

8. **Babe Ruth.** In his entire career, Babe Ruth had 1330 strikeouts and 714 home runs. What was his home-run to strikeout rate? How does it compare to Rolen's?

Source: Major League Baseball

9. **Unit Price of Mustard.** A consumer bought a 20-oz container of French's yellow mustard for $1.49. What is the unit price in cents per ounce?

10. **Meijer Brand Olives.** Complete the following table of unit prices for Meijer Brand olives. Which package has the better unit price?

Source: Meijer Stores

PACKAGE	PRICE	UNIT PRICE
7 oz	$1.69	
10 oz	$2.59	
5¾ oz	$1.39	

Answers on page A-11

a In Exercises 1–4, find the rate, or speed, as a ratio of distance to time. Round to the nearest hundredth where appropriate.

1. 120 km, 3 hr

2. 18 mi, 9 hr

3. 217 mi, 29 sec

4. 443 m, 48 sec

5. *Chevrolet Cobalt LS—City Driving.*   A 2005 Chevrolet Cobalt LS will go 300 mi on 12.5 gallons of gasoline in city driving. What is the rate in miles per gallon?
Source: *Car and Driver,* April 2005, p. 104

6. *Audi A6 4.2 Quattro—City Driving.*   A 2005 Audi A6 4.2 Quattro will go 246.5 miles on 14.5 gallons of gasoline in city driving. What is the rate in miles per gallon?
Source: *Car and Driver,* May 2005, p. 52

7. *Audi A6 4.2 Quattro—Highway Driving.*   A 2005 Audi A6 4.2 Quattro will go 448.5 miles on 19.5 gallons of gasoline in highway driving. What is the rate in miles per gallon?
Source: *Car and Driver,* May 2005, p. 52

8. *Chevrolet Cobalt LS—Highway Driving.*   A 2005 Chevrolet Cobalt LS will go 432 miles on 13.5 gallons of gasoline in highway driving. What is the rate in miles per gallon?
Source: *Car and Driver,* April 2005, p. 104

9. *Population Density of Monaco.*   Monaco is a tiny country on the Mediterranean coast of France. It has an area of 0.75 square mile and a population of 32,270 people. What is the rate of number of people per square mile? The rate per square mile is called the *population density.* Monaco has the highest population density in the world.
Source: *Time Almanac, 2005*

10. *Population Density of Australia.*   The continent of Australia, with the island state of Tasmania, has an area of 2,967,893 sq mi and a population of 19,913,144 people. What is the rate of number of people per square mile? The rate per square mile is called the *population density.* Australia has one of the lowest population densities in the world.
Source: *Time Almanac, 2005*

11. A car is driven 500 mi in 20 hr. What is the rate in miles per hour? in hours per mile?

12. A student eats 3 hamburgers in 15 min. What is the rate in hamburgers per minute? in minutes per hamburger?

13. *Points per Game.* In the 2004–2005 season, Yao Ming of the Houston Rockets scored 1465 points in 80 games. What was the rate in points per game?
Source: National Basketball Association

14. *Points per Game.* In the 2004–2005 season, Tayshaun Prince of the Detroit Pistons scored 1206 points in 82 games. What was the rate in points per game?
Source: National Basketball Association

15. *Lawn Watering.* To water a lawn adequately requires 623 gal of water for every 1000 ft^2. What is the rate in gallons per square foot?

16. A car is driven 200 km on 40 L of gasoline. What is the rate in kilometers per liter?

17. *Speed of Light.* Light travels 186,000 mi in 1 sec. What is its rate, or speed, in miles per second?
Source: The Handy Science Answer Book

18. *Speed of Sound.* Sound travels 1100 ft in 1 sec. What is its rate, or speed, in feet per second?
Source: The Handy Science Answer Book

19. Impulses in nerve fibers travel 310 km in 2.5 hr. What is the rate, or speed, in kilometers per hour?

20. A black racer snake can travel 4.6 km in 2 hr. What is its rate, or speed, in kilometers per hour?

21. *Elephant Heartbeat.* The heart of an elephant, at rest, will beat an average of 1500 beats in 60 min. What is the rate in beats per minute?

Source: *The Handy Science Answer Book*

22. *Human Heartbeat.* The heart of a human, at rest, will beat an average of 4200 beats in 60 min. What is the rate in beats per minute?

Source: *The Handy Science Answer Book*

b Find each unit price in each of Exercises 23–32. Then determine which size has the lowest unit price.

23. *Pert Plus Shampoo.*

PACKAGE	PRICE	UNIT PRICE
13.5 oz	$2.59	
25.4 oz	$3.99	

24. *Roll-on Deodorant.*

PACKAGE	PRICE	UNIT PRICE
2.25 oz	$2.49	
3.5 oz	$3.98	

25. *Miracle Whip.*

PACKAGE	PRICE	UNIT PRICE
16-oz jar	$1.84	
18-oz squeezable	$2.49	

26. *Bush's Homestyle Baked Beans.*

PACKAGE	PRICE	UNIT PRICE
16 oz	$0.99	
28 oz	$1.44	

27. *Meijer Coffee.*

PACKAGE	PRICE	UNIT PRICE
11.5 oz	$2.09	
34.5 oz	$5.27	

28. *Maxwell House Coffee.*

PACKAGE	PRICE	UNIT PRICE
13 oz	$3.74	
34.5 oz	$6.36	

29. *Jif Creamy Peanut Butter.*

PACKAGE	PRICE	UNIT PRICE
18 oz	$1.89	
28 oz	$3.25	
40 oz	$4.99	
64 oz	$7.99	

30. *Downy Fabric Softener.*

PACKAGE	PRICE	UNIT PRICE
40 oz	$3.23	
60 oz	$4.79	
80 oz	$7.99	
120 oz	$10.69	

31. *Tide Liquid Laundry Detergent.*

PACKAGE	PRICE	UNIT PRICE
50 fl oz	$4.29	
100 fl oz	$5.29	
200 fl oz	$10.49	
300 fl oz	$15.79	

32. *Del Monte Green Beans.*

PACKAGE	PRICE	UNIT PRICE
8 oz	$0.59	
14.5 oz	$0.69	
28 oz	$1.19	

Use the unit prices listed in Exercises 23–32 when doing Exercises 33 and 34.

33. **Dw** Look over the unit prices for each size package of Downy fabric softener. What seems to violate common sense about these unit prices? Why do you think the products are sold this way?

34. **Dw** Compare the prices and unit prices for the 16-oz jar of Miracle Whip and the 18-oz squeezable container of Miracle Whip. What seems unusual about these prices? Explain why you think this has happened.

SKILL MAINTENANCE

Solve.

35. There are 20.6 million people in this country who play the piano and 18.9 million who play the guitar. How many more play the piano than the guitar?   [4.7a]

36. A serving of fish steak (cross section) is generally $\frac{1}{2}$ lb. How many servings can be prepared from a cleaned $18\frac{3}{4}$-lb tuna?   [3.6c]

37. *Surf Expo.*   In a swimwear showing at Surf Expo, a trade show for retailers of beach supplies, each swimsuit test takes 8 minutes (min). If the show runs for 240 min, how many tests can be scheduled?   [1.8a]

38. *Eating Habits.*   Each year, Americans eat 24.8 billion hamburgers and 15.9 billion hot dogs. How many more hamburgers than hot dogs do Americans eat?   [4.7a]

Multiply.   [4.3a]

39.
$$\begin{array}{r} 4\ 5.6\ 7 \\ \times \qquad 2.4 \\ \hline \end{array}$$

40.
$$\begin{array}{r} 6\ 7\ 8.1\ 9 \\ \times \qquad 1\ 0\ 0 \\ \hline \end{array}$$

41. 84.3×69.2

42. 1002.56×465

SYNTHESIS

43. Recently, certain manufacturers have been changing the size of their containers in such a way that the consumer thinks the price of a product has been lowered when, in reality, a higher unit price is being charged.

Some aluminum juice cans are now concave (curved in) on the bottom. Suppose the volume of the can in the figure has been reduced from a fluid capacity of 6 oz to 5.5 oz, and the price of each can has been reduced from 65¢ to 60¢. Find the unit price of each container in cents per ounce.

$\frac{5}{16}$ in.

$1\frac{13}{16}$ in.

$2\frac{1}{16}$ in.

Objectives

a   Determine whether two pairs of numbers are proportional.

b   Solve proportions.

During the 2004 season, Peyton Manning of the Indianapolis Colts completed 336 passes out of 497 attempts. His pass-completion rate was

$$\text{Completion rate} = \frac{336 \text{ completions}}{497 \text{ attempts}} = \frac{336}{497} \frac{\text{completion}}{\text{attempt}}$$

$$\approx 0.676 \frac{\text{completion}}{\text{attempt}}.$$

The rate was 0.676 completion per attempt.

Ben Roethlisberger of the Pittsburgh Steelers completed 196 passes out of 295 attempts. His pass-completion rate was

$$\text{Completion rate} = \frac{196 \text{ completions}}{295 \text{ attempts}} = \frac{196}{295} \frac{\text{completion}}{\text{attempt}}$$

$$\approx 0.664 \frac{\text{completion}}{\text{attempt}}.$$

The rate was 0.664 completion per attempt. We can see that the rates are not equal.

Source: National Football League

Instead of comparing the rates in decimal notation, we can compare the ratios

$$\frac{336}{497} \quad \text{and} \quad \frac{196}{295}$$

using the test for equality considered in Section 2.5. We compare cross products.

$$= \frac{336 \cdot 295}{99{,}120} \qquad \frac{336}{497} \overset{?}{=} \frac{196}{295} \qquad = \frac{497 \cdot 196}{97{,}412}$$

Since

$99{,}120 \neq 97{,}412$, we know that

$$\frac{336}{497} \neq \frac{196}{295}.$$

Thus the ratios are not equal. If the ratios had been equal, we would say they are proportional.

a   Proportions

When two pairs of numbers (such as 3, 2 and 6, 4) have the same ratio, we say that they are **proportional.** The equation

$$\frac{3}{2} = \frac{6}{4}$$

states that the pairs 3, 2 and 6, 4 are proportional. Such an equation is called a **proportion.** We sometimes read $\frac{3}{2} = \frac{6}{4}$ as "3 is to 2 as 6 is to 4."

EXAMPLE 1 Determine whether 1, 2, and 3, 6 are proportional.

We can use cross products:

$$1 \cdot 6 = 6 \qquad \overset{?}{\underset{2}{\frac{1}{}} = \frac{3}{6}} \qquad 2 \cdot 3 = 6.$$

Since the cross products are the same, $6 = 6$, we know that $\frac{1}{2} = \frac{3}{6}$, so the numbers are proportional.

EXAMPLE 2 Determine whether 2, 5 and 4, 7 are proportional.

We can use cross products:

$$2 \cdot 7 = 14 \qquad \overset{?}{\frac{2}{5} = \frac{4}{7}} \qquad 5 \cdot 4 = 20.$$

Since the cross products are not the same, $14 \neq 20$, we know that $\frac{2}{5} \neq \frac{4}{7}$, so the numbers are not proportional.

Do Exercises 1–3.

EXAMPLE 3 Determine whether 3.2, 4.8 and 0.16, 0.24 are proportional.

We can use cross products:

$$3.2 \times 0.24 = 0.768 \qquad \overset{?}{\frac{3.2}{4.8} = \frac{0.16}{0.24}} \qquad 4.8 \times 0.16 = 0.768.$$

Since the cross products are the same, $0.768 = 0.768$, we know that $\frac{3.2}{4.8} = \frac{0.16}{0.24}$, so the numbers are proportional.

Do Exercises 4 and 5.

EXAMPLE 4 Determine whether $4\frac{2}{3}$, $5\frac{1}{2}$ and $8\frac{7}{8}$, $16\frac{1}{3}$ are proportional.

We can use cross products:

$$4\frac{2}{3} \cdot 16\frac{1}{3} = \frac{14}{3} \cdot \frac{49}{3} \qquad \overset{?}{\frac{4\frac{2}{3}}{5\frac{1}{2}} = \frac{8\frac{7}{8}}{16\frac{1}{3}}} \qquad 5\frac{1}{2} \cdot 8\frac{7}{8} = \frac{11}{2} \cdot \frac{71}{8}$$

$$= \frac{686}{9} \qquad\qquad\qquad\qquad\qquad = \frac{781}{16}$$

$$= 76\frac{2}{9} \qquad\qquad\qquad\qquad\qquad = 48\frac{13}{16}.$$

Since the cross products are not the same, $76\frac{2}{9} \neq 48\frac{13}{16}$, we know that the numbers are not proportional.

Do Exercise 6.

b   Solving Proportions

Let's now look at solving proportions. Consider the proportion

$$\frac{x}{3} = \frac{4}{6}.$$

One way to solve a proportion is to use cross products. Then we can divide on

---

Determine whether the two pairs of numbers are proportional.

1. 3, 4 and 6, 8

2. 1, 4 and 10, 39

3. 1, 2 and 20, 39

Determine whether the two pairs of numbers are proportional.

4. 6.4, 12.8 and 5.3, 10.6

5. 6.8, 7.4 and 3.4, 4.2

6. Determine whether $4\frac{2}{3}$, $5\frac{1}{2}$ and 14, $16\frac{1}{2}$ are proportional.

Answers on page A-11

7. Solve: $\dfrac{x}{63} = \dfrac{2}{9}$.

both sides to get the variable alone:

$$x \cdot 6 = 3 \cdot 4 \qquad$$ Equating cross products (finding cross products and setting them equal)

$$\frac{x \cdot 6}{6} = \frac{3 \cdot 4}{6} \qquad$$ Dividing by 6 on both sides

$$x = \frac{3 \cdot 4}{6} = \frac{12}{6} = 2.$$

We can check that 2 is the solution by replacing x with 2 and using cross products:

$$2 \cdot 6 = 12 \qquad \overset{?}{\frac{2}{3} = \frac{4}{6}} \qquad 3 \cdot 4 = 12.$$

Since the cross products are the same, it follows that $\frac{2}{3} = \frac{4}{6}$; so the numbers 2, 3 and 4, 6 are proportional, and 2 is the solution of the equation.

8. Solve: $\dfrac{x}{9} = \dfrac{5}{4}$.

SOLVING PROPORTIONS

To solve $\dfrac{x}{a} = \dfrac{c}{d}$, equate *cross products* and divide on both sides to get x alone.

Do Exercise 7.

Answers on page A-11

EXAMPLE 5 Solve: $\dfrac{x}{7} = \dfrac{5}{3}$. Write a mixed numeral for the answer.

We have

$$\frac{x}{7} = \frac{5}{3}$$

$$x \cdot 3 = 7 \cdot 5 \qquad$$ Equating cross products

$$\frac{x \cdot 3}{3} = \frac{7 \cdot 5}{3} \qquad$$ Dividing by 3

$$x = \frac{7 \cdot 5}{3}$$

$$x = \frac{35}{3}, \text{ or } 11\frac{2}{3}.$$

The solution is $11\frac{2}{3}$.

Do Exercise 8.

EXAMPLE 6 Solve: $\dfrac{7.7}{15.4} = \dfrac{y}{2.2}$.

We have

$$\frac{7.7}{15.4} = \frac{y}{2.2}$$

$$7.7 \times 2.2 = 15.4 \times y \qquad$$ Equating cross products

$$\frac{7.7 \times 2.2}{15.4} = \frac{15.4 \times y}{15.4}. \qquad$$ Dividing by 15.4

Then

$$\frac{7.7 \times 2.2}{15.4} = y$$

$$\frac{16.94}{15.4} = y \qquad \text{Multiplying}$$

$$1.1 = y. \qquad \text{Dividing:} \quad 1\,5.4\,\overline{)\,1\,6.9_\wedge 4}$$

$$\begin{array}{r} 1.1 \\ 1\,5.4\,\overline{)\,1\,6.9_\wedge 4} \\ \underline{1\,5\,4\,0} \\ 1\,5\,4 \\ \underline{1\,5\,4} \\ 0 \end{array}$$

The solution is 1.1.

Do Exercise 9.

EXAMPLE 7 Solve: $\dfrac{8}{x} = \dfrac{5}{3}$. Write decimal notation for the answer.

We have

$$\frac{8}{x} = \frac{5}{3}$$

$$8 \cdot 3 = x \cdot 5 \qquad \text{Equating cross products}$$

$$\frac{8 \cdot 3}{5} = \frac{x \cdot 5}{5} \qquad \text{Dividing by 5}$$

$$\frac{8 \cdot 3}{5} = x$$

$$\frac{24}{5} = x \qquad \text{Multiplying}$$

$$4.8 = x. \qquad \text{Simplifying}$$

The solution is 4.8.

Do Exercise 10.

EXAMPLE 8 Solve: $\dfrac{3.4}{4.93} = \dfrac{10}{n}$.

We have

$$\frac{3.4}{4.93} = \frac{10}{n}$$

$$3.4 \times n = 4.93 \times 10 \qquad \text{Equating cross products}$$

$$\frac{3.4 \times n}{3.4} = \frac{4.93 \times 10}{3.4} \qquad \text{Dividing by 3.4}$$

$$n = \frac{4.93 \times 10}{3.4}$$

$$n = \frac{49.3}{3.4} \qquad \text{Multiplying}$$

$$n = 14.5. \qquad \text{Dividing}$$

The solution is 14.5.

Do Exercise 11.

9. Solve: $\dfrac{21}{5} = \dfrac{n}{2.5}$.

10. Solve: $\dfrac{6}{x} = \dfrac{25}{11}$.

11. Solve: $\dfrac{0.4}{0.9} = \dfrac{4.8}{t}$.

Answers on page A-11

12. Solve:

$$\frac{8\frac{1}{3}}{x} = \frac{10\frac{1}{2}}{3\frac{3}{4}}.$$

EXAMPLE 9 Solve: $\dfrac{4\frac{2}{3}}{5\frac{1}{2}} = \dfrac{14}{x}$. Write a mixed numeral for the answer.

We have

$$\frac{4\frac{2}{3}}{5\frac{1}{2}} = \frac{14}{x}$$

$$4\frac{2}{3} \cdot x = 14 \cdot 5\frac{1}{2} \qquad \text{Equating cross products}$$

$$\frac{14}{3} \cdot x = 14 \cdot \frac{11}{2} \qquad \text{Converting to fraction notation}$$

$$\frac{\frac{14}{3} \cdot x}{\frac{14}{3}} = \frac{14 \cdot \frac{11}{2}}{\frac{14}{3}} \qquad \text{Dividing by } \frac{14}{3}$$

$$x = 14 \cdot \frac{11}{2} \div \frac{14}{3}$$

$$x = \cancel{14} \cdot \frac{11}{2} \cdot \frac{3}{\cancel{14}} \qquad \text{Multiplying by the reciprocal of the divisor}$$

$$x = \frac{11 \cdot 3}{2} \qquad \text{Simplifying by removing a factor of 1: } \frac{14}{14} = 1$$

$$x = \frac{33}{2}, \text{ or } 16\frac{1}{2}.$$

The solution is $16\frac{1}{2}$.

Answer on page A-11 *Do Exercise 12.*

CALCULATOR CORNER

Solving Proportions Note in Examples 5–9 that when we solve a proportion, we equate cross products and then we divide on both sides to isolate the variable on one side of the equation. We can use a calculator to do the calculations in this situation. In Example 8, for instance, after equating cross products and dividing by 3.4 on both sides, we have

$$n = \frac{4.93 \times 10}{3.4}.$$

To find n on a calculator, we can press [4] [.] [9] [3] [×] [1] [0] [÷] [3] [.] [4] [=]. The result is 14.5, so $n = 14.5$.

Exercises

1. Use a calculator to solve each of the proportions in Examples 5–7.

2. Use a calculator to solve each of the proportions in Margin Exercises 7–11.

Solve each proportion.

3. $\dfrac{15.75}{20} = \dfrac{a}{35}$

4. $\dfrac{32}{x} = \dfrac{25}{20}$

5. $\dfrac{t}{57} = \dfrac{17}{64}$

6. $\dfrac{71.2}{a} = \dfrac{42.5}{23.9}$

7. $\dfrac{29.6}{3.15} = \dfrac{x}{4.23}$

8. $\dfrac{a}{3.01} = \dfrac{1.7}{0.043}$

EXERCISE SET

For Extra Help

a   Determine whether the two pairs of numbers are proportional.

1. 5, 6 and 7, 9

2. 7, 5 and 6, 4

3. 1, 2 and 10, 20

4. 7, 3 and 21, 9

5. 2.4, 3.6 and 1.8, 2.7

6. 4.5, 3.8 and 6.7, 5.2

7. $5\frac{1}{3}, 8\frac{1}{4}$ and $2\frac{1}{5}, 9\frac{1}{2}$

8. $2\frac{1}{3}, 3\frac{1}{2}$ and 14, 21

Pass-Completion Rates.   The table below lists the records of four NFL quarterbacks from the 2000 season.

PLAYER	TEAM	NUMBER OF PASSES COMPLETED	NUMBER OF PASSES ATTEMPTED	NUMBER OF COMPLETIONS PER ATTEMPT (completion rate)
Tom Brady	New England Patriots	288	474	
Drew Brees	San Diego Chargers	262	400	
Daunte Culpepper	Minnesota Vikings	379	548	
Ben Roethlisberger	Pittsburgh Steelers	196	295	

Source: National Football League

9. Find each pass-completion rate rounded to the nearest hundredth. Are any the same?

10. Use cross products to determine whether any quarterback completion rates are the same.

b   Solve.

11. $\dfrac{18}{4} = \dfrac{x}{10}$

12. $\dfrac{x}{45} = \dfrac{20}{25}$

13. $\dfrac{x}{8} = \dfrac{9}{6}$

14. $\dfrac{8}{10} = \dfrac{n}{5}$

15. $\dfrac{t}{12} = \dfrac{5}{6}$

16. $\dfrac{12}{4} = \dfrac{x}{3}$

17. $\dfrac{2}{5} = \dfrac{8}{n}$

18. $\dfrac{10}{6} = \dfrac{5}{x}$

19. $\dfrac{n}{15} = \dfrac{10}{30}$

20. $\dfrac{2}{24} = \dfrac{x}{36}$

21. $\dfrac{16}{12} = \dfrac{24}{x}$

22. $\dfrac{7}{11} = \dfrac{2}{x}$

23. $\dfrac{6}{11} = \dfrac{12}{x}$

24. $\dfrac{8}{9} = \dfrac{32}{n}$

25. $\dfrac{20}{7} = \dfrac{80}{x}$

26. $\dfrac{5}{x} = \dfrac{4}{10}$

27. $\dfrac{12}{9} = \dfrac{x}{7}$

28. $\dfrac{x}{20} = \dfrac{16}{15}$

29. $\dfrac{x}{13} = \dfrac{2}{9}$

30. $\dfrac{1.2}{4} = \dfrac{x}{9}$

31. $\dfrac{t}{0.16} = \dfrac{0.15}{0.40}$

32. $\dfrac{x}{11} = \dfrac{7.1}{2}$

33. $\dfrac{100}{25} = \dfrac{20}{n}$

34. $\dfrac{35}{125} = \dfrac{7}{m}$

35. $\dfrac{7}{\frac{1}{4}} = \dfrac{28}{x}$

36. $\dfrac{x}{6} = \dfrac{1}{6}$

37. $\dfrac{\frac{1}{4}}{\frac{1}{2}} = \dfrac{\frac{1}{2}}{x}$

38. $\dfrac{1}{7} = \dfrac{x}{4\frac{1}{2}}$

39. $\dfrac{1}{2} = \dfrac{7}{x}$

40. $\dfrac{x}{3} = \dfrac{0}{9}$

41. $\dfrac{\frac{2}{7}}{\frac{3}{4}} = \dfrac{\frac{5}{6}}{y}$

42. $\dfrac{\frac{5}{4}}{\frac{5}{8}} = \dfrac{\frac{3}{2}}{Q}$

43. $\dfrac{2\frac{1}{2}}{3\frac{1}{3}} = \dfrac{x}{4\frac{1}{4}}$

44. $\dfrac{5\frac{1}{5}}{6\frac{1}{6}} = \dfrac{y}{3\frac{1}{2}}$

45. $\dfrac{1.28}{3.76} = \dfrac{4.28}{y}$

46. $\dfrac{10.4}{12.4} = \dfrac{6.76}{t}$

47. $\dfrac{10\frac{3}{8}}{12\frac{2}{3}} = \dfrac{5\frac{3}{4}}{y}$

48. $\dfrac{12\frac{7}{8}}{20\frac{3}{4}} = \dfrac{5\frac{2}{3}}{y}$

49. D_W Instead of equating cross products, a student solves $\frac{x}{7} = \frac{5}{3}$ (see Example 5) by multiplying on both sides by the least common denominator, 21. Is his approach a good one? Why or why not?

50. D_W An instructor predicts that a student's test grade will be proportional to the amount of time the student spends studying. What is meant by this? Write an example of a proportion that involves the grades of two students and their study times.

SKILL MAINTENANCE

✎ VOCABULARY REINFORCEMENT

In each of Exercises 51–58, fill in the blank with the correct term from the given list. Some of the choices may not be used.

51. A ratio is the _____ of two quantities.   [5.1a]

52. A number is divisible by 9 if the _____ of the digits is divisible by 9.   [2.2a]

53. To compute a(n) _____ of a set of numbers, we add the numbers and then divide by the number of addends.   [3.7a]

54. To convert from _____ to _____ , move the decimal point two places to the right and change the $ sign in front to the ¢ sign at the end.   [4.3b]

55. In the equation $103 - 13 = 90$, the _____ is 13.   [1.3a]

56. The decimal 0.125 is an example of a(n) _____ decimal.   [4.5a]

57. The sentence $\frac{2}{5} \cdot \frac{4}{9} = \frac{4}{9} \cdot \frac{2}{5}$ illustrates the _____ law of multiplication.   [1.5a]

58. To solve $\frac{x}{a} = \frac{c}{d}$, equate the _____ and divide on both sides to get x alone.   [5.3b]

cross products

cents

dollars

terminating

repeating

sum

difference

minuend

subtrahend

associative

commutative

average

product

quotient

SYNTHESIS

▦ Solve.

59. $\dfrac{1728}{5643} = \dfrac{836.4}{x}$

60. $\dfrac{328.56}{627.48} = \dfrac{y}{127.66}$

61. *Strikeouts per Home Run.*   Baseball Hall-of-Famer Babe Ruth had 1330 strikeouts and 714 home runs in his career. Hall-of-Famer Mike Schmidt had 1883 strikeouts and 548 home runs in his career.

   a) Find the unit rate of each player in terms of strikeouts per home run. (These rates were considered among the highest in the history of the game and yet each made the Hall of Fame.)

   b) Which player had the higher rate?

Objective

a Solve applied problems involving proportions.

1. Calories Burned. Your author generally exercises for 2 hr each day. The readout on an exercise machine tells him that if he exercises for 24 min, he will burn 356 calories. How many calories will he burn if he exercises for 30 min?

Source: Star Trac Treadmill

5.4 APPLICATIONS OF PROPORTIONS

a Applications and Problem Solving

Proportions have applications in such diverse fields as business, chemistry, health sciences, and home economics, as well as to many areas of daily life. Proportions are useful in making predictions.

EXAMPLE 1 *Predicting Total Distance.* Donna drives her delivery van 800 mi in 3 days. At this rate, how far will she drive in 15 days?

1. **Familiarize.** We let d = the distance traveled in 15 days.

2. **Translate.** We translate to a proportion. We make each side the ratio of distance to time, with distance in the numerator and time in the denominator.

$$\text{Distance in 15 days} \rightarrow \frac{d}{15} = \frac{800}{3} \leftarrow \text{Distance in 3 days}$$
$$\text{Time} \rightarrow \qquad\qquad\qquad \leftarrow \text{Time}$$

It may help to verbalize the proportion above as "the unknown distance d is to 15 days as the known distance 800 miles is to 3 days."

3. **Solve.** Next, we solve the proportion:

$$3 \cdot d = 15 \cdot 800 \qquad \text{Equating cross products}$$

$$\frac{3 \cdot d}{3} = \frac{15 \cdot 800}{3} \qquad \text{Dividing by 3 on both sides}$$

$$d = \frac{15 \cdot 800}{3}$$

$$d = 4000. \qquad \text{Multiplying and dividing}$$

4. **Check.** We substitute into the proportion and check cross products:

$$\frac{4000}{15} = \frac{800}{3};$$

$$4000 \cdot 3 = 12{,}000; \qquad 15 \cdot 800 = 12{,}000.$$

The cross products are the same.

5. **State.** Donna will drive 4000 mi in 15 days.

Do Exercise 1.

Problems involving proportion can be translated in more than one way. In Example 1, any one of the following is a correct translation:

$$\frac{800}{3} = \frac{d}{15}, \qquad \frac{15}{d} = \frac{3}{800}, \qquad \frac{15}{3} = \frac{d}{800}, \qquad \frac{800}{d} = \frac{3}{15}.$$

Equating the cross products in each equation gives us the equation $3 \cdot d = 15 \cdot 800$.

Answer on page A-12

314

CHAPTER 5: Ratio and Proportion

EXAMPLE 2 *Recommended Dosage.* To control a fever, a doctor suggests that a child who weighs 28 kg be given 320 mg of Tylenol. If the dosage is proportional to the child's weight, how much Tylenol is recommended for a child who weighs 35 kg?

1. **Familiarize.** We let t = the number of milligrams of Tylenol.

2. **Translate.** We translate to a proportion, keeping the amount of Tylenol in the numerators.

$$\text{Tylenol suggested} \rightarrow \frac{320}{28} = \frac{t}{35} \leftarrow \text{Tylenol suggested}$$
$$\text{Child's weight} \rightarrow \qquad\qquad \leftarrow \text{Child's weight}$$

3. **Solve.** Next, we solve the proportion:

$$320 \cdot 35 = 28 \cdot t \qquad \text{Equating cross products}$$

$$\frac{320 \cdot 35}{28} = \frac{28 \cdot t}{28} \qquad \text{Dividing by 28 on both sides}$$

$$\frac{320 \cdot 35}{28} = t$$

$$400 = t. \qquad \text{Multiplying and dividing}$$

4. **Check.** We substitute into the proportion and check cross products:

$$\frac{320}{28} = \frac{400}{35};$$

$$320 \cdot 35 = 11{,}200; \qquad 28 \cdot 400 = 11{,}200.$$

The cross products are the same.

5. **State.** The dosage for a child who weighs 35 kg is 400 mg.

Do Exercise 2.

EXAMPLE 3 *Purchasing Tickets.* Carey bought 8 tickets to an international food festival for $52. How many tickets could she purchase with $90?

1. **Familiarize.** We let n = the number of tickets that can be purchased with $90.

2. **Translate.** We translate to a proportion, keeping the number of tickets in the numerators.

$$\text{Tickets} \rightarrow \frac{8}{52} = \frac{n}{90} \leftarrow \text{Tickets}$$
$$\text{Cost} \rightarrow \qquad\qquad \leftarrow \text{Cost}$$

2. Determining Paint Needs.
Lowell and Chris run a summer painting company to support their college expenses. They can paint 1600 ft^2 of clapboard with 4 gal of paint. How much paint would be needed for a building with 6000 ft^2 of clapboard?

Answer on page A-12

3. Purchasing Shirts. If 2 shirts can be bought for $47, how many shirts can be bought with $200?

3. Solve. Next, we solve the proportion:

$$52 \cdot n = 8 \cdot 90 \qquad \text{Equating cross products}$$

$$\frac{52 \cdot n}{52} = \frac{8 \cdot 90}{52} \qquad \text{Dividing by 52 on both sides}$$

$$n = \frac{8 \cdot 90}{52}$$

$$n \approx 13.8. \qquad \text{Multiplying and dividing}$$

Because it is impossible to buy a fractional part of a ticket, we must round our answer *down* to 13.

4. Check. As a check, we use a different approach: We find the cost per ticket and then divide $90 by that price. Since $52 \div 8 = 6.50$ and $90 \div 6.50 \approx 13.8$, we have a check.

5. State. Carey could purchase 13 tickets with $90.

Do Exercise 3.

EXAMPLE 4 *Women's Hip Measurements.* For improved health, it is recommended that a woman's waist-to-hip ratio be 0.85 (or lower). Marta's hip measurement is 40 in. To meet the recommendation, what should Marta's waist measurement be?

Source: David Schmidt, "Lifting Weight Myths," *Nutrition Action Newsletter* 20, no. 4, October 1993

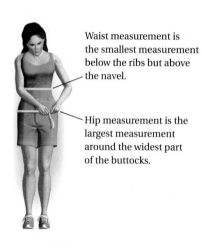

Waist measurement is the smallest measurement below the ribs but above the navel.

Hip measurement is the largest measurement around the widest part of the buttocks.

4. Men's Hip Measurements. It is recommended that a man's waist-to-hip ratio be 0.95 (or lower). Malcolm's hip measurement is 40 in. To meet the recommendation, what should Malcolm's waist measurement be?

Source: David Schmidt, "Lifting Weight Myths," *Nutrition Action Newsletter* 20, no. 4, October 1993

1. Familiarize. Note that $0.85 = \frac{85}{100}$. We let w = Marta's waist measurement.

2. Translate. We translate to a proportion as follows:

$$\begin{array}{ll} \text{Waist measurement} \rightarrow & \dfrac{w}{40} = \dfrac{85}{100}. \\ \text{Hip measurement} \rightarrow & \end{array} \quad \begin{array}{l} \text{Recommended} \\ \text{waist-to-hip ratio} \end{array}$$

3. Solve. Next, we solve the proportion:

$$100 \cdot w = 40 \cdot 85 \qquad \text{Equating cross products}$$

$$\frac{100 \cdot w}{100} = \frac{40 \cdot 85}{100} \qquad \text{Dividing by 100 on both sides}$$

$$w = \frac{40 \cdot 85}{100}$$

$$w = 34. \qquad \text{Multiplying and dividing}$$

Answers on page A-12

4. Check. As a check, we divide 34 by 40: $34 \div 40 = 0.85$. This is the desired ratio.

5. State. Marta's recommended waist measurement is 34 in. (or less).

Do Exercise 4 on the preceding page.

5. Construction Plans. In Example 5, the length of the actual deck is 28.5 ft. What is the length of the deck on the blueprints?

EXAMPLE 5 *Construction Plans.* Architects make blueprints of projects being constructed. These are scale drawings in which lengths are in proportion to actual sizes. The Hennesseys are constructing a rectangular deck just outside their house. The architectural blueprints are rendered such that $\frac{3}{4}$ in. on the drawing is actually 2.25 ft on the deck. The width of the deck on the drawing is 4.3 in. How wide is the deck in reality?

1. Familiarize. We let w = the width of the deck.

2. Translate. Then we translate to a proportion, using 0.75 for $\frac{3}{4}$ in.

Measure on drawing $\rightarrow \dfrac{0.75}{2.25} = \dfrac{4.3}{w}$ $\leftarrow$ Width of drawing
Measure on deck $\rightarrow$ $\phantom{\dfrac{0.75}{2.25}}$ $\leftarrow$ Width of deck

3. Solve. Next, we solve the proportion:

$0.75 \times w = 2.25 \times 4.3$ Equating cross products

$\dfrac{0.75 \times w}{0.75} = \dfrac{2.25 \times 4.3}{0.75}$ Dividing by 0.75 on both sides

$w = \dfrac{2.25 \times 4.3}{0.75}$

$w = 12.9.$

4. Check. We substitute into the proportion and check cross products:

$\dfrac{0.75}{2.25} = \dfrac{4.3}{12.9};$

$0.75 \times 12.9 = 9.675;$ $2.25 \times 4.3 = 9.675.$

The cross products are the same.

5. State. The width of the deck is 12.9 ft.

Do Exercise 5.

Answer on page A-12

6. Estimating a Deer Population.
To determine the number of deer in a forest, a conservationist catches 612 deer, tags them, and releases them. Later, 244 deer are caught, and it is found that 72 of them are tagged. Estimate how many deer are in the forest.

EXAMPLE 6   *Estimating a Wildlife Population.*   To determine the number of fish in a lake, a conservationist catches 225 fish, tags them, and throws them back into the lake. Later, 108 fish are caught, and it is found that 15 of them are tagged. Estimate how many fish are in the lake.

1. **Familiarize.**   We let F = the number of fish in the lake.
2. **Translate.**   We translate to a proportion as follows:

$$\text{Fish tagged originally} \rightarrow \frac{225}{F} = \frac{15}{108} \leftarrow \text{Tagged fish caught later} \atop \leftarrow \text{Fish caught later}$$
$$\text{Fish in lake} \rightarrow$$

3. **Solve.**   Next, we solve the proportion:

$$225 \cdot 108 = F \cdot 15 \qquad \text{Equating cross products}$$

$$\frac{225 \cdot 108}{15} = \frac{F \cdot 15}{15} \qquad \text{Dividing by 15 on both sides}$$

$$\frac{225 \cdot 108}{15} = F$$

$$1620 = F. \qquad \text{Multiplying and dividing}$$

4. **Check.**   We substitute into the proportion and check cross products:

$$\frac{225}{1620} = \frac{15}{108};$$

$$225 \cdot 108 = 24{,}300; \qquad 1620 \cdot 15 = 24{,}300.$$

The cross products are the same.

5. **State.**   We estimate that there are 1620 fish in the lake.

Do Exercise 6.

Answer on page A-12

CHAPTER 5: Ratio and Proportion

Translating for Success

1. *Calories in Cereal.* There are 140 calories in a $1\frac{1}{2}$-cup serving of Brand A cereal. How many calories are there in 6 cups of the cereal?

2. *Calories in Cereal.* There are 140 calories in 6 cups of Brand B cereal. How many calories are there in a $1\frac{1}{2}$-cup serving of the cereal?

3. *Gallons of Gasoline.* Jared's SUV traveled 310 miles on 15.5 gallons of gasoline. At this rate, how many gallons would be needed to travel 465 miles?

4. *Gallons of Gasoline.* Elizabeth's new fuel-efficient car traveled 465 miles on 15.5 gallons of gasoline. At this rate, how many gallons will be needed to travel 310 miles?

5. *Perimeter.* Find the perimeter of a rectangular field that measures 83.7 m by 62.4 m.

The goal of these matching questions is to practice step (2), *Translate*, of the five-step problem-solving process. Translate each word problem to an equation and select a correct translation from equations A–O.

A. $\dfrac{310}{15.5} = \dfrac{465}{x}$

B. $180 = 1\frac{1}{2} \cdot x$

C. $x = 71\frac{1}{8} - 76\frac{1}{2}$

D. $71\frac{1}{8} \cdot x = 74$

E. $74 \cdot 71\frac{1}{8} = x$

F. $x = 83.7 + 62.4$

G. $71\frac{1}{8} + x = 76\frac{1}{2}$

H. $x = 1\frac{2}{3} \cdot 180$

I. $\dfrac{140}{6} = \dfrac{x}{1\frac{1}{2}}$

J. $x = 2(83.7 + 62.4)$

K. $\dfrac{465}{15.5} = \dfrac{310}{x}$

L. $x = 83.7 \cdot 62.4$

M. $x = 180 \div 1\frac{2}{3}$

N. $\dfrac{140}{1\frac{1}{2}} = \dfrac{x}{6}$

O. $x = 1\frac{2}{3} \div 180$

Answers on page A-12

6. *Electric Bill.* Last month Todd's electric bills for his two rentals were $83.70 and $62.40. What was the total electric bill for the two properties?

7. *Package Tape.* A postal service center uses rolls of package tape that each contain 180 feet of tape. If it takes an average of $1\frac{2}{3}$ ft per package, how many packages can be taped with one roll?

8. *Online Price.* Jane spent $180 for an area rug in a department store. Later she saw the same rug for sale online and realized she had paid $1\frac{1}{2}$ times the online price. What was the online price?

9. *Heights of Sons.* Henry's three sons play basket-ball on three different college teams. Jeff, Jason, and Jared's heights are 74 in., $71\frac{1}{8}$ in., and $76\frac{1}{2}$ in., respectively. How much taller is Jared than Jason?

10. *Total Investment.* An investor bought 74 shares of stock at $71\frac{1}{8}$ per share. What was the total investment?

a   Solve.

1. *Study Time and Test Grades.* An English instructor asserted that students' test grades are directly proportional to the amount of time spent studying. Lisa studies 9 hr for a particular test and gets a score of 75. At this rate, how many hours would she have had to study to get a score of 92?

2. *Study Time and Test Grades.* A mathematics instructor asserted that students' test grades are directly proportional to the amount of time spent studying. Brent studies 15 hr for a particular test and gets a score of 85. At this rate, what score would he have received if he had studied 16 hr?

3. *Cap'n Crunch's Peanut Butter Crunch® Cereal.* The nutritional chart on the side of a box of Quaker Cap'n Crunch's Peanut Butter Crunch® Cereal states that there are 110 calories in a $\frac{3}{4}$-cup serving. How many calories are there in 6 cups of the cereal?

4. *Rice Krispies® Cereal.* The nutritional chart on the side of a box of Kellogg's Rice Krispies® Cereal states that there are 120 calories in a $1\frac{1}{4}$-cup serving. How many calories are there in 5 cups of the cereal?

5. *Overweight Americans.* A study recently confirmed that of every 100 Americans, 60 are considered overweight. There were 295 million Americans in 2005. How many would be considered overweight?
Source: U.S. Centers for Disease Control

6. *Cancer Death Rate in Illinois.* It is predicted that for every 1000 people in the state of Illinois, 130.9 will die of cancer. The population of Chicago is about 2,721,547. How many of these people will die of cancer?
Source: *2001 New York Times Almanac*

7. *Gasoline Mileage.* Nancy's van traveled 84 mi on 6.5 gal of gasoline. At this rate, how many gallons would be needed to travel 126 mi?

8. *Bicycling.* Roy bicycled 234 mi in 14 days. At this rate, how far would Roy travel in 42 days?

9. *Quality Control.* A quality-control inspector examined 100 lightbulbs and found 7 of them to be defective. At this rate, how many defective bulbs will there be in a lot of 2500?

10. *Grading.* A professor must grade 32 essays in a literature class. She can grade 5 essays in 40 min. At this rate, how long will it take her to grade all 32 essays?

11. *Painting.* Fred uses 3 gal of paint to cover 1275 ft^2 of siding. How much siding can Fred paint with 7 gal of paint?

12. *Waterproofing.* Bonnie can waterproof 450 ft^2 of decking with 2 gal of sealant. How many gallons should Bonnie buy for a 1200-ft^2 deck?

13. *Publishing.* Every 6 pages of an author's manuscript corresponds to 5 published pages. How many published pages will a 540-page manuscript become?

14. *Turkey Servings.* An 8-lb turkey breast contains 36 servings of meat. How many pounds of turkey breast would be needed for 54 servings?

15. *Exchanging Money.* On 26 April 2005, 1 U.S. dollar was worth about 0.52521 British pound.

   a) How much would 45 U.S. dollars be worth in British pounds?

   b) How much would a car cost in U.S. dollars that costs 8640 British pounds?

16. *Exchanging Money.* On 26 April 2005, 1 U.S. dollar was worth about 1.1894 Swiss francs.

   a) How much would 360 U.S. dollars be worth in Swiss francs?

   b) How much would a pair of jeans cost in U.S. dollars that costs 80 Swiss francs?

17. *Exchanging Money.* On 26 April 2005, 1 U.S. dollar was worth about 106.03 Japanese yen.

   a) How much would 200 U.S. dollars be worth in Japanese yen?

   b) Dan was traveling in Japan and bought a skateboard that cost 3180 Japanese yen. How much would it cost in U.S. dollars?

18. *Exchanging Money.* On 26 April 2005, 1 U.S. dollar was worth about 11.059 Mexican pesos.

   a) How much would 120 U.S. dollars be worth in Mexican pesos?

   b) Jackie was traveling in Mexico and bought a watch that cost 3600 Mexican pesos. How much would it cost in U.S. dollars?

19. *Gas Mileage.* A 2005 Ford Mustang GT Convertible will go 372 mi on 15.5 gal of gasoline in highway driving.

a) How many gallons of gasoline will it take to drive 2690 mi from Boston to Phoenix?
b) How far can the car be driven on 140 gal of gasoline?

Source: Ford

20. *Gas Mileage.* A 2005 Volkswagen Passat will go 462 mi on 16.5 gal of gasoline in highway driving.

a) How many gallons of gasoline will it take to drive 1650 mi from Pittsburgh to Albuquerque?
b) How far can the car be driven on 130 gal of gasoline?

Source: Volkswagen of America, Inc.

21. *Lefties.* In a class of 40 students, on average, 6 will be left-handed. If a class includes 9 "lefties," how many students would you estimate are in the class?

22. *Sugaring.* When 38 gal of maple sap are boiled down, the result is 2 gal of maple syrup. How much sap is needed to produce 9 gal of syrup?

23. *Mileage.* Jean bought a new car. In the first 8 months, it was driven 9000 mi. At this rate, how many miles will the car be driven in 1 yr?

24. *Coffee Production.* Coffee beans from 14 trees are required to produce the 17 lb of coffee that the average person in the United States drinks each year. How many trees are required to produce 375 lb of coffee?

25. *Metallurgy.* In a metal alloy, the ratio of zinc to copper is 3 to 13. If there are 520 lb of copper, how many pounds of zinc are there?

26. *Class Size.* A college advertises that its student-to-faculty ratio is 14 to 1. If 56 students register for Introductory Spanish, how many sections of the course would you expect to see offered?

27. *Painting.* Helen can paint 950 ft² with 2 gal of paint. How many 1-gal cans does she need in order to paint a 30,000-ft² wall?

28. *Snow to Water.* Under typical conditions, $1\frac{1}{2}$ ft of snow will melt to 2 in. of water. To how many inches of water will $5\frac{1}{2}$ ft of snow melt?

29. *Grass-Seed Coverage.* It takes 60 oz of grass seed to seed 3000 ft² of lawn. At this rate, how much would be needed for 5000 ft² of lawn?

30. *Grass-Seed Coverage.* In Exercise 29, how much seed would be needed for 7000 ft² of lawn?

31. *Estimating a Deer Population.* To determine the number of deer in a game preserve, a forest ranger catches 318 deer, tags them, and releases them. Later, 168 deer are caught, and it is found that 56 of them are tagged. Estimate how many deer are in the game preserve.

32. *Estimating a Trout Population.* To determine the number of trout in a lake, a conservationist catches 112 trout, tags them, and throws them back into the lake. Later, 82 trout are caught, and it is found that 32 of them are tagged. Estimate how many trout there are in the lake.

33. *Map Scaling.* On a road atlas map, 1 in. represents 16.6 mi. If two cities are 3.5-in. apart on the map, how far apart are they in reality?

34. *Map Scaling.* On a map, $\frac{1}{4}$ in. represents 50 mi. If two cities are $3\frac{1}{4}$-in. apart on the map, how far apart are they in reality?

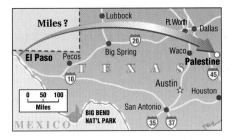

35. *Points per Game.* At one point in the 2000–2001 NBA season, Allen Iverson of the Philadelphia 76ers had scored 884 points in 33 games.

a) At this rate, how many games would it take him to score 1500 points?

b) There are 82 games in an entire NBA season. At this rate, how many points would Iverson score in the entire season?

Source: National Basketball Association

36. *Points per Game.* At one point in the 2000–2001 NBA season, Shaquille O'Neal, then of the Los Angeles Lakers, had scored 826 points in 32 games.

a) At this rate, how many games would it take him to score 2000 points?

b) There are 82 games in an entire NBA season. At this rate, how many points would O'Neal score in the entire season?

37. **D**w Can unit prices be used to solve proportions that involve money? Explain why or why not.

38. **D**w *Earned Run Average.* In baseball, the average number of runs given up by a pitcher in nine innings is his *earned run average,* or *ERA.* Set up a formula for determining a player's ERA. Then verify it using the fact that in the 2000 season, Daryl Kile of the St. Louis Cardinals gave up 101 earned runs in $232\frac{1}{3}$ innings to compile an ERA of 3.91. Then use your formula to determine the ERA of Randy Johnson of the Arizona Diamondbacks, who gave up 73 earned runs in $248\frac{2}{3}$ innings. Is a low ERA considered good or bad?

Source: Major League Baseball

> **SKILL MAINTENANCE**

Determine whether each number is prime, composite, or neither.   [2.1c]

39. 1   **40.** 28   **41.** 83   **42.** 93   **43.** 47

Find the prime factorization of each number.   [2.1d]

44. 808   **45.** 28   **46.** 866   **47.** 93   **48.** 2020

> **SYNTHESIS**

49. 🖩 Carney College is expanding from 850 to 1050 students. To avoid any rise in the student-to-faculty ratio, the faculty of 69 professors must also increase. How many new faculty positions should be created?

50. 🖩 In recognition of her outstanding work, Sheri's salary has been increased from $26,000 to $29,380. Tim is earning $23,000 and is requesting a proportional raise. How much more should he ask for?

51. *Baseball Statistics.* Cy Young, one of the greatest baseball pitchers of all time, gave up an average of 2.63 earned runs every 9 innings. Young pitched 7356 innings, more than anyone in the history of baseball. How many earned runs did he give up?

52. 🖩 *Real-Estate Values.* According to Coldwell Banker Real Estate Corporation, a home selling for $189,000 in Austin, Texas, would sell for $665,795 in San Francisco. How much would a $450,000 home in San Francisco sell for in Austin? Round to the nearest $1000.

Source: Coldwell Banker Real Estate Corporation

53. 🖩 The ratio 1:3:2 is used to estimate the relative costs of a CD player, receiver, and speakers when shopping for a stereo. That is, the receiver should cost three times the amount spent on the CD player and the speakers should cost twice as much as the amount spent on the CD player. If you had $900 to spend, how would you allocate the money, using this ratio?

324

5.5   GEOMETRIC APPLICATIONS

a   Proportions and Similar Triangles

Look at the pair of triangles below. Note that they appear to have the same shape, but their sizes are different. These are examples of **similar triangles.** By using a magnifying glass, you could imagine enlarging the smaller triangle to get the larger. This process works because the corresponding sides of each triangle have the same ratio. That is, the following proportion is true.

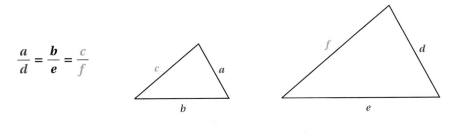

$$\frac{a}{d} = \frac{b}{e} = \frac{c}{f}$$

1. This pair of triangles is similar. Find the missing length x.

12 | 20
24

9 | x
18

$$\frac{20}{12} \quad \frac{24}{18} \quad \frac{x}{9}$$

SIMILAR TRIANGLES

Similar triangles have the same shape. The lengths of their corresponding sides have the same ratio—that is, they are proportional.

EXAMPLE 1   The triangles below are similar triangles. Find the missing length x.

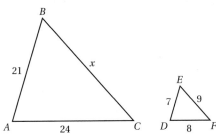

The ratio of x to 9 is the same as the ratio of 24 to 8 or 21 to 7. We get the proportions

$$\frac{x}{9} = \frac{24}{8} \quad \text{and} \quad \frac{x}{9} = \frac{21}{7}.$$

We can solve either one of these proportions. We use the first:

$$\frac{x}{9} = \frac{24}{8}$$

$x \cdot 8 = 24 \cdot 9$ Equating cross products

$$\frac{x \cdot 8}{8} = \frac{24 \cdot 9}{8}$$ Dividing by 8 on both sides

$x = 27.$ Simplifying

The missing length x is 27. Other proportions could also be used.

Answer on page A-12

Do Exercise 1.

325

2. How high is a flagpole that casts a 45-ft shadow at the same time that a 5.5-ft woman casts a 10-ft shadow?

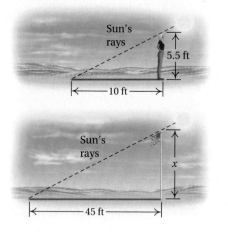

Sun's rays

5.5 ft

|← 10 ft →|

Sun's rays

x

|← 45 ft →|

Similar triangles and proportions can often be used to find lengths that would ordinarily be difficult to measure. For example, we could find the height of a flagpole without climbing it or the distance across a river without crossing it.

EXAMPLE 2 How high is a flagpole that casts a 56-ft shadow at the same time that a 6-ft man casts a 5-ft shadow?

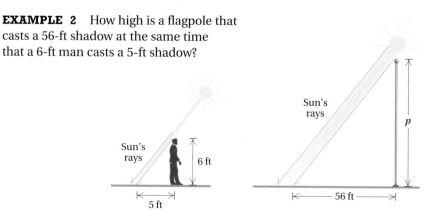

Sun's rays

6 ft

|← 5 ft →|

Sun's rays

p

|← 56 ft →|

If we use the sun's rays to represent the third side of the triangle in our drawing of the situation, we see that we have similar triangles. Let p = the height of the flagpole. The ratio of 6 to p is the same as the ratio of 5 to 56. Thus we have the proportion

Height of man → $\dfrac{6}{p} = \dfrac{5}{56}$. ← Length of shadow of man
Height of pole →   ← Length of shadow of pole

Solve: $6 \cdot 56 = 5 \cdot p$ Equating cross products

$\dfrac{6 \cdot 56}{5} = \dfrac{5 \cdot p}{5}$ Dividing by 5 on both sides

$\dfrac{6 \cdot 56}{5} = p$ Simplifying

$67.2 = p$

The height of the flagpole is 67.2 ft.

Do Exercise 2.

EXAMPLE 3 *Rafters of a House.* Carpenters use similar triangles to determine the lengths of rafters for a house. They first choose the pitch of the roof, or the ratio of the rise over the run. Then using a triangle with that ratio, they calculate the length of the rafter needed for the house. Loren is constructing rafters for a roof with a 6/12 pitch on a house that is 30 ft wide. Using a rafter guide, Loren knows that the rafter length corresponding to the 6/12 pitch is 13.4. Find the length x of the rafter of the house to the nearest tenth of a foot.

13.4

Rise: 6

Run: 12

Pitch: 6/12

x

y

15 ft

30 ft

Answer on page A-12

Thus we have the proportion

Length of rafter
in 6/12 triangle → $\dfrac{13.4}{x} = \dfrac{12}{15}$ ← Run in 6/12 triangle
Length of rafter → ← Run in similar
on the house   triangle on the house

Solve: $13.4 \cdot 15 = x \cdot 12$ Equating cross products

$\dfrac{13.4 \cdot 15}{12} = \dfrac{x \cdot 12}{12}$ Dividing by 12 on both sides

$\dfrac{13.4 \cdot 15}{12} = x$

$16.8 \text{ ft} \approx x$ Rounding to the nearest tenth of a foot

The length of the rafter x of the house is about 16.8 ft.

Do Exercise 3.

3. Rafters of a House. Referring to Example 3, find the length y in the rafter of the house to the nearest tenth of a foot.

b   Proportions and Other Geometric Shapes

When one geometric figure is a magnification of another, the figures are similar. Thus the corresponding lengths are proportional.

EXAMPLE 4 The sides in the negative and photograph below are proportional. Find the width of the photograph.

2.5 cm |←3.5 cm→| |←——— 10.5 cm ———→|

We let $x =$ the width of the photograph. Then we translate to a proportion.

Photo width → $\dfrac{x}{2.5} = \dfrac{10.5}{3.5}$ ← Photo length
Negative width →   ← Negative length

Solve: $3.5 \times x = 2.5 \times 10.5$ Equating cross products

$\dfrac{3.5 \times x}{3.5} = \dfrac{2.5 \times 10.5}{3.5}$ Dividing by 3.5 on both sides

$x = \dfrac{2.5 \times 10.5}{3.5}$ Simplifying

$x = 7.5$

Thus the width of the photograph is 7.5 cm.

Do Exercise 4.

4. The sides in the photographs below are proportional. Find the width of the larger photograph.

6 cm |←10 cm→|

35 cm

Answers on page A-12

5. Refer to the figures in Example 5. If a model skylight is 3 cm wide, how wide will the actual skylight be?

EXAMPLE 5 A scale model of an addition to an athletic facility is 12 cm wide at the base and rises to a height of 15 cm. If the actual base is to be 116 ft, what will be the actual height of the addition?

We let h = the height of the addition. Then we translate to a proportion.

$$\text{Width in model} \rightarrow \frac{12}{116} = \frac{15}{h} \leftarrow \text{Height in model} \\ \text{Actual width} \rightarrow \qquad \qquad \leftarrow \text{Actual height}$$

Solve: $12 \cdot h = 116 \cdot 15$ Equating cross products

$$\frac{12 \cdot h}{12} = \frac{116 \cdot 15}{12}$$ Dividing by 12 on both sides

$$h = \frac{116 \cdot 15}{12} = 145$$

Thus the height of the addition will be 145 ft.

Do Exercise 5.

Answer on page A-12

Study Tips

TIME MANAGEMENT (PART 3)

Here are some additional tips to help you with time management. (See also the Study Tips on time management in Sections 1.5 and 5.1.)

■ **Are you a morning or an evening person?** If you are an evening person, it might be best to avoid scheduling early-morning classes. If you are a morning person, do the opposite, but go to bed earlier to compensate. Nothing can drain your study time and effectiveness like fatigue.

■ **Keep on schedule.** Your course syllabus provides a plan for the semester's schedule. Use a write-on calendar, daily planner, laptop computer, or personal digital assistant to outline your time for the semester. Be sure to note deadlines involving term papers and exams so you can begin a task early, breaking it down into smaller segments that can be accomplished more easily.

■ **Balance your class schedule.** You may be someone who prefers large blocks of time for study on the off days. In that case, it might be advantageous for you to take courses that meet only three days a week. Keep in mind, however, that this might be a problem when tests in more than one course are scheduled for the same day.

5.5

EXERCISE SET   For Extra Help

Math XL   MyMathLab   InterAct   Math Tutor   Digital Video   Student's
                        Math   Center   Tutor CD 3   Solutions
MathXL   MyMathLab   Videotape 5   Manual

a.   The triangles in each exercise are similar. Find the missing lengths.

1.

2.

3.

4.

5.

6.

7.

8.

9. When a tree 8 m high casts a shadow 5 m long, how long a shadow is cast by a person 2 m tall?

10. How high is a flagpole that casts a 42-ft shadow at the same time that a $5\frac{1}{2}$-ft woman casts a 7-ft shadow?

11. How high is a tree that casts a 27-ft shadow at the same time that a 4-ft fence post casts a 3-ft shadow?

12. How high is a tree that casts a 32-ft shadow at the same time that an 8-ft light pole casts a 9-ft shadow?

13. Find the height h of the wall.

6 ft

h

6 ft

7 ft

14. Find the length L of the lake. Assume that the ratio of L to 120 yd is the same as the ratio of 720 yd to 30 yd.

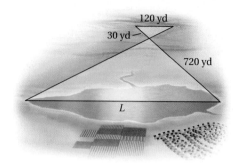

120 yd

30 yd

720 yd

L

15. Find the distance across the river. Assume that the ratio of d to 25 ft is the same as the ratio of 40 ft to 10 ft.

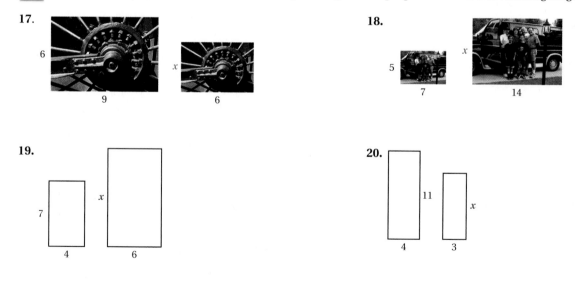

10 ft

25 ft

40 ft

d

16. To measure the height of a hill, a string is drawn tight from level ground to the top of the hill. A 3-ft stick is placed under the string, touching it at point P, a distance of 5 ft from point G, where the string touches the ground. The string is then detached and found to be 120 ft long. How high is the hill?

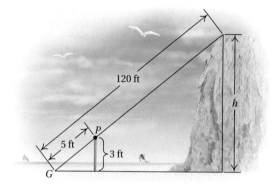

120 ft

P

5 ft

3 ft

G

h

---

b In each of Exercises 17–26, the sides in each pair of figures are proportional. Find the missing lengths.

17.

6

9

x

6

18.

5

7

x

14

19.

7

4

x

6

20.

11

4

x

3

21.

22.

23.

24.

25.

26.

27. A scale model of an addition to an athletic facility is 15 cm wide at the base and rises to a height of 19 cm. If the actual base is to be 120 ft, what will be the height of the addition?

28. Refer to the figures in Exercise 27. If a model skylight is 3 cm wide, how wide will the actual skylight be?

29. **D$_W$** Is it possible for two triangles to have two pairs of sides that are proportional without the triangles being similar? Why or why not?

30. **D$_W$** Design for a classmate a problem involving similar triangles for which

$$\frac{18}{128.95} = \frac{x}{789.89}.$$

31. *Expense Needs.* A student has $34.97 to spend for a book at $49.95, a CD at $14.88, and a sweatshirt at $29.95. How much more money does the student need to make these purchases? [4.7a]

32. Divide: 80.892 ÷ 8.4. [4.4a]

Multiply. [4.3a]

33. 8.4 × 80.892

34. 0.01 × 274.568

35. 100 × 274.568

36. 0.002 × 274.568

Find decimal notation and round to the nearest thousandth, if appropriate. [4.5a, b]

37. $\frac{17}{20}$

38. $\frac{73}{40}$

39. $\frac{10}{11}$

40. $\frac{43}{51}$

Hockey Goals. An official hockey goal is 6 ft wide. To make scoring more difficult, goalies often locate themselves far in front of the goal to "cut down the angle." In Exercises 41 and 42, suppose that a slapshot from point *A* is attempted and that the goalie is 2.7 ft wide. Determine how far from the goal the goalie should be located if point A is the given distance from the goal. (*Hint*: First find how far the goalie should be from point *A*.)

41. 🖩 25 ft

42. 🖩 35 ft

43. A miniature basketball hoop is built for the model referred to in Exercise 27. An actual hoop is 10 ft high. How high should the model hoop be?

🖩 Solve. Round the answer to the nearest thousandth.

44. $\frac{8664.3}{10,344.8} = \frac{x}{9776.2}$

45. $\frac{12.0078}{56.0115} = \frac{789.23}{y}$

🖩 The triangles in each exercise are similar triangles. Find the lengths not given.

46.

47.

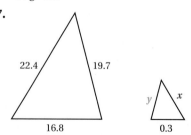

The review that follows is meant to prepare you for a chapter exam. It consists of two parts. The first part, Concept Reinforcement, is designed to increase understanding of the concepts through true/false exercises. The second part is the Review Exercises. These provide practice exercises for the exam, together with references to section objectives so you can go back and review. Before beginning, stop and look back over the skills you have obtained. What skills in mathematics do you have now that you did not have before studying this chapter?

🖑 CONCEPT REINFORCEMENT

Determine whether the statement is true or false. Answers are given at the back of the book.

_____ **1.** The proportion $\dfrac{a}{b} = \dfrac{c}{d}$ can also be written as $\dfrac{c}{a} = \dfrac{d}{b}$.

_____ **2.** Lengths of corresponding sides of similar triangles have the same ratio.

_____ **3.** The larger size of a product always has the lower unit price.

_____ **4.** If $\dfrac{x}{t} = \dfrac{y}{s}$, then $xy = ts$.

_____ **5.** A ratio is a quotient of two quantities.

Review Exercises

Write fraction notation for the ratio. Do not simplify. [5.1a]

1. 47 to 84

2. 46 to 1.27

3. 83 to 100

4. 0.72 to 197

5. _Kona Jack's Restaurants._ Kona Jack's is a seafood restaurant chain in Indianapolis. Each year they sell 12,480 lb of tuna and 16,640 lb of salmon. [5.1a]

a) Write fraction notation for the ratio of tuna sold to salmon sold.

b) Write fraction notation for the ratio of salmon sold to the total number of pounds of both kinds of fish.

Source: Kona Jack's Restaurants

Find the ratio of the first number to the second number and simplify. [5.1b]

6. 9 to 12

7. 3.6 to 6.4

8. _Gas Mileage._ The Chrysler PT Cruiser will go 377 mi on 14.5 gal of gasoline in highway driving. What is the rate in miles per gallon? [5.2a]

Source: DaimlerChrysler Corporation

9. _CD-ROM Spin Rate._ A 12x CD-ROM on a computer will spin 472,500 revolutions if left running for 75 min. What is the rate of its spin in revolutions per minute (rpm)? [5.2a]

Source: Electronic Engineering Times, June 1997

10. A lawn requires 319 gal of water for every 500 ft². What is the rate in gallons per square foot? [5.2a]

11. _Turkey Servings._ A 25-lb turkey serves 18 people. Find the rate in servings per pound. [5.2a]

12. _Calcium Supplement._ The price for a particular calcium supplement is $12.99 for 300 tablets. Find the unit price in cents per tablet. [5.2b]

13. *Pillsbury Orange Breakfast Rolls.* The price for these breakfast rolls is $1.97 for 13.9 oz. Find the unit price in cents per ounce. [5.2b]

In each of Exercises 14 and 15, find the unit prices. Then determine in each case which has the lowest unit price. [5.2b]

14. *Paper Towels.*

PACKAGE	PRICE	UNIT PRICE PER SHEET
8 rolls, 60 (2 ply) sheets per roll	$6.38	
15 rolls, 60 (2 ply) sheets per roll	$13.99	
6 big rolls, 165 (2 ply) sheets per roll	$10.99	

15. *Crisco Oil.*

PACKAGE	PRICE	UNIT PRICE
32 oz	$2.19	
48 oz	$2.49	
64 oz	$3.59	
128 oz	$7.09	

Determine whether the two pairs of numbers are proportional. [5.3a]

16. 9, 15 and 36, 59

17. 24, 37 and 40, 46.25

Solve. [5.3b]

18. $\dfrac{8}{9} = \dfrac{x}{36}$

19. $\dfrac{6}{x} = \dfrac{48}{56}$

20. $\dfrac{120}{\frac{3}{7}} = \dfrac{7}{x}$

21. $\dfrac{4.5}{120} = \dfrac{0.9}{x}$

Solve. [5.4a]

22. If 3 dozen eggs cost $2.67, how much will 5 dozen eggs cost?

23. *Quality Control.* A factory manufacturing computer circuits found 39 defective circuits in a lot of 65 circuits. At this rate, how many defective circuits can be expected in a lot of 585 circuits?

24. *Exchanging Money.* On 5 August 2005, 1 U.S. dollar was worth about 0.808 European Monetary Unit (Euro).
   a) How much would 250 U.S. dollars be worth in Euros?
   b) Jamal was traveling in France and saw a sweatshirt that cost 50 Euros. How much would it cost in U.S. dollars?

25. A train travels 448 mi in 7 hr. At this rate, how far will it travel in 13 hr?

26. Fifteen acres are required to produce 54 bushels of tomatoes. At this rate, how many acres are required to produce 97.2 bushels of tomatoes?

27. *Garbage Production.* It is known that 5 people produce 13 kg of garbage in one day. San Diego, California, has 1,266,753 people. How many kilograms of garbage are produced in San Diego in one day?

28. *Snow to Water.* Under typical conditions, $1\frac{1}{2}$ ft of snow will melt to 2 in. of water. To how many inches of water will $4\frac{1}{2}$ ft of snow melt?

29. *Lawyers in Michigan.* In Michigan, there are 2.3 lawyers for every 1000 people. The population of Detroit is 911,402. How many lawyers would you expect there to be in Detroit?

Source: U.S. Bureau of the Census

Each pair of triangles in Exercises 30 and 31 is similar. Find the missing length(s). [5.5a]

30.

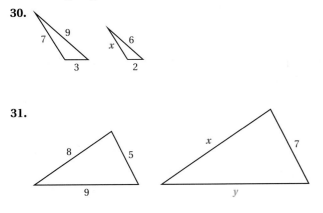

31.

32. How high is a billboard that casts a 25-ft shadow at the same time that an 8-ft sapling casts a 5-ft shadow? [5.5a]

33. The lengths in the figures below are proportional. Find the missing lengths. [5.5b]

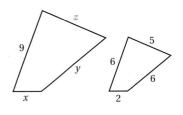

34. **D_W** If you were a college president, which would you prefer: a low or high faculty-to-student ratio? Why? [5.1a]

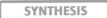

35. **D_W** Write a proportion problem for a classmate to solve. Design the problem so that the solution is "Leslie would need 16 gal of gasoline in order to travel 368 mi." [5.4a]

SYNTHESIS

36. It takes Yancy Martinez 10 min to type two-thirds of a page of his term paper. At this rate, how long will it take him to type a 7-page term paper? [5.4a]

37. ▦ The following triangles are similar. Find the missing lengths. [5.5a]

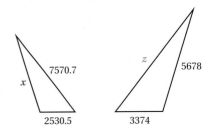

38. Shine-and-Glo Painters uses 2 gal of finishing paint for every 3 gal of primer. Each gallon of finishing paint covers 450 ft². If a surface of 4950 ft² needs both primer and finishing paint, how many gallons of each should be purchased? [5.4a]

335

Write fraction notation for the ratio. Do not simplify.

1. 85 to 97

2. 0.34 to 124

Find the ratio of the first number to the second number and simplify.

3. 18 to 20

4. 0.75 to 0.96

5. What is the rate in feet per second?

   10 feet,   16 seconds

6. *Ham Servings.* A 12-lb shankless ham contains 16 servings. What is the rate in servings per pound?

7. *Gas Mileage.* The 2005 Chevrolet Malibu Maxx will go 319 mi on 14.5 gal of gasoline in city driving. What is the rate in miles per gallon?

   Source: General Motors Corporation

8. *Laundry Detergent.* A box of Cheer laundry detergent powder sells at $6.29 for 81 oz. Find the unit price in cents per ounce.

9. The following table lists prices for various packages of Tide laundry detergent powder. Find the unit price of each package. Then determine which has the lowest unit price.

PACKAGE	PRICE	UNIT PRICE
33 oz	$3.69	
87 oz	$6.22	
131 oz	$10.99	
263 oz	$17.99	

Determine whether the two pairs of numbers are proportional.

10. 7, 8 and 63, 72

11. 1.3, 3.4 and 5.6, 15.2

Solve.

12. $\dfrac{9}{4} = \dfrac{27}{x}$

13. $\dfrac{150}{2.5} = \dfrac{x}{6}$

14. $\dfrac{x}{100} = \dfrac{27}{64}$

15. $\dfrac{68}{y} = \dfrac{17}{25}$

Solve.

16. *Distance Traveled.* An ocean liner traveled 432 km in 12 hr. At this rate, how far would the boat travel in 42 hr?

17. *Time Loss.* A watch loses 2 min in 10 hr. At this rate, how much will it lose in 24 hr?

18. *Map Scaling.* On a map, 3 in. represents 225 mi. If two cities are 7 in. apart on the map, how far are they apart in reality?

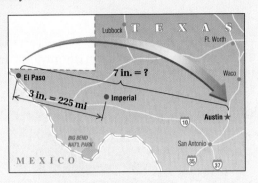

19. *Tower Height.* A birdhouse built on a pole that is 3 m high casts a shadow 5 m long. At the same time, the shadow of a tower is 110 m long. How high is the tower?

20. *Exchanging Money.* On 5 August 2005, 1 U.S. dollar was worth about 7.775 Hong Kong dollars.

a) How much would 450 U.S. dollars be worth in Hong Kong dollars?

b) Mitchell was traveling in Hong Kong and saw a DVD player that cost 795 Hong Kong dollars. How much would it cost in U.S. dollars?

21. *Automobile Violations.* In a recent year, the Indianapolis Police Department employed 1088 officers and made 37,493 arrests. At this rate, how many arrests could be made if the number of officers were increased to 2500?

Source: *Indianapolis Star,* 12-31-00

The lengths in each pair of figures are proportional. Find the missing lengths.

22.

23.

| SYNTHESIS |

24. Nancy Morano-Smith wants to win a season football ticket from the local bookstore. Her goal is to guess the number of marbles in an 8-gal jar. She knows that there are 128 oz in a gallon. She goes home and fills an 8-oz jar with 46 marbles. How many marbles should she guess are in the jar?

Cumulative Review

1. *Baseball Salaries.* Alex Rodriguez now plays with the New York Yankees under a 10-yr contract for $252 million that he originally signed with the Texas Rangers in 2000.

a) Find standard notation for the dollar amount of this contract.

b) How many billion dollars was this contract?

c) What is the average amount of money he makes each year?

d) In 2000, Rodriguez had 554 at-bats. How much money did he average for each at-bat?

Source: Major League Baseball

2. *Gas Mileage.* The 2005 Volkswagen Jetta 2.5 L will go 319 mi on 14.5 gal of gasoline in city driving. What is the rate in miles per gallon?

Source: Volkswagen of America, Inc.

Add and simplify.

3.
$$
\begin{array}{r}
2\ 7.6\ 8 \\
3.0\ 1\ 9 \\
+\ 4\ 8\ 3.2\ 9\ 7 \\
\hline
\end{array}
$$

4.
$$
\begin{array}{r}
2\frac{1}{3} \\
+4\frac{5}{12} \\
\hline
\end{array}
$$

5. $\dfrac{6}{35} + \dfrac{5}{28}$

Subtract and simplify.

6.
$$
\begin{array}{r}
4\ 0.2 \\
-\ \ \ 9.7\ 0\ 9 \\
\hline
\end{array}
$$

7. $73.82 - 0.908$

8. $\dfrac{4}{15} - \dfrac{3}{20}.$

Multiply and simplify.

9.
$$
\begin{array}{r}
3\ 7.6\ 4 \\
\times\ \ \ \ \ 5.9 \\
\hline
\end{array}
$$

10. 5.678×100

11. $2\dfrac{1}{3} \cdot 1\dfrac{2}{7}$

Divide and simplify.

12. $2.3\ \overline{)\ 9\ 8.9}$

13. $5\ 4\ \overline{)\ 4\ 8,5\ 4\ 6}$

14. $\dfrac{7}{11} \div \dfrac{14}{33}$

15. Write expanded notation: 30,074.

16. Write a word name for 120.07.

Which number is larger?

17. 0.7, 0.698

18. 0.799, 0.8

19. Find the prime factorization of 144.

20. Find the LCM of 28 and 35.

21. What part is shaded?

22. Simplify: $\dfrac{90}{144}.$

Calculate.

23. $\dfrac{3}{5} \times 9.53$

24. $\dfrac{1}{3} \times 0.645 - \dfrac{3}{4} \times 0.048$

25. Write fraction notation for the ratio 0.3 to 15.

26. Determine whether the pairs 3, 9 and 25, 75 are proportional.

27. What is the rate in meters per second?

660 meters,   12 seconds

28. The following table lists prices for various brands of liquid dish soap. Find the unit price of each brand. Then determine which has the lowest unit price.

BRAND	PACKAGE	PRICE	UNIT PRICE
Palmolive	13 oz	$1.53	
Dawn	42.7 oz	$3.99	
Dawn	14.7 oz	$1.43	
Joy	28 oz	$1.78	
Joy	42.7 oz	$2.99	

Solve.

29. $\dfrac{14}{25} = \dfrac{x}{54}$

30. $423 = 16 \cdot t$

31. $\dfrac{2}{3} \cdot y = \dfrac{16}{27}$

32. $\dfrac{7}{16} = \dfrac{56}{x}$

33. $34.56 + n = 67.9$

34. $t + \dfrac{7}{25} = \dfrac{5}{7}$

Solve.

35. A particular kind of fettuccini Alfredo has 520 calories in 1 cup. How many calories are there in $\frac{3}{4}$ cup?

36. *Exchanging Money.* On 26 April 2005, 1 U.S. dollar was worth about 6.085 South African Rand.

a) How much would 220 U.S. dollars be worth in Rand?
b) Monica was traveling in South Africa and saw a camera that cost 2050 Rand. How much would it cost in U.S. dollars?

37. *Gas Mileage.* A Greyhound tour bus traveled 347.6 mi, 249.8 mi, and 379.5 mi on three separate trips. What was the total mileage of the bus?

38. A machine can stamp out 925 washers in 5 min. The company owning the machine needs 1295 washers by the end of the morning. How long will it take to stamp them out?

39. A 46-oz juice can contains $5\frac{3}{4}$ cups of juice. A recipe calls for $3\frac{1}{2}$ cups of juice. How many cups are left over?

40. It takes a carpenter $\frac{2}{3}$ hr to hang a door. How many doors can the carpenter hang in 8 hr?

41. *The Leaning Tower of Pisa.*   At the time of this writing, the Leaning Tower of Pisa was still standing. It is 184.5 ft tall but leans about 17 ft out from its base. Each year, it leans about an additional $\frac{1}{20}$ in., or $\frac{1}{240}$ ft.
   a) After how many years will it lean the same length as it is tall?
   b) At most how many years do you think the Tower will stand?

42. *Airplane Tire Costs.*   A Boeing 747-400 jumbo jet has 2 nose tires and 16 rear tires. Each tire costs about $20,000.
   a) What is the total cost of a new set of tires for such a plane?
   b) Suppose an airline has a fleet of 400 such planes. What is the total cost of a new set of tires for all the planes?
   c) Suppose the airline has to change tires every month. What would be the total cost for tires for the airline for an entire year?
   Source: *World-Traveler*, October 2000

43. *Car Travel.*   A car travels 337.62 mi in 8 hr. How far does it travel in 1 hr?

44. *Shuttle Orbits.*   A recent space shuttle made 16 orbits a day during an 8.25-day mission. How many orbits were made during the entire mission?

For each of Exercises 45–47, choose the correct answer from the selections given.

45. How many even prime numbers are there?
   a) 5   **b)** 3   **c)** 2
   d) 1   **e)** None

46. The gas mileage of a car is 28.16 miles per gallon. How many gallons per mile is this?
   a) $\dfrac{704}{25}$ **b)** $\dfrac{25}{704}$ **c)** $\dfrac{2816}{100}$
   d) $\dfrac{250}{704}$ **e)** None

47. By what number do you multiply the side s of a square to find its perimeter?
   a) s itself   **b)** 4   **c)** 2
   d) 8   **e)** None

48. A soccer goalie wishing to block an opponent's shot moves toward the shooter to reduce the shooter's view of the goal. If the goalie can only defend a region 10 ft wide, how far in front of the goal should the goalie be? (See the figure at right.)

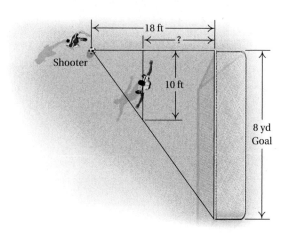

Percent Notation

6

Real-World Application

There are 1,168,195 people in the United States in active military service. The numbers in the four armed services are as follows: Air Force, 314,477; Army, 391,126; Marine Corps, 135,324; Navy, 327,268. What percent of the total does each branch represent? Round the answers to the nearest tenth of a percent.

Source: U.S. Department of Defense

This problem appears as Exercise 2 in Section 6.5.

Objectives

a Write three kinds of notation for a percent.

b Convert between percent notation and decimal notation.

a Understanding Percent Notation

Of all the surface area of the earth, 70% of it is covered by water. What does this mean? It means that of every 100 square miles of the earth's surface area, 70 square miles are covered by water. Thus, 70% is a ratio of 70 to 100, or $\frac{70}{100}$.

Source: *The Handy Geography Answer Book*

Write three kinds of notation as in Examples 1 and 2.

1. 70%

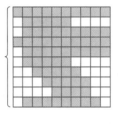

70 of 100 squares are shaded.

70% or $\frac{70}{100}$ or 0.70 of the large square is shaded.

2. 23.4%

3. 100%

Percent notation is used extensively in our everyday lives. Here are some examples:

44% of all Americans use at least one prescribed medicine.

35.5% of Massachusetts residents age 25 and over have a bachelor's degree.

18% of household personal vehicles are pickup trucks.

50.3% of all paper used in the United States is recycled.

54% of all kids prefer white bread.

0.08% blood alcohol level is a standard used by some states as the legal limit for drunk driving.

Percent notation is often represented in pie charts to show how the parts of a quantity are related. For example, the circle graph below illustrates the percentage of vehicles manufactured in the most popular car colors during 2003 in North America.

It is thought that the Roman emperor Augustus began percent notation by taxing goods sold at a rate of $\frac{1}{100}$. In time, the symbol "%" evolved by interchanging the parts of the symbol "100" to "0/0" and then to "%."

Most Popular Car Colors, 2003 (SUV / Truck / Van)

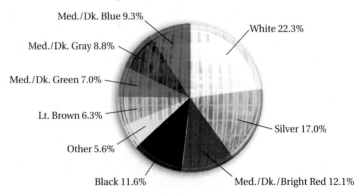

Med./Dk. Blue 9.3%
White 22.3%
Med./Dk. Gray 8.8%
Med./Dk. Green 7.0%
Lt. Brown 6.3%
Silver 17.0%
Other 5.6%
Black 11.6%
Med./Dk./Bright Red 12.1%

Source: DuPont Herberts Automotive Systems, Troy, Michigan, 2003, DuPont Automotive Color Survey Results

This definition leads us to the following equivalent ways of defining percent notation.

NOTATION FOR n%

Percent notation, n%, can be expressed using:

ratio → $n\% = $ the ratio of n to $100 = \dfrac{n}{100}$,

fraction notation → $n\% = n \times \dfrac{1}{100}$, or

decimal notation → $n\% = n \times 0.01$.

From 1998 to 2008, the number of jobs for professional chefs will increase by 13.4%.

Source: *Handbook of U.S. Labor Statistics*

EXAMPLE 1 Write three kinds of notation for 35%.

Using ratio: $35\% = \dfrac{35}{100}$ A ratio of 35 to 100

Using fraction notation: $35\% = 35 \times \dfrac{1}{100}$ Replacing % with $\times \dfrac{1}{100}$

Using decimal notation: $35\% = 35 \times 0.01$ Replacing % with $\times 0.01$

EXAMPLE 2 Write three kinds of notation for 67.8%.

Using ratio: $67.8\% = \dfrac{67.8}{100}$ A ratio of 67.8 to 100

Using fraction notation: $67.8\% = 67.8 \times \dfrac{1}{100}$ Replacing % with $\times \dfrac{1}{100}$

Using decimal notation: $67.8\% = 67.8 \times 0.01$ Replacing % with $\times 0.01$

Do Exercises 1–3 on the preceding page.

b   Converting Between Percent Notation and Decimal Notation

Consider 78%. To convert to decimal notation, we can think of percent notation as a ratio and write

$$78\% = \dfrac{78}{100} \qquad \text{Using the definition of percent as a ratio}$$

$$= 0.78. \qquad \text{Dividing}$$

Similarly,

$$4.9\% = \dfrac{4.9}{100} \qquad \text{Using the definition of percent as a ratio}$$

$$= 0.049. \qquad \text{Dividing}$$

We could also convert 78% to decimal notation by replacing "%" with "$\times 0.01$" and write

$$78\% = 78 \times 0.01 \qquad \text{Replacing % with } \times 0.01$$

$$= 0.78. \qquad \text{Multiplying}$$

CALCULATOR CORNER

Converting from Percent Notation to Decimal Notation Many calculators have a ⬚% key that can be used to convert from percent notation to decimal notation. This is often the second operation associated with a particular key and is accessed by first pressing a ⬚2nd⬚ or ⬚SHIFT⬚ key. To convert 57.6% to decimal notation, for example, you might press ⬚5⬚⬚7⬚⬚.⬚⬚6⬚⬚2nd⬚⬚%⬚ or ⬚5⬚⬚7⬚⬚.⬚⬚6⬚⬚SHIFT⬚⬚%⬚. The display would read ⬚0.576⬚, so 57.6% = 0.576. Read the user's manual to determine whether your calculator can do this conversion.

Exercises: Use a calculator to find decimal notation.

1. 14%   2. 0.069%

3. 43.8%   4. 125%

Find decimal notation.

4. 34%

5. 78.9%

6. $6\frac{5}{8}$%

Find decimal notation for the percent notation in the sentence.

7. Pickup Trucks. Of all household personal vehicles in the United States, 18% are pickup trucks.

Source: Department of Transportation, 2001 National Household Travel Survey

8. Blood Alcohol Level. A blood alcohol level of 0.08% is a standard used by some states as the legal limit for drunk driving.

Answers on page A-13

Similarly,

$$4.9\% = 4.9 \times 0.01 \qquad \text{Replacing \% with} \times 0.01$$
$$= 0.049. \qquad \text{Multiplying}$$

Dividing by 100 amounts to moving the decimal point two places to the left, which is the same as multiplying by 0.01. This leads us to a quick way to convert from percent notation to decimal notation—we drop the percent symbol and move the decimal point two places to the left.

To convert from percent notation to decimal notation,	36.5%
a) replace the percent symbol % with × 0.01, and	36.5 × 0.01
b) multiply by 0.01, which means move the decimal point two places to the left.	0.36.5 Move 2 places to the left. 36.5% = 0.365

EXAMPLE 3 Find decimal notation for 99.44%.

a) Replace the percent symbol with × 0.01.   99.44 × 0.01

b) Move the decimal point two places to the left.   0.99.44

Thus, 99.44% = 0.9944.

EXAMPLE 4 The interest rate on a $2\frac{1}{2}$-year certificate of deposit is $6\frac{3}{8}$%. Find decimal notation for $6\frac{3}{8}$%.

a) Convert $6\frac{3}{8}$ to decimal notation and replace the percent symbol with × 0.01. $6\frac{3}{8}$%
     6.375 × 0.01

b) Move the decimal point two places to the left.   0.06.375

Thus, $6\frac{3}{8}$% = 0.06375.

Do Exercises 4–8.

To convert 0.38 to percent notation, we can first write fraction notation, as follows:

$$0.38 = \frac{38}{100} \qquad \text{Converting to fraction notation}$$
$$= 38\%. \qquad \text{Using the definition of percent as a ratio}$$

Note that 100% = 100 × 0.01 = 1. Thus to convert 0.38 to percent notation, we can multiply by 1, using 100% as a symbol for 1.

$$0.38 = 0.38 \times 1$$
$$= 0.38 \times 100\%$$
$$= 0.38 \times 100 \times 0.01 \qquad \text{Replacing 100\% with 100} \times 0.01$$
$$= (0.38 \times 100) \times 0.01 \qquad \text{Using the associative law of multiplication}$$

Then

$$38 \times 0.01 = 38\%.$$ Replacing "$\times \ 0.01$" with the % symbol

Even more quickly, since $0.38 = 0.38 \times 100\%$, we can simply multiply 0.38 by 100 and write the % symbol.

To convert from decimal notation to percent notation, we multiply by 100%—that is, we move the decimal point two places to the right and write a percent symbol.

To convert from decimal notation to percent notation, multiply by 100%. That is,	$0.675 = 0.675 \times 100\%$
a) move the decimal point two places to the right, and	0.67.5 Move 2 places to the right.
b) write a % symbol.	67.5% $0.675 = 67.5\%$

EXAMPLE 5   Find percent notation for 1.27.

a) Move the decimal point two places to the right.   1.27.

b) Write a % symbol.   127%

Thus, $1.27 = 127\%$.

EXAMPLE 6   Of the time that people declare as sick leave, 0.21 is actually used for family issues. Find percent notation for 0.21.
Source: CCH Inc.

a) Move the decimal point two places to the right.   0.21.

b) Write a % symbol.   21%

Thus, $0.21 = 21\%$.

EXAMPLE 7   Find percent notation for 5.6.

a) Move the decimal point two places to the right, adding an extra zero.   5.60.

b) Write a % symbol.   560%

Thus, $5.6 = 560\%$.

EXAMPLE 8   Of those who play golf, 0.149 play 8–24 rounds per year. Find percent notation for 0.149.
Source: U.S. Golf Association

a) Move the decimal point two places to the right.   0.14.9

b) Write a % symbol.   14.9%

Thus, $0.149 = 14.9\%$.

Do Exercises 9–13.

Find percent notation.
9. 0.24

10. 3.47

11. 1

Find percent notation for the decimal notation in the sentence.

12. High School Graduate.   The highest level of education for 0.321 of persons 25 and over in the United States is high school graduation.
Source: U.S. Department of Commerce, Bureau of the Census, *Current Population Survey*

13. Golf.   Of those who play golf, 0.253 play 25–49 rounds per year.
Source: U.S. Golf Association

Answers on page A-13

a   Write three kinds of notation as in Examples 1 and 2 on p. 343.

1. 90%   **2.** 58.7%   **3.** 12.5%   **4.** 130%

b   Find decimal notation.

5. 67%   **6.** 17%   **7.** 45.6%   **8.** 76.3%

9. 59.01%   **10.** 30.02%   **11.** 10%   **12.** 80%

13. 1%   **14.** 100%   **15.** 200%   **16.** 300%

17. 0.1%   **18.** 0.4%   **19.** 0.09%   **20.** 0.12%

21. 0.18%   **22.** 5.5%   **23.** 23.19%   **24.** 87.99%

25. $14\frac{7}{8}\%$ **26.** $93\frac{1}{8}\%$ **27.** $56\frac{1}{2}\%$ **28.** $61\frac{3}{4}\%$

Find decimal notation for the percent notation(s) in the sentence.

29. *Pediatricians.*   By 2020, the population of children in the United States is expected to increase by about 9% while the number of pediatricians will leap 58%.
    Source: Study by Dr. Scott Shipman, an Oregon Health and Science University pediatrician

30. *Cancer Survival.*   In 2005, 78.9% of children with cancer survive at least 5 years compared to about 60% in 1975.
    Source: National Cancer Institute

31. *Eating Out.*   On a given day, 44% of all adults eat in a restaurant.
    Source: *AARP Bulletin*, November 2004

32. Of those who play golf, 18.6% play 100 or more rounds per year.
    Source: U.S. Golf Association

33. *Knitting and Crocheting.* Of all women ages 25 to 34, 36% know how to knit or crochet.

Source: Research Inc. for Craft Yarn Council of America

34. According to a recent survey, 95.1% of those asked to name what sports they participate in chose swimming.

Source: Sporting Goods Manufacturers

Find percent notation.

35. 0.47   **36.** 0.87   **37.** 0.03   **38.** 0.01   **39.** 8.7

40. 4   **41.** 0.334   **42.** 0.889   **43.** 0.75   **44.** 0.99

45. 0.4   **46.** 0.5   **47.** 0.006   **48.** 0.008   **49.** 0.017

50. 0.024   **51.** 0.2718   **52.** 0.8911   **53.** 0.0239   **54.** 0.00073

Find percent notation for the decimal notation(s) in the sentence.

55. *Hours of Sleep.* In 2005, only 0.26 of people get eight or more hours of sleep a night on weekdays. This rate has declined from 0.38 in 2001.

Source: National Sleep Foundation's 2005 Sleep in America Poll

56. According to a recent survey, 0.526 of those asked to name what sports they participate in chose bowling.

Source: Sporting Goods Manufacturers

57. *Bachelor's Degrees.* For 0.177 of the United States population 25 and over, the bachelor's degree is the highest level of educational attainment.

Source: U.S. Department of Commerce, Bureau of the Census, *Current Population Survey;* National Center for Educational Statistics, *Digest of Education Statistics,* 2003

58. About 0.69 of all newspapers are recycled.

Sources: American Forest and Paper Association; Newspaper Association of America

59. *65 and Over.* In Clearwater, Florida, 0.215 of the residents are 65 or over.

Source: U.S. Census Bureau

60. Of those people living in North Carolina, 0.1134 will die of heart disease.

Source: American Association of Retired Persons

61. D_W *Winning Percentage.* During the 2004 regular baseball season, the Boston Red Sox won 98 of 162 games and went on to win the World Series. Find the ratio of number of wins to total number of games played in the regular season and convert it to decimal notation. Such a rate is often called a "winning percentage." Explain why.

62. D_W Athletes sometimes speak of "giving 110%" effort. Does this make sense? Explain.

Convert to a mixed numeral. [3.4a]

63. $\dfrac{100}{3}$

64. $\dfrac{75}{2}$

65. $\dfrac{75}{8}$

66. $\dfrac{297}{16}$

67. $\dfrac{567}{98}$

68. $\dfrac{2345}{21}$

Convert to decimal notation. [4.5a]

69. $\dfrac{2}{3}$

70. $\dfrac{1}{3}$

71. $\dfrac{5}{6}$

72. $\dfrac{17}{12}$

73. $\dfrac{8}{3}$

74. $\dfrac{15}{16}$

6.2 PERCENT AND FRACTION NOTATION

Objectives

a Convert from fraction notation to percent notation.

b Convert from percent notation to fraction notation.

a Converting from Fraction Notation to Percent Notation

Consider the fraction notation $\frac{7}{8}$. To convert to percent notation, we use two skills we already have. We first find decimal notation by dividing:

$$\frac{7}{8} = 0.875$$

$$\begin{array}{r} 0.8\ 7\ 5 \\ 8\ \overline{)\ 7.0\ 0\ 0} \\ \underline{6\ 4} \\ 6\ 0 \\ \underline{5\ 6} \\ 4\ 0 \\ \underline{4\ 0} \\ 0 \end{array}$$

Then we convert the decimal notation to percent notation. We move the decimal point two places to the right

$$0.8\ 7.5$$

and write a % symbol:

$$\frac{7}{8} = 87.5\%, \text{ or } 87\frac{1}{2}\%.$$

To convert from fraction notation to percent notation,

$\frac{3}{5}$ Fraction notation

a) find decimal notation by division, and

$$\begin{array}{r} 0.6 \\ 5\ \overline{)\ 3.0} \\ \underline{3\ 0} \\ 0 \end{array}$$

b) convert the decimal notation to percent notation.

$0.6 = 0.60 = 60\%$ Percent notation

$$\frac{3}{5} = 60\%$$

EXAMPLE 1 Find percent notation for $\frac{9}{16}$.

a) We first find decimal notation by division.

$$\begin{array}{r} 0.5\ 6\ 2\ 5 \\ 1\ 6\ \overline{)\ 9.0\ 0\ 0\ 0} \\ \underline{8\ 0} \\ 1\ 0\ 0 \\ \underline{9\ 6} \\ 4\ 0 \\ \underline{3\ 2} \\ 8\ 0 \\ \underline{8\ 0} \\ 0 \end{array}$$

$$\frac{9}{16} = 0.5625$$

Study Tips

BEING A TUTOR

Try being a tutor for a fellow student. Understanding and retention of concepts can be maximized for yourself if you explain the material to someone else.

349

Find percent notation.

1. $\dfrac{1}{4}$ **2.** $\dfrac{5}{8}$

b) Next, we convert the decimal notation to percent notation. We move the decimal point two places to the right and write a % symbol.

$$0.56.25$$

$$\dfrac{9}{16} = 56.25\%, \text{ or } 56\tfrac{1}{4}\%$$

Don't forget the % symbol.

Do Exercises 1 and 2.

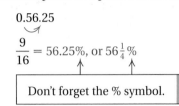

CALCULATOR CORNER

Converting from Fraction Notation to Percent Notation   A calculator can be used to convert from fraction notation to percent notation. We simply perform the division on the calculator and then use the percent key. To convert $\frac{17}{40}$ to percent notation, for example, we press ⎡1⎤ ⎡7⎤ ⎡÷⎤ ⎡4⎤ ⎡0⎤ ⎡2nd⎤ ⎡%⎤ , or ⎡1⎤ ⎡7⎤ ⎡÷⎤ ⎡4⎤ ⎡0⎤ ⎡SHIFT⎤ ⎡%⎤ . The display reads ⎡ 42.5 ⎤ , so $\frac{17}{40} = 42.5\%$. Read the user's manual to determine whether your calculator can do this conversion.

Exercises: Use a calculator to find percent notation. Round to the nearest hundredth of a percent.

1. $\dfrac{13}{25}$ **4.** $\dfrac{12}{7}$

2. $\dfrac{5}{13}$ **5.** $\dfrac{217}{364}$

3. $\dfrac{43}{39}$ **6.** $\dfrac{2378}{8401}$

EXAMPLE 2 *Without Health Insurance.*   Approximately $\frac{1}{6}$ of all people in the United States are without health insurance. Find percent notation for $\frac{1}{6}$.

Source: U.S. Bureau of the Census, *Current Population Survey*, March 2003

a) Find decimal notation by division.

$$
\begin{array}{r}
0.1\ 6\ 6 \\
6\overline{)1.0\ 0\ 0} \\
\underline{6} \\
4\ 0 \\
\underline{3\ 6} \\
4\ 0 \\
\underline{3\ 6} \\
4
\end{array}
$$

We get a repeating decimal: $0.16\overline{6}$.

b) Convert the answer to percent notation.

$$0.16.\overline{6}$$

$$\dfrac{1}{6} = 16.\overline{6}\%, \text{ or } 16\dfrac{2}{3}\%$$

Answers on page A-13

Do Exercises 3 and 4.

In some cases, division is not the fastest way to convert. The following are some optional ways in which conversion might be done.

EXAMPLE 3   Find percent notation for $\frac{69}{100}$.

We use the definition of percent as a ratio.

$$\frac{69}{100} = 69\%$$

EXAMPLE 4   Find percent notation for $\frac{17}{20}$.

We multiply by 1 to get 100 in the denominator. We think of what we have to multiply 20 by in order to get 100. That number is 5, so we multiply by 1 using $\frac{5}{5}$.

$$\frac{17}{20} \cdot \frac{5}{5} = \frac{85}{100} = 85\%$$

Note that this shortcut works only when the denominator is a factor of 100.

Do Exercises 5 and 6.

b   Converting from Percent Notation to Fraction Notation

To convert from percent notation to fraction notation,	30%	Percent notation
a) use the definition of percent as a ratio, and	$\frac{30}{100}$	
b) simplify, if possible.	$\frac{3}{10}$	Fraction notation

EXAMPLE 5   Find fraction notation for 75%.

$$75\% = \frac{75}{100} \qquad \text{Using the definition of percent}$$
$$= \frac{3 \cdot 25}{4 \cdot 25} = \frac{3}{4} \cdot \frac{25}{25} \Bigg\} \quad \text{Simplifying}$$
$$= \frac{3}{4}$$

3. Water is the single most abundant chemical in the body. The human body is about $\frac{2}{3}$ water. Find percent notation for $\frac{2}{3}$.

4. Find percent notation: $\frac{5}{6}$.

Find percent notation.

5. $\frac{57}{100}$ **6.** $\frac{19}{25}$

Answers on page A-13

Find fraction notation.

7. 60%

8. 3.25%

9. $66\frac{2}{3}\%$

10. Complete this table.

FRACTION NOTATION	$\frac{1}{5}$		
DECIMAL NOTATION		$0.83\overline{3}$	
PERCENT NOTATION			$37\frac{1}{2}\%$

Answers on page A-13

Study Tips

MEMORIZING

Memorizing is a very helpful tool in the study of mathematics. Don't underestimate its power as you memorize the table of decimal, fraction, and percent notation on the inside back cover. We will discuss memorizing more later.

EXAMPLE 6 Find fraction notation for 62.5%.

$$62.5\% = \frac{62.5}{100} \qquad \text{Using the definition of percent}$$

$$= \frac{62.5}{100} \times \frac{10}{10} \qquad \begin{array}{l}\text{Multiplying by 1 to eliminate the}\\\text{decimal point in the numerator}\end{array}$$

$$= \frac{625}{1000}$$

$$\left.\begin{array}{l}= \frac{5 \cdot 125}{8 \cdot 125} = \frac{5}{8} \cdot \frac{125}{125}\\[2mm] = \frac{5}{8}\end{array}\right\} \quad \text{Simplifying}$$

EXAMPLE 7 Find fraction notation for $16\frac{2}{3}\%$.

$$16\frac{2}{3}\% = \frac{50}{3}\% \qquad \begin{array}{l}\text{Converting from the mixed numeral}\\\text{to fraction notation}\end{array}$$

$$= \frac{50}{3} \times \frac{1}{100} \qquad \text{Using the definition of percent}$$

$$\left.\begin{array}{l}= \frac{50 \cdot 1}{3 \cdot 50 \cdot 2} = \frac{1}{6} \cdot \frac{50}{50}\\[2mm] = \frac{1}{6}\end{array}\right\} \quad \text{Simplifying}$$

The table on the inside back cover lists decimal, fraction, and percent equivalents used so often that it would speed up your work if you memorized them. For example, $\frac{1}{3} = 0.\overline{3}$, so we say that the **decimal equivalent** of $\frac{1}{3}$ is $0.\overline{3}$, or that $0.\overline{3}$ has the **fraction equivalent** $\frac{1}{3}$.

EXAMPLE 8 Find fraction notation for $16.\overline{6}\%$.

We can use the table on the inside back cover or recall that $16.\overline{6}\% = 16\frac{2}{3}\% = \frac{1}{6}$. We can also recall from our work with repeating decimals in Chapter 4 that $0.\overline{6} = \frac{2}{3}$. Then we have $16.\overline{6}\% = 16\frac{2}{3}\%$ and can proceed as in Example 7.

Do Exercises 7–10.

Applications of Ratio and Percent: The Price–Earnings Ratio and Stock Yields

The Price–Earnings Ratio   If the total earnings of a company one year were $5,000,000 and 100,000 shares of stock were issued, the earnings per share were $50. At one time, the price per share of Coca-Cola was $58.125 and the earnings per share were $0.76. The **price–earnings ratio,** P/E, is the price of the stock divided by the earnings per share. For the Coca-Cola stock, the price–earnings ratio, P/E, is given by

$$\frac{P}{E} = \frac{58.125}{0.76} \approx 76.48.$$ Dividing, using a calculator, and rounding to the nearest hundredth

Stock Yields   At one time, the price per share of Coca-Cola stock was $58.125 and the company was paying a yearly dividend of $0.68 per share. It is helpful to those interested in stocks to know what percent the dividend is of the price of the stock. The percent is called the **yield.** For the Coca-Cola stock, the yield is given by

$$\text{Yield} = \frac{\text{Dividend}}{\text{Price per share}} = \frac{0.68}{58.125} \approx 0.0117$$ Dividing and rounding to the nearest ten-thousandth

$$= 1.17\%$$ Converting to percent notation

Coca–Cola Co (Coke) (KO)
as of 24–Jan–2001

Volume (1000's)

Source: Yahoo.com

Exercises: Compute the price–earnings ratio and the yield for each stock listed below.

	STOCK	PRICE PER SHARE	EARNINGS	DIVIDEND	P/E	YIELD
1.	Pepsi (PEP)	$42.75	$1.40	$0.56		
2.	Pearson (PSO)	$25.00	$0.78	$0.30		
3.	Quaker Oats (OAT)	$92.375	$2.68	$1.10		
4.	Texas Insts (TEX)	$42.875	$1.62	$0.43		
5.	Ford Motor Co (F)	$27.5625	$2.30	$1.19		
6.	Wendy's Intl (WEN)	$25.75	$1.47	$0.23		

6.2

EXERCISE SET

For Extra Help

a Find percent notation.

1. $\dfrac{41}{100}$ 2. $\dfrac{36}{100}$ 3. $\dfrac{5}{100}$ 4. $\dfrac{1}{100}$ 5. $\dfrac{2}{10}$ 6. $\dfrac{7}{10}$

7. $\dfrac{3}{10}$ 8. $\dfrac{9}{10}$ 9. $\dfrac{1}{2}$ 10. $\dfrac{3}{4}$ 11. $\dfrac{7}{8}$ 12. $\dfrac{1}{8}$

13. $\dfrac{4}{5}$ 14. $\dfrac{2}{5}$ 15. $\dfrac{2}{3}$ 16. $\dfrac{1}{3}$ 17. $\dfrac{1}{6}$ 18. $\dfrac{5}{6}$

19. $\dfrac{3}{16}$ 20. $\dfrac{11}{16}$ 21. $\dfrac{13}{16}$ 22. $\dfrac{7}{16}$ 23. $\dfrac{4}{25}$ 24. $\dfrac{17}{25}$

25. $\dfrac{1}{20}$ 26. $\dfrac{31}{50}$ 27. $\dfrac{17}{50}$ 28. $\dfrac{3}{20}$

Find percent notation for the fraction notation in the sentence.

29. *Driving While Drowsy.* Almost half of U.S. drivers say they have driven while drowsy. Of this group, $\frac{2}{5}$ fight off sleep by opening a window, and $\frac{9}{50}$ drink a caffeinated beverage.

Source: Harris Interactive for Tylenol PM

30. *Paved Roads.* About $\frac{13}{20}$ of the roads and streets in the United States are paved.

Source: U.S. Department of Transportation, *Highway Statistics*

In Exercises 31–36, write percent notation for the fractions in this pie chart.

Time Workers Spend Sorting Unsolicited e-mail and Spam

Less than 5 minutes $\frac{59}{100}$

5–15 minutes $\frac{11}{50}$

15–30 minutes $\frac{9}{100}$

30–60 minutes $\frac{1}{20}$

More than 1 hour $\frac{1}{50}$

Did not know/ not sure $\frac{3}{100}$

Source: Data from InsightExpress

31. $\dfrac{11}{50}$ **32.** $\dfrac{3}{100}$

33. $\dfrac{1}{20}$ **34.** $\dfrac{59}{100}$

35. $\dfrac{9}{100}$ **36.** $\dfrac{1}{50}$

CHAPTER 6: Percent Notation

b Find fraction notation. Simplify.

37. 85%

38. 55%

39. 62.5%

40. 12.5%

41. $33\frac{1}{3}\%$

42. $83\frac{1}{3}\%$

43. $16.\overline{6}\%$

44. $66.\overline{6}\%$

45. 7.25%

46. 4.85%

47. 0.8%

48. 0.2%

49. $25\frac{3}{8}\%$

50. $48\frac{7}{8}\%$

51. $78\frac{2}{9}\%$

52. $16\frac{5}{9}\%$

53. $64\frac{7}{11}\%$

54. $73\frac{3}{11}\%$

55. 150%

56. 110%

57. 0.0325%

58. 0.419%

59. $33.\overline{3}\%$

60. $83.\overline{3}\%$

Find fraction notation for the percent notation in the following bar graph.

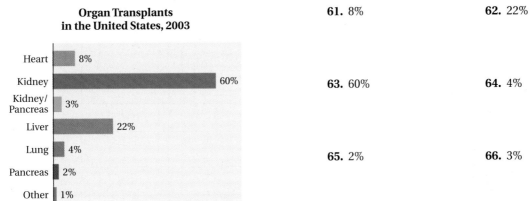

Organ Transplants in the United States, 2003

Heart 8%
Kidney 60%
Kidney/Pancreas 3%
Liver 22%
Lung 4%
Pancreas 2%
Other 1%

Source: National Organ Procurement and Transplantation Network

61. 8%

62. 22%

63. 60%

64. 4%

65. 2%

66. 3%

Find fraction notation for the percent notation in the sentence.

67. A 1.9-oz serving of Raisin Bran Crunch® cereal with $\frac{1}{2}$ cup of skim milk satisfies 35% of the minimum daily requirements for Vitamin B$_{12}$.

Source: Kellogg's USA, Inc.

68. A 1-cup serving of Wheaties® cereal with $\frac{1}{2}$ cup of skim milk satisfies 15% of the minimum daily requirements for calcium.

Source: General Mills Sales, Inc.

69. Of all those who are 85 or older, 47% have Alzheimer's disease.

Source: Alzheimer's Association

70. In 2003, 24.4% of Americans 18 and older smoked cigarettes.

Source: U.S. Centers for Disease Control and Prevention

Complete the table.

71.

FRACTION NOTATION	DECIMAL NOTATION	PERCENT NOTATION
$\frac{1}{8}$		12.5%, or $12\frac{1}{2}$%
$\frac{1}{6}$		
		20%
	0.25	
		33.$\overline{3}$%, or $33\frac{1}{3}$%
		37.5% or $37\frac{1}{2}$%
		40%
$\frac{1}{2}$		

72.

FRACTION NOTATION	DECIMAL NOTATION	PERCENT NOTATION
$\frac{3}{5}$		
	0.625	
$\frac{2}{3}$		
	0.75	75%
$\frac{4}{5}$		
$\frac{5}{6}$		83.$\overline{3}$%, or $83\frac{1}{3}$%
$\frac{7}{8}$		87.5%, or $87\frac{1}{2}$%
		100%

73.

FRACTION NOTATION	DECIMAL NOTATION	PERCENT NOTATION
	0.5	
$\frac{1}{3}$		
		25%
		16.$\overline{6}$%, or $16\frac{2}{3}$%
	0.125	
$\frac{3}{4}$		
	0.8$\overline{3}$	
$\frac{3}{8}$		

74.

FRACTION NOTATION	DECIMAL NOTATION	PERCENT NOTATION
		40%
		62.5%, or $62\frac{1}{2}$%
	0.875	
$\frac{1}{1}$		
	0.6	
	0.$\overline{6}$	
$\frac{1}{5}$		

75. D_W What do the following have in common? Explain.

$\frac{23}{16}$, $1\frac{875}{2000}$, 1.4375, $\frac{207}{144}$, $1\frac{7}{16}$, 143.75%, $1\frac{4375}{10,000}$

76. D_W Is it always best to convert from fraction notation to percent notation by first finding decimal notation? Why or why not?

Solve.

77. $13 \cdot x = 910$ [1.7b]

78. $15 \cdot y = 75$ [1.7b]

79. $0.05 \times b = 20$ [4.4b]

80. $3 = 0.16 \times b$ [4.4b]

81. $\frac{24}{37} = \frac{15}{x}$ [5.3b]

82. $\frac{17}{18} = \frac{x}{27}$ [5.3b]

83. $\frac{9}{10} = \frac{x}{5}$ [5.3b]

84. $\frac{7}{x} = \frac{4}{5}$ [5.3b]

Convert to a mixed numeral. [3.4a]

85. $\frac{100}{3}$

86. $\frac{75}{2}$

87. $\frac{250}{3}$

88. $\frac{123}{6}$

89. $\frac{345}{8}$

90. $\frac{373}{6}$

91. $\frac{75}{4}$

92. $\frac{67}{9}$

Convert from a mixed numeral to fraction notation. [3.4a]

93. $1\frac{1}{17}$

94. $20\frac{9}{10}$

95. $101\frac{1}{2}$

96. $32\frac{3}{8}$

Write percent notation.

97. $\frac{41}{369}$

98. $\frac{54}{999}$

99. $2.5\overline{74631}$

100. $3.29\overline{3847}$

Write decimal notation.

101. $\frac{14}{9}\%$

102. $\frac{19}{12}\%$

103. $\frac{729}{7}\%$

104. $\frac{637}{6}\%$

SOLVING PERCENT PROBLEMS USING PERCENT EQUATIONS

Translate to an equation. Do not solve.

1. 12% of 50 is what?

2. What is 40% of 60?

Translate to an equation. Do not solve.

3. 45 is 20% of what?

4. 120% of what is 60?

a Translating to Equations

To solve a problem involving percents, it is helpful to translate first to an equation. To distinguish the method in Section 6.3 from that of Section 6.4, we will call these *percent equations*.

EXAMPLE 1 Translate:

23%   of   5   is   what?
↓   ↓   ↓   ↓   ↓
23%   ·   5   = a This is a *percent equation*.

KEY WORDS IN PERCENT TRANSLATIONS

"**Of**" translates to "·", or "×".   "**Is**" translates to "=".

"**What**" translates to any letter.   "**%**" translates to "$\times \frac{1}{100}$" or "$\times 0.01$".

EXAMPLE 2 Translate:

What   is   11%   of   49?
↓   ↓   ↓   ↓   ↓
a =   11%   ·   49   Any letter can be used.

Do Exercises 1 and 2.

EXAMPLE 3 Translate:

3   is   10%   of   what?
↓   ↓   ↓   ↓   ↓
3   =   10%   · b

EXAMPLE 4 Translate:

45%   of   what   is   23?
↓   ↓   ↓   ↓   ↓
45%   × b =   23

Do Exercises 3 and 4.

EXAMPLE 5 Translate:

10   is   what percent   of   20?
↓   ↓   ↓   ↓   ↓
10   = p ×   20

Answers on page A-14

EXAMPLE 6 Translate:

What percent of 50 is 7?

$$p \cdot 50 = 7$$

Do Exercises 5 and 6.

b Solving Percent Problems

In solving percent problems, we use the *Translate* and *Solve* steps in the problem-solving strategy used throughout this text.

Percent problems are actually of three different types. Although the method we present does *not* require that you be able to identify which type you are solving, it is helpful to know them.

We know that

15 is 25% of 60,   or $15 = 25\% \times 60$.

We can think of this as:

> Amount = Percent number × Base.

Each of the three types of percent problems depends on which of the three pieces of information is missing.

1. **Finding the *amount* (the result of taking the percent)**

   Example: What is 25% of 60?

   Translation: $a = 25\% \cdot 60$

2. **Finding the *base* (the number you are taking the percent of)**

   Example: 15 is 25% of what number?

   Translation: $15 = 25\% \cdot b$

3. **Finding the *percent number* (the percent itself)**

   Example: 15 is what percent of 60?

   Translation: $15 = p \cdot 60$

FINDING THE AMOUNT

 EXAMPLE 7 What is 4.6% of 105,000,000?

Translate: $a = 4.6\% \times 105,000,000$.

Solve: The letter is by itself. To solve the equation, we just convert 4.6% to decimal notation and multiply:

$$a = 4.6\% \times 105,000,000$$
$$= 0.046 \times 105,000,000 = 4,830,000.$$

Thus, 4,830,000 is 4.6% of 105,000,000. The answer is 4,830,000.

Do Exercise 7.

Translate to an equation. Do not solve.

5. 16 is what percent of 40?

6. What percent of 84 is 10.5?

7. Solve:

What is 12% of 50?

Answers on page A-14

There are 105,000,000 households in the United States and approximately 4.6% own a pet bird. How many households own at least one pet bird?

8. Solve:

64% of $55 is what?

In a survey of a group of people, it was found that 5%, or 20 people, chose strawberry as their favorite ice cream. How many people were surveyed?

Source: International Ice Cream Association

Solve.

9. 20% of what is 45?

10. $60 is 120% of what?

EXAMPLE 8   120% of $42 is what?

Translate: $120\% \times 42 = a$.

Solve: The letter is by itself. To solve the equation, we carry out the calculation:

$$a = 120\% \times 42$$
$$= 1.2 \times 42 \qquad 120\% = 1.2$$
$$= 50.4.$$

Thus, 120% of $42 is $50.40. The answer is $50.40.

Do Exercise 8.

FINDING THE BASE

EXAMPLE 9   5% of what is 20?

Translate: $5\% \times b = 20$.

Solve: This time the letter is *not* by itself. To solve the equation, we divide by 5% on both sides:

$$\frac{5\% \times b}{5\%} = \frac{20}{5\%} \qquad \text{Dividing by 5\% on both sides}$$
$$b = \frac{20}{0.05} \qquad 5\% = 0.05$$
$$b = 400.$$

Thus, 5% of 400 is 20. The answer is 400.

EXAMPLE 10 $3 is 16% of what?

Translate: $3 is 16% of what?

$$3 = 16\% \times b.$$

Solve: Again, the letter is *not* by itself. To solve the equation, we divide by 16% on both sides:

$$\frac{3}{16\%} = \frac{16\% \times b}{16\%} \qquad \text{Dividing by 16\% on both sides}$$
$$\frac{3}{0.16} = b \qquad 16\% = 0.16$$
$$18.75 = b.$$

Thus, $3 is 16% of $18.75. The answer is $18.75.

Do Exercises 9 and 10.

Answers on page A-14

FINDING THE PERCENT NUMBER

In solving these problems, you *must* remember to convert to percent notation after you have solved the equation.

EXAMPLE 11 13 is what percent of 260?

Translate: 13 is what percent of 260?

$$13 = p \times 260.$$

Solve: To solve the equation, we divide by 260 on both sides and convert the result to percent notation:

$$p \cdot 260 = 13$$

$$\frac{p \cdot 260}{260} = \frac{13}{260} \qquad \text{Dividing by 260 on both sides}$$

$$p = 0.05 \qquad \text{Converting to decimal notation}$$

$$p = 5\%. \qquad \text{Converting to percent notation}$$

Thus, 13 is 5% of 260. The answer is 5%.

Do Exercise 11.

EXAMPLE 12 What percent of $50 is $16?

Translate: What percent of $50 is $16?

$$p \times 50 = 16.$$

Solve: To solve the equation, we divide by 50 on both sides and convert the answer to percent notation:

$$\frac{p \times 50}{50} = \frac{16}{50} \qquad \text{Dividing by 50 on both sides}$$

$$p = \frac{16}{50}$$

$$p = 0.32$$

$$p = 32\%. \qquad \text{Converting to percent notation}$$

Thus, 32% of $50 is $16. The answer is 32%.

Do Exercise 12.

> ### Caution!
>
> When a question asks "what percent?", be sure to give the answer in percent notation.

Workers in the United States on the average get 13 vacation days per year. What percent of the 260 work days per year are vacation days?

11. Solve:

16 is what percent of 40?

12. Solve:

What percent of $84 is $10.50?

Answers on page A-14

Using Percents in Computations Many calculators have a $\boxed{\%}$ key that can be used in computations. (See the Calculator Corner on page 343.) For example, to find 11% of 49, we press $\boxed{1}\,\boxed{1}\,\boxed{\text{2nd}}\,\boxed{\%}\,\boxed{\times}$ $\boxed{4}\,\boxed{9}\,\boxed{=}$ or $\boxed{4}\,\boxed{9}\,\boxed{\times}\,\boxed{1}\,\boxed{1}\,\boxed{\text{SHIFT}}\,\boxed{\%}$. The display reads $\boxed{\quad 5.39\quad}$, so 11% of 49 is 5.39.

In Example 9, we perform the computation 20/5%. To use the $\boxed{\%}$ key in this computation, we press $\boxed{2}\,\boxed{0}\,\boxed{\div}\,\boxed{5}\,\boxed{\text{2nd}}\,\boxed{\%}\,\boxed{=}$, or $\boxed{2}\,\boxed{0}\,\boxed{\div}\,\boxed{5}\,\boxed{\text{SHIFT}}\,\boxed{\%}$. The result is 400.

We can also use the $\boxed{\%}$ key to find the percent number in a problem. In Example 11, for instance, we answer the question "13 is what percent of 260?" On a calculator, we press $\boxed{1}\,\boxed{3}\,\boxed{\div}\,\boxed{2}\,\boxed{6}\,\boxed{0}\,\boxed{\text{2nd}}\,\boxed{\%}\,\boxed{=}$, or $\boxed{1}\,\boxed{3}\,\boxed{\div}\,\boxed{2}\,\boxed{6}\,\boxed{0}\,\boxed{\text{SHIFT}}\,\boxed{\%}$. The result is 5, so 13 is 5% of 260.

Exercises: Use a calculator to find each of the following.

1. What is 5% of 24?

2. What is 12.6% of $40?

3. What is 19% of 256?

4. 140% of $16 is what?

5. 0.04% of 28 is what?

6. 33% of $90 is what?

7. 8% of what is 36?

8. $45 is 4.5% of what?

9. 23 is what percent of 920?

10. What percent of $442 is $53.04?

6.3

EXERCISE SET

For Extra Help

MathXL   MyMathLab   InterAct   Math Tutor   Digital Video   Student's
                      Math   Center   Tutor CD 3   Solutions
                                              Videotape 6   Manual

a Translate to an equation. Do not solve.

1. What is 32% of 78?

2. 98% of 57 is what?

3. 89 is what percent of 99?

4. What percent of 25 is 8?

5. 13 is 25% of what?

6. 21.4% of what is 20?

b Translate to an equation and solve.

7. What is 85% of 276?

8. What is 74% of 53?

9. 150% of 30 is what?

10. 100% of 13 is what?

11. What is 6% of $300?

12. What is 4% of $45?

13. 3.8% of 50 is what?

14. $33\frac{1}{3}$% of 480 is what?
$\left(\textit{Hint: } 33\frac{1}{3}\% = \frac{1}{3}.\right)$

15. $39 is what percent of $50?

16. $16 is what percent of $90?

17. 20 is what percent of 10?

18. 60 is what percent of 20?

19. What percent of $300 is $150?

20. What percent of $50 is $40?

21. What percent of 80 is 100?

22. What percent of 60 is 15?

23. 20 is 50% of what?

24. 57 is 20% of what?

25. 40% of what is $16?

26. 100% of what is $74?

27. 56.32 is 64% of what?

28. 71.04 is 96% of what?

29. 70% of what is 14?

30. 70% of what is 35?

31. What is $62\frac{1}{2}$% of 10?

32. What is $35\frac{1}{4}$% of 1200?

33. What is 8.3% of $10,200?

34. What is 9.2% of $5600?

35. $\mathbf{D_W}$ Write a question that could be translated to the equation

$$25 = 4\% \times b.$$

36. $\mathbf{D_W}$ Suppose we know that 40% of 92 is 36.8. What is a quick way to find 4% of 92? 400% of 92? Explain.

SKILL MAINTENANCE

Write fraction notation.   [4.1b]

37. 0.09

38. 1.79

39. 0.875

40. 0.125

41. 0.9375

42. 0.6875

Write decimal notation.   [4.1b]

43. $\dfrac{89}{100}$

44. $\dfrac{7}{100}$

45. $\dfrac{3}{10}$

46. $\dfrac{17}{1000}$

SYNTHESIS

Solve.

47. ▦ What is 7.75% of $10,880?

Estimate _____

Calculate _____

48. ▦ 50,951.775 is what percent of 78,995?

Estimate _____

Calculate _____

49. ▦ $2496 is 24% of what amount?

Estimate _____

Calculate _____

50. ▦ What is 38.2% of $52,345.79?

Estimate _____

Calculate _____

51. 40% of $18\frac{3}{4}$% of $25,000 is what?

CHAPTER 6: Percent Notation

SOLVING PERCENT PROBLEMS USING PROPORTIONS*

Objectives

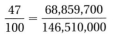 Translate percent problems to proportions.

b Solve basic percent problems.

a Translating to Proportions

A percent is a ratio of some number to 100. For example, 47% is the ratio $\frac{47}{100}$. The numbers 68,859,700 and 146,510,000 have the same ratio as 47 and 100.

$$\frac{47}{100} = \frac{68,859,700}{146,510,000}$$

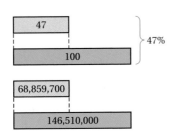

To solve a percent problem using a proportion, we translate as follows:

$$\text{Number} \rightarrow \frac{N}{100} = \frac{a}{b} \begin{array}{l} \leftarrow \text{Amount} \\ \leftarrow \text{Base} \end{array}$$

You might find it helpful to read this as "part is to whole as part is to whole."

For example, 60% of 25 is 15 translates to

$$\frac{60}{100} = \frac{15}{25}. \begin{array}{l} \leftarrow \text{Amount} \\ \leftarrow \text{Base} \end{array}$$

A clue in translating is that the base, b, corresponds to 100 and usually follows the wording "percent of." Also, $N\%$ always translates to $N/100$. Another aid in translating is to make a comparison drawing. To do this, we start with the percent side and list 0% at the top and 100% near the bottom. Then we estimate where the specified percent—in this case, 60%—is located. The corresponding quantities are then filled in. The base—in this case, 25—always corresponds to 100% and the amount—in this case, 15—corresponds to the specified percent.

In the United States, 47% of the labor force are women. In 2003, there were 146,510,000 people in the labor force. This means that 68,859,700 were women.

Sources: U.S. Department of Labor, Bureau of Labor Statistics

The proportion can then be read easily from the drawing: $\frac{60}{100} = \frac{15}{25}.$

*Note: This section presents an alternative method for solving basic percent problems. You can use either equations or proportions to solve percent problems, but you might prefer one method over the other, or your instructor may direct you to use one method over the other.

Translate to a proportion. Do not solve.

1. 12% of 50 is what?

2. What is 40% of 60?

3. 130% of 72 is what?

Translate to a proportion. Do not solve.

4. 45 is 20% of what?

5. 120% of what is 60?

EXAMPLE 1   Translate to a proportion.

23% of 5 is what?

$$\frac{23}{100} = \frac{a}{5}$$

Percents	Quantities
0%	0
23%	a
100%	5

EXAMPLE 2   Translate to a proportion.

What is 124% of 49?

$$\frac{124}{100} = \frac{a}{49}$$

Percents	Quantities
0%	0
100%	49
124%	a

Do Exercises 1–3.

EXAMPLE 3   Translate to a proportion.

3 is 10% of what?

$$\frac{10}{100} = \frac{3}{b}$$

Percents	Quantities
0%	0
10%	3
100%	b

EXAMPLE 4   Translate to a proportion.

45% of what is 23?

$$\frac{45}{100} = \frac{23}{b}$$

Percents	Quantities
0%	0
45%	23
100%	b

Do Exercises 4 and 5.

EXAMPLE 5   Translate to a proportion.

10 is what percent of 20?

$$\frac{N}{100} = \frac{10}{20}$$

Percents	Quantities
0%	0
N%	10
100%	20

Answers on page A-14

EXAMPLE 6 Translate to a proportion.

What percent of 50 is 7?

$$\frac{N}{100} = \frac{7}{50}$$

Percents	Quantities
0%	0
N%	7
100%	50

Do Exercises 6 and 7.

b Solving Percent Problems

After a percent problem has been translated to a proportion, we solve as in Section 5.3.

EXAMPLE 7 5% of what is $20?

Translate: $\dfrac{5}{100} = \dfrac{20}{b}$

Solve: $5 \cdot b = 100 \cdot 20$ Equating cross products

$\dfrac{5 \cdot b}{5} = \dfrac{100 \cdot 20}{5}$ Dividing by 5

$b = \dfrac{2000}{5}$

$b = 400$ Simplifying

Thus, 5% of $400 is $20. The answer is $400.

Percents	Quantities
0%	0
5%	20
100%	b

Do Exercise 8.

EXAMPLE 8 120% of 42 is what?

Translate: $\dfrac{120}{100} = \dfrac{a}{42}$

Solve: $120 \cdot 42 = 100 \cdot a$ Equating cross products

$\dfrac{120 \cdot 42}{100} = \dfrac{100 \cdot a}{100}$ Dividing by 100

$\dfrac{5040}{100} = a$

$50.4 = a$ Simplifying

Thus, 120% of 42 is 50.4. The answer is 50.4.

Percents	Quantities
0%	0
100%	42
120%	a

Do Exercises 9 and 10.

Translate to a proportion. Do not solve.

6. 16 is what percent of 40?

7. What percent of 84 is 10.5?

8. Solve:

20% of what is $45?

Solve.

9. 64% of 55 is what?

10. What is 12% of 50?

Answers on page A-14

11. Solve:

60 is 120% of what?

EXAMPLE 9   3 is 16% of what?

Translate: $\dfrac{3}{b} = \dfrac{16}{100}$

Solve: $3 \cdot 100 = b \cdot 16$ Equating cross products

$\dfrac{3 \cdot 100}{16} = \dfrac{b \cdot 16}{16}$ Dividing by 16

$\dfrac{300}{16} = b$ Multiplying and simplifying

$18.75 = b$ Dividing

Thus, 3 is 16% of 18.75. The answer is 18.75.

Do Exercise 11.

---

12. Solve:

$12 is what percent of $40?

EXAMPLE 10 $10 is what percent of $20?

Translate: $\dfrac{10}{20} = \dfrac{N}{100}$

Solve: $10 \cdot 100 = 20 \cdot N$ Equating cross products

$\dfrac{10 \cdot 100}{20} = \dfrac{20 \cdot N}{20}$ Dividing by 20

$\dfrac{1000}{20} = N$ Multiplying and simplifying

$50 = N$ Dividing

Thus, $10 is 50% of $20. The answer is 50%.

Do Exercise 12.

---

13. Solve:

What percent of 84 is 10.5?

EXAMPLE 11   What percent of 50 is 16?

Translate: $\dfrac{N}{100} = \dfrac{16}{50}$

Solve: $50 \cdot N = 100 \cdot 16$ Equating cross products

$\dfrac{50 \cdot N}{50} = \dfrac{100 \cdot 16}{50}$ Dividing by 50

$N = \dfrac{1600}{50}$ Multiplying and simplifying

$N = 32$ Dividing

Thus, 32% of 50 is 16. The answer is 32%.

Do Exercise 13.

Answers on page A-14

6.4

EXERCISE SET

For Extra Help

MathXL   MyMathLab   InterAct   Math Tutor   Digital Video   Student's
                     Math   Center   Tutor CD 3   Solutions
                                             Videotape 6   Manual

a   Translate to a proportion. Do not solve.

1. What is 37% of 74?

2. 66% of 74 is what?

3. 4.3 is what percent of 5.9?

4. What percent of 6.8 is 5.3?

5. 14 is 25% of what?

6. 133% of what is 40?

b   Translate to a proportion and solve.

7. What is 76% of 90?

8. What is 32% of 70?

9. 70% of 660 is what?

10. 80% of 920 is what?

11. What is 4% of 1000?

12. What is 6% of 2000?

13. 4.8% of 60 is what?

14. 63.1% of 80 is what?

15. $24 is what percent of $96?

16. $14 is what percent of $70?

17. 102 is what percent of 100?

18. 103 is what percent of 100?

19. What percent of $480 is $120?

20. What percent of $80 is $60?

21. What percent of 160 is 150?

22. What percent of 33 is 11?

23. $18 is 25% of what?

24. $75 is 20% of what?

25. 60% of what is 54?

26. 80% of what is 96?

27. 65.12 is 74% of what?

28. 63.7 is 65% of what?

29. 80% of what is 16?

30. 80% of what is 10?

31. What is $62\frac{1}{2}$% of 40?

32. What is $43\frac{1}{4}$% of 2600?

33. What is 9.4% of $8300?

34. What is 8.7% of $76,000?

35. $\mathbf{D_W}$ In your own words, list steps that a classmate could use to solve any percent problem in this section.

36. $\mathbf{D_W}$ In solving Example 10, a student simplifies $\frac{10}{20}$ before solving. Is this a good idea? Why or why not?

SKILL MAINTENANCE

Solve. [5.3b]

37. $\dfrac{x}{188} = \dfrac{2}{47}$

38. $\dfrac{15}{x} = \dfrac{3}{800}$

39. $\dfrac{4}{7} = \dfrac{x}{14}$

40. $\dfrac{612}{t} = \dfrac{72}{244}$

41. $\dfrac{5000}{t} = \dfrac{3000}{60}$

42. $\dfrac{75}{100} = \dfrac{n}{20}$

43. $\dfrac{x}{1.2} = \dfrac{36.2}{5.4}$

44. $\dfrac{y}{1\frac{1}{2}} = \dfrac{2\frac{3}{4}}{22}$

Solve.

45. A recipe for muffins calls for $\frac{1}{2}$ qt of buttermilk, $\frac{1}{3}$ qt of skim milk, and $\frac{1}{16}$ qt of oil. How many quarts of liquid ingredients does the recipe call for? [3.2b]

46. The Ferristown School District purchased $\frac{3}{4}$ ton (T) of clay. If the clay is to be shared equally among the district's 6 art departments, how much will each art department receive? [2.7d]

SYNTHESIS

Solve.

47. ▦ What is 8.85% of $12,640?
   Estimate _____
   Calculate _____

48. ▦ 78.8% of what is 9809.024?
   Estimate _____
   Calculate _____

6.5 APPLICATIONS OF PERCENT

Objectives

a Solve applied problems involving percent.

b Solve applied problems involving percent of increase or decrease.

a Applied Problems Involving Percent

Applied problems involving percent are not always stated in a manner easily translated to an equation. In such cases, it is helpful to rephrase the problem before translating. Sometimes it also helps to make a drawing.

EXAMPLE 1 *Presidential Deaths in Office.* George W. Bush was inaugurated as the 43rd President of the United States in 2001. Since Grover Cleveland was both the 22nd and the 24th presidents, there have been only 42 different presidents. Of the 42 presidents, 8 have died in office: William Henry Harrison, Zachary Taylor, Abraham Lincoln, James A. Garfield, William McKinley, Warren G. Harding, Franklin D. Roosevelt, and John F. Kennedy. What percent have died in office?

Harrison   Taylor   Garfield   McKinley

Harding   Roosevelt   Kennedy

1. **Familiarize.** The question asks for a percent of the presidents who have died in office. We note that 42 is approximately 40 and 8 is $\frac{1}{5}$, or 20%, of 40, so our answer is close to 20%. We let p = the percent who have died in office.

2. **Translate.** There are two ways in which we can translate this problem.

   Percent equation (see Section 6.3):

   8 is what percent of 42?

   8 = p · 42

   Proportion (see Section 6.4):

   $$\frac{N}{100} = \frac{8}{42}$$

   For proportions, $N\% = p$.

Percents	Quantities
0%	0
N%	8
100%	42

371

1. Presidential Assassinations in Office. Of the 42 U.S. presidents, 4 have been assassinated in office. These were Garfield, McKinley, Lincoln, and Kennedy. What percent have been assassinated in office?

3. Solve. We now have two ways in which to solve this problem.

Percent equation (see Section 6.3):

$$8 = p \cdot 42$$

$$\frac{8}{42} = \frac{p \cdot 42}{42} \qquad \text{Dividing by 42 on both sides}$$

$$\frac{8}{42} = p$$

$$0.190 \approx p \qquad \text{Finding decimal notation and rounding to the nearest thousandth}$$

$$19.0\% \approx p \qquad \text{Remember to find percent notation.}$$

Note here that the solution, p, includes the % symbol.

Proportion (see Section 6.4):

$$\frac{N}{100} = \frac{8}{42}$$

$$N \cdot 42 = 100 \cdot 8 \qquad \text{Equating cross products}$$

$$\frac{N \cdot 42}{42} = \frac{800}{42} \qquad \text{Dividing by 42 on both sides}$$

$$N = \frac{800}{42}$$

$$N \approx 19.0 \qquad \text{Dividing and rounding to the nearest tenth}$$

We use the solution of the proportion to express the answer to the problem as 19.0%. Note that in the proportion method, $N\% = p$.

4. Check. To check, we note that the answer 19.0% is close to 20%, as estimated in the *Familiarize* step.

5. State. About 19.0% of the U.S. presidents have died in office.

Do Exercise 1.

EXAMPLE 2 *Transportation to Work.* In the 15 largest cities in the United States, there are about 130,000,000 workers 16 years and over. Approximately 76.3% drive to work alone. How many workers in these 15 cities drive to work alone?

Transportation to Work (in the 15 Largest Cities in the United States)

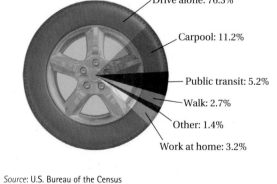

Drive alone: 76.3%
Carpool: 11.2%
Public transit: 5.2%
Walk: 2.7%
Other: 1.4%
Work at home: 3.2%

Source: U.S. Bureau of the Census

Answer on page A-14

1. Familiarize. We can make a drawing of a pie chart to help familiarize ourselves with the problem. We let b = the total number of workers who drive to work alone.

2. Transportation to Work. There are about 130,000,000 workers 16 years or over in the 15 largest cities in the United States. Approximately 11.2% carpool to work. How many workers in these 15 cities carpool to work?

Source: U.S. Bureau of the Census

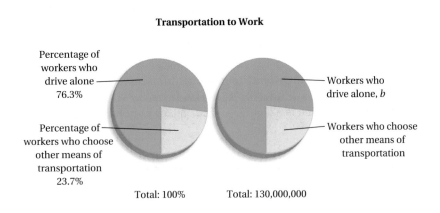

Transportation to Work

Percentage of workers who drive alone — 76.3%

Percentage of workers who choose other means of transportation 23.7%

Workers who drive alone, b

Workers who choose other means of transportation

Total: 100%   Total: 130,000,000

2. Translate. There are two ways in which we can translate this problem.

Percent equation:

$$
\begin{array}{ccccc}
\text{What number} & \text{is} & 76.3\% & \text{of} & 130{,}000{,}000? \\
\downarrow & \downarrow & \downarrow & \downarrow & \downarrow \\
b & = & 76.3\% & \cdot & 130{,}000{,}000
\end{array}
$$

Proportion:

$$\frac{76.3}{100} = \frac{b}{130{,}000{,}000}$$

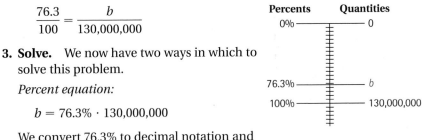

Percents   Quantities

0% —— 0

76.3% —— b

100% —— 130,000,000

3. Solve. We now have two ways in which to solve this problem.

Percent equation:

$$b = 76.3\% \cdot 130{,}000{,}000$$

We convert 76.3% to decimal notation and multiply:

$$b = 0.763 \times 130{,}000{,}000 = 99{,}190{,}000.$$

Proportion:

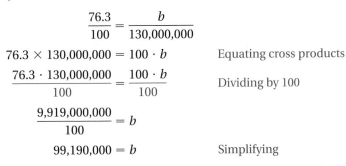

$$\frac{76.3}{100} = \frac{b}{130{,}000{,}000}$$

$76.3 \times 130{,}000{,}000 = 100 \cdot b$ Equating cross products

$\dfrac{76.3 \cdot 130{,}000{,}000}{100} = \dfrac{100 \cdot b}{100}$ Dividing by 100

$\dfrac{9{,}919{,}000{,}000}{100} = b$

$99{,}190{,}000 = b$ Simplifying

4. Check. To check, we can repeat the calculations. We can also do a partial check by estimating. Since 76.3% is about 80%, or $\frac{4}{5}$, and $\frac{4}{5}$ of 130,000,000 is 104,000,000 and 104,000,000 is close to 99,190,000, our answer is reasonable.

5. State. The number of workers in the 15 largest cities who drive to work alone is 99,190,000.

Answer on page A-14

Do Exercise 2.

3. Percent of Increase. The value of a car is $36,875. The price is increased by 4%.

   a) How much is the increase?

   b) What is the new price?

b Percent of Increase or Decrease

Percent is often used to state increase or decrease. Let's consider an example of each, using the price of a car as the original number.

PERCENT OF INCREASE

One year a car sold for $20,455. The manufacturer decides to raise the price of the following year's model by 6%. The increase is $0.06 \times \$20,455$, or $1227.30. The new price is $20,455 + $1227.30, or $21,682.30. The *percent of increase* is 6%.

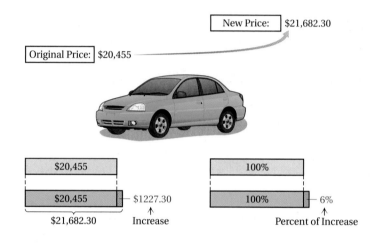

4. Percent of Decrease. The value of a car is $36,875. The car depreciates in value by 25% after one year.

   a) How much is the decrease?

   b) What is the depreciated value of the car?

PERCENT OF DECREASE

Lisa buys the car listed above for $20,455. After one year, the car depreciates in value by 25%. This is $0.25 \times \$20,455$, or $5113.75. This lowers the value of the car to

$$\$20,455 - \$5113.75, \quad \text{or} \quad \$15,341.25.$$

Note that the new price is thus 75% of the original price. If Lisa decides to sell the car after a year, $15,341.25 might be the most she could expect to get for it. The *percent of decrease* is 25%, and the decrease is $5113.75.

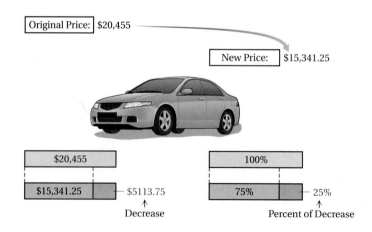

Answers on page A-14

Do Exercises 3 and 4.

When a quantity is decreased by a certain percent, we say we have **percent of decrease.**

EXAMPLE 3 *Chain Link Fence.* For one week only, Sam's Farm Supply had 4 ft × 50 ft rolls of galvanized chain link fence on sale for $39.99. The regular price was $49.99 per roll. What was the percent of decrease?

1. Familiarize. We find the amount of decrease and then make a drawing.

4 9 . 9 9	Retail price
− 3 9 . 9 9	Sale price
1 0 . 0 0	Decrease

2. Translate. There are two ways in which we can translate this problem.

Percent equation:

10.00 is what percent of 49.99?

$$10.00 = p \times 49.99$$

Proportion:

$$\frac{N}{100} = \frac{10.00}{49.99}$$

For proportions, $N\% = p$.

3. Solve. We have two ways in which to solve this problem.

Percent equation:

$$10.00 = p \times 49.99$$

$$\frac{10.00}{49.99} = \frac{p \times 49.99}{49.99} \qquad \text{Dividing by 49.99 on both sides}$$

$$\frac{10.00}{49.99} = p$$

$$0.20 \approx p$$

$$20\% \approx p \qquad \text{Converting to percent notation}$$

Proportion:

$$\frac{N}{100} = \frac{10.00}{49.99}$$

$$49.99 \times N = 100 \times 10 \qquad \text{Equating cross products}$$

$$\frac{49.99 \times N}{49.99} = \frac{100 \times 10}{49.99} \qquad \text{Dividing by 49.99 on both sides}$$

$$N = \frac{1000}{49.99}$$

$$N \approx 20$$

We use the solution of the proportion to express the answer to the problem as 20%.

Percents   Quantities
0% ——————— 0
N% ——————— 10.00

100% ——————— 49.99

5. Volume of Mail. The volume of U.S. mail decreased from about 208 billion pieces of mail in 2000 to 202 billion pieces in 2003. What was the percent of decrease?

Source: U.S. Postal Service

Answer on page A-14

4. Check. To check, we note that, with a 20% decrease, the reduced (or sale) price should be 80% of the retail (or original) price. Since

$$80\% \times 49.99 = 0.80 \times 49.99 = 39.992 \approx 39.99,$$

our answer checks.

5. State. The percent of decrease in the price of the roll of fence was 20%.

Do Exercise 5 on the preceding page.

When a quantity is increased by a certain percent, we say we have **percent of increase.**

EXAMPLE 4 *Motor Vehicle Production.* The number of motor vehicles produced worldwide increased from approximately 57.5 million in 2000 to 60.3 million in 2003. What was the percent of increase in motor vehicle production?

Source: Ward's Communications, *Ward's Motor Vehicle Facts & Figures,* 2004

1. Familiarize. We first note that the increase in the number of vehicles produced was $60.3 - 57.5$ million, or 2.8 million. A drawing can help us visualize the situation.

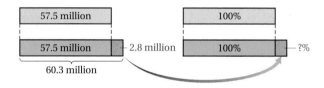

We are asking this question: The increase is what percent of the *original* amount? We let p = the percent of increase.

2. Translate. There are two ways in which we can translate this problem.

Percent equation:

2.8 million   is   what percent   of   57.5 million?

$$2.8 \quad = \quad p \quad \cdot \quad 57.5$$

Proportion:

$$\frac{N}{100} = \frac{2.8}{57.5}$$

For proportions, $N\% = p$.

Percents	Quantities
0%	0
N%	2.8
100%	57.5

3. Solve. We have two ways in which to solve this problem.

Percent equation:

$$2.8 = p \cdot 57.5$$

$$\frac{2.8}{57.5} = \frac{p \times 57.5}{57.5} \qquad \text{Dividing by 57.5 on both sides}$$

$$\frac{2.8}{57.5} = p$$

$$0.049 \approx p$$

$$4.9\% \approx p \qquad \text{Converting to percent notation}$$

Proportion:

$$\frac{N}{100} = \frac{2.8}{57.5}$$

$$57.5 \times N = 100 \times 2.8 \qquad \text{Equating cross products}$$

$$\frac{57.5 \times N}{57.5} = \frac{100 \times 2.8}{57.5} \qquad \text{Dividing by 57.5 on both sides}$$

$$N = \frac{280}{57.5}$$

$$N \approx 4.9$$

We use the solution of the proportion to express the answer to the problem as 4.9%.

4. Check. To check, we take 4.9% of 57.5:

$$4.9\% \times 57.5 = 0.049 \times 57.5 = 2.8175.$$

Since we rounded the percent, this approximation is close enough to 2.8 to be a good check.

5. State. The percent of increase in the number of motor vehicles produced is 4.9%.

Do Exercise 6.

6. Patents Issued. The number of patents issued per year by the U.S. government increased from 107,332 in 1993 to 189,597 in 2003. What was the percent of increase over the ten-year period?

Source: U.S. Patent and Trademark Office

Answer on page A-14

377

6.5 Applications of Percent

Translating for Success

1. *Distance Walked.*   After knee replacement, Alex walked $\frac{1}{8}$ mi each morning and $\frac{1}{5}$ mi each afternoon. How much farther did he walk in the afternoon?

2. *Stock Prices.*   A stock sold for $5 per share on Monday and only $2\frac{1}{8}$ per share on Friday. What was the percent of decrease from Monday to Friday?

3. *SAT Score.*   After attending a class titled *Improving Your SAT Scores,* Jacob raised his total score from 884 to 1040. What was the percent of increase?

4. *Change in Population.*   The population of a small farming community decreased from 1040 to 884. What was the percent of decrease?

5. *Lawn Mowing.*   During the summer, brothers Steve and Rob earned money for college by mowing lawns. The largest lawn they mowed was $2\frac{1}{8}$ acres. Steve can mow $\frac{1}{5}$ acre per hour, and Rob can mow only $\frac{1}{8}$ acre per hour. Working together, how many acres did they mow per hour?

The goal of these matching questions is to practice step (2), *Translate,* of the five-step problem-solving process. Translate each word problem to an equation and select a correct translation from equations A–O.

A. $x + \dfrac{1}{5} = \dfrac{1}{8}$

B. $250 = x \cdot 1040$

C. $884 = x \cdot 1040$

D. $\dfrac{250}{16.25} = \dfrac{1000}{x}$

E. $156 = x \cdot 1040$

F. $16.25 = 250 \cdot x$

G. $\dfrac{1}{5} + \dfrac{1}{8} = x$

H. $2\dfrac{1}{8} = x \cdot 5$

I. $5 = 2\dfrac{7}{8} \cdot x$

J. $\dfrac{1}{8} + x = \dfrac{1}{5}$

K. $1040 = x \cdot 884$

L. $\dfrac{250}{16.25} = \dfrac{x}{1000}$

M. $2\dfrac{7}{8} = x \cdot 5$

N. $x \cdot 884 = 156$

O. $x = 16.25 \cdot 250$

Answers on page A-14

---

6. *Land Sale.*   Cole sold $2\frac{1}{8}$ acres from the 5 acres he inherited from his uncle. What percent did he sell?

7. *Travel Expenses.*   A magazine photographer is reimbursed 16.25¢ per mile for business travel up to 1000 miles per week. In a recent week, he traveled only 250 miles. What was the total reimbursement for travel?

8. *Trip Expenses.*   The total expenses for Claire's recent business trip were $1040. She put $884 on her charge card and paid the balance in cash. What percent did she place on her charge card?

9. *Cost of Copies.*   During the first summer session at a community college, the campus copy center advertised 250 copies for $16.25. At this rate, what is the cost of 1000 copies?

10. *Cost of Insurance.*   Following a raise in the cost of health insurance, 250 of a company's 1040 employees dropped their health coverage. What percent of the employees canceled their insurance?

6.5

EXERCISE SET

For Extra Help

MathXL   MyMathLab   InterAct   Math Tutor   Digital Video   Student's
                        Math   Center   Tutor CD 3   Solutions
                                                 Videotape 6   Manual

a Solve.

1. *Wild Horses.* There are 27,369 wild horses on land managed by the Federal Bureau of Land Management. It is estimated that 48.4% of this total is in Nevada. How many wild horses are in Nevada?

Source: Bureau of Land Management, 2005

2. *U.S. Armed Forces* There are 1,168,195 people in the United States in active military service. The numbers in the four armed services are listed in the table below. What percent of the total does each branch represent? Round the answers to the nearest tenth of a percent.

U.S. ARMED FORCES WORLDWIDE, 2004	
Total	1,168,195
Air Force	314,477
Army	391,126
Marine Corps	135,324
Navy	327,268

Source: U.S. Department of Defense

3. *Car Value.* The base price of a Nissan 350Z is $34,000. This vehicle is expected to retain 62% of its value at the end of three years and 52% at the end of five years. What is the value of a Nissan 350Z after three years? after five years?

Source: November/December *Kelley Blue Book Residual Values Guide*

4. *Panda Survival.* Breeding the much-loved panda bear in captivity has been quite difficult for zookeepers.

  a) From 1964 to 1997, of 133 panda cubs born in captivity, only 90 lived to be one month old. What percent lived to be one month old?

  b) In 1999, Mark Edwards of the San Diego Zoo developed a nutritional formula on which 18 of 20 newborns lived to be one month old. What percent lived to be one month old?

5. *Overweight and Obese.* Of the 294 million people in the United States, 60% are considered overweight and 25% are considered obese. How many are overweight? How many are obese?

Source: U.S. Centers for Disease Control

6. *Smoking and Diabetes.* Of the 294 million people in the United States, 26% are smokers. How many are smokers?

Source: SAMHSA, Office of Applied Studies, National Survey on Drug Use and Health

7. A lab technician has 680 mL of a solution of water and acid; 3% is acid. How many milliliters are acid? water?

8. A lab technician has 540 mL of a solution of alcohol and water; 8% is alcohol. How many milliliters are alcohol? water?

9. *Mississippi River.* The Mississippi River, which extends from Minneapolis, Minnesota, to the Gulf of Mexico, is 2348 miles long. Approximately 77% of the river is navigable. How many miles of the river are navigable?

Source: National Oceanic and Atmospheric Administration

Mississippi River

10. *Immigrants.* In 2003, 705,827 immigrants entered the United States. Of this total, 16.4% were from Mexico and 7.1% were from India. How many immigrants came from Mexico? from India?

Source: U.S. Department of Justice, *2003 Yearbook of Immigration Statistics*

11. *Hispanic Population.* The Hispanic population is growing rapidly in the United States. In 2003, the population of the United States was about 291,000,000 and 13.7% of this total was Hispanic. How many Hispanic people lived in the United States in 2003?

Source: U.S. Bureau of the Census

12. *Age 65 and Over.* By 2010, it is predicted that 13.2% of the U.S. population will be 65 and over. If the population of the United States in 2010 is 307,000,000, how many people will be 65 and over?

Source: U.S. Bureau of the Census

13. *Test Results.* On a test of 40 items, Christina got 91% correct. (There was partial credit on some items.) How many items did she get correct? incorrect?

14. *Test Results.* On a test of 80 items, Pedro got 93% correct. (There was partial credit on some items.) How many items did he get correct? incorrect?

15. *Test Results.* On a test, Maj Ling got 86%, or 81.7, of the items correct. (There was partial credit on some items.) How many items were on the test?

16. *Test Results.* On a test, Juan got 85%, or 119, of the items correct. How many items were on the test?

17. *TV Usage.* Of the 8760 hr in a year, most television sets are on for 2190 hr. What percent is this?

18. *Colds from Kissing.* In a medical study, it was determined that if 800 people kiss someone who has a cold, only 56 will actually catch a cold. What percent is this?

Source: U.S. Centers for Disease Control

19. *Maximum Heart Rate.* Treadmill tests are often administered to diagnose heart ailments. A guideline in such a test is to try to get you to reach your *maximum heart rate,* in beats per minute. The maximum heart rate is found by subtracting your age from 220 and then multiplying by 85%. What is the maximum heart rate of someone whose age is 25? 36? 48? 55? 76? Round to the nearest one.

20. It costs an oil company $40,000 a day to operate two refineries. Refinery A accounts for 37.5% of the cost, and refinery B for the rest of the cost.
  a) What percent of the cost does it take to run refinery B?
  b) What is the cost of operating refinery A? refinery B?

b Solve.

21. *Savings Increase.* The amount in a savings account increased from $200 to $216. What was the percent of increase?

22. *Population Increase.* The population of a small mountain town increased from 840 to 882. What was the percent of increase?

23. During a sale, a dress decreased in price from $90 to $72. What was the percent of decrease?

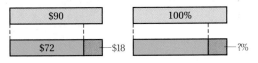

24. A person on a diet goes from a weight of 125 lb to a weight of 110 lb. What is the percent of decrease?

25. *Population Increase.* The population of the state of Nevada increased from 1,201,833 in 1990 to 2,241,154 in 2003. What is the percent of increase?
Source: U.S. Bureau of the Census

26. *Population Increase.* The population of the state of Utah increased from 1,722,850 in 1990 to 2,351,467 in 2003. What is the percent of increase?
Source: U.S. Bureau of the Census

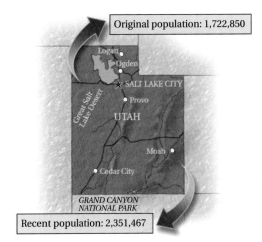

27. A person earns $28,600 one year and receives a 5% raise in salary. What is the new salary?

28. A person earns $43,200 one year and receives an 8% raise in salary. What is the new salary?

29. *Car Depreciation.* Irwin buys a car for $21,566. It depreciates 25% each year that he owns it. What is the depreciated value of the car after 1 yr? after 2 yr?

30. *Car Depreciation.* Janice buys a car for $22,688. It depreciates 25% each year that she owns it. What is the depreciated value of the car after 1 yr? after 2 yr?

31. *Doormat.* A bordered coir doormat 42 in. × 24 in. × $1\frac{1}{2}$ in. has a retail price of $89.95. Over the holidays it is on sale for $65.49. What is the percent of decrease?

32. *Business Tote.* A leather business tote retails for $239.99. An insurance company bought 30 totes for its sales staff at a reduced price of $184.95. What was the percent of decrease in the price?

Sale $65⁴⁹

JONES

COIR DOORMAT

Leather Business Tote

$239⁹⁹

33. *Two-by-Four.* A cross-section of a standard or nominal "two-by-four" board actually measures $1\frac{1}{2}$ in. by $3\frac{1}{2}$ in. The rough board is 2 in. by 4 in. but is planed and dried to the finished size. What percent of the wood is removed in planing and drying?

34. *Tipping.* Diners frequently add a 15% tip when charging a meal to a credit card. What is the total amount charged if the cost of the meal, without tip, is $18? $34? $49?

4 in.

2 in.

$3\frac{1}{2}$ in.

$1\frac{1}{2}$ in.

35. 📱 *Population Decrease.*   Between 1990 and 2000, the population of Detroit, Michigan, decreased from 1,028,000 to 951,000.

a) What is the percent of decrease?

b) If this percent of decrease over a 10-yr period repeated itself in the following decade, what would the population be in 2010?

Source: U.S. Bureau of the Census

36. 📱 *World Population.*   World population is increasing by 1.14% each year. In 2004, it was 6.39 billion. How much will it be in 2006? 2010? 2015?

Source: *The World Factbook,* 2004

Life Insurance Rates for Smokers and Nonsmokers.   The following table provides data showing how yearly rates (premiums) for a $500,000 term life insurance policy are increased for smokers. Complete the missing numbers in the table.

TYPICAL INSURANCE PREMIUMS (DOLLARS)

	AGE	RATE FOR NONSMOKER	RATE FOR SMOKER	PERCENT INCREASE FOR SMOKER
	35	$ 345	$ 630	83%
37.	40	$ 430	$ 735	
38.	45	$ 565		84%
39.	50	$ 780		100%
40.	55	$ 985		117%
41.	60	$1645	$2955	
42.	65	$2943	$5445	

Source: Pacific Life PL Protector Term Life Portfolio, OYT Rates

Population Increase.   The following table provides data showing how the populations of various states increased from 1990 to 2003. Complete the missing numbers in the table.

	STATE	POPULATION IN 1990	POPULATION IN 2003	CHANGE	PERCENT CHANGE
43.	Alaska	550,043	648,818		
44.	Connecticut	3,287,116		196,256	
45.	Montana		917,621	118,556	
46.	Texas		22,118,509	5,131,999	
47.	Colorado	3,294,394		1,256,294	
48.	Pennsylvania	11,881,643	12,365,455		

Source: U.S. Bureau of the Census

49. *Car Depreciation.* A car generally depreciates 25% of its original value in the first year. A car is worth $27,300 after the first year. What was its original cost?

50. *Car Depreciation.* Given normal use, an American-made car will depreciate 25% of its original cost the first year and 14% of its remaining value in the second year. What is the value of a car at the end of the second year if its original cost was $36,400? $28,400? $26,800?

51. *Strike Zone.* In baseball, the *strike zone* is normally a 17-in. by 30-in. rectangle. Some batters give the pitcher an advantage by swinging at pitches thrown out of the strike zone. By what percent is the area of the strike zone increased if a 2-in. border is added to the outside?

Source: Major League Baseball

52. Tony has planted grass seed on a 24-ft by 36-ft area in his back yard. He has also installed a 6-ft by 8-ft garden. By what percent does the garden reduce the area he will have to mow?

53. D_W Which is better for a wage earner, and why: a 10% raise followed by a 5% raise a year later, or a 5% raise followed by a 10% raise a year later?

54. D_W A worker receives raises of 3%, 6%, and then 9%. By what percent has the original salary increased? Explain.

SKILL MAINTENANCE

Convert to decimal notation. [4.1b], [4.5a]

55. $\dfrac{25}{11}$

56. $\dfrac{11}{25}$

57. $\dfrac{27}{8}$

58. $\dfrac{43}{9}$

59. $\dfrac{23}{25}$

60. $\dfrac{20}{24}$

61. $\dfrac{14}{32}$

62. $\dfrac{2317}{1000}$

63. $\dfrac{34,809}{10,000}$

64. $\dfrac{27}{40}$

SYNTHESIS

65. *Adult Height.* It has been determined that at the age of 10, a girl has reached 84.4% of her final adult growth. Cynthia is 4 ft, 8 in. at the age of 10. What will be her final adult height?

Source: *Dunlop Illustrated Encyclopedia of Facts.* New York: Sterling Publishing, 1970.

67. If p is 120% of q, then q is what percent of p?

66. *Adult Height.* It has been determined that at the age of 15, a boy has reached 96.1% of his final adult height. Claude is 6 ft, 4 in. at the age of 15. What will be his final adult height?

Source: *Dunlop Illustrated Encyclopedia of Facts.* New York: Sterling Publishing, 1970.

68. A coupon allows a couple to have dinner and then have $10 subtracted from the bill. Before subtracting $10, however, the restaurant adds a tip of 15%. If the couple is presented with a bill for $44.05, how much would the dinner (without tip) have cost without the coupon?

6.6 SALES TAX, COMMISSION, AND DISCOUNT

Objectives

a Solve applied problems involving sales tax and percent.

b Solve applied problems involving commission and percent.

c Solve applied problems involving discount and percent.

a Sales Tax

Sales tax computations represent a special type of percent of increase problem. The sales tax rate in Maryland is 5%. This means that the tax is 5% of the purchase price. Suppose the purchase price on a coat is $124.95. The sales tax is then 5% of $124.95, or 0.05×124.95, or 6.2475, or about $6.25.

$124.95
+ 5% sales tax

Baltimore

Annapolis

BILL:		
Purchase price	=	$124.95
Sales tax (5% of $124.95)	=	+ 6.25
Total price		$131.20

The total that you pay is the price plus the sales tax:

$124.95 + $6.25,   or $131.20.

SALES TAX

Sales tax = Sales tax rate × Purchase price

Total price = Purchase price + Sales tax

EXAMPLE 1 *Florida Sales Tax.* The sales tax rate in Florida is 6%. How much tax is charged on the purchase of 4 inflatable rafts at $89.99 each? What is the total price?

a) We first find the cost of the rafts. It is

$4 \times \$89.99 = \$359.96.$

b) The sales tax on items costing $359.96 is

$$\underbrace{\text{Sales tax rate}}_{6\%} \times \underbrace{\text{Purchase price}}_{\$359.96}$$

or 0.06×359.96, or 21.5976. Thus the tax is $21.60 (rounded to the nearest cent).

$89⁹⁹ plus 6% each

c) The total price is given by the purchase price plus the sales tax:

$359.96 + $21.60,   or $381.56.

To check, note that the total price is the purchase price plus 6% of the purchase price. Thus the total price is 106% of the purchase price. Since $1.06 \times 359.96 \approx 381.56$, we have a check. The sales tax is $21.60 and the total price is $381.56.

Do Exercises 1 and 2.

1. **California Sales Tax.** The sales tax rate in California is 7.25%. How much tax is charged on the purchase of a refrigerator that sells for $668.95? What is the total price?

2. **Louisiana Sales Tax.** Sam buys 5 hardcover copies of Dean Koontz's novel *From the Corner of His Eye* for $26.95 each. The sales tax rate in Louisiana is 4%. How much sales tax will be charged? What is the total price?

Answers on page A-14

3. The sales tax is $50.94 on the purchase of a night table that costs $849. What is the sales tax rate?

4. The sales tax on a portable navigation system is $59.94 and the sales tax rate is 6%. Find the purchase price (the price before taxes are added).

Portable
Navigation
System
$?
Tax at 6% =
$59.94

NEW LOW
PRICE
$?

$8.93 Tax @ 5%

EXAMPLE 2   The sales tax is $43.96 on the purchase of this wooden play center, which costs $1099. What is the sales tax rate?

Wooden Play Center
27 sq ft play deck
with slide and
swing set

$1099

+ $43.96 sales tax at ?% sales tax rate

Rephrase: Sales tax   is   what percent   of   purchase price?

Translate: $43.96   =   r   × $1099

To solve the equation, we divide by 1099 on both sides:

$$\frac{43.96}{1099} = \frac{r \times 1099}{1099}$$

$$\frac{43.96}{1099} = r$$

$$0.04 = r$$

$$4\% = r.$$

The sales tax rate is 4%.

Do Exercise 3.

EXAMPLE 3   The sales tax on a Tiffany torchiere lamp is $8.93 and the sales tax rate is 5%. Find the purchase price (the price before taxes are added).

Rephrase: Sales tax   is   5%   of   what?

Translate: $8.93   =   5%   ×   b,   or 8.93 = 0.05 × b.

To solve, we divide by 0.05 on both sides:

$$\frac{8.93}{0.05} = \frac{0.05 \times b}{0.05}$$

$$\frac{8.93}{0.05} = b$$

$$\$178.60 = b.$$

The purchase price is $178.60.

Do Exercise 4.

Answers on page A-14

b Commission

When you work for a **salary,** you receive the same amount of money each week or month. When you work for a **commission,** you are paid a percentage of the total sales for which you are responsible.

5. Raul's commission rate is 30%. What is the commission from the sale of $18,760 worth of air conditioners?

COMMISSION	
Commission = Commission rate × Sales	

EXAMPLE 4 *Exercise Equipment Sales.* A salesperson's commission rate is 16%. What is the commission from the sale of $9700 worth of exercise equipment?

16% commission

$$
\begin{aligned}
Commission &= Commission\ rate \times Sales \\
C &= 16\% \times 9700 \\
C &= 0.16 \times 9700 \\
C &= 1552
\end{aligned}
$$

The commission is $1552.

Do Exercise 5.

EXAMPLE 5 *Farm Machinery Sales.* Dawn earns a commission of $30,000 selling $600,000 worth of farm machinery. What is the commission rate?

FARM SUPPLIES

COMMISSION $30,000
TOTAL SALES $600,000

$$
\begin{aligned}
Commission &= Commission\ rate \times Sales \\
30,000 &= r \times 600,000
\end{aligned}
$$

Answer on page A-14

6. Liz earns a commission of $3000 selling $24,000 worth of *NSYNC concert tickets. What is the commission rate?

To solve this equation, we divide by 600,000 on both sides:

$$\frac{30,000}{600,000} = \frac{r \times 600,000}{600,000}$$

$$\frac{1}{20} = r$$

$$0.05 = r$$

$$5\% = r.$$

The commission rate is 5%.

Do Exercise 6.

EXAMPLE 6 *Motorcycle Sales.* Joyce's commission rate is 12%. She receives a commission of $936 on the sale of a motorcycle. How much did the motorcycle cost?

7. Ben's commission rate is 16%. He receives a commission of $268 from sales of clothing. How many dollars worth of clothing were sold?

$$
\begin{array}{ccccc}
Commission & = & Commission\ rate & \times & Sales \\
936 & = & 12\% & \times & S, \quad \text{or } 936 = 0.12 \times S
\end{array}
$$

To solve this equation, we divide by 0.12 on both sides:

$$\frac{936}{0.12} = \frac{0.12 \times S}{0.12}$$

$$\frac{936}{0.12} = S$$

$$7800 = S.$$

The motorcycle cost $7800.

Do Exercise 7.

c Discount

Suppose that the regular price of a rug is $60, and the rug is on sale at 25% off. Since 25% of $60 is $15, the sale price is $60 − $15, or $45. We call $60 the **original,** or **marked price,** 25% the **rate of discount,** $15 the **discount,** and $45 the **sale price.** Note that discount problems are a type of percent of decrease problem.

DISCOUNT AND SALE PRICE
Discount = Rate of discount × Original price
Sale price = Original price − Discount

Answers on page A-14

EXAMPLE 7 A masonite door marked $1389 is on sale at $33\frac{1}{3}\%$ off. What is the discount? the sale price?

DOOR
$1389 orig.

Save

$33\frac{1}{3}\%$

Sale price = ?

a) *Discount* = *Rate of discount* × *Original price*

$$D = 33\frac{1}{3}\% \times 1389$$

$$D = \frac{1}{3} \times 1389$$

$$D = \frac{1389}{3} = 463$$

b) *Sale price* = *Original price* − *Discount*

$$S = 1389 - 463$$

$$S = 926$$

The discount is $463 and the sale price is $926.

Do Exercise 8.

EXAMPLE 8 *Antique Pricing.* An antique table is marked down from $620 to $527. What is the rate of discount?

We first find the discount by subtracting the sale price from the original price:

$$620 - 527 = 93.$$

The discount is $93.

Next, we use the equation for discount:

Discount = *Rate of discount* × *Original price*

$$93 = r \times 620.$$

To solve, we divide by 620 on both sides:

$$\frac{93}{620} = \frac{r \times 620}{620}$$

$$\frac{93}{620} = r$$

$$0.15 = r$$

$$15\% = r.$$

The discount rate is 15%.

> To check, note that a 15% discount rate means that 85% of the original price is paid:
> $$0.85 \times 620 = 527.$$

Do Exercise 9.

8. A suit marked $540 is on sale at $33\frac{1}{3}\%$ off. What is the discount? the sale price?

9. A pair of hiking boots is reduced from $75 to $60. Find the rate of discount.

Reduced to
$60
Original price
$75

6.6

EXERCISE SET

For Extra Help

MathXL   MyMathLab   InterAct   Math Tutor   Digital Video   Student's
                        Math   Center   Tutor CD 3   Solutions
                                                Videotape 6   Manual

a   Solve.

1. *Tennessee Sales Tax.*   The sales tax rate in Tennessee is 7%. How much sales tax would be charged on a lawn mower that costs $279?

2. *Arizona Sales Tax.*   The sales tax rate in Arizona is 5.6%. How much sales tax would be charged on a lawn mower that costs $279?

3. *Kansas Sales Tax.*   The sales tax rate in Kansas is 5.3%. How much sales tax would be charged on a video game, Quest of the Planets, which sells for $49.99?

NEW ARRIVAL
QUEST OF THE PLANETS

VIDEO $49.99

4. *New Jersey Sales Tax.*   The sales tax rate in New Jersey is 6%. How much sales tax would be charged on a copy of John Grisham's novel *A Painted House,* which sells for $27.95?

Source: Borders Bookstore; Andrea Sutcliffe, *Numbers*

5. *Utah Sales Tax.*   The sales tax rate in Utah is 4.75%. How much tax is charged on a purchase of 5 telephones at $69 apiece? What is the total price?

6. *New York Sales Tax.*   The sales tax rate in New York is 4.25%. How much tax is charged on a purchase of 5 teapots at $37.99 apiece? What is the total price?

7. The sales tax is $48 on the purchase of a dining room set that sells for $960. What is the sales tax rate?

8. The sales tax is $15 on the purchase of a diamond ring that sells for $500. What is the sales tax rate?

9. The sales tax is $35.80 on the purchase of a refrigerator–freezer that sells for $895. What is the sales tax rate?

10. The sales tax is $9.12 on the purchase of a patio set that sells for $456. What is the sales tax rate?

11. The sales tax on a used car is $100 and the sales tax rate is 5%. Find the purchase price (the price before taxes are added).

12. The sales tax on the purchase of a new boat is $112 and the sales tax rate is 2%. Find the purchase price.

13. The sales tax on a dining room set is $28 and the sales tax rate is 3.5%. Find the purchase price.

14. The sales tax on a portable DVD player is $24.75 and the sales tax rate is 5.5%. Find the purchase price.

15. The sales tax rate in Austin is 2% for the city and county and 6.25% for the state. Find the total amount paid for 2 shower units at $332.50 apiece.

16. The sales tax rate in Omaha is 1.5% for the city and 5% for the state. Find the total amount paid for 3 air conditioners at $260 apiece.

17. The sales tax is $1030.40 on an automobile purchase of $18,400. What is the sales tax rate?

18. The sales tax is $979.60 on an automobile purchase of $15,800. What is the sales tax rate?

b   Solve.

19. Katrina's commission rate is 6%. What is the commission from the sale of $45,000 worth of furnaces?

20. Jose's commission rate is 32%. What is the commission from the sale of $12,500 worth of sailboards?

21. Mitchell earns $120 selling $2400 worth of television sets in a consignment shop. What is the commission rate?

22. Donna earns $408 selling $3400 worth of shoes. What is the commission rate?

23. An art gallery's commission rate is 40%. They receive a commission of $392. How many dollars worth of artwork were sold?

24. A real estate agent's commission rate is 7%. She receives a commission of $5600 on the sale of a home. How much did the home sell for?

25. A real estate commission is 6%. What is the commission on the sale of a $98,000 home?

26. A real estate commission is 8%. What is the commission on the sale of a piece of land for $68,000?

27. Bonnie earns $280.80 selling $2340 worth of tee shirts. What is the commission rate?

28. Chuck earns $1147.50 selling $7650 worth of ski passes. What is the commission rate?

29. Miguel's commission is increased according to how much he sells. He receives a commission of 5% for the first $2000 and 8% on the amount over $2000. What is the total commission on sales of $6000?

30. Lucinda earns a salary of $500 a month, plus a 2% commission on sales. One month, she sold $990 worth of encyclopedias. What were her wages that month?

Find what is missing.

	MARKED PRICE	RATE OF DISCOUNT	DISCOUNT	SALE PRICE
31.	$300	10%		
32.	$2000	40%		
33.	$17	15%		
34.	$20	25%		
35.		10%	$12.50	
36.		15%	$65.70	
37.	$600		$240	
38.	$12,800		$1920	

39. Find the discount and the rate of discount for the car seat in this ad.

40. Find the discount and the rate of discount for the wicker chair in this ad.

41. Find the marked price and the rate of discount for the stacked tool storage units in this ad.

42. Find the marked price and the rate of discount for the basketball system in this ad.

392

CHAPTER 6: Percent Notation

Copyright © 2007 Pearson Education, Inc.

43. **Dw** Is the following ad mathematically correct? Why or why not?

44. **Dw** An item that is no longer on sale at "25% off" receives a price tag that is $33\frac{1}{3}\%$ more than the sale price. Has the item price been restored to its original price? Why or why not?

45. **Dw** Which is better, a discount of 40% or a discount of 20% followed by another of 20%? Explain.

46. **Dw** You take 40% of 50% of a number. What percent of the number could you take to obtain the same result making only one multiplication? Explain your answer.

SKILL MAINTENANCE

Solve.   [5.3b]

47. $\dfrac{x}{12} = \dfrac{24}{16}$

48. $\dfrac{7}{2} = \dfrac{11}{x}$

Solve.   [4.4b]

49. $0.64 \cdot x = 170$

50. $28.5 = 25.6 \times y$

Find decimal notation.   [4.5a]

51. $\dfrac{5}{9}$

52. $\dfrac{23}{11}$

53. $\dfrac{11}{12}$

54. $\dfrac{13}{7}$

55. $\dfrac{15}{7}$

56. $\dfrac{19}{12}$

Convert to standard notation.   [4.3b]

57. $4.03 trillion

58. 5.8 million

59. 42.7 million

60. 6.09 trillion

SYNTHESIS

61. ▦ *Magazine Subscriptions.* In a recent subscription drive, *People* offered a subscription of 52 weekly issues for a price of $1.89 per issue. They advertised that this was a savings of 29.7% off the newsstand price. What was the newsstand price?

Source: *People Magazine*

62. ▦ Gordon receives a 10% commission on the first $5000 in sales and 15% on all sales beyond $5000. If Gordon receives a commission of $2405, how much did he sell? Use a calculator and trial and error if you wish.

63. Herb collects baseball memorabilia. He bought two autographed plaques, but became short of funds and had to sell them quickly for $200 each. On one, he made a 20% profit and on the other, he lost 20%. Did he make or lose money on the sale?

64. Tee shirts are being sold at the mall for $5 each, or 3 for $10. If you buy three tee shirts, what is the rate of discount?

Objectives

6.7 SIMPLE AND COMPOUND INTEREST

a Solve applied problems involving simple interest.

b Solve applied problems involving compound interest.

1. What is the simple interest on $4300 invested at an interest rate of 7% for 1 year?

a Simple Interest

Suppose you put $1000 into an investment for 1 year. The $1000 is called the **principal.** If the **interest rate** is 8%, in addition to the principal, you get back 8% of the principal, which is

8% of $1000,   or   0.08 × 1000,   or $80.00.

The $80.00 is called the **simple interest.** It is, in effect, the price that a financial institution pays for the use of the money over time.

> **SIMPLE INTEREST FORMULA**
>
> The **simple interest** I on principal P, invested for t years at interest rate r, is given by
>
> $$I = P \cdot r \cdot t.$$

EXAMPLE 1   What is the simple interest on $2500 invested at an interest rate of 6% for 1 year?

We use the formula $I = P \cdot r \cdot t$:

$$I = P \cdot r \cdot t = \$2500 \times 6\% \times 1$$
$$= \$2500 \times 0.06$$
$$= \$150.$$

The simple interest for 1 year is $150.

Do Exercise 1.

2. What is the simple interest on a principal of $4300 invested at an interest rate of 7% for $\frac{3}{4}$ year?

EXAMPLE 2   What is the simple interest on a principal of $2500 invested at an interest rate of 6% for $\frac{1}{4}$ year?

We use the formula $I = P \cdot r \cdot t$:

$$I = P \cdot r \cdot t = \$2500 \times 6\% \times \frac{1}{4}$$
$$= \frac{\$2500 \times 0.06}{4}$$
$$= \$37.50.$$

> We could instead have found $\frac{1}{4}$ of 6% and then multiplied by 2500.

The simple interest for $\frac{1}{4}$ year is $37.50.

Do Exercise 2.

Answers on page A-15

When time is given in days, we generally divide it by 365 to express the time as a fractional part of a year.

EXAMPLE 3 To pay for a shipment of tee shirts, New Wave Designs borrows $8000 at $9\frac{3}{4}$% for 60 days. Find (a) the amount of simple interest that is due and (b) the total amount that must be paid after 60 days.

a) We express 60 days as a fractional part of a year:

$$I = P \cdot r \cdot t = \$8000 \times 9\frac{3}{4}\% \times \frac{60}{365}$$

$$= \$8000 \times 0.0975 \times \frac{60}{365}$$

$$\approx \$128.22.$$

The interest due for 60 days is $128.22.

b) The total amount to be paid after 60 days is the principal plus the interest:

$$\$8000 + \$128.22 = \$8128.22.$$

The total amount due is $8128.22.

Do Exercise 3.

3. The Glass Nook borrows $4800 at $8\frac{1}{2}$% for 30 days. Find (a) the amount of simple interest due and (b) the total amount that must be paid after 30 days.

b Compound Interest

When interest is paid *on interest,* we call it **compound interest.** This is the type of interest usually paid on investments. Suppose you have $5000 in a savings account at 6%. In 1 year, the account will contain the original $5000 plus 6% of $5000. Thus the total in the account after 1 year will be

106% of $5000,   or   1.06 × $5000,   or $5300.

Now suppose that the total of $5300 remains in the account for another year. At the end of this second year, the account will contain the $5300 plus 6% of $5300. The total in the account would thus be

106% of $5300,   or   1.06 × $5300,   or $5618.

Note that in the second year, interest is earned on the first year's interest. When this happens, we say that interest is **compounded annually.**

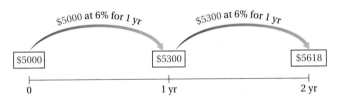

Answer on page A-15

4. Find the amount in an account if $2000 is invested at 9%, compounded annually, for 2 years.

EXAMPLE 4 Find the amount in an account if $2000 is invested at 8%, compounded annually, for 2 years.

a) After 1 year, the account will contain 108% of $2000:

$$1.08 \times \$2000 = \$2160.$$

b) At the end of the second year, the account will contain 108% of $2160:

$$1.08 \times \$2160 = \$2332.80.$$

The amount in the account after 2 years is $2332.80.

Do Exercise 4.

Suppose that the interest in Example 4 were **compounded semi-annually**—that is, every half year. Interest would then be calculated twice a year at a rate of 8% ÷ 2, or 4% each time. The approach used in Example 4 can then be adapted, as follows.

After the first $\frac{1}{2}$ year, the account will contain 104% of $2000:

$$1.04 \times \$2000 = \$2080.$$

After a second $\frac{1}{2}$ year (1 full year), the account will contain 104% of $2080:

$$1.04 \times \$2080 = \$2163.20.$$

After a third $\frac{1}{2}$ year $\left(1\frac{1}{2}\text{ full years}\right)$, the account will contain 104% of $2163.20:

$$1.04 \times \$2163.20 = \$2249.728$$
$$\approx \$2249.73. \qquad \text{Rounding to the nearest cent}$$

Finally, after a fourth $\frac{1}{2}$ year (2 full years), the account will contain 104% of $2249.73:

$$1.04 \times \$2249.73 = \$2339.7192$$
$$\approx \$2339.72. \qquad \text{Rounding to the nearest cent}$$

Let's summarize our results and look at them another way:

End of 1st $\frac{1}{2}$ year $\rightarrow 1.04 \times 2000 = 2000 \times (1.04)^1$;

End of 2nd $\frac{1}{2}$ year $\rightarrow 1.04 \times (1.04 \times 2000) = 2000 \times (1.04)^2$;

End of 3rd $\frac{1}{2}$ year $\rightarrow 1.04 \times (1.04 \times 1.04 \times 2000) = 2000 \times (1.04)^3$;

End of 4th $\frac{1}{2}$ year $\rightarrow 1.04 \times (1.04 \times 1.04 \times 1.04 \times 2000) = 2000 \times (1.04)^4$.

Note that each multiplication was by 1.04 and that

$$\$2000 \times 1.04^4 \approx \$2339.72. \qquad \text{Using a calculator and rounding to the nearest cent}$$

We have illustrated the following result.

COMPOUND INTEREST FORMULA

If a principal P has been invested at interest rate r, compounded n times a year, in t years it will grow to an amount A given by

$$A = P \cdot \left(1 + \frac{r}{n}\right)^{n \cdot t}.$$

Answer on page A-15

CHAPTER 6: Percent Notation

Let's apply this formula to confirm our preceding discussion, where the amount invested is $P = \$2000$, the number of years is $t = 2$, and the number of compounding periods each year is $n = 2$. Substituting into the compound interest formula, we have

$$A = P \cdot \left(1 + \frac{r}{n}\right)^{n \cdot t} = 2000 \cdot \left(1 + \frac{8\%}{2}\right)^{2 \cdot 2}$$

$$= 2000 \cdot \left(1 + \frac{0.08}{2}\right)^4 = 2000(1.04)^4$$

$$= 2000 \times 1.16985856 \approx \$2339.72.$$

If you are using a calculator, you could perform this computation in one step.

EXAMPLE 5 The Ibsens invest \$4000 in an account paying $5\frac{5}{8}\%$, compounded quarterly. Find the amount in the account after $2\frac{1}{2}$ years.

The compounding is quarterly, so n is 4. We substitute \$4000 for P, $5\frac{5}{8}\%$, or 0.05625, for r, 4 for n, and $2\frac{1}{2}$, or $\frac{5}{2}$, for t and compute A:

$$A = P \cdot \left(1 + \frac{r}{n}\right)^{n \cdot t} = \$4000 \cdot \left(1 + \frac{5\frac{5}{8}\%}{4}\right)^{4 \cdot 5/2}$$

$$= \$4000 \cdot \left(1 + \frac{0.05625}{4}\right)^{10}$$

$$= \$4000(1.0140625)^{10}$$

$$\approx \$4599.46.$$

The amount in the account after $2\frac{1}{2}$ years is \$4599.46.

Do Exercise 5.

5. A couple invests \$7000 in an account paying $6\frac{3}{8}\%$, compounded semiannually. Find the amount in the account after $1\frac{1}{2}$ years.

Answer on page A-15

CALCULATOR CORNER

Compound Interest   A calculator is useful in computing compound interest. Not only does it do computations quickly but it also eliminates the need to round until the computation is completed. This minimizes "round-off errors" that occur when rounding is done at each stage of the computation. We must keep order of operations in mind when computing compound interest.

To find the amount due on a \$20,000 loan made for 25 days at 11% interest, compounded daily, we would compute $20{,}000\left(1 + \frac{0.11}{365}\right)^{25}$. To do this on a calculator, we press $\boxed{2}\,\boxed{0}\,\boxed{0}\,\boxed{0}\,\boxed{0}\,\boxed{\times}\,\boxed{(}\,\boxed{1}\,\boxed{+}\,\boxed{.}\,\boxed{1}\,\boxed{1}\,\boxed{\div}$
$\boxed{3}\,\boxed{6}\,\boxed{5}\,\boxed{)}\,\boxed{y^x}$ (or $\boxed{x^y}$) $\boxed{2}\,\boxed{5}\,\boxed{=}$. Without parentheses, we would first find $1 + \frac{0.11}{365}$, raise this result to the
25th power, and then multiply by 20,000. To do this, we press $\boxed{1}\,\boxed{+}\,\boxed{.}\,\boxed{1}\,\boxed{1}\,\boxed{\div}\,\boxed{3}\,\boxed{6}\,\boxed{5}\,\boxed{=}\,\boxed{y^x}$ (or $\boxed{x^y}$)
$\boxed{2}\,\boxed{5}\,\boxed{=}\,\boxed{\times}\,\boxed{2}\,\boxed{0}\,\boxed{0}\,\boxed{0}\,\boxed{0}$. In either case, the result is 20,151.23, rounded to the nearest cent.
Some calculators have business keys that allow such computations to be done more quickly.

Exercises:

1. Find the amount due on a \$16,000 loan made for 62 days at 13% interest, compounded daily.

2. An investment of \$12,500 is made for 90 days at 8.5% interest, compounded daily. How much is the investment worth after 90 days?

6.7

EXERCISE SET

For Extra Help

MathXL   MyMathLab   InterAct   Math Tutor   Digital Video   Student's
                        Math   Center   Tutor CD 3   Solutions
                                                 Videotape 6   Manual

a   Find the simple interest.

	PRINCIPAL	RATE OF INTEREST	TIME	SIMPLE INTEREST
1.	$200	4%	1 year	
2.	$450	2%	1 year	
3.	$2000	8.4%	$\frac{1}{2}$ year	
4.	$200	7.7%	$\frac{1}{2}$ year	
5.	$4300	10.56%	$\frac{1}{4}$ year	
6.	$8000	9.42%	$\frac{1}{6}$ year	
7.	$20,000	$4\frac{5}{8}\%$	1 year	
8.	$100,000	$3\frac{7}{8}\%$	1 year	
9.	$50,000	$5\frac{3}{8}\%$	$\frac{1}{4}$ year	
10.	$80,000	$6\frac{3}{4}\%$	$\frac{1}{12}$ year	

Solve. Assume that simple interest is being calculated in each case.

11. CopiPix, Inc. borrows $10,000 at 9% for 60 days. Find (a) the amount of interest due and (b) the total amount that must be paid after 60 days.

12. Sal's Laundry borrows $8000 at 10% for 90 days. Find (a) the amount of interest due and (b) the total amount that must be paid after 90 days.

13. Animal Instinct, a pet supply shop, borrows $6500 at 5% for 90 days. Find (a) the amount of interest due and (b) the total amount that must be paid after 90 days.

14. Andante's Cafe borrows $4500 at 12% for 60 days. Find (a) the amount of interest due and (b) the total amount that must be paid after 60 days.

15. Jean's Garage borrows $5600 at 10% for 30 days. Find (a) the amount of interest due and (b) the total amount that must be paid after 30 days.

16. Shear Delights, a hair salon, borrows $3600 at 4% for 30 days. Find (a) the amount of interest due and (b) the total amount that must be paid after 30 days.

b Interest is compounded annually. Find the amount in the account after the given length of time. Round to the nearest cent.

	PRINCIPAL	RATE OF INTEREST	TIME	AMOUNT IN THE ACCOUNT
17.	$400	5%	2 years	
18.	$450	4%	2 years	
19.	$2000	8.8%	4 years	
20.	$4000	7.7%	4 years	
21.	$4300	10.56%	6 years	
22.	$8000	9.42%	6 years	
23.	$20,000	$6\frac{5}{8}\%$	25 years	
24.	$100,000	$5\frac{7}{8}\%$	30 years	

Interest is compounded semiannually. Find the amount in the account after the given length of time. Round to the nearest cent.

	PRINCIPAL	RATE OF INTEREST	TIME	AMOUNT IN THE ACCOUNT
25.	$4000	6%	1 year	
26.	$1000	5%	1 year	
27.	$20,000	8.8%	4 years	
28.	$40,000	7.7%	4 years	
29.	$5000	10.56%	6 years	
30.	$8000	9.42%	8 years	
31.	$20,000	$7\frac{5}{8}\%$	25 years	
32.	$100,000	$4\frac{7}{8}\%$	30 years	

Solve.

33. A family invests $4000 in an account paying 6%, compounded monthly. How much is in the account after 5 months?

34. A couple invests $2500 in an account paying 3%, compounded monthly. How much is in the account after 6 months?

35. A couple invests $1200 in an account paying 10%, compounded quarterly. How much is in the account after 1 year?

36. The O'Hares invest $6000 in an account paying 8%, compounded quarterly. How much is in the account after 18 months?

37. D_W Which is a better investment and why: $1000 invested at $7\frac{3}{4}$% simple interest for 1 year, or $1000 invested at 7% compounded monthly for 1 year?

38. D_W A firm must choose between borrowing $5000 at 10% for 30 days and borrowing $10,000 at 8% for 60 days. Give arguments in favor of and against each option.

VOCABULARY REINFORCEMENT

In each of Exercises 39–46, fill in the blank with the correct term from the given list. Some of the choices may not be used.

39. If the product of two numbers is 1, they are _____ of each other.   [2.7a]

40. A number is _____ if its ones digit is even and the sum of its digits is divisible by 3.   [2.2a]

41. The number 0 is the _____ identity.   [1.2a]

42. A(n) _____ is the ratio of price to the number of units.   [5.2b]

43. The distance around an object is its _____ .   [1.2b]

44. A number is _____ if the sum of its digits is divisible by 3.   [2.2a]

45. A natural number that has exactly two different factors, only itself and 1, is called a _____ number.   [2.1c]

46. When two pairs of numbers have the same ratio, they are _____ .   [5.3a]

divisible by 3

divisible by 4

divisible by 6

divisible by 9

perimeter

area

unit rate

reciprocals

proportional

composite

prime

additive

multiplicative

Effective Yield. The *effective yield* is the yearly rate of simple interest that corresponds to a rate for which interest is compounded two or more times a year. For example, if P is invested at 12%, compounded quarterly, we would multiply P by $(1 + 0.12/4)^4$, or 1.03^4. Since $1.03^4 \approx 1.126$, the 12% compounded quarterly corresponds to an effective yield of approximately 12.6%. In Exercises 47 and 48, find the effective yield for the indicated account.

47. 🖩 The account pays 9% compounded monthly.

48. 🖩 The account pays 10% compounded daily.

400

6.8 INTEREST RATES ON CREDIT CARDS AND LOANS

a Credit Cards and Loans

Objective

a Solve applied problems involving interest rates on credit cards and loans.

Look at the following graphs. They offer good reason for a study of the real-world applications of percent, interest, loans, and credit cards.

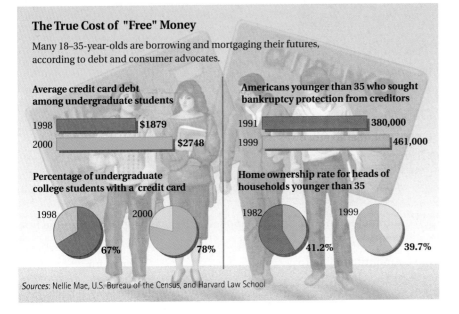

The True Cost of "Free" Money

Many 18–35-year-olds are borrowing and mortgaging their futures, according to debt and consumer advocates.

Average credit card debt among undergraduate students

1998 $1879
2000 $2748

Americans younger than 35 who sought bankruptcy protection from creditors

1991 380,000
1999 461,000

Percentage of undergraduate college students with a credit card

1998 67% 2000 78%

Home ownership rate for heads of households younger than 35

1982 41.2% 1999 39.7%

Sources: Nellie Mae, U.S. Bureau of the Census, and Harvard Law School

Comparing interest rates is essential if one is to become financially responsible. A small change in an interest rate can make a *large* difference in the cost of a loan. When you make a payment on a loan, do you know how much of that payment is interest and how much is applied to reducing the principal?

We begin with an example involving credit cards. A balance carried on a credit card is a type of loan. In a recent year in the United States, 100,000 young adults declared bankruptcy because of excessive credit card debt. The money you obtain through the use of a credit card is not "free" money. There is a price (interest) to be paid for the privilege.

EXAMPLE 1 *Credit Cards.* After the holidays, Sarah has a balance of $3216.28 on a credit card with an annual percentage rate (APR) of 19.7%. She decides not to make additional purchases with this card until she has paid off the balance.

a) Many credit cards require a minimum monthly payment of 2% of the balance. What is Sarah's minimum payment on a balance of $3216.28? Round the answer to the nearest dollar.

b) Find the amount of interest and the amount applied to reduce the principal in the minimum payment found in part (a).

c) If Sarah had transferred her balance to a card with an APR of 12.5%, how much of her first payment would be interest and how much would be applied to reduce the principal?

d) Compare the amounts for 12.5% from part (c) with the amounts for 19.7% from part (b).

1. Credit Cards. After the holidays, Jamal has a balance of $4867.59 on a credit card with an annual percentage rate (APR) of 21.3%. He decides not to make additional purchases with this card until he has paid off the balance.

a) Many credit cards require a minimum monthly payment of 2% of the balance. What is Jamal's minimum payment on a balance of $4867.59? Round the answer to the nearest dollar.

b) Find the amount of interest and the amount applied to reduce the principal in the minimum payment found in part (a).

c) If Jamal had transferred his balance to a card with an APR of 13.6%, how much of his first payment would be interest and how much would be applied to reduce the principal?

We solve as follows.

a) We multiply the balance of $3216.28 by 2%:

$$0.02 \times \$3216.28 = \$64.3256.$$ Sarah's minimum payment, rounded to the nearest dollar, is $64.

b) The amount of interest on $3216.28 at 19.7% for one month* is given by

$$I = P \cdot r \cdot t = \$3216.28 \times 0.197 \times \frac{1}{12} \approx \$52.80.$$

We subtract to find the amount applied to reduce the principal in the first payment:

$$\text{Amount applied to reduce the principal} = \text{Minimum payment} - \text{Interest for the month}$$

$$= \$64 - \$52.80$$

$$= \$11.20.$$

Thus the principal of $3216.28 is decreased by only $11.20 with the first payment. (Sarah still owes $3205.08.)

c) The amount of interest on $3216.28 at 12.5% for one month is

$$I = P \cdot r \cdot t = \$3216.28 \times 0.125 \times \frac{1}{12} \approx \$33.50.$$

We subtract to find the amount applied to reduce the principal in the first payment:

$$\text{Amount applied to reduce the principal} = \text{Minimum payment} - \text{Interest for the month}$$

$$= \$64 - \$33.50$$

$$= \$30.50.$$

Thus the principal of $3216.28 is decreased by $30.50 with the first payment. (Sarah still owes $3185.78.)

d) Let's organize the information for both rates in the following table.

BALANCE BEFORE FIRST PAYMENT	FIRST MONTH'S PAYMENT	% APR	AMOUNT OF INTEREST	AMOUNT APPLIED TO PRINCIPAL	BALANCE AFTER FIRST PAYMENT
$3216.28	$64	19.7%	$52.80	$11.20	$3205.08
3216.28	64	12.5	33.50	30.50	3185.78

Difference in balance after first payment → $19.30

d) Compare the amounts for 13.6% from part (c) with the amounts for 21.3% from part (b).

At 19.7%, the interest is $52.80 and the principal is decreased by $11.20. At 12.5%, the interest is $33.50 and the principal is decreased by $30.50. Thus the principal is decreased by $30.50 − $11.20, or $19.30 more with the 12.5% rate than with the 19.7% rate. Thus the interest at 19.7% is $19.30 greater than the interest at 12.5%.

---

*Actually, the interest on a credit card is computed daily with a rate called a daily percentage rate (DPR). The DPR for Example 1 would be 19.7%/365 ≈ 0.054%. When no payments or additional purchases are made during the month, the difference in total interest for the month is minimal and we will not deal with it here.

Answers on page A-15

Do Exercise 1 on the preceding page.

Even though the mathematics of the information in the chart below is beyond the scope of this text, it is interesting to compare how long it takes to pay off the balance of Example 1 if Sarah continues to pay $64 for each payment with how long it takes if she pays double that amount, $128, for each payment. Financial consultants frequently tell clients that if they want to take control of their debt, they should pay double the minimum payment.

RATE	PAYMENT PER MONTH	NUMBER OF PAYMENTS TO PAY OFF DEBT	TOTAL PAID BACK	ADDITIONAL COST OF PURCHASES
19.7%	$64	107, or 8 yr 11 mo	$6848	$3631.72
19.7	128	33, or 2 yr 9 mo	4224	1007.72
12.5	64	72, or 6 yr	4608	1391.72
12.5	128	29, or 2 yr 5 mo	3712	495.72

As with most loans, if you pay an extra amount toward the principal with each payment, the length of the loan can be greatly reduced. Note that at the rate of 19.7%, it will take Sarah almost 9 yr to pay off her debt if she pays only $64 per month and does not make additional purchases. If she transfers her balance to a card with a 12.5% rate and pays $128 per month, she could eliminate her debt in approximately $2\frac{1}{2}$ yr. You can see how debt can get out of control if you continue to make purchases and pay only the minimum payment. The debt will never be eliminated.

The Federal Stafford Loan program provides educational loans to students at interest rates that are much lower than those on credit cards. Payments on a loan do not begin until 6 months after graduation. At that time, the student has 10 years, or 120 monthly payments, to pay off the loan.

EXAMPLE 2 *Federal Stafford Loans.* After graduation, the balance on Taylor's Stafford loan is $28,650. If the rate on his loan is 3.37%, he will make 120 payments of approximately $282 each to pay off the loan.

a) Find the amount of interest and the amount of principal in the first payment.

b) If the interest rate were 5.25%, he would make 120 monthly payments of approximately $307 each. How much more of the first payment is interest if the loan is 5.25% rather than 3.37%?

c) Compare the total amount of interest on the loan at 3.37% with the amount on the loan at 5.25%. How much more would Taylor pay in interest on the 5.25% loan than on the 3.37% loan?

We solve as follows.

a) We use the formula $I = P \cdot r \cdot t$, substituting $28,650 for P, 0.0337 for r, and 1/12 for t:

$$I = \$28{,}650 \times 0.0337 \times \frac{1}{12}$$

$$\approx \$80.46.$$

The amount of interest in the first payment is $80.46. The payment is $282. We subtract to determine the amount applied to the principal:

$$\$282 - \$80.46 = \$201.54.$$

With the first payment, the principal will be reduced by $201.54.

403

2. Federal Stafford Loans. After graduation, the balance on Maggie's Stafford loan is $32,680. To pay off the loan at 3.37%, she will make 120 payments of approximately $321 each.

a) Find the amount of interest and the amount of principal in the first payment.

b) If the interest rate were 5.5%, she would make 120 payments of approximately $355 each. How much more of the first payment is interest if the loan is 5.5% rather than 3.37%?

c) Compare the total amount of interest on the loan at 3.37% with the amount of interest on the loan at 5.5%. How much more would Maggie pay in interest on the 5.5% loan than on the 3.37% loan?

Answers on page A-15

b) The interest at 5.25% would be

$$I = \$28{,}650 \times 0.0525 \times \frac{1}{12}$$
$$\approx \$125.34.$$

At the rate of 5.25%, the additional interest in the first payment is

$$\$125.34 - \$80.46 = \$44.88.$$

The higher interest rate results in an additional $44.88 in interest in the first payment.

c) For the 3.37% loan, there will be 120 payments of $282 each:

$$120 \times \$282 = \$33{,}840.$$

The total amount of interest at this rate is

$$\$33{,}840 - \$28{,}650 = \$5190.$$

For the 5.25% loan, there will be 120 payments of $307 each:

$$120 \times \$307 = \$36{,}840.$$

The total amount of interest at this rate is

$$\$36{,}840 - \$28{,}650 = \$8190.$$

At the rate of 5.25%, Taylor would pay

$$\$8190 - \$5190 = \$3000$$

more in interest than at the rate of 3.37%.

Do Exercise 2.

EXAMPLE 3 *Home Loans.* The Sawyers recently purchased their first home. They borrowed $153,000 at $6\frac{5}{8}$% for 30 years (360 payments). Their monthly payment (excluding insurance and taxes) is $979.68.

a) How much of the first payment is interest and how much is applied to reduce the principal?

b) If the Sawyers pay the entire 360 payments, how much interest will be paid on the loan?

We solve as follows.

a) To find the amount of interest paid in the first payment, we use the formula $I = P \cdot r \cdot t$:

$$I = P \cdot r \cdot t = \$153{,}000 \times 0.06625 \times \frac{1}{12} \approx \$844.69.$$

The amount applied to the principal is

$$\$979.68 - \$844.69, \text{ or } \$134.99.$$

b) Over the 30-year period, the total paid will be

$$360 \times \$979.68, \text{ or } \$352{,}684.80.$$

The total amount of interest paid over the lifetime of the loan is

$$\$352{,}684.80 - \$153{,}000, \text{ or } \$199{,}684.80.$$

Do Exercises 3 and 4 on the following page.

AMORTIZATION TABLES

If we make 360 calculations as in Example 3(a) and continue with a decreased principal as in Margin Exercise 3, we can create an *amortization table,* part of which is shown below. Such tables are also found in reference books. The beginning, middle, and last part of the loan described are shown. Look over the table and note how small a portion of the payment reduces the principal at the beginning of the loan and how that portion increases throughout the lifetime of the loan. Do you see again why a loan is not "free"?

MORTGAGE AMORTIZATION PROGRAM

MORTGAGE AMOUNT:	$153,000
INTEREST RATE:	6.625%
NUMBER OF YEARS:	30
MONTHLY PAYMENTS ARE:	$979.68

PAYMENT	PRINCIPAL	INTEREST	BALANCE	
1	$134.99	$844.69	$152,865.01	← Example 3
2	$135.73	$843.94	$152,729.28	← Margin
3	$136.48	$843.19	$152,592.80	Exercise 3
4	$137.24	$842.44	$152,455.56	
5	$137.99	$841.68	$152,317.56	
6	$138.76	$840.92	$152,178.81	
7	$139.52	$840.15	$152,039.29	
8	$140.29	$839.38	$151,898.99	
9	$141.07	$838.61	$151,757.93	
10	$141.85	$837.83	$151,616.08	
11	$142.63	$837.05	$151,473.45	
12	$143.42	$836.26	$151,330.04	
⋮	⋮			

Interest for 12 periods = $10,086.14

174	$349.91	$629.77	$113,721.56
175	$351.84	$627.84	$113,369.72
176	$353.78	$625.90	$113,015.94
177	$355.73	$623.94	$112,660.20
178	$357.70	$621.98	$112,302.51
179	$359.67	$620.00	$111,942.83
180	$361.66	$618.02	$111,581.18
181	$363.65	$616.02	$111,217.52
182	$365.66	$614.01	$110,851.86
183	$367.68	$611.99	$110,484.18
184	$369.71	$609.96	$110,114.47
185	$371.75	$607.92	$109,742.71
186	$373.80	$605.87	$109,368.91
⋮	⋮		

Interest for 12 periods = $8033.22

349	$917.04	$62.63	$10,427.84
350	$922.11	$57.57	$9,505.74
351	$927.20	$52.48	$8,578.54
352	$932.32	$47.36	$7,646.23
353	$937.46	$42.21	$6,708.76
354	$942.64	$37.04	$5,766.13
355	$947.84	$31.83	$4,818.28
356	$953.07	$26.60	$3,865.21
357	$958.34	$21.34	$2,906.87
358	$963.63	$16.05	$1,943.24
359	$968.95	$10.73	$974.30
360	$974.30	$0.00	$0.00

Interest for 12 periods = $405.84

Total interest for 360 periods = $199,684.80

Refer to Example 3 for Margin Exercises 3 and 4.

3. **Home Loans.** Since the principal has been reduced by the first payment, at the time of the second payment of the Sawyers' 30-year loan, the new principal is the decreased principal

$153,000 − $134.99,

or

$152,865.01.

Use $152,865.01 as the principal, and determine how much of the second payment is interest and how much is applied to reduce the principal. (In effect, repeat Example 3(a) using the new principal.)

4. **Home Loans.** The Sawyers decide to change the period of their home loan from 30 years to 15 years. Their monthly payment increases to $1343.33.

a) How much of the first payment is interest and how much is applied to reduce the principal?

b) If the Sawyers pay the entire 180 payments, how much interest will be paid on this loan?

c) Compare the amount of interest to pay off the 15-yr loan with the amount of interest to pay off the 30-yr loan.

Answers on page A-15

5. Refinancing a Home Loan.
Consider Example 4 for a 15-yr loan. The new monthly payment is $1250.14.

a) How much of the first payment is interest and how much is applied to reduce the principal?

b) If the Sawyers pay the entire 180 payments, how much interest will be paid on this loan?

c) Compare the amount of interest to pay off the 15-yr loan at $5\frac{1}{2}$% with the amount of interest to pay off the 15-yr loan at $6\frac{5}{8}$% in Margin Exercise 4.

EXAMPLE 4 *Refinancing a Home Loan.* Refer to Example 3. Ten months after the Sawyers buy their home financed at a rate of $6\frac{5}{8}$%, the rates drop to $5\frac{1}{2}$%. After much consideration, they decide to refinance even though the new loan will cost them $1200 in refinance charges. They have reduced the principal a small amount in the 10 payments they have made, but they decide to again borrow $153,000 for 30 years at the new rate. Their new monthly payment is $868.72.

a) How much of the first payment is interest and how much is applied to the principal?

b) Compare the amounts at $5\frac{1}{2}$% found in part (a) with the amounts at $6\frac{5}{8}$% found in Example 3(a).

c) With the lower house payment, how long will it take the Sawyers to recoup the refinance charge of $1200?

d) If the Sawyers pay the entire 360 payments, how much interest will be paid on this loan? How much less is the total interest at $5\frac{1}{2}$% than at $6\frac{5}{8}$%?

We solve as follows.

a) To find the interest paid in the first payment, we use the formula $I = P \cdot r \cdot t$:

$$I = P \cdot r \cdot t = \$153{,}000 \times 0.055 \times \frac{1}{12} = \$701.25.$$

The amount applied to the principal is

$$\$868.72 - \$701.25, \text{ or } \$167.47.$$

b) We compare the amounts found in part (a) with the amounts found in Example 3(a):

Rate	Monthly payment	Interest in first payment	Amount applied to principal
$6\frac{5}{8}$%	$979.68	$844.69	$134.99
$5\frac{1}{2}$%	$868.72	$701.25	$167.47

At $5\frac{1}{2}$%, the amount of interest in the first payment is $844.69 − $701.25, or $143.44, less than at $6\frac{5}{8}$%. The amount applied to the principal is $167.47 − $134.99, or $32.48, more.

c) The monthly payment at $5\frac{1}{2}$% is $979.68 − $868.72, or $110.96, less than the payment at $6\frac{5}{8}$%. The total savings each month is approximately $111. We can divide the cost of the refinancing by this monthly savings to determine the number of months it will take to recoup the $1200 refinancing charge: $1200 ÷ $111 ≈ 11. It will take the Sawyers approximately 11 months to break even.

d) Over the 30-year period, the total paid will be

$$360 \times \$868.72, \quad \text{or} \quad \$312{,}739.20.$$

The total amount of interest paid over the lifetime of the loan is

$$\$312{,}739.20 - \$153{,}000, \quad \text{or} \quad \$159{,}739.20.$$

The total interest paid at $5\frac{1}{2}$% is

$$\$199{,}684.80 \text{ (see Example 3)} - \$159{,}739.20, \quad \text{or} \quad \$39{,}945.60,$$

less than the total interest paid at $6\frac{5}{8}$%. Thus the $5\frac{1}{2}$% loan saves the Sawyers approximately $40,000 in interest charges over the 30 years.

Do Exercise 5.

Answers on page A-15

19. *Used-Car Loan.*   Twin brothers, Jerry and Terry, each take a job at the college cafeteria in order to have the money to make payments on the purchase of a 2002 Chrysler PT Cruiser for $11,900. They make a down payment of 5% and finance the remainder at 9.3% for 3 years. (Used-car loan rates are generally higher than new-car loan rates.) Their monthly payment is $361.08.

a) What is the down payment? the amount borrowed?
b) How much of the first payment is interest and how much is applied to reduce the principal?
c) If they pay all 36 payments, how much interest will they pay for the loan?

20. *Used-Car Loan.*   For his construction job, Clint buys a 2003 Dodge Ram 1500 truck for $13,800. He makes a down payment of $1380 and finances the remainder for 4 years at 8.8%. The monthly payment is $307.89.

a) How much is financed?
b) How much of the first payment is interest and how much is applied to reduce the principal?
c) If he pays all 48 payments, how much interest will he pay for the loan?

21. D_W Based on the skills of mathematics you have obtained in this section, discuss the significant new ideas you now have about interest rates and credit cards that you didn't have before.

22. D_W Examine the information in the graphs at the beginning of the section. Discuss how a knowledge of this section might have been of help to some of these students.

23. D_W Compare the following two purchases and describe a situation in which each purchase is the best choice.

Purchase A: A new car for $15,145. The loan is for $14,500 at 6.9% for 4 years. The monthly payment is $346.55.

Purchase B: A used car for $10,600. The loan is for $9300 at 12.5% for 3 years. The monthly payment is $311.12.

24. D_W Look over the examples and exercises in this section. What seems to happen to the monthly payment on a loan if the time of payment changes from 30 years to 15 years, assuming the interest rate stays the same? Discuss the pros and cons of both time periods.

> SKILL MAINTENANCE

Solve. Round the answer to the nearest hundredth where appropriate.   [5.3b]

25. $\dfrac{5}{8} = \dfrac{x}{28}$

26. $\dfrac{5}{8} = \dfrac{17.5}{y}$

27. $\dfrac{13}{16} = \dfrac{81.25}{N}$

28. $\dfrac{9}{16} = \dfrac{p}{100}$

29. $\dfrac{1284}{t} = \dfrac{3456}{5000}$

30. $\dfrac{12.8}{32.5} = \dfrac{x}{2000}$

31. $\dfrac{56.3}{78.4} = \dfrac{t}{100}$

32. $\dfrac{28}{x} = \dfrac{8}{5}$

33. $\dfrac{16}{9} = \dfrac{100}{p}$

34. $\dfrac{t}{1284} = \dfrac{5000}{3456}$

Summary and Review

The review that follows is meant to prepare you for a chapter exam. It consists of three parts. The first part, Concept Reinforcement, is designed to increase understanding through true/false exercises. The second part is a list of important properties and formulas. The third part is the Review Exercises. These provide practice exercises for the exam, together with references to section objectives so you can go back and review. Before beginning, stop and look back over the skills you have obtained. What skills in mathematics do you have now that you did not have before studying this chapter?

CONCEPT REINFORCEMENT

Determine whether the statement is true or false. Answers are given at the back of the book.

_____ **1.** A fixed principal invested for four years will earn more interest when interest is compounded quarterly than when interest is compounded semi-annually.

_____ **2.** Of the numbers 0.5%, $\frac{5}{1000}$%, $\frac{1}{2}$%, $\frac{1}{5}$, and $0.\overline{1}$, the largest number is $0.\overline{1}$.

_____ **3.** If principal A equals principal B and principal A is invested for 2 years at 4% compounded quarterly while principal B is invested for 4 years at 2% compounded semi-annually, the interest earned from each investment is the same.

_____ **4.** The symbol % is equivalent to $\times \frac{1}{10}$.

IMPORTANT PROPERTIES AND FORMULAS

Commission = Commission rate × Sales

Discount = Rate of discount × Original price

Sale price = Original price − Discount

Simple Interest: $I = P \cdot r \cdot t$

Compound Interest: $A = P \cdot \left(1 + \frac{r}{n}\right)^{n \cdot t}$

Review Exercises

Find percent notation for the decimal notation in the sentence in Exercises 1 and 2. [6.1b]

1. Of all the snacks eaten on Super Bowl Sunday, 0.56 of them are chips and salsa.

Source: Korbel Research and Pace Foods

2. Of all the vehicles in Mexico City, 0.017 of them are taxis.

Source: The Handy Geography Answer Book

Find percent notation. [6.2a]

3. $\frac{3}{8}$

4. $\frac{1}{3}$

Find decimal notation. [6.1b]

5. 73.5%

6. $6\frac{1}{2}\%$

Find fraction notation. [6.2b]

7. 24%

8. 6.3%

Translate to a percent equation. Then solve. [6.3a, b]

9. 30.6 is what percent of 90?

10. 63 is 84 percent of what?

11. What is $38\frac{1}{2}\%$ of 168?

Translate to a proportion. Then solve. [6.4a, b]

12. 24 percent of what is 16.8?

13. 42 is what percent of 30?

14. What is 10.5% of 84?

Solve. [6.5a, b]

15. *Favorite Ice Creams.* According to a recent survey, 8.9% of those interviewed chose chocolate as their favorite ice cream flavor and 4.2% chose butter pecan. Of the 2500 students in a freshman class, how many would choose chocolate as their favorite ice cream? butter pecan?

Source: International Ice Cream Association

16. *Prescriptions.* Of the 295 million people in the United States, 123.64 million take at least one kind of prescription drug per day. What percent take at least one kind of prescription drug per day?

Source: American Society of Health-System Pharmacies

17. *Water Output.* The average person expels 200 mL of water per day by sweating. This is 8% of the total output of water from the body. How much is the total output of water?

Source: Elaine N. Marieb, *Essentials of Human Anatomy and Physiology,* 6th ed. Boston: Addison Wesley Longman, Inc., 2000

18. *Test Scores.* Jason got a 75 on a math test. He was allowed to go to the math lab and take a retest. He increased his score to 84. What was the percent of increase?

19. *Test Scores.* Jenny got an 81 on a math test. By taking a retest in the math lab, she increased her score by 15%. What was her new score?

Solve. [6.6a, b, c]

20. A state charges a meals tax of $4\frac{1}{2}\%$. What is the meals tax charged on a dinner party costing $320?

21. In a certain state, a sales tax of $378 is collected on the purchase of a used car for $7560. What is the sales tax rate?

22. Kim earns $753.50 selling $6850 worth of televisions. What is the commission rate?

23. An air conditioner has a marked price of $350. It is placed on sale at 12% off. What are the discount and the sale price?

24. A fax machine priced at $305 is discounted at the rate of 14%. What are the discount and the sale price?

25. An insurance salesperson receives a 7% commission. If $42,000 worth of life insurance is sold, what is the commission?

26. Find the rate of discount.

Our lowest price of the season!
18-ft metal frame
$399⁶⁹
Reg. 489.99
Family-size metal-frame pool

Solve. [6.7a, b]

27. What is the simple interest on $1800 at 6% for $\frac{1}{3}$ year?

28. The Dress Shack borrows $24,000 at 10% simple interest for 60 days. Find (a) the amount of interest due and (b) the total amount that must be paid after 60 days.

29. What is the simple interest on $2200 principal at the interest rate of 5.5% for 1 year?

30. The Kleins invest $7500 in an investment account paying an annual interest rate of 12%, compounded monthly. How much is in the account after 3 months?

31. Find the amount in an investment account if $8000 is invested at 9%, compounded annually, for 2 years.

Solve. [6.8a]

32. *Credit Cards.* At the end of her junior year of college, Judy has a balance of $6428.74 on a credit card with an annual percentage rate (APR) of 18.7%. She decides not to make additional purchases with this card until she has paid off the balance.

   a) Many credit cards require a minimum payment of 2% of the balance. What is Judy's minimum payment on a balance of $6428.74? Round the answer to the nearest dollar.
   b) Find the amount of interest and the amount applied to reduce the principal in the minimum payment found in part (a).
   c) If Judy had transferred her balance to a card with an APR of 13.2%, how much of her first payment would be interest and how much would be applied to reduce the principal?
   d) Compare the amounts for 13.2% from part (c) with the amounts for 18.7% from part (b).

33. **D_W** Ollie buys a microwave oven during a 10%-off sale. The sale price that Ollie paid was $162. To find the original price, Ollie calculates 10% of $162 and adds that to $162. Is this correct? Why or why not? [6.6c]

34. **D_W** Which is the better deal for a consumer and why: a discount of 40% or a discount of 20% followed by another of 22%? [6.6c]

SYNTHESIS

35. ▦ *Land Area of the United States.* After Hawaii and Alaska became states, the total land area of the United States increased from 2,963,681 mi² to 3,540,939 mi². What was the percent of increase? [6.5b]

36. Rhonda's Dress Shop reduces the price of a dress by 40% during a sale. By what percent must the store increase the sale price, after the sale, to get back to the original price? [6.6c]

37. A $200 coat is marked up 20%. After 30 days, it is marked down 30% and sold. What was the final selling price of the coat? [6.6c]

1. *Bookmobiles.* Since 1991, the number of bookmobiles has decreased by approximately 6.4%. Find decimal notation for 6.4%.
   Source: American Library Association

2. *Gravity.* The gravity of Mars is 0.38 as strong as Earth's. Find percent notation for 0.38.
   Source: www.marsinstitute.info/epo/mermarsfacts.html

3. Find percent notation for $\frac{11}{8}$.

4. Find fraction notation for 65%.

5. Translate to a percent equation. Then solve.

   What is 40% of 55?

6. Translate to a proportion. Then solve.

   What percent of 80 is 65?

Solve.

7. *Cruise Ship Passengers.* Of the passengers on a typical cruise ship, 16% are in the 25–34 age group and 23% are in the 35–44 age group. A cruise ship has 2500 passengers. How many are in the 25–34 age group? the 35–44 age group?
   Source: Polk

8. *Batting Averages.* Luis Castillo, second baseman for the Florida Marlins, got 180 hits during the 2000 baseball season. This was about 33.4% of his at-bats. How many at-bats did he have?
   Source: Major League Baseball

9. *Airline Profits.* Profits of the entire U. S. Airline industry decreased from $5.5 billion in 1999 to $2.7 billion in 2000. Find the percent of decrease.
   Source: Air Transport Association

10. There are 6.6 billion people living in the world today. It is estimated that the total number who have ever lived is about 120 billion. What percent of people who have ever lived are alive today?
    Source: *The Handy Geography Answer Book*

11. *Maine Sales Tax.* The sales tax rate in Maine is 5%. How much tax is charged on a purchase of $324? What is the total price?

12. Gwen's commission rate is 15%. What is the commission from the sale of $4200 worth of merchandise?

13. The marked price of a DVD player is $200 and the item is on sale at 20% off. What are the discount and the sale price?

14. What is the simple interest on a principal of $120 at the interest rate of 7.1% for 1 year?

15. The Burnham Parents–Teachers Association invests $5200 at 6% simple interest. How much is in the account after $\frac{1}{2}$ year?

16. Find the amount in an account if $1000 is invested at $5\frac{3}{8}$%, compounded annually, for 2 years.

17. The Suarez family invests $10,000 at an annual interest rate of 4.9%, compounded monthly. How much is in the account after 3 years?

18. *Job Opportunities.* The table below lists job opportunities, in millions, in 2002 and projected increases to 2012. Find the missing numbers.

OCCUPATION	NUMBER OF JOBS IN 2002 (in millions)	NUMBER OF JOBS IN 2012 (in millions)	CHANGE	PERCENT OF INCREASE
Retail salespersons	4.1	4.8	0.7	17.1%
Registered nurses	2.3		0.6	
Post-secondary teachers	1.6	2.2		
Food preparation and service workers		2.4	0.4	
Restaurant servers		2.5		19.0%

Source: Department of Labor

19. Find the discount and the discount rate of the washer-dryer duet in this ad.

SHORT TIME ONLY

$1675

Was $1950

No interest for 18 months

20. *Home Loan.* Complete the following table, assuming the monthly payment as given.

Interest Rate	7.4%
Mortgage	$120,000
Time of Loan	360 mos
Monthly Payment	$830.86
Principal after First Payment	
Principal after Second Payment	

21. By selling a home without using a realtor, Juan and Marie can avoid paying a 7.5% commission. They receive an offer of $180,000 from a potential buyer. In order to give a comparable offer, for what price would a realtor need to sell the house? Round to the nearest hundred.

22. Karen's commission rate is 16%. She invests her commission from the sale of $15,000 worth of merchandise at the interest rate of 12%, compounded quarterly. How much is Karen's investment worth after 6 months?

Data, Graphs, and Statistics

Real-World Application

The rhinoceros is considered one of the world's most endangered animals. The worldwide total number of rhinoceroses is approximately 17,800. Using the pictograph shown in Example 3 on page 429, determine how many more black rhinos there are than Indian/Nepalese rhinos.

This problem appears as Example 3 in Section 7.2.

AVERAGES, MEDIANS, AND MODES

Objectives

a Find the average of a set of numbers and solve applied problems involving averages.

b Find the median of a set of numbers and solve applied problems involving medians.

c Find the mode of a set of numbers and solve applied problems involving modes.

d Compare two sets of data using their means.

In real-world applications, we can use tables and graphs of various kinds to show information about data and to extract information from data that can lead us to make analyses and predictions. Graphs allow us to communicate a message from the data.

For example, the following two pages show data regarding the number of alcohol-related traffic deaths in recent years. Examine each method of presentation. Which method, if any, do you like the best and why?

Paragraph The National Highway Traffic Safety Administration has released data regarding the number of alcohol-related traffic deaths for various years. In 1982, there were 26,173 deaths; in 1984, there were 24,762 deaths; in 1986, there were 25,017 deaths; in 1988, there were 23,833 deaths; in 1990, there were 22,587 deaths; in 1992, there were 18,290 deaths; in 1994, there were 17,308 deaths; in 1996, there were 17,749 deaths; in 1998, there were 16,673 deaths; in 2000, there were 17,380 deaths; and finally, in 2002, there were 17,419 deaths.

Table

YEAR	ALCOHOL-RELATED TRAFFIC DEATHS
1982	26,173
1984	24,762
1986	25,017
1988	23,833
1990	22,587
1992	18,290
1994	17,308
1996	17,749
1998	16,673
2000	17,380
2002	17,419

Source: 2002 FARS Annual Report File, FHWA's Highway Statistics Annual Series, from the National Center for Statistics and Analysis

Pictograph

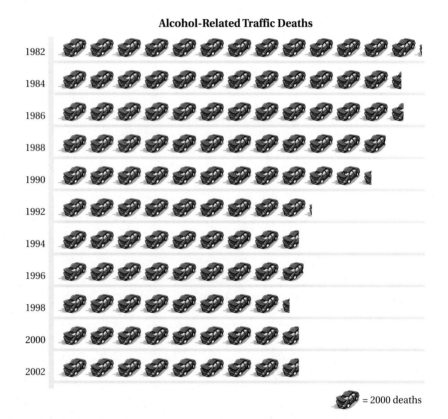

Alcohol-Related Traffic Deaths

= 2000 deaths

Bar Graph

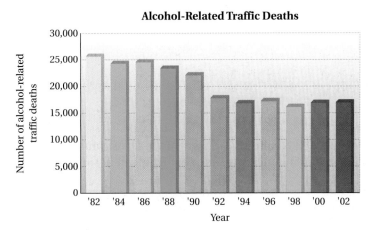

Alcohol-Related Traffic Deaths

Line Graph

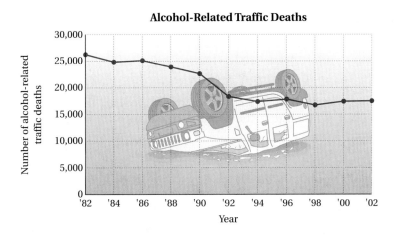

Alcohol-Related Traffic Deaths

Most people would not find the paragraph method for displaying the data most useful. It takes time to read, and it is hard to look for a trend. The bar and line graphs are more helpful if we want to see the overall trend of decreased deaths and to make a prediction about years to come. The exact data values are most easily read in tables.

In this chapter, we will learn not only how to extract information from various kinds of tables and graphs, but also how to create various kinds of graphs, such as bar, line, and circle.

a   Averages

A **statistic** is a number describing a set of data. One statistic is a *center point*, or *measure of central tendency*, that characterizes the data. The most common kind of center point is the *mean*, or *average*, of a set of numbers. We first considered averages in Section 1.9 and extended the coverage in Sections 3.7 and 4.4.

Let's consider the data of the number of alcohol-related traffic deaths:

26,173,   24,762,   25,017,   23,833,   22,587,   18,290,   17,308, 17,749,   16,673,   17,380,   17,419.

What is the average of this set of numbers? First, we add the numbers:

$$26{,}173 + 24{,}762 + 25{,}017 + 23{,}833 + 22{,}587 + 18{,}290 + 17{,}308$$
$$+ 17{,}749 + 16{,}673 + 17{,}380 + 17{,}419 = 227{,}191.$$

417

Find the average.

1. 14, 175, 36

2. 75, 36.8, 95.7, 12.1

3. A student scored the following on five tests; 68, 85, 82, 74, 96. What was the average score?

4. In the first five games, a basketball player scored points as follows: 26, 21, 13, 14, 23. Find the average number of points scored per game.

5. Home-Run Batting Average. At the end of the 2004 baseball season, Barry Bonds had the most career home runs of any active player in the major leagues, 703 in 19 seasons. What was his average number of home runs per season?

Source: Major League Baseball

Next, we divide by the number of data items, 11:

$$\frac{227{,}191}{11} \approx 20{,}654 \qquad \text{Rounding to the nearest one}$$

Note that if the traffic deaths had been the average (same) for 11 years, we would have:

$$20{,}654 + 20{,}654 + 20{,}654 + 20{,}654 + 20{,}654 + 20{,}654 + 20{,}654$$
$$+ \; 20{,}654 + 20{,}654 + 20{,}654 + 20{,}654$$
$$= 227{,}194 \approx 227{,}191.$$

The number 20,654 is called the *average* of the set of numbers. It is also called the **arithmetic** (pronounced ăr´ ĭth-mĕt´-ĭk) **mean** or simply the **mean.**

> **AVERAGE**
>
> To find the **average** of a set of numbers, add the numbers and then divide by the number of items of data.

EXAMPLE 1 On a 4-day trip, a car was driven the following number of miles each day: 240, 302, 280, 320. What was the average number of miles per day?

$$\frac{240 + 302 + 280 + 320}{4} = \frac{1142}{4}, \quad \text{or} \quad 285.5$$

The car was driven an average of 285.5 mi per day. Had the car been driven exactly 285.5 mi each day, the same total distance (1142 mi) would have been traveled.

Do Exercises 1–4.

EXAMPLE 2 *Scoring Average.* Kareem Abdul-Jabbar is the all-time leading scorer in the history of the National Basketball Association. He scored 38,387 points in 1560 games. What was the average number of points scored per game? Round to the nearest tenth.

Source: National Basketball Association

We already know the total number of points, 38,387, and the number of games, 1560. We divide and round to the nearest tenth:

$$\frac{38{,}387}{1560} = 24.60705\ldots \approx 24.6.$$

Abdul-Jabbar's average was 24.6 points per game.

Do Exercise 5.

Answers on page A-16

EXAMPLE 3 *Gas Mileage.* The 2005 Chevrolet Cobalt LS, shown here, is estimated to go 450 miles of city driving on 18 gallons of gasoline. What is the expected average number of miles per gallon—that is, what is its gas mileage for city driving?
Source: *Motor Trend,* July 2005

6. Gas Mileage. The 2005 Chevrolet Cobalt LS gets 612 miles of highway driving on 18 gallons of gasoline. What is the average number of miles per gallon—that is, what is its gas mileage for highway driving?
Source: *Motor Trend,* July 2005

We divide the total number of miles, 450, by the total number of gallons, 18:

$$\frac{450}{18} = 25 \text{ mpg.}$$

The Chevrolet Cobalt's expected average is 25 miles per gallon for city driving.

Do Exercise 6.

EXAMPLE 4 *Grade Point Average.* In most colleges, students are assigned grade point values for grades obtained. The **grade point average,** or **GPA,** is the average of the grade point values for each credit hour taken. At most colleges, grade point values are assigned as follows:

A: 4.0   B: 3.0   C: 2.0   D: 1.0   F: 0.0

Meg earned the following grades for one semester. What was her grade point average?

COURSE	GRADE	NUMBER OF CREDIT HOURS IN COURSE
Colonial History	B	3
Basic Mathematics	A	4
English Literature	A	3
French	C	4
Time Management	D	1

To find the GPA, we first multiply the grade point value (in color below) by the number of credit hours in the course to determine the number of *quality points,* and then add, as follows:

Colonial History	$3.0 \cdot 3 =$	9
Basic Mathematics	$4.0 \cdot 4 =$	16
English Literature	$4.0 \cdot 3 =$	12
French	$2.0 \cdot 4 =$	8
Time Management	$1.0 \cdot 1 =$	1

46 (Total)

Answer on page A-16

7. Grade Point Average. Alex earned the following grades one semester.

GRADE	NUMBER OF CREDIT HOURS IN COURSE
B	3
C	4
C	4
A	2

What was Alex's grade point average? Assume that the grade point values are 4.0 for an A, 3.0 for a B, and so on. Round to the nearest tenth.

The total number of credit hours taken is $3 + 4 + 3 + 4 + 1$, or 15. We divide the number of quality points, 46, by the number of credits, 15, and round to the nearest tenth:

$$\text{GPA} = \frac{46}{15} \approx 3.1.$$

Meg's grade point average was 3.1.

Do Exercise 7.

EXAMPLE 5 *Grading.* To get a B in math, Geraldo must score an average of 80 on the tests. On the first four tests, his scores were 79, 88, 64, and 78. What is the lowest score that Geraldo can get on the last test and still get a B?

We can find the total of the five scores needed as follows:

$$80 + 80 + 80 + 80 + 80 = 5 \cdot 80, \quad \text{or} \quad 400.$$

The total of the scores on the first four tests is

$$79 + 88 + 64 + 78 = 309.$$

Thus Geraldo needs to get at least

$$400 - 309, \quad \text{or} \quad 91$$

in order to get a B. We can check this as follows:

$$\frac{79 + 88 + 64 + 78 + 91}{5} = \frac{400}{5}, \quad \text{or} \quad 80.$$

Do Exercise 8.

8. Grading. To get an A in math, Rosa must score an average of 90 on the tests. On the first three tests, her scores were 80, 100, and 86. What is the lowest score that Rosa can get on the last test and still get an A?

CALCULATOR CORNER

Computing Averages Averages can be computed easily on a calculator. We must keep the rules for order of operations in mind when doing this. For example, to calculate

$$\frac{84 + 92 + 79}{3}$$

on a calculator with parenthesis keys, we press (8 4 + 9 2 + 7 9) ÷ 3 = . On a calculator without parenthesis keys, we first add the numbers in the numerator and then divide that result by 3. To do this, we press 8 4 + 9 2 + 7 9 = ÷ 3 = . In either case, the result is 85.

Exercises:

1. Use a calculator to perform the computation in Example 1.

2. Use a calculator to perform the computations in Margin Exercises 1–4.

3. What would the result have been if we had not used parentheses in the first set of keystrokes above? (Keep the rules for order of operations in mind.)

Answers on page A-16

b Medians

Another type of center-point statistic is the *median*. Medians are useful when we wish to de-emphasize unusually extreme scores. For example, suppose a small class scored as follows on an exam.

Phil:	78	Pat:	56
Jill:	81	Olga:	84
Matt:	82		

Let's first list the scores in order from smallest to largest:

56, 78, 81, 82, 84.
↑
Middle score

The middle score—in this case, 81—is called the **median.** Note that because of the extremely low score of 56, the average of the scores is 76.2. In this example, the median may be a more appropriate center-point statistic.

EXAMPLE 6 What is the median of this set of numbers?

99, 870, 91, 98, 106, 90, 98

We first rearrange the numbers in order from smallest to largest. Then we locate the middle number, 98.

90, 91, 98, 98, 99, 106, 870
↑
Middle number

The median is 98.

Do Exercises 9–11.

> **MEDIAN**
>
> Once a set of data is listed in order, from smallest to largest, the **median** is the middle number if there is an odd number of data items. If there is an even number of items, the median is the number that is the average of the two middle numbers.

EXAMPLE 7 What is the median of this set of numbers?

69, 80, 61, 63, 62, 65

We first rearrange the numbers in order from smallest to largest. There is an even number of numbers. We look for the middle two, which are 63 and 65. The median is halfway between 63 and 65, the number 64.

61, 62, 63, 65, 69, 80   The average of the middle numbers is
↑ $\dfrac{63 + 65}{2}$, or 64.
└── The median is 64.

Find the median.

9. 17, 13, 18, 14, 19

10. 20, 14, 13, 19, 16, 18, 17

11. 78, 81, 83, 91, 103, 102, 122, 119, 88

Answers on page A-16

Find the median.

12. Salaries of Part-Time Typists.
$3300, $4000, $3900, $3600,
$3800, $3400

13. 68, 34, 67, 69, 34, 70

Find the modes of these data.

14. 23, 45, 45, 45, 78

15. 34, 34, 67, 67, 68, 70

16. 13, 24, 27, 28, 67, 89

17. In a lab, Gina determined the
mass, in grams, of each of five
eggs:

  15 g, 19 g, 19 g, 14 g, 18 g.

a) What is the mean?
b) What is the median?
c) What is the mode?

Answers on page A-16

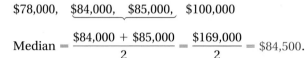 **EXAMPLE 8** *Salaries.* The following are the salaries of the top four employees of Verducci's Dress Company. What is the median of the salaries?

  $85,000, $100,000, $78,000, $84,000

We rearrange the numbers in order from smallest to largest. The two middle numbers are $84,000 and $85,000. Thus the median is halfway between $84,000 and $85,000 (the average of $84,000 and $85,000):

  $78,000, $84,000, $85,000, $100,000

$$\text{Median} = \frac{\$84,000 + \$85,000}{2} = \frac{\$169,000}{2} = \$84,500.$$

Do Exercises 12 and 13.

 Modes

The final type of center-point statistic is the *mode*.

MODE

The **mode** of a set of data is the number or numbers that occur most often. If each number occurs the same number of times, there is *no* mode.

EXAMPLE 9 Find the mode of these data.

   13, 14, 17, 17, 18, 19

The number that occurs most often is 17. Thus the mode is 17.

A set of data has just one average (mean) and just one median, but it can have more than one mode. It is also possible for a set of data to have no mode—when all numbers are equally represented. For example, the set of data 5, 7, 11, 13, 19 has no mode.

EXAMPLE 10 Find the modes of these data.

   33, 34, 34, 34, 35, 36, 37, 37, 37, 38, 39, 40

There are two numbers that occur most often, 34 and 37. Thus the modes are 34 and 37.

Do Exercises 14–17.

Which statistic is best for a particular situation? If someone is bowling, the *average* from several games is a good indicator of that person's ability. If someone is applying for a job, the *median* salary at that business is often most indicative of what people are earning there because although executives tend to make a lot more money, there are fewer of them. Finally, if someone is re-ordering for a clothing store, the *mode* of the sizes sold is probably the most important statistic.

d | Comparing Two Sets of Data

We have seen how to calculate averages, medians, and modes from data. A way to analyze two sets of data is to make a determination about which of two groups is "better." One way to do so is by comparing the averages.

EXAMPLE 11 *Battery Testing.* An experiment is performed to compare battery quality. Two kinds of battery were tested to see how long, in hours, they kept a portable CD player running. On the basis of this test, which battery is better?

BATTERY A: ETERNREADY TIMES (in hours)			BATTERY B: STURDYCELL TIMES (in hours)		
27.9	28.3	27.4	28.3	27.6	27.8
27.6	27.9	28.0	27.4	27.6	27.9
26.8	27.7	28.1	26.9	27.8	28.1
28.2	26.9	27.4	27.9	28.7	27.6

Note that it is difficult to analyze the data at a glance because the numbers are close together. We need a way to compare the two groups. Let's compute the average of each set of data.

Battery A: Average

$$= \frac{27.9 + 28.3 + 27.4 + 27.6 + 27.9 + 28.0 + 26.8 + 27.7 + 28.1 + 28.2 + 26.9 + 27.4}{12}$$

$$= \frac{332.2}{12} \approx 27.68$$

Battery B: Average

$$= \frac{28.3 + 27.6 + 27.8 + 27.4 + 27.6 + 27.9 + 26.9 + 27.8 + 28.1 + 27.9 + 28.7 + 27.6}{12}$$

$$= \frac{333.6}{12} = 27.8$$

We see that the average time of battery B is higher than that of battery A and thus conclude that battery B is "better." (It should be noted that statisticians might question whether these differences are what they call "significant." The answer to that question belongs to a later math course.)

Do Exercise 18.

18. Growth of Wheat. Rudy experiments to see which of two kinds of wheat is better. (In this situation, the shorter wheat is considered "better.") He grows both kinds under similar conditions and measures stalk heights, in inches, as follows. Which kind is better?

WHEAT A STALK HEIGHTS (in inches)			
16.2	42.3	19.5	25.7
25.6	18.0	15.6	41.7
22.6	26.4	18.4	12.6
41.5	13.7	42.0	21.6

WHEAT B STALK HEIGHTS (in inches)			
19.7	18.4	19.7	17.2
19.7	14.6	32.0	25.7
14.0	21.6	42.5	32.6
22.6	10.9	26.7	22.8

Answer on page A-16

7.1

EXERCISE SET

For Extra Help

MathXL | MyMathLab | InterAct Math | Tutor Center Math Tutor Center | Digital Video Tutor CD 4 Videotape 7 | Student's Solutions Manual

a , b , c For each set of numbers, find the average, the median, and any modes that exist.

1. 17, 19, 29, 18, 14, 29

2. 72, 83, 85, 88, 92

3. 5, 37, 20, 20, 35, 5, 25

4. 13, 32, 25, 27, 13

5. 4.3, 7.4, 1.2, 5.7, 7.4

6. 13.4, 13.4, 12.6, 42.9

7. 234, 228, 234, 229, 234, 278

8. $29.95, $28.79, $30.95, $29.95

9. *Atlantic Storms and Hurricanes.* The following bar graph shows the number of Atlantic storms or hurricanes that formed in various months from 1980 to 2000. What is the average number for the 9 months given? the median? the mode?

Atlantic Storms and Hurricanes
Tropical storm and hurricane formation in 1980–2000, by month

April 1 May 1 June 11 July 25 Aug. 60 Sept. 72 Oct. 29 Nov. 15 Dec. 1

Source: Colorado State University

10. *Tornadoes.* The following bar graph shows the average number of tornado deaths by month for years 1950–1999. What is the average number of tornado deaths for the 12 months? the median? the mode?

Average Number of Deaths by Tornado by Month

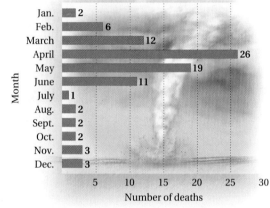

Jan. 2 Feb. 6 March 12 April 26 May 19 June 11 July 1 Aug. 2 Sept. 2 Oct. 2 Nov. 3 Dec. 3

Number of deaths

Source: National Weather Service's Storm Prediction Center

11. *Gas Mileage.* The Acura RSX Type-S gets 279 miles of highway driving on 9 gallons of gasoline. What is the average number of miles expected per gallon—that is, what is its gas mileage?
Source: EPA; *Car and Driver,* September 2005

12. *Gas Mileage.* The Chevrolet Cobalt SS Supercharged gets 322 miles of city driving on 14 gallons of gasoline. What is the average number of miles expected per gallon—that is, what is its gas mileage?
Source: EPA; *Car and Driver,* September 2005

Grade Point Average. The tables in Exercises 13 and 14 show the grades of a student for one semester. In each case, find the grade point average. Assume that the grade point values are 4.0 for an A, 3.0 for a B, and so on. Round to the nearest tenth.

13.

GRADE	NUMBER OF CREDIT HOURS IN COURSE
B	4
A	5
D	3
C	4

14.

GRADE	NUMBER OF CREDIT HOURS IN COURSE
A	5
C	4
F	3
B	5

15. *Salmon Prices.* The following prices per pound of Atlantic salmon were found at five fish markets:

$6.99, $8.49, $8.99, $6.99, $9.49.

What was the average price per pound? the median price? the mode?

16. *Cheddar Cheese Prices.* The following prices per pound of sharp cheddar cheese were found at five supermarkets:

$5.99, $6.79, $5.99, $6.99, $6.79.

What was the average price per pound? the median price? the mode?

17. *Grading.* To get a B in math, Rich must score an average of 80 on five tests. Scores on the first four tests were 80, 74, 81, and 75. What is the lowest score that Rich can get on the last test and still receive a B?

18. *Grading.* To get an A in math, Cybil must score an average of 90 on five tests. Scores on the first four tests were 90, 91, 81, and 92. What is the lowest score that Cybil can get on the last test and still receive an A?

19. *Length of Pregnancy.* Marta was pregnant 270 days, 259 days, and 272 days for her first three pregnancies. In order for Marta's average pregnancy to equal the worldwide average of 266 days, how long must her fourth pregnancy last?

Source: David Crystal (ed.), *The Cambridge Factfinder.* Cambridge CB2 1RP: Cambridge University Press, 1993, p. 84.

20. *Male Height.* Jason's brothers are 174 cm, 180 cm, 179 cm, and 172 cm tall. The average male is 176.5 cm tall. How tall is Jason if he and his brothers have an average height of 176.5 cm?

d Solve.

21. *Light-Bulb Testing.* An experiment is performed to compare the lives of two types of light bulb. Several bulbs of each type were tested and the results are listed in the following table. On the basis of this test, which bulb is better?

BULB A: HOTLIGHT TIMES (in hours)			BULB B: BRIGHTBULB TIMES (in hours)		
983	964	1214	979	1083	1344
1417	1211	1521	984	1445	975
1084	1075	892	1492	1325	1283
1423	949	1322	1325	1352	1432

22. *Cola Testing.* An experiment is conducted to determine which of two colas tastes better. Students drank each cola and gave it a rating from 1 to 10. The results are given in the following table. On the basis of this test, which cola tastes better?

COLA A: VERVCOLA				COLA B: COLA-COLA			
6	8	10	7	10	9	9	6
7	9	9	8	8	8	10	7
5	10	9	10	8	7	4	3
9	4	7	6	7	8	10	9

23. $\mathbf{D}_\mathbf{W}$ You are applying for an entry-level job at a large firm. You can be informed of the mean, median, or mode salary. Which of the three figures would you request? Why?

24. $\mathbf{D}_\mathbf{W}$ Is it possible for a driver to average 20 mph on a 30-mi trip and still receive a ticket for driving 75 mph? Why or why not?

Multiply.

25. $14 \cdot 14$ [1.5a]

26. $\dfrac{2}{3} \cdot \dfrac{2}{3}$ [2.6a]

27. 1.4×1.4 [4.3a]

28. 1.414×1.414 [4.3a]

29. 12.86×17.5 [4.3a]

30. 222×0.5678 [4.3a]

31. $\dfrac{4}{5} \cdot \dfrac{3}{28}$ [2.6a]

32. $\dfrac{28}{45} \cdot \dfrac{3}{2}$ [2.6a]

Convert to percent notation.   [6.2a]

33. $\dfrac{19}{16}$

34. $\dfrac{11}{16}$

35. $\dfrac{64}{125}$

36. $\dfrac{9781}{10,000}$

Bowling Averages.   Bowling averages are always computed by rounding down to the nearest integer. For example, suppose a bowler gets a total of 599 for 3 games. To find the average, we divide 599 by 3 and drop the amount to the right of the decimal point:

$$\frac{599}{3} \approx 199.67 \qquad \text{The bowler's average is 199.}$$

In each case, find the bowling average.

37. 🔢 547 in 3 games

38. 🔢 4621 in 27 games

39. *Hank Aaron.*   Hank Aaron averaged $34\frac{7}{22}$ home runs per year over a 22-yr career. After 21 yr, Aaron had averaged $35\frac{10}{21}$ home runs per year. How many home runs did Aaron hit in his final year?

40. The ordered set of data 18, 21, 24, a, 36, 37, b has a median of 30 and an average of 32. Find a and b.

41. *Grades.*   Because of a poor grade on the fifth of five tests, Chris's average test grade fell from 90.5 to 84.0. What did Chris score on the fifth test? Assume that all tests are equally important.

42. *Price Negotiations.*   Amy offers $3200 for a used Ford Taurus advertised at $4000. The first offer from Jim, the car's owner, is to "split the difference" and sell the car for $(3200 + 4000) \div 2$, or $3600. Amy's second offer is to split the difference between Jim's offer and her first offer. Jim's second offer is to split the difference between Amy's second offer and his first offer. If this pattern continues and Amy accepts Jim's third (and final) offer, how much will she pay for the car?

7.2 TABLES AND PICTOGRAPHS

Objectives

a Extract and interpret data from tables.

b Extract and interpret data from pictographs.

a Reading and Interpreting Tables

A **table** is often used to present data in rows and columns.

EXAMPLE 1 *Nutrition Information.* The following table lists daily nutrition information for eleven diets.

Diet	Average daily calories	Grams of fat	Grams of saturated fat	Grams of carbohydrates	Grams of protein	Grams of fiber*	Fruits and vegetables†
Atkins (Induction)	1640	61	19	8	31	8	6
Atkins (Ongoing Weight Loss)	1520	60	20	11	29	12	6
eDiets	1450	23	5	53	24	19	12
Jenny Craig	1520	18	7	62	20	16	6
Ornish (Eat More, Weigh Less)	1520	6	1	77	16	31	17
Slim-Fast	1540	22	6	57	21	21	12
South Beach (Phase One)	1530	51	14	15	34	9	12
South Beach (Phase Two)	1340	39	9	38	22	19	13
Volumetrics	1500	23	7	55	22	20	14
Weight Watchers	1450	24	7	56	20	20	11
Zone (Men's Menu)	1660	27	7	42	30	21	17

Source: Consumer Reports   *Per 1000 calories †Daily servings

a) Which diet has the least number of average daily calories?

b) Which diet has the greatest number of grams of carbohydrates?

c) Which diet has the least number of grams of fiber?

d) Find the average number of grams of fat in the diets listed.

Careful examination of the table will give the answers.

a) To determine which diet has the least number of average daily calories, look down the column headed "Average Daily Calories" and find the smallest number. That number is 1340. Then look across that row to find the name of the diet, South Beach (Phase Two).

b) To determine which diet has the greatest amount of carbohydrates, look down the column headed "Grams of Carbohydrates" and find the largest number. That number is 77 g. Then look across that row to find the diet, Ornish (Eat More, Weigh Less).

c) To determine which diet has the least amount of fiber, look down the column headed "Grams of Fiber" and find the smallest number. That number is 8g. Then look across that row to find the diet, Atkins (Induction).

d) Find the average of all the numbers in the column headed "Grams of Fat":

$$\frac{61 + 60 + 23 + 18 + 6 + 22 + 51 + 39 + 23 + 24 + 27}{11} = \frac{354}{11} \approx 32.2 \text{ g.}$$

Use the table in Example 1 to answer Margin Exercises 1–7.

1. Which diet has the greatest number of average daily calories?

2. Which diet has the greatest number of servings of fruit and vegetables?

3. Which diet has the least number of grams of fat?

4. Find the average number of grams of carbohydrates in the diets listed.

Answers on page A-16

427

7.2 Tables and Pictographs

5. Find the average amount of protein in the eleven diets.

The average daily number of grams of fat in these eleven diets is about 32.2 g.

Do Exercises 1–7. (Exercises 1–4 are on the preceding page.)

6. Find the median of the amount of protein in the diets.

EXAMPLE 2 *Wheaties Nutrition Facts.* Most foods are required by law to provide factual information regarding nutrition, as shown in the following table of Nutrition Facts from a box of Wheaties cereal. Although this can be very helpful to the consumer, one must be careful in interpreting the data. The % Daily Value figures shown here are based on a 2000-calorie diet. Your daily values may be higher or lower, depending on your calorie needs or intake.

7. Find the average, the median, and the mode of the amount of saturated fat in the diets listed.

Source: General Mills

Use the Nutrition Facts data from the Wheaties box and the bowl of cereal described in Example 2 to answer Margin Exercises 8–12.

8. How many calories from fat are in your bowl of cereal?

Suppose your morning bowl of cereal consists of 2 cups of Wheaties with 1 cup of skim milk.

a) How many calories have you consumed?

b) What percent of the daily value of total fat have you consumed?

9. A nutritionist recommends that you look for foods that provide 10% or more of the daily value for iron. Do you get that with your bowl of Wheaties?

c) A nutritionist recommends that you look for foods that provide 10% or more of the daily value for vitamin C. Do you get that with your bowl of Wheaties?

d) Suppose you are trying to limit your daily caloric intake to 2500 calories. How many bowls of cereal would it take to exceed the 2500 calories?

Careful examination of the table of nutrition facts will give the answers.

10. How much sodium have you consumed? *Hint:* Use the data listed in the first footnote below the table of nutrition facts.

a) Look at the column marked "with 1/2 cup skim milk" and note that 1 cup of cereal with 1/2 cup skim milk contains 150 calories. Since you are having twice that amount, you are consuming

$$2 \times 150, \quad \text{or} \quad 300 \text{ calories.}$$

11. What daily value of sodium have you consumed?

b) Read across from "Total Fat" and note that in 1 cup of cereal with 1/2 cup skim milk, you get 2% of the daily value of fat. Since you are doubling that, you get 4% of the daily value of fat.

c) Find the row labeled "Vitamin C" on the left and look under the column labeled "with 1/2 cup skim milk." Note that you get 10% of the daily value for "1 cup with 1/2 cup of skim milk," and since you are doubling that, you are more than satisfying the 10% requirement.

12. How much protein have you consumed? *Hint:* Use the data listed in the first footnote below the table of nutrition facts.

d) From part (a), we know that you are consuming 300 calories per bowl. Dividing 2500 by 300 gives $\frac{2500}{300} \approx 8.33$. Thus if you eat 9 bowls of cereal in this manner, you will exceed the 2500 calories.

Answers on page A-16

Do Exercises 8–12.

b Reading and Interpreting Pictographs

Pictographs (or *picture graphs*) are another way to show information. Instead of actually listing the amounts to be considered, a **pictograph** uses symbols to represent the amounts. In addition, a *key* is given telling what each symbol represents.

EXAMPLE 3 *Rhino Population.* The rhinoceros is considered one of the world's most endangered animals. The worldwide total number of rhinoceroses is approximately 17,800. The following pictograph shows the population of the five remaining rhino species. Located in the graph is a key that tells you that each symbol represents 300 rhinos.

Source: International Rhino Foundation, 2005

a) Which species has the greatest number of rhinos?

b) Which species has the least number of rhinos?

c) How many more black rhinos are there than Indian/Nepalese rhinos?

We can compute the answers by first reading the pictograph.

a) The species with the most symbols has the greatest number of rhinos: white rhino, with $37\frac{3}{4}$ (symbols) $\times$ 300, or 11,325 rhinos.

b) The species with the fewest symbols has the least number of rhinos: Javan rhino, with about $\frac{1}{5}$ (of a symbol) $\times$ 300, or 60 rhinos.

c) From the graph we see that there are 12 $\times$ 300, or 3600, black rhinos, and 8 $\times$ 300, or 2400, Indian/Nepalese rhinos. Thus, there are 3600 $-$ 2400, or 1200, more black rhinos than Indian/Nepalese rhinos.

Do Exercises 13–15.

You have probably noticed that, although they seem to be very easy to read, pictographs are difficult to draw accurately because whole symbols reflect loose approximations due to significant rounding. In pictographs, you also need to use some mathematics to find the actual amounts.

On July 5, 2004, at the Cincinnati Zoo and Botanical Gardens, Emi, a critically endangered Sumatran rhinoceros, became the first rhino to produce two calves in captivity.

Use the pictograph in Example 3 to answer Margin Exercises 13–15.

13. How many more Indian/Nepalese rhinos are there than Sumatran rhinos?

14. How does the black rhino population compare with the Sumatran rhino population?

15. What is the average number of rhinos in these five species?

Answers on page A-16

Use the pictograph in Example 4 to answer Margin Exercises 16–18.

16. Determine the approximate coffee consumption per capita of France.

17. Determine the approximate coffee consumption per capita of Italy.

18. The approximate coffee consumption of Finland is about the same as the combined coffee consumptions of Switzerland and the United States. What is the approximate coffee consumption of Finland?

EXAMPLE 4 *Coffee Consumption.* For selected countries, the following pictograph shows approximately how many cups of coffee each person (per capita) drinks annually.

Coffee Consumption

Germany			
United States			
Switzerland			
France			
Italy			

🍵 = 100 cups

Source: Beverage Marketing Corporation

a) Determine the approximate annual coffee consumption per capita of Germany.

b) Which two countries have the greatest difference in coffee consumption? Estimate that difference.

We use the data from the pictograph as follows.

a) Germany's consumption is represented by 11 whole symbols (1100 cups) and, though it is visually debatable, about $\frac{1}{8}$ of another symbol (about 13 cups), for a total of 1113 cups.

b) Visually, we see that Switzerland has the most consumption and that the United States has the least consumption. Switzerland's annual coffee consumption per capita is represented by 12 whole symbols (1200 cups) and about $\frac{1}{5}$ of another symbol (20 cups), for a total of 1220 cups. U.S. consumption is represented by 6 whole symbols (600 cups) and about $\frac{1}{10}$ of another symbol (10 cups), for a total of 610 cups. The difference between these amounts is 1220 − 610, or 610 cups.

One advantage of pictographs is that the appropriate choice of a symbol will tell you, at a glance, the kind of measurement being made. Another advantage is that the comparison of amounts represented in the graph can be expressed more easily by just counting symbols. For instance, in Example 3, the ratio of Indian/Nepalese rhinoceroses to black rhinoceroses is 8 : 12.

There are at least three disadvantages of pictographs:

1. To make a pictograph easy to read, the amounts must be rounded significantly to the unit that a symbol represents. This makes it difficult to accurately represent an amount.

2. It is difficult to determine very accurately how much a partial symbol represents.

3. Some mathematics is required to finally compute the amount represented, since there is usually no explicit statement of the amount.

Do Exercises 16–18.

Answers on page A-16

7.2

EXERCISE SET

For Extra Help

Math XL   MyMathLab   InterAct Math   Math Tutor Center   Digital Video Tutor CD 4 Videotape 7   Student's Solutions Manual

a   *Planets.*   Use the following table, which lists information about the planets, for Exercises 1–10.

PLANET	AVERAGE DISTANCE FROM SUN (in miles)	DIAMETER (in miles)	LENGTH OF PLANET'S DAY IN EARTH TIME (in days)	TIME OF REVOLUTION IN EARTH TIME (in years)
Mercury	35,983,000	3,031	58.82	0.24
Venus	67,237,700	7,520	224.59	0.62
Earth	92,955,900	7,926	1.00	1.00
Mars	141,634,800	4,221	1.03	1.88
Jupiter	483,612,200	88,846	0.41	11.86
Saturn	888,184,000	74,898	0.43	29.46
Uranus	1,782,000,000	31,763	0.45	84.01
Neptune	2,794,000,000	31,329	0.66	164.78
Pluto	3,666,000,000	1,423	6.41	248.53

Source: The Handy Science Answer Book, Gale Research, Inc.

1. Find the average distance from the sun to Jupiter.

2. How long is a day on Venus?

3. Which planet has a time of revolution of 164.78 yr?

4. Which planet has a diameter of 4221 mi?

5. Which planets have an average distance from the sun that is greater than 1,000,000 mi?

6. Which planets have a diameter that is less than 100,000 mi?

7. About how many Earth diameters would it take to equal one Jupiter diameter?

8. How much longer is the longest time of revolution than the shortest?

9. What are the average, the median, and the mode of the diameters of the planets?

10. What are the average, the median, and the mode of the average distances from the sun of the planets?

431

Heat Index.   In warm weather, a person can feel hotter due to reduced heat loss from the skin caused by higher humidity. The **temperature–humidity index,** or **apparent temperature,** is what the temperature would have to be with no humidity in order to give the same heat effect. The following table lists the apparent temperatures for various actual temperatures and relative humidities. Use this table for Exercises 11–22.

ACTUAL TEMPERATURE (°F)	RELATIVE HUMIDITY									
	10%	20%	30%	40%	50%	60%	70%	80%	90%	100%
	APPARENT TEMPERATURE (°F)									
75°	75	77	79	80	82	84	86	88	90	92
80°	80	82	85	87	90	92	94	97	99	102
85°	85	88	91	94	97	100	103	106	108	111
90°	90	93	97	100	104	107	111	114	118	121
95°	95	99	103	107	111	115	119	123	127	131
100°	100	105	109	114	118	123	127	132	137	141
105°	105	110	115	120	125	131	136	141	146	151

In Exercises 11–14, find the apparent temperature for the given actual temperature and humidity combinations.

11. 80°, 60%   **12.** 90°, 70%   **13.** 85°, 90%   **14.** 95°, 80%

15. How many listed temperature–humidity combinations give an apparent temperature of 100°?

16. How many listed temperature–humidity combinations given an apparent temperature of 111°?

17. At a relative humidity of 50%, what actual temperatures give an apparent temperature above 100°?

18. At a relative humidity of 90%, what actual temperatures give an apparent temperature above 100°?

19. At an actual temperature of 95°, what relative humidities give an apparent temperature above 100°?

20. At an actual temperature of 85°, what relative humidities give an apparent temperature above 100°?

21. At an actual temperature of 85°, by how much would the humidity have to increase in order to raise the apparent temperature from 94° to 108°?

22. At an actual temperature of 80°, by how much would the humidity have to increase in order to raise the apparent temperature from 87° to 102°?

Cigarette Consumption.   The May 2004 U.S. Surgeon General's report, *The Health Consequences of Smoking,* reports that smoking causes many diseases and reduces the health of smokers. The per capita cigarette consumption in the United States for specific years from 1920 to 2003 is listed in the table below. Use this data for Exercises 23–26.

U.S. Cigarette Consumption (1920 – 2003) (Per capita among persons 18 and older)

YEAR	1920	1930	1940	1950	1960	1970	1980	1990	2000	2003
Cigarette Consumption	665	1485	1976	3552	4171	3985	3849	2817	2092	1903

Sources: U.S. Department of Health and Human Services, Centers for Disease Control and Prevention; USDA, Economic Research Service

23. Find the cigarette consumption in 1940 and 1980. What was the percent of increase in the consumption from 1940 to 1980?

24. Find the cigarette consumption in 1960 and 2000. What was the percent of decrease in the consumption from 1960 to 2000?

25. Find the average of the cigarette consumption for the years 1920 to 1950. Find the average of the cigarette consumption for the years 1970 to 2000. By how many cigarettes per capita does the latter average exceed the former?

26. Find the average of the cigarette consumption for the years 2000 and 2003. Find the average of the cigarette consumption for the years 1920 to 2000. By how many cigarettes per capita does the average for 1920 to 2000 exceed the average of the years 2000 and 2003?

b *World Population Growth.*   The following pictograph shows world population in various years. Use the pictograph for Exercises 27–34.

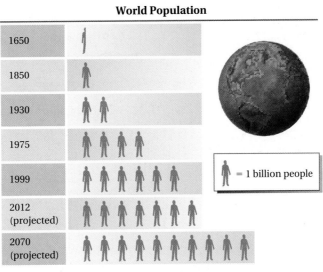

Source: U.S. Bureau of the Census, International Data Base

27. What was the world population in 1850?

28. What was the world population in 1975?

29. In which year will the population be the greatest?

30. In which year was the population the least?

31. Between which two years was the amount of growth the least?

32. Between which two years was the amount of growth the greatest?

33. How much greater will the world population in 2012 be than in 1975? What is the percent of increase?

34. How much greater was the world population in 1999 than in 1930? What is the percent of increase?

Water Consumption. The following pictograph shows water consumption, per person, in different regions of the world in a recent year. Use the pictograph for Exercises 35–40.

Region	Water Consumption
North America	
Central America	
Europe	
Oceania	
Asia	
South America	
Africa	

= 100,000 gallons

Sources: World Resources Institute; U.S. Energy Information Administration

35. What region consumes the least water?

36. Which region consumes the most water?

37. About how many gallons are consumed per person in North America?

38. About how many gallons are consumed per person in Europe?

39. Approximately how many more gallons are consumed per person in North America than in Asia?

40. Approximately how many more gallons are consumed per person in Central America than in Africa?

41. D_W Loreena is drawing a pictograph in which dollar bills are used as symbols to represent the tuition at various private colleges. Should each dollar bill represent $10,000, $1000, or $100? Why?

42. D_W What advantage(s) does a table have over a pictograph?

Solve. [6.5a]

43. *Kitchen Costs.* The average cost of a kitchen is $26,888. Some of the cost percentages are as follows.

Cabinets:	50%
Countertops:	15%
Appliances:	8%
Fixtures:	3%

Find the costs for each part of a kitchen.

Source: National Kitchen and Bath Association

44. *Bathroom Costs.* The average cost of a bathroom is $11,605. Some of the cost percentages are as follows.

Cabinets:	31%
Countertops:	11%
Labor:	25%
Flooring:	6%

Find the costs for each part of a bathroom.

Source: National Kitchen and Bath Association

Convert to fraction notation and simplify. [6.2b]

45. 24%

46. 45%

47. 4.8%

48. 6.4%

49. 53.1%

50. 87.3%

51. 100%

52. 2%

53. Redraw the pictograph appearing in Example 4 as one in which each symbol represents 150 cups of coffee.

Objectives

Use the bar graph in Example 1 to answer Margin Exercises 1–3.

1. About how many fans does the NHL (National Hockey League) have?

2. Which sport has the fewest number of fans?

3. Which sports have 100 million or more fans?

7.3 BAR GRAPHS AND LINE GRAPHS

A **bar graph** is convenient for showing comparisons because you can tell at a glance which amount represents the largest or smallest quantity. Of course, since a bar graph is a more abstract form of pictograph, this is true of pictographs as well. However, with bar graphs, a *second scale* is usually included so that a more accurate determination of the amount can be made.

a Reading and Interpreting Bar Graphs

EXAMPLE 1 *Major League Sports Fans.* As NASCAR racing increases in popularity, its fan base is growing rapidly. The following horizontal bar graph shows the number of fans of various major league sports.

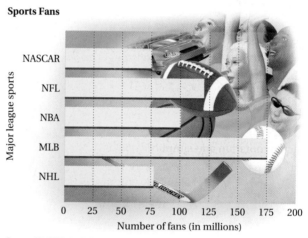

Sports Fans

Sources: NASCAR; the individual leagues; ESPN

a) Which of the major league sports shown in the graph above has the greatest number of fans?

b) About how many fans does the NFL (National Football League) have?

c) Which of the major league sports has about 100 million fans?

We look at the graph to answer the questions.

a) The longest bar is for MLB (Major League Baseball). Thus, that sport has the greatest number of fans.

b) We move to the right along the bar representing the NFL. We can read, fairly accurately, that this sport has approximately 120 million fans.

c) We locate the line representing 100 million and then go up until we reach a bar that ends at 100 million. We then go to the left and read the name of the sport, NBA (National Basketball Association).

Do Exercises 1–3.

Answers on page A-16

Bar graphs are often drawn vertically and sometimes a double bar graph is used to make comparisons.

EXAMPLE 2 *Breast Cancer.* The following graph indicates the incidence and mortality rates of breast cancer for women of various age groups.

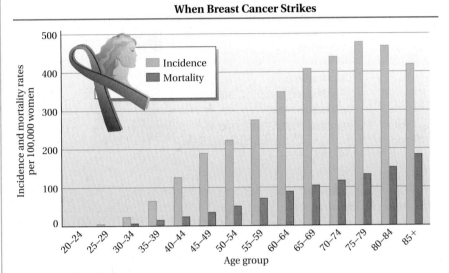

When Breast Cancer Strikes

Source: National Cancer Institute

a) Approximately how many women, per 100,000, develop breast cancer between the ages of 40 and 44?

b) In what age range is the mortality rate for breast cancer approximately 100 for every 100,000 women?

c) In what age range is the incidence of breast cancer the highest?

d) Does the incidence of breast cancer seem to increase from the youngest to the oldest age group?

We look at the graph to answer the questions.

a) We go to the right, across the bottom, to the green bar above the age group 40–44. Next, we go up to the top of that bar and, from there, back to the left to read approximately 130 on the vertical scale. About 130 out of every 100,000 women develop breast cancer between the ages of 40 and 44.

b) We read up the vertical scale to the number 100. From there we move to the right until we come to the top of a red bar. Moving down that bar, we find that in the 65–69 age group, about 100 out of every 100,000 women die of breast cancer.

c) We look for the tallest green bar and read the age range below it. The incidence of breast cancer is highest for women in the 75–79 age group.

d) Looking at the heights of the bars, we see that the incidence of breast cancer increases to a high point in the 75–79 age group and then decreases.

Do Exercises 4–7.

Use the bar graph in Example 2 to answer Margin Exercises 4–7.

4. Approximately how many women, per 100,000, develop breast cancer between the ages of 35 and 39?

5. In what age group is the mortality rate the highest?

6. In what age group do about 350 out of every 100,000 women develop breast cancer?

7. Does the breast cancer mortality rate seem to increase from the youngest to the oldest age group?

Answers on page A-16

8. Planetary Moons. Make a horizontal bar graph to show the number of moons orbiting the various planets.

PLANET	MOONS
Earth	1
Mars	2
Jupiter	17
Saturn	28
Uranus	21
Neptune	8
Pluto	1

Source: National Aeronautics and Space Administration

b   Drawing Bar Graphs

EXAMPLE 3 *Police Officers.* Listed below are the numbers of police officers per 10,000 people in various cities. Make a vertical graph of the data.

CITY	POLICE OFFICERS, PER 10,000 PEOPLE
Cincinnati	31
Cleveland	37
Minneapolis	26
Pittsburgh	30
Columbus	26
St. Louis	44

Sources: U.S. Bureau of the Census; FBI Uniform Crime Reports

First, we indicate the different names of the cities in six equally spaced intervals on the horizontal scale and give the horizontal scale the title "Cities." (See the figure on the left below.)

Next, we scale the vertical axis. To do so, we look over the data and note that it ranges from 26 to 44. We start the vertical scaling at 0, labeling the marks by 5's from 0 to 50. We give the vertical scale the title "Number of Officers (per 10,000 people)."

Finally, we draw vertical bars to show the various numbers, as shown in the figure at the right. We give the graph an overall title, "Police Officers."

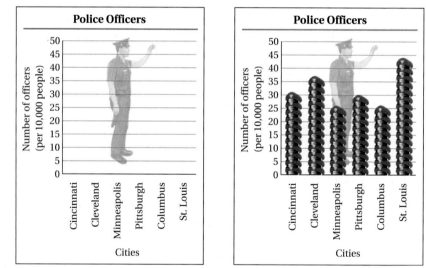

Sources: U.S. Bureau of the Census; FBI Uniform Crime Reports

Do Exercise 8.

c   Reading and Interpreting Line Graphs

Line graphs are often used to show a change over time as well as to indicate patterns or trends.

EXAMPLE 4 *New Home Sales.* The following line graph shows the number of new home sales, in thousands, over a twelve-month period. The jagged

Answer on page A-16

line at the base of the vertical scale indicates an unnecessary portion of the scale. Note that the vertical scale differs from the horizontal scale so that the data can be shown reasonably.

New Home Sales

Source: U.S. Department of Commerce

a) For which month were new home sales the greatest?

b) Between which months did new home sales increase?

c) For which months were new home sales about 700 thousand?

We look at the graph to answer the questions.

a) The greatest number of new home sales was about 825 thousand in month 1, December.

b) Reading the graph from left to right, we see that new home sales increased from month 2 to month 3, from month 3 to month 4, from month 5 to month 6, from month 7 to month 8, from month 8 to month 9, from month 9 to month 10, and from month 10 to month 11.

c) We look from left to right along the line at 700.

New Home Sales

We see that points are close to 700 thousand at months 3, 6, 10, 11, and 12.

Do Exercises 9–11.

EXAMPLE 5 *Monthly Loan Payment.* Suppose that you borrow $110,000 at an interest rate of 9% to buy a home. The following graph shows the monthly payment required to pay off the loan, depending on the length of

Use the line graph in Example 4 to answer Margin Exercises 9–11.

9. For which month were new home sales lowest?

10. Between which months did new home sales decrease?

11. For which months were new home sales about 650 thousand?

Answers on page A-16

Use the line graph in Example 5 to answer Margin Exercises 12–14.

12. Estimate the monthly payment for a loan of 25 yr.

13. What time period corresponds to a monthly payment of about $850?

14. By how much does the monthly payment decrease when the loan period is increased from 5 yr to 20 yr?

the loan. (*Caution:* A low monthly payment means that you will pay more interest over the duration of the loan.)

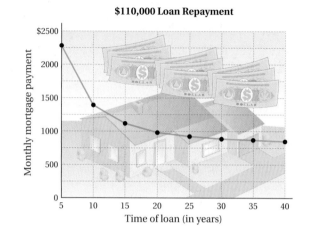

$110,000 Loan Repayment

a) Estimate the monthly payment for a loan of 15 yr.

b) What time period corresponds to a monthly payment of about $1400?

c) By how much does the monthly payment decrease when the loan period is increased from 10 yr to 20 yr?

We look at the graph to answer the questions.

a) We find the time period labeled "15" on the bottom scale and move up from that point to the line. We then go straight across to the left and find that the monthly payment is about $1100.

b) We locate $1400 on the vertical axis. Then we move to the right until we hit the line. The point $1400 is on the line at the 10-yr time period.

c) The graph shows that the monthly payment for 10 yr is about $1400; for 20 yr, it is about $990. Thus the monthly payment is decreased by $1400 − $990, or $410. (It should be noted that you will pay back $990 · 20 · 12 − $1400 · 10 · 12, or $69,600, more in interest for a 20-yr loan.)

Do Exercises 12–14.

d Drawing Line Graphs

EXAMPLE 6 *Cell Phones with Internet Access.* Listed below are projections on the use of cell phones with access to the Internet. Make a line graph of the data.

YEAR	CELL PHONES WITH WEB ACCESS (in millions)
2001	29.4
2002	69.6
2003	120.1
2004	152.4
2005	171.1

Source: Forrester Research

Answers on page A-16

First, we indicate the different years on the horizontal scale and give the horizontal scale the title "Year." (See the figure on the left below.) Next, we scale the vertical axis by 25's to show the number of phones, in millions, and give the vertical scale the title "Number of cell phones (in millions)". We also give the graph the overall title "Cell Phones with Web Access."

15. Cell Phones. Listed below are projections on the use of cell phones with or without access to the Internet. Make a line graph of the data.

YEAR	NUMBER OF CELL PHONES (in millions)
2001	119.8
2002	135.3
2003	150.2
2004	163.8
2005	176.9

Source: Forrester Research

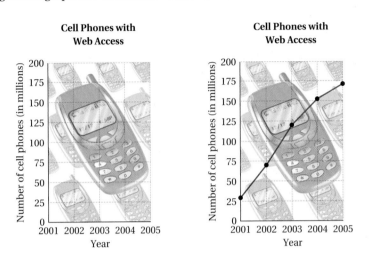

Next, we mark the number of phones at the appropriate level above each year. Then we draw line segments connecting the points. The dramatic change over time can now be observed easily from the graph.

Do Exercise 15.

Answer on page A-16

Study Tips

ATTITUDE AND THE POWER OF YOUR CHOICES

Making the right choices can give you the power to succeed in learning mathematics.

You can choose to improve your attitude and raise the academic goals that you have set for yourself. Projecting a positive attitude toward your study of mathematics and expecting a positive outcome can make it easier for you to learn and to perform well in this course.

Here are some positive choices you can make:

■ Choose to allocate the proper amount of time to learn.

■ Choose to place the primary responsibility for learning on yourself.

■ Choose to make a strong commitment to learning.

Well-known American psychologist William James once said, "The one thing that will guarantee the successful conclusion of a doubtful undertaking is faith in the beginning that you can do it." Having a positive attitude and making the right choices will add to your confidence in this course and multiply your successes.

7.3
EXERCISE SET
For Extra Help

MathXL · MyMathLab · InterAct Math · Math Tutor Center · Digital Video Tutor CD 4 Videotape 7 · Student's Solutions Manual

a *Chocolate Desserts.*   The following horizontal bar graph shows the average caloric content of various kinds of chocolate desserts. Use the bar graph for Exercises 1–12.

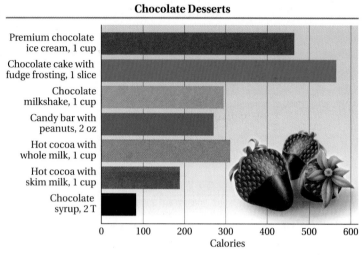

Chocolate Desserts

Source: *Better Homes and Gardens*, December 1996

1. Estimate how many calories there are in 1 cup of hot cocoa with skim milk.

2. Estimate how many calories there are in 1 cup of premium chocolate ice cream.

3. Which dessert has the highest caloric content?

4. Which dessert has the lowest caloric content?

5. Which dessert contains about 460 calories?

6. Which desserts contain about 300 calories?

7. How many more calories are there in 1 cup of hot cocoa made with whole milk than in 1 cup of hot cocoa made with skim milk?

8. Fred generally drinks a 4-cup chocolate milkshake. How many calories does he consume?

9. Kristin likes to eat 2 cups of premium chocolate ice cream at bedtime. How many calories does she consume?

10. Barney likes to eat a 6-oz chocolate bar with peanuts for lunch. How many calories does he consume?

11. Paul adds a 2-oz chocolate bar with peanuts to his diet each day for 1 yr (365 days) and makes no other changes in his eating or exercise habits. Consumption of 3500 extra calories will add about 1 lb to his body weight. How many pounds will he gain?

12. Tricia adds one slice of chocolate cake with fudge frosting to her diet each day for one year (365 days) and makes no other changes in her eating or exercise habits. The consumption of 3500 extra calories will add about 1 lb to her body weight. How many pounds will she have gained at the end of the year?

Education and Earnings. Side-by-side bar graphs allow for comparisons. The one shown at right provides data on the effect of education on earning power for men and women from 1970 to 2002. Use the bar graph in Exercises 13–20.

Mean earnings for full-time, year-round workers— high school graduates vs. bachelor's degree recipients, ages 25 to 34

H.S. Diploma
Bachelor's Degree

Source: USA Group Foundation

13. How much were the mean earnings for men with bachelor's degrees in 1970? in 2002? How much had they increased? What was the percent of increase?

14. How much were the mean earnings for women with bachelor's degrees in 1970? in 2002? How much had they increased? What was the percent of increase?

15. How much were the mean earnings for women who had ended their education at high school graduation in 1970? in 2002? How much had they increased? What was the percent of increase?

16. How much were the mean earnings for men who had ended their education at high school graduation in 1970? in 2002? How much had they increased? What was the percent of increase?

17. In 1970, how much more did men with bachelor's degrees earn than men who ended their education at high school graduation?

18. In 2002, how much more did men with bachelor's degrees earn than men who ended their education at high school graduation?

19. In 2002, how much more did women with bachelor's degrees earn than men who ended their education at high school graduation?

20. In 1970, how much more did men with bachelor's degrees earn than women who ended their education at high school graduation?

443

21. *Commuting Time.* The following table lists the average commuting time in six metropolitan areas with more than 1 million people. Make a vertical bar graph to illustrate the data.

CITY	COMMUTING TIME (in minutes)
New York	39.0
Los Angeles	28.1
Phoenix	24.7
Houston	25.9
Indianapolis	21.6
Chicago	33.1

Source: U.S. Bureau of the Census

Use the data and the bar graph in Exercise 21 to do Exercises 22–25.

22. Which city has the greatest commuting time?

23. Which city has the least commuting time?

24. What was the median commuting time for all six cities?

25. What was the average commuting time for the six cities?

26. *Airline Net Profits.* The net profits (the amount remaining after all deductions like expenses have been made) of U.S. airlines in various years are listed in the table below. Make a horizontal bar graph illustrating the data.

YEAR	NET PROFIT (in billions)
1995	$2.3
1996	2.8
1997	5.2
1998	4.9
1999	5.5
2000	2.7

Source: Air Transportation Association

Use the data and the bar graph in Exercise 26 to do Exercises 27–32.

27. Between which pairs of years was there an increase in profit?

28. Between which pairs of years was there a decrease in profit?

29. What was the percent of decrease between 1999 and 2000?

30. What was the percent of increase between 1996 and 1997?

31. What was the average net profit for all 6 yr?

32. What was the median net profit for all 6 yr?

C *Golf Distances.* In recent years, new equipment and technology have had a tremendous impact on the distance a golfer can hit a golf ball. The line graph below shows the average driving distances for years from 1980 to 2004. Use the graph for Exercises 33–36.

Average Driving Distance on the PGA Tour

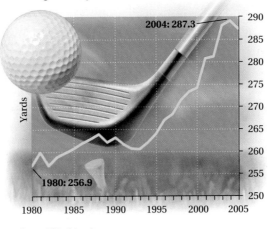

2004: 287.3

1980: 256.9

Yards

290
285
280
275
270
265
260
255
250

1980   1985   1990   1995   2000   2005

Source: PGA of America

33. How much farther was the driving distance in 2004 than in 1980?

34. What was the percent of increase in the driving distance from 1980 to 2004?

35. In what years was the average driving distance about 264 yd?

36. In what year was the average driving distance about 270 yd?

d Make a line graph of the data in the following tables (Exercises 37 and 42), using the horizontal axis to scale "Year."

37. *Longevity Beyond Age 65.* These data tell us the number of years a 65-year-old male in the given year can expect to live. Draw a line graph.

YEAR	AVERAGE NUMBER OF YEARS MEN ARE ESTIMATED TO LIVE BEYOND AGE 65
1980	14
1990	15
2000	15.9
2010	16.4
2020	16.9
2030	17.5

Source: 2000 Social Security Report

38. What was the percent of increase in longevity (years beyond 65) between 1980 and 2000?

39. What is the expected percent of increase in longevity between 1980 and 2030?

40. What will be the percent of increase in longevity between 2020 and 2030?

41. What will be the percent of increase in longevity between 2000 and 2030?

42. *Homicide Rate in Baltimore.* These data indicate the number of homicides in Baltimore, Maryland, for several years. Draw a line graph.

YEAR	NUMBER OF MURDERS COMMITTED
1995	325
1996	331
1997	311
1998	313
1999	305
2000	262

Source: Baltimore Police Department

43. Between which two years was the increase in murders the greatest?

44. Between which two years was the increase in murders the least?

45. What is the average number of murders committed per year from 1995 to 2000?

46. What is the median number of murders committed?

47. What was the percent of decrease in the murders from 1999 to 2000?

48. What was the percent of increase in the number of murders from 1995 to 1996?

49. D_W Can bar graphs always, sometimes, or never be converted to line graphs? Why?

50. D_W Compare bar graphs and line graphs. Discuss why you might use one rather than the other to graph a particular set of data.

VOCABULARY REINFORCEMENT

In each of Exercises 51–58, fill in the blank with the correct term from the given list. Some of the choices may not be used.

51. The set of numbers 1, 2, 3, 4, 5,… is called the set of _____ numbers. [1.1b]

52. To find the average of a set of numbers, _____ the numbers and then _____ by the number of items of data. [7.1a]

53. The _____ interest I on principal P, invested for t years at interest rate r, is given by $I = P \cdot r \cdot t$. [6.7a]

54. If an item has a regular price of $100 and is on sale for 25% off, $100 is called the _____ , 25% the _____ , 25% of $100, or $25, the _____ , and $100 − $25, or $75, the _____ . [6.6c]

55. When interest is paid on interest, it is called _____ interest. [6.7b]

56. The decimal $0.\overline{1518}$ is an example of a(n) _____ decimal. [4.5a]

57. The statement $a(b + c) = ab + ac$ illustrates the _____ law. [1.5a]

58. The statement $x + t = t + x$ illustrates the _____ law. [1.5a]

add

subtract

multiply

divide

simple

compound

discount

rate of discount

sale price

marked price

repeating

terminating

natural

whole

distributive

associative

commutative

---

59. *Movie Theater Screens.* Draw a line graph of the data presented in the following bar graph. Use the line graph and the data to make the best estimate you can of the number of screens in 2000, 2001, and 2003. Explain.

Source: Motion Picture Association of America

Objectives

 a Extract and interpret data from circle graphs.

b Draw circle graphs.

Use the circle graph in Example 1 to answer Margin Exercises 1–4.

1. Which item costs the least?

2. What percent is not spent on either toys or supplies?

3. How much of the $6600 lifetime cost of owning a dog is for grooming?

4. What part of the expense is for supplies and for buying the dog?

 7.4 **CIRCLE GRAPHS**

We often use **circle graphs,** also called **pie charts,** to show the percent of a quantity used in different categories. Circle graphs can also be used very effectively to show visually the *ratio* of one category to another. In either case, it is quite often necessary to use mathematics to find the actual amounts represented for each specific category.

a **Reading and Interpreting Circle Graphs**

EXAMPLE 1 *Costs of Owning a Dog.* The following circle graph shows the relative costs of raising a dog from birth to death.

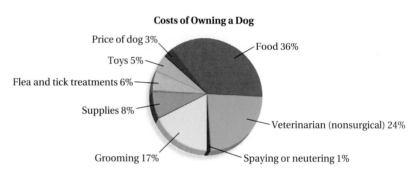

Costs of Owning a Dog

Price of dog 3%
Toys 5%
Flea and tick treatments 6%
Supplies 8%
Grooming 17%
Food 36%
Veterinarian (nonsurgical) 24%
Spaying or neutering 1%

Source: The American Pet Products Manufacturers Association

a) Which item costs the most?

b) What percent of the total cost is spent on grooming?

c) Which item involves 24% of the cost?

d) The American Pet Products Manufacturers Association estimates that the total cost of owning a dog for its lifetime is $6600. How much of that amount is spent for food?

e) What percent of the expense is for grooming and flea and tick treatments?

We look at the sections of the graph to find the answers.

a) The largest section (or sector) of the graph, 36%, is for food.

b) We see that grooming is 17% of the cost.

c) Nonsurgical veterinarian bills account for 24% of the cost.

d) The section of the graph representing food costs is 36%; 36% of $6600 is $2376.

e) We add the percents corresponding to grooming and flea and tick treatments. We have

17% (grooming) + 6% (flea and tick treatments) = 23%.

Do Exercises 1–4.

Answers on page A-17

b Drawing Circle Graphs

To draw a circle graph, or pie chart, like the one in Example 1, think of a pie cut into 100 equally sized pieces. We would then shade in a wedge equal in size to 36 of these pieces to represent 36% for food. We shade a wedge equal in size to 5 of these pieces to represent 5% for toys, and so on.

EXAMPLE 2 *Fruit Juice Sales.* The percents of various kinds of fruit juice sold are given in the list at right. Use this information to draw a circle graph.

Source: Beverage Marketing Corporation

Apple:	14%
Orange:	56%
Blends:	6%
Grape:	5%
Grapefruit:	4%
Prune:	1%
Other:	14%

Using a circle with 100 equally spaced tick marks, we start with the 14% given for apple juice. We draw a line from the center to any tick mark. Then we count off 14 ticks and draw another line. We shade the wedge with a color—in this case, red—and label the wedge as shown in the figure on the left below.

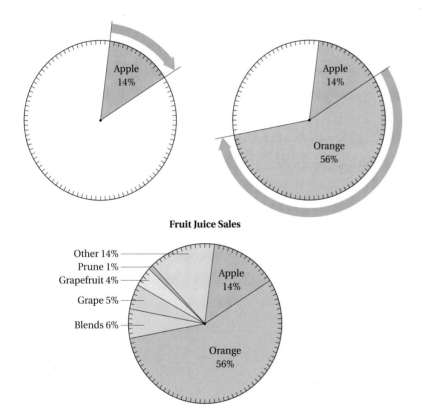

Fruit Juice Sales

To shade a wedge for orange juice, at 56%, we start at one side of the apple wedge, count off 56 ticks, and draw another line. We shade the wedge with a different color—in this case, orange—and label the wedge as shown in the figure on the right above. Continuing in this manner and choosing different colors, we obtain the graph shown above. Finally, we give the graph the overall title "Fruit Juice Sales."

Do Exercise 5.

5. Lengths of Engagement of Married Couples. The data below relate the percent of married couples who were engaged for a certain time period before marriage. Use this information to draw a circle graph.

ENGAGEMENT PERIOD	PERCENT
Less than 1 year	24
1–2 years	21
More than 2 years	35
Never engaged	20

Source: Bruskin Goldring Research

Answer on page A-17

Translating
for Success

1. *Vacation Miles.* The Saenz family drove their new van 13,640.8 mi in the first year. Of this total, 2018.2 miles were driven while on vacation. How many nonvacation miles did they drive?

2. *Rail Miles.* Of the recent $15\frac{1}{2}$ million passenger miles on a rail passenger line, 80% were transportation-to-work miles. How many rail miles, in millions, were to and from work?

3. *Sales Tax Rate.* The sales tax is $8.42 on the purchase of 10 bath towels which cost $129.50. What is the sales tax rate?

4. *Water Level.* During heavy rains in early spring, the water level in a pond rose 0.5 in. every 35 minutes. How much did the water rise in 90 minutes?

5. *Marathon Training.* When training for a marathon, Vebjoern ran $15\frac{1}{2}$ miles per day six days a week. This distance is 80% of the distance Yuriy ran per day. How far did Yuriy run each day?

The goal of these matching questions is to practice step (2), *Translate*, of the five-step problem-solving process. Translate each word problem to an equation and select a correct translation from equations A–O.

A. $8.42 \cdot x = 129.50$

B. $x = 80\% \cdot 15\frac{1}{2}$

C. $x = \dfrac{84 - 68}{84}$

D. $2018.2 + x = 13,640.8$

E. $\dfrac{5}{100} = \dfrac{x}{3875}$

F. $2018.2 = x \cdot 13,640.8$

G. $4\frac{1}{6} \cdot 73 = x$

H. $\dfrac{x}{5} = \dfrac{100}{3875}$

I. $15\frac{1}{2} = 80\% \cdot x$

J. $8.42 = x \cdot 129.50$

K. $\dfrac{0.5}{35} = \dfrac{x}{90}$

L. $x \cdot 4\frac{1}{6} = 73$

M. $x = \dfrac{84 - 68}{68}$

N. $x = 8.42\% \cdot 129.50$

O. $0.5 \times 35 = 90 \cdot x$

Answers on page A-17

6. *Vacation Miles.* The Ning family drove 2018.2 miles on their summer vacation. If they put a total of 13,640.8 miles on their new van during that year, what percent were vacation miles?

7. *Sales Tax.* The sales tax rate is 8.42%. Salena purchased 10 pillows at $12.95 each. How much tax is charged on this purchase?

8. *Charity Donations.* Rachel donated $5 to her favorite charity for each $100 she earned. One month, she earned $3875. How much did she donate that month?

9. *Tuxedos.* Emil Tailoring Company purchased 73 yards of fabric for a new line of tuxedos. How many tuxedos can be produced if it takes $4\frac{1}{6}$ yd of fabric for each tuxedo?

10. *Percent of Increase.* In a calculus-based physics course, Mime got 68% on the first exam and 84% on the second. What was the percent of increase in her score?

7.4

EXERCISE SET For Extra Help

| MathXL | MyMathLab | InterAct Math | Math Tutor Center | Digital Video Tutor CD 4 Videotape 7 | Student's Solutions Manual |

a *Popular Majors.* This circle graph shows the most popular majors among incoming freshmen. Use this graph for Exercises 1–6.

Majors Among Incoming Freshmen

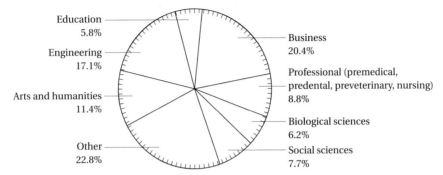

Education 5.8%

Engineering 17.1%

Arts and humanities 11.4%

Other 22.8%

Business 20.4%

Professional (premedical, predental, preveterinary, nursing) 8.8%

Biological sciences 6.2%

Social sciences 7.7%

1. What percent of freshmen major in engineering?

2. Together, what percent of freshmen major in biological science or professional?

3. In the fall of 2004, the University of Minnesota had approximately 10,562 freshmen. According to the data in the graph above, how many majored in education?

4. Texas Tech had 5766 freshmen in the fall of 2004. How many freshmen would be expected to major in the arts and humanities?

5. What percent of all freshmen do not major in biological or social sciences?

6. What percent of freshmen do not major in business or education?

Family Expenses. This circle graph shows expenses as a percent of income for a family of four. Use the graph for Exercises 7–10.

Family Expenses

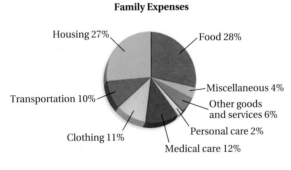

Housing 27%

Food 28%

Transportation 10%

Miscellaneous 4%

Other goods and services 6%

Personal care 2%

Clothing 11%

Medical care 12%

Source: U.S. Bureau of Labor Statistics

7. Which item accounts for the greatest expense?

8. In a family with a $4000 monthly income, how much is spent for transportation?

9. Some surveys combine medical care with personal care. What percent would be spent on those two items combined?

10. In a family with a $2000 monthly income, what is the ratio of the amount spent on medical care to the amount spent on personal care?

b Use the given information to complete a circle graph. Note that each circle is divided into 100 sections.

11. *Holiday Baking.*   The table below lists the percentages of the times when people do their holiday baking.

PREFERENCE	PERCENT
Pre-dawn, before 6 A.M.	28.7%
Late night, 9 P.M. to midnight	11.7%
Overnight, midnight to 5 A.M.	11.5%
Marathon style, all day/night	10.5%
Take day off work	3.4%
Don't bake for the holidays	23.4%
Don't know	10.8%

Source: Land O'Lakes Holiday Bakeline

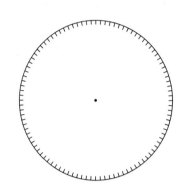

12. *Wealthy Givers.*   The table below lists the amounts that people with a net worth of more than $1 million make in charitable contributions.

DONATIONS	PERCENT
Under $1000	18%
$1000–$2499	21%
$2500–$4999	20%
$5000–$9999	17%
$10,000 or more	18%
Don't contribute/no answer	6%

Source: Yankelovich Partners

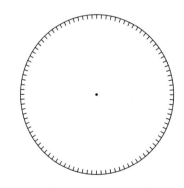

13. *Pregnancy Weight Gain.*   The table below lists the amounts of weight gain during pregnancy.

WEIGHT GAIN (in pounds)	PERCENT
Less than 20	22%
21–30	32%
31–40	27%
41 or more	19%

Source: National Vital Statistics Report

14. *e-mail.*   The table below lists the numbers of e-mails people get per day at work.

NUMBER OF E-MAILS PER DAY	PERCENT
Less than 1	28%
1–5	20%
6–10	12%
11–20	9%
21 or more	31%

Source: John J. Heldrich Center for Workforce Development

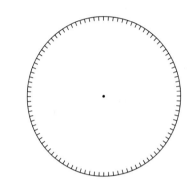

15. *Causes of Spinal Cord Injuries.*   The table below lists the causes of spinal cord injury.

CAUSES	PERCENT
Motor vehicle accidents	44%
Acts of violence	24%
Falls	22%
Sports	8%
Other	2%

Source: National Spinal Cord Injury Association

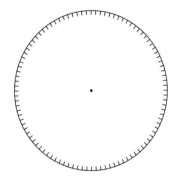

16. *Kids in Foster Care.*   There are approximately one-half million children in foster care in the United States. Most of these children are under the age of 10. The table below lists the percentages by ages of children in foster care.

AGE GROUP	PERCENT
Under 1	3%
1–5	25%
6–10	27%
11–15	27%
16+	18%

Source: The Administration for Children and Families

17. **D**w Discuss the advantages of being able to read a circle graph.

18. **D**w Compare circle graphs to bar graphs.

The review that follows is meant to prepare you for a chapter exam. It consists of two parts. The first part, Concept Reinforcement, is designed to increase understanding of the concepts through true/false exercises. The second part is the Review Exercises. These provide practice exercises for the exam, together with references to section objectives so you can go back and review. Before beginning, stop and look back over the skills you have obtained. What skills in mathematics do you have now that you did not have before studying this chapter?

✍ CONCEPT REINFORCEMENT

Determine whether the statement is true or false. Answers are given at the back of the book.

_____ **1.** A set of data has just one average and just one median, but it can have more than one mode.

_____ **2.** To find the average of a set of numbers, add the numbers and then multiply by the number of items of data.

_____ **3.** It is possible for the average, median, and mode of a set of data to be the same number.

_____ **4.** If each number in a set of data occurs the same number of times, there is no mode.

Review Exercises

FedEx Mailing Costs. There are three types of Federal Express delivery service for packages of various weights within Zone 5 (shipments from 601 to 1000 miles from origin), as shown in the following table. Use this table for Exercises 1–6. [7.2a]

1. Find the cost of a 3-lb FedEx Priority Overnight delivery.

2. Find the cost of a 10-lb FedEx Standard Overnight delivery.

3. How much would you save by sending the package listed in Exercise 1 by FedEx 2Day delivery?

4. How much would you save by sending the package in Exercise 2 by FedEx 2Day delivery?

Delivery by 10:00 a.m. next business day	Delivery by 3:00 p.m. next business day	Delivery by 4:30 p.m. second business day

All other packaging / weight in lbs.	FEDEX LETTER	FEDEX PRIORITY OVERNIGHT®	FEDEX STANDARD OVERNIGHT®	FEDEX 2DAY®
up to 8 oz.		$ 17.55	$ 15.95	$ n/a
1 lb.		$ 26.50	$ 23.25	$ 9.40
2 lbs.		29.75	26.25	10.10
3		32.25	28.50	11.30
4		36.00	31.50	13.10
5		38.75	33.25	14.50
6		41.50	36.25	15.80
7		45.25	39.75	17.30
8		47.50	41.75	18.80
9		51.00	45.00	20.20
10		53.50	47.00	21.60
11		56.50	49.50	23.00

Source: Federal Express Corporation

5. Is there any difference in price between sending a 5-oz package FedEx Priority Overnight and sending an 8-oz package in the same way?

6. An author has a 4-lb manuscript to send by FedEx Standard Overnight delivery to her publisher. She calls and the package is picked up. Later that day she completes work on another part of her manuscript that weighs 5 lb. She calls and sends it by FedEx Standard Overnight delivery to the same address. How much could she have saved if she had waited and sent both packages as one?

U.S. Police Forces. This pictograph shows the number of officers in the largest U.S. police forces. Use the graph for Exercises 7–10.

America's Largest Police Forces

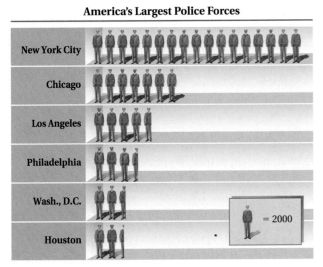

Source: International Association of Chiefs of Police

7. About how many officers are in the Chicago police force? [7.2b]

8. Which city has about 9000 officers on its force? [7.2b]

9. Of the cities listed, which has the smallest police force? [7.2b]

10. Estimate the average size of these six police forces. [7.1a], [7.2b]

Find the mode. [7.1c]

11. 26, 34, 43, 26, 51

12. 17, 7, 11, 11, 14, 17, 18

13. 0.2, 0.2, 1.7, 1.9, 2.4, 0.2

14. 700, 700, 800, 2700, 800

15. $14, $17, $21, $29, $17, $2

16. 20, 20, 20, 20, 20, 500

17. One summer, a student earned the following amounts over a four-week period: $102, $112, $130, and $98. What was the average amount earned per week? the median? [7.1a, b]

18. *Gas Mileage.* A 2001 Ford Focus gets 528 miles of highway driving on 16 gallons of gasoline. What is the gas mileage? [7.1a]
Source: Ford Motor Company

19. To get an A in math, a student must score an average of 90 on four tests. Scores on the first three tests were 94, 78, and 92. What is the lowest score that the student can make on the last test and still get an A? [7.1a]

Calorie Content in Fast Foods. Wendy's Hamburgers is a national food franchise. The following bar graph shows the caloric content of various sandwiches sold by Wendy's. Use the graph for Exercises 20–27. [7.3a]

Wendy's Sandwiches

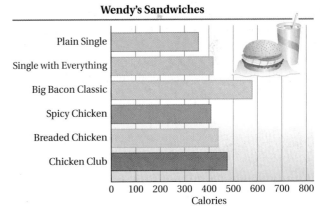

Source: Wendy's International

20. How many calories are in a Single with Everything?

21. How many calories are in a Breaded Chicken sandwich?

22. Which sandwich has the highest caloric content?

23. Which sandwich has the lowest caloric content?

24. Which sandwich contains about 360 calories?

25. Which sandwich contains about 470 calories?

26. How many more calories are in a Chicken Club than in a Single with Everything?

27. How many more calories are in a Big Bacon Classic than in a Plain Single?

Accidents by Driver Age. The following line graph shows the number of accidents per 100 drivers, by age. Use the graph for Exercises 28–33. [7.3c]

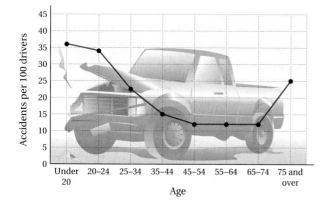

28. Which age group has the most accidents per 100 drivers?

29. What is the fewest number of accidents per 100 in any age group?

30. How many more accidents do people over 75 yr of age have than those in the age range of 65–74?

31. Between what ages does the number of accidents stay basically the same?

32. How many fewer accidents do people 25–34 yr of age have than those 20–24 yr of age?

33. Which age group has accidents right around three times as often as people 55–64 yr of age?

Hotel Preferences. This circle graph shows hotel preferences for travelers. Use the graph for Exercises 34–37. [7.4a]

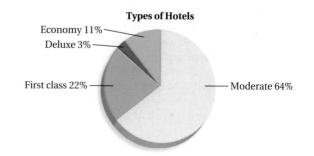

34. What percent of travelers prefer a first-class hotel?

35. What percent of travelers prefer an economy hotel?

36. Suppose 2500 travelers arrive in a city one day. How many of them might seek a moderate room?

37. What percent of travelers prefer either a first-class or a deluxe hotel?

First-Class Postage. [*] The following table shows the cost of first-class postage in various years. Use the table for Exercises 38 and 39.

YEAR	FIRST-CLASS POSTAGE
1983	20¢
1989	25¢
1991	29¢
1995	32¢
1999	33¢
2001	34¢
2002	37¢

Source: U.S. Postal Service

38. Make a vertical bar graph of the data. [7.3b]

39. Make a line graph of the data. [7.3d]

40. *Battery Testing.*   An experiment is performed to compare battery quality. Two kinds of battery were tested to see how long, in hours, they kept a hand radio running. On the basis of this test, which battery is better?   [7.1d]

BATTERY A: TIMES (in hours)	BATTERY B: TIMES (in hours)
38.9 39.3 40.4	39.3 38.6 38.8
53.1 41.7 38.0	37.4 47.6 37.9
36.8 47.7 48.1	46.9 37.8 38.1
38.2 46.9 47.4	47.9 50.1 38.2

Find the average.   [7.1a]

41. 26, 34, 43, 51

42. 11, 14, 17, 18, 7

43. 0.2, 1.7, 1.9, 2.4

44. 700, 2700, 3000, 900, 1900

45. $2, $14, $17, $17, $21, $29

46. 20, 190, 280, 470, 470, 500

Find the median.   [7.1b]

47. 26, 34, 43, 51

48. 7, 11, 14, 17, 18

49. 0.2, 1.7, 1.9, 2.4

50. 700, 900, 1900, 2700, 3000

51. $2, $17, $21, $29, $14, $17

52. 470, 20, 190, 280, 470, 500

53. *Grade Point Average.*   Find the grade point average for one semester given the following grades. Assume the grade point values are 4.0 for A, 3.0 for B, and so on. Round to the nearest tenth.   [7.1a]

COURSE	GRADE	NUMBER OF CREDIT HOURS IN COURSE
Basic math	A	5
English	B	3
Computer applications	C	4
Russian	B	3
College skills	B	1

54. **D**w Compare and contrast averages, medians, and modes. Discuss why you might use one over the others to analyze a set of data.   [7.1a, b, c]

55. **D**w Find a real-world situation that fits this equation: [7.1a]

$$T = \frac{(20{,}500 + 22{,}800 + 23{,}400 + 26{,}000)}{4}.$$

56. The ordered set of data 298, 301, 305, *a*, 323, *b*, 390 has a median of 316 and an average of 326. Find *a* and *b*. [7.1a, b]

Desirable Body Weights.   The following tables list the desirable body weights for men and women over age 25. Use the tables for Exercises 1–4.

DESIRABLE WEIGHT OF MEN			
HEIGHT	SMALL FRAME (in pounds)	MEDIUM FRAME (in pounds)	LARGE FRAME (in pounds)
5 ft, 7 in.	138	152	166
5 ft, 9 in.	146	160	174
5 ft, 11 in.	154	169	184
6 ft, 1 in.	163	179	194
6 ft, 3 in.	172	188	204

DESIRABLE WEIGHT OF WOMEN			
HEIGHT	SMALL FRAME (in pounds)	MEDIUM FRAME (in pounds)	LARGE FRAME (in pounds)
5 ft, 1 in.	105	113	122
5 ft, 3 in.	111	120	130
5 ft, 5 in.	118	128	139
5 ft, 7 in.	126	137	147
5 ft, 9 in.	134	144	155

Source: U.S. Department of Agriculture

1. What is the desirable weight for a 6 ft, 1 in. man with a medium frame?

2. What is the desirable weight for a 5 ft, 3 in. woman with a small frame?

3. What size woman has a desirable weight of 120 lb?

4. What size man has a desirable weight of 169 lb?

Waste Generated.   The number of pounds of waste generated per person per year varies greatly among countries around the world. In the pictograph at right, each symbol represents approximately 100 pounds of waste. Use the pictograph for Exercises 5–8.

5. In which country does each person generate 600 pounds of waste per year?

6. In which country does each person generate 1000 pounds of waste per year?

7. How many pounds of waste per person per year are generated in France?

8. How many pounds of waste per person per year are generated in Finland?

**Amount of Waste Generated 2000
(per person per year)**

Australia Finland France Ireland Japan Netherlands United States

Source: Data from OECD Environmental Data Compendium: 2002

Find the average.

9. 45, 49, 52, 52

10. 1, 1, 3, 5, 3

11. 3, 17, 17, 18, 18, 20

Find the median and the mode.

12. 45, 49, 52, 52

13. 1, 1, 3, 5, 3

14. 3, 17, 17, 18, 18, 20

15. *Gas Mileage.* A 2006 Mitsubishi Eclipse GT V-6 gets 432 miles of highway driving on 16 gallons of gasoline. What is the gas mileage?
Source: Mitsubishi Motors

16. *Grades.* To get a C in chemistry, a student must score an average of 70 on four tests. Scores on the first three tests were 68, 71, and 65. What is the lowest score that the student can make on the last test and still get a C?

Food Dollars Spent Away from Home. The line graph below shows the percentage of food dollars spent away from home for various years, and projected to 2010. Use the graph for Exercises 17–20.

Food dollars spent away from home

Sources: U.S. Bureau of Labor Statistics; National Restaurant Association

17. What percent of meals will be eaten away from home in 2010?

18. What percent of meals were eaten away from home in 1985?

19. In what year was the percent of meals eaten away from home about 30%?

20. In what year will the percent of meals eaten away from home be about 50%?

21. *Animal Speeds.* The following table lists maximum speeds of movement for various animals, in miles per hour, compared to the speed of the fastest human. Make a vertical bar graph of the data.

ANIMAL	SPEED (in miles per hour)
Antelope	61
Peregrine falcon	225
Cheetah	70
Fastest human	28
Greyhound	42
Golden eagle	150
Grant's gazelle	47

Source: Barbara Ann Kipfer, *The Order of Things.* New York: Random House, 1998.

Refer to the table and the graph in Exercise 21 for Exercises 22–25.

22. By how much does the fastest speed exceed the slowest speed?

23. Does a human have a chance of outrunning a greyhound? Explain.

24. Find the average of all the speeds.

25. Find the median of all the speeds.

26. *Shoplifting and Employee Theft.* The following table lists ways in which American retailers lost money recently. Construct a circle graph representing these data.

TYPE OF LOSS	PERCENT
Employee theft	44
Shoplifting	32.7
Administrative error	17.5
Vendor fraud	5.1
Other	0.7

Source: University of Florida, Department of Sociology for Sensormatic Electronics Corporation

27. In reference to Exercise 26, it is known that retailers lost $23 billion. Using the percents from the table and the circle graph, find the amount of money lost from each type of loss.

Porsche Sales. The table below lists the number of Porsche sales in the United States for various years. Use the table for Exercises 28 and 29.

YEAR	PORSCHE U.S. SALES
1996	7,152
1997	12,980
1998	17,239
1999	20,877
2000	23,000

Sources: Autodata, Bridge Information Systems

28. Make a bar graph of the data.

29. Make a line graph of the data.

30. *Chocolate Bars.* An experiment is performed to compare the quality of new Swiss chocolate bars being introduced in the United States. People were asked to taste the candies and rate them on a scale of 1 to 10. On the basis of this test, which chocolate bar is better?

BAR A: SWISS PECAN			BAR B: SWISS HAZELNUT		
9	10	8	10	6	8
10	9	7	9	10	10
6	9	10	8	7	6
7	8	8	9	10	8

31. *Grade Point Average.* Find the grade point average for one semester given the following grades. Assume the grade point values are 4.0 for A, 3.0 for B, and so on. Round to the nearest tenth.

COURSE	GRADE	NUMBER OF CREDIT HOURS IN COURSE
Introductory algebra	B	3
English	A	3
Business	C	4
Spanish	B	3
Typing	B	2

SYNTHESIS

32. The ordered set of data 69, 71, 73, a, 78, 98, b has a median of 74 and a mean of 82. Find a and b.

Cumulative Review

1. *Net Worth.* In 2004, Bill Gates, of Microsoft fame, was worth $46.6 billion. Write standard notation for $46.6 billion.
   Source: Forbes

2. *Gas Mileage.* A 2005 Subaru Impreza WRX gets 324 miles of highway driving on 12 gallons of gasoline. What is the gas mileage?
   Source: EPA; *Car and Driver,* September 2005

3. In 402,513, what does the digit 5 mean?

4. Evaluate: $3 + 5^3$.

5. Find all the factors of 60.

6. Round 52.045 to the nearest tenth.

7. Convert to fraction notation: $3\frac{3}{10}$.

8. Convert from cents to dollars: 210¢.

9. Convert to standard notation: $3.25 trillion.

10. Determine whether 11, 30 and 4, 12 are proportional.

Compute and simplify.

11. $2\frac{2}{5} + 4\frac{3}{10}$

12. $41.063 + 3.5721$

13. $\frac{14}{15} - \frac{3}{5}$

14. $350 - 24.57$

15. $3\frac{3}{7} \cdot 4\frac{3}{8}$

16. $12,456 \times 220$

17. $\frac{13}{15} \div \frac{26}{27}$

18. $104,676 \div 24$

Solve.

19. $\frac{5}{8} = \frac{6}{x}$

20. $\frac{2}{5} \cdot y = \frac{3}{10}$

21. $21.5 \cdot y = 146.2$

22. $x = 398,112 \div 26$

Solve.

23. Tortilla chips cost $2.99 for 14.5 oz. Find the unit price rounded to the nearest tenth of a cent, in cents per ounce.

24. A college has a student body of 6000 students. Of these, 55.4% own a car. How many students own a car?

25. A piece of fabric $1\frac{3}{4}$ yd long is cut into 7 equal strips. What is the length of each strip?

26. A recipe calls for $\frac{3}{4}$ cup of sugar. How much sugar should be used for $\frac{1}{2}$ of the recipe?

27. *Peanut Products.* In any given year, the average American eats 2.7 lb of peanut butter, 1.5 lb of salted peanuts, 1.2 lb of peanut candy, 0.7 lb of in-shell peanuts, and 0.1 lb of peanuts in other forms. How many pounds of peanuts and products containing peanuts does the average American eat in one year?

28. *Energy Consumption.* In a recent year, American utility companies generated 1464 billion kilowatt-hours of electricity using coal, 455 billion using nuclear power, 273 billion using natural gas, 250 billion using hydroelectric plants, 118 billion using petroleum, and 12 billion using geothermal technology and other methods. How many kilowatt-hours of electricity were produced that year?

29. *Heart Disease.* Of the 295 million people in the United States, about 7.4 million have coronary heart disease and about 500,000 die of heart attacks each year. What percent have coronary heart disease? What percent die of heart attacks? Round your answers to the nearest tenth of a percent.
Source: U.S. Centers for Disease Control

30. *Billionaires.* In 2003, the mean net worth of U.S. billionaires was $3.18 billion. By 2004, this figure had increased to $3.31 billion. What was the percent of increase?
Source: Forbes

31. *Football Fields.* The Arena Football League (AFL) is a professional league playing indoors, mostly on converted basketball and/or hockey rinks. The figure shows the AFL field compared with the larger field of the National Football League (NFL).

a) Find the area of an AFL field. Include the end zones in your calculation.
b) Find the area of an NFL field. Include the end zones in your calculation.
c) How much larger is an NFL field than an AFL field?

FedEx. The following table lists the cost of delivering a package by FedEx Priority Overnight shipping in Zone 4. Use the table for Questions 32–34.

WEIGHT (in pounds)	COST
1	$24.25
2	27.25
3	30.25
4	33.00
5	35.75
6	38.25
7	41.00
8	43.25
9	46.50
10	49.25
11	52.00

Source: Federal Express Corporation

AFL field

8 yd

$28\frac{1}{3}$ yd

NFL field

10 yd

50 yd

66 yd total

$53\frac{1}{3}$ yd

100 yd

120 yd total

32. Find the average and the median of these costs.

33. Make a vertical bar graph of the data.

34. Make a line graph of the data.

35. A business is owned by four people. One owns $\frac{1}{3}$, the second owns $\frac{1}{4}$, and the third owns $\frac{1}{6}$. How much does the fourth person own?

36. In manufacturing valves for engines, a factory was discovered to have made 4 defective valves in a lot of 18 valves. At this rate, how many defective valves can be expected in a lot of 5049 valves?

37. A landscaper bought 22 evergreen trees for $210. What was the cost of each tree? Round to the nearest cent.

38. A salesperson earns $182 selling $2600 worth of electronic equipment. What is the commission rate?

Teen Spending. Teenagers are big spenders. More and more retailers are catering to the 13–19 year-old crowd. Those in this group who shop regularly spend an average of $381 per month. Use the table to answer Exercises 39–42.

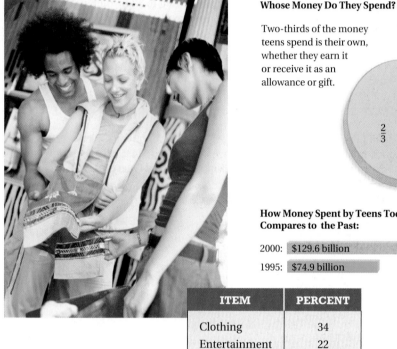

Whose Money Do They Spend?

Two-thirds of the money teens spend is their own, whether they earn it or receive it as an allowance or gift.

$\frac{1}{3}$

$\frac{2}{3}$

Parents provide the rest.

How Money Spent by Teens Today Compares to the Past:

2000: $129.6 billion

1995: $74.9 billion

ITEM	PERCENT
Clothing	34
Entertainment	22
Food	22
Other	22

Source: Rand Youth Poll, eMarketer

39. How is their monthly spending allocated?

40. How much of their monthly spending, whether they earn it or receive it as an allowance or gift, is their own? The rest of their spending money comes from their parents. How much of the monthly spending money comes from their parents?

41. In 2000, teenagers spent $129.6 billion. How was it allocated?

42. By what percent has teenage spending increased from 1995 to 2000?

SYNTHESIS

43. A photography club meets four times a month. In September, the attendance figures were 28, 23, 26, and 23. In October, the attendance figures were 26, 20, 14, and 28. What was the percent of increase or decrease in average attendance from September to October?

Measurement

Real-World Application

"The Gates," designed by artists Christo and Jeanne-Claude, is the largest art project in New York City's history. For the 16-day exhibit in February 2005, 7500 gates with fabric panels were placed at 12-ft intervals along 23 miles of pedestrian paths in Central Park. The panels contained 46 miles of saffron-colored fabric. Convert 46 miles to yards.

Sources: Indianapolis Star, February 13, 2005; Associated Press, Verena Dobnik

This problem appears as Example 6 in Section 8.1.

Objective

a Convert from one American unit of length to another.

Use the unit below to measure the length of each segment or object.

$A \vdash\!\!\!\underbrace{}_{1}\!\!\!\dashv B$

1. $\vdash\!\!\!\rule{3cm}{0pt}\!\!\!\dashv$

2.

3.

4.

Length, or distance, is one kind of measure. To find lengths, we start with some **unit segment** and assign to it a measure of 1. Suppose $\overline{AB}$ below is a unit segment.

$A \vdash\!\!\!\underbrace{}_{1}\!\!\!\dashv B$

Let's measure segment $\overline{CD}$ below, using $\overline{AB}$ as our unit segment.

Since we can place 4 unit segments end to end along $\overline{CD}$, the measure of $\overline{CD}$ is 4.

Sometimes we have to use parts of units, called **subunits.** For example, the measure of the segment $\overline{MN}$ below is $1\frac{1}{2}$. We place one unit segment and one half-unit segment end to end.

Do Exercises 1–4.

a American Measures

American units of length are related as follows.

(Actual size, in inches)

AMERICAN UNITS OF LENGTH	
12 inches (in.) = 1 foot (ft)	3 feet = 1 yard (yd)
36 inches = 1 yard	5280 feet = 1 mile (mi)

We can visualize comparisons of the units as follows:

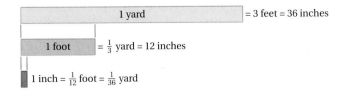

The symbolism 13 in. = 13″ and 27 ft = 27′ is also used for inches and feet. American units have also been called "English," or "British–American," because at one time they were used by both countries. Today, both Canada and England have officially converted to the metric system. However, if you travel in England, you will still see units such as "miles" on road signs.

To change from certain American units to others, we make substitutions. Such a substitution is usually helpful when we are converting from a *larger* unit to a *smaller* one.

EXAMPLE 1 Complete: $7\frac{1}{3}$ yd = _____ in.

$$7\frac{1}{3} \text{ yd} = 7\frac{1}{3} \times 1 \text{ yd} \qquad \text{We think of } 7\frac{1}{3} \text{ yd as } 7\frac{1}{3} \times \text{yd, or } 7\frac{1}{3} \times 1 \text{ yd.}$$

$$= 7\frac{1}{3} \times 36 \text{ in.} \qquad \text{Substituting 36 in. for 1 yd}$$

$$= \frac{22}{3} \times 36 \text{ in.}$$

$$= 264 \text{ in.}$$

Do Exercises 5–7.

Sometimes it helps to use multiplying by 1 in making conversions. For example, 12 in. = 1 ft, so

$$\frac{12 \text{ in.}}{1 \text{ ft}} = 1 \quad \text{and} \quad \frac{1 \text{ ft}}{12 \text{ in.}} = 1.$$

If we divide 12 in. by 1 ft or 1 ft by 12 in., we get 1 because the lengths are the same. Let's first convert from *smaller* to *larger* units.

EXAMPLE 2 Complete: 48 in. = _____ ft.

We want to convert from "in." to "ft." We multiply by 1 using a symbol for 1 with "in." on the bottom and "ft" on the top to eliminate inches and to convert to feet:

$$48 \text{ in.} = \frac{48 \text{ in.}}{1} \times \frac{1 \text{ ft}}{12 \text{ in.}} \qquad \text{Multiplying by 1 using } \frac{1 \text{ ft}}{12 \text{ in.}} \text{ to eliminate in.}$$

$$= \frac{48 \text{ in.}}{12 \text{ in.}} \times 1 \text{ ft}$$

$$= \frac{48}{12} \times \frac{\text{in.}}{\text{in.}} \times 1 \text{ ft}$$

$$= 4 \times 1 \text{ ft} \qquad \text{The } \frac{\text{in.}}{\text{in.}} \text{ acts like 1, so we can omit it.}$$

$$= 4 \text{ ft.}$$

Complete.

5. 8 yd = _____ in.

6. $2\frac{5}{6}$ yd = _____ ft

7. 3.8 mi = _____ in.

Answers on page A-19

Complete.

8. 72 in. = _____ ft

We can also look at this conversion as "canceling" units:

$$48 \text{ in.} = \frac{48 \text{ in.}}{1} \times \frac{1 \text{ ft}}{12 \text{ in.}} = \frac{48}{12} \times 1 \text{ ft} = 4 \text{ ft.}$$

This method is not used only in mathematics, as here, but also in fields such as medicine, chemistry, and physics.

Do Exercises 8 and 9.

9. 17 in. = _____ ft

EXAMPLE 3 Complete: 25 ft = _____ yd.

Since we are converting from "ft" to "yd," we choose a symbol for 1 with "yd" on the top and "ft" on the bottom:

$$25 \text{ ft} = 25 \text{ ft} \times \frac{1 \text{ yd}}{3 \text{ ft}}$$

3 ft = 1 yd, so $\frac{3 \text{ ft}}{1 \text{ yd}} = 1$, and $\frac{1 \text{ yd}}{3 \text{ ft}} = 1$. We use $\frac{1 \text{ yd}}{3 \text{ ft}}$ to eliminate ft.

$$= \frac{25}{3} \times \frac{\text{ft}}{\text{ft}} \times 1 \text{ yd}$$

$$= 8\frac{1}{3} \times 1 \text{ yd}$$ The $\frac{\text{ft}}{\text{ft}}$ acts like 1, so we can omit it.

$$= 8\frac{1}{3} \text{ yd, or } 8.\overline{3} \text{ yd.}$$

Again, in this example, we can consider conversion from the point of view of canceling:

$$25 \text{ ft} = 25 \text{ ft} \times \frac{1 \text{ yd}}{3 \text{ ft}} = \frac{25}{3} \times 1 \text{ yd} = 8\frac{1}{3} \text{ yd, or } 8.\overline{3} \text{ yd.}$$

Complete.

10. 24 ft = _____ yd

Do Exercises 10 and 11.

11. 35 ft = _____ yd

EXAMPLE 4 Complete: 23,760 ft = _____ mi.

We choose a symbol for 1 with "mi" on the top and "ft" on the bottom:

$$23,760 \text{ ft} = 23,760 \text{ ft} \times \frac{1 \text{ mi}}{5280 \text{ ft}}$$

5280 ft = 1 mi, so $\frac{1 \text{ mi}}{5280 \text{ ft}} = 1$.

$$= \frac{23,760}{5280} \times \frac{\text{ft}}{\text{ft}} \times 1 \text{ mi}$$

$$= 4.5 \times 1 \text{ mi}$$ Dividing

$$= 4.5 \text{ mi.}$$

Let's also consider this example using canceling:

$$23,760 \text{ ft} = 23,760 \text{ ft} \times \frac{1 \text{ mi}}{5280 \text{ ft}}$$

$$= \frac{23,760}{5280} \times 1 \text{ mi}$$

$$= 4.5 \times 1 \text{ mi} = 4.5 \text{ mi.}$$

Answers on page A-19

Do Exercises 12 and 13 on the following page.

CHAPTER 8: Measurement

We can also use multiplying by 1 to convert from larger to smaller units. Let's redo Example 1.

■ **EXAMPLE 5** Complete: $7\frac{1}{3}$ yd = _____ in.

$$7\frac{1}{3} \text{ yd} = \frac{22 \text{ yd}}{3} \times \frac{36 \text{ in.}}{1 \text{ yd}} = \frac{22 \times 36}{3} \times 1 \text{ in.} = 264 \text{ in.}$$

Do Exercise 14.

■ **EXAMPLE 6** *Art Project in Central Park.* "The Gates," designed by artists Christo and Jeanne-Claude, is the largest art project in New York City's history. For the 16-day exhibit in February 2005, 7500 gates with fabric panels were placed at 12-ft intervals along 23 miles of pedestrian paths in Central Park. The panels contained 46 miles of saffron-colored fabric. Convert 46 miles to yards.

Sources: *Indianapolis Star,* February 13, 2005; Associated Press, Verena Dobnik

$$46 \text{ mi} = 46 \text{ mi} \times \frac{5280 \text{ ft}}{1 \text{ mi}} \times \frac{1 \text{ yd}}{3 \text{ ft}}$$

$$= \frac{46 \times 5280}{1 \times 3} \times 1 \text{ yd}$$

$$= 80{,}960 \text{ yd}$$

"The Gates" project contained 80,960 yd of fabric.

Do Exercise 15.

Complete.

12. 26,400 ft = _____ mi

13. 2640 ft = _____ mi

14. Complete. Use multiplying by 1.

8 yd = _____ in.

15. Pedestrian Paths. There are 23 miles of pedestrian paths in Central Park in New York City. Convert 23 miles to yards.

Answers on page A-19

Study Tips

TAKING NOTES

Effective note-taking can greatly enhance your learning of mathematics. Here are some suggestions.

■ This textbook has been written and designed so that it represents a quality set of notes at the same time that it teaches. Thus you might not need to take as many notes in class. Just watch, listen, and ask questions as the class moves along, rather than racing to keep up your note-taking.

■ If you have any difficulty keeping up with the instructor, use abbreviations to speed up your note-taking. Consider standard abbreviations like "Ex" for "Example," "≈" for "is approximately equal to," or "∴" for "therefore." Create your own abbreviations as well.

EXERCISE SET

For Extra Help

MathXL   MyMathLab   InterAct Math   Math Tutor Center   Digital Video Tutor CD 4 Videotape 8   Student's Solutions Manual

a Complete.

1. 1 ft = _____ in.

2. 1 yd = _____ ft

3. 1 in. = _____ ft

4. 1 mi = _____ yd

5. 1 mi = _____ ft

6. 1 ft = _____ yd

7. 3 yd = _____ in.

8. 10 yd = _____ ft

9. 84 in. = _____ ft

10. 48 ft = _____ yd

11. 18 in. = _____ ft

12. 29 ft = _____ yd

13. 5 mi = _____ ft

14. 5 mi = _____ yd

15. 63 in. = _____ ft

16. 11,616 ft = _____ mi

17. 10 ft = _____ yd

18. 9.6 yd = _____ ft

19. 7.1 mi = _____ ft

20. 31,680 ft = _____ mi

21. $4\frac{1}{2}$ ft = _____ yd

22. 48 in. = _____ ft

23. 45 in. = _____ yd

24. $6\frac{1}{3}$ yd = _____ in.

25. 330 ft = _____ yd

26. 5280 yd = _____ mi

27. 3520 yd = _____ mi

28. 25 mi = _____ ft

29. 100 yd = _____ ft

30. 480 in. = _____ ft

31. 360 in. = _____ ft

32. 720 in. = _____ yd

33. 1 in. = _____ yd

34. 25 in. = _____ ft

35. 2 mi = _____ in.

36. 63,360 in. = _____ mi

37. D_W A student makes the following error:
$$23 \text{ in.} = 23 \cdot (12 \text{ ft}) = 276 \text{ ft.}$$
Explain the error in at least two ways.

38. D_W Describe two methods of making unit conversions discussed in this section.

SKILL MAINTENANCE

Convert to fraction notation.   [6.2b]

39. 9.25%

40. $87\frac{1}{2}\%$

Find fraction notation for the percent notation in the sentence.   [6.2b]

41. Of all 18-year olds, 27.5% are registered to vote.

42. Of all those who buy CDs, 57% are in the 20–39 age group.

Convert to percent notation.   [6.2a]

43. $\frac{11}{8}$

44. $\frac{2}{3}$

45. $\frac{1}{4}$

46. $\frac{7}{16}$

Find fraction notation for the ratio.   [5.1a]

47. *Heart Disease.*   In one year, in California, there are 373 deaths due to heart disease for every 100,000 women. What is the ratio of number of deaths due to heart disease to number of women? What is the ratio of number of women to number of deaths due to heart disease?
Source: U.S. Centers for Disease Control

48. *Wildfires.*   In 2003, 58,000 forest fires burned 4.0 million acres in the United States. What was the ratio of number of fires to number of acres? What was the ratio of number of acres to number of fires?
Source: National Forest Service

SYNTHESIS

49. *Noah's Ark.*   In biblical measures, it is thought that 1 cubit ≈ 18 in. The dimensions of Noah's ark are given as follows: "The length of the ark shall be three hundred cubits, the breadth of it fifty cubits, and the height of it thirty cubits." What were the dimensions of Noah's ark in inches? in feet?
Source: *Holy Bible, King James Version*, Gen. 6:15

50. *Goliath's Height.*   In biblical measures, a span was considered to be half of a cubit (1 cubit ≈ 18 in.; see Exercise 49). The giant Goliath's height "was six cubits and a span." What was the height of Goliath in inches? in feet?
Source: *Holy Bible, King James Version*, 1 Sam. 17:4

Objective

a Convert from one metric unit of length to another.

The **metric system** is used in most countries of the world, but very little in the United States. The metric system does not use inches, feet, pounds, and so on, although units for time and electricity are the same as those used now in the United States.

An advantage of the metric system is that it is easier to convert from one unit to another. That is because the metric system is based on the number 10.

The basic unit of length is the **meter.** It is just over a yard. In fact, 1 meter ≈ 1.1 yd.

(Comparative sizes are shown.)

1 Meter

1 Yard

The other units of length are multiples of the length of a meter:

10 times a meter,   100 times a meter,   1000 times a meter,   and so on,

or fractions of a meter:

$\frac{1}{10}$ of a meter, $\frac{1}{100}$ of a meter, $\frac{1}{1000}$ of a meter,   and so on.

METRIC UNITS OF LENGTH

1 *kilo*meter (km) = 1000 meters (m)

1 *hecto*meter (hm) = 100 meters (m)

1 *deka*meter (dam) = 10 meters (m)

1 meter (m)

1 *deci*meter (dm) = $\frac{1}{10}$ meter (m)

1 *centi*meter (cm) = $\frac{1}{100}$ meter (m)

1 *milli*meter (mm) = $\frac{1}{1000}$ meter (m)

dam and *dm* are not used often.

You should memorize these names and abbreviations. Think of *kilo-* for 1000, *hecto-* for 100, *deka-* for 10, *deci-* for $\frac{1}{10}$, *centi-* for $\frac{1}{100}$, and *milli-* for $\frac{1}{1000}$. We will also use these prefixes when considering units of area, capacity, and mass.

Study Tips

WORKING WITH A CLASSMATE

If you are finding it difficult to master a particular topic or concept, try talking about it with a classmate. Verbalizing your questions about the material might help clarify it. If your classmate is also finding the material difficult, it is possible that the majority of the people in your class are confused and you can ask your instructor to explain the concept again.

THINKING METRIC

To familiarize yourself with metric units, consider the following.

1 kilometer (1000 meters)	is slightly more than $\frac{1}{2}$ mile (0.6 mi).
1 meter	is just over a yard (1.1 yd).
1 centimeter (0.01 meter)	is a little more than the width of a paperclip (about 0.3937 inch).

1 cm

} 1 cm

1 inch is about 2.54 centimeters.

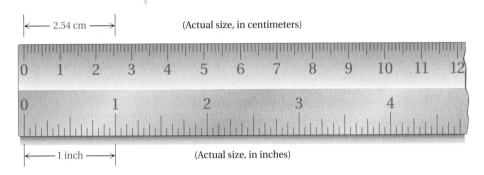

2.54 cm ──→ (Actual size, in centimeters)

1 inch ──→ (Actual size, in inches)

1 millimeter is about the diameter of a paperclip wire.

1 mm

The millimeter (mm) is used to measure small distances, especially in industry.

1 mm

3 mm

Use a centimeter ruler. Measure each object.

1.

2.

3.

Answers on page A-19

Complete with mm, cm, m, or km.

4. A stick of gum is 7 _____ long.

In many countries, the centimeter (cm) is used for body dimensions and clothing sizes.

163 cm
(64.2 in.)
5 ft 4 in.

RELAXED FIT
97cm/81cm
(38 in./32 in.)

5. Minneapolis is 3213 _____ from San Francisco.

Do Exercises 1–3 on the preceding page.

6. A penny is 1 _____ thick.

The meter (m) is used for expressing dimensions of larger objects—say, the height of a diving board—and for shorter distances, like the length of a rug.

10 m
(33 ft)

3.7 m
(12 ft)

2.7 m
(9 ft)

7. The halfback ran 7 _____ .

8. The book is 3 _____ thick.

The kilometer (km) is used for longer distances, mostly in cases where miles are now being used.

1 mile is about 1.6 km.

1 km

1 mi

9. The desk is 2 _____ long.

ST. LOUIS
30 MI

ST. LOUIS
48 KM

Do Exercises 4–9.

a Changing Metric Units

As with American units, when changing from a *larger* unit to a *smaller* unit, we usually make substitutions.

EXAMPLE 1 Complete: 4 km = _____ m.

We want to convert from "km" to "m." Since we are converting from a *larger* to a *smaller* unit, we use substitution.

$$4 \text{ km} = 4 \times 1 \text{ km}$$
$$= 4 \times 1000 \text{ m} \qquad \text{Substituting 1000 m for 1 km}$$
$$= 4000 \text{ m}$$

Do Exercises 10 and 11.

Since

$$\frac{1}{10}\text{ m} = 1 \text{ dm}, \qquad \frac{1}{100}\text{ m} = 1 \text{ cm}, \quad \text{and} \quad \frac{1}{1000}\text{ m} = 1 \text{ mm},$$

it follows that

$$1 \text{ m} = 10 \text{ dm}, \qquad 1 \text{ m} = 100 \text{ cm}, \quad \text{and} \quad 1 \text{ m} = 1000 \text{ mm}.$$

EXAMPLE 2 Complete: 93.4 m = _____ cm.

We want to convert from "m" to "cm." Since we are converting from a *larger* to a *smaller* unit, we use substitution.

$$93.4 \text{ m} = 93.4 \times 1 \text{ m}$$
$$= 93.4 \times 100 \text{ cm} \qquad \text{Substituting 100 cm for 1 m}$$
$$= 9340 \text{ cm}$$

EXAMPLE 3 Complete: 0.248 m = _____ mm.

We want to convert from "m" to "mm." Since we are again converting from a *larger* to a *smaller* unit, we use substitution.

$$0.248 \text{ m} = 0.248 \times 1 \text{ m}$$
$$= 0.248 \times 1000 \text{ mm} \qquad \text{Substituting 1000 mm for 1 m}$$
$$= 248 \text{ mm}$$

Do Exercises 12 and 13.

Complete.

10. 23 km = _____ m

11. 4 hm = _____ m

Complete.

12. 1.78 m = _____ cm

13. 9.04 m = _____ mm

Answers on page A-19

Complete.

14. 7814 m = _____ km

15. 7814 m = _____ dam

Complete.

16. 9.67 mm = _____ cm

17. 89 km = _____ cm

We now convert from "m" to "km." Since we are converting from a *smaller* unit to a *larger* unit, we use multiplying by 1. We choose a symbol for 1 with "km" in the numerator and "m" in the denominator.

EXAMPLE 4 Complete: 2347 m = _____ km.

$$2347 \text{ m} = 2347 \text{ m} \times \frac{1 \text{ km}}{1000 \text{ m}} \qquad \text{Multiplying by 1 using } \frac{1 \text{ km}}{1000 \text{ m}}$$

$$= \frac{2347}{1000} \times \frac{\text{m}}{\text{m}} \times 1 \text{ km} \qquad \text{The } \frac{\text{m}}{\text{m}} \text{ acts like 1, so we omit it.}$$

$$= 2.347 \text{ km} \qquad \text{Dividing by 1000 moves the decimal point three places to the left.}$$

Using canceling, we can work this example as follows:

$$2347 \text{ m} = 2347 \text{ m̸} \times \frac{1 \text{ km}}{1000 \text{ m̸}}$$

$$= \frac{2347}{1000} \times 1 \text{ km} = 2.347 \text{ km}.$$

Do Exercises 14 and 15.

Sometimes we multiply by 1 more than once.

EXAMPLE 5 Complete: 8.42 mm = _____ cm.

$$8.42 \text{ mm} = 8.42 \text{ mm} \times \frac{1 \text{ m}}{1000 \text{ mm}} \times \frac{100 \text{ cm}}{1 \text{ m}} \qquad \begin{array}{l}\text{Multiplying by 1 using}\\ \frac{1 \text{ m}}{1000 \text{ mm}} \text{ and } \frac{100 \text{ cm}}{1 \text{ m}}\end{array}$$

$$= \frac{8.42 \times 100}{1000} \times \frac{\text{mm}}{\text{mm}} \times \frac{\text{m}}{\text{m}} \times 1 \text{ cm}$$

$$= \frac{842}{1000} \text{ cm} = 0.842 \text{ cm}$$

Do Exercises 16 and 17.

MENTAL CONVERSION

Look back over the examples and exercises done so far and you will see that changing from one unit to another in the metric system amounts to only the movement of a decimal point. That is because the metric system is based on 10. Let's find a faster way to convert. Look at the following table.

1000 m	100 m	10 m	1 m	0.1 m	0.01 m	0.001 m
1 km	1 hm	1 dam	1 m	1 dm	1 cm	1 mm

Each place in the table has a value $\frac{1}{10}$ that to the left or 10 times that to the right. Thus moving one place in the table corresponds to moving one decimal place.

Answers on page A-19

Let's convert mentally.

Complete. Try to do this mentally using the table.

18. 6780 m = _____ km

EXAMPLE 6 Complete: 8.42 mm = _____ cm.

Think: To go from mm to cm in the table is a move of one place to the left. Thus we move the decimal point one place to the left.

1000 m	100 m	10 m	1 m	0.1 m	0.01 m	0.001 m
1 km	1 hm	1 dam	1 m	1 dm	1 cm	1 mm

1 place to the left

8.42   0.8.42   8.42 mm = 0.842 cm

19. 9.74 cm = _____ mm

EXAMPLE 7 Complete: 1.886 km = _____ cm.

Think: To go from km to cm is a move of five places to the right. Thus we move the decimal point five places to the right.

1000 m	100 m	10 m	1 m	0.1 m	0.01 m	0.001 m
1 km	1 hm	1 dam	1 m	1 dm	1 cm	1 mm

5 places to the right

1.886   1.88600.   1.886 km = 188,600 cm

20. 1 mm = _____ cm

EXAMPLE 8 Complete: 3 m = _____ cm.

Think: To go from m to cm in the table is a move of two places to the right. Thus we move the decimal point two places to the right.

1000 m	100 m	10 m	1 m	0.1 m	0.01 m	0.001 m
1 km	1 hm	1 dam	1 m	1 dm	1 cm	1 mm

2 places to the right

3   3.00.   3 m = 300 cm

21. 845.1 mm = _____ dm

You should try to make metric conversions mentally as much as possible.
The fact that conversions can be done so easily is an important advantage of the metric system. The most commonly used metric units of length are km, m, cm, and mm. We have purposely used these more often than the others in the exercises.

Do Exercises 18–21.

Answers on page A-19

8.2

EXERCISE SET

For Extra Help

MathXL · MyMathLab · InterAct Math · Math Tutor Center · Digital Video Tutor CD 4 Videotape 8 · Student's Solutions Manual

a Complete. Do as much as possible mentally.

1. **a)** 1 km = _____ m
   b) 1 m = _____ km

2. **a)** 1 hm = _____ m
   b) 1 m = _____ hm

3. **a)** 1 dam = _____ m
   b) 1 m = _____ dam

4. **a)** 1 dm = _____ m
   b) 1 m = _____ dm

5. **a)** 1 cm = _____ m
   b) 1 m = _____ cm

6. **a)** 1 mm = _____ m
   b) 1 m = _____ mm

7. **a)** 6.7 km = _____ m
   b) This conversion is from a larger unit to a smaller. Did you substitute or multiply by 1?

8. 27 km = _____ m

9. **a)** 98 cm = _____ m
   b) This conversion is from a smaller unit to a larger. Did you substitute or multiply by 1?

10. 0.789 cm = _____ m

11. 8921 m = _____ km

12. 8664 m = _____ km

13. 56.66 m = _____ km

14. 4.733 m = _____ km

15. 5666 m = _____ cm

16. 869 m = _____ cm

17. 477 cm = _____ m

18. 6.27 mm = _____ m

19. 6.88 m = _____ cm

20. 6.88 m = _____ dm

21. 1 mm = _____ cm

22. 1 cm = _____ km

23. 1 km = _____ cm

24. 2 km = _____ cm

25. 14.2 cm = _____ mm

26. 25.3 cm = _____ mm

27. 8.2 mm = _____ cm

28. 9.7 mm = _____ cm

29. 4500 mm = _____ cm

30. 8,000,000 m = _____ km

31. 0.024 mm = _____ m

32. 60,000 mm = _____ dam

33. 6.88 m = _____ dam

34. 7.44 m = _____ hm

35. 2.3 dam = _____ dm

36. 9 km = _____ hm

37. 392 dam = _____ km

38. 0.056 mm = _____ dm

Complete the following table.

	OBJECT	MILLIMETERS (mm)	CENTIMETERS (cm)	METERS (m)
39.	Length of a calculator		18	
40.	Width of a calculator	85		
41.	Length of a piece of typing paper			0.278
42.	Length of a football field			109.09
43.	Width of a football field		4844	
44.	Width of credit card	56		
45.	Length of 4 meter sticks			4
46.	Length of 3 meter sticks		300	
47.	Thickness of an index card	0.27		
48.	Thickness of a piece of cardboard		0.23	
49.	Height of the Sears Tower			442
50.	Height of the CN Tower (Toronto)	553,000		

51. D_W Recall the guidelines for conversion: (1) "If the conversion is from a larger unit to a smaller unit, substitute." (2) "If the conversion is from a smaller unit to a larger unit, multiply by 1." Explain why each is the easier way to convert in that situation.

52. D_W Explain in your own words why metric units are easier to work with than American units.

SKILL MAINTENANCE

Divide. [4.4a]

53. $23.4 \div 100$

54. $23.4 \div 1000$

Multiply.

55. 3.14×4.41 [4.3a]

56. $4 \times 20\frac{1}{8}$ [3.6a]

Convert to percent notation. [6.2a]

57. $\frac{3}{8}$

58. $\frac{5}{8}$

59. $\frac{2}{3}$

60. $\frac{1}{5}$

Find decimal notation for the percent notation in the sentence. [6.1b]

61. Blood is 90% water.

62. Of those accidents requiring medical attention, 10.8% of them occur on roads.

SYNTHESIS

Each sentence is incorrect. Insert or alter a decimal point to make the sentence correct.

63. When my right arm is extended, the distance from my left shoulder to the end of my right hand is 10 m.

64. The height of the Empire State Building is 38.1 m.

65. A stack of ten quarters is 140 cm high.

66. The width of an adult's hand is 112 cm.

Objective

 a Convert between American units of length and metric units of length.

Complete.

1. 100 yd = _____ m
(The length of a football field excluding the end zones)

2. 500 mi = _____ km
(The Indianapolis 500-mile race)

3. 2383 km = _____ mi
(The distance from St. Louis to Phoenix)

a Converting Units

We can make conversions between American and metric units by using the following table. These listings are rounded approximations. Again, we either make a substitution or multiply by 1 appropriately.

AMERICAN	METRIC
1 in.	2.540 cm
1 ft	0.305 m
1 yd	0.914 m
1 mi	1.609 km
0.621 mi	1 km
1.094 yd	1 m
3.281 ft	1 m
39.370 in.	1 m

This table is set up to enable us to make conversions by substitution.

EXAMPLE 1 Complete: 26.2 mi = _____ km.
(The length of the Olympic marathon)

$$26.2 \text{ mi} = 26.2 \times 1 \text{ mi}$$
$$\approx 26.2 \times 1.609 \text{ km} \quad \text{Substituting 1.609 km for 1 mi}$$
$$= 42.1558 \text{ km}$$

EXAMPLE 2 Complete: 2.16 m = _____ in.
(The height of Shaquille O'Neal of the Miami Heat)

$$2.16 \text{ m} = 2.16 \times 1 \text{ m}$$
$$\approx 2.16 \times 39.37 \text{ in.} \quad \text{Substituting 39.37 in. for 1 m}$$
$$= 85.0392 \text{ in.}$$

In an application like this one, the answer would probably be rounded to the nearest one, as 85 in.

EXAMPLE 3 Complete: 100 m = _____ ft.
(The length of the 100-meter dash)

$$100 \text{ m} = 100 \times 1 \text{ m}$$
$$\approx 100 \times 3.281 \text{ ft} \quad \text{Substituting 3.281 ft for 1 m}$$
$$= 328.1 \text{ ft}$$

EXAMPLE 4 Complete: 4544 km = _____ mi.
(The distance from New York to Los Angeles)

$$4544 = 4544 \times 1 \text{ km}$$
$$\approx 4544 \times 0.621 \text{ mi} \quad \text{Substituting 0.621 mi for 1 km}$$
$$= 2821.824 \text{ mi}$$

In practical situations, we would probably round this answer to 2822 mi.

Answers on page A-19

Do Exercises 1–3 on the preceding page.

EXAMPLE 5 *Petronas Towers.* The height of the Petronas Towers in Kuala Lumpur, Malaysia, is 1483 ft. Find the height in meters.

Source: *The New York Times Almanac*

1483 ft

The height *H*, in meters, is given by

$$H = 1483 \text{ ft}$$
$$= 1483 \times 1 \text{ ft}$$
$$\approx 1483 \times 0.305 \text{ m} \qquad \text{Substituting 0.305 m for 1 ft}$$
$$= 452.315 \text{ m}.$$

Do Exercises 4 and 5.

EXAMPLE 6 Complete: 0.10414 mm = _____ in.
(The thickness of a $1 bill)

In this case, we must make two substitutions or multiply by two forms of 1 since the chart on the preceding page does not provide an easy way to convert from millimeters to inches. Here we choose to multiply by forms of 1.

$$0.10414 \text{ mm} = 0.10414 \times 1 \text{ mm} \times \frac{1 \text{ cm}}{10 \text{ mm}}$$

$$= 0.010414 \text{ cm}$$

$$\approx 0.010414 \times 1 \text{ cm} \times \frac{1 \text{ in.}}{2.54 \text{ cm}}$$

$$= 0.0041 \text{ in.}$$

Do Exercise 6.

Complete.

4. 568 mi = _____ km
(The distance from San Francisco to Las Vegas)

5. The height of the John Hancock Building in Chicago is 1127 ft. Find the height in meters.

6. Complete: 3.175 mm = _____ in.
(The thickness of a quarter)

Answers on page A-19

8.3

EXERCISE SET

For Extra Help

MathXL   MyMathLab   InterAct   Math Tutor   Digital Video   Student's
                       Math   Center   Tutor CD 4   Solutions
                                                 Videotape 8   Manual

a Complete.

1. 330 ft = _____ m
   (The length of most baseball foul lines)

2. 12 in. = _____ cm
   (The length of a common ruler)

3. 1171.4 km = _____ mi
   (The distance from Cleveland to Atlanta)

4. 2 m = _____ ft
   (The length of a desk)

5. 65 mph = _____ km/h
   (The common speed limit in the United States)

6. 100 km/h = _____ mph
   (A common speed limit in Canada)

7. 180 mi = _____ km
   (The distance from Indianapolis to Chicago)

8. 141,600,000 mi = _____ km
   (The farthest distance of Mars from the sun)

9. 70 mph = _____ km/h
   (An interstate speed limit in Arizona)

10. 60 km/h = _____ mph
    (A city speed limit in Canada)

11. 10 yd = _____ m
    (The length needed for a first down in football)

12. 450 ft = _____ m
    (The length of a long home run in baseball)

13. 2.08 m = _____ in.
    (The height of Amare Stoudemire of the Phoenix Suns)

14. 76 in. = _____ m
    (The height of Jason Kidd of the New Jersey Nets)

15. 381 m = _____ ft
    (The height of the Empire State Building)

16. 1127 ft = _____ m
    (The height of the John Hancock Center)

17. 7.5 in. = _____ cm
    (The length of a pencil)

18. 15.7 cm = _____ in.
    (The length of a $1 bill)

19. 2216 km = _____ mi
    (The distance from Chicago to Miami)

20. 1862 mi = _____ km
    (The distance from Seattle to Kansas City)

21. 13 mm = _____ in.
    (The thickness of a plastic case for a DVD)

22. 0.25 in. = _____ mm
    (The thickness of an eraser on a pencil)

Complete the following table.

	OBJECT	YARDS (yd)	CENTIMETERS (cm)	INCHES (in.)	METERS (m)	MILLIMETERS (mm)
23.	Length of a mousepad		23.8			
24.	Width of a mousepad		20.3			
25.	Width of a piece of typing paper			$8\frac{1}{2}$		
26.	Length of a football field	120 yd				
27.	Width of a football field		4844			
28.	Width of a credit card					56
29.	Length of 4 yardsticks	4				
30.	Length of 3 meter sticks		300			
31.	Thickness of an index card				0.00027	
32.	Thickness of a piece of cardboard		0.23			
33.	Height of the Sears Tower				442	
34.	Height of Central Plaza, Hong Kong	409				

35. **D_W** Do some research in a library or on the Internet about the metric system versus the American system. Why do you think the United States has not converted to the metric system?

36. **D_W** List all the memory devices you know for comparing metric and American units of measure.

SKILL MAINTENANCE

Convert the number in the sentence to standard notation. [4.3b]

37. *Electricity Usage.* In 2025, it is expected that the world will be using 23.072 trillion kilowatt-hours of electricity per year.
Source: Energy Information Administration, *International Energy Outlook,* 2004

38. *Rail Travel.* From October 2003 to September 2004, 25.1 million passengers traveled by rail on Amtrak.
Source: Amtrak

39. *Cases of Diabetes.* By 2030, it is estimated that there will be 366 million cases of diabetes in the world.
Source: *Diabetes Care*

40. *Domestic and Foreign Travel.* It is estimated that domestic and foreign travelers spent 585 billion dollars in 2004 while traveling in the United States.
Source: Travel Industry Association of America

SYNTHESIS

41. Develop a formula to convert from inches to millimeters.

42. Develop a formula to convert from millimeters to inches. How does it relate to the answer for Exercise 47?

43. As of September 2004, the women's world record for the 100-m dash was 10.49 sec. How fast is this in miles per hour? Round to the nearest tenth of a mile per hour.

44. As of September 2004, the men's world record for the 100-m dash was 9.78 sec. How fast is this in miles per hour? Round to the nearest tenth of a mile per hour.

8.4 WEIGHT AND MASS; MEDICAL APPLICATIONS

Objectives

a Convert from one American unit of weight to another.

b Convert from one metric unit of mass to another.

c Make conversions and solve applied problems concerning medical dosages.

There is a difference between **mass** and **weight,** but the terms are often used interchangeably. People sometimes use the word "weight" instead of "mass." Weight is related to the force of gravity. The farther you are from the center of the earth, the less you weigh. Your mass stays the same no matter where you are.

a Weight: The American System

AMERICAN UNITS OF WEIGHT	
1 ton (T) = 2000 pounds (lb)	1 lb = 16 ounces (oz)

The term "ounce" used here for weight is different from the "ounce" we will use for capacity in Section 8.5. We convert units using the same techniques that we use with linear measure.

Complete.

1. 5 lb = _____ oz

EXAMPLE 1 A well-known hamburger is called a "quarter-pounder." Find its name in ounces: a "_____ ouncer."

$$\frac{1}{4} \text{ lb} = \frac{1}{4} \cdot 1 \text{ lb} = \frac{1}{4} \cdot 16 \text{ oz}$$

Substituting 16 oz for 1 lb. Since we are converting from a larger unit to a smaller unit, we use substitution.

$$= 4 \text{ oz}$$

A "quarter-pounder" can also be called a "four-ouncer."

2. 8640 lb = _____ T

EXAMPLE 2 Complete: 15,360 lb = _____ T.

$$15{,}360 \text{ lb} = 15{,}360 \text{ lb} \times \frac{1 \text{ T}}{2000 \text{ lb}}$$

Multiplying by 1. Since we are converting from a smaller unit to a larger unit, we use multiplying by 1.

$$= \frac{15{,}360}{2000} \text{ T} = 7.68 \text{ T}$$

Do Exercises 1–3.

3. 1 T = _____ oz

b Mass: The Metric System

The basic unit of mass is the **gram** (g), which is the mass of 1 cubic centimeter (1 cm³) of water. Since a cubic centimeter is small, a gram is a small unit of mass.

$$1 \text{ g} = 1 \text{ gram} = \text{the mass of 1 cm}^3 \text{ of water}$$

The following table lists the metric units of mass. The prefixes are the same as those for length.

METRIC UNITS OF MASS	
1 metric ton (t)	= 1000 kilograms (kg)
1 *kilo*gram (kg)	= 1000 grams (g)
1 *hecto*gram (hg)	= 100 grams (g)
1 *deka*gram (dag)	= 10 grams (g)
1 gram (g)	
1 *deci*gram (dg)	$= \dfrac{1}{10}$ gram (g)
1 *centi*gram (cg)	$= \dfrac{1}{100}$ gram (g)
1 *milli*gram (mg)	$= \dfrac{1}{1000}$ gram (g)

THINKING METRIC

One gram is about the mass of 1 raisin or 1 paperclip or 1 package of "Splenda" sweetener. Since 1 kg is about 2.2 lb, 1000 kg is about 2200 lb, or 1 metric ton (t), which is just a little more than 1 American ton (T), which is 2000 lb.

1 g   1 gram

1 kilogram   1 pound

Small masses, such as dosages of medicine and vitamins, may be measured in milligrams (mg). The gram (g) is used for objects ordinarily measured in ounces, such as the mass of a letter, a piece of candy, a coin, or a small package of food.

Complete with mg, g, kg, or t.

4. A laptop computer has a mass of 6 _____ .

Each 2.5 mg

2 g   2 kg   90 kg

15 g

5. Eric has a body mass of 85.4 _____ .

The kilogram (kg) is used for larger food packages, such as fruit, or for human body mass. The metric ton (t) is used for very large masses, such as the mass of an automobile, a truckload of gravel, or an airplane.

Do Exercises 4–8.

6. This is a 3-_____ vitamin.

CHANGING UNITS MENTALLY

As before, changing from one metric unit to another amounts to only the movement of a decimal point. We use this table.

7. A pen has a mass of 12 _____ .

1000 g	100 g	10 g	1 g	0.1 g	0.01 g	0.001 g
1 kg	1 hg	1 dag	1 g	1 dg	1 cg	1 mg

8. A minivan has a mass of 3 _____ .

EXAMPLE 3 Complete: 8 kg = _____ g.

Think: To go from kg to g in the table is a move of three places to the right. Thus we move the decimal point three places to the right.

1000 g	100 g	10 g	1 g	0.1 g	0.01 g	0.001 g
1 kg	1 hg	1 dag	1 g	1 dg	1 cg	1 mg

3 places to the right

Complete.

9. 6.2 kg = _____ g

8.0   8.000.   8 kg = 8000 g

EXAMPLE 4 Complete: 4235 g = _____ kg.

Think: To go from g to kg in the table is a move of three places to the left. Thus we move the decimal point three places to the left.

10. 304.8 cg = _____ g

1000 g	100 g	10 g	1 g	0.1 g	0.01 g	0.001 g
1 kg	1 hg	1 dag	1 g	1 dg	1 cg	1 mg

3 places to the left

4235.0   4.235.0   4235 g = 4.235 kg

Do Exercises 9 and 10.

EXAMPLE 5   Complete: 6.98 cg = _____ mg.

Think: To go from cg to mg is a move of one place to the right. Thus we move the decimal point one place to the right.

1000 g	100 g	10 g	1 g	0.1 g	0.01 g	0.001 g
1 kg	1 hg	1 dag	1 g	1 dg	1 cg	1 mg

1 place to the right

6.98   6.9.8   6.98 cg = 69.8 mg

The most commonly used metric units of mass are kg, g, cg, and mg. We have purposely used those more than the others in the exercises.

EXAMPLE 6   Complete: 89.21 mg = _____ g.

Think: To go from mg to g is a move of three places to the left. Thus we move the decimal point three places to the left.

1000 g	100 g	10 g	1 g	0.1 g	0.01 g	0.001 g
1 kg	1 hg	1 dag	1 g	1 dg	1 cg	1 mg

3 places to the left

89.21   0.089.21   89.21 mg = 0.08921 g

Do Exercises 11–13.

C   Medical Applications

Another metric unit that is used in medicine is the microgram (mcg). It is de-fined as follows.

> **MICROGRAM**
>
> $$1 \text{ microgram} = 1 \text{ mcg} = \frac{1}{1{,}000{,}000} \text{ g} = 0.000001 \text{ g}$$
>
> $$1{,}000{,}000 \text{ mcg} = 1 \text{ g}$$

Complete.

11. 7.7 cg = _____ mg

12. 2344 mg = _____ cg

13. 67 dg = _____ mg

14. Complete:

   1 mcg = _____ mg.

Answers on page A-19

15. Medical Dosage. A physician prescribes 500 mcg of alprazolam, an antianxiety medication. How many milligrams is this dosage?

Source: Steven R. Smith, M.D.

Answer on page A-19

EXAMPLE 7 Complete: 1 mg = _____ mcg.

We convert to grams and then to micrograms:

$$1 \text{ mg} = 0.001 \text{ g}$$
$$= 0.001 \times 1 \text{ g}$$
$$= 0.001 \times 1{,}000{,}000 \text{ mcg} \qquad \text{Substituting } 1{,}000{,}000 \text{ mcg for } 1 \text{ g}$$
$$= 1000 \text{ mcg.}$$

Do Exercise 14 on the preceding page.

EXAMPLE 8 *Medical Dosage.* Nitroglycerin sublingual tablets come in 0.4-mg tablets. How many micrograms are in each tablet?

Source: Steven R. Smith, M.D.

We are to complete: 0.4 mg = _____ mcg. Thus,

$$0.4 \text{ mg} = 0.4 \times 1 \text{ mg}$$
$$= 0.4 \times 1000 \text{ mcg} \qquad \text{From Example 7, substituting 1000 mcg for 1 mg}$$
$$= 400 \text{ mcg.}$$

We can also do this problem in a manner similar to Example 7.

Do Exercise 15.

Study Tips

THE 20 BEST JOBS

Although this does not qualify as a Study Tip, you can use the information to motivate your study of mathematics. The book Best Jobs for the 21st Century, *3rd edition, developed by Michael Farr with database work by Laurence Shatkin, Ph.D, lists 500 jobs, based on overall scores for pay, growth rate through 2010, and number of annual openings. Note that the use of mathematics is significant in most of the 20 top jobs.*

JOB	ANNUAL EARNINGS	GROWTH RATE THROUGH 2010	ANNUAL OPENING
1. Computer Software Engineers, Applications	$70,210	100.0%	28,000
2. Computer Systems Analysts	$61,990	59.7%	34,000
3. Computer and Information Systems Managers	$82,480	47.9%	28,000
4. Teachers, Postsecondary	$52,115	23.5%	184,000
5. Management Analysts	$57,970	28.9%	50,000
6. Registered Nurses	$46,670	25.6%	140,000
7. Computer Software Engineers, Systems Software	$73,280	89.7%	23,000
8. Medical and Health Services Managers	$59,220	32.3%	27,000
9. Sales Agents, Financial Services	$59,690	22.3%	55,000
10. Sales Agents, Securities and Commodities	$59,690	22.3%	55,000
11. Securities, Commodities, and Financial Services Sales Agents	$59,690	22.3%	55,000
12. Computer Support Specialists	$38,560	97.0%	40,000
13. Sales Managers	$71,620	32.8%	21,000
14. Computer Security Specialists	$53,770	81.9%	18,000
15. Network and Computer Systems Administrators	$53,770	81.9%	18,000
16. Financial Managers	$70,210	18.5%	53,000
17. Financial Managers, Branch or Department	$70,210	18.5%	53,000
18. Treasurers, Controllers, and Chief Financial Officers	$70,210	18.5%	53,000
19. Accountants	$45,380	18.5%	100,000
20. Accountants and Auditors	$45,380	18.5%	100,000

8.4

EXERCISE SET

For Extra Help

MathXL · MyMathLab · InterAct Math · Math Tutor Center · Digital Video Tutor CD 4 Videotape 8 · Student's Solutions Manual

a Complete.

1. 1 T = _____ lb

2. 1 lb = _____ oz

3. 6000 lb = _____ T

4. 8 T = _____ lb

5. 4 lb = _____ oz

6. 10 lb = _____ oz

7. 6.32 T = _____ lb

8. 8.07 T = _____ lb

9. 3200 oz = _____ T

10. 6400 oz = _____ T

11. 80 oz = _____ lb

12. 960 oz = _____ lb

13. *Excelsior.* Western Excelsior is a company that makes a packing material called excelsior. Excelsior is produced from the wood of aspen trees. The largest grove of aspen trees on record contained 13,000,000 tons of aspen. How many pounds of aspen were there?

Source: Western Excelsior Corporation, Mancos CO

14. *Excelsior.* Western Excelsior buys 44,800 tons of aspen each year to make excelsior. How many pounds of aspen does it buy each year?

Source: Western Excelsior Corporation, Mancos CO

b Complete.

15. 1 kg = _____ g

16. 1 hg = _____ g

17. 1 dag = _____ g

18. 1 dg = _____ g

19. 1 cg = _____ g

20. 1 mg = _____ g

21. 1 g = _____ mg

22. 1 g = _____ cg

23. 1 g = _____ dg

24. 25 kg = _____ g

25. 234 kg = _____ g

26. 9403 g = _____ kg

27. 5200 g = _____ kg

28. 1.506 kg = _____ g

29. 67 hg = _____ kg

30. 45 cg = _____ g

31. 0.502 dg = _____ g

32. 0.0025 cg = _____ mg

33. 8492 g = _____ kg

34. 9466 g = _____ kg

35. 585 mg = _____ cg

36. 96.1 mg = _____ cg

37. 8 kg = _____ cg

38. 0.06 kg = _____ mg

39. 1 t = _____ kg

40. 2 t = _____ kg

41. 3.4 cg = _____ dag

42. 115 mg = _____ g

C   Complete.

43. 1 mg = _____ mcg

44. 1 mcg = _____ mg

45. 325 mcg = _____ mg

46. 0.45 mg = _____ mcg

Medical Dosage.   Solve each of the following. (None of these medications should be taken without consulting your own physician.)

Source: Steven R. Smith, M.D.

47. Digoxin is a medication used to treat heart problems. A physician orders 0.125 mg of digoxin to be taken once daily. How many micrograms of digoxin are there in the daily dosage?

48. Digoxin is a medication used to treat heart problems. A physician orders 0.25 mg of digoxin to be taken once a day. How many micrograms of digoxin are there in the daily dosage?

49. Triazolam is a medication used for the short-term treatment of insomnia. A physician advises her patient to take one of the 0.125-mg tablets each night for 7 nights. How many milligrams of triazolam will the patient have ingested over that 7-day period? How many micrograms?

50. Clonidine is a medication used to treat high blood pressure. The usual starting dose of clonidine is one 0.1-mg tablet twice a day. If a patient is started on this dose by his physician, how many total milligrams of clonidine will the patient have taken before he returns to see his physician 14 days later? How many micrograms?

51. Cephalexin is an antibiotic that frequently is prescribed in a 500-mg tablet form. A physician prescribes 2 grams of cephalexin per day for a patient with a skin abscess. How many 500-mg tablets would have to be taken in order to achieve this daily dosage?

52. Quinidine gluconate is a liquid mixture, part medicine and part water, which is administered intravenously. There are 80 mg of quinidine gluconate in each cubic centimeter (cc) of the liquid mixture. A physician orders 900 mg of quinidine gluconate to be administered daily to a patient with malaria. How much of the solution would have to be administered in order to achieve the recommended daily dosage?

53. Amoxicillin is a common antibiotic prescribed for children. It is a liquid suspension composed of part amoxicillin and part water. In one formulation of amoxicillin suspension, there is 250 mg of amoxicillin in 5 cc of the liquid suspension. A physician prescribes 400 mg per day for a 2-year-old child with an ear infection. How much of the amoxicillin liquid suspension would the child's parent need to administer in order to achieve the recommended daily dosage of amoxicillin?

54. Albuterol is a medication used for the treatment of asthma. It comes in an inhaler that contains 17 mg albuterol mixed with a liquid. One actuation (inhalation) from the mouthpiece delivers a 90-mcg dose of albuterol.

  a) A physician orders 2 inhalations 4 times per day. How many micrograms of albuterol does the patient inhale per day?

  b) How many actuations/inhalations are contained in one inhaler?

  c) Danielle is going away for 4 months of college and wants to take enough albuterol to last for that time. Her physician has prescribed 2 inhalations 4 times per day. Estimate how many inhalers Danielle will need to take with her for the 4-month period.

55. D_W Give at least two reasons why someone might prefer the use of grams to the use of ounces.

56. D_W Describe a situation in which one object weighs 70 kg, another weighs 3 g, and another weighs 125 mg.

Convert to fraction notation.   [6.2b]

57. 35%

58. 99%

59. 85.5%

60. 34.2%

61. $37\frac{1}{2}\%$

62. $66.\overline{6}\%$

63. $83.\overline{3}\%$

64. $16\frac{2}{3}\%$

Solve.

65. A country has a population that is increasing by 4% each year. This year the population is 180,000. What will it be next year? [6.5b]

66. A college has a student body of 1850 students. Of these, 17.5% are seniors. How many students are seniors? [6.5a]

67. A state charges a meals tax of $4\frac{1}{2}$%. What is the meals tax charged on a dinner party costing $540? [6.6a]

68. The price of a microwave oven was reduced from $350 to $308. Find the percent of decrease in price. [6.5b]

69. A ream of paper contains 500 sheets. How many sheets are there in 15 reams? [1.8a]

70. A lab technician separates a vial containing 140 cc of blood into test tubes, each of which contains 3 cc of blood. How many test tubes can be filled? How much blood is left over? [1.8a]

---

71. A box of gelatin-mix packages weighs $15\frac{3}{4}$ lb. Each package weighs $1\frac{3}{4}$ oz. How many packages are in the box?

$1\frac{3}{4}$ oz

$15\frac{3}{4}$ lb

72. At $0.90 a dozen, the cost of eggs is $0.60 per pound. How much does an egg weigh?

73. *Large Diamonds.* A **carat** (also spelled **karat**) is a unit of weight for precious stones; 1 carat = 200 mg. The Golden Jubilee Diamond weighs 545.67 carats and is the largest cut diamond in the world. The Hope Diamond, located at the Smithsonian Institution Museum of Natural History, weighs 45.52 carats.

Source: *National Geographic,* February 2001

a) How many grams does the Golden Jubilee Diamond weigh?
b) How many grams does the Hope Diamond weigh?
c) Given that 1 lb = 453.6 g, how many ounces does each diamond weigh?

492

CHAPTER 8: Measurement

8.5

CAPACITY; MEDICAL APPLICATIONS

a Capacity

AMERICAN UNITS

To answer a question like "How much soda is in the can?" we need measures of **capacity.** American units of capacity are fluid ounces, cups, pints, quarts, and gallons. These units are related as follows.

AMERICAN UNITS OF CAPACITY	
1 gallon (gal) = 4 quarts (qt)	1 pt = 2 cups = 16 fluid ounces (fl oz)
1 qt = 2 pints (pt)	1 cup = 8 fluid oz

Fluid ounces, abbreviated fl oz, are often referred to as ounces, or oz.

EXAMPLE 1 Complete: 9 gal = _____ oz.

Since we are converting from a *larger* unit to a *smaller* unit, we use substitution.

$$9 \text{ gal} = 9 \cdot 1 \text{ gal} = 9 \cdot 4 \text{ qt} \qquad \text{Substituting 4 qt for 1 gal}$$
$$= 9 \cdot 4 \cdot 1 \text{ qt} = 9 \cdot 4 \cdot 2 \text{ pt} \qquad \text{Substituting 2 pt for 1 qt}$$
$$= 9 \cdot 4 \cdot 2 \cdot 1 \text{ pt} = 9 \cdot 4 \cdot 2 \cdot 16 \text{ oz} \qquad \text{Substituting 16 oz for 1 pt}$$
$$= 1152 \text{ oz}$$

EXAMPLE 2 Complete: 24 qt = _____ gal.

Since we are converting from a *smaller* unit to a *larger* unit, we multiply by 1 using 1 gal in the numerator and 4 qt in the denominator.

$$24 \text{ qt} = 24 \text{ qt} \cdot \frac{1 \text{ gal}}{4 \text{ qt}} = \frac{24}{4} \cdot 1 \text{ gal} = 6 \text{ gal}$$

Do Exercises 1 and 2.

METRIC UNITS

One unit of capacity in the metric system is a **liter.** A liter is just a bit more than a quart. It is defined as follows.

1 liter ≈ 1.06 quarts

1 liter   1 quart

Complete.

1. 5 gal = _____ pt

2. 80 qt = _____ gal

Answers on page A-20

Complete with mL or L.

3. The patient received an injection of 2 _____ of penicillin.

METRIC UNITS OF CAPACITY

1 liter (L) = 1000 cubic centimeters (1000 cm³)

The script letter ℓ is also used for "liter."

The metric prefixes are also used with liters. The most common is **milli-.** The milliliter (mL) is, then, $\frac{1}{1000}$ liter. Thus,

$$1 \text{ L} = 1000 \text{ mL} = 1000 \text{ cm}^3;$$
$$0.001 \text{ L} = 1 \text{ mL} = 1 \text{ cm}^3.$$

Although the other metric prefixes are rarely used for capacity, we display them in the following table as we did for linear measure.

4. There are 250 _____ in a coffee cup.

1000 L	100 L	10 L	1 L	0.1 L	0.01 L	0.001 L
1 kL	1 hL	1 daL	1 L	1 dL	1 cL	1 mL (cc)

A preferred unit for drug dosage is the milliliter (mL) or the cubic centimeter (cm³). The notation "cc" is also used for cubic centimeter, especially in medicine. The milliliter and the cubic centimeter represent the same measure of capacity. A milliliter is about $\frac{1}{5}$ of a teaspoon.

5. The gas tank holds 80 _____ .

5 mL

3 cm³

$$1 \text{ mL} = 1 \text{ cm}^3 = 1 \text{ cc}$$

Volumes for which quarts and gallons are used are expressed in liters. Large volumes in business and industry are expressed using measures of cubic meters (m³).

6. Bring home 8 _____ of milk.

Do Exercises 3–6.

EXAMPLE 3 Complete: 4.5 L = _____ mL.

$$4.5 \text{ L} = 4.5 \times 1 \text{ L} = 4.5 \times 1000 \text{ mL} \qquad \text{Substituting 1000 mL for 1 L}$$
$$= 4500 \text{ mL}$$

1000 L	100 L	10 L	1 L	0.1 L	0.01 L	0.001 L
1 kL	1 hL	1 daL	1 L	1 dL	1 cL	1 mL (cc)

Answers on page A-20

3 places to the right

EXAMPLE 4 Complete: 280 mL = _____ L.

$$280 \text{ mL} = 280 \times 1 \text{ mL}$$
$$= 280 \times 0.001 \text{ L} \qquad \text{Substituting 0.001 L for 1 mL}$$
$$= 0.28 \text{ L}$$

1000 L	100 L	10 L	1 L	0.1 L	0.01 L	0.001 L
1 kL	1 hL	1 daL	1 L	1 dL	1 cL	1 mL (cc)

3 places to the left

We do find metric units of capacity in frequent use in the United States—for example, in sizes of soda bottles and automobile engines.

Do Exercises 7 and 8.

 b Medical Applications

The metric system is used extensively in medicine.

EXAMPLE 5 *Medical Dosage.* A physician orders 3.5 L of 5% dextrose in water (abbrev. D5W) to be administered over a 24-hr period. How many milliliters were ordered?

We convert 3.5 L to milliliters:

$$3.5 \text{ L} = 3.5 \times 1 \text{ L} = 3.5 \times 1000 \text{ mL} = 3500 \text{ mL}.$$

The physician had ordered 3500 mL of D5W.

Do Exercise 9.

EXAMPLE 6 *Medical Dosage.* Liquids at a pharmacy are often labeled in liters or milliliters. Thus if a physician's prescription is given in ounces, it must be converted. For conversion, a pharmacist knows that 1 oz ≈ 29.57 mL.* A prescription calls for 3 oz of theophylline. For how many milliliters is the prescription?

We convert as follows:

$$3 \text{ oz} = 3 \times 1 \text{ oz} \approx 3 \times 29.57 \text{ mL} = 88.71 \text{ mL}.$$

The prescription calls for 88.71 mL of theophylline.

Do Exercise 10.

Complete.

7. 0.97 L = _____ mL

8. 8990 mL = _____ L

9. Medical Dosage. A physician orders 2400 mL of 0.9% saline solution to be administered intravenously over a 24-hr period. How many liters were ordered?

10. Medical Dosage. A prescription calls for 2 oz of theophylline.
   a) For how many milliliters is the prescription?
   b) For how many liters is the prescription?

---

*In practice, most physicians use 30 mL as an approximation to 1 oz.

Answers on page A-20

8.5

EXERCISE SET

For Extra Help

MathXL · MyMathLab · InterAct Math · Math Tutor Center · Digital Video Tutor CD 4 Videotape 8 · Student's Solutions Manual

a   Complete.

1. 1 L = _____ mL = _____ cm^3

2. _____ L = 1 mL = _____ cm^3

3. 87 L = _____ mL

4. 806 L = _____ mL

5. 49 mL = _____ L

6. 19 mL = _____ L

7. 0.401 mL = _____ L

8. 0.816 mL = _____ L

9. 78.1 L = _____ cm^3

10. 99.6 L = _____ cm^3

11. 10 qt = _____ oz

12. 9.6 oz = _____ pt

13. 20 cups = _____ pt

14. 1 gal = _____ oz

15. 8 gal = _____ qt

16. 1 gal = _____ cups

17. 5 gal = _____ qt

18. 11 gal = _____ qt

19. 56 qt = _____ gal

20. 84 qt = _____ gal

21. 11 gal = _____ pt

22. 5 gal = _____ pt

Complete.

	OBJECT	GALLONS (gal)	QUARTS (qt)	PINTS (pt)	CUPS	OUNCES (oz)
23.	12-can package of 12-oz sodas					144
24.	6-bottle package of 16-oz sodas		3			
25.	Full tank of gasoline	16				
26.	Container of milk			8		
27.	Minute Maid Orange Juice				4	
28.	Scope Mouthwash					33
29.	Revlon Flex Shampoo					15
30.	Jhirmack Moisturizing Shampoo					20

Complete.

	OBJECT	LITERS (l)	MILLILITERS (ml)	CUBIC CENTIMETERS (cc)	CUBIC CENTIMETERS (cm³)
31.	2-L bottle of soda	2			
32.	Heinz Vinegar		3755		
33.	Full tank of gasoline in Europe	64			
34.	Old Spice Aftershave				125
35.	Revlon Flex Shampoo			443	
36.	Jhirmack Moisturizing Shampoo		591		

Medical Dosage. Solve each of the following.
Source: Steven R. Smith, M.D.

37. An emergency-room physician orders 2.0 L of Ringer's lactate to be administered over 2 hr for a patient in shock. How many milliliters is this?

38. An emergency-room physician orders 2.5 L of 0.9% saline solution over 4 hr for a patient suffering from dehydration. How many milliliters is this?

39. A physician orders 320 mL of 5% dextrose in water (D5W) solution to be administered intravenously over 4 hr. How many liters of D5W is this?

40. A physician orders 40 mL of 5% dextrose in water (D5W) solution to be administered intravenously over 2 hr to an elderly patient. How many liters of D5W is this?

41. A physician orders 0.5 oz of magnesia and alumina oral suspension antacid 4 times per day for a patient with indigestion. How many milliliters of the antacid is the patient to ingest in a day?

42. A physician orders 0.25 oz of magnesia and alumina oral suspension antacid 3 times per day for a child with upper abdominal discomfort. How many milliliters of the antacid is the child to ingest in a day?

43. A physician orders 0.5 L of normal saline solution. How many milliliters are ordered?

44. A physician has ordered that his patient receive 60 mL per hour of normal saline solution intravenously. How many liters of the saline solution is the patient to receive in a 24-hr period?

45. A physician wants her patient to receive 3.0 L of normal saline intravenously over a 24-hr period. How many milliliters per hour must the nurse administer?

46. A physician tells a patient to purchase 0.5 L of hydrogen peroxide. Commercially, hydrogen peroxide is found on the shelf in bottles that hold 4 oz, 8 oz, and 16 oz. Which bottle comes closest to filling the prescription?

Medical Dosage. Because patients do not always have a working knowledge of the metric system, physicians often prescribe dosages in teaspoons (t or tsp) and tablespoons (T or tbs). The units are related to each other and to the metric system as follows:

$$5 \text{ mL} \approx 1 \text{ tsp}, \qquad 3 \text{ tsp} = 1 \text{ T}.$$

Complete.

47. 45 mL = _____ tsp

48. 3 T = _____ tsp

49. 1 mL = _____ tsp

50. 18.5 mL = _____ tsp

51. 2 T = _____ tsp

52. 8.5 tsp = _____ T

53. 1 T = _____ mL

54. 18.5 mL = _____ T

55. D_W What advantages does the use of metric units of capacity have over that of American units?

56. D_W Why do you think most liquid containers list both metric and American units of measure?

SKILL MAINTENANCE

Convert to percent notation. [6.1b], [6.2a]

57. 0.452

58. 0.999

59. $\frac{1}{3}$

60. $\frac{2}{3}$

61. $\frac{11}{20}$

62. $\frac{21}{20}$

63. $\frac{22}{25}$

64. $\frac{2}{25}$

65. *Tourist Spending.* Foreign tourists spend $79.3 billion in this country annually. The most money, $18.2 billion, is spent in Florida. What is the ratio of amount spent in Florida to total amount spent? What is the ratio of total amount spent to amount spent in Florida? [5.1a]

Source: Travel Industry Association of America, 2000

66. One person in four plays a musical instrument. In a given group of people, what is the ratio of those who play an instrument to total number of people? What is the ratio of those who do not play an instrument to total number of people? [5.1a]

SYNTHESIS

67. *Wasting Water.* Many people leave the water running while they are brushing their teeth. Suppose that 32 oz of water is wasted in such a way each day by one person. How much water, in gallons, is wasted in a week? in a month (30 days)? in a year? Assuming each of the 294 million people in this country wastes water in this way, estimate how much water is wasted in a day; in a year.

68. *Cost of Gasoline.* Suppose that premium gasoline is selling for about $2.54/gallon. Using the fact that 1 L = 1.057 qt, determine the price of the gasoline in liters.

69. *Bees and Honey.* The average bee produces only $\frac{1}{8}$ teaspoon of honey in its lifetime. It takes 60,000 honeybees to produce 100 lb of honey. How much does a teaspoon of honey weight?

498

8.6 TIME AND TEMPERATURE

Objectives

a Convert from one unit of time to another.

b Convert between Celsius and Fahrenheit temperatures using the formulas

$$F = \frac{9}{5} \cdot C + 32$$

and

$$C = \frac{5}{9} \cdot (F - 32).$$

a Time

A table of units of time is shown below. The metric system sometimes uses "h" for hour and "s" for second, but we will use the more familiar "hr" and "sec."

UNITS OF TIME

1 day = 24 hours (hr)	1 year (yr) = $365\frac{1}{4}$ days
1 hr = 60 minutes (min)	1 week (wk) = 7 days
1 min = 60 seconds (sec)	

Since we cannot have $\frac{1}{4}$ day on the calendar, we give each year 365 days and every fourth year 366 days (a leap year), unless it is a year at the beginning of a century not divisible by 400.

EXAMPLE 1   Complete: 1 hr = _____ sec.

$$
\begin{aligned}
1 \text{ hr} &= 60 \text{ min} \\
&= 60 \cdot 1 \text{ min} \\
&= 60 \cdot 60 \text{ sec} \qquad \text{Substituting 60 sec for 1 min} \\
&= 3600 \text{ sec}
\end{aligned}
$$

EXAMPLE 2   Complete: 5 yr = _____ days.

$$
\begin{aligned}
5 \text{ yr} &= 5 \cdot 1 \text{ yr} \\
&= 5 \cdot 365\frac{1}{4} \text{ days} \qquad \text{Substituting } 365\frac{1}{4} \text{ days for 1 yr} \\
&= 1826\frac{1}{4} \text{ days}
\end{aligned}
$$

EXAMPLE 3   Complete: 4320 min = _____ days.

$$4320 \text{ min} = 4320 \text{ min} \cdot \frac{1 \text{ hr}}{60 \text{ min}} \cdot \frac{1 \text{ day}}{24 \text{ hr}} = \frac{4320}{60 \cdot 24} \text{ days} = 3 \text{ days}$$

Do Exercises 1–4.

Complete.

1. 2 hr = _____ min

2. 4 yr = _____ days

3. 1 day = _____ min

4. 168 hr = _____ wk

Answers on page A-20

Convert to Celsius. Approximate to the nearest ten degrees.

5. 180°F (Brewing coffee)

6. 25°F (Cold day)

7. −10°F (Miserably cold day)

Convert to Fahrenheit. Approximate to the nearest ten degrees.

8. 25°C (Warm day at the beach)

9. 40°C (Temperature of a patient with a high fever)

10. 10°C (A cold bath)

Answers on page A-20

b Temperature

Below are two temperature scales: **Fahrenheit** for American measure and **Celsius** for metric measure.

By laying a ruler or a piece of paper horizontally between the scales, we can make an approximate conversion from one measure of temperature to another and get an idea of how the temperature scales compare.

EXAMPLES Convert to Celsius (using the scales shown above). Approximate to the nearest ten degrees.

4. 212°F (Boiling point of water)   100°C   This is exact.
5. 32°F (Freezing point of water)   0°C   This is exact.
6. 105°F   40°C   This is approximate.

Do Exercises 5–7.

EXAMPLES Make an approximate conversion to Fahrenheit.

7. 44°C (Hot bath)   110°F   This is approximate.
8. 20°C (Room temperature)   68°F   This is exact.
9. 83°C   180°F   This is approximate.

Do Exercises 8–10.

The following formula allows us to make exact conversions from Celsius to Fahrenheit.

CELSIUS TO FAHRENHEIT

$$F = \frac{9}{5} \cdot C + 32, \quad \text{or} \quad F = 1.8 \cdot C + 32$$

$\left(\text{Multiply the Celsius temperature by } \dfrac{9}{5}, \text{ or } 1.8, \text{ and add } 32.\right)$

EXAMPLES Convert to Fahrenheit.

10. 0°C (Freezing point of water)

$$F = \frac{9}{5} \cdot C + 32 = \frac{9}{5} \cdot 0 + 32 = 0 + 32 = 32°$$

Thus, 0°C = 32°F.

11. 37°C (Normal body temperature)

$$F = 1.8 \cdot C + 32 = 1.8 \cdot 37 + 32 = 66.6 + 32 = 98.6°$$

Thus, 37°C = 98.6°F.

Check the answers to Examples 10 and 11 using the scales on p. 500.

Do Exercises 11 and 12.

The following formula allows us to make exact conversions from Fahrenheit to Celsius.

FAHRENHEIT TO CELSIUS

$$C = \frac{5}{9} \cdot (F - 32), \quad \text{or} \quad C = \frac{F - 32}{1.8}$$

$\left(\text{Subtract } 32 \text{ from the Fahrenheit temperature and multiply by } \dfrac{5}{9} \text{ or divide by } 1.8.\right)$

EXAMPLES Convert to Celsius.

12. 212°F (Boiling point of water)

$$C = \frac{5}{9} \cdot (F - 32)$$

$$= \frac{5}{9} \cdot (212 - 32)$$

$$= \frac{5}{9} \cdot 180 = 100°$$

Thus, 212°F = 100°C.

13. 77°F

$$C = \frac{F - 32}{1.8}$$

$$= \frac{77 - 32}{1.8}$$

$$= \frac{45}{1.8} = 25°$$

Thus, 77°F = 25°C.

Check the answers to Examples 12 and 13 using the scales on p. 500.

Do Exercises 13 and 14.

Convert to Fahrenheit.

11. 80°C

12. 35°C

Convert to Celsius.

13. 95°F

14. 113°F

Answers on page A-20

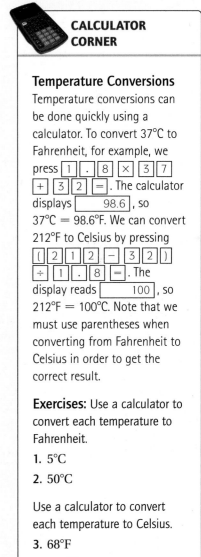

CALCULATOR CORNER

Temperature Conversions
Temperature conversions can be done quickly using a calculator. To convert 37°C to Fahrenheit, for example, we press [1] [.] [8] [×] [3] [7] [+] [3] [2] [=]. The calculator displays [98.6], so 37°C = 98.6°F. We can convert 212°F to Celsius by pressing [(] [2] [1] [2] [–] [3] [2] [)] [÷] [1] [.] [8] [=]. The display reads [100], so 212°F = 100°C. Note that we must use parentheses when converting from Fahrenheit to Celsius in order to get the correct result.

Exercises: Use a calculator to convert each temperature to Fahrenheit.

1. 5°C
2. 50°C

Use a calculator to convert each temperature to Celsius.

3. 68°F
4. 113°F

a   Complete.

1. 1 day = _____ hr

2. 1 hr = _____ min

3. 1 min = _____ sec

4. 1 wk = _____ days

5. 1 yr = _____ days

6. 2 yr = _____ days

7. 180 sec = _____ hr

8. 60 sec = _____ hr

9. 492 sec = _____ min
(The amount of time it takes for the rays of the sun to reach the earth)

10. 18,000 sec = _____ hr

11. 156 hr = _____ days

12. 444 hr = _____ days

13. 645 min = _____ hr

14. 375 min = _____ hr

15. 2 wk = _____ hr

16. 4 hr = _____ sec

17. 756 hr = _____ wk

18. 166,320 min = _____ wk

19. 2922 wk = _____ yr

20. 623 days = _____ wk

21. *Actual Time in a Day.*   Although we round it to 24 hr, the actual length of a day is 23 hr, 56 min, and 4.2 sec. How many seconds are there in an actual day?
Source: *The Handy Geography Answer Book*

22. *Time Length.*   What length of time is 86,400 sec? Is it 1 hr, 1 day, 1 week, or 1 month? (This was a question on the TV game show "Who Wants to Be a Millionaire?".)

b   Convert to Fahrenheit. Use the formula $F = \frac{9}{5} \cdot C + 32$ or $F = 1.8 \cdot C + 32$.

23. 25°C

24. 85°C

25. 40°C

26. 90°C

27. 86°C

28. 93°C

29. 58°C

30. 35°C

31. 2°C

32. 78°C

33. 5°C

34. 15°C

35. 3000°C
(The melting point of iron)

36. 1000°C
(The melting point of gold)

Convert to Celsius. Use the formula $C = \dfrac{5}{9} \cdot (F - 32)$ or $C = \dfrac{F - 32}{1.8}$.

37. 86°F

38. 59°F

39. 131°F

40. 140°F

41. 178°F

42. 195°F

43. 140°F

44. 107°F

45. 68°F

46. 50°F

47. 44°F

48. 120°F

49. 98.6°F
(Normal body temperature)

50. 104°F
(High-fevered body temperature)

51. *Highest Temperatures.* The highest temperature ever recorded in the world is 136°F in the desert of Libya in 1922. The highest temperature ever recorded in the United States is $56\frac{2}{3}$°C in California's Death Valley in 1913.

Source: *The Handy Geography Answer Book*

a) Convert each temperature to the other scale.
b) How much higher in degrees Fahrenheit was the world record than the U. S. record?

52. *Boiling Point and Altitude.* The boiling point of water actually changes with altitude. The boiling point is 212°F at sea level, but lowers about 1°F for every 500 ft that the altitude increases above sea level.

Sources: *The Handy Geography Answer Book; The New York Times Almanac*

a) What is the boiling point at an elevation of 1500 ft above sea level?
b) The elevation of Tucson is 2564 ft above sea level and that of Phoenix is 1117 ft. What is the boiling point in each city?
c) How much lower is the boiling point in Denver, whose elevation is 5280 ft, than in Tucson?
d) What is the boiling point at the top of Mt. McKinley in Alaska, the highest point in the United States, at 20,320 ft?

53. D_W Write a report on the origin of the words *Fahrenheit* and *Celsius*. How is the word *centigrade* related to temperature?

54. D_W **a)** The temperature is 23°C. Would you want to play golf? Explain.
b) Your bathwater has a temperature of 10°C. Would you want to take a bath?
c) The nearby lake has a temperature of −10°C. Would it be safe to go ice skating?

SKILL MAINTENANCE

VOCABULARY REINFORCEMENT

In each of Exercises 55–62, fill in the blank with the correct term from the given list. Some of the choices may not be used.

55. When interest is paid on interest, it is called _____ interest.   [6.7b]

56. A _____ is the quotient of two quantities.   [5.1a]

57. The _____ of a set of data is the middle number if there is an odd number of data items.   [7.1b]

58. When you work for a _____ , you are paid a percentage of the total sales for which you are responsible. [6.6b]

59. In _____ triangles, the lengths of their corresponding sides have the same ratio.   [5.5a]

60. To find the _____ of a set of data, add the numbers and then divide by the number of items of data. [7.1a]

61. A natural number, other than 1, that is not prime is _____ .   [2.1c]

62. In the sentence $35 - 11 = 24$, 11 is called the _____ .   [1.3a]

mean

median

mode

complementary

composite

commission

salary

ratio

minuend

subtrahend

simple

compound

isosceles

similar

SYNTHESIS

63. Estimate the number of years in one million seconds.

64. Estimate the number of years in one billion seconds.

65. Estimate the number of years in one trillion seconds.

Complete.

66. $88 \dfrac{ft}{sec} = \underline{\hspace{1cm}} \dfrac{mi}{hr}$

67. $0.9 \dfrac{L}{hr} = \underline{\hspace{1cm}} \dfrac{mL}{sec}$

8.7 CONVERTING UNITS OF AREA

Objectives

a Convert from one American unit of area to another.

b Convert from one metric unit of area to another.

a   American Units

Let's do some conversions from one American unit of area to another.

EXAMPLE 1   Complete: $1 \text{ ft}^2 =$ _____ in^2.

$1 \text{ ft}^2 = 1 \cdot (12 \text{ in.})^2$ Substituting 12 in. for 1 ft

$= 12 \text{ in.} \cdot 12 \text{ in.} = 144 \text{ in}^2$

EXAMPLE 2   Complete: $8 \text{ yd}^2 =$ _____ ft^2.

$8 \text{ yd}^2 = 8 \cdot (3 \text{ ft})^2$ Substituting 3 ft for 1 yd

$= 8 \cdot 3 \text{ ft} \cdot 3 \text{ ft} = 8 \cdot 3 \cdot 3 \cdot \text{ft} \cdot \text{ft} = 72 \text{ ft}^2$

Do Exercises 1–3.

Complete.

1. $1 \text{ yd}^2 =$ _____ ft^2

2. $5 \text{ yd}^2 =$ _____ ft^2

3. $20 \text{ ft}^2 =$ _____ in^2

AMERICAN UNITS OF AREA

1 square yard (yd^2) = 9 square feet (ft^2)

1 square foot (ft^2) = 144 square inches (in^2)

1 square mile (mi^2) = 640 acres

1 acre = 43,560 ft^2

EXAMPLE 3   Complete: $36 \text{ ft}^2 =$ _____ yd^2.

We are converting from "ft^2" to "yd^2". Thus we choose a symbol for 1 with yd^2 on top and ft^2 on the bottom.

$36 \text{ ft}^2 = 36 \text{ ft}^2 \times \dfrac{1 \text{ yd}^2}{9 \text{ ft}^2}$ Multiplying by 1 using $\dfrac{1 \text{ yd}^2}{9 \text{ ft}^2}$

$= \dfrac{36}{9} \times \dfrac{\text{ft}^2}{\text{ft}^2} \times 1 \text{ yd}^2 = 4 \text{ yd}^2$

EXAMPLE 4   Complete: $7 \text{ mi}^2 =$ _____ acres.

$7 \text{ mi}^2 = 7 \cdot 1 \text{ mi}^2$

$= 7 \cdot 640 \text{ acres}$ Substituting 640 acres for 1 mi^2

$= 4480 \text{ acres}$

Do Exercises 4 and 5.

Complete.

4. $360 \text{ in}^2 =$ _____ ft^2

5. $5 \text{ mi}^2 =$ _____ acres

b   Metric Units

Let's now convert from one metric unit of area to another.

EXAMPLE 5   Complete: $1 \text{ km}^2 =$ _____ m^2.

$1 \text{ km}^2 = 1 \cdot (1000 \text{ m})^2$ Substituting 1000 m for 1 km

$= 1000 \text{ m} \cdot 1000 \text{ m} = 1{,}000{,}000 \text{ m}^2$

Answers on page A-20

Complete.

6. $1 \text{ m}^2 =$ _____ mm^2

EXAMPLE 6 Complete: $10{,}000 \text{ cm}^2 =$ _____ m^2.

$$10{,}000 \text{ cm}^2 = 10{,}000 \text{ cm}^2 \cdot \frac{1 \text{ m}}{100 \text{ cm}} \cdot \frac{1 \text{ m}}{100 \text{ cm}}$$

$$= 10{,}000 \text{ cm}^2 \cdot \frac{1 \text{ m}^2}{10{,}000 \text{ cm}^2}$$

$$= 1 \text{ m}^2$$

Do Exercises 6 and 7.

7. $100 \text{ mm}^2 =$ _____ cm^2

MENTAL CONVERSION

To convert mentally, we first note that $10^2 = 100$, $100^2 = 10{,}000$, and $0.1^2 = 0.01$. We use the diagram as before and multiply the number of moves by 2 to determine the number of moves of the decimal point.

1000 m	100 m	10 m	1 m	0.1 m	0.01 m	0.001 m
1 km	1 hm	1 dam	1 m	1 dm	1 cm	1 mm

Complete.

8. $2.88 \text{ m}^2 =$ _____ cm^2

EXAMPLE 7 Complete: $3.48 \text{ km}^2 =$ _____ m^2.

Think: To go from km to m in the table is a move of 3 places to the right.

1000 m	100 m	10 m	1 m	0.1 m	0.01 m	0.001 m
1 km	1 hm	1 dam	1 m	1 dm	1 cm	1 mm

3 moves to the right

So we move the decimal point $2 \cdot 3$, or 6 places to the right.

$$3.48 \qquad 3.480000. \qquad 3.48 \text{ km}^2 = 3{,}480{,}000 \text{ m}^2$$

6 places to the right

9. $4.3 \text{ mm}^2 =$ _____ cm^2

EXAMPLE 8 Complete: $586.78 \text{ cm}^2 =$ _____ m^2.

Think: To go from cm to m in the table is a move of 2 places to the left.

1000 m	100 m	10 m	1 m	0.1 m	0.01 m	0.001 m
1 km	1 hm	1 dam	1 m	1 dm	1 cm	1 mm

10. $678{,}000 \text{ m}^2 =$ _____ km^2

2 moves to the left

So we move the decimal point $2 \cdot 2$, or 4 places to the left.

$$586.78 \qquad 0.0586.78 \qquad 586.78 \text{ cm}^2 = 0.058678 \text{ m}^2$$

4 places to the left

Answers on page A-20

Do Exercises 8–10.

Translating
for Success

1. *Test Items.* On a test of 90 items, Sally got 80% correct. How many items did she get correct?

2. *Suspension Bridge.* The San Francisco/Oakland Bay suspension bridge is 0.4375 mi long. Convert this distance to yards.

3. *Population Growth.* City A's growth rate per year is 0.9%. If the population was 1,500,000 in 2005, what is the population in 2006?

4. *Roller Coaster Drop.* The Manhattan Express Roller Coaster at the New York-New York Hotel and Casino, Las Vegas, Nevada, has a 144-ft drop. The California Screamin' Roller Coaster at Disney's California Adventure, Anaheim, California, has a 32.635-m drop. How much larger in meters is the drop of the Manhattan Express than the drop of the California Screamin'?

5. *Driving Distance.* Nate drives the company car 675 mi in 15 days. At this rate, how far will he drive in 20 days?

The goal of these matching questions is to practice step (2), *Translate,* of the five-step problem-solving process. Translate each word problem to an equation and select a correct translation from equations A–O.

A. $32.635 \text{ m} + x = 144 \text{ ft} \cdot \dfrac{0.305 \text{ m}}{1 \text{ ft}}$

B. $x = 0.4375 \text{ mi} \times \dfrac{5280 \text{ ft}}{1 \text{ mi}} \times \dfrac{12 \text{ in.}}{1 \text{ ft}}$

C. $80\% \cdot x = 90$

D. $x = 0.89 \text{ km} \cdot \dfrac{1000 \text{ m}}{1 \text{ km}} \cdot \dfrac{1 \text{ m}}{3.281 \text{ ft}}$

E. $x = 80\% \cdot 90$

F. $x = 420 \text{ m} + 75 \text{ ft} \cdot \dfrac{0.305 \text{ m}}{1 \text{ ft}}$

G. $\dfrac{x}{20} = \dfrac{675}{15}$

H. $x = 0.4375 \text{ mi} \times \dfrac{5280 \text{ ft}}{1 \text{ mi}} \times \dfrac{1 \text{ yd}}{3 \text{ ft}}$

I. $x = 0.89 \text{ km} \cdot \dfrac{0.621 \text{ mi}}{1 \text{ km}} \cdot \dfrac{5280 \text{ ft}}{1 \text{ mi}}$

J. $\dfrac{x}{15} = \dfrac{675}{20}$

K. $x = 420 \text{ m} \cdot \dfrac{3.281 \text{ ft}}{1 \text{ m}} + 75 \text{ ft}$

L. $144 \text{ ft} + x = 32.635 \text{ m} \cdot \dfrac{1 \text{ ft}}{0.305 \text{ m}}$

M. $x = 1,500,000 - 0.9\%(1,500,000)$

N. $20 \cdot x = 675$

O. $x = 1,500,000 + 0.9\%(1,500,000)$

Answers on page A-20

6. *Test Items.* Jason answered correctly 90 items on a recent test. These items represented 80% of the total number of questions. How many items were on the test?

7. *Population Decline.* City B's growth rate per year is −0.9%. If the population was 1,500,000 in 2005, what is the population in 2006?

8. *Bridge Length.* The Tatara Bridge in Onomichi-Imabari, Japan, is 0.89 km long. Convert this distance to feet.

9. *Height of Tower.* The Sears Tower in Chicago is 75 ft taller than the Jin Mao Building in Shanghai. The height of the Jin Mao Building is 420 m. What is the height of the Sears Tower in feet?

10. *Gasoline Usage.* Nate's company car gets 20 miles to the gallon in city driving. How many gallons will it use in 675 miles of city driving?

a   Complete.

1. $1 \text{ ft}^2 = $ _____ in^2

2. $1 \text{ yd}^2 = $ _____ ft^2

3. $1 \text{ mi}^2 = $ _____ acres

4. $1 \text{ acre} = $ _____ ft^2

5. $1 \text{ in}^2 = $ _____ ft^2

6. $1 \text{ ft}^2 = $ _____ yd^2

7. $22 \text{ yd}^2 = $ _____ ft^2

8. $40 \text{ ft}^2 = $ _____ in^2

9. $44 \text{ yd}^2 = $ _____ ft^2

10. $144 \text{ ft}^2 = $ _____ yd^2

11. $20 \text{ mi}^2 = $ _____ acres

12. $576 \text{ in}^2 = $ _____ ft^2

13. $1 \text{ mi}^2 = $ _____ ft^2

14. $1 \text{ mi}^2 = $ _____ yd^2

15. $720 \text{ in}^2 = $ _____ ft^2

16. $27 \text{ ft}^2 = $ _____ yd^2

17. $144 \text{ in}^2 = $ _____ ft^2

18. $72 \text{ in}^2 = $ _____ ft^2

19. $1 \text{ acre} = $ _____ mi^2

20. $4 \text{ acres} = $ _____ ft^2

b   Complete.

21. $5.21 \text{ km}^2 = $ _____ m^2

22. $65 \text{ km}^2 = $ _____ m^2

23. $0.014 \text{ m}^2 = $ _____ cm^2

24. $0.028 \text{ m}^2 = $ _____ mm^2

25. $2345.6 \text{ mm}^2 = $ _____ cm^2

26. $8.38 \text{ cm}^2 = $ _____ mm^2

27. $852.14 \text{ cm}^2 = $ _____ m^2

28. $125 \text{ mm}^2 = $ _____ m^2

29. $250{,}000 \text{ mm}^2 = $ _____ cm^2

30. $2400 \text{ mm}^2 = $ _____ cm^2

31. $472{,}800 \text{ m}^2 = $ _____ km^2

32. $1.37 \text{ cm}^2 = $ _____ mm^2

33. D_W Explain the difference between the way we move the decimal point for area conversion and the way we do so for length conversion.

34. D_W Which is larger and why: one square meter or nine square feet?

In Exercises 35 and 36, find the simple interest. [6.7a]

35. On $2000 at an interest rate of 8% for 1.5 yr

36. On $2000 at an interest rate of 5.3% for 2 yr

In each of Exercises 37–40, find (a) the amount of simple interest due and (b) the total amount that must be paid back. [6.7a]

37. A firm borrows $15,500 at 9.5% for 120 days.

38. A firm borrows $8500 at 10% for 90 days.

39. A firm borrows $6400 at 8.4% for 150 days.

40. A firm borrows $4200 at 11% for 30 days.

Complete.

41. $1 \text{ m}^2 = $ _____ ft^2

42. $1 \text{ in}^2 = $ _____ cm^2

43. $2 \text{ yd}^2 = $ _____ m^2

44. $1 \text{ acre} = $ _____ m^2

45. *The White House.*   The president's family has about 20,175 ft^2 of living area in the White House. Estimate the amount of living area in square meters.

The review that follows is meant to prepare you for a chapter exam. It consists of three parts. The first part, Concept Reinforcement, is designed to increase understanding of the concepts through true/false exercises. The second part is a list of important properties and formulas. The third part is the Review Exercises. These provide practice exercises for the exam, together with references to section objectives so you can go back and review. Before beginning, stop and look back over the skills you have obtained. What skills in mathematics do you have now that you did not have before studying this chapter?

↪ CONCEPT REINFORCEMENT

Determine whether the statement is true or false. Answers are given at the back of the book.

_____ **1.** Distances that are measured in miles in the American system would probably be measured in meters in the metric system.

_____ **2.** One meter is slightly more than one yard.

_____ **3.** When converting from meters to centimeters we move the decimal point to the right.

_____ **4.** To convert mm^2 to cm^2, move the decimal point 2 places to the left.

_____ **5.** Since 1 yd = 3 ft, we multiply by 3 to convert square yards to square feet.

_____ **6.** You would probably use your furnace when the temperature is 40°C.

IMPORTANT PROPERTIES AND FORMULAS

American Units of Length: 12 in. = 1 ft; 3 ft = 1 yd; 36 in. = 1 yd; 5280 ft = 1 mi

Metric Units of Length: 1 km = 1000 m; 1 hm = 100 m; 1 dam = 10 m; 1 dm = 0.1 m; 1 cm = 0.01 m;
                                 1 mm = 0.001 m

American–Metric Conversion: 1 m = 39.370 in.; 1 m = 3.281 ft; 0.305 m = 1 ft; 2.540 cm = 1 in.;
                                 1 km = 0.621 mi; 1.609 km = 1 mi; 0.914 m = 1 yd; 1 m = 1.094 yd

American System of Weights: 1 T = 2000 lb; 1 lb = 16 oz

Metric System of Mass: 1 t = 1000 kg; 1 kg = 1000 g; 1 hg = 100 g; 1 dag = 10 g; 1 dg = 0.1 g;
                                 1 cg = 0.01 g; 1 mg = 0.001 g; 1 mcg = 0.000001 g

American Units of Capacity: 1 gal = 4 qt; 1 qt = 2 pt; 1 pt = 16 oz; 1 pt = 2 cups; 1 cup = 8 oz

Metric Units of Capacity: 1 L = 1000 mL = 1000 cm^3 = 1000 cc

American–Metric Conversion: 1 oz = 29.57 mL; 1 L = 1.06 qt

Units of Time: 1 min = 60 sec; 1 hr = 60 min; 1 day = 24 hr; 1 wk = 7 days; 1 yr = $365\frac{1}{4}$ days

Temperature Conversion: $F = \dfrac{9}{5} \cdot C + 32$, or $F = 1.8 \cdot C + 32$;

$C = \dfrac{5}{9} \cdot (F - 32)$, or $C = \dfrac{F - 32}{1.8}$

Review Exercises

Complete. [8.1a], [8.2a], [8.3a]

1. 8 ft = _____ yd

2. $\frac{5}{6}$ yd = _____ in.

3. 0.3 mm = _____ cm

4. 4 m = _____ km

5. 2 yd = _____ in.

6. 4 km = _____ cm

7. 14 in. = _____ ft

8. 15 cm = _____ m

9. 200 m = _____ yd

10. 20 mi = _____ km

Complete. [8.4a, b], [8.5a], [8.6a]

13. 7 lb = _____ oz

14. 4 g = _____ kg

15. 16 min = _____ hr

16. 464 mL = _____ L

17. 3 min = _____ sec

18. 4.7 kg = _____ g

19. 8.07 T = _____ lb

20. 0.83 L = _____ mL

21. 6 hr = _____ days

22. 4 cg = _____ g

23. 0.2 g = _____ mg

24. 0.0003 kg = _____ cg

25. 0.7 mL = _____ L

26. 60 mL = _____ L

27. 0.8 T = _____ lb

28. 0.4 L = _____ mL

29. 20 oz = _____ lb

30. $\frac{5}{6}$ min = _____ sec

Complete the following table [8.2a]

	OBJECT	MILLIMETERS (mm)	CENTIMETERS (cm)	METERS (m)
11.	Length of a key on a calculator		1	
12.	Height of Texas Commerce Center, Houston			305

31. 20 gal = _____ pt

32. 960 oz = _____ gal

33. 54 qt = _____ gal

34. 2.5 days = _____ hr

35. 3020 cg = _____ kg

36. 10,500 lb = _____ T

Medical Dosage. Solve.

37. Amoxicillin is an antibiotic obtainable in a liquid suspension form, part medication and part water, and is frequently used to treat infections in infants. One formulation of the drug contains 125 mg of amoxicillin per 5 mL of liquid. A pediatrician orders 150 mg per day for a 4-month-old child with an ear infection. How much of the amoxicillin suspension would the parent need to administer to the infant in order to achieve the recommended daily dose? [8.4c]

38. An emergency-room physician orders 3 L of Ringer's lactate to be administered over 4 hr for a patient suffering from shock and severe low blood pressure. How many milliliters is this? [8.5b]

39. A physician prescribes 0.25 mg of alprazolam, an antianxiety medication. How many micrograms are in this dose? [8.4c]

40. Convert 27°C to Fahrenheit. [8.6b]

41. Convert −5°C to Fahrenheit. [8.6b]

42. Convert 68°F to Celsius. [8.6b]

43. Convert 10°F to Celsius. [8.6b]

Complete. [8.7a, b]

44. $4 \text{ yd}^2 =$ _____ ft^2

45. $0.3 \text{ km}^2 =$ _____ m^2

46. $2070 \text{ in}^2 =$ _____ ft^2

47. $600 \text{ cm}^2 =$ _____ m^2

48. $\mathbf{D_W}$ Napoleon is credited with influencing the use of the metric system. Research this possibility and make a report. [8.2a]

49. $\mathbf{D_W}$ It is known that 1 gal of water weighs 8.3453 lb. Which weighs more, an ounce of pennies or an ounce (as capacity) of water? Explain. [8.4a], [8.5a]

50. *Running Record.* The world's record for running the 200-m dash is 19.32 sec, set by Michael Johnson of the United States in the 1996 Olympics in Atlanta. How should the record be changed if it were a 200-yd dash? [8.3a]

Source: *The Guinness Book of Records*

Complete.

1. 4 ft = _____ in.

2. 4 in. = _____ ft

3. 6 km = _____ m

4. 8.7 mm = _____ cm

5. 200 yd = _____ m

6. 2400 km = _____ mi

Complete the following table.

	OBJECT	MILLIMETERS (mm)	CENTIMETERS (cm)	METERS (m)
7.	Width of a key on a calculator		0.5	
8.	Height of your author			1.8542

Complete.

9. 3080 mL = _____ L

10. 0.24 L = _____ mL

11. 4 lb = _____ oz

12. 4.11 T = _____ lb

13. 3.8 kg = _____ g

14. 4.325 mg = _____ cg

15. 2200 mg = _____ g

16. 5 hr = _____ min

17. 15 days = _____ hr

18. 64 pt = _____ qt

19. 10 gal = _____ oz

20. 5 cups = _____ oz

21. 0.37 mg = _____ mcg

22. Convert 95°F to Celsius.

23. Convert 59°C to Fahrenheit.

Complete the following table.

	OBJECT	YARDS (yd)	CENTIMETERS (cm)	INCHES (in.)	METERS (m)	MILLIMETERS (mm)
24.	Length of a meter stick				1	
25.	Height of Xiamen Posts and Telecommunications Building, Xiamen, China	398				

Medical Dosage. Solve each of the following.

26. An emergency-room physician prescribes 2.5 L of normal saline intravenously over 8 hr for a patient who is severely dehydrated. How many milliliters is this?

27. A physician prescribes 0.5 mg of alprazolam to be taken 3 times a day by a patient suffering from anxiety. How many micrograms of alprazolam is the patient to ingest each day?

28. A prescription calls for 4 oz of dextromethorphan, a cough-suppressant medication. For how many milliliters is the prescription? (Use 1 oz = 29.57 mL.)

Complete.

29. $12 \text{ ft}^2 = $ _____ in^2

30. $3 \text{ cm}^2 = $ _____ m^2

31. *Running Record.* The world's record for running the 400-m run is 43.18 sec, set by Michael Johnson of the United States in Seville, Spain, on August 26, 1999. How should the record be changed if it were a 400-yd run?

Source: *The Guinness Book of Records 2004*

Geometry

9

Real-World Application

The average American eats 54 quarts of popcorn annually. This quantity of popcorn would fill a bag measuring 10.5 in. by 10.5 in. by 28 in. Find the volume of such a bag.

Source: The Popcorn Board

This problem appears as Margin Exercise 2 in Section 9.4.

Objectives

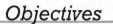

a Find the perimeter of a polygon.

b Solve applied problems involving perimeter.

Find the perimeter of the polygon.

1.

11 cm
4 cm
2 cm
7 cm
2 cm

2.

9.2 in.   9.2 in.

9.2 in.   9.2 in.

9.2 in.

3. Find the perimeter of a rectangle that is 2 cm by 4 cm.

2 cm

4 cm

Answers on page A-21

a Finding Perimeters

PERIMETER OF A POLYGON
A **polygon** is a geometric figure with three or more sides. The **perimeter** of a **polygon** is the distance around it, or the sum of the lengths of its sides.

EXAMPLE 1   Find the perimeter of this polygon.

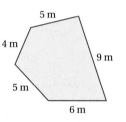

5 m
4 m
9 m
5 m
6 m

We add the lengths of the sides. Since all units are the same, we add the numbers, keeping meters (m) as the unit.

Perimeter = 6 m + 5 m + 4 m + 5 m + 9 m
          = (6 + 5 + 4 + 5 + 9) m
          = 29 m

Do Exercises 1 and 2.

A **rectangle** is a figure with four sides and four 90° angles, like the one shown in Example 2.

EXAMPLE 2   Find the perimeter of a rectangle that is 3 cm by 4 cm.

4 cm
90°
3 cm

Perimeter = 3 cm + 3 cm + 4 cm + 4 cm
          = (3 + 3 + 4 + 4) cm
          = 14 cm

Do Exercise 3.

PERIMETER OF A RECTANGLE

The **perimeter of a rectangle** is twice the sum of the length and the width, or 2 times the length plus 2 times the width:

$$P = 2 \cdot (l + w), \quad \text{or} \quad P = 2 \cdot l + 2 \cdot w.$$

EXAMPLE 3 Find the perimeter of a rectangle that is 4.3 ft by 7.8 ft.

$$
\begin{aligned}
P &= 2 \cdot (l + w) \\
&= 2 \cdot (4.3 \text{ ft} + 7.8 \text{ ft}) \\
&= 2 \cdot (12.1 \text{ ft}) \\
&= 24.2 \text{ ft}
\end{aligned}
$$

Do Exercises 4 and 5.

A square is a rectangle with all sides the same length.

EXAMPLE 4 Find the perimeter of a square whose sides are 9 mm long.

$$
\begin{aligned}
P &= 9 \text{ mm} + 9 \text{ mm} + 9 \text{ mm} + 9 \text{ mm} \\
&= (9 + 9 + 9 + 9) \text{ mm} \\
&= 36 \text{ mm}
\end{aligned}
$$

Do Exercise 6.

PERIMETER OF A SQUARE

The **perimeter of a square** is four times the length of a side:

$$P = 4 \cdot s.$$

---

4. Find the perimeter of a rectangle that is 5.25 yd by 3.5 yd.

5. Find the perimeter of a rectangle that is $8\frac{1}{4}$ in. by $5\frac{2}{3}$ in.

6. Find the perimeter of a square with sides of length 10 km.

Answers on page A-21

7. Find the perimeter of a square with sides of length $5\frac{1}{4}$ yd.

EXAMPLE 5 Find the perimeter of a square whose sides are $20\frac{1}{8}$ in. long.

$$P = 4 \cdot s = 4 \cdot 20\frac{1}{8} \text{ in.}$$

$$= 4 \cdot \frac{161}{8} \text{ in.} = \frac{4 \cdot 161}{4 \cdot 2} \text{ in.}$$

$$= \frac{161}{2} \cdot \frac{4}{4} \text{ in.} = 80\frac{1}{2} \text{ in.}$$

8. Find the perimeter of a square with sides of length 7.8 km.

Do Exercises 7 and 8.

b Solving Applied Problems

EXAMPLE 6 A vegetable garden is 20 ft by 15 ft. A fence is to be built around the garden. How many feet of fence will be needed? If fencing sells for $2.95 per foot, what will the fencing cost?

1. Familiarize. We make a drawing and let P = the perimeter.

9. A play area is 25 ft by 10 ft. A fence is to be built around the play area. How many feet of fencing will be needed? If fencing costs $4.95 per foot, what will the fencing cost?

2. Translate. The perimeter of the garden is given by

$$P = 2 \cdot (l + w) = 2 \cdot (20 \text{ ft} + 15 \text{ ft}).$$

3. Solve. We calculate the perimeter as follows:

$$P = 2 \cdot (20 \text{ ft} + 15 \text{ ft}) = 2 \cdot (35 \text{ ft}) = 70 \text{ ft}$$

Then we multiply by $2.95 to find the cost of the fencing:

Cost = $2.95 \times$ Perimeter = 2.95×70 ft = $206.50.

4. Check. The check is left to the student.

5. State. The 70 ft of fencing that is needed will cost $206.50.

Answers on page A-21

Do Exercise 9.

a   Find the perimeter of the polygon.

1.
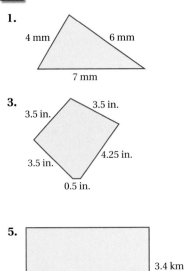

4 mm   6 mm

7 mm

2.

3 yd

1.2 yd

1.2 yd

3 yd

3.

3.5 in.

3.5 in.

3.5 in.

4.25 in.

3.5 in.

0.5 in.

4.

46 in.

18 in.

14 in.

22 in.

4 in.

13 in.

8 in.

19 in.

5.

3.4 km

5.6 km

6.

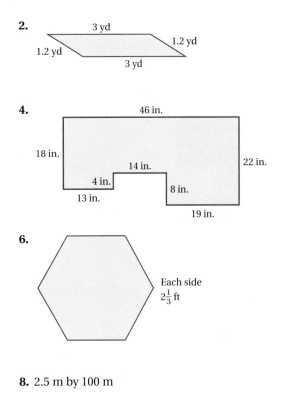

Each side $2\frac{1}{3}$ ft

Find the perimeter of the rectangle.

7. 5 ft by 10 ft

8. 2.5 m by 100 m

9. 34.67 cm by 4.9 cm

10. $3\frac{1}{2}$ yd by $4\frac{1}{2}$ yd

Find the perimeter of the square.

11. 22 ft on a side

12. 56.9 km on a side

13. 45.5 mm on a side

14. $3\frac{1}{8}$ yd on a side

 b   Solve.

15. A security fence is to be built around a 173-m by 240-m rectangular field. What is the perimeter of the field? If 5-ft–high galvanized fence wire costs $7.29 per meter, what will the fencing cost?

16. *Softball Diamond.* A standard-sized slow-pitch softball diamond is a square with sides of length 65 ft. What is the perimeter of this softball diamond? (This is the distance you would have to run if you hit a home run.)

Source: American Softball Association

65 ft

65 ft

17. A piece of flooring tile is a square with sides of length 30.5 cm. What is the perimeter of a piece of tile?

18. A rectangular posterboard is 61.8 cm by 87.9 cm. What is the perimeter of the board?

19. A rain gutter is to be installed around the office building shown in the figure.

   a) Find the perimeter of the office building.
   b) If the gutter costs $4.59 per foot, what is the total cost of the gutter?

20. A carpenter is to build a fence around a 9-m by 12-m garden.

   a) The posts are 3 m apart. How many posts will be needed?
   b) The posts cost $2.40 each. How much will the posts cost?
   c) The fence will surround all but 3 m of the garden, which will be a gate. How long will the fence be?
   d) The fence costs $2.85 per meter. What will the cost of the fence be?
   e) The gate costs $9.95. What is the total cost of the materials?

21. **D**_W Create for a fellow student a development of the formula

$$P = 2 \cdot (l + w) = 2 \cdot l + 2 \cdot w$$

for the perimeter of a rectangle.

22. **D**_W Create for a fellow student a development of the formula

$$P = 4 \cdot s$$

for the perimeter of a square.

23. Find the simple interest on $600 at 6.4% for $\frac{1}{2}$ yr. [6.7a]

24. Find the simple interest on $600 at 8% for 2 yr. [6.7a]

Evaluate. [1.9b]

25. 10^3

26. 11^3

27. 15^2

28. 22^2

29. 7^2

30. 4^3

Solve.

31. *Sales Tax.* In a certain state, a sales tax of $878 is collected on the purchase of a car for $17,560. What is the sales tax rate? [6.6a]

32. *Commission Rate.* Rich earns $1854.60 selling $16,860 worth of cellular phones. What is the commission rate? [6.6b]

Find the perimeter, in feet, of the figure.

33.

18 in.

3 ft

34.

78 in.

5.5 yd

9.2 AREA

Objectives

a Find the area of a rectangle and a square.

b Find the area of a parallelogram, a triangle, and a trapezoid.

c Solve applied problems involving areas of rectangles, squares, parallelograms, triangles, and trapezoids.

a Rectangles and Squares

A polygon and its interior form a plane region. We can find the area of a *rectangular region*, or *rectangle*, by filling it in with square units. Two such units, a *square inch* and a *square centimeter*, are shown below.

Square inch

Square centimeter

1 cm

1 in.

EXAMPLE 1 What is the area of this region?

We have a rectangular array. Since the region is filled with 12 square centimeters, its area is 12 square centimeters (sq cm), or 12 cm². The number of units is 3 × 4, or 12.

3 cm

4 cm

Do Exercise 1.

1. What is the area of this region? Count the number of square centimeters.

2 cm

4 cm

AREA OF A RECTANGLE

The **area of a rectangle** is the product of the length *l* and the width *w*:

$$A = l \cdot w.$$

w

l

EXAMPLE 2 Find the area of a rectangle that is 7 yd by 4 yd.

$$A = l \cdot w = 7 \text{ yd} \cdot 4 \text{ yd}$$
$$= 7 \cdot 4 \cdot \text{yd} \cdot \text{yd} = 28 \text{ yd}^2$$

We think of yd · yd as (yd)² and denote it yd². Thus we read "28 yd²" as "28 square yards."

Do Exercises 2 and 3.

2. Find the area of a rectangle that is 7 km by 8 km.

3. Find the area of a rectangle that is $5\frac{1}{4}$ yd by $3\frac{1}{2}$ yd.

Answers on page A-21

4. Find the area of a square with sides of length 12 km.

12 km

12 km

EXAMPLE 3 Find the area of a square with sides of length 9 mm.

$$A = (9 \text{ mm}) \cdot (9 \text{ mm})$$
$$= 9 \cdot 9 \cdot \text{mm} \cdot \text{mm}$$
$$= 81 \text{ mm}^2$$

9 mm

9 mm

Do Exercise 4.

AREA OF A SQUARE

The **area of a square** is the square of the length of a side:

$$A = s \cdot s, \quad \text{or} \quad A = s^2.$$

s

s

---

5. Find the area of a square with sides of length 10.9 m.

EXAMPLE 4 Find the area of a square with sides of length 20.3 m.

$$A = s \cdot s = 20.3 \text{ m} \times 20.3 \text{ m} = 20.3 \times 20.3 \times \text{m} \times \text{m} = 412.09 \text{ m}^2$$

Do Exercises 5 and 6.

6. Find the area of a square with sides of length $3\frac{1}{2}$ yd.

b   Finding Other Areas

PARALLELOGRAMS

A **parallelogram** is a four-sided figure with two pairs of parallel sides, as shown below.

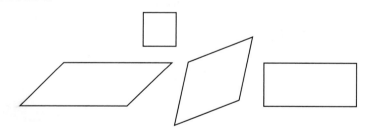

Answers on page A-21

To find the area of a parallelogram, consider the one below.

If we cut off a piece and move it to the other end, we get a rectangle.

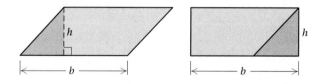

h

b

h

b

We can find the area by multiplying the length b, called a **base,** by h, called the **height.**

AREA OF A PARALLELOGRAM

The **area of a parallelogram** is the product of the length of a base b and the height h:

$$A = b \cdot h.$$

Find the area.

7.

6 cm

7.3 cm

EXAMPLE 5 Find the area of this parallelogram.

$A = b \cdot h$
$\quad = 7 \text{ km} \cdot 5 \text{ km}$
$\quad = 35 \text{ km}^2$

5 km

7 km

EXAMPLE 6 Find the area of this parallelogram.

$A = b \cdot h$
$\quad = 1.2 \text{ m} \times 6 \text{ m}$
$\quad = 7.2 \text{ m}^2$

6 m

1.2 m

Do Exercises 7 and 8.

TRIANGLES

To find the area of a triangle like the one shown on the left below, think of cutting out another just like it and placing it as shown on the right below.

8.

5.5 km

2.25 km

h

b

The resulting figure is a parallelogram whose area is

$$b \cdot h.$$

The triangle we started with has half the area of the parallelogram, or

$$\frac{1}{2} \cdot b \cdot h.$$

AREA OF A TRIANGLE

The **area of a triangle** is half the length of the base times the height:

$$A = \frac{1}{2} \cdot b \cdot h.$$

h

b

Answers on page A-21

Find the area.

9.

12 m

|←——16 m——→|

EXAMPLE 7 Find the area of this triangle.

$$A = \frac{1}{2} \cdot b \cdot h$$

$$= \frac{1}{2} \cdot 9 \text{ m} \cdot 6 \text{ m}$$

$$= \frac{9 \cdot 6}{2} \text{ m}^2$$

$$= 27 \text{ m}^2$$

6 m

|←——9 m——→|

EXAMPLE 8 Find the area of this triangle.

$$A = \frac{1}{2} \cdot b \cdot h$$

$$= \frac{1}{2} \times 6.25 \text{ cm} \times 5.5 \text{ cm}$$

$$= 0.5 \times 6.25 \times 5.5 \text{ cm}^2$$

$$= 17.1875 \text{ cm}^2$$

5.5 cm

|←——6.25 cm——→|

Do Exercises 9 and 10.

TRAPEZOIDS

A **trapezoid** is a polygon with four sides, two of which, the **bases,** are parallel to each other.

10.

3.4 cm

|←———11 cm———→|

To find the area of a trapezoid, think of cutting out another just like it.

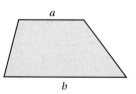

a

b

Then place the second one like this.

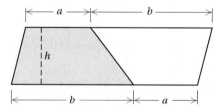

|←— *a* —→|←—— *b* ——→|

h

|←——— *b* ———→|←— *a* —→|

The resulting figure is a parallelogram whose area is

$$h \cdot (a + b). \qquad \text{The base is } a + b.$$

The trapezoid we started with has half the area of the parallelogram, or

$$\frac{1}{2} \cdot h \cdot (a + b).$$

Answers on page A-21

The **area of a trapezoid** is half the product of the height and the sum of the lengths of the parallel sides (bases):

$$A = \frac{1}{2} \cdot h \cdot (a + b), \quad \text{or} \quad A = \frac{a + b}{2} \cdot h.$$

Find the area.

11.

EXAMPLE 9 Find the area of this trapezoid.

$$A = \frac{1}{2} \cdot h \cdot (a + b)$$

$$= \frac{1}{2} \cdot 7 \text{ cm} \cdot (12 + 18) \text{ cm}$$

$$= \frac{7 \cdot 30}{2} \cdot \text{cm}^2 = \frac{7 \cdot 15 \cdot 2}{1 \cdot 2} \text{ cm}^2$$

$$= \frac{7 \cdot 15}{1} \cdot \frac{2}{2} \text{ cm}^2 = 105 \text{ cm}^2$$

Do Exercises 11 and 12.

C Solving Applied Problems

EXAMPLE 10 *Baseball's Strike Zones.* Major League Baseball underwent a rule change between the 2000 and 2001 seasons. Over the years before 2001, the strike zone evolved to something other than what was defined in the rule book. The zone in the rule book is described by rectangle *ABCD* below. The zone used before 2001 is described by the region *AQRST*. The figure shown here represents the zones for a normal-sized player, but they vary depending on the height of the player.

12.

Answers on page A-21

13. Find the area of this kite.

←8 in.→←8 in.→

28.5 in.

a) Find the area of the new zone.

b) Find the area of the former zone.

c) How much larger is the rule-book zone than the former zone?

d) By what percent has the area of the strike zone been increased by the change?

Sources: *The Cincinnati Inquirer;* Major League Baseball; Gannett News Service; *The Sporting News Official Baseball Rules Book*

This is a multistep problem that makes use of many of the skills we have learned in this book.

a) The area of rectangle *ABCD* (the rule-book zone), denoted A_1, is the length times width:

$A_1 = l \cdot w = (17.75$ in. $+ 12.25$ in.$) \cdot (17$ in.$) = (30$ in.$) \cdot (17$ in.$) = 510$ in^2.

b) The area of the region *AQRST* (the former zone), denoted A_2, is shown shaded in the figure. To find that area, we add the area of rectangle *AQRD* to the area of triangle *SDT*. We first determine

Area of *AQRD* $= l \cdot w = (17.75$ in.$) \cdot (17$ in.$) = 301.75$ in^2.

To find the area of triangle *SDT*, we first note that the length of the base is given as 9.7 in. To find the height that is the length of segment *SD*, we subtract 11 in. from 17.75 in.: 17.75 in. $-$ 11 in. $= 6.75$ in. Then

Area of triangle $SDT = \frac{1}{2} \cdot b \cdot h = \frac{1}{2}(9.7$ in.$) \cdot (6.75$ in.$) = 32.7375$ in^2.

The area of the former zone A_2 is the sum of the areas of the triangle *SDT* and the rectangle *AQRD*:

$A_2 = 301.75$ in$^2 + 32.7375$ in$^2 = 334.4875$ in$^2 \approx 334.5$ in^2.

c) To find the increase in the area, we subtract A_2 from A_1:

510 in$^2 - 334.5$ in$^2 = 175.5$ in^2.

d) To determine the percent of increase, note that we are asking "What percent of the former area is the increase?" We translate this to an equation as follows:

$\underbrace{\text{What percent}}$ of 334.5 is 175.5?

↓ ↓ ↓ ↓ ↓

p $\cdot$ 334.5 $=$ 175.5

We solve the equation:

$p \cdot 334.5 = 175.5$

$\dfrac{p \cdot 334.5}{334.5} = \dfrac{175.5}{334.5}$

$p = \dfrac{175.5}{334.5} \approx 0.5247 \approx 52\%.$

There was an increase of approximately 52% in the strike zone.

Do Exercise 13.

Answer on page A-21

9.2

EXERCISE SET

For Extra Help

Math XL   MyMathLab   InterAct   Math Tutor   Digital Video   Student's
                        Math   Center   Tutor CD 4   Solutions
MathXL   MyMathLab   InterAct   Math Tutor   Videotape 9   Manual
                      Math   Center

a   Find the area.

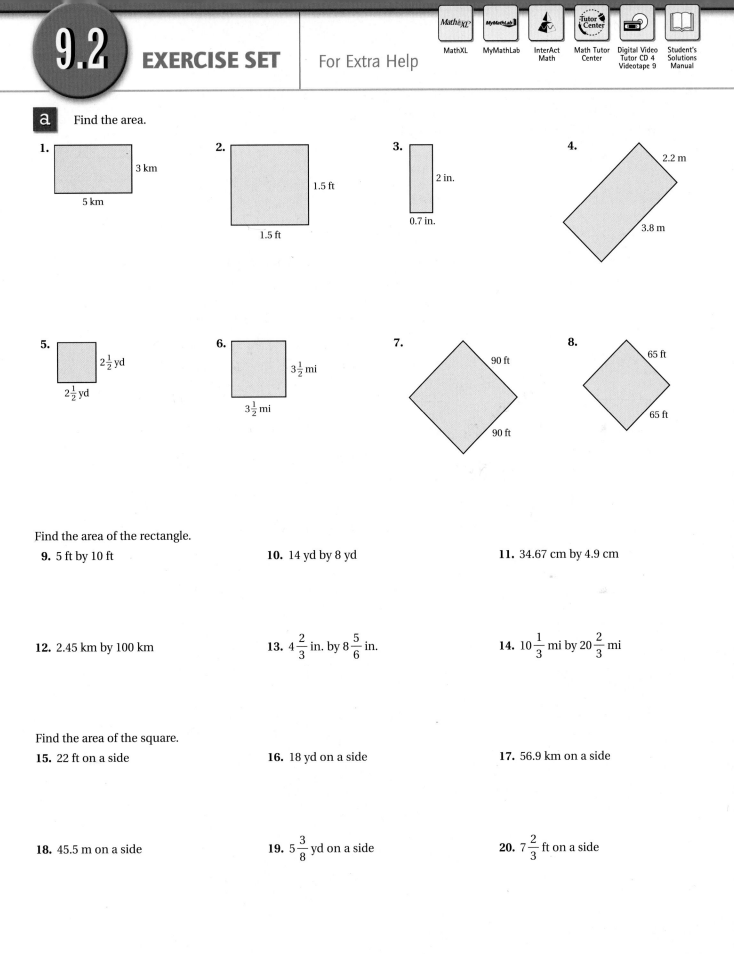

1. 3 km, 5 km

2. 1.5 ft, 1.5 ft

3. 2 in., 0.7 in.

4. 2.2 m, 3.8 m

5. $2\frac{1}{2}$ yd, $2\frac{1}{2}$ yd

6. $3\frac{1}{2}$ mi, $3\frac{1}{2}$ mi

7. 90 ft, 90 ft

8. 65 ft, 65 ft

Find the area of the rectangle.

9. 5 ft by 10 ft

10. 14 yd by 8 yd

11. 34.67 cm by 4.9 cm

12. 2.45 km by 100 km

13. $4\frac{2}{3}$ in. by $8\frac{5}{6}$ in.

14. $10\frac{1}{3}$ mi by $20\frac{2}{3}$ mi

Find the area of the square.

15. 22 ft on a side

16. 18 yd on a side

17. 56.9 km on a side

18. 45.5 m on a side

19. $5\frac{3}{8}$ yd on a side

20. $7\frac{2}{3}$ ft on a side

b Find the area.

21.

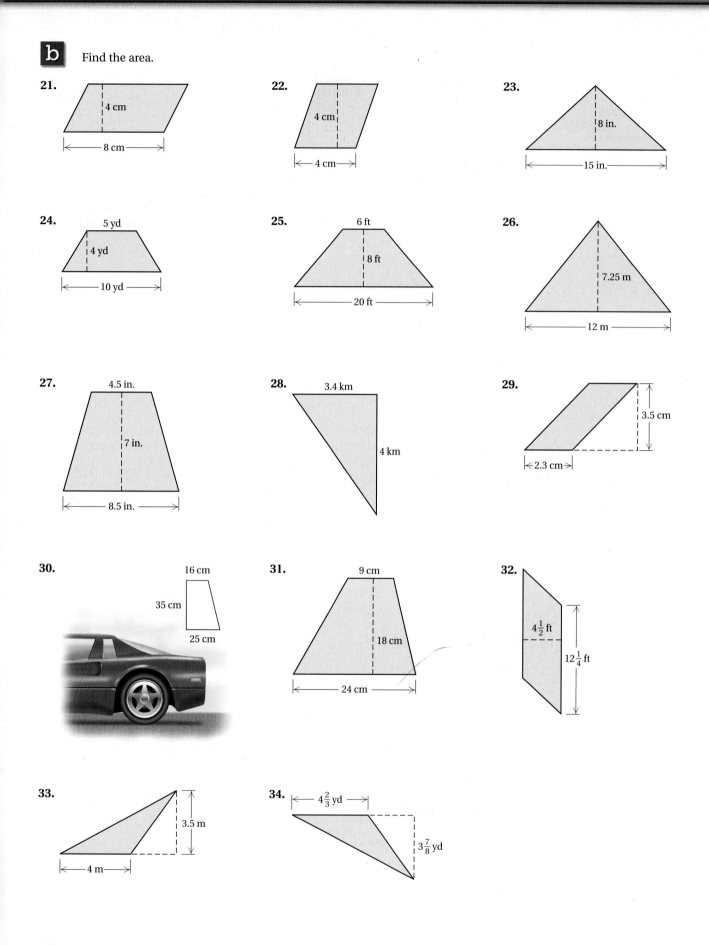

4 cm

8 cm

22.

4 cm

4 cm

23.

8 in.

15 in.

24.

5 yd

4 yd

10 yd

25.

6 ft

8 ft

20 ft

26.

7.25 m

12 m

27.

4.5 in.

7 in.

8.5 in.

28.

3.4 km

4 km

29.

3.5 cm

2.3 cm

30.

16 cm

35 cm

25 cm

31.

9 cm

18 cm

24 cm

32.

$4\frac{1}{2}$ ft

$12\frac{1}{4}$ ft

33.

3.5 m

4 m

34.

$4\frac{2}{3}$ yd

$3\frac{7}{8}$ yd

Solve.

35. *Area of a Lawn.* A lot is 40 m by 36 m. A house 27 m by 9 m is built on the lot. How much area is left over for a lawn?

36. *Area of a Field.* A field is 240.8 m by 450.2 m. Part of the field, 160.4 m by 90.6 m, is paved for a parking lot. How much area is unpaved?

37. *Mowing Expense.* A square basketball court $19\frac{1}{2}$ ft on a side is placed on an 80-ft by $110\frac{2}{3}$-ft lawn.

   a) Find the area of the lawn.
   b) It costs $0.012 per square foot to have the lawn mowed. What is the total cost of the mowing?

38. *Mowing Expense.* A square flower bed 10.5 ft on a side is dug on a 90-ft by $67\frac{1}{4}$-ft lawn.

   a) Find the area of the lawn.
   b) It costs $0.03 per square foot to have the lawn mowed. What is the total cost of the mowing?

39. *Area of a Sidewalk.* Franklin Construction Company builds a sidewalk around two sides of the Municipal Trust Bank building, as shown in the figure. What is the area of the sidewalk?

40. *Margin Area.* A standard sheet of typewriter paper is $8\frac{1}{2}$ in. by 11 in. We generally type on a $7\frac{1}{2}$-in. by 9-in. area of the paper. What is the area of the margin?

41. *Painting Costs.* A room is 15 ft by 20 ft. The ceiling is 8 ft above the floor. There are two windows in the room, each 3 ft by 4 ft. The door is $2\frac{1}{2}$ ft by $6\frac{1}{2}$ ft.

   a) What is the total area of the walls and the ceiling?
   b) A gallon of paint will cover 86.625 ft^2. How many gallons of paint are needed for the room, including the ceiling?
   c) Paint costs $17.95 a gallon. How much will it cost to paint the room?

42. *Carpeting Costs.* A restaurant owner wants to carpet a 15-yd by 20-yd room.

   a) How many square yards of carpeting are needed?
   b) The carpeting she wants is $18.50 per square yard. How much will it cost to carpet the room?

Find the area of the shaded region.

43.

4 cm

Each side 4 cm

44.

11 mm   3 mm
←5 mm→
2 mm
←12.5 mm→

45.

15 cm   30 cm

←30 cm→

46.

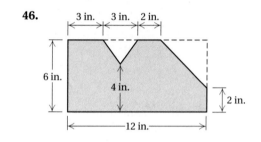

3 in.   3 in.   2 in.

6 in.   4 in.   2 in.

←12 in.→

47.

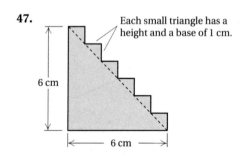

Each small triangle has a
height and a base of 1 cm.

6 cm

←6 cm→

48.

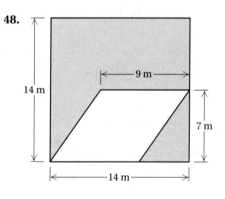

14 m   ←9 m→   7 m

←14 m→

49. *Triangular Sail.* A rectangular piece of sailcloth is 36 ft by 24 ft. A triangular area with a height of 4.6 ft and a base of 5.2 ft is cut from the sailcloth. How much area is left over?

50. *Building Area.* Find the total area of the sides and ends of the building.

51. **D**_W The length and the width of one rectangle are each three times the length and the width of another rectangle. Is the area of the first rectangle three times the area of the other rectangle? Why or why not?

52. **D**_W Explain how the area of a triangle can be found by considering the area of a parallelogram.

Complete.   [8.1a], [8.2a]

53. 23.4 cm = _____ mm

54. 0.23 km = _____ m

55. 28 ft = _____ in.

56. 72 ft = _____ yd

57. 72.4 cm = _____ m

58. 72.4 m = _____ km

59. 70 yd = _____ in.

60. 31,680 ft = _____ mi

61. 84 ft = _____ yd

62. $7\frac{1}{2}$ yd = _____ ft

63. 144 in. = _____ ft

64. 0.73 mi = _____ in.

Interest is compounded semiannually. Find the amount in the account after the given length of time. Round to the nearest cent.   [6.7b]

	PRINCIPAL	RATE OF INTEREST	TIME	AMOUNT IN THE ACCOUNT
65.	$25,000	6%	5 years	
66.	$150,000	$6\frac{7}{8}$%	15 years	
67.	$150,000	7.4%	20 years	
68.	$160,000	7.4%	20 years	

69. Find the area, in square inches, of the shaded region.

11 in.

2 ft, 2 in.

10 in.

11 ft

12.5 ft

70. Find the area, in square feet, of the shaded region.

20 in.
20 in.

8 yd

15 in.
15 in.

8 yd

531

Objectives

a Find the length of a radius of a circle given the length of a diameter, and find the length of a diameter given the length of a radius.

b Find the circumference of a circle given the length of a diameter or a radius.

c Find the area of a circle given the length of a radius.

d Solve applied problems involving circles.

a Radius and Diameter

Shown below is a circle with center *O*. Segment $\overline{AC}$ is a *diameter*. A **diameter** is a segment that passes through the center of the circle and has endpoints on the circle. Segment $\overline{OB}$ is called a *radius*. A **radius** is a segment with one end-point on the center and the other endpoint on the circle.

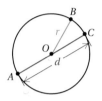

1. Find the length of a radius.

2. Find the length of a diameter.

DIAMETER AND RADIUS

Suppose that *d* is the length of the diameter of a circle and *r* is the length of the radius. Then

$$d = 2 \cdot r \quad \text{and} \quad r = \frac{d}{2}.$$

EXAMPLE 1   Find the length of a radius of this circle.

$$r = \frac{d}{2}$$

$$= \frac{12 \text{ m}}{2}$$

$$= 6 \text{ m}$$

The radius is 6 m.

EXAMPLE 2   Find the length of a diameter of this circle.

$$d = 2 \cdot r$$

$$= 2 \cdot \frac{1}{4} \text{ ft}$$

$$= \frac{1}{2} \text{ ft}$$

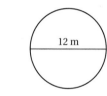

The diameter is $\frac{1}{2}$ ft.

Do Exercises 1 and 2.

Answers on page A-21

b Circumference

The **circumference** of a circle is the distance around it. Calculating circumference is similar to finding the perimeter of a polygon.

To find a formula for the circumference of any circle given its diameter, we first need to consider the ratio C/d. Take a 12-oz soda can and measure the circumference C with a tape measure. Also measure the diameter d. The results are shown in the figure. Then

$C \approx 7.8$ in.

$\leftarrow d \approx 2.5$ in. $\rightarrow$

$$\frac{C}{d} = \frac{7.8 \text{ in.}}{2.5 \text{ in.}} \approx 3.1.$$

Suppose we did this with cans and circles of several sizes. We would get a number close to 3.1. For any circle, if we divide the circumference C by the diameter d, we get the same number. We call this number π (pi). The *exact* value of the ratio C/d is π; 3.14 and 22/7 are approximations of π.

CIRCUMFERENCE AND DIAMETER

The circumference C of a circle of diameter d is given by

$$C = \pi \cdot d.$$

The number π is about 3.14, or about $\frac{22}{7}$.

EXAMPLE 3 Find the circumference of this circle. Use 3.14 for π.

$C = \pi \cdot d$

$\approx 3.14 \times 6$ cm

$= 18.84$ cm

6 cm

The circumference is about 18.84 cm.

Do Exercise 3.

3. Find the circumference of this circle. Use 3.14 for π.

20 m

Answer on page A-21

4. Find the circumference of this circle. Use $\frac{22}{7}$ for π.

14 m

5. Find the perimeter of this figure. Use 3.14 for π.

3.2 yd

7.1 yd

Since $d = 2 \cdot r$, where r is the length of a radius, it follows that

$$C = \pi \cdot d = \pi \cdot (2 \cdot r).$$

CIRCUMFERENCE AND RADIUS

The circumference C of a circle of radius r is given by

$$C = 2 \cdot \pi \cdot r.$$

EXAMPLE 4 Find the circumference of this circle. Use $\frac{22}{7}$ for π.

$$C = 2 \cdot \pi \cdot r$$

$$\approx 2 \cdot \frac{22}{7} \cdot 70 \text{ in.}$$

$$= 2 \cdot 22 \cdot \frac{70}{7} \text{ in.}$$

$$= 44 \cdot 10 \text{ in.}$$

$$= 440 \text{ in.}$$

The circumference is about 440 in.

70 in.

EXAMPLE 5 Find the perimeter of this figure. Use 3.14 for π.

We let P = the perimeter. We see that we have half a circle attached to a square. Thus we add half the circumference of the circle to the lengths of the three line segments.

9.4 km   4.7 km

9.4 km

$$P = \begin{array}{c} \text{Length of} \\ \text{three sides} \\ \text{of the square} \end{array} + \begin{array}{c} \text{Half of the} \\ \text{circumference} \\ \text{of the circle} \end{array}$$

$$= 3 \times 9.4 \text{ km} + \frac{1}{2} \times 2 \times \pi \times 4.7 \text{ km}$$

$$= 28.2 \text{ km} + 3.14 \times 4.7 \text{ km}$$

$$= 28.2 \text{ km} + 14.758 \text{ km}$$

$$= 42.958 \text{ km}$$

The perimeter is about 42.958 km.

Do Exercises 4 and 5.

Answers on page A-21

C Area

Below is a circle of radius r.

5 km

Think of cutting half the circular region into small pieces and arranging them as shown below.

$\pi \cdot r$

Then imagine cutting the other half of the circular region and arranging the pieces in with the others as shown below.

$\pi \cdot r$

This is almost a parallelogram. The base has length $\frac{1}{2} \cdot 2 \cdot \pi \cdot r$, or $\pi \cdot r$ (half the circumference) and the height is r. Thus the area is

$$(\pi \cdot r) \cdot r.$$

This is the area of a circle.

AREA OF A CIRCLE

The **area of a circle** with radius of length r is given by

$$A = \pi \cdot r \cdot r, \quad \text{or} \quad A = \pi \cdot r^2.$$

r

EXAMPLE 6   Find the area of this circle. Use $\frac{22}{7}$ for π.

$A = \pi \cdot r \cdot r$

$\approx \dfrac{22}{7} \cdot 14 \text{ cm} \cdot 14 \text{ cm}$

$= \dfrac{22}{7} \cdot 196 \text{ cm}^2$

$= 616 \text{ cm}^2$

14 cm

The area is about 616 cm^2.

Do Exercise 6.

Answer on page A-21

535

7. Find the area of this circle. Use 3.14 for π. Round to the nearest hundredth.

10.4 cm

EXAMPLE 7 Find the area of this circle. Use 3.14 for π. Round to the nearest hundredth.

$$A = \pi \cdot r \cdot r$$
$$\approx 3.14 \times 2.1 \text{ m} \times 2.1 \text{ m}$$
$$= 3.14 \times 4.41 \text{ m}^2$$
$$= 13.8474 \text{ m}^2$$
$$\approx 13.85 \text{ m}^2$$

The area is about 13.85 m^2.

2.1 m

> **Caution!**
>
> Remember that circumference is always measured in linear units like ft, m, cm, yd, and so on. But area is measured in square units like ft^2, m^2, cm^2, yd^2, and so on.

Do Exercise 7.

d Solving Applied Problems

8. Which is larger and by how much: a 10-ft square flower bed or a 12-ft–diameter flower bed?

EXAMPLE 8 *Area of Pizza Pans.* Which is larger and by how much: a pizza made in a 16-in. square pizza pan or a pizza made in a 16-in.–diameter circular pan?

First, we make a drawing of each.

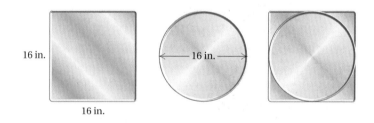

16 in.

16 in.

16 in.

Then we compute areas.
    The area of the square is

$$A = s \cdot s$$
$$= 16 \text{ in.} \times 16 \text{ in.} = 256 \text{ in}^2.$$

The diameter of the circle is 16 in., so the radius is 16 in./2, or 8 in. The area of the circle is

$$A = \pi \cdot r \cdot r$$
$$\approx 3.14 \times 8 \text{ in.} \times 8 \text{ in.} = 200.96 \text{ in}^2.$$

We see that the square pizza is larger by about

$$256 \text{ in}^2 - 200.96 \text{ in}^2, \quad \text{or} \quad 55.04 \text{ in}^2.$$

Thus the pizza made in the square pan is larger, by about 55.04 in^2.

Do Exercise 8.

Answers on page A-21

a , **b** , **c**   For each circle, find the length of a diameter, the circumference, and the area. Use $\frac{22}{7}$ for π.

1.

7 cm

2.

8 m

3.

$\frac{3}{4}$ in.

4.

$8\frac{2}{3}$ mi

For each circle, find the length of a radius, the circumference, and the area. Use 3.14 for π.

5.

32 ft

6.

24 in.

7.

1.4 cm

8.

60.9 km

d   Solve. Use 3.14 for π.

9. *Soda-Can Top.*   The top of a soda can has a 6-cm diameter. What is its radius? its circumference? its area?

6 cm

10. *Penny.*   A penny has a 1-cm radius. What is its diameter? its circumference? its area?

1 cm

11. *Trampoline.*   The standard backyard trampoline has a diameter of 14 ft. What is its area?

Source: International Trampoline Industry Association, Inc.

14 ft

Frame height: 36 in.

12. *Area of Pizza Pans.*   Which is larger and by how much: a pizza made in a 12-in square pizza pan or a pizza made in a 12-in–diameter circular pan?

13. *Dimensions of a Quarter.*   The circumference of a quarter is 7.85 cm. What is the diameter? the radius? the area?

14. *Dimensions of a Dime.*   The circumference of a dime is 2.23 in. What is the diameter? the radius? the area?

15. *Gypsy-Moth Tape.*   To protect an elm tree in your backyard, you need to attach gypsy moth caterpillar tape around the trunk. The tree has a 1.1-ft diameter. What length of tape is needed?

16. *Earth.*   The diameter of the earth at the equator is 7926.41 mi. What is the circumference of the earth at the equator?

Source: *The Handy Geography Answer Book*

17. *Swimming-Pool Walk.*   You want to install a 1-yd–wide walk around a circular swimming pool. The diameter of the pool is 20 yd. What is the area of the walk?

18. *Roller-Rink Floor.*   A roller-rink floor is shown below. Each end is a semicircle. What is its area? If hardwood flooring costs $32.50 per square meter, how much will the flooring cost?

Find the perimeter. Use 3.14 for π.

19.

20.

21.

22.

23.

24.

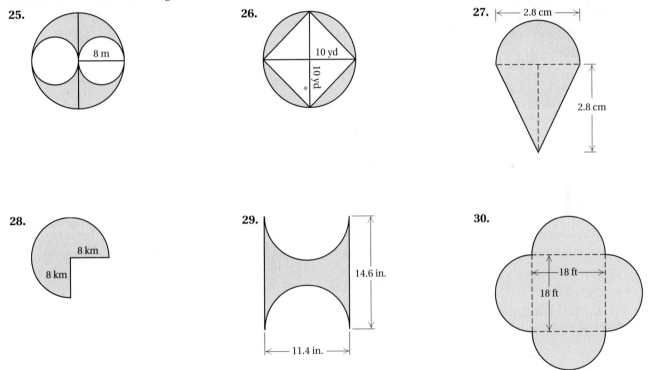

Find the area of the shaded region. Use 3.14 for π.

25.

8 m

26.

10 yd

10 yd

27. 2.8 cm

2.8 cm

28.

8 km

8 km

29.

14.6 in.

11.4 in.

30.

18 ft

18 ft

31. D_W Explain why a 16-in.–diameter pizza that costs $16.25 is a better buy than a 10-in.–diameter pizza that costs $7.85.

32. D_W The radius of one circle is twice the length of another circle's radius. Is the area of the first circle twice the area of the other circle? Why or why not?

Evaluate. [1.9b]

33. 2^4 **34.** 17^2 **35.** 5^3 **36.** 8^2

Complete. [8.2a]

37. 5.43 m = _____ cm **38.** 5.43 m = _____ km

39. Find the rate of discount. [6.6c]

Cookware Set 58^{99}

Lowest Price of the Year
Original price $100

40. Jack's commission is increased according to how much he sells. He receives a commission of 6% for the first $3000 and 10% on the amount over $3000. What is the total commission on sales of $8500? [6.6b]

41. If 2 cans of tomato paste cost $1.49, how many cans of tomato paste can you buy for $7.45? [5.4a]

42. The weight of a human brain is 2.5% of total body weight. A person weighs 200 lb. What does the brain weigh? [6.5a]

43. ▦ $\pi \approx \frac{3927}{1250}$ is another approximation for π. Find decimal notation using a calculator.

44. ▦ The distance from Kansas City to Indianapolis is 500 mi. A car was driven this distance using tires with a radius of 14 in. How many revolutions of each tire occurred on the trip? Use $\frac{22}{7}$ for π.

45. *Tennis Balls.* Tennis balls are generally packed vertically three in a can, one on top of another. Suppose the diameter of a tennis ball is d. Find the height of the stack of balls. Find the circumference of one ball. Which is greater? Explain.

B unce
OFFICIAL BALL

3 tennis balls

540

9.4 VOLUME

a Rectangular Solids

Objectives

a Find the volume of a rectangular solid using the formula $V = l \cdot w \cdot h$.

b Given the radius and the height, find the volume of a circular cylinder.

c Given the radius, find the volume of a sphere.

d Given the radius and the height, find the volume of a circular cone.

e Solve applied problems involving volume of rectangular solids, circular cylinders, spheres, and cones.

The **volume** of a **rectangular solid** is the number of unit cubes needed to fill it.

Unit cube

Volume = 18

Two other units are shown below.

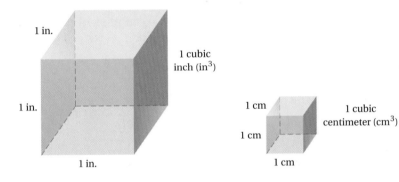

1 in.

1 in.

1 in.

1 cubic inch (in³)

1 cm

1 cm

1 cm

1 cubic centimeter (cm³)

EXAMPLE 1 Find the volume.

2 cm

4 cm

3 cm

The figure is made up of 2 layers of 12 cubes each, so its volume is 24 cubic centimeters (cm³).

Do Exercise 1.

---

1. Find the volume.

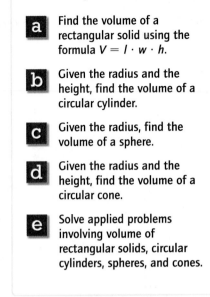

2 cm

3 cm

2 cm

VOLUME OF A RECTANGULAR SOLID

The **volume of a rectangular solid** is found by multiplying length by width by height:

$$V = l \cdot w \cdot h.$$

h

w l

Answer on page A-21

541

2. Popcorn. The average American eats 54 quarts of popcorn annually. This quantity of popcorn would fill a bag measuring 10.5 in. by 10.5 in. by 28 in. Find the volume of such a bag.

Source: The Popcorn Board

EXAMPLE 2 *Carry-on Luggage.* The largest piece of luggage that you can carry on an airplane measures 23 in. by 10 in. by 13 in. Find the volume of this solid.

$$V = l \cdot w \cdot h$$
$$= 23 \text{ in.} \cdot 10 \text{ in.} \cdot 13 \text{ in.}$$
$$= 230 \cdot 13 \text{ in}^3$$
$$= 2990 \text{ in}^3$$

Do Exercises 2 and 3.

b Cylinders

A rectangular solid is shown below. Note that we can think of the volume as the product of the area of the base times the height:

$$V = l \cdot w \cdot h$$
$$= (l \cdot w) \cdot h$$
$$= (\text{Area of the base}) \cdot h$$
$$= B \cdot h,$$

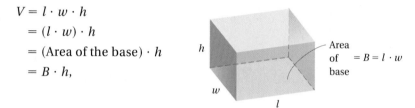

where B represents the area of the base.

Like rectangular solids, **circular cylinders** have bases of equal area that lie in parallel planes. The bases of circular cylinders are circular regions.

3. Cord of Wood. A cord of wood measures 4 ft by 4 ft by 8 ft. What is the volume of a cord of wood?

Answers on page A-21

The volume of a circular cylinder is found in a manner similar to finding the volume of a rectangular solid. The volume is the product of the area of the base times the height. The height is always measured perpendicular to the base.

VOLUME OF A CIRCULAR CYLINDER

The **volume of a circular cylinder** is the product of the area of the base B and the height h:

$$V = B \cdot h, \quad \text{or} \quad V = \pi \cdot r^2 \cdot h.$$

EXAMPLE 3 Find the volume of this circular cylinder. Use 3.14 for π.

12 cm

4 cm

$$V = Bh = \pi \cdot r^2 \cdot h$$
$$\approx 3.14 \times 4 \text{ cm} \times 4 \text{ cm} \times 12 \text{ cm}$$
$$= 602.88 \text{ cm}^3$$

Do Exercises 4 and 5.

C ## Spheres

A **sphere** is the three-dimensional counterpart of a circle. It is the set of all points in space that are a given distance (the radius) from a given point (the center).

We find the volume of a sphere as follows.

VOLUME OF A SPHERE

The **volume of a sphere** of radius r is given by

$$V = \frac{4}{3} \cdot \pi \cdot r^3.$$

EXAMPLE 4 *Bowling Ball.* The radius of a standard-sized bowling ball is 4.2915 in. Find the volume of a standard-sized bowling ball. Round to the nearest hundredth of a cubic inch. Use 3.14 for π.

$r = 4.2915$ in.

4. Find the volume of the cylinder. Use 3.14 for π.

10 ft

5 ft

5. Find the volume of the cylinder. Use $\frac{22}{7}$ for π.

49 m

21 m

6. Find the volume of the sphere. Use $\frac{22}{7}$ for π.

28 ft

7. The radius of a standard-sized golf ball is 2.1 cm. Find its volume. Use 3.14 for π.

Answers on page A-21

8. Find the volume of this cone. Use 3.14 for π.

20 m

9 m

We have

$$V = \frac{4}{3} \cdot \pi \cdot r^3 \approx \frac{4}{3} \times 3.14 \times (4.2915 \text{ in.})^3$$
$$\approx 330.90 \text{ in}^3 \qquad \text{Using a calculator}$$

Do Exercises 6 and 7 on the preceding page.

d | Cones

Consider a circle in a plane and choose any point P not in the plane. The circular region, together with the set of all segments connecting P to a point on the circle, is called a **circular cone.**

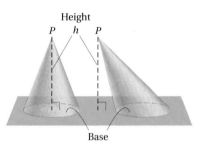

Height

P h P

Base

We find the volume of a cone as follows.

9. Find the volume of this cone. Use $\frac{22}{7}$ for π.

14 in.

6 in.

> **VOLUME OF A CIRCULAR CONE**
>
> The **volume of a circular cone** with base radius r is one-third the product of the base area and the height:
>
> $$V = \frac{1}{3} \cdot B \cdot h = \frac{1}{3}\pi \cdot r^2 \cdot h.$$

EXAMPLE 5   Find the volume of this circular cone. Use 3.14 for π.

$$V = \frac{1}{3}\pi \cdot r^2 \cdot h$$
$$\approx \frac{1}{3} \times 3.14 \times 3 \text{ cm} \times 3 \text{ cm} \times 7 \text{ cm}$$
$$= 65.94 \text{ cm}^3$$

7 cm

3 cm

Answers on page A-21

Do Exercises 8 and 9.

CALCULATOR CORNER

The [π] Key   Many calculators have a [π] key that can be used to enter the value of π in a computation. It might be necessary to press a [2nd] or [SHIFT] key before pressing the [π] key on some calculators. Since 3.14 is a rounded value for π, results obtained using the [π] key might be slightly different from those obtained when 3.14 is used for the value of π in a computation.

To find the volume of the circular cylinder in Example 3, we press [2nd] [π] [×] [4] [×] [4] [×] [1] [2] [=] or [SHIFT] [π] [×] [4] [×] [4] [×] [1] [2] [=]. The result is approximately 603.19. Note that this is slightly different from the result found using 3.14 for the value of π.

Exercises

1. Use a calculator with a [π] key to perform the computations in Examples 4 and 5.
2. Use a calculator with a [π] key to perform the computations in Margin Exercises 5–10.

e | Solving Applied Problems

EXAMPLE 6 *Propane Gas Tank.* A propane gas tank is shaped like a circular cylinder with half of a sphere at each end. Find the volume of the tank if the cylindrical section is 5 ft long with a 4-ft diameter. Use 3.14 for π.

1. **Familiarize.** We first make a drawing.

2. **Translate.** This is a two-step problem. We first find the volume of the cylindrical portion. Then we find the volume of the two ends and add. Note that the radius is 2 ft and that together the two ends make a sphere. We let

$$V = \pi \cdot r^2 \cdot h + \frac{4}{3} \cdot \pi \cdot r^3,$$

where V is the total volume. Then

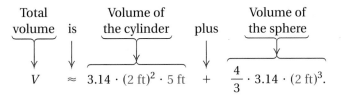

3. **Solve.** The volume of the cylinder is approximately

$$3.14 \cdot (2 \text{ ft})^2 \cdot 5 \text{ ft} = 3.14 \cdot 2 \text{ ft} \cdot 2 \text{ ft} \cdot 5 \text{ ft}$$
$$= 62.8 \text{ ft}^3.$$

The volume of the two ends is approximately

$$\frac{4}{3} \cdot 3.14 \cdot (2 \text{ ft})^3 = \frac{4}{3} \cdot 3.14 \cdot 2 \text{ ft} \cdot 2 \text{ ft} \cdot 2 \text{ ft}$$
$$\approx 33.5 \text{ ft}^3.$$

The total volume is about

$$62.8 \text{ ft}^3 + 33.5 \text{ ft}^3 = 96.3 \text{ ft}^3.$$

4. **Check.** The check is left to the student.

5. **State.** The volume of the tank is about 96.3 ft³.

Do Exercise 10.

---

10. Medicine Capsule. A cold capsule is 8 mm long and 4 mm in diameter. Find the volume of the capsule. Use 3.14 for π. (*Hint*: First find the length of the cylindrical section.)

Answer on page A-21

Study Tips

BEGINNING TO STUDY FOR THE FINAL EXAM (PART 1)

It is never too soon to begin to study for the final examination. Take a few minutes each week to review the highlighted information, such as formulas, properties, and procedures. Make special use of the Summary and Reviews, Chapter Tests, and Cumulative Reviews, as well as supplements such as the Work It Out! Chapter Test Video on CD, MyMathLab, MathXL, or the MathXL Tutorials on CD. The Cumulative Review and Final Examination for Chapters 1–11 is a sample final exam.

"Practice does not make perfect; practice makes permanent."

Dr. Richard Chase, former president, Wheaton College

9.4

EXERCISE SET

For Extra Help

MathXL MyMathLab InterAct Math Tutor Digital Video Student's
                    Math   Center   Tutor CD 4   Solutions
                                         Videotape 9   Manual

a Find the volume.

1.

8 cm

12 cm   8 cm

2.

0.6 m

0.6 m   0.6 m

3.

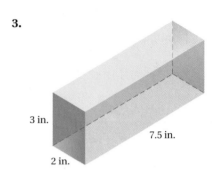

3 in.   7.5 in.

2 in.

4.

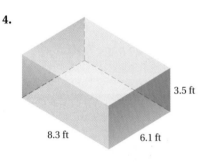

3.5 ft

8.3 ft   6.1 ft

5.

1.5 m

10 m   5 m

6.

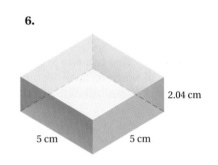

2.04 cm

5 cm   5 cm

7.

10 yd

$5\frac{1}{2}$ yd $6\frac{1}{2}$ yd

8.

$6\frac{1}{4}$ ft

$2\frac{1}{2}$ ft $1\frac{1}{2}$ ft

b   Find the volume of the circular cylinder. Use 3.14 for π in Exercises 9–12. Use $\frac{22}{7}$ for π in Exercises 13 and 14.

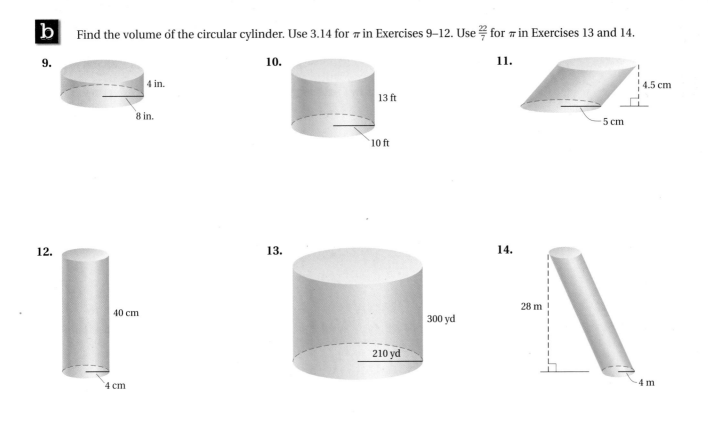

9.   4 in.   8 in.

10.   13 ft   10 ft

11.   4.5 cm   5 cm

12.   40 cm   4 cm

13.   300 yd   210 yd

14.   28 m   4 m

c   Find the volume of the sphere. Use 3.14 for π in Exercises 15–18. Use $\frac{22}{7}$ for π in Exercises 19 and 20.

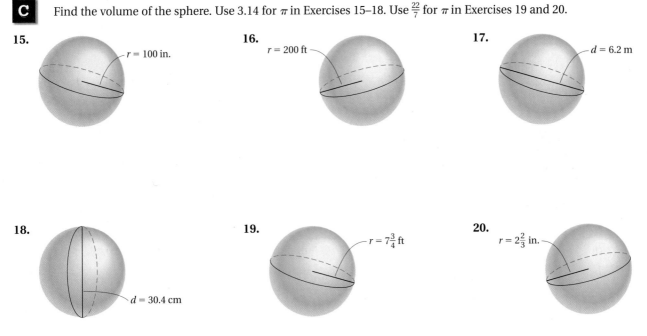

15. $r = 100$ in.

16. $r = 200$ ft

17. $d = 6.2$ m

18. $d = 30.4$ cm

19. $r = 7\frac{3}{4}$ ft

20. $r = 2\frac{2}{3}$ in.

d. Find the volume of the circular cone. Use 3.14 for π in Exercises 21 and 22. Use $\frac{22}{7}$ for π in Exercises 23 and 24.

21.

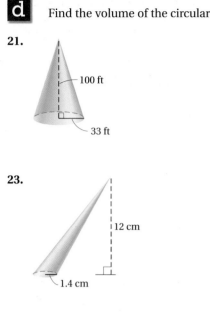

100 ft

33 ft

22.

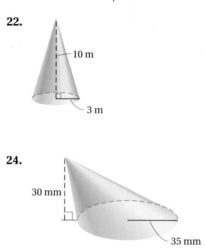

10 m

3 m

23.

12 cm

1.4 cm

24.

30 mm

35 mm

e Solve.

25. *Oak Log.* An oak log has a diameter of 12 cm and a length (height) of 42 cm. Find the volume. Use 3.14 for π.

26. *Ladder Rung.* A rung of a ladder is 2 in. in diameter and 16 in. long. Find the volume. Use 3.14 for π.

27. *Barn Silo.* A barn silo, excluding the top, is a circular cylinder. The silo is 6 m in diameter and the height is 13 m. Find the volume of the silo excluding the top. Use 3.14 for π.

6 m

13 m

28. *Volume of a Trash Can.* The diameter of the base of a cylindrical trash can is 0.7 yd. The height is 1.1 yd. Find the volume. Use 3.14 for π.

29. *Tennis Ball.* The diameter of a tennis ball is 6.5 cm. Find the volume. Use 3.14 for π.

30. *Spherical Gas Tank.* The diameter of a spherical gas tank is 6 m. Find the volume. Use 3.14 for π.

31. *Volume of Earth.* The diameter of the earth is about 3980 mi. Find the volume of the earth. Use 3.14 for π. Round to the nearest ten thousand cubic miles.

32. *Astronomy.* The radius of Pluto's moon is about 500 km. Find the volume of this satellite. Use $\frac{22}{7}$ for π.

33. *Tennis-Ball Packaging.* Tennis balls are generally packaged in circular cylinders that hold 3 balls each. The diameter of a tennis ball is 6.5 cm. Find the volume of a can of tennis balls. Use 3.14 for π.

34. *Golf-Ball Packaging.* The box shown is just big enough to hold 3 golf balls. If the radius of a golf ball is 2.1 cm, how much air surrounds the three balls? Use 3.14 for π.

35. *Water Storage.* A water storage tank is a right circular cylinder with a radius of 14 cm and a height of 100 cm. What is the tank's volume? Use $\frac{22}{7}$ for π.

36. *Oceanography.* A research submarine is capsule-shaped. Find the volume of the submarine if it has a length of 10 m and a diameter of 8 m. Use 3.14 for π. (*Hint*: First find the length of the cylindrical section.)

37. *Metallurgy.* If all the gold in the world could be gathered together, it would form a cube 18 yd on a side. Find the volume of the world's gold.

38. The volume of a ball is 36π cm^3. Find the dimensions of a rectangular box that is just large enough to hold the ball.

39. **D**_W How could you use the volume formulas given in this section to help estimate the volume of an egg?

40. **D**_W The design of a modern home includes a cylindrical tower that will be capped with either a 10-ft–high dome or a 10-ft–high cone. Which type of cap will be more energy-efficient and why?

Complete. [8.1a]

41. 11 yd = _____ in.

42. 15,840 ft = _____ mi

43. 42 ft = _____ yd

44. 48 mi = _____ ft

45. 144 in. = _____ ft

46. 5.3 mi = _____ in.

Complete. [8.5a]

47. 6 gal = _____ qt

48. 56 qt = _____ gal

49. 566 mL = _____ L

50. 13.4 L = _____ cm^3

SYNTHESIS

51. ▦ The width of a dollar bill is 2.3125 in., the length is 6.0625 in., and the thickness is 0.0041 in. Find the volume occupied by one million one-dollar bills.

© 1998 AL SATTERWHITE

52. ▦ Audio-cassette cases are typically 7 cm by 10.75 cm by 1.5 cm and contain 90 min of music. Compact-disc cases are typically 12.4 cm by 14.1 cm by 1 cm and contain 50 min of music. Which container holds the most music per cubic centimeter?

53. ▦ A 2-cm–wide stream of water passes through a 30-m–long garden hose. At the instant that the water is turned off, how many liters of water are in the hose? Use 3.141593 for π.

54. ▦ The volume of a basketball is 2304π cm³. Find the volume of a cube-shaped box that is just large enough to hold a ball.

55. *Circumference of Earth.*   The circumference of the earth at the equator is about 24,901.55 mi. Due to the irregular shape of the earth, the circumference of a circle of longitude wrapped around the earth between the north and south poles is about 24,859.82 mi. Describe and carry out a procedure for estimating the volume of the earth.

Source: *The Handy Geography Answer Book*

56. ▦ A sphere with diameter 1 m is circumscribed by a cube. How much more volume is in the cube?

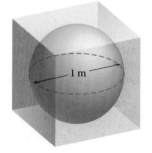

57. ▦ A cube is circumscribed by a sphere with a 1-m diameter. How much more volume is in the sphere?

550

9.5 ANGLES AND TRIANGLES

a Measuring Angles

We see a real-world application of *angles* of various types in the spokes of these bicycles and the different back postures of the riders.

Style of Biking Determines Cycling Posture

Road
About 180° flat

Mountain
About 45°

Comfort
About 90°

Riders prefer a more aerodynamic flat-back position.

Riders prefer a semi-upright position to help lift the front wheel over obstacles.

Riders prefer an upright position that lessens stress on the lower back and neck.

Source: USA TODAY research

An **angle** is a set of points consisting of two **rays,** or half-lines, with a common endpoint. The endpoint is called the **vertex.**

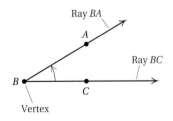

The rays are called the *sides*. The angle above can be named

angle *ABC*,   angle *CBA*,   ∠*ABC*,   ∠*CBA*,   or   ∠*B*.

Note that the name of the vertex is either in the middle or, if no confusion results, listed by itself.

Do Exercises 1 and 2.

Measuring angles is similar to measuring segments. To measure angles, we start with some arbitrary angle and assign to it a measure of 1. We call it a *unit angle*. Suppose that ∠*U*, shown on the following page, is a unit angle. Let's measure ∠*DEF*. If we made 3 copies of ∠*U*, they would "fill up" ∠*DEF*. Thus the measure of ∠*DEF* would be 3.

Objectives

a Name an angle in five different ways and given an angle, measure it with a protractor.

b Classify an angle as right, straight, acute, or obtuse.

c Identify complementary and supplementary angles and find the measure of a complement or a supplement of a given angle.

d Classify a triangle as equilateral, isosceles, or scalene, and as right, obtuse, or acute.

e Given two of the angle measures of a triangle, find the third.

Name the angle in five different ways.

1.

2.

Answers on page A-21

551

3. Use a protractor to measure this angle.

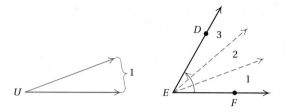

The unit most commonly used for angle measure is the degree. Below is such a unit. Its measure is 1 degree, or 1°.

A 1° angle:

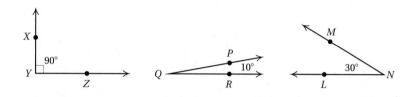

Here are some other angles with their degree measures.

To indicate the *measure* of ∠*XYZ*, we write *m*∠*XYZ* = 90°. The symbol ⌐ is sometimes drawn on a figure to indicate a 90° angle.

A device called a **protractor** is used to measure angles. Protractors have two scales. To measure an angle like ∠*Q* below, we place the protractor's ▲ at the vertex and line up one of the angle's sides at 0°. Then we check where the angle's other side crosses the scale. In the figure below, 0° is on the inside scale, so we check where the angle's other side crosses the inside scale. We see that *m*∠*Q* = 145°. The notation *m*∠*Q* is read "the measure of angle *Q*."

Do Exercise 3.

Answer on page A-21

Let's find the measure of $\angle ABC$. This time we will use the 0° on the outside scale. We see that $m \angle ABC = 42°$.

Do Exercise 4.

b   Classifying Angles

The following are ways in which we classify angles.

TYPES OF ANGLES

Right angle: An angle whose measure is 90°.

Straight angle: An angle whose measure is 180°.

Acute angle: An angle whose measure is greater than 0° and less than 90°.

Obtuse angle: An angle whose measure is greater than 90° and less than 180°.

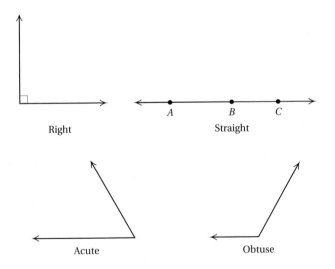

Do Exercises 5–8.

4. Use a protractor to measure this angle.

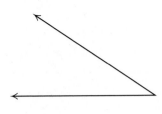

Classify the angle as right, straight, acute, or obtuse. Use a protractor if necessary.

5.

6.

7.

8.

Answers on page A-21

9. Identify each pair of complementary angles.

Find the measure of a complement of the angle.

10.

11.

12.

Answers on page A-21

C Complementary and Supplementary Angles

$\angle 1$ and $\angle 2$ above are **complementary** angles.

$$m\angle 1 + m\angle 2 = 90°$$
$$75° + 15° = 90°$$

COMPLEMENTARY ANGLES

Two angles are **complementary** if the sum of their measures is 90°. Each angle is called a **complement** of the other.

If two angles are complementary, each is an acute angle. When complementary angles are adjacent to each other, they form a right angle.

EXAMPLE 1 Identify each pair of complementary angles.

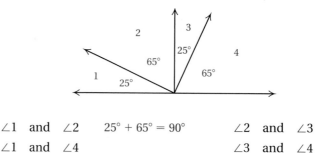

| $\angle 1$ | and | $\angle 2$ | $25° + 65° = 90°$ | $\angle 2$ | and | $\angle 3$ |
| $\angle 1$ | and | $\angle 4$ | | $\angle 3$ | and | $\angle 4$ |

EXAMPLE 2 Find the measure of a complement of an angle of 39°.

$$90° - 39° = 51°$$

The measure of a complement is 51°.

Do Exercises 9–12.

Next, consider $\angle 1$ and $\angle 2$ as shown below. Because the sum of their measures is 180°, $\angle 1$ and $\angle 2$ are said to be **supplementary.** Note that when supplementary angles are adjacent, they form a straight angle.

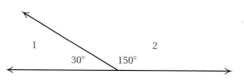

$$m\angle 1 + m\angle 2 = 180°$$
$$30° + 150° = 180°$$

SUPPLEMENTARY ANGLES

Two angles are **supplementary** if the sum of their measures is 180°. Each angle is called a **supplement** of the other.

EXAMPLE 3 Identify each pair of supplementary angles.

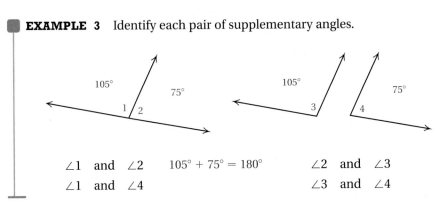

∠1 and ∠2   105° + 75° = 180°   ∠2 and ∠3
∠1 and ∠4   ∠3 and ∠4

EXAMPLE 4 Find the measure of a supplement of an angle of 112°.

(180 − 112)°   112°

180° − 112° = 68°

The measure of a supplement is 68°.

Do Exercises 13–16.

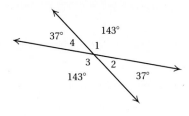

Find the measure of a supplement of an angle with the given measure.

14. 38°

d Triangles

A **triangle** is a polygon made up of three segments, or sides. Consider these triangles. The triangle with vertices *A*, *B*, and *C* can be named △*ABC*.

15. 157°

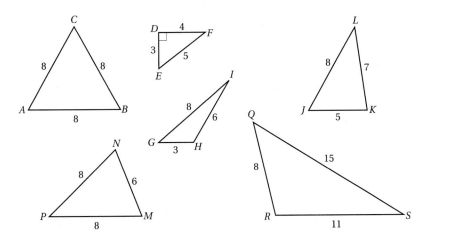

16. 90°

Answers on page A-21

17. Which triangles on p. 555 are:

a) equilateral?

b) isosceles?

c) scalene?

18. Are all equilateral triangles isosceles?

19. Are all isosceles triangles equilateral?

20. Which triangles on p. 555 are:

a) right triangles?

b) obtuse triangles?

c) acute triangles?

21. Find $m(\angle P) + m(\angle Q) + m(\angle R)$.

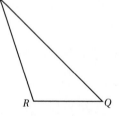

22. Find the missing angle measure.

Answers on page A-21

We can classify triangles according to sides and according to angles.

TYPES OF TRIANGLES

Equilateral triangle: All sides are the same length.

Isosceles triangle: Two or more sides are the same length.

Scalene triangle: All sides are of different lengths.

Right triangle: One angle is a right angle.

Obtuse triangle: One angle is an obtuse angle.

Acute triangle: All three angles are acute.

Do Exercises 17–20.

e Sum of the Angle Measures of a Triangle

The sum of the angle measures of a triangle is 180°. To see this, note that we can think of cutting apart a triangle as shown on the left below. If we re-assemble the pieces, we see that a straight angle is formed.

$$64° + 32° + 84° = 180°$$

SUM OF THE ANGLE MEASURES OF A TRIANGLE

In any triangle ABC, the sum of the measures of the angles is 180°:

$$m(\angle A) + m(\angle B) + m(\angle C) = 180°.$$

Do Exercise 21.

If we know the measures of two angles of a triangle, we can calculate the third.

EXAMPLE 5 Find the missing angle measure.

$$m(\angle A) + m(\angle B) + m(\angle C) = 180°$$
$$x + 65° + 24° = 180°$$
$$x + 89° = 180°$$
$$x = 180° - 89°$$
$$x = 91°$$

Do Exercise 22.

9.5

EXERCISE SET

For Extra Help

MathXL   MyMathLab   InterAct   Math Tutor   Digital Video   Student's
                       Math   Center   Tutor CD 4   Solutions
                                                 Videotape 9   Manual

a Name the angle in five different ways.

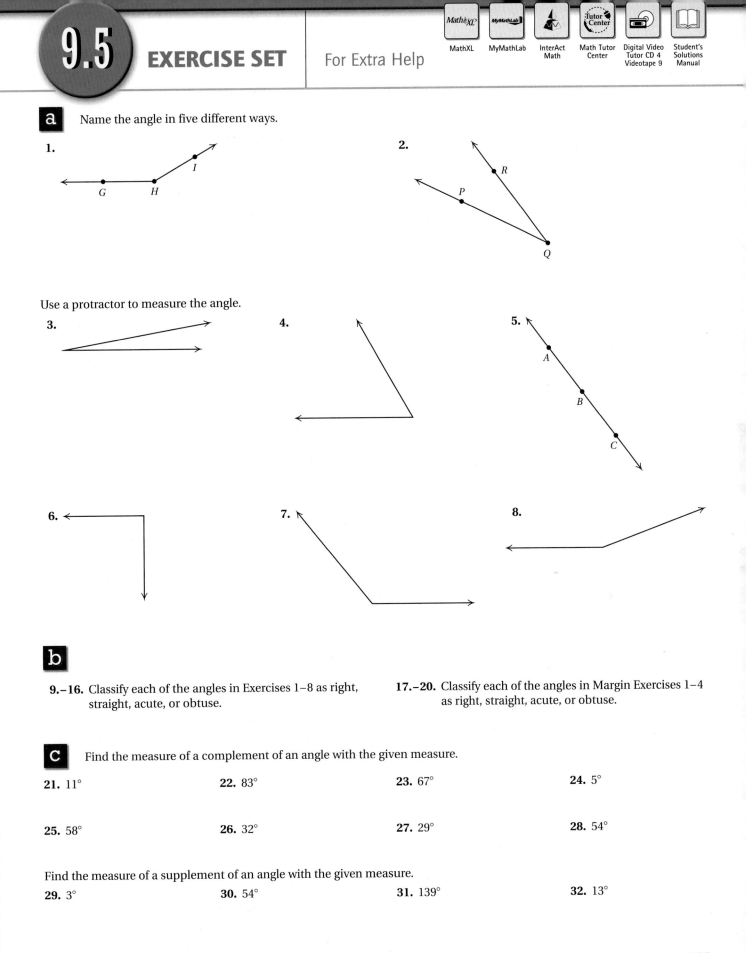

1.

2.

Use a protractor to measure the angle.

3.

4.

5.

6.

7.

8.

b

9.–16. Classify each of the angles in Exercises 1–8 as right, straight, acute, or obtuse.

17.–20. Classify each of the angles in Margin Exercises 1–4 as right, straight, acute, or obtuse.

c Find the measure of a complement of an angle with the given measure.

21. 11°

22. 83°

23. 67°

24. 5°

25. 58°

26. 32°

27. 29°

28. 54°

Find the measure of a supplement of an angle with the given measure.

29. 3°

30. 54°

31. 139°

32. 13°

33. 85°   **34.** 129°   **35.** 102°   **36.** 45°

d Classify the triangle as equilateral, isosceles, or scalene. Then classify it as right, obtuse, or acute.

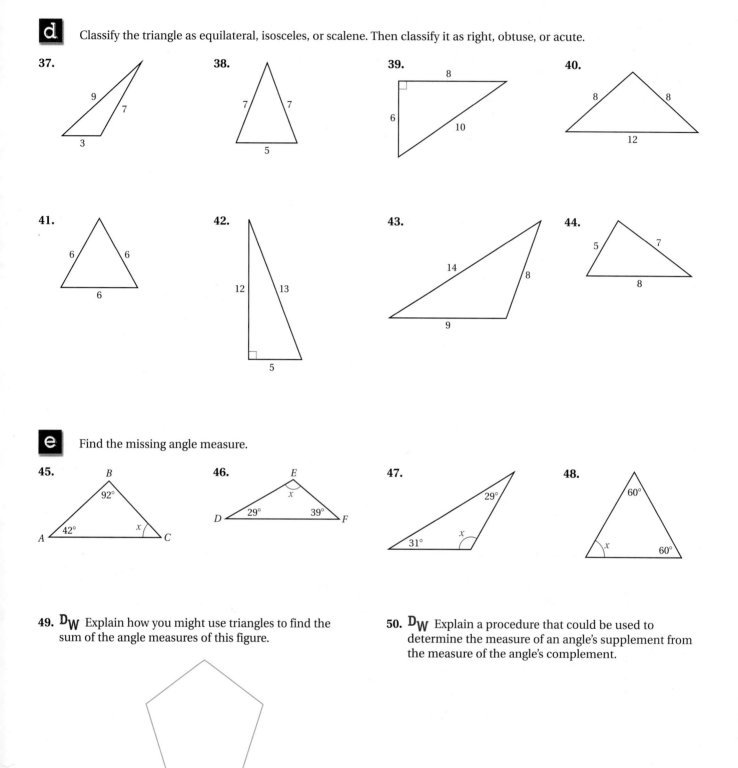

37.

9

7

3

38.

7   7

5

39.

8

6

10

40.

8   8

12

41.

6   6

6

42.

12   13

5

43.

14

8

9

44.

5   7

8

e Find the missing angle measure.

45.
B
92°
42°   x
A   C

46.
E
x
D   29°   39°   F

47.
29°
x
31°

48.
60°
x
60°

49. **DW** Explain how you might use triangles to find the sum of the angle measures of this figure.

50. **DW** Explain a procedure that could be used to determine the measure of an angle's supplement from the measure of the angle's complement.

VOCABULARY REINFORCEMENT

In each of Exercises 51–58, fill in the blank with the correct term from the given list. Some of the choices may not be used.

51. A number is divisible by 8 if the number named by the last _____ digits is divisible by 8.   [2.2a]

52. A parallelogram is a four-sided figure with two pairs of _____ sides.   [9.2b]

53. In the metric system, the _____ is the basic unit of mass.   [8.4b]

54. A natural number, other than 1, that is not _____ is composite.   [2.1c]

55. A(n) _____ is a set of points consisting of two rays.   [9.5a]

56. To convert from _____ to _____ , move the decimal point two places to the left and change the ¢ sign at the end to the $ sign in front.   [4.3b]

57. The _____ of a polygon is the sum of the lengths of its sides.   [9.1a]

58. The number 1 is known as the _____ identity, and the number 0 is known as the _____ identity.   [1.2a], [1.5a]

prime
composite
two
three
dollars
cents
perimeter
parallel
perpendicular
additive
multiplicative
meter
gram
vertex
angle
area

SYNTHESIS

59. ▦ In the figure, $m\angle 1 = 79.8°$ and $m\angle 6 = 33.07°$. Find $m\angle 2$, $m\angle 3$, $m\angle 4$, and $m\angle 5$.

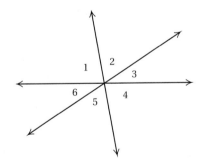

60. ▦ In the figure, $m\angle 2 = 42.17°$ and $m\angle 3 = 81.9°$. Find $m\angle 1$, $m\angle 4$, $m\angle 5$, and $m\angle 6$.

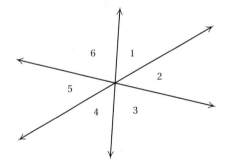

61. Find $m\angle ACB$, $m\angle CAB$, $m\angle EBC$, $m\angle EBA$, $m\angle AEB$, and $m\angle ADB$ in the rectangle shown below.

Objectives

a Simplify square roots of squares such as $\sqrt{25}$.

b Approximate square roots.

c Given the lengths of any two sides of a right triangle, find the length of the third side.

d Solve applied problems involving right triangles.

Find the square. (See Section 1.9.)

1. 9^2 **2.** 10^2

3. 11^2 **4.** 12^2

> It would be helpful to memorize the squares of numbers from 1 to 25. We call these "perfect" squares.

5. 13^2 **6.** 14^2

7. 15^2 **8.** 16^2

9. 17^2 **10.** 18^2

11. 20^2 **12.** 25^2

Simplify. The results of Exercises 1–12 above may be helpful here.

13. $\sqrt{9}$ **14.** $\sqrt{16}$

15. $\sqrt{121}$ **16.** $\sqrt{100}$

17. $\sqrt{81}$ **18.** $\sqrt{64}$

19. $\sqrt{324}$ **20.** $\sqrt{400}$

21. $\sqrt{225}$ **22.** $\sqrt{169}$

23. $\sqrt{1}$ **24.** $\sqrt{0}$

Answers on page A-22

9.6 SQUARE ROOTS AND THE PYTHAGOREAN THEOREM

a Square Roots

SQUARE ROOT

If a number is a product of two identical factors, then either factor is called a **square root** of the number. (If $a = c^2$, then c is a square root of a.) The symbol $\sqrt{}$ (called a **radical sign**) is used in naming square roots.

For example, $\sqrt{36}$ is the square root of 36. It follows that

$$\sqrt{36} = \sqrt{6 \cdot 6} = 6 \qquad \text{The square root of 36 is 6.}$$

because $6^2 = 36$.

EXAMPLE 1 Simplify: $\sqrt{25}$.

$$\sqrt{25} = \sqrt{5 \cdot 5} = 5 \qquad \text{The square root of 25 is 5 because } 5^2 = 25.$$

EXAMPLE 2 Simplify: $\sqrt{144}$.

$$\sqrt{144} = \sqrt{12 \cdot 12} = 12 \qquad \text{The square root of 144 is 12 because } 12^2 = 144.$$

Caution!

It is common to confuse squares and square roots. A number squared is that number multiplied by itself. For example, $16^2 = 16 \cdot 16 = 256$. A square root of a number is a number that when multiplied by itself gives the original number. For example, $\sqrt{16} = 4$, because $4 \cdot 4 = 16$.

EXAMPLES Simplify.

3. $\sqrt{4} = 2$ **4.** $\sqrt{256} = 16$ **5.** $\sqrt{361} = 19$

Do Exercises 1–24.

b Approximating Square Roots

Many square roots can't be written as whole numbers or fractions. For example,

$$\sqrt{2}, \qquad \sqrt{3}, \qquad \sqrt{39}, \quad \text{and} \quad \sqrt{70}$$

cannot be precisely represented in decimal notation. To see this, consider the following decimal approximations for $\sqrt{2}$. Each gives a closer approximation, but none is exactly $\sqrt{2}$:

$$\sqrt{2} \approx 1.4 \quad \text{because} \quad (1.4)^2 = 1.96;$$
$$\sqrt{2} \approx 1.41 \quad \text{because} \quad (1.41)^2 = 1.9881;$$
$$\sqrt{2} \approx 1.414 \quad \text{because} \quad (1.414)^2 = 1.999396;$$
$$\sqrt{2} \approx 1.4142 \quad \text{because} \quad (1.4142)^2 = 1.99996164.$$

Decimal approximations like these are commonly found by using a calculator.

EXAMPLE 6 Approximate $\sqrt{3}$, $\sqrt{27}$, and $\sqrt{180}$ to three decimal places. Use a calculator.

We use a calculator to find each square root. Since more than three decimal places are given, we round back to three places.

$$\sqrt{3} \approx 1.732,$$
$$\sqrt{27} \approx 5.196,$$
$$\sqrt{180} \approx 13.416$$

As a check, note that $1 \cdot 1 = 1$ and $2 \cdot 2 = 4$, so we expect $\sqrt{3}$ to be between 1 and 2. Similarly, we expect $\sqrt{27}$ to be between 5 and 6 and $\sqrt{180}$ to be between 13 and 14.

Do Exercises 25–27 on the following page.

C   The Pythagorean Theorem

A **right triangle** is a triangle with a 90° angle, as shown here.

In a right triangle, the longest side is called the **hypotenuse.** It is also the side opposite the right angle. The other two sides are called **legs.** We generally use the letters a and b for the lengths of the legs and c for the length of the hypotenuse. They are related as follows.

THE PYTHAGOREAN THEOREM

In any right triangle, if a and b are the lengths of the legs and c is the length of the hypotenuse, then

$$a^2 + b^2 = c^2, \quad \text{or}$$
$$(\text{Leg})^2 + (\text{Other leg})^2 = (\text{Hypotenuse})^2.$$

The equation $a^2 + b^2 = c^2$ is called the **Pythagorean equation.***

---

*The *converse* of the Pythagorean theorem is also true. That is, if $a^2 + b^2 = c^2$, then the triangle is a right triangle.

Approximate to three decimal places.

25. $\sqrt{5}$

The Pythagorean theorem is named for the Greek mathematician Pythagoras (569?–500? B.C.). We can think of this relationship as adding areas.

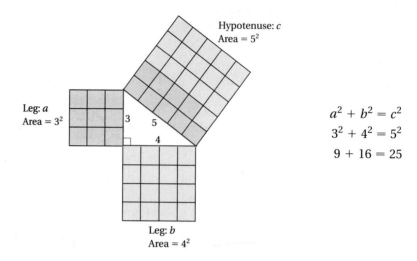

$$a^2 + b^2 = c^2$$
$$3^2 + 4^2 = 5^2$$
$$9 + 16 = 25$$

26. $\sqrt{78}$

If we know the lengths of any two sides of a right triangle, we can use the Pythagorean equation to determine the length of the third side.

EXAMPLE 7 Find the length of the hypotenuse of this right triangle.

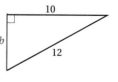

We substitute in the Pythagorean equation:

$$a^2 + b^2 = c^2$$
$$6^2 + 8^2 = c^2 \qquad \text{Substituting}$$
$$36 + 64 = c^2$$
$$100 = c^2.$$

27. $\sqrt{168}$

The solution of this equation is the square root of 100, which is 10:

$$c = \sqrt{100} = 10.$$

Do Exercise 28.

28. Find the length of the hypotenuse of this right triangle.

EXAMPLE 8 Find the length b for the right triangle shown. Give an exact answer and an approximation to three decimal places.

We substitute in the Pythagorean equation. Next, we solve for b^2 and then b, as follows:

$$a^2 + b^2 = c^2$$
$$10^2 + b^2 = 12^2 \qquad \text{Substituting}$$
$$100 + b^2 = 144.$$

Answers on page A-22

Then

$$100 + b^2 - 100 = 144 - 100$$ Subtracting 100 on both sides

$$b^2 = 144 - 100$$

$$b^2 = 44$$

Exact answer: $b = \sqrt{44}$

Approximation: $b \approx 6.633.$ Using a calculator

Do Exercises 29–31.

d | Applications

EXAMPLE 9   *Height of Ladder.*   A 12-ft ladder leans against a building. The bottom of the ladder is 7 ft from the building. How high is the top of the ladder? Give an exact answer and an approximation to the nearest tenth of a foot.

1. **Familiarize.**   We first make a drawing. In it we see a right triangle. We let h = the unknown height.

2. **Translate.**   We substitute 7 for a, h for b, and 12 for c in the Pythagorean equation:

$$a^2 + b^2 = c^2$$ Pythagorean equation

$$7^2 + h^2 = 12^2.$$

3. **Solve.**   We solve for h^2 and then h:

$$49 + h^2 = 144$$

$$49 + h^2 - 49 = 144 - 49$$

$$h^2 = 144 - 49$$

$$h^2 = 95$$

Exact answer: $h = \sqrt{95}$

Approximation: $h \approx 9.7$ ft.

4. **Check.** $7^2 + \left(\sqrt{95}\right)^2 = 49 + 95 = 144 = 12^2.$

5. **State.**   The top of the ladder is $\sqrt{95}$, or about 9.7 ft from the ground.

Do Exercise 32.

Find the length of the leg of the right triangle. Give an exact answer and an approximation to three decimal places.

29.

30.

31.

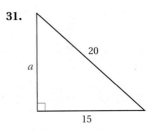

32. How long is a guy wire reaching from the top of an 18-ft pole to a point on the ground 10 ft from the pole? Give an exact answer and an approximation to the nearest tenth of a foot.

Answers on page A-22

Translating for Success

1. *Servings of Pork.* An 8-lb pork roast contains 37 servings of meat. How many pounds of pork would be needed for 55 servings?

2. *Height of a Ladder.* A 14.5-ft ladder leans against a house. The bottom of the ladder is 9.4 ft from the building. How high is the top of the ladder?

3. *Cruise Cost.* A group of 6 college students pays $4608 for a spring break cruise. What is each person's share?

4. *Sales Tax Rate.* The sales tax is $14.95 on the purchase of a new ladder which costs $299. What is the sales tax rate?

5. *Volume of a Sphere.* Find the volume of a sphere whose radius is 7.2 cm.

The goal of these matching questions is to practice step (2), *Translate*, of the five-step problem-solving process. Translate each word problem to an equation and select a correct translation from equations A–O.

A. $x = \dfrac{1}{3}\pi \cdot 6^2 \cdot (7.2)$

B. $6 \cdot x = \$4608$

C. $x = \dfrac{4}{3} \cdot \pi \cdot 6^2 \cdot (7.2)$

D. $x = \pi \cdot \left(5\dfrac{1}{2} \div 2\right)^2 \cdot 7$

E. $x = 6\% \times 5 \times \$14.95$

F. $x = \dfrac{1}{3}\pi\left(5\dfrac{1}{2}\right)^2$

G. $(9.4)^2 + x^2 = (14.5)^2$

H. $\$14.95 = x \cdot \$299$

I. $x = 2(14.5 + 9.4)$

J. $(9.4 + 14.5)^2 = x$

K. $\dfrac{8}{37} = \dfrac{x}{55}$

L. $x = 4(14.5 + 9.4)$

M. $x = 6 \cdot \$4608$

N. $8 \cdot 37 = 55 \cdot x$

O. $x = \dfrac{4}{3}\pi(7.2)^3$

Answers on page A-22

6. *Inheritance.* Each of 6 children inherit $4608 from their mother's estate. What is the total inheritance?

7. *Sales Tax.* Erica buys 5 pairs of earrings at $14.95 each. The sales tax rate in Indiana is 6%. How much sales tax will be charged?

8. *Volume of a Cone.* Find the volume of a circular cone with a 6-cm base radius and a height of 7.2 cm.

9. *Volume of a Storage Tank.* The diameter of a cylindrical grain-storage tank is $5\frac{1}{2}$ yd. Its height is 7 yd. Find its volume.

10. *Perimeter of a Photo.* A rectangular photo is 14.5 cm by 9.4 cm. What is the perimeter of the photo?

a Simplify.

1. $\sqrt{100}$

2. $\sqrt{25}$

3. $\sqrt{441}$

4. $\sqrt{225}$

5. $\sqrt{625}$

6. $\sqrt{576}$

7. $\sqrt{361}$

8. $\sqrt{484}$

9. $\sqrt{529}$

10. $\sqrt{169}$

11. $\sqrt{10,000}$

12. $\sqrt{4,000,000}$

b Approximate to three decimal places.

13. $\sqrt{48}$

14. $\sqrt{17}$

15. $\sqrt{8}$

16. $\sqrt{3}$

17. $\sqrt{18}$

18. $\sqrt{7}$

19. $\sqrt{6}$

20. $\sqrt{61}$

21. $\sqrt{10}$

22. $\sqrt{21}$

23. $\sqrt{75}$

24. $\sqrt{220}$

25. $\sqrt{196}$

26. $\sqrt{123}$

27. $\sqrt{183}$

28. $\sqrt{300}$

c Find the length of the third side of the right triangle. Give an exact answer and, where appropriate, an approximation to three decimal places.

29.

30.

31.

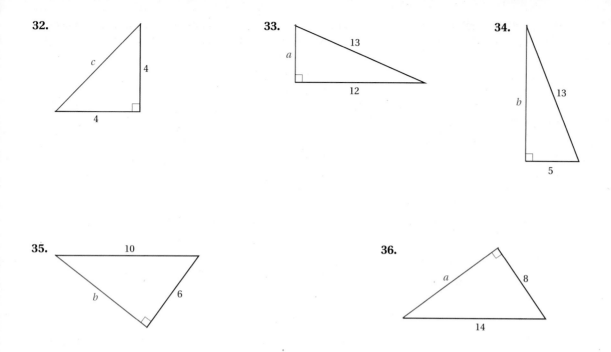

32.

c 4 4

33.

a 13 12

34.

b 13 5

35.

10 b 6

36.

a 8 14

For each right triangle, find the length of the side not given. Assume that c represents the length of the hypotenuse. Give an exact answer and, when appropriate, an approximation to three decimal places.

37. $a = 10, b = 24$

38. $a = 5, b = 12$

39. $a = 9, c = 15$

40. $a = 18, c = 30$

41. $a = 1, c = 32$

42. $b = 1, c = 20$

43. $a = 4, b = 3$

44. $a = 1, c = 15$

d   In Exercises 45–54, give an exact answer and an approximation to the nearest tenth.

45. How long must a wire be in order to reach from the top of a 13-m telephone pole to a point on the ground 9 m from the base of the pole?

46. How long is a light cord reaching from the top of a 12-ft pole to a point on the ground 8 ft from the base of the pole?

47. *Softball Diamond.* A slow-pitch softball diamond is actually a square 65 ft on a side. How far is it from home plate to second base?

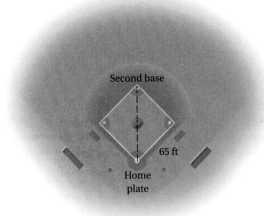

Second base

65 ft

Home plate

48. *Baseball Diamond.* A baseball diamond is actually a square 90 ft on a side. How far is it from home plate to second base?

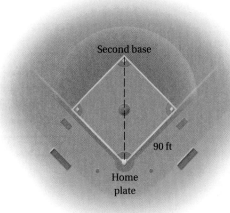

Second base

90 ft

Home plate

49. How tall is this tree?

30 ft

h

20 ft

50. How far is the base of the fence post from point A?

8 ft

5 ft

A

a

51. An airplane is flying at an altitude of 4100 ft. The slanted distance directly to the airport is 15,100 ft. How far is the airplane horizontally from the airport?

4100 ft

15,100 ft

?

52. A surveyor had poles located at points P, Q, and R around a lake. The distances that the surveyor was able to measure are marked on the drawing. What is the approximate distance from P to R across the lake?

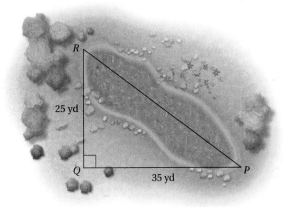

R

25 yd

Q

35 yd

P

53. How long is a string of lights reaching from the top of a 24-ft pole to a point on the ground 16 ft from the base of the pole?

54. How long must a wire be in order to reach from the top of a 26-m telephone pole to a point on the ground 18 m from the base of the pole?

55. **D**_W Explain how the Pythagorean theorem can be used to prove that a triangle is a *right* triangle.

56. **D**_W Write a problem similar to Exercises 45–54 for a classmate to solve. Design the problem so that its solution involves the length $\sqrt{58}$ m.

SKILL MAINTENANCE

Evaluate.   [1.9b]

57. 10^3 **58.** 10^2 **59.** 10^5 **60.** 10^4

Find the simple interest.   [6.7a]

	PRINCIPAL	RATE OF INTEREST	TIME	SIMPLE INTEREST
61.	$2,600	8%	1 year	
62.	$750	6%	$\frac{1}{4}$ year	
63.	$20,600	6.7%	$\frac{1}{12}$ year	

Interest is compounded annually. Find the amount in the account after the given length of time. Round to the nearest cent. [6.7b]

	PRINCIPAL	RATE OF INTEREST	TIME	AMOUNT IN THE ACCOUNT
64.	$200,000	$6\frac{5}{8}$%	25 years	
65.	$150,000	8.4%	30 years	

SYNTHESIS

66. 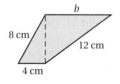 Find the area of the trapezoid shown. Round to the nearest hundredth.

67. Which of the triangles below has the larger area?

68. *Television Screen.*   A 19-in. television set has a rectangular screen that measures 19 in. diagonally. The ratio of length to width in a conventional television set is 4 to 3. Find the length and the width of the screen.

69. *Television Screen.*   A Philips 42-in. plasma television has a rectangular screen that measures 42 in. diagonally. The ratio of length to width is 16 to 9. Find the length and the width of the screen.

The review that follows is meant to prepare you for a chapter exam. It consists of three parts. The first part, Concept Reinforcement, is designed to increase understanding of the concepts through true/false exercises. The second part is a list of important properties and formulas. The third part is the Review Exercises. These provide practice exercises for the exam, together with references to section objectives so you can go back and review. Before beginning, stop and look back over the skills you have obtained. What skills in mathematics do you have now that you did not have before studying this chapter?

CONCEPT REINFORCEMENT

Determine whether the statement is true or false. Answers are given at the back of the book.

_____ **1.** The area of a square is four times the length of a side.

_____ **2.** The area of a parallelogram with base 8 cm and height 5 cm is the same as the area of a rectangle with length 8 cm and width 5 cm.

_____ **3.** The acute angles of a right triangle are complementary.

_____ **4.** Two angles are supplementary if the sum of their measures is between 90° and 180°.

_____ **5.** The measure of any obtuse angle is larger than the measure of any acute angle.

_____ **6.** Some right triangles are equilateral triangles.

IMPORTANT PROPERTIES AND FORMULAS

Perimeter of a Rectangle:	$P = 2 \cdot (l + w)$, or $P = 2 \cdot l + 2 \cdot w$
Perimeter of a Square:	$P = 4 \cdot s$
Area of a Rectangle:	$A = l \cdot w$
Area of a Square:	$A = s \cdot s$, or $A = s^2$
Area of a Parallelogram:	$A = b \cdot h$
Area of a Triangle:	$A = \frac{1}{2} \cdot b \cdot h$
Area of a Trapezoid:	$A = \frac{1}{2} \cdot h \cdot (a + b)$
Radius and Diameter of a Circle:	$d = 2 \cdot r$, or $r = \frac{d}{2}$
Circumference of a Circle:	$C = \pi \cdot d$, or $C = 2 \cdot \pi \cdot r$
Area of a Circle:	$A = \pi \cdot r \cdot r$, or $A = \pi \cdot r^2$
Volume of a Rectangular Solid:	$V = l \cdot w \cdot h$
Volume of a Circular Cylinder:	$V = \pi \cdot r^2 \cdot h$
Volume of a Sphere:	$V = \frac{4}{3} \cdot \pi \cdot r^3$
Volume of a Cone:	$V = \frac{1}{3} \cdot \pi \cdot r^2 \cdot h$
Sum of Angle Measures of a Triangle:	$m(\angle A) + m(\angle B) + m(\angle C) = 180°$
Pythagorean Equation:	$a^2 + b^2 = c^2$

Review Exercises

Find the perimeter.   [9.1a]

1.
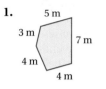
5 m
3 m
7 m
4 m
4 m

2.
0.5 m   1.9 m
0.8 m
1.2 m

3. *Tennis Court.*   The dimensions of a standard-sized tennis court are 78 ft by 36 ft. Find the perimeter and the area of the tennis court.   [9.1b], [9.2c]

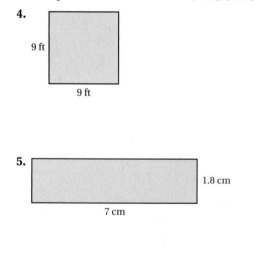

36 ft

78 ft

Find the perimeter and the area.   [9.1a], [9.2a]

4.

9 ft

9 ft

5.

1.8 cm

7 cm

Find the area.   [9.2b]

6.

5 cm

12 cm

7.

4 mm

5 mm

10 mm

8.

3 m

15 m

9.

5.2 cm

11.4 cm

10.

5 m

8 m

17 m

11.

$6\frac{2}{3}$ in.

$21\frac{5}{6}$ in.

12. *Seeded Area.* A grassy area is to be seeded around three sides of a building and has equal width on the three sides, as shown below. What is the seeded area? [9.2c]

Find the length of a radius of the circle. [9.3a]

13.

16 m

14.

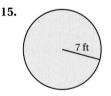

$\frac{28}{11}$ in.

Find the length of a diameter of the circle. [9.3a]

15.

7 ft

16.

10 cm

17. Find the circumference of the circle in Exercise 13. Use 3.14 for π. [9.3b]

18. Find the circumference of the circle in Exercise 14. Use $\frac{22}{7}$ for π. [9.3b]

19. Find the area of the circle in Exercise 13. Use 3.14 for π. [9.3c]

20. Find the area of the circle in Exercise 14. Use $\frac{22}{7}$ for π. [9.3c]

21. Find the area of the shaded region. Use 3.14 for π. [9.3d]

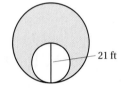

21 ft

Find the volume. [9.4a]

22.

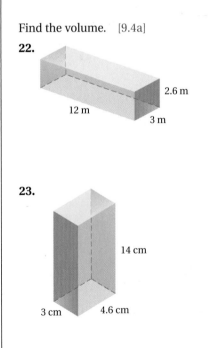

2.6 m

12 m

3 m

23.

14 cm

3 cm   4.6 cm

Find the volume. Use 3.14 for π.

24.   [9.4b]

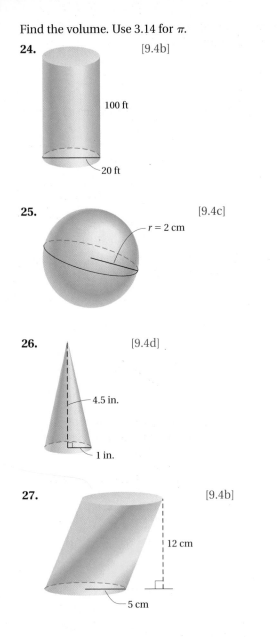

100 ft

20 ft

25.   [9.4c]

$r = 2$ cm

26.   [9.4d]

4.5 in.

1 in.

27.   [9.4b]

12 cm

5 cm

28. A "Norman" window is designed with dimensions as shown. Find its area and its perimeter. Use 3.14 for π. [9.3d]

2 ft

5 ft

Use a protractor to measure each angle.   [9.5a]

29.   **30.**

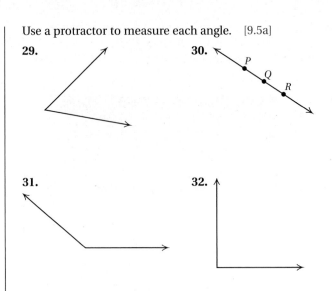

P

Q

R

31.   **32.**

33.–36. Classify each of the angles in Exercises 29–32 as right, straight, acute, or obtuse.   [9.5b]

37. Find the measure of a complement of $\angle BAC$.   [9.5c]

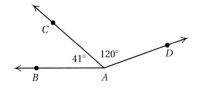

C

$41°$ $120°$ D

B A

38. Find the measure of a supplement of a 44° angle. [9.5c]

Use the following triangle for Exercises 39–41.

3.5

$30°$

4

x

2

39. Find the missing angle measure.   [9.5e]

40. Classify the triangle as equilateral, isosceles, or scalene. [9.5d]

41. Classify the triangle as right, obtuse, or acute.   [9.5d]

42. Simplify: $\sqrt{64}$. [9.6a]

43. Approximate to three decimal places: $\sqrt{83}$. [9.6b]

In a right triangle, find the length of the side not given. Give an exact answer and an approximation to three decimal places. [9.6c]

44. $a = 15, b = 25$

45. $a = 7, c = 10$

Find the length of the side not given. Give an exact answer and an approximation to three decimal places. [9.6c]

46.

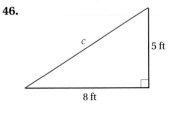

c

5 ft

8 ft

47.

20 cm

a

18 cm

Solve. [9.6d]

48. How long is a wire reaching from the top of a 24-ft pole to a point on the ground 16 ft from the base of the pole?

49. How tall is this tree?

60 ft

h

40 ft

50. Find the length of a diagonal from one corner to another of the tennis court in Exercise 3.

51. **D**_W List and describe all the volume formulas that you have learned in this chapter. [9.4a, b, c, d]

52. **D**_W Which occupies more volume: two spheres, each with radius r, or one sphere with radius $2r$? Explain why. [9.4c]

SYNTHESIS

53. A square is cut in half so that the perimeter of the resulting rectangle is 30 ft. Find the area of the original square. [9.1a], [9.2a]

54. Find the area, in square meters, of the shaded region. [8.2a], [9.2c]

2.8 m

1.8 mm

2.8 m

1.8 mm

55. Find the area, in square centimeters, of the shaded region. [8.2a], [9.2c]

Each small triangle has a height and base of 1.3 mm.

10 cm

8.4 cm

573

Find the perimeter and the area.

1.

7.01 cm

9.4 cm

2.

$4\frac{7}{8}$ in.

$4\frac{7}{8}$ in.

Find the area.

3.

2.5 cm

10 cm

4.

3 m

8 m

5.

4 ft

3 ft

8 ft

6. Find the length of a diameter of this circle.

$\frac{1}{8}$ in.

7. Find the length of a radius of this circle.

18 cm

8. Find the circumference of the circle in Question 6. Use $\frac{22}{7}$ for π.

9. Find the area of the circle in Question 7. Use 3.14 for π.

10. Find the perimeter and the area of the shaded region. Use 3.14 for π.

18.6 km

9.0 km

11. Find the volume.

10.5 cm

4 cm

2 cm

12. A twelve-box carton of 12-oz juice boxes comes in a rectangular box $10\frac{1}{2}$ in. by 8 in. by 5 in. What is the volume of the carton?

Find the volume. Use 3.14 for π.

13.

15 ft

5 ft

14.

$d = 20$ yd

15.

12 cm

3 cm

Use a protractor to measure each angle.

16.

17.

18.

Q R S

19.

20.–23. Classify each of the angles in Questions 16–19 as right, straight, acute, or obtuse.

Use the following triangle for Questions 24–26.

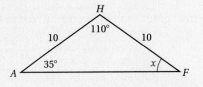

H

110°

10   10

35° x

A F

24. Find the missing angle measure.

25. Classify the triangle as equilateral, isosceles, or scalene.

26. Classify the triangle as right, obtuse, or acute.

27. Find the measure of a complement and a supplement of $\angle CAD$.

C

D

65°

75°   30°

B A E

28. Simplify: $\sqrt{225}$.

29. Approximate to three decimal places: $\sqrt{87}$

In a right triangle, find the length of the side not given. Give an exact answer and, where appropriate, an approximation to three decimal places.

30. $a = 24$, $b = 32$

31. $a = 2$, $c = 8$

32.

33.

34. How long must a wire be in order to reach from the top of a 13-m antenna to a point on the ground 9 m from the base of the antenna?

SYNTHESIS

Find the area of the shaded region. (Note that the figures are not drawn in perfect proportion.) Give the answer in square feet.

35.

36.

Find the volume of the solid. (Note that the solids are not drawn in perfect proportion.) Give the answer in cubic feet. Use 3.14 for π.

37.

38.

39.

576

Cumulative Review

Solve.

1. *Quito, Ecuador.* In Quito, Ecuador, there are 1.5 million people living under threat of ash clouds and mudflows from a giant volcano, Mt. Antisana. Find standard notation for 1.5 million.

   Source: *National Geographic Magazine,* February 2001

Egg Consumption. As shown in the line graph below, egg consumption per person in the United States has been increasing in recent years. Use the graph for Exercises 3–8.

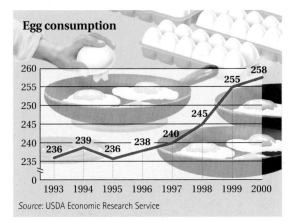

9. *Medical Dosage.* A doctor suggests that a child who weighs 24 kg be given 42 mg of Phenytoin. If the dosage is proportional to the child's weight, how much Phenytoin is recommended for a child who weighs 32 kg?

2. *Dead Sea.* The lowest point in the world is the Dead Sea on the border of Israel and Jordan. It is 1312 ft below sea level. Convert 1312 ft to yards; to meters.

   Source: *The Handy Geography Answer Book*

3. Find the lowest egg consumption and the year(s) in which it occurred.

4. Find the highest egg consumption and the year(s) in which it occurred.

5. Find the average, the median, and the mode of the egg consumptions.

6. Find the average egg consumption over the years 1997 to 2000.

7. Find the average egg consumption over the years 1993 to 1996. How does it compare to the answer to Exercise 6?

8. What was the percent of increase in egg consumption from 1996 to 2000?

10. *Firefighting.* During a fire, the firefighters get a 1-ft layer of water on the 25-ft by 60-ft first floor of a 5-floor building. Water weighs $62\frac{1}{2}$ lb per cubic foot. What is the total weight of the water on the floor?

Calculate.

11. $1\dfrac{1}{2} + 2\dfrac{2}{3}$

12. $\left(\dfrac{1}{4}\right)^2 \div \left(\dfrac{1}{2}\right)^3 \times 2^4 + (10.3)(4)$

13. $120.5 - 32.98$

14. $22\overline{)27{,}148}$

15. $14 \div [33 \div 11 + 8 \times 2 - (15 - 3)]$

16. $8^3 + 45 \cdot 24 - 9^2 \div 3$

Find fraction notation.

17. 1.209

18. 17%

Use $<$, $>$, or $=$ for $\square$ to write a true sentence.

19. $\dfrac{5}{6} \ \square \ \dfrac{7}{8}$

20. $\dfrac{15}{18} \ \square \ \dfrac{10}{12}$

Complete.

21. $6 \text{ oz} = \underline{\hspace{1cm}} \text{ lb}$

22. $15°\text{C} = \underline{\hspace{1cm}} °\text{F}$

23. $0.087 \text{ L} = \underline{\hspace{1cm}} \text{ mL}$

24. $9 \text{ sec} = \underline{\hspace{1cm}} \text{ min}$

25. $3 \text{ yd}^2 = \underline{\hspace{1cm}} \text{ ft}^2$

26. $17 \text{ cm} = \underline{\hspace{1cm}} \text{ m}$

27. $2437 \text{ mcg} = \underline{\hspace{1cm}} \text{ mg}$

28. $29{,}082 \text{ ft} = \underline{\hspace{1cm}} \text{ m}$
(The height of Mt. Everest)

29. $9 \text{ L} = \underline{\hspace{1cm}} \text{ qt}$

Solve.

30. $\dfrac{12}{15} = \dfrac{x}{18}$

31. $\dfrac{3}{x} = \dfrac{7}{10}$

32. $25 \cdot x = 2835$

33. $x + \dfrac{3}{4} = \dfrac{7}{8}$

Find the perimeter and the area.

34.

35.

36. Find the diameter and the area of this circle. Use $\frac{22}{7}$ for π.

37. Find the volume of this sphere. Use $\frac{22}{7}$ for π.

Solve.

38. To get an A in math, a student must score an average of 90 on five tests. On the first four tests, the scores were 85, 92, 79, and 95. What is the lowest score that the student can get on the last test and still get an A?

39. Americans own 52 million dogs, 56 million cats, 45 million birds, 250 million fish, and 125 million other creatures as house pets. How many pets do Americans own altogether?

40. What is the simple interest on $8000 at 4.2% for $\frac{1}{4}$ year?

41. What is the amount in an account after 25 years if $8000 is invested at 4.2%, compounded annually?

42. How long must a rope be in order to reach from the top of an 8-m tree to a point on the ground 15 m from the bottom of the tree?

43. The sales tax on an office supply purchase of $5.50 is $0.33. What is the sales tax rate?

44. A bolt of fabric in a fabric store has $10\frac{3}{4}$ yd on it. A customer purchases $8\frac{5}{8}$ yd. How many yards remain on the bolt?

45. What is the cost, in dollars, of 15.6 gal of gasoline at 239.9¢ per gallon? Round to the nearest cent.

46. A box of powdered milk that makes 20 qt costs $4.99. A box that makes 8 qt costs $1.99. Which size has the lower unit price?

47. It is $\frac{7}{10}$ km from a student's dormitory to the library. Maria starts to walk there, changes her mind after going $\frac{1}{4}$ of the distance, and returns home. How far did she walk?

48. Find the missing angle measure.

49. Classify the triangle as equilateral, isosceles, or scalene.

12 in.   12 in.

7 in.

50. Classify the triangle in Question 48 as right, obtuse, or acute.

51. *Matching.* Match each item in the first column with the appropriate item in the second column by drawing connecting lines. Some expressions in the second column will be used more than once. Some expressions will not be used.

Area of a circle of radius 4 ft	24 ft
Area of a square of side 4 ft	16 ft
Circumference of a circle of radius 4 ft	$16 \cdot \pi \, \text{ft}^2$
Volume of a cone with radius of the base 4 ft and height 8 ft	$32 \, \text{ft}^2$
Area of a rectangle of length 8 ft and width 4 ft	$\frac{4^4}{3} \cdot \pi \, \text{ft}^3$
Area of a triangle with base 4 ft and height 8 ft	$128 \cdot \pi \, \text{ft}^3$
Volume of a sphere of radius 4 ft	$\left(21\frac{1}{3}\right) \cdot \pi \, \text{ft}^3$
Volume of a right circular cylinder with radius of the base 4 ft and height 8 ft	$8 \cdot \pi \, \text{ft}$
	$64 \, \text{ft}$
Perimeter of a square of side 4 ft	$128 \cdot \pi \, \text{ft}^2$
Perimeter of a rectangle of length 8 ft and width 4 ft	$16 \, \text{ft}^2$
	$\frac{128}{3} \cdot \pi \, \text{ft}^3$

SYNTHESIS

52. Your house sits on a lot measuring 75 ft by 200 ft. The lot is at the intersection of two streets, so there are sidewalks on the lot on two sides of the lot. In the winter, you have to shovel the snow off the sidewalks. If the sidewalks are 3 ft wide and the snow is 4 in. deep, what volume of snow must you shovel?

Find the volume in cubic feet. Use 3.14 for π.

53.

100 yd

10 ft

54.

14 ft

3 in.   4.6 in.

55. J. C. Payne, a 71-year-old rancher at the time, recently asked *The Guinness Book of Records* to accept a world record for constructing a ball of string. The ball was 13.2 ft in diameter. What was the volume of the ball? Assuming the diameter of the string was 0.1 in., how long was the string in feet? in miles? Use the π key on your calculator.

Source: *The Guinness Book of Records*

Real Numbers

Real-World Application

Surface temperatures on Mars vary from $-128°C$ during polar night to $27°C$ at the equator during midday at the closest point in orbit to the sun. Find the difference between the highest value and the lowest value in this temperature range.

Source: Mars Institute

This problem appears as Example 13 in Section 10.3.

Objectives

a State the integer that corresponds to a real-world situation.

b Graph rational numbers on a number line.

c Convert from fraction notation for a rational number to decimal notation.

d Determine which of two real numbers is greater and indicate which, using $<$ or $>$.

e Find the absolute value of a real number.

10.1 THE REAL NUMBERS

In this section, we introduce the *real numbers*. We begin with numbers called *integers* and build up to the real numbers. To describe integers, we start with the whole numbers, 0, 1, 2, 3, and so on. For each number 1, 2, 3, and so on, we obtain a new number to the left of zero on the number line:

For the number 1, there will be an *opposite* number -1 (negative 1).

For the number 2, there will be an *opposite* number -2 (negative 2).

For the number 3, there will be an *opposite* number -3 (negative 3), and so on.

The **integers** consist of the whole numbers and these new numbers. We picture them on a number line as follows.

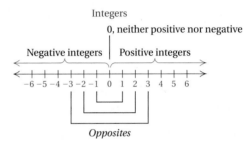

Opposites

We call the numbers to the left of zero on the number line **negative integers.** The natural numbers are called **positive integers.** Zero is neither positive nor negative. We call -1 and 1 **opposites** of each other. Similarly, -2 and 2 are opposites, -3 and 3 are opposites, -100 and 100 are opposites, and 0 is its own opposite. Opposite pairs of numbers like -3 and 3 are the same distance from 0. The integers extend infinitely on the number line to the left and right of zero.

INTEGERS

The **integers:** $\ldots, -5, -4, -3, -2, -1, 0, 1, 2, 3, 4, 5, \ldots$

---

a Integers and the Real World

Integers correspond to many real-world problems and situations. The following examples will help you get ready to translate problem situations that involve integers to mathematical language.

EXAMPLE 1 Tell which integer corresponds to this situation: The temperature is 4 degrees below zero.

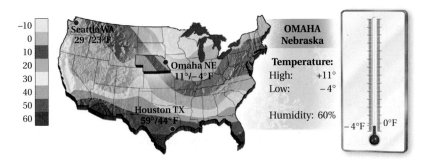

The integer −4 corresponds to the situation. The temperature is −4°.

EXAMPLE 2 *Elevation.* Tell which integer corresponds to this situation: The shores of California's largest lake, the Salton Sea, are 227 ft below sea level.

Source: *National Geographic,* February 2005, p. 88. "Salton Sea," by Joel K. Bourne, Jr., Senior Writer

The integer −227 corresponds to the situation. The elevation is −227 ft.

EXAMPLE 3 *Stock Price Change.* Tell which integers correspond to the situation: The stock price of Pearson Education decreased from $27 per share to $11 per share over a recent time period. The price of Safeway stock increased from $20 per share to $22 per share over a recent time period.

Source: The New York Stock Exchange

The integer −16 corresponds to the decrease in the value of the stock. The integer 2 represents the increase in the value of the stock.

Do Exercises 1–5.

b The Rational Numbers

Fractions such as $\frac{1}{2}$ are not integers. A larger system called the **rational numbers** contains integers and fractions. The rational numbers consist of quotients of integers with nonzero divisors.

Tell which integers correspond to the situation.

1. The halfback gained 8 yd on first down. The quarterback was sacked for a 5-yd loss on second down.

2. **Temperature High and Low.** The highest recorded temperature in Nevada is 125°F on June 29, 1994, in Laughlin. The lowest recorded temperature in Nevada is 50°F below zero on June 8, 1937, in San Jacinto.

   Source: National Climatic Data Center, Asheville, NC, and Storm Phillips, STORMFAX, INC.

3. **Stock Decrease.** The stock of Wendy's decreased from $41 per share to $38 per share over a recent period.

   Source: The New York Stock Exchange

4. At 10 sec before liftoff, ignition occurs. At 148 sec after liftoff, the first stage is detached from the rocket.

5. A submarine dove 120 ft, rose 50 ft, and then dove 80 ft.

Answers on page A-23

Graph on a number line.

6. $-\dfrac{7}{2}$

7. 1.4

8. $-\dfrac{11}{4}$

The following are rational numbers:

$$\frac{2}{3}, \quad -\frac{2}{3}, \quad \frac{7}{1}, \quad 4, \quad -3, \quad 0, \quad \frac{23}{-8}, \quad 2.4, \quad -0.17, \quad 10\frac{1}{2}.$$

The number $-\frac{2}{3}$ (read "negative two-thirds") can also be named $\frac{-2}{3}$ or $\frac{2}{-3}$. That is,

$$-\frac{a}{b} = \frac{-a}{b} = \frac{a}{-b}.$$

The number 2.4 can be named $\frac{24}{10}$, or $\frac{12}{5}$, and -0.17 can be named $-\frac{17}{100}$.

Note that the rational numbers contain the natural numbers, the whole numbers, the integers, the arithmetic numbers (also called the nonnegative rational numbers), and the negative rational numbers.

RATIONAL NUMBERS

The **rational numbers** consist of all numbers that can be named in the form $\dfrac{a}{b}$, where a and b are integers and b is not 0.

We picture the rational numbers on a number line, as follows.

To **graph** a number means to find and mark its point on the number line. Some rational numbers are graphed in the preceding figure.

EXAMPLE 4 Graph: $\frac{5}{2}$.

The number $\frac{5}{2}$ can be named $2\frac{1}{2}$, or 2.5. Its graph is halfway between 2 and 3.

EXAMPLE 5 Graph: -3.2.

The graph of -3.2 is $\frac{2}{10}$ of the way from -3 to -4.

EXAMPLE 6 Graph: $\frac{13}{8}$.

The number $\frac{13}{8}$ can be named $1\frac{5}{8}$, or 1.625. The graph is $\frac{5}{8}$ of the way from 1 to 2.

Answers on page A-23

Do Exercises 6–8.

c | Notation for Rational Numbers

Each rational number can be named using fraction or decimal notation.

EXAMPLE 7 Find decimal notation for $-\frac{5}{8}$.

We first find decimal notation for $\frac{5}{8}$. Since $\frac{5}{8}$ means $5 \div 8$, we divide,

```
       0.6 2 5
  8 ) 5.0 0 0
       4 8
         2 0
         1 6
           4 0
           4 0
             0
```

Thus, $\frac{5}{8} = 0.625$, so $-\frac{5}{8} = -0.625$.

Decimal notation for $-\frac{5}{8}$ is -0.625. We consider -0.625 to be a **terminating decimal.** Decimal notation for some numbers repeats.

EXAMPLE 8 Find decimal notation for $-\frac{7}{11}$.

We divide to find decimal notation for $\frac{7}{11}$.

```
        0.6 3 6 3...
  1 1 ) 7.0 0 0 0
        6 6
          4 0
          3 3
            7 0
            6 6
              4 0
              3 3
                7
```

Thus, $\frac{7}{11} = 0.6363\ldots$, so $-\frac{7}{11} = -0.6363\ldots$. Repeating decimal notation can be abbreviated by writing a bar over the repeating part; in this case, we write $-0.\overline{63}$.

The following are other examples to show how each rational number can be named using fraction or decimal notation:

$$0 = \frac{0}{6}, \qquad \frac{27}{100} = 0.27, \qquad -8\frac{3}{4} = -8.75, \qquad -\frac{13}{6} = -2.1\overline{6}.$$

Do Exercises 9–11.

d | The Real Numbers and Order

Every rational number has a point on the number line. However, there are some points on the line for which there are no rational numbers. These points correspond to what are called **irrational numbers.** Some examples of irrational numbers are π and $\sqrt{2}$.

Find decimal notation.

9. $-\frac{3}{8}$

10. $-\frac{6}{11}$

11. $\frac{4}{3}$

Answers on page A-23

Use either < or > for ☐ to write a true sentence.

12. −3 ☐ 7

13. −8 ☐ −5

14. 7 ☐ −10

15. 3.1 ☐ −9.5

16. $-\dfrac{2}{3}$ ☐ −1

17. $-\dfrac{11}{8}$ ☐ $\dfrac{23}{15}$

18. $-\dfrac{2}{3}$ ☐ $-\dfrac{5}{9}$

19. −4.78 ☐ −5.01

Decimal notation for rational numbers *either* terminates *or* repeats. Decimal notation for irrational numbers *neither* terminates *nor* repeats. Some other examples of irrational numbers are $\sqrt{3}$, $-\sqrt{8}$, $\sqrt{11}$, and 0.121221222122221 .... *Whenever we take the square root of a number that is not a perfect square (see Section 9.6), we will get an irrational number.*

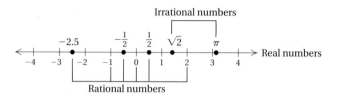

THE REAL-NUMBER SYSTEM

The rational numbers and the irrational numbers together correspond to all the points on a number line and make up what is called the **real-number system.**

The **real numbers** consist of the rational numbers and the irrational numbers. The following figure shows the relationship among various kinds of numbers.

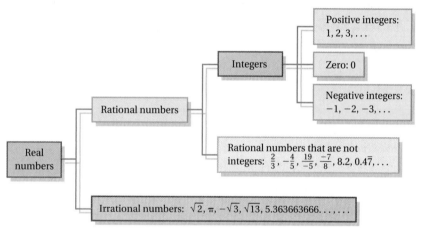

ORDER

Real numbers are named in order on the number line, with larger numbers named farther to the right. (See Section 1.4.) For any two numbers on the line, the one to the left is less than the one to the right.

We use the symbol **<** to mean "**is less than.**" The sentence −8 < 6 means "−8 is less than 6." The symbol **>** means "**is greater than.**" The sentence −3 > −7 means "−3 is greater than −7."

EXAMPLES Use either < or > for ☐ to write a true sentence.

9. −7 ☐ 3 Since −7 is to the left of 3, we have −7 < 3.

10. 6 ☐ −12 Since 6 is to the right of −12, then 6 > −12.

11. −18 ☐ −5 Since −18 is to the left of −5, we have −18 < −5.

Answers on page A-23

12. $-2.7 \;\square\; -\dfrac{3}{2}$ The answer is $-2.7 < -\dfrac{3}{2}$.

13. $1.5 \;\square\; -2.7$ The answer is $1.5 > -2.7$.

14. $-3.45 \;\square\; 1.32$ The answer is $-3.45 < 1.32$.

15. $\dfrac{5}{8} \;\square\; \dfrac{7}{11}$ We convert to decimal notation: $\dfrac{5}{8} = 0.625$, and $\dfrac{7}{11} = 0.6363\ldots$. Thus, $\dfrac{5}{8} < \dfrac{7}{11}$.

16. $-4 \;\square\; 0$ The answer is $-4 < 0$.

17. $5.8 \;\square\; 0$ The answer is $5.8 > 0$.

Do Exercises 12–19 on the preceding page.

e   Absolute Value

From the number line, we see that numbers like 4 and -4 are the same distance from zero. Distance is always a nonnegative number. We call the distance of a number from zero the **absolute value** of the number.

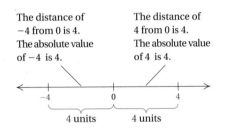

The distance of -4 from 0 is 4.
The absolute value of -4 is 4.

The distance of 4 from 0 is 4.
The absolute value of 4 is 4.

4 units   4 units

ABSOLUTE VALUE

The **absolute value** of a number is its distance from zero on a number line. We use the symbol $|x|$ to represent the absolute value of a number x.

To find the absolute value of a number:

a) If a number is negative, its absolute value is positive.
b) If a number is positive or zero, its absolute value is the same as the number.

EXAMPLES   Find the absolute value.

18. $|-7|$ The distance of -7 from 0 is 7, so $|-7| = 7$.

19. $|12|$ The distance of 12 from 0 is 12, so $|12| = 12$.

20. $|0|$ The distance of 0 from 0 is 0, so $|0| = 0$.

21. $\left|\dfrac{3}{2}\right| = \dfrac{3}{2}$

22. $|-2.73| = 2.73$

Do Exercises 20–24.

Find the absolute value.

20. $|8|$

21. $|0|$

22. $|-9|$

23. $\left|-\dfrac{2}{3}\right|$

24. $|5.6|$

Answers on page A-23

10.1

EXERCISE SET

For Extra Help

MathXL MyMathLab InterAct Math Math Tutor Center Digital Video Tutor CD 5 Videotape 10 Student's Solutions Manual

a Tell which integers correspond to the situation.

1. *Pollution Fine.* In 2003, The Colonial Pipeline Company was fined a record $34 million for pollution.
Source: Green Consumer Guide.com

2. *Lake Powell.* The water level of Lake Powell, a desert reservoir behind Glen Canyon Dam in northern Arizona and southeastern Utah, has dropped 130 ft since 2000.

3. On Wednesday, the temperature was 24° above zero. On Thursday, it was 2° below zero.

4. A student deposited her tax refund of $750 in a savings account. Two weeks later, she withdrew $125 to pay sorority fees.

5. *Highest and Lowest Temperatures* The highest temperature ever created on earth was 950,000,000°F. The lowest temperature ever created was approximately 460°F below zero.
Source: *The Guinness Book of Records,* 2004

6. *Extreme Climate.* Verkhoyansk, a river port in northeast Siberia, has the most extreme climate on the planet. Its average monthly winter temperature is 58.5°F below zero, and its average monthly summer temperature is 56.5°F.
Source: *The Guinness Book of Records,* 2004

7. In bowling, after the game, team A is 34 pins behind team B, and team B is 15 pins ahead of team C.

8. During a video game, Sara intercepted a missile worth 20 points, lost a starship worth 150 points, and captured a base worth 300 points.

b Graph the number on the number line.

9. $\dfrac{10}{3}$

10. $-\dfrac{17}{4}$

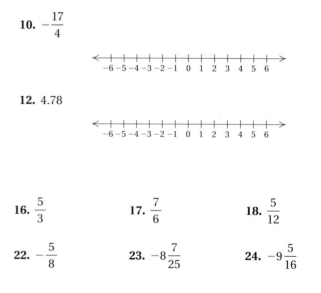

11. -5.2

12. 4.78

c Convert to decimal notation.

13. $-\dfrac{7}{8}$ **14.** $-\dfrac{1}{8}$ **15.** $\dfrac{5}{6}$ **16.** $\dfrac{5}{3}$ **17.** $\dfrac{7}{6}$ **18.** $\dfrac{5}{12}$

19. $\dfrac{2}{3}$ **20.** $\dfrac{1}{4}$ **21.** $-\dfrac{1}{2}$ **22.** $-\dfrac{5}{8}$ **23.** $-8\dfrac{7}{25}$ **24.** $-9\dfrac{5}{16}$

588

CHAPTER 10: Real Numbers

Copyright © 2007 Pearson Education, Inc.

Use either $<$ or $>$ for ☐ to write a true sentence.

25. 8 ☐ 0   **26.** 3 ☐ 0   **27.** -8 ☐ 3   **28.** 6 ☐ -6 **29.** -8 ☐ 8

30. 0 ☐ -9 **31.** -8 ☐ -5 **32.** -4 ☐ -3 **33.** -5 ☐ -11 **34.** -3 ☐ -4

35. -6 ☐ -5 **36.** -10 ☐ -14 **37.** 2.14 ☐ 1.24   **38.** -3.3 ☐ -2.2 **39.** -14.5 ☐ 0.011

40. 17.2 ☐ -1.67 **41.** $-12\frac{5}{8}$ ☐ $-6\frac{3}{8}$ **42.** $-7\frac{5}{16}$ ☐ $-3\frac{11}{16}$ **43.** $\frac{5}{12}$ ☐ $\frac{11}{25}$ **44.** $-\frac{13}{16}$ ☐ $-\frac{5}{9}$

Find the absolute value.

45. $|-3|$ **46.** $|-7|$ **47.** $|18|$ **48.** $|0|$ **49.** $|325|$ **50.** $|-4|$

51. $|-3.625|$ **52.** $\left|-7\frac{4}{5}\right|$ **53.** $\left|-\frac{2}{3}\right|$ **54.** $\left|-\frac{10}{7}\right|$ **55.** $\left|\frac{0}{4}\right|$ **56.** $|14.8|$

57. D_W Give three examples of rational numbers that are not integers. Explain.

58. D_W Give three examples of irrational numbers. Explain the difference between an irrational number and a rational number.

Find the prime factorization.   [2.1d]

59. 54   **60.** 192   **61.** 102

62. 260   **63.** 864   **64.** 468

Find the LCM.   [3.1a]

65. 6,   18   **66.** 18,   24   **67.** 6,   24,   32

68. 12,   24,   36   **69.** 48,   56,   64   **70.** 12,   36,   84

Use either $<$, $>$, or $=$ for ☐ to write a true sentence.

71. $|-5|$ ☐ $|-2|$ **72.** $|4|$ ☐ $|-7|$ **73.** $|-8|$ ☐ $|8|$

List in order from the least to the greatest.

74. $-\frac{2}{3}, \frac{1}{2}, -\frac{3}{4}, -\frac{5}{6}, \frac{3}{8}, \frac{1}{6}$

75. $-8\frac{7}{8}, 7, -5, |-6|, 4, |3|, -8\frac{5}{8}, -100, 0, 1^7, \frac{14}{4}, -\frac{67}{8}$

Objectives

a Add real numbers without using a number line.

b Find the opposite, or additive inverse, of a real number.

Add using a number line.

1. $0 + (-3)$

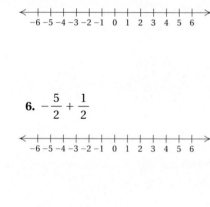

2. $1 + (-4)$

3. $-3 + (-2)$

4. $-3 + 7$

5. $-2.4 + 2.4$

6. $-\dfrac{5}{2} + \dfrac{1}{2}$

Answers on page A-23

10.2 ADDITION OF REAL NUMBERS

We now consider addition of real numbers. First, to gain an understanding, we add using a number line. Then we consider rules for addition.

ADDITION ON A NUMBER LINE

Addition of numbers can be illustrated on a number line. To do the addition $a + b$, we start at 0. Then move to a and then move according to b.

a) If b is positive, we move from a to the right.

b) If b is negative, we move from a to the left.

c) If b is 0, we stay at a.

EXAMPLE 1 Add: $3 + (-5)$.

We start at 0 and move to 3. Then we move 5 units left since -5 is negative.

$3 + (-5) = -2$

EXAMPLE 2 Add: $-4 + (-3)$.

We start at 0 and move to -4. Then we move 3 units left since -3 is negative.

$-4 + (-3) = -7$

EXAMPLE 3 Add: $-4 + 9$.

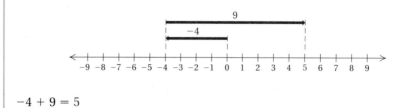

$-4 + 9 = 5$

EXAMPLE 4 Add: $-5.2 + 0$.

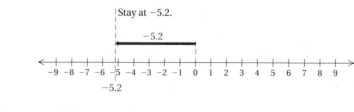

$-5.2 + 0 = -5.2$

a Addition Without a Number Line

You may have noticed some patterns in the preceding examples. These lead us to rules for adding without using a number line that are more efficient for adding larger or more complicated numbers.

> **RULES FOR ADDITION OF REAL NUMBERS**
>
> 1. *Positive numbers*: Add the same as arithmetic numbers. The answer is positive.
> 2. *Negative numbers*: Add absolute values. The answer is negative.
> 3. *A positive and a negative number*: Subtract the smaller absolute value from the larger. Then:
>
>   a) If the positive number has the greater absolute value, the answer is positive.
>   b) If the negative number has the greater absolute value, the answer is negative.
>   c) If the numbers have the same absolute value, the answer is 0.
>
> 4. *One number is zero*: The sum is the other number.

Rule 4 is known as the **identity property of 0.** It says that for any real number a, $a + 0 = a$.

EXAMPLES   Add without using a number line.

5. $-12 + (-7) = -19$ Add the absolute values, 12 and 7, getting 19. Make the answer *negative*, -19.

6. $-1.4 + 8.5 = 7.1$ The absolute values are 1.4 and 8.5. The difference is 7.1. The positive number has the larger absolute value, so the answer is *positive*, 7.1.

7. $-36 + 21 = -15$ The absolute values are 36 and 21. The difference is 15. The negative number has the larger absolute value, so the answer is *negative*, -15.

8. $1.5 + (-1.5) = 0$ The numbers have the same absolute value. The sum is 0.

9. $-\dfrac{7}{8} + 0 = -\dfrac{7}{8}$ One number is zero. The sum is $-\dfrac{7}{8}$.

10. $-9.2 + 3.1 = -6.1$

11. $-\dfrac{3}{2} + \dfrac{9}{2} = \dfrac{6}{2} = 3$

12. $-\dfrac{2}{3} + \dfrac{5}{8} = -\dfrac{16}{24} + \dfrac{15}{24} = -\dfrac{1}{24}$

Do Exercises 7–20.

Suppose we wish to add several numbers, some positive and some negative, as follows. How can we proceed?

$$15 + (-2) + 7 + 14 + (-5) + (-12)$$

Add without using a number line.

7. $-5 + (-6)$ -11

8. $-9 + (-3)$ -12

9. $-4 + 6$ 2

10. $-7 + 3$ -4

11. $5 + (-7)$ -2

12. $-20 + 20$ 0

13. $-11 + (-11)$ -22

14. $10 + (-7)$

15. $-0.17 + 0.7$

16. $-6.4 + 8.7$

17. $-4.5 + (-3.2)$

18. $-8.6 + 2.4$

19. $\dfrac{5}{9} + \left(-\dfrac{7}{9}\right)$

20. $-\dfrac{1}{5} + \left(-\dfrac{3}{4}\right)$

Answers on page A-23

Add.

21. $(-15) + (-37) + 25 + 42 + (-59) + (-14)$

22. $42 + (-81) + (-28) + 24 + 18 + (-31)$

23. $-2.5 + (-10) + 6 + (-7.5)$

24. $-35 + 17 + 14 + (-27) + 31 + (-12)$

Find the opposite, or additive inverse, of each of the following.

25. -4

26. 8.7

27. -7.74

28. $-\dfrac{8}{9}$

29. 0

30. 12

Answers on page A-23

The commutative and associative laws hold for real numbers. Thus we can change grouping and order as we please when adding. For instance, we can group the positive numbers together and the negative numbers together and add them separately. Then we add the two results.

EXAMPLE 13 Add: $15 + (-2) + 7 + 14 + (-5) + (-12)$.

a) $15 + 7 + 14 = 36$ Adding the positive numbers
b) $-2 + (-5) + (-12) = -19$ Adding the negative numbers
c) $36 + (-19) = 17$ Adding the results of (a) and (b)

We can also add the numbers in any other order we wish, say, from left to right as follows:

$$
\begin{aligned}
15 + (-2) + 7 + 14 + (-5) + (-12) &= 13 + 7 + 14 + (-5) + (-12) \\
&= 20 + 14 + (-5) + (-12) \\
&= 34 + (-5) + (-12) \\
&= 29 + (-12) \\
&= 17
\end{aligned}
$$

Do Exercises 21–24.

b   Opposites, or Additive Inverses

Suppose we add two numbers that are **opposites,** such as 6 and -6. The result is 0. When opposites are added, the result is always 0. Such numbers are also called **additive inverses.** Every real number has an opposite, or additive inverse.

OPPOSITES, OR ADDITIVE INVERSES

Two numbers whose sum is 0 are called **opposites,** or **additive inverses,** of each other.

EXAMPLES Find the opposite, or additive inverse, of each number.

14. 34   The opposite of 34 is -34 because $34 + (-34) = 0$.
15. -8 The opposite of -8 is 8 because $-8 + 8 = 0$.
16. 0   The opposite of 0 is 0 because $0 + 0 = 0$.
17. $-\dfrac{7}{8}$ The opposite of $-\dfrac{7}{8}$ is $\dfrac{7}{8}$ because $-\dfrac{7}{8} + \dfrac{7}{8} = 0$.

Do Exercises 25–30.

To name the opposite, we use the symbol $-$, as follows.

SYMBOLIZING OPPOSITES

The opposite, or additive inverse, of a number a can be named $-a$ (read "the opposite of a," or "the additive inverse of a").

Note that if we take a number, say 8, and find its opposite, -8, and then find the opposite of the result, we will have the original number, 8, again.

THE OPPOSITE OF THE OPPOSITE

The opposite of the opposite of a number is the number itself. (The additive inverse of the additive inverse of a number is the number itself.) That is, for any number a,

$$-(-a) = a.$$

EXAMPLE 18 Evaluate $-x$ and $-(-x)$ when $x = 16$.

We replace x in each case with 16.

a) If $x = 16$, then $-x = -16 = -16$.   The opposite of 16 is -16.
b) If $x = 16$, then $-(-x) = -(-16) = 16$.   The opposite of the opposite of 16 is 16.

EXAMPLE 19 Evaluate $-x$ and $-(-x)$ when $x = -3$.

We replace x in each case with -3.

a) If $x = -3$, then $-x = -(-3) = 3$.
b) If $x = -3$, then $-(-x) = -(-(-3)) = -(3) = -3$.

Note that in Example 19 we used an extra set of parentheses to show that we are substituting the negative number -3 for x. Symbolism like $--x$ is not considered meaningful.

Do Exercises 31–36.

A symbol such as -8 is usually read "negative 8." It could be read "the additive inverse of 8," because the additive inverse of 8 is negative 8. It could also be read "the opposite of 8," because the opposite of 8 is -8. Thus a symbol like -8 can be read in more than one way. A symbol like $-x$, which has a variable, should be read "the opposite of x" or "the additive inverse of x" and *not* "negative x," because we do not know whether x represents a positive number, a negative number, or 0.

We can use the symbolism $-a$ to restate the definition of opposite, or additive inverse.

THE SUM OF OPPOSITES

For any real number a, the opposite, or additive inverse, of a, expressed as $-a$, is such that

$$a + (-a) = (-a) + a = 0.$$

SIGNS OF NUMBERS

A negative number is sometimes said to have a "negative sign." A positive number is said to have a "positive sign." When we replace a number with its opposite, we can say that we have "changed its sign."

EXAMPLES Change the sign. (Find the opposite.)

20. -3 $-(-3) = 3$

21. $-\dfrac{2}{13}$ $-\left(-\dfrac{2}{13}\right) = \dfrac{2}{13}$

22. 0 $-(0) = 0$

23. 14 $-(14) = -14$

Do Exercises 37–40.

Evaluate $-x$ and $-(-x)$ when:

31. $x = 14$.   **32.** $x = 1$.

33. $x = -19$.   **34.** $x = -1.6$.

35. $x = \dfrac{2}{3}$.   **36.** $x = -\dfrac{9}{8}$.

Find the opposite. (Change the sign.)

37. -4 **38.** -13.4

39. 0 **40.** $\dfrac{1}{4}$

Answers on page A-23

a Add. Do not use a number line except as a check.

1. $-9 + 2$

2. $-5 + 2$

3. $-10 + 6$

4. $4 + (-3)$

5. $-8 + 8$

6. $4 + (-4)$

7. $-3 + (-5)$

8. $-6 + (-8)$

9. $-7 + 0$

10. $-10 + 0$

11. $0 + (-27)$

12. $0 + (-36)$

13. $17 + (-17)$

14. $-20 + 20$

15. $-17 + (-25)$

16. $-23 + (-14)$

17. $18 + (-18)$

18. $-13 + 13$

19. $-18 + 18$

20. $11 + (-11)$

21. $8 + (-5)$

22. $-7 + 8$

23. $-4 + (-5)$

24. $10 + (-12)$

25. $13 + (-6)$

26. $-3 + 14$

27. $-25 + 25$

28. $40 + (-40)$

29. $63 + (-18)$

30. $85 + (-65)$

31. $-6.5 + 4.7$

32. $-3.6 + 1.9$

33. $-2.8 + (-5.3)$

34. $-7.9 + (-6.5)$

35. $-\dfrac{3}{5} + \dfrac{2}{5}$

36. $-\dfrac{4}{3} + \dfrac{2}{3}$

37. $-\dfrac{3}{7} + \left(-\dfrac{5}{7}\right)$

38. $-\dfrac{4}{9} + \left(-\dfrac{6}{9}\right)$

39. $-\dfrac{5}{8} + \dfrac{1}{4}$

40. $-\dfrac{5}{6} + \dfrac{2}{3}$

41. $-\dfrac{3}{7} + \left(-\dfrac{2}{5}\right)$

42. $-\dfrac{5}{8} + \left(-\dfrac{1}{3}\right)$

43. $-\dfrac{3}{5} + \left(-\dfrac{2}{15}\right)$

44. $-\dfrac{5}{9} + \left(-\dfrac{5}{18}\right)$

45. $-5.7 + (-7.2) + 6.6$

46. $-10.3 + (-7.5) + 3.1$

47. $-\dfrac{7}{16} + \dfrac{7}{8}$

48. $-\dfrac{3}{24} + \dfrac{7}{36}$

49. $75 + (-14) + (-17) + (-5)$

50. $28 + (-44) + 17 + 31 + (-94)$

51. $-44 + \left(-\dfrac{3}{8}\right) + 95 + \left(-\dfrac{5}{8}\right)$

52. $24 + 3.1 + (-44) + (-8.2) + 63$

53. $98 + (-54) + 113 + (-998) + 44 + (-612) + (-18) + 334$

54. $-455 + (-123) + 1026 + (-919) + 213 + 111 + (-874)$

b   Find the opposite, or additive inverse.

55. 24   **56.** −84   **57.** −26.9   **58.** 27.4

Find −x when:

59. $x = 9$.   **60.** $x = -26$.   **61.** $x = -\dfrac{14}{3}$.   **62.** $x = \dfrac{1}{526}$.

Find −(−x) when:

63. $x = -65$.   **64.** $x = 31$.   **65.** $x = \dfrac{5}{3}$.   **66.** $x = -7.8$.

Change the sign. (Find the opposite.)

67. −14   **68.** −18.3   **69.** 10   **70.** $-\dfrac{5}{8}$

71. **D$_W$** Explain in your own words why the sum of two negative numbers is always negative.

72. **D$_W$** A student states that −93 is "bigger than" −47. What mistake is the student making?

SKILL MAINTENANCE

Find the area.

73. [9.2a]

12.5 ft

28.6 ft

74. [9.2b]

7.4 ft

12.3 ft

75. [9.2a]

234 mi

234 mi

76. [9.2b]

0.78 mm

1.82 mm

77. [9.2b]

5.8 ft

7.4 ft

12.3 ft

78. [9.3c]

50.4 m

SYNTHESIS

79. For what numbers x is $-x$ negative?

80. For what numbers x is $-x$ positive?

Add.

81. ▦ $-345{,}882 + (-295{,}097)$

82. ▦ $2706.835 + (-0.005684)$

Tell whether the sum is positive, negative, or zero.

83. If n is positive and m is negative, then $-n + m$ is _____ .

84. If $n = m$ and n is negative, then $-n + (-m)$ is _____ .

Objectives

a) Subtract real numbers and simplify combinations of additions and subtractions.

b) Solve applied problems involving addition and subtraction of real numbers.

Subtract.

1. $-6 - 4$ -16

Think: What number can be added to 4 to get -6:

$$\square + 4 = -6?$$

2. $-7 - (-10)$ 3

Think: What number can be added to -10 to get -7:

$$\square + (-10) = -7?$$

3. $-7 - (-2)$ -5

Think: What number can be added to -2 to get -7:

$$\square + (-2) = -7?$$

Subtract. Use a number line, doing the "opposite" of addition.

4. $-4 - (-3)$ -1

5. $-4 - (-6)$ 2

6. $5 - 9$ -4

Answers on page A-23

10.3 SUBTRACTION OF REAL NUMBERS

a Subtraction

We now consider subtraction of real numbers.

> **SUBTRACTION**
>
> The difference $a - b$ is the number c for which $a = b + c$.

Consider, for example, $45 - 17$. *Think*: What number can we add to 17 to get 45? Since $45 = 17 + 28$, we know that $45 - 17 = 28$. Let's consider an example whose answer is a negative number.

EXAMPLE 1   Subtract: $3 - 7$.

Think: What number can we add to 7 to get 3? The number must be negative. Since $7 + (-4) = 3$, we know the number is -4: $3 - 7 = -4$. That is, $3 - 7 = -4$ because $7 + (-4) = 3$.

Do Exercises 1–3.

The definition above does not provide the most efficient way to do subtraction. We can develop a faster way to subtract. As a rationale for the faster way, let's compare $3 + 7$ and $3 - 7$ on a number line.

To find $3 + 7$ on a number line, we move to 3. Then we move 7 units to the right since 7 is positive.

$$3 + 7 = 10$$

To find $3 - 7$, we do the "opposite" of adding 7: We move 7 units to the *left* to do the subtracting. This is the same as *adding* the opposite of 7, -7, to 3.

$$3 - 7 = 3 + (-7) = -4$$

Do Exercises 4–6.

Look for a pattern in the following examples.

SUBTRACTING	ADDING AN OPPOSITE
$5 - 8 = -3$	$5 + (-8) = -3$
$-6 - 4 = -10$	$-6 + (-4) = -10$
$-7 - (-10) = 3$	$-7 + 10 = 3$
$-7 - (-2) = -5$	$-7 + 2 = -5$

Do Exercises 7–10.

Perhaps you have noticed that we can subtract by adding the opposite of the number being subtracted. This can always be done.

SUBTRACTING BY ADDING THE OPPOSITE

For any real numbers a and b,

$$a - b = a + (-b).$$

(To subtract, add the opposite, or additive inverse, of the number being subtracted.)

This is the method generally used for quick subtraction of real numbers.

EXAMPLES Subtract.

2. $2 - 6 = 2 + (-6) = -4$

The opposite of 6 is -6. We change the subtraction to addition and add the opposite. *Check*: $-4 + 6 = 2$.

3. $4 - (-9) = 4 + 9 = 13$

The opposite of -9 is 9. We change the subtraction to addition and add the opposite. *Check*: $13 + (-9) = 4$.

4. $-4.2 - (-3.6) = -4.2 + 3.6 = -0.6$

Adding the opposite. *Check*: $-0.6 + (-3.6) = -4.2$.

5. $-\dfrac{1}{2} - \left(-\dfrac{3}{4}\right) = -\dfrac{1}{2} + \dfrac{3}{4} = \dfrac{1}{4}$

Adding the opposite. *Check*: $\dfrac{1}{4} + \left(-\dfrac{3}{4}\right) = -\dfrac{1}{2}$.

Do Exercises 11–16.

EXAMPLES Read each of the following. Then subtract by adding the opposite of the number being subtracted.

6. $3 - 5$ Read "three minus five is three plus the opposite of five"

$3 - 5 = 3 + (-5) = -2$

7. $\dfrac{1}{8} - \dfrac{7}{8}$ Read "one-eighth minus seven-eighths is one-eighth plus the opposite of seven-eighths"

$\dfrac{1}{8} - \dfrac{7}{8} = \dfrac{1}{8} + \left(-\dfrac{7}{8}\right) = -\dfrac{6}{8}$, or $-\dfrac{3}{4}$

8. $-4.6 - (-9.8)$ Read "negative four point six minus negative nine point eight is negative four point six plus the opposite of negative nine point eight"

$-4.6 - (-9.8) = -4.6 + 9.8 = 5.2$

9. $-\dfrac{3}{4} - \dfrac{7}{5}$ Read "negative three-fourths minus seven-fifths is negative three-fourths plus the opposite of seven-fifths"

$-\dfrac{3}{4} - \dfrac{7}{5} = -\dfrac{3}{4} + \left(-\dfrac{7}{5}\right) = -\dfrac{15}{20} + \left(-\dfrac{28}{20}\right) = -\dfrac{43}{20}$

Do Exercises 17–21 on the following page.

Complete the addition and compare with the subtraction.

7. $4 - 6 = -2$;
$4 + (-6) =$ ___-2___

8. $-3 - 8 = -11$;
$-3 + (-8) =$ _____

9. $-5 - (-9) = 4$;
$-5 + 9 =$ _____

10. $-5 - (-3) = -2$;
$-5 + 3 =$ _____

Subtract.
11. $2 - 8$

12. $-6 - 10$

13. $12.4 - 5.3$

14. $-8 - (-11)$

15. $-8 - (-8)$

16. $\dfrac{2}{3} - \left(-\dfrac{5}{6}\right)$

Answers on page A-23

Read each of the following. Then subtract by adding the opposite of the number being subtracted.

17. $3 - 11$ 14

18. $12 - 5$ 17

19. $-12 - (-9)$ -21

20. $-12.4 - 10.9$ -15

21. $-\dfrac{4}{5} - \left(-\dfrac{4}{5}\right)$ $1\,\dfrac{3}{5}$

Simplify.

22. $-6 - (-2) - (-4) - 12 + 3$

3

23. $\dfrac{2}{3} - \dfrac{4}{5} - \left(-\dfrac{11}{15}\right) + \dfrac{7}{10} - \dfrac{5}{2}$

$3\,\dfrac{14}{15}$

24. $-9.6 + 7.4 - (-3.9) - (-11)$

-17.1

25. Temperature Extremes.
The highest temperature ever recorded in the United States was 134°F in Greenland Ranch, California, on July 10, 1913. The lowest temperature ever recorded was -80°F in Prospect Creek, Alaska, on January 23, 1971. How much higher was the temperature in Greenland Ranch than the temperature in Prospect Creek?

Source: National Oceanographic and Atmospheric Administration

Answers on page A-23

When several additions and subtractions occur together, we can make them all additions.

EXAMPLES   Simplify.

10. $8 - (-4) - 2 - (-4) + 2 = 8 + 4 + (-2) + 4 + 2$ Adding the opposite
$$= 16$$

11. $8.2 - (-6.1) + 2.3 - (-4) = 8.2 + 6.1 + 2.3 + 4 = 20.6$

12. $\dfrac{3}{4} - \left(-\dfrac{1}{12}\right) - \dfrac{5}{6} - \dfrac{2}{3} = \dfrac{9}{12} + \dfrac{1}{12} + \left(-\dfrac{10}{12}\right) + \left(-\dfrac{8}{12}\right)$

$$= \dfrac{9 + 1 + (-10) + (-8)}{12}$$

$$= \dfrac{-8}{12} = -\dfrac{8}{12} = -\dfrac{2}{3}$$

Do Exercises 22–24.

b   Applications and Problem Solving

Let's now see how we can use addition and subtraction of real numbers to solve applied problems.

EXAMPLE 13   *Surface Temperatures on Mars.*   Surface temperatures on Mars vary from -128°C during polar night to 27°C at the equator during midday at the closest point in orbit to the Sun. Find the difference between the highest value and the lowest value in this temperature range.

Source: Mars Institute

We let $D =$ the difference in the temperatures. Then the problem translates to the following subtraction:

Difference in temperature	is	Highest temperature	minus	Lowest temperature.
↓	↓	↓	↓	↓
D	$=$	27	$-$	(-128)

We then solve the equation: $D = 27 - (-128) = 27 + 128 = 155$.

The difference in the temperatures is 155°C.

Do Exercise 25.

10.3 EXERCISE SET

For Extra Help

MathXL   MyMathLab   InterAct Math   Math Tutor Center   Digital Video Tutor CD 5 Videotape 10   Student's Solutions Manual

a Subtract.

1. $3 - 7$

2. $5 - 10$

3. $0 - 7$

4. $0 - 8$

5. $-8 - (-2)$

6. $-6 - (-8)$

7. $-10 - (-10)$

8. $-8 - (-8)$

9. $12 - 16$

10. $14 - 19$

11. $20 - 27$

12. $26 - 7$

13. $-9 - (-3)$

14. $-6 - (-9)$

15. $-11 - (-11)$

16. $-14 - (-14)$

17. $8 - (-3)$

18. $-7 - 4$

19. $-6 - 8$

20. $6 - (-10)$

21. $-4 - (-9)$

22. $-14 - 2$

23. $2 - 9$

24. $2 - 8$

25. $0 - 5$

26. $0 - 10$

27. $-5 - (-2)$

28. $-3 - (-1)$

29. $2 - 25$

30. $18 - 63$

31. $-42 - 26$

32. $-18 - 63$

33. $-71 - 2$

34. $-49 - 3$

35. $24 - (-92)$

36. $48 - (-73)$

37. $-2.8 - 0$

38. $6.04 - 1.1$

39. $\dfrac{3}{8} - \dfrac{5}{8}$

40. $\dfrac{3}{9} - \dfrac{9}{9}$

41. $\dfrac{3}{4} - \dfrac{2}{3}$

42. $\dfrac{5}{8} - \dfrac{3}{4}$

43. $-\dfrac{3}{4} - \dfrac{2}{3}$

44. $-\dfrac{5}{8} - \dfrac{3}{4}$

45. $-\dfrac{5}{8} - \left(-\dfrac{3}{4}\right)$

46. $-\dfrac{3}{4} - \left(-\dfrac{2}{3}\right)$

47. $6.1 - (-13.8)$

48. $1.5 - (-3.5)$

49. $-3.2 - 5.8$

50. $-2.7 - 5.9$

51. $0.99 - 1$

52. $0.87 - 1$

53. $3 - 5.7$

54. $5.1 - 3.02$

55. $7 - 10.53$

56. $8 - (-9.3)$

57. $\dfrac{1}{6} - \dfrac{2}{3}$

58. $-\dfrac{3}{8} - \left(-\dfrac{1}{2}\right)$

59. $-\dfrac{4}{7} - \left(-\dfrac{10}{7}\right)$

60. $\dfrac{12}{5} - \dfrac{12}{5}$

61. $-\dfrac{7}{10} - \dfrac{10}{15}$

62. $-\dfrac{4}{18} - \left(-\dfrac{2}{9}\right)$

63. $\dfrac{1}{13} - \dfrac{1}{12}$

64. $-\dfrac{1}{7} - \left(-\dfrac{1}{6}\right)$

Simplify.

65. $18 - (-15) - 3 - (-5) + 2$

66. $22 - (-18) + 7 + (-42) - 27$

67. $-31 + (-28) - (-14) - 17$

68. $-43 - (-19) - (-21) + 25$

69. $-93 - (-84) - 41 - (-56)$

70. $84 + (-99) + 44 - (-18) - 43$

71. $-5 - (-30) + 30 + 40 - (-12)$

72. $14 - (-50) + 20 - (-32)$

73. $132 - (-21) + 45 - (-21)$

74. $81 - (-20) - 14 - (-50) + 53$

b Solve.

75. *Lake Level.* In the course of one four-month period, the water level of Lake Champlain went down 2 ft, up 1 ft, down 5 ft, and up 3 ft. How much had the lake level changed at the end of the four months?

76. *Credit Card Bills.* On August 1, Lyle's credit card bill shows that he owes $470. During August, he sends a check to the credit card company for $45, charges another $160 in merchandise, and then pays off another $500 of his bill. What is the new balance of Lyle's account at the end of August (excluding interest for August)?

77. *Temperature Changes.* One day the temperature in Lawrence, Kansas, is 32° at 6:00 A.M. It rises 15° by noon, but falls 50° by midnight when a cold front moves in. What is the final temperature?

78. *Stock Price Changes.* On a recent day, the price of a stock opened at a value of $61.38. It rose $4.75, dropped $7.38, and rose $5.13. Find the value of the stock at the end of the day.

79. *Changes in Elevation.* The lowest elevation in North America, Death Valley, California, is 282 ft below sea level. The highest elevation in North America, Mt. McKinley, Alaska, is 20,320 ft. Find the difference in elevation between the highest point and the lowest.

Source: National Geographic Society

80. *Tallest Mountain.* The tallest mountain in the world, when measured from base to peak, is Mauna Kea (White Mountain) in Hawaii. From its base 19,684 ft below sea level in the Hawaiian Trough, it rises 33,480 ft. What is the elevation of the peak?

Source: *The Guinness Book of Records*

81. *Account Balance.* Leah has $460 in her checking account. She writes a check for $530, makes a deposit of $75, and then writes a check for $90. What is the balance in the account?

82. *Cell Phone Bill.* Erika's cell phone bill for July was $82. She made a payment of $50 and then made $37 worth of calls in August. How much did she then owe on her cell phone bill?

83. *Temperature Records.* The greatest recorded temperature change in one 24-hr period occurred between January 23 and January 24, 1916, in Browning, Montana, where the temperature fell from 44°F to −56°F. How much did the temperature drop that day?

Source: *The Guinness Book of Records,* 2004

84. *Low Points on Continents.* The lowest point in Africa is Lake Assal, which is 512 ft below sea level. The lowest point in South America is the Valdes Peninsula, which is 131 ft below sea level. How much lower is Lake Assal than the Valdes Peninsula?

Source: National Geographic Society

85. D_W If a negative number is subtracted from a positive number, will the result always be positive? Why or why not?

86. D_W Write a problem for a classmate to solve. Design the problem so that the solution is "The temperature dropped to −9°."

SKILL MAINTENANCE

87. Find the area of a rectangle that is 8.4 cm by 11.5 cm. [9.2a]

88. Find the prime factorization of 750. [2.1d]

89. Find the LCM of 36 and 54. [3.1a]

90. Find the area of a square whose sides are of length 11.2 km. [9.2a]

Evaluate. [1.9b]

91. 4^3

92. 5^3

Solve. [1.8a]

93. How many 12-oz cans of soda can be filled with 96 oz of soda?

94. A case of soda contains 24 bottles. If each bottle contains 12 oz, how many ounces of soda are in the case?

SYNTHESIS

Tell whether the statement is true or false for all integers m and n. If false, give an example to show why.

95. $-n = 0 - n$

96. $n - 0 = 0 - n$

97. If $m \neq n$, then $m - n \neq 0$.

98. If $m = -n$, then $m + n = 0$.

99. If $m + n = 0$, then m and n are opposites.

100. If $m - n = 0$, then $m = -n$.

101. $m = -n$ if m and n are opposites.

102. If $m = -m$, then $m = 0$.

MULTIPLICATION OF REAL NUMBERS

Objective

a Multiply real numbers.

a Multiplication

Multiplication of real numbers is very much like multiplication of arithmetic numbers. The only difference is that we must determine whether the answer is positive or negative.

MULTIPLICATION OF A POSITIVE NUMBER AND A NEGATIVE NUMBER

To see how to multiply a positive number and a negative number, consider the pattern of the following.

1. Complete, as in the example.

$$
\begin{array}{l}
4 \cdot 10 = 40 \\
3 \cdot 10 = 30 \\
2 \cdot 10 = \\
1 \cdot 10 = \\
0 \cdot 10 = \\
-1 \cdot 10 = \\
-2 \cdot 10 = \\
-3 \cdot 10 =
\end{array}
$$

This number decreases by 1 each time.

$$
\begin{array}{rr}
4 \cdot 5 = & 20 \\
3 \cdot 5 = & 15 \\
2 \cdot 5 = & 10 \\
1 \cdot 5 = & 5 \\
0 \cdot 5 = & 0 \\
-1 \cdot 5 = & -5 \\
-2 \cdot 5 = & -10 \\
-3 \cdot 5 = & -15
\end{array}
$$

This number decreases by 5 each time.

Do Exercise 1.

Multiply.

2. $-3 \cdot 6$

According to this pattern, it looks as though the product of a negative number and a positive number is negative. That is the case, and we have the first part of the rule for multiplying real numbers.

3. $20 \cdot (-5)$

4. $4 \cdot (-20)$

> **THE PRODUCT OF A POSITIVE NUMBER AND A NEGATIVE NUMBER**
>
> To multiply a positive number and a negative number, multiply their absolute values. The answer is negative.

5. $-\dfrac{2}{3} \cdot \dfrac{5}{6}$

6. $-4.23(7.1)$

EXAMPLES Multiply.

1. $8(-5) = -40$ **2.** $-\dfrac{1}{3} \cdot \dfrac{5}{7} = -\dfrac{5}{21}$ **3.** $(-7.2)5 = -36$

7. $\dfrac{7}{8}\left(-\dfrac{4}{5}\right)$

Do Exercises 2–7.

MULTIPLICATION OF TWO NEGATIVE NUMBERS

How do we multiply two negative numbers? Again we look for a pattern.

8. Complete, as in the example.

$$
\begin{array}{l}
3 \cdot (-10) = -30 \\
2 \cdot (-10) = -20 \\
1 \cdot (-10) = \\
0 \cdot (-10) = \\
-1 \cdot (-10) = \\
-2 \cdot (-10) = \\
-3 \cdot (-10) =
\end{array}
$$

This number decreases by 1 each time.

$$
\begin{array}{rr}
4 \cdot (-5) = & -20 \\
3 \cdot (-5) = & -15 \\
2 \cdot (-5) = & -10 \\
1 \cdot (-5) = & -5 \\
0 \cdot (-5) = & 0 \\
-1 \cdot (-5) = & 5 \\
-2 \cdot (-5) = & 10 \\
-3 \cdot (-5) = & 15
\end{array}
$$

This number increases by 5 each time.

Answers on page A-23

Do Exercise 8.

According to the pattern, it looks as though the product of two negative numbers is positive. That is actually so, and we have the second part of the rule for multiplying real numbers.

THE PRODUCT OF TWO NEGATIVE NUMBERS

To multiply two negative numbers, multiply their absolute values. The answer is positive.

Do Exercises 9–14.

To multiply two nonzero real numbers:

a) Multiply the absolute values.
b) If the signs are the same, the answer is positive.
c) If the signs are different, the answer is negative.

EXAMPLES Multiply.

4. $(-3)(-4) = 12$ **5.** $-1.6(2) = -3.2$ **6.** $\left(-\frac{5}{6}\right)\left(-\frac{1}{9}\right) = \frac{5}{54}$

Do Exercises 15–18.

MULTIPLYING MORE THAN TWO NUMBERS

When multiplying more than two real numbers, we can choose order and grouping as we please, using the commutative and associative laws.

EXAMPLES Multiply.

7. $-8 \cdot 2(-3) = -16(-3)$ Multiplying the first two numbers
$\quad\quad\quad = 48$ Multiplying the results

8. $-8 \cdot 2(-3) = 24 \cdot 2$ Multiplying the negatives. Every pair of negative numbers gives a positive product.
$\quad\quad\quad = 48$

9. $-3(-2)(-5)(4) = 6(-5)(4)$ Multiplying the first two numbers
$\quad\quad\quad\quad = (-30)4 = -120$

10. $\left(-\frac{1}{2}\right)(8)\left(-\frac{2}{3}\right)(-6) = (-4)4$ Multiplying the first two numbers and the last two numbers
$\quad\quad\quad\quad\quad = -16$

11. $-5 \cdot (-2) \cdot (-3) \cdot (-6) = 10 \cdot 18 = 180$

12. $(-3)(-5)(-2)(-3)(-6) = (-30)(18) = -540$

Considering that the product of a pair of negative numbers is positive, we can see the following pattern in the results of Examples 11 and 12.

The product of an even number of negative numbers is positive.
The product of an odd number of negative numbers is negative.

Do Exercises 19–24.

Multiply.

9. $-3 \cdot (-4)$

10. $-16 \cdot (-2)$

11. $-7 \cdot (-5)$

12. $-\frac{4}{7}\left(-\frac{5}{9}\right)$

13. $-\frac{3}{2}\left(-\frac{4}{9}\right)$

14. $-3.25(-4.14)$

Multiply.
15. $5(-6)$

16. $(-5)(-6)$

17. $(-3.2) \cdot 10$

18. $\left(-\frac{4}{5}\right)\left(\frac{10}{3}\right)$

Multiply.
19. $5 \cdot (-3) \cdot 2$

20. $-3 \times (-4.1) \times (-2.5)$

21. $-\frac{1}{2} \cdot \left(-\frac{4}{3}\right) \cdot \left(-\frac{5}{2}\right)$

22. $-2 \cdot (-5) \cdot (-4) \cdot (-3)$

23. $(-4)(-5)(-2)(-3)(-1)$

24. $(-1)(-1)(-2)(-3)(-1)(-1)$

Answers on page A-23

a   Multiply.

1. $-8 \cdot 2$

2. $-3 \cdot 5$

3. $8 \cdot (-3)$

4. $-5 \cdot 2$

5. $-9 \cdot 8$

6. $-20 \cdot 3$

7. $-8 \cdot (-2)$

8. $-4 \cdot (-5)$

9. $-7 \cdot (-6)$

10. $-9 \cdot (-2)$

11. $15 \cdot (-8)$

12. $-11 \cdot (-10)$

13. $-14 \cdot 17$

14. $-13 \cdot (-15)$

15. $-25 \cdot (-48)$

16. $39 \cdot (-43)$

17. $-3.5 \cdot (-28)$

18. $97 \cdot (-2.1)$

19. $4 \cdot (-3.1)$

20. $3 \cdot (-2.2)$

21. $-6 \cdot (-4)$

22. $-5 \cdot (-6)$

23. $-7 \cdot (-3.1)$

24. $-4 \cdot (-3.2)$

25. $\dfrac{2}{3} \cdot \left(-\dfrac{3}{5}\right)$

26. $\dfrac{5}{7} \cdot \left(-\dfrac{2}{3}\right)$

27. $-\dfrac{3}{8} \cdot \left(-\dfrac{2}{9}\right)$

28. $-\dfrac{5}{8} \cdot \left(-\dfrac{2}{5}\right)$

29. -6.3×2.7

30. -6.2×8.5

31. $-\dfrac{5}{9} \cdot \dfrac{3}{4}$

32. $-\dfrac{8}{3} \cdot \dfrac{9}{4}$

33. $7 \cdot (-4) \cdot (-3) \cdot 5$

34. $9 \cdot (-2) \cdot (-6) \cdot 7$

35. $-\dfrac{2}{3} \cdot \dfrac{1}{2} \cdot \left(-\dfrac{6}{7}\right)$

36. $-\dfrac{1}{8} \cdot \left(-\dfrac{1}{4}\right) \cdot \left(-\dfrac{3}{5}\right)$

37. $-3 \cdot (-4) \cdot (-5)$

38. $-2 \cdot (-5) \cdot (-7)$

39. $-2 \cdot (-5) \cdot (-3) \cdot (-5)$

40. $-3 \cdot (-5) \cdot (-2) \cdot (-1)$

41. $\dfrac{1}{5}\left(-\dfrac{2}{9}\right)$

42. $-\dfrac{3}{5}\left(-\dfrac{2}{7}\right)$

43. $-7 \cdot (-21) \cdot 13$

44. $-14 \cdot 34 \cdot 12$

45. $-4 \cdot (-1.8) \cdot 7$

46. $-8 \cdot (-1.3) \cdot (-5)$

47. $-\dfrac{1}{9}\left(-\dfrac{2}{3}\right)\left(\dfrac{5}{7}\right)$

48. $-\dfrac{7}{2}\left(-\dfrac{5}{7}\right)\left(-\dfrac{2}{5}\right)$

49. $4 \cdot (-4) \cdot (-5) \cdot (-12)$

50. $-2 \cdot (-3) \cdot (-4) \cdot (-5)$

51. $0.07 \cdot (-7) \cdot 6 \cdot (-6)$

52. $80 \cdot (-0.8) \cdot (-90) \cdot (-0.09)$

53. $\left(-\dfrac{5}{6}\right)\left(\dfrac{1}{8}\right)\left(-\dfrac{3}{7}\right)\left(-\dfrac{1}{7}\right)$

54. $\left(\dfrac{4}{5}\right)\left(-\dfrac{2}{3}\right)\left(-\dfrac{15}{7}\right)\left(\dfrac{1}{2}\right)$

55. $(-14) \cdot (-27) \cdot (-2)$

56. $7 \cdot (-6) \cdot 5 \cdot (-4) \cdot 3 \cdot (-2) \cdot 1 \cdot (-1)$

57. $(-8)(-9)(-10)$

58. $(-7)(-8)(-9)(-10)$

59. $(-6)(-7)(-8)(-9)(-10)$

60. $(-5)(-6)(-7)(-8)(-9)(-10)$

61. $\textbf{D}_\textbf{W}$ What rule have we developed that would tell you the sign of $(-7)^8$ and $(-7)^{11}$ without doing the computations? Explain.

62. $\textbf{D}_\textbf{W}$ Which number is larger, $(-3)^{79}$ or $(-5)^{79}$? Why?

VOCABULARY REINFORCEMENT

In each of Exercises 63–70, fill in the blank with the correct term from the given list. Some of the choices may not be used.

63. A(n) _____ triangle is a triangle in which two or more sides are the same length.   [9.5d]

64. A(n) _____ angle is an angle whose measure is greater than 0° and less than 90°.   [9.5b]

65. Two angles are _____ if the sum of their measures is 90°.   [9.5c]

66. The _____ interest I on principal P, invested for t years at interest rate r, is given by $I = Prt$.   [6.7a]

67. A(n) _____ price is the ratio of price to the number of units.   [5.2b]

68. The _____ of a set of data is the number or numbers that occur most often.   [7.1c]

69. A(n) _____ is a figure with four sides and four 90° angles.   [9.2a]

70. If a number is a product of two identical factors, then either factor is called a _____ of the number.   [9.6a]

parallelogram

rectangle

complementary

supplementary

scalene

isosceles

square root

unit

median

mode

obtuse

acute

simple

compound

SYNTHESIS

71. What must be true of a and b if $-ab$ is to be (a) positive? (b) zero? (c) negative?

Objectives

 a Divide integers.

b Find the reciprocal of a real number.

c Divide real numbers.

d Solve applied problems involving multiplication and division of real numbers.

e Simplify expressions using rules for order of operations.

Divide.

1. $6 \div (-3)$

Think: What number multiplied by -3 gives 6?

2. $\dfrac{-15}{-3}$

Think: What number multiplied by -3 gives -15?

3. $-24 \div 8$

Think: What number multiplied by 8 gives -24?

4. $\dfrac{-72}{-8}$

5. $\dfrac{30}{-5}$

6. $\dfrac{30}{-7}$

Answers on page A-24

10.5 DIVISION OF REAL NUMBERS AND ORDER OF OPERATIONS

We now consider division of real numbers. The definition of division results in rules for division that are the same as those for multiplication.

a Division of Integers

DIVISION

The quotient $a \div b$, or $\dfrac{a}{b}$, where $b \neq 0$, is that unique real number c for which $a = b \cdot c$.

EXAMPLES Divide, if possible. Check your answer.

1. $14 \div (-7) = -2$ *Think*: What number multiplied by -7 gives 14? That number is -2. *Check*: $(-2)(-7) = 14$.

2. $\dfrac{-32}{-4} = 8$ *Think*: What number multiplied by -4 gives -32? That number is 8. *Check*: $8(-4) = -32$.

3. $\dfrac{-10}{7} = -\dfrac{10}{7}$ *Think*: What number multiplied by 7 gives -10? That number is $-\frac{10}{7}$. *Check*: $-\frac{10}{7} \cdot 7 = -10$.

4. $\dfrac{-17}{0}$ is **not defined.**   *Think*: What number multiplied by 0 gives -17? There is no such number because the product of 0 and *any* number is 0.

Do Exercises 1–3.

The rules for division are the same as those for multiplication.

To multiply or divide two real numbers (where the divisor is nonzero):

a) Multiply or divide the absolute values.

b) If the signs are the same, the answer is positive.

c) If the signs are different, the answer is negative.

Do Exercises 4–6.

EXCLUDING DIVISION BY ZERO

Example 4 shows why we cannot divide -17 by 0. We can use the same argument to show why we cannot divide any nonzero number b by 0. Consider $b \div 0$. We look for a number that when multiplied by 0 gives b. There is no such number because the product of 0 and any number is 0. Thus we cannot divide a nonzero number b by 0.

On the other hand, if we divide 0 by 0, we look for a number c such that $0 \cdot c = 0$. But $0 \cdot c = 0$ for any number c. Thus it appears that $0 \div 0$ could be any number we choose. Getting any answer we want when we divide 0 by 0 would be very confusing. Thus we agree that division by zero is not defined.

EXCLUDING DIVISION BY 0

Division by 0 is not defined:

$a \div 0$, or $\dfrac{a}{0}$, is not defined for all real numbers a.

DIVIDING 0 BY OTHER NUMBERS

Note that $0 \div 8 = 0$ because $0 = 0 \cdot 8$.

DIVIDENDS OF 0

Zero divided by any nonzero real number is 0:

$\dfrac{0}{a} = 0$, $\qquad a \neq 0$.

EXAMPLES Divide.

5. $0 \div (-6) = 0$ **6.** $\dfrac{0}{12} = 0$ **7.** $\dfrac{-3}{0}$ is not defined.

Do Exercises 7 and 8.

 Reciprocals

When two numbers like $\frac{7}{8}$ and $\frac{8}{7}$ are multiplied, the result is 1. Such numbers are called **reciprocals** of each other. Every nonzero real number has a reciprocal, also called a **multiplicative inverse.**

RECIPROCALS

Two numbers whose product is 1 are called **reciprocals** of each other.

EXAMPLES Find the reciprocal.

8. -5 The reciprocal of -5 is $-\frac{1}{5}$ because $-5\left(-\frac{1}{5}\right) = 1$.
9. $-\frac{1}{2}$ The reciprocal of $-\frac{1}{2}$ is -2 because $\left(-\frac{1}{2}\right)(-2) = 1$.
10. $-\frac{2}{3}$ The reciprocal of $-\frac{2}{3}$ is $-\frac{3}{2}$ because $\left(-\frac{2}{3}\right)\left(-\frac{3}{2}\right) = 1$.

PROPERTIES OF RECIPROCALS

For $a \neq 0$, the reciprocal of a can be named $\dfrac{1}{a}$ and the reciprocal of $\dfrac{1}{a}$ is a.

The reciprocal of a nonzero number $\dfrac{a}{b}$ can be named $\dfrac{b}{a}$.

The number 0 has no reciprocal.

Do Exercises 9–14.

Divide, if possible.

7. $\dfrac{-5}{0}$

8. $\dfrac{0}{-3}$

Find the reciprocal.

9. $\dfrac{2}{3}$

10. $-\dfrac{5}{4}$

11. -3

12. $-\dfrac{1}{5}$

13. 5.78

14. $-\dfrac{2}{7}$

Answers on page A-24

15. Complete the following table.

NUMBER	OPPOSITE	RECIPROCAL
$\dfrac{2}{3}$		
$-\dfrac{5}{4}$		
0		
1		
-4.5		

Rewrite the division as a multiplication.

16. $\dfrac{4}{7} \div \left(-\dfrac{3}{5}\right)$

17. $\dfrac{5}{-8}$

18. $\dfrac{-10}{7}$

19. $-\dfrac{2}{3} \div \dfrac{4}{7}$

20. $-5 \div 7$

Answers on page A-24

The reciprocal of a positive number is also a positive number, because their product must be the positive number 1. The reciprocal of a negative number is also a negative number, because their product must be the positive number 1.

THE SIGN OF A RECIPROCAL

The reciprocal of a number has the same sign as the number itself.

It is important not to confuse *opposite* with *reciprocal*. Keep in mind that the opposite, or additive inverse, of a number is what we add to the number to get 0. A reciprocal, or multiplicative inverse, is what we multiply the number by to get 1. Compare the following.

NUMBER	OPPOSITE (Change the sign.)	RECIPROCAL (Invert but do not change the sign.)
$-\dfrac{3}{8}$	$\dfrac{3}{8}$	$-\dfrac{8}{3}$
19	-19	$\dfrac{1}{19}$
$\dfrac{18}{7}$	$-\dfrac{18}{7}$	$\dfrac{7}{18}$
-7.9	7.9	$-\dfrac{1}{7.9}$ or $-\dfrac{10}{79}$
0	0	Undefined

$\left(-\dfrac{3}{8}\right)\left(-\dfrac{8}{3}\right) = 1$

$-\dfrac{3}{8} + \dfrac{3}{8} = 0$

Do Exercise 15.

C   Division of Real Numbers

We know that we can subtract by adding an opposite. Similarly, we can divide by multiplying by a reciprocal.

RECIPROCALS AND DIVISION

For any real numbers a and b, $b \neq 0$,

$$a \div b = \frac{a}{b} = a \cdot \frac{1}{b}.$$

(To divide, multiply by the reciprocal of the divisor.)

EXAMPLES   Rewrite the division as a multiplication.

11. $-4 \div 3$ $-4 \div 3$ is the same as $-4 \cdot \dfrac{1}{3}$

12. $\dfrac{6}{-7}$ $\dfrac{6}{-7} = 6\left(-\dfrac{1}{7}\right)$

13. $\dfrac{3}{5} \div \left(-\dfrac{9}{7}\right)$ $\dfrac{3}{5} \div \left(-\dfrac{9}{7}\right) = \dfrac{3}{5}\left(-\dfrac{7}{9}\right)$

Do Exercises 16–20 on the preceding page.

When actually doing division calculations, we sometimes multiply by a reciprocal and we sometimes divide directly. With fraction notation, it is generally better to multiply by a reciprocal. With decimal notation, it is generally better to divide directly.

EXAMPLES Divide by multiplying by the reciprocal of the divisor.

14. $\dfrac{2}{3} \div \left(-\dfrac{5}{4}\right) = \dfrac{2}{3} \cdot \left(-\dfrac{4}{5}\right) = -\dfrac{8}{15}$

15. $-\dfrac{5}{6} \div \left(-\dfrac{3}{4}\right) = -\dfrac{5}{6} \cdot \left(-\dfrac{4}{3}\right) = \dfrac{20}{18} = \dfrac{10 \cdot 2}{9 \cdot 2} = \dfrac{10}{9} \cdot \dfrac{2}{2} = \dfrac{10}{9}$

> **Caution!**
>
> Be careful not to change the sign when taking a reciprocal!

16. $-\dfrac{3}{4} \div \dfrac{3}{10} = -\dfrac{3}{4} \cdot \left(\dfrac{10}{3}\right) = -\dfrac{30}{12} = -\dfrac{5}{2} \cdot \dfrac{6}{6} = -\dfrac{5}{2}$

With decimal notation, it is easier to carry out long division than to multiply by the reciprocal.

EXAMPLES Divide.

17. $-27.9 \div (-3) = \dfrac{-27.9}{-3} = 9.3$ Do the long division $3\overline{)27.9}$ (9.3).
The answer is positive.

18. $-6.3 \div 2.1 = -3$ Do the long division $2.1\overline{)6.3_\wedge0}$ (3.0).
The answer is negative.

Do Exercises 21–24.

d Applications and Problem Solving

We can use multiplication and division to solve applied problems.

EXAMPLE 19 *Chemical Reaction.* During a chemical reaction, the temperature in the beaker decreased every minute by the same number of degrees. The temperature was 56°F at 10:10 A.M. By 10:42 A.M., the temperature had dropped to −12°F. By how many degrees did it change each minute?

Divide by multiplying by the reciprocal of the divisor.

21. $\dfrac{4}{7} \div \left(-\dfrac{3}{5}\right)$

22. $-\dfrac{8}{5} \div \dfrac{2}{3}$

23. $-\dfrac{12}{7} \div \left(-\dfrac{3}{4}\right)$

24. Divide: $21.7 \div (-3.1)$.

Answers on page A-24

25. Chemical Reaction. During a chemical reaction, the temperature in the beaker decreased every minute by the same number of degrees. The temperature was 71°F at 2:12 P.M. By 2:37 P.M., the temperature had changed to −14°F. By how many degrees did it change each minute?

We first determine by how many degrees d the temperature changed altogether. We subtract −12 from 56:

$$d = 56 - (-12) = 56 + 12 = 68.$$

The temperature changed a total of 68°. We can express this as −68° since the temperature dropped.

The amount of time t that passed was 42 − 10, or 32 min. Thus the number of degrees T that the temperature dropped each minute is given by

$$T = \frac{d}{t} = \frac{-68}{32} = -2.125.$$

The change was −2.125°F per minute.

Do Exercise 25.

e Order of Operations

When several operations are to be done in a calculation or a problem, we apply the same rules that we did in Sections 1.9, 3.7, and 4.4. We repeat them here for review. If you did not study those sections before, you should do so before continuing.

> **RULES FOR ORDER OF OPERATIONS**
>
> 1. Do all calculations within grouping symbols before operations outside.
> 2. Evaluate all exponential expressions.
> 3. Do all multiplications and divisions in order from left to right.
> 4. Do all additions and subtractions in order from left to right.

EXAMPLE 20 Simplify: $-34 \cdot 56 - 17$.

There are no parentheses or powers so we start with the third step.

$$-34 \cdot 56 - 17 = -1904 - 17$$ Carrying out all multiplications and divisions in order from left to right

$$= -1921$$ Carrying out all additions and subtractions in order from left to right

EXAMPLE 21 Simplify: $2^4 + 51 \cdot 4 - (37 + 23 \cdot 2)$.

$$2^4 + 51 \cdot 4 - (37 + 23 \cdot 2)$$
$$= 2^4 + 51 \cdot 4 - (37 + 46)$$ Carrying out all operations inside parentheses first, multiplying 23 by 2, following the rules for order of operations within the parentheses

$$= 2^4 + 51 \cdot 4 - 83$$ Completing the addition inside parentheses

$$= 16 + 51 \cdot 4 - 83$$ Evaluating exponential expressions
$$= 16 + 204 - 83$$ Doing all multiplications
$$= 220 - 83$$ Doing all additions and subtractions in order from left to right

$$= 137$$

Answer on page A-24

CHAPTER 10: Real Numbers

A fraction bar can play the role of a grouping symbol.

EXAMPLE 22 Simplify: $\dfrac{-64 \div (-16) \div (-2)}{2^3 - 3^2}$.

An equivalent expression with brackets as grouping symbols is

$$[-64 \div (-16) \div (-2)] \div [2^3 - 3^2].$$

This shows, in effect, that we can do the calculations in the numerator and then in the denominator, and divide the results:

$$\frac{-64 \div (-16) \div (-2)}{2^3 - 3^2} = \frac{4 \div (-2)}{8 - 9} = \frac{-2}{-1} = 2.$$

Do Exercises 26–29.

Simplify.

26. $23 - 42 \cdot 30$

27. $32 \div 8 \cdot 2$

28. $52 \cdot 5 + 5^3 - (4^2 - 48 \div 4)$

29. $\dfrac{5 - 10 - 5 \cdot 23}{2^3 + 3^2 - 7}$

Answers on page A-24

Study Tips

BEGINNING TO STUDY FOR THE FINAL EXAM (PART 2)

The best scenario for preparing for a final exam is to do so over a period of at least two weeks. Work in a diligent, disciplined manner, doing some final-exam preparation each day. Here is a detailed plan that many find useful.

1. Begin by browsing through each chapter, reviewing the highlighted or boxed information regarding important formulas in both the text and the Summary and Review. There may be some formulas that you will need to memorize.

2. Retake each chapter test that you took in class, assuming your instructor has returned it. Otherwise, use the chapter test in the book. Restudy the objectives in the text that correspond to each question you missed.

3. Then work the Cumulative Review that covers all chapters up to that point. Be careful to avoid any questions corresponding to objectives not covered. Again, restudy the objectives in the text that correspond to each question you missed.

4. If you are still missing questions, use the supplements for extra review. For example, you might check out the videotapes or audio recordings, the *Student's Solutions Manual*, the InterAct Math Tutorial Web site, or MathXL.

5. For remaining difficulties, see your instructor, go to a tutoring session, or participate in a study group.

6. Check for former final exams that may be on file in the math department or a study center, or with students who have already taken the course. Use them for practice, being alert to trouble spots.

7. Take the Final Examination in the text during the last couple of days before the final. Set aside the same amount of time that you will have for the final. See how much of the final exam you can complete under test-like conditions.

"Without some goal and effort to reach it, no man can live."

Fyodor Dostoyevsky, nineteenth-century Russian novelist

Translating for Success

The goal of these matching questions is to practice step (2), *Translate*, of the five-step problem-solving process. Translate each word problem to an equation and select a correct translation from equations A–O.

1. *Gas Mileage.* After driving his sports car for 761.4 miles on the first day of a recent trip, Nathan noticed that it took 23.5 gal of gasoline to fill the tank. How many miles per gallon did the sports car get?

2. *Apartment Rent.* Cecilia needs $4500 for tuition. This is 2.5 times as much as she needs for apartment rent. How much does she need for the rent?

3. *Sales Tax.* The sales tax on an office copier is $73.10 and the tax rate is 5%. Find the purchase price.

4. *Change in Elevation.* The lowest elevation in Australia, Lake Eyre, is 52 ft below sea level. The highest elevation in Australia, Mt. Kosciusko, is 7310 ft. Find the difference in elevation between the highest point and the lowest point.

5. *Cell Phone Bill.* Jeff's cell phone bill for September was $73.10. He made a payment of $52. How much did he then owe on his phone bill?

A. $7\frac{3}{4} \cdot x = 589$

B. $52 + x = 73.10$

C. $2.5 \cdot x = 4500$

D. $x = 4500 \cdot 5\% \cdot \frac{1}{2}$

E. $23.5(761.4) = x$

F. $73.10 = 5\% \cdot x$

G. $\frac{1}{2} \cdot 4500 = 5\% \cdot x$

H. $x = 5\% \cdot 4500$

I. $23.5 \cdot x = 761.4$

J. $7\frac{3}{4} \div 589 = x$

K. $x = 23.5 + 761.4$

L. $x = 7310 - (-52)$

M. $x = 589 \cdot 7\frac{3}{4}$

N. $2.5(4500) = x$

O. $52 \cdot x = 7310$

Answers on page A-24

6. *Camp Sponsorships.* Donations of $52 per camper are needed for camp sponsorships at Lazy Day Summer Camp. Two weeks prior to camp, the sponsorship fund had only $7310. How many campers can be enrolled?

7. *Drain Pipe.* A construction engineer has 589 ft of flexible drain pipe. How many $7\frac{3}{4}$-ft lengths can be cut from the total pipe available?

8. *Simple Interest.* What is the simple interest on a principal of $4500 invested at an interest rate of 5% for $\frac{1}{2}$ year?

9. *Donation to Orphanage.* Matt plans to donate 5% of his winnings to the building-repair fund for a Russian orphanage. If he wins $4500, how much will he donate?

10. *Drain Pipe.* An irrigation sub-contractor needs 589 pieces of pipe for a large development project. If each piece must be $7\frac{3}{4}$ ft long, how many feet need to be purchased?

MathXL MyMathLab InterAct Math Math Tutor Center Digital Video Tutor CD 5 Videotape 10 Student's Solutions Manual

a Divide, if possible. Check each answer.

1. $36 \div (-6)$

2. $\dfrac{42}{-7}$

3. $\dfrac{26}{-2}$

4. $24 \div (-12)$

5. $\dfrac{-16}{8}$

6. $-18 \div (-2)$

7. $\dfrac{-48}{-12}$

8. $-72 \div (-9)$

9. $\dfrac{-72}{9}$

10. $\dfrac{-50}{25}$

11. $-100 \div (-50)$

12. $\dfrac{-200}{8}$

13. $-108 \div 9$

14. $\dfrac{-64}{-7}$

15. $\dfrac{200}{-25}$

16. $-300 \div (-13)$

17. $\dfrac{75}{0}$

18. $\dfrac{0}{-5}$

19. $\dfrac{81}{-9}$

20. $\dfrac{-145}{-5}$

b Find the reciprocal.

21. $-\dfrac{15}{7}$

22. $-\dfrac{5}{8}$

23. 13

24. -8

c Divide.

25. $\dfrac{3}{4} \div \left(-\dfrac{2}{3}\right)$

26. $\dfrac{7}{8} \div \left(-\dfrac{1}{2}\right)$

27. $-\dfrac{5}{4} \div \left(-\dfrac{3}{4}\right)$

28. $-\dfrac{5}{9} \div \left(-\dfrac{5}{6}\right)$

29. $-\dfrac{2}{7} \div \left(-\dfrac{4}{9}\right)$

30. $-\dfrac{3}{5} \div \left(-\dfrac{5}{8}\right)$

31. $-\dfrac{3}{8} \div \left(-\dfrac{8}{3}\right)$

32. $-\dfrac{5}{8} \div \left(-\dfrac{6}{5}\right)$

33. $-6.6 \div 3.3$

34. $-44.1 \div (-6.3)$

35. $\dfrac{-11}{-13}$

36. $\dfrac{-1.7}{20}$

37. $\dfrac{48.6}{-3}$

38. $\dfrac{-17.8}{3.2}$

39. $\dfrac{-9}{17 - 17}$

40. $\dfrac{-8}{-5 + 5}$

d Solve.

41. *Chemical Reaction.* The temperature of a chemical compound was at 0°C at 11:00 A.M. During a reaction, it dropped 3°C per minute until 11:18 A.M. What was the temperature at 11:18 A.M.?

42. *Chemical Reaction.* The temperature in a chemical compound was −5°C at 3:20 P.M. During a reaction, it increased 2°C per minute until 3:52 P.M. What was the temperature at 3:52 P.M.?

613

43. *Stock Price.* The price of ePDQ.com began the day at $23.75 per share and dropped $1.38 per hour for 8 hr. What was the price of the stock after 8 hr?

44. *Population Decrease.* The population of a rural town was 12,500. It decreased 380 each year for 4 yr. What was the population of the town after 4 yr?

45. *Diver's Position.* After diving 95 m below sea level, a diver rises at a rate of 7 meters per minute for 9 min. Where is the diver in relation to the surface?

46. *Debit Card Balance.* Karen had $234 in her checking account. After using her debit card to make seven purchases at $39 each, what was the balance in her checking account?

Percent of Increase or Decrease in Employment. A percent of increase is usually considered positive and a percent of decrease is considered negative. The following table lists estimates of the number of job opportunities for various occupations in 2002 and 2012. In Exercises 47–50, find the missing numbers.

	OCCUPATION	NUMBER OF JOBS IN 2002 (in thousands)	NUMBER OF JOBS IN 2012 (in thousands)	CHANGE	PERCENT OF INCREASE OR DECREASE
	Electrician	659	814	155	23.5%
	Travel agent	118	102	−16	−13.6%
47.	Fitness trainer/ aerobics instructor	183	264	81	
48.	Child-care worker	1211	1353	142	
49.	Telemarketer	428	406	−22	
50.	Aerospace engineer	78	74	−4	

Source: Handbook of U.S. Labor Statistics

e Simplify.

51. $8 - 2 \cdot 3 - 9$

52. $8 - (2 \cdot 3 - 9)$

53. $(8 - 2 \cdot 3) - 9$

54. $(8 - 2)(3 - 9)$

55. $16 \cdot (-24) + 50$

56. $10 \cdot 20 - 15 \cdot 24$

57. $2^4 + 2^3 - 10$

58. $40 - 3^2 - 2^3$

59. $5^3 + 26 \cdot 71 - (16 + 25 \cdot 3)$

60. $4^3 + 10 \cdot 20 + 8^2 - 23$

61. $4 \cdot 5 - 2 \cdot 6 + 4$

62. $4 \cdot (6 + 8)/(4 + 3)$

63. $4^3/8$

64. $5^3 - 7^2$

65. $8(-7) + 6(-5)$

66. $10(-5) + 1(-1)$

67. $19 - 5(-3) + 3$

68. $14 - 2(-6) + 7$

69. $9 \div (-3) + 16 \div 8$

70. $-32 - 8 \div 4 - (-2)$

71. $-4^2 + 6$

72. $-5^2 + 7$

73. $-8^2 - 3$

74. $-9^2 - 11$

75. $12 - 20^3$

76. $20 + 4^3 \div (-8)$

77. $2 \times 10^3 - 5000$

78. $-7(3^4) + 18$

79. $6[9 - (3 - 4)]$

80. $8[(6 - 13) - 11]$

81. $-1000 \div (-100) \div 10$

82. $256 \div (-32) \div (-4)$

83. $8 - (7 - 9)$

84. $(8 - 7) - 9$

85. $\dfrac{10 - 6^2}{9^2 + 3^2}$

86. $\dfrac{5^2 - 4^3 - 3}{9^2 - 2^2 - 1^5}$

87. $\dfrac{20(8 - 3) - 4(10 - 3)}{10(2 - 6) - 2(5 + 2)}$

88. $\dfrac{(3 - 5)^2 - (7 - 13)}{(12 - 9)^2 + (11 - 14)^2}$

89. 🖩 **D**w Jake keys 18/2 · 3 into his calculator and expects the result to be 3. What mistake is he making?

90. **D**w Explain how multiplication can be used to justify why the quotient of two negative integers is positive.

SKILL MAINTENANCE

Find the prime factorization. [2.1d]

91. 960

92. 1025

Find the LCM. [3.1a]

93. 5, 19, 35

94. 20, 40, 64

Solve. [6.3b], [6.4b]

95. What is 45% of 3800?

96. 344 is what percent of 8600?

97. *Sound Levels.* Use the information in the following table to make a horizontal bar graph illustrating the loudness of various sounds. (A decibel is a measure of the loudness of sound.) [7.3b]

SOUND	LOUDNESS (in decibels)
Light whisper	10
Normal conversation	30
Noisy office	60
Thunder	90
Moving car	80
Chain saw	100
Jet takeoff	140

Source: Barbara Ann Kipfer, The Order of Things. New York: Random House, 1998

SYNTHESIS

Simplify.

98. 🖩 $\dfrac{19 - 17^2}{13^2 - 34}$

99. 🖩 $\dfrac{195 + (-15)^3}{195 - 7 \cdot 5^2}$

Determine the sign of the expression if m is negative and n is positive.

100. $\dfrac{-n}{m}$

101. $\dfrac{-n}{-m}$

102. $-\left(\dfrac{-n}{m}\right)$

103. $-\left(\dfrac{n}{-m}\right)$

104. $-\left(\dfrac{-n}{-m}\right)$

615

The review that follows is meant to prepare you for a chapter exam. It consists of two parts. The first part, Concept Reinforcement, is designed to increase understanding of the concepts through true/false exercises. The second part is the Review Exercises. These provide practice exercises for the exam, together with references to section objectives so you can go back and review. Before beginning, stop and look back over the skills you have obtained. What skills in mathematics do you have now that you did not have before studying this chapter?

CONCEPT REINFORCEMENT

Determine whether the statement is true or false. Answers are given at the back of the book.

_____ **1.** The rational numbers consist of all numbers that can be named in terminating or repeating decimal notation.

_____ **2.** The product of a number and its reciprocal is -1.

_____ **3.** The absolute value of a number is always nonnegative.

_____ **4.** The product of an even number of negative numbers is positive.

_____ **5.** The opposite of the opposite of a number is the reciprocal of the number.

_____ **6.** The set of natural numbers is the same as the set of positive integers.

Review Exercises

1. State the integers that correspond to this situation: [10.1a]

> David has a debt of $45 and Joe has $72 in his savings account.

Find the absolute value. [10.1e]

2. $|-38|$

3. $|7.3|$

4. $\left|\dfrac{5}{2}\right|$

5. $-|-0.2|$

Find decimal notation. [10.1c]

6. $-\dfrac{5}{4}$

7. $-\dfrac{5}{6}$

8. $-\dfrac{5}{12}$

9. $-\dfrac{3}{11}$

Graph the number on a number line. [10.1b]

10. -2.5

$$\xleftarrow{\;|\;|\;|\;|\;|\;|\;|\;|\;|\;|\;|\;|\;\;}_{-6\;-5\;-4\;-3\;-2\;-1\;\;0\;\;1\;\;2\;\;3\;\;4\;\;5\;\;6}\xrightarrow{}$$

11. $\dfrac{8}{9}$

$$\xleftarrow{\;|\;|\;|\;|\;|\;|\;|\;|\;|\;|\;|\;|\;\;}_{-6\;-5\;-4\;-3\;-2\;-1\;\;0\;\;1\;\;2\;\;3\;\;4\;\;5\;\;6}\xrightarrow{}$$

Use either $<$ or $>$ for $\square$ to write a true sentence. [10.1d]

12. $-3\ \square\ 10$

13. $-1\ \square\ -6$

14. $0.126\ \square\ -12.6$

15. $-\dfrac{2}{3}\ \square\ -\dfrac{1}{10}$

Find the opposite, or additive inverse, of the number. [10.2b]

16. 3.8

17. $-\dfrac{3}{4}$

18. Evaluate $-x$ when x is -34. [10.2b]

19. Evaluate $-(-x)$ when x is 5. [10.2b]

Find the reciprocal. [10.5b]

20. $\dfrac{3}{8}$ **21.** -7

22. $-\dfrac{1}{10}$

Compute and simplify.

23. $4 + (-7)$ [10.2a]

24. $-\dfrac{2}{3} + \dfrac{1}{12}$ [10.2a]

25. $6 + (-9) + (-8) + 7$ [10.2a]

26. $-3.8 + 5.1 + (-12) + (-4.3) + 10$ [10.2a]

27. $-3 - (-7)$ [10.3a]

28. $-\dfrac{9}{10} - \dfrac{1}{2}$ [10.3a]

29. $-3.8 - 4.1$ [10.3a]

30. $-9 \cdot (-6)$ [10.4a]

31. $-2.7(3.4)$ [10.4a]

32. $\dfrac{2}{3} \cdot \left(-\dfrac{3}{7}\right)$ [10.4a]

33. $3 \cdot (-7) \cdot (-2) \cdot (-5)$ [10.4a]

34. $35 \div (-5)$ [10.5a]

35. $-5.1 \div 1.7$ [10.5c]

36. $-\dfrac{3}{11} \div \left(-\dfrac{4}{11}\right)$ [10.5c]

37. $(-3.4 - 12.2) - 8(-7)$ [10.5e]

Simplify. [10.5e]

38. $[-12(-3) - 2^3] - (-9)(-10)$

39. $625 \div (-25) \div 5$

40. $-16 \div 4 - 30 \div (-5)$

41. $9[(7 - 14) - 13]$

Solve.

42. On the first, second, and third downs, a football team had these gains and losses: 5-yd gain, 12-yd loss, and 15-yd gain, respectively. Find the total gain (or loss). [10.3b]

43. Kaleb's total assets are $170. He borrows $300. What are his total assets now? [10.3b]

44. *Stock Price.* The price of a stock opened at a value of $17.68 per share and dropped by $1.63 per hour for 8 hr. What was the price of the stock after 8 hr? [10.5d]

45. *Checking Account Balance.* Yuri had $68 in his checking account. After writing checks to make seven purchases of DVDs at the same price for each, the balance in his account was −$64.65. What was the price of each DVD? [10.5d]

46. D_W Is it possible for a number to be its own reciprocal? Explain. [10.5b]

47. D_W Write as many arguments as you can to convince a fellow classmate that $-(-a) = a$ for all real numbers a. [10.2b]

48. The sum of two numbers is 800. The difference is 6. Find the numbers. [10.3b]

49. The sum of two numbers is 5. The product is −84. Find the numbers. [10.3b], [10.5d]

50. The following are examples of consecutive integers: 4, 5, 6, 7, 8: and −13, −12, −11, −10. [10.3b], [10.5d]
 a) Express the number 8 as the sum of 16 consecutive integers.
 b) Find the product of the 16 consecutive integers in part (a).

51. Describe how you might find the following product quickly: [10.4a]
$$\left(-\tfrac{1}{11}\right)\left(-\tfrac{1}{9}\right)\left(-\tfrac{1}{7}\right)\left(-\tfrac{1}{5}\right)\left(-\tfrac{1}{3}\right)(-1)(-3)(-5)(-7)(-9)(-11).$$

52. Simplify: $-\left|\dfrac{7}{8} - \left(-\dfrac{1}{2}\right) - \dfrac{3}{4}\right|.$ [10.1e], [10.3a]

53. Simplify: $\left(|2.7 - 3| + 3^2 - |-3|\right) \div (-3).$ [10.1e], [10.5e]

Use either < or > for ☐ to write a true sentence.

1. -4 ☐ 0

2. -3 ☐ -8

3. -0.78 ☐ -0.87

4. $-\dfrac{1}{8}$ ☐ $\dfrac{1}{2}$

Find decimal notation.

5. $-\dfrac{1}{8}$

6. $-\dfrac{4}{9}$

7. $-\dfrac{2}{11}$

Find the absolute value.

8. $|-7|$

9. $\left|\dfrac{9}{4}\right|$

10. $-|-2.7|$

Find the opposite, or additive inverse.

11. $\dfrac{2}{3}$

12. -1.4

13. Evaluate $-x$ when x is -8.

14. Graph -0.2 on a number line.

Find the reciprocal.

15. -2

16. $\dfrac{4}{7}$

Compute and simplify.

17. $3.1 - (-4.7)$

18. $-8 + 4 + (-7) + 3$

19. $-\dfrac{1}{5} + \dfrac{3}{8}$

20. $2 - (-8)$

21. $3.2 - 5.7$

22. $\dfrac{1}{8} - \left(-\dfrac{3}{4}\right)$

23. $4 \cdot (-12)$

24. $-\dfrac{1}{2} \cdot \left(-\dfrac{3}{8}\right)$

25. $-45 \div 5$

26. $-\dfrac{3}{5} \div \left(-\dfrac{4}{5}\right)$

27. $4.864 \div (-0.5)$

28. $-2(16) - [2(-8) - 5^3]$

29. *Antarctica Highs and Lows.* The continent of Antarctica, which lies in the southern hemisphere, experiences winter in July. The average high temperature is $-67°$F and the average low temperature is $-81°$F. How much higher is the average high than the average low?

Source: National Climatic Data Center

30. Maureen is a stockbroker. She kept track of the changes in the stock market over a period of 5 weeks. By how many points had the market risen or fallen over this time?

WEEK 1	WEEK 2	WEEK 3	WEEK 4	WEEK 5
Down 13 pts	Down 16 pts	Up 36 pts	Down 11 pts	Up 19 pts

31. *Population Decrease.* The population of a city was 18,600. It decreased by 420 each year for 6 yr. What was the population of the city after 6 yr?

32. *Chemical Reaction.* During a chemical reaction, the temperature in the beaker decreased every minute by the same number of degrees. The temperature was $16°$C at 11:08 A.M. By 11:43 A.M., the temperature had dropped to $-17°$C. By how many degrees did it drop each minute?

SYNTHESIS

33. Simplify: $|-27 - 3(4)| - |-36| + |-12|$.

34. The deepest point in the Pacific Ocean is the Marianas Trench with a depth of 11,033 m. The deepest point in the Atlantic Ocean is the Puerto Rico Trench with a depth of 8648 m. How much higher is the Puerto Rico Trench than the Marianas Trench?

Source: Defense Mapping Agency, Hydrographic/Topographic Center

35. Find the next three numbers in each sequence.

a) 6, 5, 3, 0, ___ , ___ , ___
b) 14, 10, 6, 2, ___ , ___ , ___
c) $-4, -6, -9, -13,$ ___ , ___ , ___
d) $8, -4, 2, -1, 0.5,$ ___ , ___ , ___

Algebra: Solving Equations and Problems

Real-World Application

The average top speed of the three fastest roller coasters in the world is 109 mph. The third-fastest roller coaster, Superman the Escape (located at Six Flags Magic Mountain, Los Angeles, California), reaches a top speed of 20 mph less than the fastest roller coaster, Top Thrill Dragster (located in Cedar Point, Sandusky, Ohio). The second-fastest roller coaster, Dodonpa (located in Fujikyu Highlands, Japan), has a top speed of 107 mph. What is the top speed of the fastest roller coaster?

Source: Fortune Small Business, June 2004, p. 48

This problem appears as Example 10 in Section 11.5.

Objectives

a Evaluate an algebraic expression by substitution.

b Use the distributive laws to multiply expressions like 8 and $x - y$.

c Use the distributive laws to factor expressions like $4x - 12 + 24y$.

d Collect like terms.

1. Evaluate $a + b$ when $a = 38$ and $b = 26$.

Many types of problems require the use of equations in order to be solved efficiently. The study of algebra involves the use of equations to solve problems. Equations are constructed from algebraic expressions.

a Evaluating Algebraic Expressions

In arithmetic, you have worked with expressions such as

$$37 + 86, \quad 7 \times 8, \quad 19 - 7, \quad \text{and} \quad \frac{3}{8}.$$

In algebra, we use letters for numbers and work with *algebraic expressions* such as

$$x + 86, \quad 7 \times t, \quad 19 - y, \quad \text{and} \quad \frac{a}{b}.$$

Expressions like these should be familiar from the equation and problem solving that we have already done.

Sometimes a letter can stand for various numbers. In that case, we call the letter a **variable.** Let $a = $ your age. Then a is a variable since a changes from year to year. Sometimes a letter can stand for just one number. In that case, we call the letter a **constant.** Let $b = $ your date of birth. Then b is a constant.

An **algebraic expression** consists of variables, constants, numerals, and operation signs. When we replace a variable with a number, we say that we are **substituting** for the variable. This process is called **evaluating the expression.**

EXAMPLE 1 Evaluate $x + y$ for $x = 37$ and $y = 29$.

We substitute 37 for x and 29 for y and carry out the addition:

$$x + y = 37 + 29 = 66.$$

The number 66 is called the **value** of the expression.

2. Evaluate $x - y$ when $x = 57$ and $y = 29$.

Do Exercises 1 and 2.

Algebraic expressions involving multiplication can be written in several ways. For example, "8 times a" can be written as

$$8 \times a, \quad 8 \cdot a, \quad 8(a), \quad \text{or} \quad \text{simply } 8a.$$

Two letters written together without an operation sign, such as ab, also indicates a multiplication.

EXAMPLE 2 Evaluate $3y$ for $y = 14$.

$$3y = 3(14) = 42$$

Answers on page A-25

EXAMPLE 3 *Area of a Rectangle.* The area A of a rectangle of length l and width w is given by the formula $A = lw$. Find the area when l is 24.5 in. and w is 16 in.

We substitute 24.5 in. for l and 16 in. for w and carry out the multiplication:

$$A = lw = (24.5 \text{ in.})(16 \text{ in.})$$
$$= (24.5)(16)(\text{in.})(\text{in.})$$
$$= 392 \text{ in}^2, \text{ or } 392 \text{ square inches.}$$

Do Exercises 3 and 4.

Algebraic expressions involving division can also be written in several ways. For example, "8 divided by t" can be written as

$$8 \div t, \quad \frac{8}{t}, \quad 8/t, \quad \text{or} \quad 8 \cdot \frac{1}{t},$$

where the fraction bar is a division symbol.

EXAMPLE 4 Evaluate $\dfrac{a}{b}$ for $a = 63$ and $b = 9$.

We substitute 63 for a and 9 for b and carry out the division:

$$\frac{a}{b} = \frac{63}{9} = 7.$$

3. Evaluate $4t$ when $t = 15$ and when $t = -6.8$.

4. Find the area of a rectangle when l is 24 ft and w is 8 ft.

Answers on page A-25

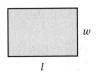

CALCULATOR CORNER

Evaluating Algebraic Expressions We can use a calculator to evaluate algebraic expressions. To evaluate $x - y$ when $x = 48$ and $y = 19$, for example, we press $\boxed{4}\boxed{8}\boxed{-}\boxed{1}\boxed{9}\boxed{=}$. The calculator displays the result, 29. To evaluate $3x$ when $x = -27$, we press $\boxed{3}\boxed{\times}\boxed{2}\boxed{7}\boxed{+/-}\boxed{=}$. The result is -81. When we evaluate an expression like $\dfrac{x+y}{4}$, we must enclose the numerator in parentheses. To evaluate this expression when $x = 29$ and $y = 15$, we press $\boxed{(}\boxed{2}\boxed{9}\boxed{+}\boxed{1}\boxed{5}\boxed{)}\boxed{\div}\boxed{4}\boxed{=}$. The calculator displays the result, 11.

A calculator can also be used to demonstrate the distributive laws. We can evaluate $5(x + y)$ and $5x + 5y$ when $x = 3$ and $y = 9$, for example, and see that the results are the same. To evaluate $5(x + y)$ when $x = 3$ and $y = 9$, we press $\boxed{5}\boxed{\times}\boxed{(}\boxed{3}\boxed{+}\boxed{9}\boxed{)}\boxed{=}$. The calculator displays the result, 60. Next, we evaluate $5x + 5y$ when $x = 3$ and $y = 9$. We press $\boxed{5}\boxed{\times}\boxed{3}\boxed{+}\boxed{5}\boxed{\times}\boxed{9}\boxed{=}$. Again the calculator displays 60. This verifies that $5(3 + 9) = 5 \cdot 3 + 5 \cdot 9$.

Exercises: Evaluate.

1. $x + y$, when $x = 35$ and $y = 16$

2. $7t$, when $t = 14$

3. $\dfrac{a}{b}$, when $a = 54$ and $b = -9$

4. $\dfrac{2m}{n}$, when $m = 38$ and $n = -4$

5. $\dfrac{x - y}{7}$, when $x = 94$ and $y = 31$

6. $\dfrac{p + q}{12}$, when $p = 47$ and $q = 97$

7. $4(x + y)$ and $4x + 4y$, when $x = 6$ and $y = 7$

8. $3(x + y)$ and $3x + 3y$, when $x = 34$ and $y = 18$

9. $6(a - b)$ and $6a - 6b$, when $a = 67$ and $b = 29$

10. $7(a - b)$ and $7a - 7b$, when $a = 18$ and $b = 6$

5. Evaluate $\dfrac{a}{b}$ when $a = -200$ and $b = 8$.

EXAMPLE 5 Evaluate $\dfrac{12m}{n}$ for $m = 8$ and $n = 16$.

$$\frac{12m}{n} = \frac{12 \cdot 8}{16} = \frac{96}{16} = 6$$

Do Exercises 5 and 6.

Equivalent Expressions and the Distributive Laws

In solving and doing other kinds of work in algebra, we manipulate expressions in various ways. To see how to do this, we consider some examples in which we evaluate expressions.

6. Evaluate $\dfrac{10p}{q}$ when $p = 40$ and $q = 25$.

EXAMPLE 6 Evaluate $1 \cdot x$ when $x = 5$ and $x = -8$ and compare the results to x.

We substitute 5 for x:

$$1 \cdot x = 1 \cdot 5 = 5.$$

Then we substitute -8 for x:

$$1 \cdot x = 1 \cdot (-8) = -8.$$

We see that $1 \cdot x$ and x represent the same number.

Complete the table by evaluating each expression for the given values.

7.

VALUE OF x	$1 \cdot x$	x
$x = 3$		
$x = -6$		
$x = 4.8$		

Do Exercises 7 and 8.

We see in Example 6 and Margin Exercise 7 that the expressions represent the same number for any allowable replacement of x. In that sense, the expressions $1 \cdot x$ and x are **equivalent.**

> **EQUIVALENT EXPRESSIONS**
>
> Two expressions that have the same value for all allowable replacements are called **equivalent.**

In the expression $3/x$, the number 0 is not allowable because $3/0$ is not defined. Even so, the expressions $6/2x$ and $3/x$ are *equivalent* because they represent the same number for any allowable (not 0) replacement of x. For example, when $x = 5$,

$$\frac{6}{2x} = \frac{6}{2 \cdot 5} = \frac{6}{10} = \frac{3}{5} \quad \text{and} \quad \frac{3}{x} = \frac{3}{5}.$$

8.

VALUE OF x	$2x$	$5x$
$x = 2$		
$x = -6$		
$x = 4.8$		

We see in Margin Exercise 8 that the expressions $2x$ and $5x$ are *not* equivalent.

The fact that $1 \cdot x$ and x are equivalent is a law of real numbers. It is called the **identity property of 1.** We often refer to the use of the identity property of 1 as "multiplying by 1." We have used multiplying by 1 for understanding many times in this text.

> **THE IDENTITY PROPERTY OF 1 (MULTIPLICATIVE IDENTITY)**
>
> For any real number a,
>
> $$a \cdot 1 = 1 \cdot a = a.$$

Answers on page A-25

We now consider two other laws of real numbers called the **distributive laws.** They are the basis of many procedures in both arithmetic and algebra and are probably the most important laws that we use to manipulate algebraic expressions. The first distributive law involves two operations: addition and multiplication.

Let's begin by considering a multiplication problem from arithmetic:

$$
\begin{array}{r}
4\ 5 \\
\times\ \ \ \ 7 \\
\hline
3\ 5 \\
2\ 8\ 0 \\
3\ 1\ 5 \\
\end{array}
$$

3 5 ← This is $7 \cdot 5$.
2 8 0 ← This is $7 \cdot 40$.
3 1 5 ← This is the sum $7 \cdot 40 + 7 \cdot 5$.

To carry out the multiplication, we actually added two products. That is,

$$7 \cdot 45 = 7(40 + 5) = 7 \cdot 40 + 7 \cdot 5.$$

Let's examine this further. If we wish to multiply a sum of several numbers by a factor, we can either add and then multiply or multiply and then add.

EXAMPLE 7 Evaluate $5(x + y)$ and $5x + 5y$ when $x = 2$ and $y = 8$ and compare the results.

We substitute 2 for x and 8 for y in each expression. Then we use the rules for order of operations to calculate.

a) $5(x + y) = 5(2 + 8)$
$\qquad\qquad = 5(10)$ Adding within parentheses first, and then multiplying
$\qquad\qquad = 50$

b) $5x + 5y = 5 \cdot 2 + 5 \cdot 8$
$\qquad\qquad = 10 + 40$ Multiplying first and then adding
$\qquad\qquad = 50$

The results of (a) and (b) are the same.

Do Exercises 9–11.

The expressions $5(x + y)$ and $5x + 5y$, in Example 7 and Margin Exercise 9, are equivalent. They illustrate the distributive law of multiplication over addition.

> **THE DISTRIBUTIVE LAW OF MULTIPLICATION OVER ADDITION**
>
> For any numbers a, b, and c,
>
> $$a(b + c) = ab + ac.$$

Margin Exercises 10 and 11 also illustrate the distributive law.

In the statement of the distributive law, we know that in an expression such as $ab + ac$, the multiplications are to be done first according to the rules for order of operations. So, instead of writing $(4 \cdot 5) + (4 \cdot 7)$, we can write $4 \cdot 5 + 4 \cdot 7$. However, in $a(b + c)$, we cannot omit the parentheses. If we did, we would have $ab + c$, which means $(ab) + c$. For example, $3(4 + 2) = 18$, but $3 \cdot 4 + 2 = 14$.

9. Complete this table.

VALUES OF x AND y	$5(x + y)$	$5x + 5y$
$x = 6, y = 7$		
$x = -3, y = 4$		
$x = -10, y = 5$		

10. Evaluate $6x + 6y$ and $6(x + y)$ when $x = 10$ and $y = 5$.

11. Evaluate $4(x + y)$ and $4x + 4y$ when $x = 11$ and $y = 5$.

12. Evaluate $7(x - y)$ and $7x - 7y$ when $x = 9$ and $y = 7$.

13. Evaluate $6x - 6y$ and $6(x - y)$ when $x = 10$ and $y = 5$.

14. Evaluate $2(x - y)$ and $2x - 2y$ when $x = 11$ and $y = 5$.

Answers on page A-25

What are the terms of the expression?

15. $5x - 4y + 3$

16. $-4y - 2x + 3z$

Multiply.

17. $3(x - 5)$

18. $5(x + 1)$

19. $\dfrac{5}{4}(x - y + 4)$

20. $-2(x - 3)$

21. $-5(x - 2y + 4z)$

The second distributive law relates multiplication and subtraction. This law says that to multiply by a difference, we can either subtract and then multiply or multiply and then subtract.

THE DISTRIBUTIVE LAW OF MULTIPLICATION OVER SUBTRACTION

For any numbers a, b, and c,

$$a(b - c) = ab - ac.$$

We often refer to "*the* distributive law" when we mean *either or both* of these laws.

Do Exercises 12–14 on the preceding page.

What do we mean by the *terms* of an expression? **Terms** are separated by addition signs. If there are subtraction signs, we can find an equivalent expression that uses addition signs.

EXAMPLE 8 What are the terms of $3x - 4y + 2z$?

$$3x - 4y + 2z = 3x + (-4y) + 2z \qquad \text{Separating parts with + signs}$$

The terms are $3x$, $-4y$, and $2z$.

Do Exercises 15 and 16.

The distributive laws are the basis for a procedure in algebra called **multiplying.** In an expression such as $8(a + 2b - 7)$, we multiply each term inside the parentheses by 8:

$$8(a + 2b - 7) = 8 \cdot a + 8 \cdot 2b - 8 \cdot 7 = 8a + 16b - 56.$$

EXAMPLES Multiply.

9. $9(x - 5) = 9x - 9(5)$ Using the distributive law of multiplication over subtraction

$$= 9x - 45$$

10. $\dfrac{2}{3}(w + 1) = \dfrac{2}{3} \cdot w + \dfrac{2}{3} \cdot 1$ Using the distributive law of multiplication over addition

$$= \dfrac{2}{3}w + \dfrac{2}{3}$$

Answers on page A-25

EXAMPLE 11 Multiply: $-4(x - 2y + 3z)$.

$$-4(x - 2y + 3z) = -4 \cdot x - (-4)(2y) + (-4)(3z) \qquad \text{Using both distributive laws}$$

$$= -4x - (-8y) + (-12z) \qquad \text{Multiplying}$$

$$= -4x + 8y - 12z.$$

We can also do this problem by first finding an equivalent expression with all plus signs and then multiplying:

$$-4(x - 2y + 3z) = -4[x + (-2y) + 3z]$$

$$= -4 \cdot x + (-4)(-2y) + (-4)(3z) = -4x + 8y - 12z.$$

Do Exercises 17–21 on the preceding page.

C Factoring

Factoring is the reverse of multiplying. To factor, we can use the distributive laws in reverse:

$$ab + ac = a(b + c) \quad \text{and} \quad ab - ac = a(b - c).$$

> **FACTOR**
>
> To **factor** an expression is to find an equivalent expression that is a product.

Look at Example 9. To *factor* $9x - 45$, we find an equivalent expression that is a product, $9(x - 5)$. When all the terms of an expression have a factor in common, we can "factor it out" using the distributive laws. Note the following.

$9x$ has the factors $9, -9, 3, -3, 1, -1, x, -x, 3x, -3x, 9x, -9x$;

-45 has the factors $1, -1, 3, -3, 5, -5, 9, -9, 15, -15, 45, -45$.

We remove the greatest common factor. In this case, that factor is 9. Thus,

$$9x - 45 = 9 \cdot x - 9 \cdot 5 = 9(x - 5).$$

Remember that an expression is factored when we find an equivalent expression that is a product.

EXAMPLES Factor.

12. $5x - 10 = 5 \cdot x - 5 \cdot 2 \qquad$ Try to do this step mentally.

$\qquad\quad = 5(x - 2) \longleftarrow$ You can check by multiplying.

13. $9x + 27y - 9 = 9 \cdot x + 9 \cdot 3y - 9 \cdot 1 = 9(x + 3y - 1)$

> **Caution!**
>
> Note that although $3(3x + 9y - 3)$ is also equivalent to $9x + 27y - 9$, it is *not* the desired form. However, we can complete the process by factoring out another factor of 3:
>
> $$9x + 27y - 9 = 3(3x + 9y - 3) = 3 \cdot 3(x + 3y - 1) = 9(x + 3y - 1).$$
>
> Remember to factor out the *greatest common factor*.

Factor.

22. $6z - 12$

23. $3x - 6y + 9$

24. $16a - 36b + 42$

25. $-12x + 32y - 16z$

Answers on page A-25

Collect like terms.

26. $6x - 3x$

27. $7x - x$

28. $x - 9x$

29. $x - 0.41x$

30. $5x + 4y - 2x - y$

31. $3x - 7x - 11 + 8y + 4 - 13y$

32. $-\dfrac{2}{3} - \dfrac{3}{5}x + y + \dfrac{7}{10}x - \dfrac{2}{9}y$

EXAMPLES Factor. Try to write just the answer, if you can.

14. $5x - 5y = 5(x - y)$

15. $-3x + 6y - 9z = -3 \cdot x - 3(-2y) - 3(3z) = -3(x - 2y + 3z)$

We usually factor out a negative factor when the first term is negative. The way we factor can depend on the situation in which we are working. We might also factor the expression in Example 15 as follows:

$$-3x + 6y - 9z = 3(-x + 2y - 3z).$$

16. $18z - 12x - 24 = 6(3z - 2x - 4)$

Remember that you can always check such factoring by multiplying. Keep in mind that an expression is factored when it is written as a product.

Do Exercises 22–25 on the preceding page.

d Collecting Like Terms

Terms such as $5x$ and $-4x$, whose variable factors are exactly the same, are called **like terms.** Similarly, numbers, such as -7 and 13, are like terms. Also, $3y^2$ and $9y^2$ are like terms because the variables are raised to the same power. Terms such as $4y$ and $5y^2$ are not like terms, and $7x$ and $2y$ are not like terms.

The process of **collecting like terms** is based on the distributive laws. We can also apply the distributive law when a factor is on the right.

EXAMPLES Collect like terms. Try to write just the answer, if you can.

17. $4x + 2x = (4 + 2)x = 6x$ Factoring out the x using a distributive law

18. $2x + 3y - 5x - 2y = 2x - 5x + 3y - 2y$

$$= (2 - 5)x + (3 - 2)y = -3x + y$$

19. $3x - x = 3x - 1x = (3 - 1)x = 2x$

20. $x - 0.24x = 1 \cdot x - 0.24x = (1 - 0.24)x = 0.76x$

21. $x - 6x = 1 \cdot x - 6 \cdot x = (1 - 6)x = -5x$

22. $4x - 7y + 9x - 5 + 3y - 8 = 13x - 4y - 13$

23. $\dfrac{2}{3}a - b + \dfrac{4}{5}a + \dfrac{1}{4}b - 10 = \dfrac{2}{3}a - 1 \cdot b + \dfrac{4}{5}a + \dfrac{1}{4}b - 10$

$$= \left(\dfrac{2}{3} + \dfrac{4}{5}\right)a + \left(-1 + \dfrac{1}{4}\right)b - 10$$

$$= \left(\dfrac{10}{15} + \dfrac{12}{15}\right)a + \left(-\dfrac{4}{4} + \dfrac{1}{4}\right)b - 10$$

$$= \dfrac{22}{15}a - \dfrac{3}{4}b - 10$$

Do Exercises 26–32.

Answers on page A-25

CHAPTER 11: Algebra: Solving Equations and Problems

11.1 EXERCISE SET

For Extra Help

MathXL   MyMathLab   InterAct Math   Math Tutor Center   Digital Video Tutor CD 5 Videotape 11   Student's Solutions Manual

a Evaluate.

1. $6x$, when $x = 7$

2. $9t$, when $t = 8$

3. $\dfrac{x}{y}$, when $x = 9$ and $y = 3$

4. $\dfrac{m}{n}$, when $m = 18$ and $n = 3$

5. $\dfrac{3p}{q}$, when $p = -2$ and $q = 6$

6. $\dfrac{5y}{z}$, when $y = -15$ and $z = -25$

7. $\dfrac{x + y}{5}$, when $x = 10$ and $y = 20$

8. $\dfrac{p - q}{2}$, when $p = 17$ and $q = 3$

9. ab, when $a = -5$ and $b = 4$

10. ba, when $a = -5$ and $b = 4$

b Evaluate.

11. $10(x + y)$ and $10x + 10y$, when $x = 20$ and $y = 4$

12. $5(a + b)$ and $5a + 5b$, when $a = 16$ and $b = 6$

13. $10(x - y)$ and $10x - 10y$, when $x = 20$ and $y = 4$

14. $5(a - b)$ and $5a - 5b$, when $a = 16$ and $b = 6$

Multiply.

15. $2(b + 5)$

16. $4(x + 3)$

17. $7(1 - t)$

18. $4(1 - y)$

19. $6(5x + 2)$

20. $9(6m + 7)$

21. $7(x + 4 + 6y)$

22. $4(5x + 8 + 3p)$

23. $-7(y - 2)$

24. $-9(y - 7)$

25. $-9(-5x - 6y + 8)$

26. $-7(-2x - 5y + 9)$

27. $\dfrac{3}{4}(x - 3y - 2z)$

28. $\dfrac{2}{5}(2x - 5y - 8z)$

29. $3.1(-1.2x + 3.2y - 1.1)$

30. $-2.1(-4.2x - 4.3y - 2.2)$

c Factor. Check by multiplying.

31. $2x + 4$

32. $5y + 20$

33. $30 + 5y$

34. $7x + 28$

35. $14x + 21y$

36. $18a + 24b$

37. $5x + 10 + 15y$

38. $9a + 27b + 81$

39. $8x - 24$

40. $10x - 50$

41. $32 - 4y$

42. $24 - 6m$

629

Exercise Set 11.1

43. $8x + 10y - 22$ **44.** $9a + 6b - 15$ **45.** $-18x - 12y + 6$ **46.** $-14x + 21y + 7$

d   Collect like terms.

47. $9a + 10a$ **48.** $14x + 3x$ **49.** $10a - a$

50. $-10x + x$ **51.** $2x + 9z + 6x$ **52.** $3a - 5b + 4a$

53. $41a + 90 - 60a - 2$ **54.** $42x - 6 - 4x + 20$ **55.** $23 + 5t + 7y - t - y - 27$

56. $95 - 90d - 87 - 9d + 3 + 7d$ **57.** $11x - 3x$ **58.** $9t - 13t$

59. $6n - n$ **60.** $10t - t$ **61.** $y - 17y$

62. $5m - 8m + 4$ **63.** $-8 + 11a - 5b + 6a - 7b + 7$ **64.** $8x - 5x + 6 + 3y - 2y - 4$

65. $9x + 2y - 5x$ **66.** $8y - 3z + 4y$ **67.** $\dfrac{11}{4}x + \dfrac{2}{3}y - \dfrac{4}{5}x - \dfrac{1}{6}y + 12$

68. $\dfrac{13}{2}a + \dfrac{9}{5}b - \dfrac{2}{3}a - \dfrac{3}{10}b - 42$ **69.** $2.7x + 2.3y - 1.9x - 1.8y$ **70.** $6.7a + 4.3b - 4.1a - 2.9b$

71. $\mathbf{D_W}$ Determine whether $(a + b)^2$ and $a^2 + b^2$ are equivalent for all real numbers. Explain.

72. $\mathbf{D_W}$ The distributive law is introduced before the material on collecting like terms. Why do you think this is?

SKILL MAINTENANCE

For a circle with the given radius, find the diameter, the circumference, and the area. Use 3.14 for π.   [9.3a, b, c]

73. $r = 15$ yd   **74.** $r = 8.2$ m   **75.** $r = 9\frac{1}{2}$ mi   **76.** $r = 2400$ cm

For a circle with the given diameter, find the radius, the circumference, and the area. Use 3.14 for π.   [9.3a, b, c]

77. $d = 20$ mm   **78.** $d = 264$ km   **79.** $d = 4.6$ ft   **80.** $d = 10.3$ m

SYNTHESIS

Collect like terms, if possible, and factor the result.

81. $q + qr + qrs + qrst$ **82.** $21x + 44xy + 15y - 16x - 8y - 38xy + 2y + xy$

CHAPTER 11: Algebra: Solving Equations and Problems

11.2 SOLVING EQUATIONS: THE ADDITION PRINCIPLE

Objective

a | Solve equations using the addition principle.

a | Using the Addition Principle

Consider the equation $x = 7$. We can easily "see" that the solution of this equation is 7. If we replace x with 7, we get $7 = 7$, which is true. Now consider the equation $x + 6 = 13$. The solution of this equation is also 7, but the fact that 7 is the solution is not as obvious. We now begin to consider principles that allow us to start with an equation like $x + 6 = 13$ and end up with an equation like $x = 7$, in which the variable is alone on one side and for which the solution is easier to find. The equations $x + 6 = 13$ and $x = 7$ are **equivalent.**

> ### EQUIVALENT EQUATIONS
>
> Equations with the same solutions are called **equivalent equations.**

One principle that we use to solve equations involves the addition principle, which we have used throughout this text.

> ### THE ADDITION PRINCIPLE
>
> For any real numbers a, b, and c,
>
> $a = b$ is equivalent to $a + c = b + c$.

Let's solve $x + 6 = 13$ using the addition principle. We want to get x alone on one side. To do so, we use the addition principle, choosing to add -6 on both sides because $6 + (-6) = 0$:

$$x + 6 = 13$$
$$x + 6 + (-6) = 13 + (-6) \quad \text{Using the addition principle; adding } -6 \text{ on both sides}$$
$$x + 0 = 7 \quad \text{Simplifying}$$
$$x = 7. \quad \text{Identity property of 0}$$

The solution of $x + 6 = 13$ is 7.

Do Exercise 1.

To visualize the addition principle, think of a jeweler's balance. When both sides of the balance hold equal amounts of weight, the balance is level. If weight is added or removed, equally, on both sides, the balance remains level.

1. Solve $x + 2 = 11$ using the addition principle.

   a) First, complete this sentence:
   $$2 + \boxed{} = 0.$$

   b) Then solve the equation.

Answers on page A-25

2. Solve using the addition principle:

$$x + 7 = 2.$$

Solve.

3. $8.7 = n - 4.5$

4. $y + 17.4 = 10.9$

When we use the addition principle, we sometimes say that we "add the same number on both sides of the equation." This is also true for subtraction, since we can express every subtraction as an addition. That is, since

$$a - c = b - c \quad \text{is equivalent to} \quad a + (-c) = b + (-c),$$

the addition principle tells us that we can "subtract the same number on both sides of an equation."

EXAMPLE 1 Solve: $x + 5 = -7$.

We have

$x + 5 = -7$	
$x + 5 - 5 = -7 - 5$	Using the addition principle: adding -5 on both sides or subtracting 5 on both sides
$x + 0 = -12$	Simplifying
$x = -12.$	Identity property of 0

To check the answer, we substitute -12 in the original equation.

Check:
$$\frac{x + 5 = -7}{-12 + 5 \; ? \; -7}$$
$$-7 \; | \qquad \text{TRUE}$$

The solution of the original equation is -12.

In Example 1, to get x alone, we used the addition principle and subtracted 5 on both sides. This eliminated the 5 on the left. We started with $x + 5 = -7$, and using the addition principle we found a simpler equation $x = -12$, for which it was easier to "*see*" the solution. The equations $x + 5 = -7$ and $x = -12$ are *equivalent*.

Do Exercise 2.

Now we solve an equation that involves a subtraction by using the addition principle.

EXAMPLE 2 Solve: $-6.5 = y - 8.4$.

We have

$-6.5 = y - 8.4$	
$-6.5 + 8.4 = y - 8.4 + 8.4$	Using the addition principle: adding 8.4 to eliminate -8.4 on the right
$1.9 = y.$	

Check:
$$\frac{-6.5 = y - 8.4}{-6.5 \; ? \; 1.9 - 8.4}$$
$$| \; -6.5 \qquad \text{TRUE}$$

The solution is 1.9.

Note that equations are reversible. That is, if $a = b$ is true, then $b = a$ is true. Thus, when we solve $-6.5 = y - 8.4$, we can reverse it and solve $y - 8.4 = -6.5$ if we wish.

Do Exercises 3 and 4.

Answers on page A-25

a   Solve using the addition principle. Don't forget to check!

1. $x + 2 = 6$

Check: $x + 2 = 6$
         ?

2. $y + 4 = 11$

Check: $y + 4 = 11$
         ?

3. $x + 15 = -5$

Check: $x + 15 = -5$
         ?

4. $t + 10 = 44$

Check: $t + 10 = 44$
         ?

5. $x + 6 = -8$

6. $y + 8 = 37$

7. $x + 5 = 12$

8. $x + 3 = 7$

9. $-22 = t + 4$

10. $t + 8 = -14$

11. $x + 16 = -2$

12. $y + 34 = -8$

13. $x - 9 = 6$

14. $x - 9 = 2$

15. $x - 7 = -21$

16. $x - 5 = -16$

17. $5 + t = 7$

18. $6 + y = 22$

19. $-7 + y = 13$

20. $-8 + z = 16$

21. $-3 + t = -9$

22. $-8 + y = -23$

23. $r + \dfrac{1}{3} = \dfrac{8}{3}$

24. $t + \dfrac{3}{8} = \dfrac{5}{8}$

25. $m + \dfrac{5}{6} = -\dfrac{11}{12}$

26. $x + \dfrac{2}{3} = -\dfrac{5}{6}$

27. $x - \dfrac{5}{6} = \dfrac{7}{8}$

28. $y - \dfrac{3}{4} = \dfrac{5}{6}$

29. $-\dfrac{1}{5} + z = -\dfrac{1}{4}$

30. $-\dfrac{1}{8} + y = -\dfrac{3}{4}$

31. $7.4 = x + 2.3$

32. $9.3 = 4.6 + x$

33. $7.6 = x - 4.8$

34. $9.5 = y - 8.3$

35. $-9.7 = -4.7 + y$

36. $-7.8 = 2.8 + x$

37. $5\frac{1}{6} + x = 7$

38. $5\frac{1}{4} = 4\frac{2}{3} + x$

39. $q + \frac{1}{3} = -\frac{1}{7}$

40. $47\frac{1}{8} = -76 + z$

41. $\mathbf{D_W}$ Explain the following mistake made by a fellow student.

$$x + \frac{1}{3} = -\frac{5}{3}$$
$$x = -\frac{4}{3}$$

42. $\mathbf{D_W}$ Explain the role of the opposite of a number when using the addition principle.

SKILL MAINTENANCE

Add. [10.2a]

43. $-3 + (-8)$

44. $-\frac{2}{3} + \frac{5}{8}$

45. $-14.3 + (-19.8)$

46. $3.2 + (-4.9)$

Subtract. [10.3a]

47. $-3 - (-8)$

48. $-\frac{2}{3} - \frac{5}{8}$

49. $-14.3 - (-19.8)$

50. $3.2 - (-4.9)$

Multiply. [10.4a]

51. $-3(-8)$

52. $-\frac{2}{3} \cdot \frac{5}{8}$

53. $-14.3 \times (-19.8)$

54. $3.2(-4.9)$

Divide. [10.5c]

55. $\dfrac{-24}{-3}$

56. $-\frac{2}{3} \div \frac{5}{8}$

57. $\dfrac{283.14}{-19.8}$

58. $\dfrac{-15.68}{3.2}$

SYNTHESIS

Solve.

59. ▦ $-356.788 = -699.034 + t$

60. $-\frac{4}{5} + \frac{7}{10} = x - \frac{3}{4}$

61. $x + \frac{4}{5} = -\frac{2}{3} - \frac{4}{15}$

62. $8 - 25 = 8 + x - 21$

63. $16 + x - 22 = -16$

64. $x + x = x$

65. $-\frac{3}{2} + x = -\frac{5}{17} - \frac{3}{2}$

66. $|x| = 5$

634

11.3

SOLVING EQUATIONS: THE MULTIPLICATION PRINCIPLE

a Using the Multiplication Principle

Suppose that $a = b$ is true, and we multiply a by some number c. We get the same answer if we multiply b by c, because a and b are the same number.

THE MULTIPLICATION PRINCIPLE

For any real numbers a, b, and c, $c \neq 0$,

$$a = b \quad \text{is equivalent to} \quad a \cdot c = b \cdot c.$$

When using the multiplication principle, we sometimes say that we "multiply on both sides of the equation by the same number."

EXAMPLE 1 Solve: $5x = 70$.

To get x alone, we multiply by the *multiplicative inverse*, or *reciprocal*, of 5. Then we get the *multiplicative identity* 1 times x, or $1 \cdot x$, which simplifies to x. This allows us to eliminate 5 on the left.

$5x = 70$	The reciprocal of 5 is $\frac{1}{5}$.
$\frac{1}{5} \cdot 5x = \frac{1}{5} \cdot 70$	Multiplying by $\frac{1}{5}$ to get $1 \cdot x$ and eliminate 5 on the left
$1 \cdot x = 14$	Simplifying
$x = 14$	Identity property of 1: $1 \cdot x = x$

Check:
$$\begin{array}{c} 5x = 70 \\ \hline 5 \cdot 14 \ ? \ 70 \\ 70 \ | \quad \text{TRUE} \end{array}$$

The solution is 14.

The multiplication principle also tells us that we can "divide on both sides of the equation by a nonzero number." This is because division is the same as multiplying by a reciprocal. That is,

$$\frac{a}{c} = \frac{b}{c} \quad \text{is equivalent to} \quad a \cdot \frac{1}{c} = b \cdot \frac{1}{c}, \quad \text{when } c \neq 0.$$

In an expression like $5x$ in Example 1, the number 5 is called the **coefficient.** Example 1 could be done as follows, dividing on both sides by 5, the coefficient of x.

EXAMPLE 2 Solve: $5x = 70$.

$5x = 70$	
$\frac{5x}{5} = \frac{70}{5}$	Dividing by 5 on both sides
$1 \cdot x = 14$	Simplifying
$x = 14$	Identity property of 1

Do Exercises 1 and 2.

1. Solve. Multiply on both sides.

$$6x = 90$$

2. Solve. Divide on both sides.

$$4x = -7$$

Answers on page A-25

3. Solve: $-6x = 108$.

4. Solve: $-x = -10$.

EXAMPLE 3 Solve: $-4x = 92$.

We have

$$-4x = 92$$

$$\frac{-4x}{-4} = \frac{92}{-4}$$ Using the multiplication principle. Dividing by -4 on both sides is the same as multiplying by $-\frac{1}{4}$.

$$1 \cdot x = -23$$ Simplifying

$$x = -23.$$ Identity property of 1

Check: $$\frac{-4x = 92}{-4(-23) \ ? \ 92}$$
$$92 \ | \qquad \text{TRUE}$$

The solution is -23.

Do Exercise 3.

EXAMPLE 4 Solve: $-x = 9$.

$$-x = 9$$

$$-1(-x) = -1 \cdot 9$$ Multiplying by -1 on both sides

$$-1 \cdot (-1) \cdot x = -9 \qquad -x = (-1) \cdot x$$

$$1 \cdot x = -9$$

$$x = -9.$$

Check: $$\frac{-x = 9}{-(-9) \ ? \ 9}$$
$$9 \ | \qquad \text{TRUE}$$

The solution is -9.

Do Exercise 4.

In practice, it is generally more convenient to "divide" on both sides of the equation if the coefficient of the variable is in decimal notation or is an integer. If the coefficient is in fraction notation, it is more convenient to "multiply" by a reciprocal.

EXAMPLE 5 Solve: $\dfrac{3}{8} = -\dfrac{5}{4}x$.

$$\frac{3}{8} = -\frac{5}{4}x$$

The reciprocal of $-\frac{5}{4}$ is $-\frac{4}{5}$. There is no sign change.

$$-\frac{4}{5} \cdot \frac{3}{8} = -\frac{4}{5} \cdot \left(-\frac{5}{4}x\right)$$ Multiplying by $-\frac{4}{5}$ to get $1 \cdot x$ and eliminate $-\frac{5}{4}$ on the right

$$-\frac{12}{40} = 1 \cdot x$$

$$-\frac{3}{10} = 1 \cdot x$$ Simplifying

$$-\frac{3}{10} = x$$ Identity property of 1

Check: $\dfrac{3}{8} = -\dfrac{5}{4}x$

$\dfrac{3}{8}$? $-\dfrac{5}{4}\left(-\dfrac{3}{10}\right)$

$\dfrac{3}{8}$ TRUE

The solution is $-\dfrac{3}{10}$.

If $a = b$ is true, then $b = a$ is true. Thus when we solve $\frac{3}{8} = -\frac{5}{4}x$, we can reverse it and solve $-\frac{5}{4}x = \frac{3}{8}$ if we wish.

Do Exercise 5.

EXAMPLE 6 Solve: $1.16y = 9744$.

$\dfrac{1.16y}{1.16} = \dfrac{9744}{1.16}$ Dividing by 1.16 on both sides

$y = \dfrac{9744}{1.16}$

$y = 8400$ Using a calculator to divide

Check: $1.16y = 9744$

$1.16(8400)$? 9744
9744 TRUE

The solution is 8400.

Do Exercises 6 and 7.

Now we use the multiplication principle to solve an equation that involves division.

EXAMPLE 7 Solve: $\dfrac{-y}{9} = 14$.

$9 \cdot \dfrac{-y}{9} = 9 \cdot 14$ Multiplying by 9 on both sides

$-y = 126$

$-1 \cdot (-y) = -1 \cdot 126$ Multiplying by -1 on both sides

$y = -126$

Check: $\dfrac{-y}{9} = 14$

$\dfrac{-(-126)}{9}$? 14

$\dfrac{126}{9}$

14 TRUE

The solution is -126.

Do Exercise 8.

5. Solve: $\dfrac{2}{3} = -\dfrac{5}{6}y$.

Solve.
6. $1.12x = 8736$

7. $6.3 = -2.1y$

8. Solve: $-14 = \dfrac{-y}{2}$.

Answers on page A-25

a Solve using the multiplication principle. Don't forget to check!

1. $6x = 36$

2. $4x = 52$

3. $5x = 45$

4. $8x = 56$

5. $84 = 7x$

6. $63 = 7x$

7. $-x = 40$

8. $50 = -x$

9. $-2x = -10$

10. $-78 = -39p$

11. $7x = -49$

12. $9x = -54$

13. $-12x = 72$

14. $-15x = 105$

15. $-21x = -126$

16. $-13x = -104$

17. $\frac{1}{7}t = -9$

18. $-\frac{1}{8}y = 11$

19. $\frac{3}{4}x = 27$

20. $\frac{4}{5}x = 16$

21. $-\frac{1}{3}t = 7$

22. $-\frac{1}{6}x = 9$

23. $-\frac{1}{3}m = \frac{1}{5}$

24. $\frac{1}{5} = -\frac{1}{8}z$

25. $-\frac{3}{5}r = \frac{9}{10}$

26. $\frac{2}{5}y = -\frac{4}{15}$

27. $-\frac{3}{2}r = -\frac{27}{4}$

28. $-\frac{5}{7}x = -\frac{10}{14}$

CHAPTER 11: Algebra: Solving Equations and Problems

29. $6.3x = 44.1$

30. $2.7y = 54$

31. $-3.1y = 21.7$

32. $-3.3y = 6.6$

33. $38.7m = 309.6$

34. $29.4m = 235.2$

35. $-\dfrac{2}{3}y = -10.6$

36. $-\dfrac{9}{7}y = 12.06$

37. $\dfrac{-x}{5} = 10$

38. $\dfrac{-x}{8} = -16$

39. $\dfrac{t}{-2} = 7$

40. $\dfrac{m}{-3} = 10$

41. $\mathbf{D_W}$ Explain the following mistake made by a fellow student.

$$\frac{2}{3}x = -\frac{5}{3}$$
$$x = \frac{10}{9}$$

42. $\mathbf{D_W}$ Explain the role of the reciprocal of a number when using the multiplication principle.

SKILL MAINTENANCE

43. Find the circumference, the diameter, and the area of a circle painted on a driveway. The radius is 10 ft. Use 3.14 for π.   [9.3a, b, c]

44. Find the circumference, the radius, and the area of a circle whose diameter is 24 cm. Use 3.14 for π.   [9.3a, b, c]

45. Find the volume of a rectangular block of granite of length 25 ft, width 10 ft, and height 32 ft.   [9.4a]

46. Find the volume of a rectangular solid of length 1.3 cm, width 10 cm, and height 2.4 cm.   [9.4a]

Find the area of the figure.   [9.2b]

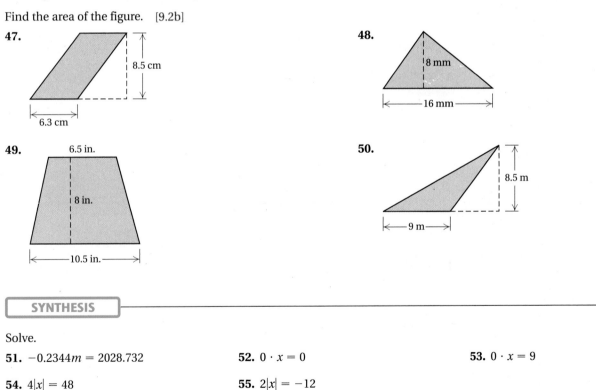

47. 8.5 cm   6.3 cm

48. 8 mm   16 mm

49. 6.5 in.   8 in.   10.5 in.

50. 8.5 m   9 m

SYNTHESIS

Solve.

51. $-0.2344m = 2028.732$

52. $0 \cdot x = 0$

53. $0 \cdot x = 9$

54. $4|x| = 48$

55. $2|x| = -12$

56. A student makes a calculation and gets an answer of 22.5. On the last step, the student multiplies by 0.3 when she should have divided by 0.3. What should the correct answer be?

Objectives

a Solve equations using both the addition and the multiplication principles.

b Solve equations in which like terms may need to be collected.

c Solve equations by first removing parentheses and collecting like terms.

1. Solve: $9x + 6 = 51$.

Solve.

2. $8x - 4 = 28$

3. $-\dfrac{1}{2}x + 3 = 1$

a Applying Both Principles

Consider the equation $3x + 4 = 13$. It is more complicated than those in the preceding two sections. In order to solve such an equation, we first isolate the x-term, $3x$, using the addition principle. Then we apply the multiplication principle to get x by itself.

EXAMPLE 1 Solve: $3x + 4 = 13$.

$$3x + 4 = 13$$

$$3x + 4 - 4 = 13 - 4 \qquad \text{Using the addition principle: adding } -4 \text{ or subtracting 4 on both sides}$$

$$3x = 9 \qquad \text{Simplifying}$$

$$\frac{3x}{3} = \frac{9}{3} \qquad \text{Using the multiplication principle: multiplying by } \frac{1}{3} \text{ or dividing by 3 on both sides}$$

$$x = 3 \qquad \text{Simplifying}$$

Check:

$$\begin{array}{r} 3x + 4 = 13 \\ \hline 3 \cdot 3 + 4 \; ? \; 13 \\ 9 + 4 \; \Big| \\ 13 \; \Big| \qquad \text{TRUE} \end{array}$$

The solution is 3.

Do Exercise 1.

EXAMPLE 2 Solve: $-5x - 6 = 16$.

$$-5x - 6 = 16$$

$$-5x - 6 + 6 = 16 + 6 \qquad \text{Adding 6 on both sides}$$

$$-5x = 22$$

$$\frac{-5x}{-5} = \frac{22}{-5} \qquad \text{Dividing by } -5 \text{ on both sides}$$

$$x = -\frac{22}{5}, \text{ or } -4\frac{2}{5} \qquad \text{Simplifying}$$

Check:

$$\begin{array}{r} -5x - 6 = 16 \\ \hline -5\left(-\dfrac{22}{5}\right) - 6 \; ? \; 16 \\ 22 - 6 \; \Big| \\ 16 \; \Big| \qquad \text{TRUE} \end{array}$$

The solution is $-\dfrac{22}{5}$.

Do Exercises 2 and 3.

EXAMPLE 3   Solve: $45 - x = 13$.

$$45 - x = 13$$
$$-45 + 45 - x = -45 + 13 \qquad \text{Adding } -45 \text{ on both sides}$$
$$-x = -32$$
$$-1 \cdot (-x) = -1 \cdot (-32) \qquad \text{Multiplying by } -1 \text{ on both sides}$$
$$-1 \cdot (-1) \cdot x = 32$$
$$x = 32$$

Check:
$$\begin{array}{c} 45 - x = 13 \\ \hline 45 - 32 \; ? \; 13 \\ 13 \; | \quad \text{TRUE} \end{array}$$

The solution is 32.

Do Exercise 4.

EXAMPLE 4   Solve: $16.3 - 7.2y = -8.18$.

$$16.3 - 7.2y = -8.18$$
$$-16.3 + 16.3 - 7.2y = -16.3 + (-8.18) \qquad \text{Adding } -16.3 \text{ on both sides}$$
$$-7.2y = -24.48$$
$$\frac{-7.2y}{-7.2} = \frac{-24.48}{-7.2} \qquad \text{Dividing by } -7.2 \text{ on both sides}$$
$$y = 3.4$$

Check:
$$\begin{array}{c} 16.3 - 7.2y = -8.18 \\ \hline 16.3 - 7.2(3.4) \; ? \; -8.18 \\ 16.3 - 24.48 \; | \\ -8.18 \; | \qquad \text{TRUE} \end{array}$$

The solution is 3.4.

Do Exercises 5 and 6.

b   Collecting Like Terms

If there are like terms on one side of the equation, we collect them before using the addition or multiplication principle.

EXAMPLE 5   Solve: $3x + 4x = -14$.

$$3x + 4x = -14$$
$$7x = -14 \qquad \text{Collecting like terms}$$
$$\frac{7x}{7} = \frac{-14}{7} \qquad \text{Dividing by 7 on both sides}$$
$$x = -2.$$

The number -2 checks, so the solution is -2.

Do Exercises 7 and 8.

4. Solve: $-18 - x = -57$.

Solve.

5. $-4 - 8x = 8$

6. $41.68 = 4.7 - 8.6y$

Solve.

7. $4x + 3x = -21$

8. $x - 0.09x = 728$

Answers on page A-25

Solve.

9. $7y + 5 = 2y + 10$

10. $5 - 2y = 3y - 5$

Solve.

11. $7x - 17 + 2x = 2 - 8x + 15$

12. $3x - 15 = 5x + 2 - 4x$

If there are like terms on opposite sides of the equation, we get them on the same side by using the addition principle. Then we collect them. In other words, we get all terms with a variable on one side and all numbers on the other.

EXAMPLE 6 Solve: $2x - 2 = -3x + 3$.

$$2x - 2 = -3x + 3$$
$$2x - 2 + 2 = -3x + 3 + 2 \qquad \text{Adding 2}$$
$$2x = -3x + 5 \qquad \text{Collecting like terms}$$
$$2x + 3x = -3x + 5 + 3x \qquad \text{Adding } 3x$$
$$5x = 5 \qquad \text{Simplifying}$$
$$\frac{5x}{5} = \frac{5}{5} \qquad \text{Dividing by 5}$$
$$x = 1 \qquad \text{Simplifying}$$

Check:

$$\begin{array}{c|c} \multicolumn{2}{c}{2x - 2 = -3x + 3} \\ \hline 2 \cdot 1 - 2 \;?\; -3 \cdot 1 + 3 & \text{Substituting in the original equation} \\ 2 - 2 & -3 + 3 \\ 0 & 0 \qquad \text{TRUE} \end{array}$$

The solution is 1.

Do Exercises 9 and 10.

In Example 6, we used the addition principle to get all terms with a variable on one side and all numbers on the other side. Then we collected like terms and proceeded as before. If there are like terms on one side at the outset, they should be collected first.

EXAMPLE 7 Solve: $6x + 5 - 7x = 10 - 4x + 3$.

$$6x + 5 - 7x = 10 - 4x + 3$$
$$-x + 5 = 13 - 4x \qquad \text{Collecting like terms}$$
$$4x - x + 5 = 13 - 4x + 4x \qquad \text{Adding } 4x$$
$$3x + 5 = 13 \qquad \text{Simplifying}$$
$$3x + 5 - 5 = 13 - 5 \qquad \text{Subtracting 5}$$
$$3x = 8 \qquad \text{Simplifying}$$
$$\frac{3x}{3} = \frac{8}{3} \qquad \text{Dividing by 3}$$
$$x = \frac{8}{3} \qquad \text{Simplifying}$$

The number $\frac{8}{3}$ checks, so $\frac{8}{3}$ is the solution.

Do Exercises 11 and 12.

Answers on page A-25

CLEARING FRACTIONS AND DECIMALS

For the equations considered thus far, we generally use the addition principle first. There are, however, some situations in which it is to our advantage to use the multiplication principle first. Consider, for example,

$$\frac{1}{2}x = \frac{3}{4}.$$

The LCM of the denominators is 4. If we multiply by 4 on both sides, we get $2x = 3$, which has no fractions. We have "cleared fractions." Now consider

$$2.3x = 4.78.$$

If we multiply by 100 on both sides, we get $230x = 478$, which has no decimal points. We have "cleared decimals." The equations are then easier to solve. It is your choice whether to clear fractions or decimals, but doing so often eases computations.

In what follows, we use the multiplication principle first to "clear," or "eliminate," fractions or decimals. For fractions, the number by which we multiply is the **least common multiple of all the denominators.**

EXAMPLE 8 Solve:

$$\frac{2}{3}x - \frac{1}{6} + \frac{1}{2}x = \frac{7}{6} + 2x.$$

The number 6 is the least common multiple of all the denominators. We multiply by 6 on both sides:

$$6\left(\frac{2}{3}x - \frac{1}{6} + \frac{1}{2}x\right) = 6\left(\frac{7}{6} + 2x\right) \quad \text{Multiplying by 6 on both sides}$$

$$6 \cdot \frac{2}{3}x - 6 \cdot \frac{1}{6} + 6 \cdot \frac{1}{2}x = 6 \cdot \frac{7}{6} + 6 \cdot 2x \quad \text{Using the distributive laws. (Caution! Be sure to multiply all terms by 6.)}$$

$$4x - 1 + 3x = 7 + 12x \quad \text{Simplifying. Note that the fractions are cleared.}$$

$$7x - 1 = 7 + 12x \quad \text{Collecting like terms}$$

$$7x - 1 - 12x = 7 + 12x - 12x \quad \text{Subtracting } 12x$$

$$-5x - 1 = 7 \quad \text{Simplifying}$$

$$-5x - 1 + 1 = 7 + 1 \quad \text{Adding 1}$$

$$-5x = 8 \quad \text{Collecting like terms}$$

$$\frac{-5x}{-5} = \frac{8}{-5} \quad \text{Dividing by } -5$$

$$x = -\frac{8}{5}.$$

13. Solve: $\dfrac{7}{8}x - \dfrac{1}{4} + \dfrac{1}{2}x = \dfrac{3}{4} + x.$

Check:

$$\dfrac{2}{3}x - \dfrac{1}{6} + \dfrac{1}{2}x = \dfrac{7}{6} + 2x$$

$$\dfrac{2}{3}\left(-\dfrac{8}{5}\right) - \dfrac{1}{6} + \dfrac{1}{2}\left(-\dfrac{8}{5}\right) \;?\; \dfrac{7}{6} + 2\left(-\dfrac{8}{5}\right)$$

$$-\dfrac{16}{15} - \dfrac{1}{6} - \dfrac{8}{10} \;\Big|\; \dfrac{7}{6} - \dfrac{16}{5}$$

$$-\dfrac{32}{30} - \dfrac{5}{30} - \dfrac{24}{30} \;\Big|\; \dfrac{35}{30} - \dfrac{96}{30}$$

$$\dfrac{-32 - 5 - 24}{30} \;\Big|\; -\dfrac{61}{30}$$

$$-\dfrac{61}{30} \qquad\qquad \text{TRUE}$$

The solution is $-\dfrac{8}{5}$.

Do Exercise 13.

To illustrate clearing decimals, we repeat Example 4, but this time we clear the decimals first.

EXAMPLE 9 Solve: $16.3 - 7.2y = -8.18.$

The greatest number of decimal places in any one number is *two*. Multiplying by 100, which has *two* 0's, will clear the decimals.

14. Solve: $41.68 = 4.7 - 8.6y.$

$$100(16.3 - 7.2y) = 100(-8.18) \qquad \text{Multiplying by 100 on both sides}$$

$$100(16.3) - 100(7.2y) = 100(-8.18) \qquad \text{Using a distributive law}$$

$$1630 - 720y = -818 \qquad \text{Simplifying}$$

$$1630 - 720y - 1630 = -818 - 1630 \qquad \text{Subtracting 1630 on both sides}$$

$$-720y = -2448 \qquad \text{Collecting like terms}$$

$$\dfrac{-720y}{-720} = \dfrac{-2448}{-720} \qquad \text{Dividing by } -720 \text{ on both sides}$$

$$y = \dfrac{17}{5}, \text{ or } 3.4$$

The number $\dfrac{17}{5}$, or 3.4, checks, as shown in Example 4, so it is the solution.

Do Exercise 14.

Answers on page A-25

C Equations Containing Parentheses

To solve certain kinds of equations that contain parentheses, we first use the distributive laws to remove the parentheses. Then we proceed as before.

EXAMPLE 10 Solve: $8x = 2(12 - 2x)$.

$$8x = 2(12 - 2x)$$

$$8x = 24 - 4x \qquad \text{Using the distributive law to multiply and remove parentheses}$$

$$8x + 4x = 24 - 4x + 4x \qquad \text{Adding } 4x \text{ to get all } x\text{-terms on one side}$$

$$12x = 24 \qquad \text{Collecting like terms}$$

$$\frac{12x}{12} = \frac{24}{12} \qquad \text{Dividing by 12}$$

$$x = 2$$

Check:

$$8x = 2(12 - 2x)$$

$$\begin{array}{c|c} 8 \cdot 2 \; ? \; 2(12 - 2 \cdot 2) & \text{We use the rules for order of operations to} \\ 16 \quad \Big| \quad 2(12 - 4) & \text{carry out the calculations on each side of} \\ \quad \Big| \quad 2 \cdot 8 & \text{the equation.} \\ \quad \Big| \quad 16 \qquad \text{TRUE} \end{array}$$

The solution is 2.

Do Exercises 15 and 16.

Here is a procedure for solving the types of equations discussed in this section.

> ### AN EQUATION-SOLVING PROCEDURE
>
> 1. Multiply on both sides to clear the equation of fractions or decimals. (This is optional, but it can ease computations.)
> 2. If parentheses occur, multiply using the *distributive laws* to remove them.
> 3. Collect like terms on each side, if necessary.
> 4. Get all terms with variables on one side and all constant terms on the other side, using the *addition principle*.
> 5. Collect like terms again, if necessary.
> 6. Multiply or divide to solve for the variable, using the *multiplication principle*.
> 7. Check all possible solutions in the original equation.

Solve.

15. $2(2y + 3) = 14$

16. $5(3x - 2) = 35$

Answers on page A-25

Solve.

17. $3(7 + 2x) = 30 + 7(x - 1)$

18. $4(3 + 5x) - 4 = 3 + 2(x - 2)$

■ **EXAMPLE 11**   Solve: $2 - 5(x + 5) = 3(x - 2) - 1$.

$$2 - 5(x + 5) = 3(x - 2) - 1$$

$$2 - 5x - 25 = 3x - 6 - 1 \qquad \text{Using the distributive laws to}$$
$$\text{multiply and remove parentheses}$$

$$-5x - 23 = 3x - 7 \qquad \text{Collecting like terms}$$

$$-5x - 23 + 5x = 3x - 7 + 5x \qquad \text{Adding } 5x$$

$$-23 = 8x - 7 \qquad \text{Collecting like terms}$$

$$-23 + 7 = 8x - 7 + 7 \qquad \text{Adding } 7$$

$$-16 = 8x \qquad \text{Collecting like terms}$$

$$\frac{-16}{8} = \frac{8x}{8} \qquad \text{Dividing by 8}$$

$$-2 = x$$

Check:
$$\frac{2 - 5(x + 5) = 3(x - 2) - 1}{2 - 5(-2 + 5) \;?\; 3(-2 - 2) - 1}$$
$$\begin{array}{c|c} 2 - 5(3) & 3(-4) - 1 \\ 2 - 15 & -12 - 1 \\ -13 & -13 \end{array} \quad \text{TRUE}$$

The solution is -2.

Note that the solution of $-2 = x$ is -2, which is also the solution of $x = -2$.

Do Exercises 17 and 18.

Answers on page A-25

CALCULATOR CORNER

Checking Solutions of Equations   We can use a calculator to check solutions of equations by substituting the possible solution for each occurrence of the variable on the left side of the equation and then on the right side of the equation. If both sides have the same value, then the number that was substituted for the variable is the solution of the equation. In Example 5, for instance, to check if -2 is the solution of the equation $3x + 4x = -14$, we substitute -2 for x on the left side of the equation. To do this, we press $\boxed{3}\;\boxed{\times}\;\boxed{2}\;\boxed{+/-}\;\boxed{+}\;\boxed{4}\;\boxed{\times}\;\boxed{2}\;\boxed{+/-}\;\boxed{=}$. The result is -14. Note that the right side of the equation does not contain a variable. It is -14, the value that was obtained when -2 was substituted for x on the left side. Thus, the number -2 is the solution of the equation.

   In Example 6, we can check if 1 is the solution of the equation $2x - 2 = -3x + 3$. First, we substitute 1 for x on the left side of the equation by pressing $\boxed{2}\;\boxed{\times}\;\boxed{1}\;\boxed{-}\;\boxed{2}\;\boxed{=}$. We get 0. Next, we substitute 1 for x on the right side of the equation by pressing $\boxed{3}\;\boxed{+/-}\;\boxed{\times}\;\boxed{1}\;\boxed{+}\;\boxed{3}\;\boxed{=}$. We get 0 again. Since the same value was obtained on both sides of the equation, the number 1 is the solution.

Exercises:

1. Use a calculator to check the solutions of the equations in Examples 1–4 and 7–11.

2. Use a calculator to check the solutions of the equations in Margin Exercises 1–18.

For Extra Help

MathXL   MyMathLab   InterAct   Math Tutor   Digital Video   Student's
                       Math   Center   Tutor CD 5   Solutions
                                              Videotape 11   Manual

a   Solve. Don't forget to check!

1. $5x + 6 = 31$

2. $8x + 6 = 30$

3. $8x + 4 = 68$

4. $8z + 7 = 79$

5. $4x - 6 = 34$

6. $4x - 11 = 21$

7. $3x - 9 = 33$

8. $6x - 9 = 57$

9. $7x + 2 = -54$

10. $5x + 4 = -41$

11. $-45 = 3 + 6y$

12. $-91 = 9t + 8$

13. $-4x + 7 = 35$

14. $-5x - 7 = 108$

15. $-7x - 24 = -129$

16. $-6z - 18 = -132$

b   Solve.

17. $5x + 7x = 72$

18. $4x + 5x = 45$

19. $8x + 7x = 60$

20. $3x + 9x = 96$

21. $4x + 3x = 42$

22. $6x + 19x = 100$

23. $-6y - 3y = 27$

24. $-4y - 8y = 48$

25. $-7y - 8y = -15$

26. $-10y - 3y = -39$

27. $10.2y - 7.3y = -58$

28. $6.8y - 2.4y = -88$

29. $x + \dfrac{1}{3}x = 8$

30. $x + \dfrac{1}{4}x = 10$

31. $8y - 35 = 3y$

32. $4x - 6 = 6x$

33. $8x - 1 = 23 - 4x$

34. $5y - 2 = 28 - y$

35. $2x - 1 = 4 + x$

36. $5x - 2 = 6 + x$

37. $6x + 3 = 2x + 11$

38. $5y + 3 = 2y + 15$

39. $5 - 2x = 3x - 7x + 25$

40. $10 - 3x = 2x - 8x + 40$

41. $4 + 3x - 6 = 3x + 2 - x$

42. $5 + 4x - 7 = 4x - 2 - x$

43. $4y - 4 + y + 24 = 6y + 20 - 4y$

44. $5y - 7 + y = 7y + 21 - 5y$

Solve. Clear fractions or decimals first.

45. $\dfrac{7}{2}x + \dfrac{1}{2}x = 3x + \dfrac{3}{2} + \dfrac{5}{2}x$

46. $\dfrac{7}{8}x - \dfrac{1}{4} + \dfrac{3}{4}x = \dfrac{1}{16} + x$

47. $\dfrac{2}{3} + \dfrac{1}{4}t = \dfrac{1}{3}$

48. $-\dfrac{3}{2} + x = -\dfrac{5}{6} - \dfrac{4}{3}$

49. $\dfrac{2}{3} + 3y = 5y - \dfrac{2}{15}$

50. $\dfrac{1}{2} + 4m = 3m - \dfrac{5}{2}$

51. $\dfrac{5}{3} + \dfrac{2}{3}x = \dfrac{25}{12} + \dfrac{5}{4}x + \dfrac{3}{4}$

52. $1 - \dfrac{2}{3}y = \dfrac{9}{5} - \dfrac{y}{5} + \dfrac{3}{5}$

53. $2.1x + 45.2 = 3.2 - 8.4x$

54. $0.96y - 0.79 = 0.21y + 0.46$

55. $1.03 - 0.62x = 0.71 - 0.22x$

56. $1.7t + 8 - 1.62t = 0.4t - 0.32 + 8$

57. $\dfrac{2}{7}x - \dfrac{1}{2}x = \dfrac{3}{4}x + 1$

58. $\dfrac{5}{16}y + \dfrac{3}{8}y = 2 + \dfrac{1}{4}y$

C   Solve.

59. $3(2y - 3) = 27$

60. $4(2y - 3) = 28$

61. $40 = 5(3x + 2)$

62. $9 = 3(5x - 2)$

63. $2(3 + 4m) - 9 = 45$

64. $3(5 + 3m) - 8 = 88$

65. $5r - (2r + 8) = 16$

66. $6b - (3b + 8) = 16$

67. $6 - 2(3x - 1) = 2$

68. $10 - 3(2x - 1) = 1$

69. $5(d + 4) = 7(d - 2)$

70. $3(t - 2) = 9(t + 2)$

71. $8(2t + 1) = 4(7t + 7)$

72. $7(5x - 2) = 6(6x - 1)$

73. $3(r - 6) + 2 = 4(r + 2) - 21$

74. $5(t + 3) + 9 = 3(t - 2) + 6$

75. $19 - (2x + 3) = 2(x + 3) + x$

76. $13 - (2c + 2) = 2(c + 2) + 3c$

77. $0.7(3x + 6) = 1.1 - (x + 2)$

78. $0.9(2x + 8) = 20 - (x + 5)$

79. $a + (a - 3) = (a + 2) - (a + 1)$

80. $0.8 - 4(b - 1) = 0.2 + 3(4 - b)$

81. **D**_W A student begins solving the equation $\frac{2}{3}x + 1 = \frac{5}{6}$ by multiplying by 6 on both sides. Is this a wise thing to do? Why or why not?

82. **D**_W Describe a procedure that a classmate could use to solve the equation $ax + b = c$ for x.

🖎 VOCABULARY REINFORCEMENT

In each of Exercises 83–90, fill in the blank with the correct term from the given list. Some of the choices may not be used.

83. The _____ numbers consist of all numbers that can be named in the form $\frac{a}{b}$, where a and b are integers and b is not _____ .   [10.1b]

84. The symbol > means _____ .   [1.4c]

85. Two numbers whose product is _____ are called reciprocals of each other.   [2.7a]

86. The _____ of a number is its distance from zero on a number line.   [10.1e]

87. A(n) _____ angle is an angle whose measure is greater than 90° and less than 180°.   [9.5b]

88. A(n) _____ triangle is a triangle in which all sides are the same length.   [9.5d]

89. The basic unit of length in the metric system is the _____ .   [8.2a]

90. To _____ an expression is to find an equivalent expression that is a product.   [11.1c]

is greater than	absolute value
is less than	opposite
acute	equilateral
obtuse	scalene
negative	meter
one	gram
zero	rational
factor	irrational
multiply	

Solve.

91. $\dfrac{y-2}{3} = \dfrac{2-y}{5}$

92. $3x = 4x$

93. $\dfrac{5+2y}{3} = \dfrac{25}{12} + \dfrac{5y+3}{4}$

94. ▦ $0.05y - 1.82 = 0.708y - 0.504$

95. $\dfrac{2}{3}(2x - 1) = 10$

96. $\dfrac{2}{3}\left(\dfrac{7}{8} - 4x\right) - \dfrac{5}{8} = \dfrac{3}{8}$

97. The perimeter of the figure shown is 15 cm. Solve for x.

650

11.5

APPLICATIONS AND PROBLEM SOLVING

a   Translating to Algebraic Expressions

In algebra, we translate problems to equations. The different parts of an equation are translations of word phrases to algebraic expressions. To translate, it helps to learn which words translate to certain operation symbols.

KEY WORDS

ADDITION (+)	SUBTRACTION (−)	MULTIPLICATION (·)	DIVISION (÷)
add	subtract	multiply	divide
added to	subtracted from	multiplied by	quotient
sum	difference	product	divided by
total	minus	times	
plus	less than	of	
more than	decreased by		
increased by	take away		

EXAMPLE 1   Translate to an algebraic expression:

   Twice (or two times) some number.

   Think of some number—say, 8. We can write 2 times 8 as 2×8, or $2 \cdot 8$. We multiplied by 2. Do the same thing using a variable. We can use any variable we wish, such as x, y, m, or n. Let's use y to stand for some number. If we multiply by 2, we get an expression

$$y \times 2, \quad 2 \times y, \quad 2 \cdot y, \quad \text{or} \quad 2y.$$

In algebra, $2y$ is the expression used most often.

EXAMPLE 2   Translate to an algebraic expression:

   Seven less than some number.

   We let

   $x =$ the number.

   If the number were 23, then 7 less than 23 is 16, $(23 - 7)$ not $(7 - 23)$. If we knew the number to be 345, then the translation would be $345 - 7$. Thus if the number is x, then the translation is

   $x - 7$.

Caution!

Note that $7 - x$ is *not* a correct translation of the expression in Example 2. The expression $7 - x$ is a translation of "seven minus some number" or "some number less than seven."

Translate to an algebraic expression.

1. Twelve less than some number

2. Twelve more than some number

3. Four less than some number

4. Half of some number

5. Six more than eight times some number

6. The difference of two numbers

7. Fifty-nine percent of some number

8. Two hundred less than the product of two numbers

9. The sum of two numbers

EXAMPLE 3 Translate to an algebraic expression:

Eighteen more than a number.

We let

t = the number.

Now if the number were 26, then the translation would be $18 + 26$, or $26 + 18$. If we knew the number to be 174, then the translation would be $18 + 174$, or $174 + 18$. The translation is

$$18 + t, \quad \text{or} \quad t + 18.$$

EXAMPLE 4 Translate to an algebraic expression:

A number divided by 5.

We let

m = the number.

If the number were 76, then the translation would be $76 \div 5$, or $76/5$, or $\frac{76}{5}$. If the number were 213, then the translation would be $213 \div 5$, or $213/5$, or $\frac{213}{5}$. The translation is found as follows:

A number   divided by   5

$$m \div 5, \quad \text{or} \quad m/5, \quad \text{or} \quad \frac{m}{5}.$$

EXAMPLE 5 Translate to an algebraic expression.

PHRASE	ALGEBRAIC EXPRESSION
Five more than some number	$5 + n$, or $n + 5$
Half of a number	$\frac{1}{2}t$, or $\frac{t}{2}$
Five more than three times some number	$5 + 3p$, or $3p + 5$
The difference of two numbers	$x - y$
Six less than the product of two numbers	$mn - 6$
Seventy-six percent of some number	$76\%z$, or $0.76z$

Do Exercises 1–9.

Answers on page A-26

b Five Steps for Solving Problems

We have studied many new equation-solving tools in this chapter. We now apply them to problem solving. We have purposely used the following strategy throughout this text in order to introduce you to algebra.

> **FIVE STEPS FOR PROBLEM SOLVING IN ALGEBRA**
>
> 1. *Familiarize* yourself with the problem situation.
> 2. *Translate* to an equation.
> 3. *Solve* the equation.
> 4. *Check* your possible answer in the original problem.
> 5. *State* the answer clearly.

Of the five steps, the most important is probably the first one: becoming familiar with the problem situation. The table below lists some hints for familiarization.

To familiarize yourself with a problem:

- If a problem is given in words, read it carefully. Reread the problem, perhaps aloud. Try to verbalize the problem as if you were explaining it to someone else.

- Choose a variable (or variables) to represent the unknown and clearly state what the variable represents. Be descriptive! For example, let L = the length, d = the distance, and so on.

- Make a drawing and label it with known information, using specific units if given. Also, indicate unknown information.

- Find further information. Look up formulas or definitions with which you are not familiar. (Geometric formulas appear on the inside back cover of this text.) Consult a reference librarian or an expert in the field.

- Create a table that lists all the information you have available. Look for patterns that may help in the translation to an equation.

- Think of a possible answer and check the guess. Observe the manner in which the guess is checked.

EXAMPLE 6 *Hiking.* In 1998, at age 79, Earl Shaffer became the oldest person to thru-hike all 2100 miles of the Appalachian Trail—from Springer Mountain, Georgia, to Mount Katahdin, Maine. Shaffer thru-hiked the trail three times, in 1948 (Georgia to Maine), in 1965 (Maine to Georgia), and in 1998 (Georgia to Maine) near the 50th anniversary of his first hike. At one point in 1998, Shaffer stood atop Big Walker Mountain, Virginia, which is three times as far from the northern end as from the southern end. How far was Shaffer from each end of the trail?

Source: Appalachian Trail Conference; Earl Shaffer Foundation

10. Running. In 1997, Yiannis Kouros of Australia set the record for the greatest distance run in 24 hr by running 188 mi. After 8 hr, he was approximately twice as far from the finish line as he was from the start. How far had he run?

Source: *Guinness World Records 2004 Millennium Edition*

1. Familiarize. Let's consider a drawing.

To become familiar with the problem, let's guess a possible distance that Shaffer stood from Springer Mountain—say, 600 mi. Three times 600 mi is 1800 mi. Since 600 mi + 1800 mi = 2400 mi and 2400 mi is greater than 2100 mi, we see that our guess is too large. Rather than guess again, let's use the skills we have obtained in the ability to solve equations. We let

d = the distance, in miles, to the southern end, and

$3d$ = the distance, in miles, to the northern end.

(We could also let x = the distance to the northern end and $\frac{1}{3}x$ = the distance to the southern end.)

2. Translate. From the drawing, we see that the lengths of the two parts of the trail must add up to 2100 mi. This leads to our translation.

$$
\underbrace{\text{Distance to southern end}}_{d} \quad \underset{+}{\text{plus}} \quad \underbrace{\text{Distance to northern end}}_{3d} \quad \underset{=}{\text{is}} \quad \underset{2100}{\text{2100 mi}}
$$

3. Solve. We solve the equation:

$$d + 3d = 2100$$
$$4d = 2100 \qquad \text{Collecting like terms}$$
$$\frac{4d}{4} = \frac{2100}{4} \qquad \text{Dividing by 4 on both sides}$$
$$d = 525.$$

4. Check. As expected, d is less than 600 mi. If $d = 525$ mi, then $3d = 1575$ mi. Since 525 mi + 1575 mi = 2100 mi, we have a check.

5. State. Atop Big Walker Mountain, Shaffer stood 525 mi from Springer Mountain and 1575 mi from Mount Katahdin.

Answer on page A-26

CHAPTER 11: Algebra: Solving Equations and Problems

Do Exercise 10.

EXAMPLE 7 *Rocket Sections.* A rocket is divided into three sections: the payload and navigation section at the top, the fuel section in the middle, and the rocket engine section at the bottom. The top section is one-sixth the length of the bottom section. The middle section is one-half the length of the bottom section. The total length of the rocket is 240 ft. Find the length of each section.

1. **Familiarize.** We first make a drawing. Noting that the lengths of the top and middle sections are expressed in terms of the bottom section, we let

$$x = \text{the length of the bottom section.}$$
Then $\frac{1}{2}x$ = the length of the middle section
and $\frac{1}{6}x$ = the length of the top section.

2. **Translate.** From the statement of the problem and the drawing, we see that the lengths add up to 240 ft. This gives us our translation.

Length of top	plus	Length of middle	plus	Length of bottom	is	Total length
$\frac{1}{6}x$	$+$	$\frac{1}{2}x$	$+$	x	$=$	240

3. **Solve.** We solve the equation by clearing fractions, as follows:

$$\frac{1}{6}x + \frac{1}{2}x + x = 240 \qquad \text{The LCM of all the denominators is 6.}$$

$$6 \cdot \left(\frac{1}{6}x + \frac{1}{2}x + x \right) = 6 \cdot 240 \qquad \text{Multiplying by the LCM, 6}$$

$$6 \cdot \frac{1}{6}x + 6 \cdot \frac{1}{2}x + 6 \cdot x = 6 \cdot 240 \qquad \text{Using the distributive law}$$

$$x + 3x + 6x = 1440 \qquad \text{Simplifying}$$

$$10x = 1440 \qquad \text{Adding}$$

$$\frac{10x}{10} = \frac{1440}{10} \qquad \text{Dividing by 10}$$

$$x = 144.$$

4. **Check.** Do we have an answer *to the problem*? If the rocket engine (bottom section) is 144 ft, then the middle section is $\frac{1}{2} \cdot 144$, or 72 ft, and the top section is $\frac{1}{6} \cdot 144$, or 24 ft. Since these lengths total 240 ft, our answer checks.

5. **State.** The top section is 24 ft, the middle 72 ft, and the bottom 144 ft.

Do Exercise 11.

11. **Gourmet Sandwiches.** A gourmet sandwich shop located near a college campus specializes in sandwiches prepared in buns of length 18 in. Suppose Jenny, Emma, and Sarah buy one of these sandwiches and take it back to their apartment. Since they have different appetites, Jenny cuts the sandwich in such a way that Emma gets half of what Jenny gets and Sarah gets three-fourths of what Jenny gets. Find the length of each person's sandwich.

Answer on page A-26

12. Van Rental. Truck-Rite also rents vans at a daily rate of $64.95 plus 57 cents per mile. What mileage will allow a salesperson to stay within a daily budget of $400?

Answer on page A-26

656

EXAMPLE 8 *Truck Rental.* Truck-Rite Rentals rents trucks at a daily rate of $59.95 plus 49¢ per mile. Concert Productions has budgeted $400 per day for renting a truck to haul equipment to an upcoming concert. How many miles can a rental truck be driven on a $400 daily budget?

1. **Familiarize.** Suppose that Concert Productions drives 75 mi. Then the cost is

$$
\begin{array}{ccccc}
\text{Daily charge} & \text{plus} & & \text{Mileage charge} & \\
(\$59.95) & \text{plus} & (\text{Cost per mile}) & \text{times} & (\text{Number of miles driven}) \\
\$59.95 & + & \$0.49 & \cdot & 75,
\end{array}
$$

which is $59.95 + $36.75, or $96.70. This familiarizes us with the way in which a calculation is made. Note that we convert 49 cents to $0.49 so that we are using the same units, dollars, throughout the problem.

Let m = the number of miles that can be driven for $400.

2. **Translate.** We reword the problem and translate as follows.

$$
\begin{array}{ccccccc}
\text{Daily rate} & \text{plus} & \begin{array}{c}\text{Cost}\\\text{per mile}\end{array} & \text{times} & \begin{array}{c}\text{Number of}\\\text{miles driven}\end{array} & \text{is} & \begin{array}{c}\text{Total}\\\text{cost}\end{array} \\
\$59.95 & + & \$0.49 & \cdot & m & = & \$400
\end{array}
$$

3. **Solve.** We solve the equation:

$$\$59.95 + 0.49m = 400$$
$$100(59.95 + 0.49m) = 100(400) \qquad \text{Multiplying by 100 on both sides to clear decimals}$$
$$100(59.95) + 100(0.49m) = 40{,}000 \qquad \text{Using the distributive law}$$
$$5995 + 49m = 40{,}000$$
$$49m = 34{,}005 \qquad \text{Subtracting 5995}$$
$$\frac{49m}{49} = \frac{34{,}005}{49} \qquad \text{Dividing by 49}$$
$$m \approx 694.0. \qquad \text{Rounding to the nearest tenth}$$

4. **Check.** We check in the original problem. We multiply 694 by $0.49, getting $340.06. Then we add $340.06 to $59.95 and get $400.01, which is just about the $400 allotted.

5. **State.** The truck can be driven about 694 mi on the rental allotment of $400.

Do Exercise 12.

EXAMPLE 9 *Angles of a Triangle.* The second angle of a triangle is twice as large as the first. The measure of the third angle is 20° greater than that of the first angle. How large are the angles?

1. **Familiarize.** We first make a drawing, letting

   the measure of the first angle = x.

   Then the measure of the second angle = $2x$

   and the measure of the third angle = $x + 20$.

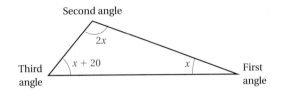

2. **Translate.** To translate, we recall from Section 9.5 that the measures of the angles of any triangle add up to 180°.

$$
\underbrace{\text{Measure of first angle}}_{x} + \underbrace{\text{Measure of second angle}}_{2x} + \underbrace{\text{Measure of third angle}}_{(x + 20)} = \underbrace{180°}_{180}
$$

3. **Solve.** We solve the equation:

$$x + 2x + (x + 20) = 180$$
$$4x + 20 = 180$$
$$4x + 20 - 20 = 180 - 20$$
$$4x = 160$$
$$\frac{4x}{4} = \frac{160}{4}$$
$$x = 40.$$

Possible measures for the angles are as follows:

First angle: $x = 40°$;
Second angle: $2x = 2(40) = 80°$;
Third angle: $x + 20 = 40 + 20 = 60°$.

4. **Check.** Consider 40°, 80°, and 60°. The second is twice the first, and the third is 20° greater than the first. The sum is 180°. These numbers check.

5. **State.** The measures of the angles are 40°, 80°, and 60°.

Caution!

Units are important in answers. Remember to include them, where appropriate.

Do Exercise 13.

13. **Angles of a Triangle.** The second angle of a triangle is three times as large as the first. The third angle measures 30° more than the first angle. Find the measures of the angles.

Answer on page A-26

EXAMPLE 10 *Top Speeds of Roller Coasters.* The average top speed of the three fastest roller coasters in the world is 109 mph. The third-fastest roller coaster, Superman the Escape (located at Six Flags Magic Mountain, Los Angeles, California), reaches a top speed of 20 mph less than the fastest roller coaster, Top Thrill Dragster (located in Cedar Point, Sandusky, Ohio). The second-fastest roller coaster, Dodonpa (located in Fujikyu Highlands, Japan), has a top speed of 107 mph. What is the top speed of the fastest roller coaster?

Source: *Fortune Small Business,* June 2004, p. 48

1. **Familiarize.** The **average** of a set of numbers is the sum of the numbers divided by the number of addends.

   We are given that the second-fastest speed is 107 mph. Suppose the three top speeds are 90, 107, and 112. The average is then

   $$\frac{90 + 107 + 112}{3} = \frac{309}{3} = 103,$$

   which is too low. Instead of continuing to guess, let's use the equation-solving skills we have learned in this chapter. We let x represent the top speed of the fastest roller coaster. Then $x - 20$ is the top speed of the third-fastest roller coaster.

2. **Translate.** We reword the problem and translate as follows.

   $$\frac{\text{Speed of fastest coaster} + \text{Speed of second-fastest coaster} + \text{Speed of third-fastest coaster}}{\text{Number of roller coasters}} = \begin{array}{c}\text{Average speed of three fastest roller coasters}\end{array}$$

   $$\frac{x + 107 + (x - 20)}{3} = 109$$

3. **Solve.** We solve as follows.

   $$\frac{x + 107 + (x - 20)}{3} = 109$$

   $$3 \cdot \frac{x + 107 + (x - 20)}{3} = 3 \cdot 109 \qquad \text{Multiplying on both sides by 3 to clear the fraction}$$

   $$x + 107 + (x - 20) = 327$$

   $$2x + 87 = 327 \qquad \text{Collecting like terms}$$

   $$2x = 240 \qquad \text{Subtracting 87}$$

   $$x = 120 \qquad \text{Dividing by 2}$$

4. Check. If the top speed of the fastest roller coaster is 120 mph, then the top speed of the third-fastest is $120 - 20$, or 100 mph. The average of the top speeds of the three fastest is $(120 + 107 + 100) \div 3 = 327 \div 3$, or 109 mph. The answer checks.

5. State. The top speed of the fastest roller coaster in the world is 120 mph.

Do Exercise 14.

EXAMPLE 11 *Savings Investment.* An investment is made at 6% simple interest for 1 year. It grows to $768.50. How much was originally invested (the principal)?

1. Familiarize. Suppose that $100 was invested. Recalling the formula for simple interest, $I = Prt$, we know that the interest for 1 year on $100 at 6% simple interest is given by $I = \$100 \cdot 0.06 \cdot 1 = \$6$. Then, at the end of the year, the amount in the account is found by adding the principal and the interest:

$$
\begin{array}{ccccc}
\text{Principal} & + & \text{Interest} & = & \text{Amount} \\
\downarrow & & \downarrow & & \downarrow \\
\$100 & + & \$6 & = & \$106.
\end{array}
$$

In this problem, we are working backward. We are trying to find the principal, which is the original investment. We let x = the principal. Then the interest earned is $6\%x$.

2. Translate. We reword the problem and then translate.

$$
\begin{array}{ccccc}
\text{Principal} & + & \text{Interest} & = & \text{Amount} \\
\downarrow & & \downarrow & & \downarrow \\
x & + & 6\%x & = & 768.50
\end{array}
$$
Interest is 6% of the principal.

3. Solve. We solve the equation:

$$
\begin{aligned}
x + 6\%x &= 768.50 \\
x + 0.06x &= 768.50 \qquad \text{Converting to decimal notation} \\
1x + 0.06x &= 768.50 \qquad \text{Identity property of 1} \\
1.06x &= 768.50 \qquad \text{Collecting like terms} \\
\frac{1.06x}{1.06} &= \frac{768.50}{1.06} \qquad \text{Dividing by 1.06} \\
x &= 725.
\end{aligned}
$$

4. Check. We check by taking 6% of $725 and adding it to $725:

$$6\% \times \$725 = 0.06 \times 725 = \$43.50.$$

Then $\$725 + \$43.50 = \$768.50$, so $725 checks.

5. State. The original investment was $725.

Do Exercise 15.

---

14. Average Test Score. Sam's average score on his first three math tests is 77. He scored 62 on the first test. On the third test, he scored nine points more than he scored on his second test. What did he score on the second and third tests?

15. Savings Investment. An investment is made at 7% simple interest for 1 year. It grows to $8988. How much was originally invested (the principal)?

Answers on page A-26

Study Tips

FINAL STUDY TIP

You are arriving at the end of your course in Basic Mathematics. If you have not begun to prepare for the final examination, be sure to read the comments in the Study Tips on pp. 545, 611, 643, and 656.

"We make a living by what we get, but we make a life by what we give."

Winston Churchill

659

Translating for Success

1. *Angle Measures.* The measure of the second angle of a triangle is 51° more than that of the first angle. The measure of the third angle is 3° less than twice the first angle. Find the measures of the angles.

2. *Sales Tax.* Tina paid $3976 for a used car. This amount included 5% for sales tax. How much did the car cost before tax?

3. *Perimeter.* The perimeter of a rectangle is 2347 ft. The length is 28 ft greater than the width. Find the length and the width.

4. *Fraternity or Sorority Membership.* At Arches Tech University, 3976 students belong to a fraternity or a sorority. This is 35% of the total enrollment. What is the total enrollment at Arches Tech?

5. *Fraternity or Sorority Membership.* At Moab Tech University, 35% of the students belong to a fraternity or a sorority. The total enrollment of the university is 11,360 students. How many students belong to either a fraternity or a sorority?

The goal of these matching questions is to practice step (2), *Translate,* of the five-step problem-solving process. Translate each word problem to an equation and select a correct translation from equations A–O.

A. $x + (x - 3) + \dfrac{4}{5}x = 384$

B. $x + (x + 51) + (2x - 3) = 180$

C. $x + (x + 96{,}000) = 180{,}000$

D. $2 \cdot 96 + 2x = 3976$

E. $x + (x + 1) + (x + 2) = 384$

F. $3976 = x \cdot 11{,}360$

G. $2x + 2(x + 28) = 2347$

H. $3976 = x + 5\%x$

I. $x + (x + 28) = 2347$

J. $x = 35\% \cdot 11{,}360$

K. $x + 96 = 3976$

L. $x + (x + 3) + \dfrac{4}{5}x = 384$

M. $x + (x + 2) + (x + 4) = 384$

N. $35\% \cdot x = 3976$

O. $x + (x + 28) = 2347$

Answers on page A-26

6. *Island Population.* There are 180,000 people living on a small Caribbean island. The women outnumber the men by 96,000. How many men live on the island?

7. *Wire Cutting.* A 384-m wire is cut into three pieces. The second piece is 3 m longer than the first. The third is four-fifths as long as the first. How long is each piece?

8. *Locker Numbers.* The numbers on three adjoining lockers are consecutive integers whose sum is 384. Find the integers.

9. *Fraternity or Sorority Membership.* The total enrollment at Canyonlands Tech University is 11,360 students. Of these, 3976 students belong to a fraternity or a sorority. What percent of the students belong to a fraternity or a sorority?

10. *Width of a Rectangle.* The length of a rectangle is 96 ft. The perimeter of the rectangle is 3976 ft. Find the width.

11.5 EXERCISE SET

For Extra Help

a Translate to an algebraic expression.

1. Three less than twice a number

2. Three times a number divided by a

3. The product of 97% and some number

4. 43% of some number

5. Four more than five times some number

6. Seventy-five less than eight times a number

7. *Pipe Cutting.* A 240-in. pipe is cut into two pieces. One piece is three times the length of the other. Let $x =$ the length of the longer piece. Write an expression for the length of the shorter piece.

8. The price of a book is decreased by 30% during a sale. Let $b =$ the price of the book before the reduction. Write an expression for the sale price.

b Solve.

9. What number added to 85 is 117?

10. Eight times what number is 2552?

11. *Statue of Liberty.* The height of the Eiffel Tower is 974 ft, which is about 669 ft higher than the Statue of Liberty. What is the height of the Statue of Liberty?

12. *Area of Lake Ontario.* The area of Lake Superior is about four times the area of Lake Ontario. The area of Lake Superior is 30,172 mi². What is the area of Lake Ontario?

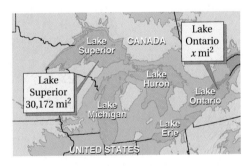

13. *Cinnamon Life* Recently, the cost of four 21-oz boxes of Cinnamon Life cereal was $17.16. What was the cost of one box?

Source: General Mills

14. *Women's Dresses.* In a recent year, the total amount spent on women's blouses was $6.5 billion. This was $0.2 billion more than what was spent on women's dresses. How much was spent on women's dresses?

15. When 17 is subtracted from four times a certain number, the result is 211. What is the number?

16. When 36 is subtracted from five times a certain number, the result is 374. What is the number?

17. If you double a number and then add 16, you get $\frac{2}{3}$ of the original number. What is the original number?

18. If you double a number and then add 85, you get $\frac{3}{4}$ of the original number. What is the original number?

19. *Iditarod Race.* The Iditarod sled dog race extends for 1049 mi from Anchorage to Nome. If a musher is twice as far from Anchorage as from Nome, how many miles has the musher traveled?

Source: Iditarod Trail Commission

20. *Home Remodeling.* In a recent year, Americans spent a total of approximately $32.8 billion to remodel bathrooms and kitchens. Twice as much was spent on kitchens as on bathrooms. How much was spent on each?

Source: *Remodeling Magazine*

21. *Pipe Cutting.* A 480-m pipe is cut into three pieces. The second piece is three times as long as the first. The third piece is four times as long as the second. How long is each piece?

22. *Rope Cutting.* A 180-ft rope is cut into three pieces. The second piece is twice as long as the first. The third piece is three times as long as the second. How long is each piece of rope?

23. *Perimeter of NBA Court.* The perimeter of an NBA basketball court is 288 ft. The length is 44 ft longer than the width. Find the dimensions of the court.

Source: National Basketball Association

w $w + 44$

24. *Perimeter of High School Court.* The perimeter of a standard high school basketball court is 268 ft. The length is 34 ft longer than the width. Find the dimensions of the court.

Source: Indiana High School Athletic Association

25. *Hancock Building Dimensions.* The ground floor of the John Hancock Building in Chicago is in the shape of a rectangle whose length is 100 ft more than the width. The perimeter is 860 ft. Find the length, the width, and the area of the ground floor.

26. *Hancock Building Dimensions.* The top floor of the John Hancock Building in Chicago is in the shape of a rectangle whose length is 60 ft more than the width. The perimeter is 520 ft. Find the length, the width, and the area of the top floor.

$w + 60$ 　 w

$w + 100$ 　 w

27. *Car Rental.* Value Rent-A-Car rents a family-sized car at a daily rate of $74.95 plus 40 cents per mile. Rick Moneycutt is allotted a daily budget of $240. How many miles can he drive per day and stay within his budget?

28. *Van Rental.* Value Rent-A-Car rents a van at a daily rate of $84.95 plus 60 cents per mile. Molly Rickert rents a van to deliver electrical parts to her customers. She is allotted a daily budget of $320. How many miles can she drive per day and stay within her budget?

29. *Angles of a Triangle.*   The second angle of a triangular parking lot is four times as large as the first. The third angle is 45° less than the sum of the other two angles. How large are the angles?

30. *Angles of a Triangle.*   The second angle of a triangular field is three times as large as the first. The third angle is 40° greater than the first. How large are the angles?

31. *Stock Prices.*   Sarah's investment in AOL/Time Warner stock grew 28% to $448. How much did she invest?

32. *Savings Interest.*   Sharon invested money in a savings account at a rate of 6% simple interest. After 1 yr, she has $6996 in the account. How much did Sharon originally invest?

33. *Credit Cards.*   The balance in Will's Mastercard® account grew 2%, to $870, in one month. What was his balance at the beginning of the month?

34. *Loan Interest.*   Alvin borrowed money from a cousin at a rate of 10% simple interest. After 1 yr, $7194 paid off the loan. How much did Alvin borrow?

35. *Taxi Fares.*   In Beniford, taxis charge $3 plus 75¢ per mile for an airport pickup. How far from the airport can Courtney travel for $12?

36. *Taxi Fares.*   In Cranston, taxis charge $4 plus 90¢ per mile for an airport pickup. How far from the airport can Ralph travel for $17.50?

37. *Tipping.*   Leon left a 15% tip for a meal. The total cost of the meal, including the tip, was $41.40. What was the cost of the meal before the tip was added?

38. *Tipping.*   Selena left an 18% tip for a meal. The total cost of the meal, including the tip, was $40.71. What was the cost of the meal before the tip was added?

CHAPTER 11: Algebra: Solving Equations and Problems

39. *Standard Billboard Sign.* A standard rectangular highway billboard sign has a perimeter of 124 ft. The length is 6 ft more than three times the width. Find the dimensions.

40. *Two-by-Four.* The perimeter of a cross section of a "two-by-four" piece of lumber is 10 in. The length is 2 in. more than the width. Find the actual dimensions of the cross section of a two-by-four.

41. D_W A fellow student claims to be able to solve most of the problems in this section by guessing. Is there anything wrong with this approach? Why or why not?

42. D_W Write a problem for a classmate to solve so that it can be translated to the equation

$$\tfrac{2}{3}x + (x + 5) + x = 375.$$

SKILL MAINTENANCE

Calculate.

43. $-\dfrac{4}{5} - \left(\dfrac{3}{8}\right)$ [10.3a]

44. $-\dfrac{4}{5} + \dfrac{3}{8}$ [10.2a]

45. $-\dfrac{4}{5} \cdot \dfrac{3}{8}$ [10.4a]

46. $-\dfrac{4}{5} \div \left(\dfrac{3}{8}\right)$ [10.5c]

47. $-25.6 \div (-16)$ [10.5c]

48. $-25.6(-16)$ [10.4a]

49. $-25.6 - (-16)$ [10.3a]

50. $-25.6 + (-16)$ [10.2a]

SYNTHESIS

51. Abraham Lincoln's 1863 Gettysburg Address refers to the year 1776 as "Four *score* and seven years ago." Write an equation to find out what a score is.

52. *Test Questions.* A student scored 78 on a test that had 4 seven-point fill-ins and 24 three-point multiple-choice questions. The student had 1 fill-in wrong. How many multiple-choice questions did the student get right?

53. The width of a rectangle is $\frac{3}{4}$ of the length. The perimeter of the rectangle becomes 50 cm when the length and the width are each increased by 2 cm. Find the length and the width.

54. Cookies are set out on a tray for six people to take home. One-third, one-fourth, one-eighth, and one-fifth are given to four people, respectively. The fifth person eats ten cookies on the spot, with one cookie remaining for the sixth person. Find the original number of cookies on the tray.

55. A storekeeper goes to the bank to get $10 worth of change. She requests twice as many quarters as half dollars, twice as many dimes as quarters, three times as many nickels as dimes, and no pennies or dollars. How many of each coin did the storekeeper receive?

56. A student has an average score of 82 on three tests. His average score on the first two tests is 85. What was the score on the third test?

11 Summary and Review

The review that follows is meant to prepare you for a chapter exam. It consists of three parts. The first part, Concept Reinforcement, is designed to increase understanding of the concepts through true/false exercises. The second part is a list of important properties and formulas. The third part is the Review Exercises. These provide practice exercises for the exam, together with references to section objectives so you can go back and review. Before beginning, stop and look back over the skills you have obtained. What skills in mathematics do you have now that you did not have before studying this chapter?

CONCEPT REINFORCEMENT

Determine whether the statement is true or false. Answers are given at the back of the book.

_____ **1.** The expression $2(x + 3)$ is equivalent to the expression $2 \cdot x + 3$.

_____ **2.** The expression $x - 7$ is not equivalent to the expression $7 - x$.

_____ **3.** To factor an expression is to find an equivalent expression that is a product.

_____ **4.** $3 - x = 4x$ and $5x = -3$ are equivalent equations.

_____ **5.** Collecting like terms is based on the distributive laws.

IMPORTANT PROPERTIES AND FORMULAS

The Addition Principle: For any real numbers a, b, and c, $a = b$ is equivalent to $a + c = b + c$.

The Multiplication Principle: For any real numbers a, b, and c, $c \neq 0$, $a = b$ is equivalent to $a \cdot c = b \cdot c$.

The Distributive Laws: $a(b + c) = ab + ac$, $a(b - c) = ab - ac$

Review Exercises

1. Evaluate $\dfrac{x - y}{3}$ when $x = 17$ and $y = 5$.   [11.1a]

Multiply.   [11.1b]

2. $5(3x - 7)$

3. $-2(4x - 5)$

4. $10(0.4x + 1.5)$

5. $-8(3 - 6x)$

Factor.   [11.1c]

6. $2x - 14$

7. $6x - 6$

8. $5x + 10$

9. $12 - 3x$

Collect like terms.   [11.1d]

10. $11a + 2b - 4a - 5b$

11. $7x - 3y - 9x + 8y$

12. $6x + 3y - x - 4y$

13. $-3a + 9b + 2a - b$

Solve. [11.2a], [11.3a]

14. $x + 5 = -17$

15. $-8x = -56$

16. $-\dfrac{x}{4} = 48$

17. $n - 7 = -6$

18. $15x = -35$

19. $x - 11 = 14$

20. $-\dfrac{2}{3} + x = -\dfrac{1}{6}$

21. $\dfrac{4}{5}y = -\dfrac{3}{16}$

22. $y - 0.9 = 9.09$

23. $5 - x = 13$

Solve. [11.4a, b, c]

24. $5t + 9 = 3t - 1$

25. $7x - 6 = 25x$

26. $\dfrac{1}{4}x - \dfrac{5}{8} = \dfrac{3}{8}$

27. $14y = 23y - 17 - 10$

28. $0.22y - 0.6 = 0.12y + 3 - 0.8y$

29. $\dfrac{1}{4}x - \dfrac{1}{8}x = 3 - \dfrac{1}{16}x$

30. $4(x + 3) = 36$

31. $3(5x - 7) = -66$

32. $8(x - 2) - 5(x + 4) = 20x + x$

33. $-5x + 3(x + 8) = 16$

34. Translate to an algebraic expression: [11.5a]
   Nineteen percent of some number.

Solve. [11.5b]

35. *Dimensions of Wyoming.* The state of Wyoming is roughly in the shape of a rectangle whose perimeter is 1280 mi. The length is 90 mi more than the width. Find the dimensions.

36. An entertainment center sold for $2449 in June. This was $332 more than the cost in February. Find the cost in February.

37. Ty is paid a commission of $8 for each appliance he sells. One week, he received $216 in commissions. How many appliances did he sell?

38. The measure of the second angle of a triangle is 50° more than that of the first angle. The measure of the third angle is 10° less than twice the first angle. Find the measures of the angles.

39. After a 30% reduction, a bread maker is on sale for $154. What was the marked price (the price before the reduction)?

40. A hotel manager's salary is $90,000, which is a 20% increase over the previous year's salary. What was the previous salary?

41. A tax-exempt charity received a bill of $145.90 for a sump pump. The bill incorrectly included sales tax of 5%. How much does the charity actually owe?

42. *Nile and Amazon Rivers.* The total length of the Nile and Amazon Rivers is 13,108 km. If the Amazon were 234 km longer, it would be as long as the Nile. Find the length of each river.

Source: *The Handy Geography Answer Book*

43. *HDTV Costs.* An HDTV television sold for $829 in May. This was $38 more than the cost in January. Find the cost in January.

44. *Writing Pad.* The perimeter of a rectangular writing pad is 56 cm. The width is 6 cm less than the length. Find the width and the length.

45. D_W Explain at least three uses of the distributive laws considered in this chapter. [11.1b, c, d]

46. D_W Explain all the errors in each of the following. [11.4a, c]

a) $4 - 3x = 9$
$3x = 9$
$x = 3$

b) $2(x - 5) = 7$
$2x - 5 = 7$
$2x = 12$
$x = 6$

SYNTHESIS

Solve. [10.1e], [11.4a, b]

47. $2|n| + 4 = 50$

48. $|3n| = 60$

1. Evaluate $\dfrac{3x}{y}$ when $x = 10$ and $y = 5$.

Multiply.

2. $3(6 - x)$

3. $-5(y - 1)$

Factor.

4. $12 - 22x$

5. $7x + 21 + 14y$

Collect like terms.

6. $9x - 2y - 14x + y$

7. $-a + 6b + 5a - b$

Solve.

8. $x + 7 = 15$

9. $t - 9 = 17$

10. $3x = -18$

11. $-\dfrac{4}{7}x = -28$

12. $3t + 7 = 2t - 5$

13. $\dfrac{1}{2}x - \dfrac{3}{5} = \dfrac{2}{5}$

14. $8 - y = 16$

15. $-\dfrac{2}{5} + x = -\dfrac{3}{4}$

16. $0.4p + 0.2 = 4.2p - 7.8 - 0.6p$

17. $3(x + 2) = 27$

18. $-3x - 6(x - 4) = 9$

19. Translate to an algebraic expression:
Nine less than some number.

Solve.

20. *Perimeter of a Photograph.* The perimeter of a rectangular photograph is 36 cm. The length is 4 cm greater than the width. Find the width and the length.

21. *Charitable Contributions.* About 35.9% of all charitable contributions are made to religious organizations. In 2003, about $86.4 billion was given to religious organizations. How much was given to charities in general?

Source: AAFRC Trust for Philanthropy/Giving USA 2004

22. *Board Cutting.* An 8-m board is cut into two pieces. One piece is 2 m longer than the other. How long are the pieces?

23. *Savings Account.* Money is invested in a savings account at 5% simple interest. After 1 year, there is $924 in the account. How much was originally invested?

24. If you triple a number and then subtract 14, you get $\frac{2}{3}$ of the original number. What is the original number?

25. The second angle of a triangle is three times as large as the first. The third angle is 25° less than the sum of the other two angles. Find the measure of the first angle.

SYNTHESIS

26. Solve: $3|w| - 8 = 37$.

27. A movie theater had a certain number of tickets to give away. Five people got the tickets. The first got $\frac{1}{3}$ of the tickets, the second got $\frac{1}{4}$ of the tickets, and the third got $\frac{1}{5}$ of the tickets. The fourth person got 8 tickets, and there were 5 tickets left for the fifth person. Find the total number of tickets given away.

CHAPTER 11: Algebra: Solving Equations and Problems

Cumulative Review/ Final Examination

This cumulative review also serves as a final examination for the entire book. A question that may occur at this point is what notation to use for a particular problem or exercise. Although there is no particular rule, especially as you use mathematics outside the classroom, here is the guideline that we follow: Use the notation given in the problem. That is, if the problem is given using mixed numerals, give the answer in mixed numerals. If the problem is given in decimal notation, give the answer in decimal notation.

1. *Pears.* Find the unit price of each brand. Then determine which package has the lower unit price.

BRAND	PACKAGE	PRICE	UNIT PRICE
A	$8\frac{1}{2}$ oz	$0.65	
B	15 oz	$1.07	
C	$15\frac{1}{4}$ oz	$1.07	
D	29 oz	$1.69	
E	29 oz	$1.32	

2. Write a word name for 7.463.

3. *Pluto.* The planet Pluto has a diameter of 1400 mi. Use $\frac{22}{7}$ for π.

a) Find the circumference of Pluto.

b) Find the volume of Pluto.

Kitchen Costs. The average cost of a remodeled kitchen is $26,888, according to the National Kitchen and Bath Association. Complete the table below, which relates percents and costs of certain items.

	ITEM	PERCENT OF COST	COST
4.	Cabinets	50%	
5.	Countertops		$4033.20
6.	Appliances	8%	
7.	Fixtures		$8066.40
8.	Flooring	2%	

Source: National Kitchen and Bath Association

9. In 47,201, what digit tells the number of thousands?

10. Write expanded notation for 7405.

Add and simplify, if appropriate.

11.
```
  7 4 1
+ 2 7 1
```

12.
```
  4 9 0 3
  5 2 7 8
  6 3 9 1
+ 4 5 1 3
```

13. $\dfrac{2}{13} + \dfrac{1}{26}$

14.
$$2\frac{4}{9}$$
$$+ 3\frac{1}{3}$$

15.
$$\begin{array}{r} 2.0\,4\,8 \\ 6\,3.9\,1\,4 \\ +\ 4\,2\,8.0\,0\,9 \\ \hline \end{array}$$

16. $34.56 + 2.783 + 0.433 + 765.1$

Subtract and simplify, if possible.

17.
$$\begin{array}{r} 6\,7\,4 \\ -\ 5\,2\,2 \\ \hline \end{array}$$

18.
$$\begin{array}{r} 9\,4\,6\,5 \\ -\ 8\,7\,9\,1 \\ \hline \end{array}$$

19. $\dfrac{7}{8} - \dfrac{2}{3}$

20.
$$\begin{array}{r} 4\dfrac{1}{3} \\ -\ 1\dfrac{5}{8} \\ \hline \end{array}$$

21.
$$\begin{array}{r} 2\,0.0 \\ -\ 0.0\,0\,2\,7 \\ \hline \end{array}$$

22. $40.03 - 5.789$

Multiply and simplify, if possible.

23.
$$\begin{array}{r} 2\,9\,7 \\ \times\ 1\,6 \\ \hline \end{array}$$

24.
$$\begin{array}{r} 3\,4\,9 \\ \times\ 7\,6\,3 \\ \hline \end{array}$$

25. $1\dfrac{3}{4} \cdot 2\dfrac{1}{3}$

26. $\dfrac{9}{7} \cdot \dfrac{14}{15}$

27. $12 \cdot \dfrac{5}{6}$

28.
$$\begin{array}{r} 3\,4.0\,9 \\ \times\ 7.6 \\ \hline \end{array}$$

Divide and simplify. State the answer using a whole-number quotient and remainder.

29. $6\,\overline{)\,3\,4\,3\,8}$

30. $3\,4\,\overline{)\,1\,9\,1\,4}$

31. Give a mixed numeral for the quotient in Question 30.

Divide and simplify, if possible.

32. $\dfrac{4}{5} \div \dfrac{8}{15}$

33. $2\dfrac{1}{3} \div 30$

34. $2.7\,\overline{)\,1\,0\,5.3}$

35. Round 68,489 to the nearest thousand.

36. Round 0.4275 to the nearest thousandth.

37. Round $21.\overline{83}$ to the nearest hundredth.

38. Determine whether 1368 is divisible by 8.

39. Find all the factors of 15.

40. Find the LCM of 16, 25, and 32.

Simplify.

41. $\dfrac{21}{30}$

42. $\dfrac{275}{5}$

43. Convert to a mixed numeral: $\dfrac{18}{5}$.

44. Use = or ≠ for ☐ to write a true sentence:
$$\dfrac{4}{7}\ \square\ \dfrac{3}{5}.$$

45. Use < or > for ☐ to write a true sentence:
$$\dfrac{4}{7}\ \square\ \dfrac{3}{5}.$$

46. Which number is greater, 1.001 or 0.9976?

CHAPTER 11: Algebra: Solving Equations
and Problems

47. Use < or > for ☐ to write a true sentence: 987 ☐ 879.

48. What part is shaded?

Convert to decimal notation.

49. $\dfrac{37}{1000}$

50. $\dfrac{13}{25}$

51. $\dfrac{8}{9}$

52. 7%

Convert to fraction notation.

53. 4.63

54. $7\dfrac{1}{4}$

55. 40%

Convert to percent notation.

56. $\dfrac{17}{20}$

57. 1.5

Solve.

58. $234 + y = 789$

59. $3.9 \times y = 249.6$

60. $\dfrac{2}{3} \cdot t = \dfrac{5}{6}$

61. $\dfrac{8}{17} = \dfrac{36}{x}$

Golf Rounds. The data in the following table show the percent of golfers who play a certain number of rounds of golf per year.

NUMBER OF ROUNDS PER YEAR	PERCENT
1–7	2.2
8–24	14.9
25–49	25.3
50–99	39.0
100 or more	18.6

Source: U.S. Golf Foundation

62. Make a circle graph of the data.

63. Make a bar graph of the data.

64. Find the missing angle measure.

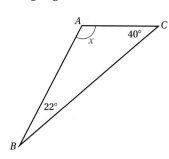

65. Classify the triangle in Exercise 64 as right, obtuse, or acute.

Solve.

66. *Donations.* Lorenzo makes donations of $627 and $48. What was the total donation?

67. *Candy Bars.* A machine wraps 134 candy bars per minute. How long does it take this machine to wrap 8710 bars?

68. *Stock Prices.* A share of stock bought for $29.63 dropped $3.88 before it was resold. What was the price when it was resold?

69. *Length of Trip.* At the start of a trip, a car's odometer read 27,428.6 mi, and at the end of the trip, the reading was 27,914.5 mi. How long was the trip?

70. *Taxes.* From an income of $12,000, amounts of $2300 and $1600 are paid for federal and state taxes. How much remains after these taxes have been paid?

71. *Teacher Salary.* A substitute teacher is paid $87 a day for 9 days. How much was received altogether?

72. *Walking Distance.* A person walks $\frac{3}{5}$ km per hour. At this rate, how far would the person walk in $\frac{1}{2}$ hr?

73. *Sweater Costs.* Eight identical sweaters cost a total of $679.68. What is the cost of each sweater?

74. *Paint Needs.* Eight gallons of exterior paint covers 400 ft^2. How much paint is needed to cover 650 ft^2?

75. *Simple Interest.* What is the simple interest on $4000 principal at 5% for $\frac{3}{4}$ year?

76. *Commission Rate.* A real estate agent received $5880 commission on the sale of an $84,000 home. What was the rate of commission?

77. *Population Growth.* The population of a city is 29,000 this year and is increasing at 4% per year. What will the population be next year?

78. *Student Ages.* The ages of students in a math class at a community college are as follows:

18, 21, 26, 31, 32, 18, 50.

Find the average, the median, and the mode of their ages.

Evaluate.

79. 18^2

80. 20^2

Simplify.

81. $\sqrt{9}$

82. $\sqrt{121}$

83. Approximate to three decimal places: $\sqrt{20}$.

Complete.

84. $\frac{1}{3}$ yd = _____ in.

85. 4280 mm = _____ cm

86. 3 days = _____ hr

87. 20,000 g = _____ kg

88. 5 lb = _____ oz

89. 0.008 cg = _____ mg

90. 8190 mL = _____ L

91. 20 qt = _____ gal

92. Find the length of the third side of this right triangle. Give an exact answer and an approximation to three decimal places.

93. Find the perimeter and the area.

Find the area.

94.

95.

96.

97. Find the diameter, the circumference, and the area of this circle. Use 3.14 for π.

98. Find the volume.

Find the volume. Use 3.14 for π.

99.

100.

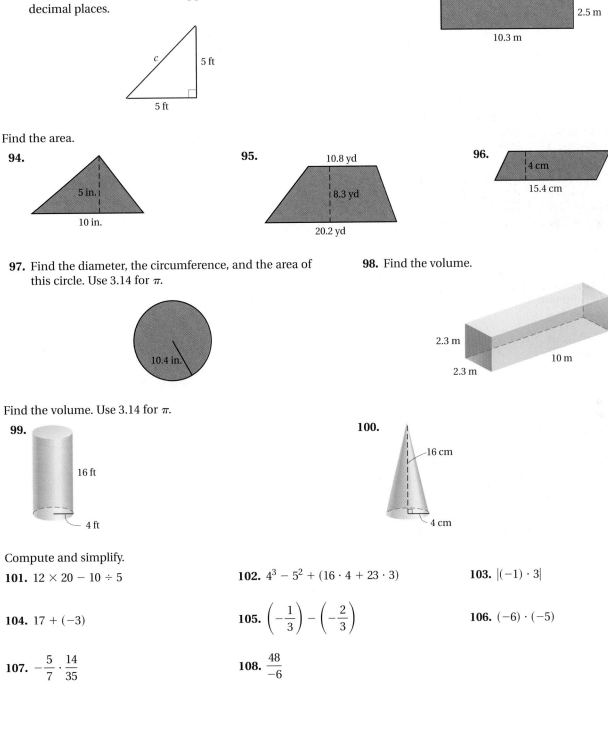

Compute and simplify.

101. $12 \times 20 - 10 \div 5$

102. $4^3 - 5^2 + (16 \cdot 4 + 23 \cdot 3)$

103. $|(-1) \cdot 3|$

104. $17 + (-3)$

105. $\left(-\frac{1}{3}\right) - \left(-\frac{2}{3}\right)$

106. $(-6) \cdot (-5)$

107. $-\frac{5}{7} \cdot \frac{14}{35}$

108. $\frac{48}{-6}$

Solve.

109. $7 - x = 12$

110. $-4.3x = -17.2$

111. $5x + 7 = 3x - 9$

112. $5(x - 2) - 8(x - 4) = 20$

Translate to an algebraic expression.

113. 17 more than some number

114. 38 percent of some number

Solve.

115. *Rollerblade Costs.*   Melinda and Susan purchased rollerblades for a total of $107. Melinda paid $17 more for her rollerblades than Susan did. What did Susan pay?

116. *Savings Investment.*   Money is invested in a savings account at 8% simple interest. After 1 year, there is $1134 in the account. How much was originally invested?

117. *Wire Cutting.*   A 143-m wire is cut into three pieces. The second piece is 3 m longer than the first. The third is four-fifths as long as the first. How long is each piece?

Solve.

118. $\dfrac{2}{3}x + \dfrac{1}{6} - \dfrac{1}{2}x = \dfrac{1}{6} - 3x$

119. $29.966 - 8.673y = -8.18 + 10.4y$

For each of Questions 120–123, choose the correct answer from the selection given.

120. Collect like terms: $\dfrac{1}{4}x - \dfrac{3}{4}y + \dfrac{1}{4}x - \dfrac{3}{4}y$.

   a) 0   **b)** $-\dfrac{2}{3}y$ **c)** $\dfrac{1}{2}x - \dfrac{3}{2}y$

   d) $\dfrac{1}{2}x$ **e)** None

121. Factor: $8x + 4y - 12z$.

   a) $2x + y - 3z$ **b)** $4x + 2y - 6z$ **c)** $2x - 3z$
   d) $2x + 4y - 3z$ **e)** None

122. Divide: $-\dfrac{13}{25} \div \left(-\dfrac{13}{5}\right)$.

   a) $\dfrac{169}{125}$ **b)** 5   **c)** $\dfrac{125}{169}$

   d) $\dfrac{1}{5}$ **e)** None

123. Add: $-27 + (-11)$.

   a) -38 **b)** -16 **c)** 16
   d) 38   **e)** None

Answers

CHAPTER 1

Margin Exercises, Section 1.1, pp. 2–5

1. 2 ten thousands   **2.** 2 hundred thousands
3. 2 millions   **4.** 2 ten millions
5. 2 tens   **6.** 2 hundreds
7. 2 hundred thousands; 8 ten thousands; 0 thousands;
2 hundreds; 1 ten; 9 ones
8. 1 thousand + 8 hundreds + 9 tens + 5 ones
9. 2 ten thousands + 3 thousands + 4 hundreds +
1 ten + 6 ones
10. 3 thousands + 0 hundreds + 3 tens + 1 one, or
3 thousands + 3 tens + 1 one
11. 4 thousands + 1 hundred + 8 tens + 0 ones, or
4 thousands + 1 hundred + 8 tens
12. 1 hundred thousand + 5 ten thousands +
4 thousands + 6 hundreds + 1 ten + 6 ones
13. Forty-nine   **14.** Sixteen   **15.** Thirty-eight
16. Two hundred four
17. Forty-four thousand, one hundred fifty-five
18. One million, eight hundred seventy-nine thousand,
two hundred four   **19.** Six billion, four hundred forty-
nine million   **20.** 213,105,329

Exercise Set 1.1, p. 6

1. 5 thousands   **3.** 5 hundreds   **5.** 2   **7.** 1
9. 5 thousands + 7 hundreds + 0 tens + 2 ones, or
5 thousands + 7 hundreds + 2 ones
11. 9 ten thousands + 3 thousands + 9 hundreds +
8 tens + 6 ones
13. 2 thousands + 0 hundreds + 5 tens + 8 ones, or
2 thousands + 5 tens + 8 ones
15. 1 thousand + 2 hundreds + 6 tens + 8 ones
17. 4 hundred thousands + 0 ten thousands +
5 thousands + 6 hundreds + 9 tens + 8 ones, or 4 hundred
thousands + 5 thousands + 6 hundreds + 9 tens + 8 ones
19. 2 hundred thousands + 7 ten thousands +
2 thousands + 1 hundred + 6 tens + 1 one
21. 1 million + 1 hundred thousand + 8 ten thousands +
0 thousands + 2 hundreds + 1 ten + 2 ones, or 1 million +
1 hundred thousand + 8 ten thousands + 2 hundreds +
1 ten + 2 ones   **23.** Eighty-five

25. Eighty-eight thousand   **27.** One hundred twenty-
three thousand, seven hundred sixty-five   **29.** Seven
billion, seven hundred fifty-four million, two hundred
eleven thousand, five hundred seventy-seven
31. 2,233,812   **33.** 8,000,000,000   **35.** Five hundred
sixty-six thousand, two hundred eighty   **37.** seventy-six
million, eighty-six thousand, seven hundred ninety-two
39. 9,460,000,000,000   **41.** 64,186,000   **43.** D_W
45. 138

Margin Exercises, Section 1.2, pp. 10–11

1. 13,465   **2.** 9745   **3.** 16,182   **4.** 27,474   **5.** 29 in.
6. 62 ft   **7.** 16 in.; 26 in.

Calculator Corner, p. 12

1. 55   **2.** 121   **3.** 1602   **4.** 734   **5.** 1932   **6.** 864

Exercise Set 1.2, p. 13

1. 387   **3.** 5198   **5.** 164   **7.** 100   **9.** 8503   **11.** 5266
13. 4466   **15.** 6608   **17.** 34,432   **19.** 101,310
21. 230   **23.** 18,424   **25.** 31,685   **27.** 114 mi
29. 570 ft   **31.** D_W **33.** 8 ten thousands
34. Five billion, two hundred ninety-four million,
two hundred forty-seven thousand
35. $1 + 99 = 100, 2 + 98 = 100, \ldots, 49 + 51 = 100$. Then
$49 \cdot 100 = 4900$ and $4900 + 50 + 100 = 5050$.

Margin Exercises, Section 1.3, pp. 16–19

1. $7 = 2 + 5$, or $7 = 5 + 2$ **2.** $17 = 9 + 8$, or $17 = 8 + 9$
3. $5 = 13 - 8; 8 = 13 - 5$ **4.** $11 = 14 - 3; 3 = 14 - 11$
5. 3801   **6.** 6328   **7.** 4747   **8.** 56   **9.** 205
10. 658   **11.** 2851   **12.** 1546

Calculator Corner, p. 18

1. 28   **2.** 47   **3.** 67   **4.** 119   **5.** 2128   **6.** 2593

Exercise Set 1.3, p. 20

1. $7 = 3 + 4$, or $7 = 4 + 3$ **3.** $13 = 5 + 8$, or $13 = 8 + 5$
5. $23 = 14 + 9$, or $23 = 9 + 14$ **7.** $43 = 27 + 16$, or
$43 = 16 + 27$ **9.** $6 = 15 - 9; 9 = 15 - 6$
11. $8 = 15 - 7; 7 = 15 - 8$ **13.** $17 = 23 - 6; 6 = 23 - 17$
15. $23 = 32 - 9; 9 = 32 - 23$ **17.** 44   **19.** 533
21. 39   **23.** 234   **25.** 5382   **27.** 3831   **29.** 7748
31. 43,028   **33.** 56   **35.** 454   **37.** 3749   **39.** 2191
41. 95,974   **43.** 9989   **45.** 4206   **47.** 10,305   **49.** D_W
51. 1024   **52.** 12,732   **53.** 90,283   **54.** 29,364
55. 1345   **56.** 924   **57.** 22,692   **58.** 10,920
59. Six million, three hundred seventy-five thousand, six
hundred two   **60.** 7 ten thousands   **61.** 3; 4

Margin Exercises, Section 1.4, pp. 22–26

1. 40   **2.** 50   **3.** 70   **4.** 100   **5.** 40   **6.** 80   **7.** 90
8. 140   **9.** 470   **10.** 240   **11.** 290   **12.** 600   **13.** 800
14. 800   **15.** 9300   **16.** 8000   **17.** 8000   **18.** 19,000
19. 69,000   **20.** 48,970; 49,000; 49,000   **21.** 269,580;
269,600; 270,000   **22.** Eliminate the power sunroof and
the power package. Answers may vary.   **23.** (a) $18,300;
(b) yes   **24.** $70 + 20 + 40 + 70 = 200$
25. $700 + 700 + 200 + 200 = 1800$
26. $9300 - 6700 = 2600$ **27.** $23,000 - 12,000 = 11,000$
28. <   **29.** >   **30.** >   **31.** <   **32.** <   **33.** >

Exercise Set 1.4, p. 27

1. 50   **3.** 470   **5.** 730   **7.** 900   **9.** 100   **11.** 1000
13. 9100   **15.** 32,900   **17.** 6000   **19.** 8000
21. 45,000   **23.** 373,000   **25.** 180   **27.** 5720
29. 220; incorrect   **31.** 890; incorrect   **33.** 16,500
35. 5200   **37.** $11,200   **39.** $18,900; no   **41.** Answers
will vary depending on the options chosen.   **43.** 1600
45. 1500   **47.** 31,000   **49.** 69,000   **51.** <   **53.** >
55. <   **57.** >   **59.** >   **61.** >
63. $1,800,607 < 2,136,068$, or $2,136,068 > 1,800,607$
65. $6482 > 4827$, or $4827 < 6482$ **67.** D_W
69. 7 thousands + 9 hundreds + 9 tens + 2 ones
70. 2 ten millions + 3 millions   **71.** Two hundred forty-
six billion, six hundred five million, four thousand, thirty-
two   **72.** One million, five thousand, one hundred
73. 86,754   **74.** 13,589   **75.** 48,824   **76.** 4415
77. Left to the student   **79.** Left to the student

Margin Exercises, Section 1.5, pp. 32–36

1. 116   **2.** 148   **3.** 4938   **4.** 6740   **5.** 1035   **6.** 3024
7. 46,252   **8.** 205,065   **9.** 144,432   **10.** 287,232
11. 14,075,720   **12.** 391,760   **13.** 17,345,600
14. 56,200   **15.** 562,000   **16.** (a) 1081; (b) 1081; (c) same
17. 40   **18.** 15   **19.** $15,300   **20.** $840 \times 250 = 210,000$;
$800 \times 200 = 160,000$ **21.** 45 sq ft

Calculator Corner, p. 35

1. 448   **2.** 21,970   **3.** 6380   **4.** 39,564   **5.** 180,480
6. 2,363,754

Exercise Set 1.5, p. 37

1. 870   **3.** 2,340,000   **5.** 520   **7.** 564   **9.** 1527
11. 64,603   **13.** 4770   **15.** 3995   **17.** 46,080
19. 14,652   **21.** 207,672   **23.** 798,408   **25.** 20,723,872
27. 362,128   **29.** 20,064,048   **31.** 25,236,000
33. 302,220   **35.** 49,101,136   **37.** $50 \cdot 70 = 3500$
39. $30 \cdot 30 = 900$ **41.** $900 \cdot 300 = 270,000$
43. $400 \cdot 200 = 80,000$ **45.** (a) $2,840,000; (b) $2,850,000
47. 529,984 sq mi   **49.** 8100 sq ft   **51.** D_W **53.** 12,685
54. 10,834   **55.** 427,477   **56.** 111,110   **57.** 1241
58. 8889   **59.** 254,119   **60.** 66,444   **61.** 6,376,000
62. 6,375,600   **63.** 247,464 sq ft

Margin Exercises, Section 1.6, pp. 40–46

1. $\frac{54}{6} = 9; 6\overline{)54}^{\,9}$ **2.** $15 = 5 \cdot 3$, or $15 = 3 \cdot 5$
3. $72 = 9 \cdot 8$, or $72 = 8 \cdot 9$ **4.** $6 = 12 \div 2; 2 = 12 \div 6$
5. $6 = 42 \div 7; 7 = 42 \div 6$ **6.** $6; 6 \cdot 9 = 54$ **7.** 6 R 7;
$6 \cdot 9 = 54, 54 + 7 = 61$ **8.** 4 R 5; $4 \cdot 12 = 48, 48 + 5 = 53$
9. 6 R 13; $6 \cdot 24 = 144, 144 + 13 = 157$ **10.** 59 R 3
11. 1475 R 5   **12.** 1015   **13.** 134   **14.** 63 R 12
15. 807 R 4   **16.** 1088   **17.** 360 R 4   **18.** 800 R 47

Calculator Corner, p. 47

1. 3 R 11   **2.** 28   **3.** 124 R 2   **4.** 131 R 18   **5.** 283 R 57
6. 843 R 187

Exercise Set 1.6, p. 48

1. $18 = 3 \cdot 6$, or $18 = 6 \cdot 3$ **3.** $22 = 22 \cdot 1$, or $22 = 1 \cdot 22$
5. $54 = 6 \cdot 9$, or $54 = 9 \cdot 6$ **7.** $37 = 1 \cdot 37$, or $37 = 37 \cdot 1$
9. $9 = 45 \div 5; 5 = 45 \div 9$ **11.** $37 = 37 \div 1; 1 = 37 \div 37$
13. $8 = 64 \div 8$ **15.** $11 = 66 \div 6; 6 = 66 \div 11$ **17.** 12
19. 1   **21.** 22   **23.** Not defined   **25.** 55 R 2   **27.** 108
29. 307   **31.** 753 R 3   **33.** 74 R 1   **35.** 92 R 2
37. 1703   **39.** 987 R 5   **41.** 12,700   **43.** 127
45. 52 R 52   **47.** 29 R 5   **49.** 40 R 12   **51.** 90 R 22
53. 29   **55.** 105 R 3   **57.** 1609 R 2   **59.** 1007 R 1
61. 23   **63.** 107 R 1   **65.** 370   **67.** 609 R 15   **69.** 304
71. 3508 R 219   **73.** 8070   **75.** D_W **77.** perimeter
78. equation, inequality   **79.** digits, periods
80. additive   **81.** dividend   **82.** factors, product
83. minuend   **84.** associative   **85.** 54, 122; 33, 2772;
4, 8   **87.** 30 buses

Margin Exercises, Section 1.7, pp. 52–55

1. 7   **2.** 5   **3.** No   **4.** Yes   **5.** 5   **6.** 10   **7.** 5
8. 22   **9.** 22,490   **10.** 9022   **11.** 570   **12.** 3661
13. 8   **14.** 45   **15.** 77   **16.** 3311   **17.** 6114   **18.** 8
19. 16   **20.** 644   **21.** 96   **22.** 94

Exercise Set 1.7, p. 56

1. 14   **3.** 0   **5.** 29   **7.** 0   **9.** 8   **11.** 14   **13.** 1035
15. 25   **17.** 450   **19.** 90,900   **21.** 32   **23.** 143
25. 79   **27.** 45   **29.** 324   **31.** 743   **33.** 37   **35.** 66

37. 15 **39.** 48 **41.** 175 **43.** 335 **45.** 104 **47.** 45 **49.** 4056 **51.** 17,603 **53.** 18,252 **55.** 205 **57.** $\mathbf{D_W}$ **59.** $7 = 15 - 8; 8 = 15 - 7$ **60.** $6 = 48 \div 8; 8 = 48 \div 6$ **61.** $<$ **62.** $>$ **63.** $>$ **64.** $<$ **65.** 142 R 5 **66.** 142 **67.** 334 **68.** 334 R 11 **69.** 347

Margin Exercises, Section 1.8, pp. 59–66

1. 11,277 adoptions **2.** 8441 adoptions **3.** 19,089 adoptions **4.** $1874 **5.** $171 **6.** $5572 **7.** 9180 sq in. **8.** 378 cartons with 1 can left over **9.** 79 gal **10.** 181 seats

Translating for Success, p. 67

1. E **2.** M **3.** D **4.** G **5.** A **6.** O **7.** F **8.** K **9.** J **10.** H

Exercise Set 1.8, p. 68

1. 33,042 performances **3.** 704 performances **5.** 2054 miles **7.** 18 rows **9.** 792,316 degrees; 1,291,900 degrees **11.** 192,268 degrees **13.** $69,277 **15.** 240 mi **17.** 168 hr **19.** 225 squares **21.** $24,456 **23.** 35 weeks; 2 episodes left over **25.** 236 gal **27.** 3616 mi **29.** 12,804 gal **31.** $14,445 **33.** $247 **35.** (a) 4200 sq ft; (b) 268 ft **37.** $29,105,000,000 **39.** 151,500 **41.** 563 packages; 7 bars left over **43.** 384 mi; 27 in. **45.** 21 columns **47.** 56 full cartons; 11 books left over. If 1355 books are shipped, it will take 57 cartons. **49.** 32 $10 bills **51.** $400 **53.** 280 min; or 4 hr 40 min **55.** 525 min, or 8 hr 45 min **57.** 106 bones **59.** 3000 sq in. **61.** $\mathbf{D_W}$ **63.** 234,600 **64.** 234,560 **65.** 235,000 **66.** 22,000 **67.** 16,000 **68.** 8000 **69.** 4000 **70.** 320,000 **71.** 720,000 **72.** 46,800,000 **73.** 792,000 mi; 1,386,000 mi

Margin Exercises, Section 1.9, pp. 75–80

1. 5^4 **2.** 5^5 **3.** 10^2 **4.** 10^4 **5.** 10,000 **6.** 100 **7.** 512 **8.** 32 **9.** 51 **10.** 30 **11.** 584 **12.** 84 **13.** 4; 1 **14.** 52; 52 **15.** 29 **16.** 1880 **17.** 253 **18.** 93 **19.** 1880 **20.** 305 **21.** 75 **22.** 4 **23.** 1496 ft **24.** 46 **25.** 4

Calculator Corner, p. 76

1. 243 **2.** 15,625 **3.** 20,736 **4.** 2048

Calculator Corner, p. 78

1. 49 **2.** 85 **3.** 36 **4.** 0 **5.** 73 **6.** 49

Exercise Set 1.9, p. 81

1. 3^4 **3.** 5^2 **5.** 7^5 **7.** 10^3 **9.** 49 **11.** 729 **13.** 20,736 **15.** 121 **17.** 22 **19.** 20 **21.** 100 **23.** 1 **25.** 49 **27.** 5 **29.** 434 **31.** 41 **33.** 88

35. 4 **37.** 303 **39.** 20 **41.** 70 **43.** 295 **45.** 32 **47.** 906 **49.** 62 **51.** 102 **53.** 32 **55.** $94 **57.** 401 **59.** 110 **61.** 7 **63.** 544 **65.** 708 **67.** 27 **69.** $\mathbf{D_W}$ **71.** 452 **72.** 835 **73.** 13 **74.** 37 **75.** 2342 **76.** 4898 **77.** 25 **78.** 100 **79.** 104,286 mi^2 **80.** 98 gal **81.** 24; $1 + 5 \cdot (4 + 3) = 36$ **83.** 7; $12 \div (4 + 2) \cdot 3 - 2 = 4$

Concept Reinforcement, p. 84

1. False **2.** True **3.** True **4.** True **5.** False **6.** False

Summary and Review: Chapter 1, p. 84

1. 8 thousands **2.** 3 **3.** 2 thousands + 7 hundreds + 9 tens + 3 ones **4.** 5 ten thousands + 6 thousands + 0 hundreds + 7 tens + 8 ones, or 5 ten thousands + 6 thousands + 7 tens + 8 ones **5.** 4 millions + 0 hundred thousands + 0 ten thousands + 7 thousands + 1 hundred + 0 tens + 1 one, or 4 millions + 7 thousands + 1 hundred + 1 one **6.** Sixty-seven thousand, eight hundred nineteen **7.** Two million, seven hundred eighty-one thousand, four hundred twenty-seven **8.** 1 billion, sixty-five million, seventy thousand, six hundred seven **9.** 476,588 **10.** 2,000,400,000 **11.** 14,272 **12.** 66,024 **13.** 21,788 **14.** 98,921 **15.** $10 = 6 + 4$, or $10 = 4 + 6$ **16.** $8 = 11 - 3; 3 = 11 - 8$ **17.** 5148 **18.** 1689 **19.** 2274 **20.** 17,757 **21.** 345,800 **22.** 345,760 **23.** 346,000 **24.** 300,000 **25.** $41,300 + 19,700 = 61,000$ **26.** $38,700 - 24,500 = 14,200$ **27.** $400 \cdot 700 = 280,000$ **28.** $>$ **29.** $<$ **30.** 5,100,000 **31.** 6,276,800 **32.** 506,748 **33.** 27,589 **34.** 5,331,810 **35.** $56 = 8 \cdot 7$, or $56 = 7 \cdot 8$ **36.** $4 = 52 \div 13; 13 = 52 \div 4$ **37.** 12 R 3 **38.** 5 **39.** 913 R 3 **40.** 384 R 1 **41.** 4 R 46 **42.** 54 **43.** 452 **44.** 5008 **45.** 4389 **46.** 8 **47.** 45 **48.** 58 **49.** 0 **50.** 4^3 **51.** 10,000 **52.** 36 **53.** 65 **54.** 233 **55.** 56 **56.** 32 **57.** 260 **58.** 165 **59.** $502 **60.** $484 **61.** 1982 **62.** 19 cartons **63.** 14 beehives **64.** $13,585 **65.** $27,598 **66.** 137 beakers filled; 13 mL left over **67.** 98 ft^2; 42 ft **68.** $\mathbf{D_W}$ A vat contains 1152 oz of hot sauce. If 144 bottles are to be filled equally, how much will each bottle contain? Answers may vary. **69.** $\mathbf{D_W}$ No; if subtraction were associative, then $a - (b - c) = (a - b) - c$ for any a, b, and c. But, for example,
$$12 - (8 - 4) = 12 - 4 = 8,$$
whereas
$$(12 - 8) - 4 = 4 - 4 = 0.$$
Since $8 \neq 0$, this example shows that subtraction is not associative. **70.** $d = 8$ **71.** $a = 8, b = 4$ **72.** 6 days

Test: Chapter 1, p. 87

1. [1.1a] 5 **2.** [1.1b] 8 thousands + 8 hundreds + 4 tens + 3 ones **3.** [1.1c] Thirty-eight million, four hundred three thousand, two hundred seventy-seven

4. [1.2a] 9989   **5.** [1.2a] 63,791   **6.** [1.2a] 34
7. [1.2a] 10,515   **8.** [1.3b] 3630   **9.** [1.3b] 1039
10. [1.3b] 6848   **11.** [1.3b] 5175   **12.** [1.5a] 41,112
13. [1.5a] 5,325,600   **14.** [1.5a] 2405
15. [1.5a] 534,264   **16.** [1.6b] 3 R 3   **17.** [1.6b] 70
18. [1.6b] 97   **19.** [1.6b] 805 R 8
20. [1.8a] 1852 12-packs; 7 cakes left over
21. [1.8a] 1,256,615 mi^2 **22. (a)** [1.2b], [1.5c] 300 in.,
5000 in^2; 264 in., 3872 in^2; 228 in., 2888 in^2;
(b) [1.8a] 2112 in^2 **23.** [1.8a] 206,330 voters
24. [1.8a] 1808 lb   **25.** [1.8a] 20 staplers   **26.** [1.7b] 46
27. [1.7b] 13   **28.** [1.7b] 14   **29.** [1.7b] 381
30. [1.4a] 35,000   **31.** [1.4a] 34,580   **32.** [1.4a] 34,600
33. [1.4b] 23,600 + 54,700 = 78,300
34. [1.4b] 54,800 − 23,600 = 31,200
35. [1.5b] 800 · 500 = 400,000   **36.** [1.4c] >
37. [1.4c] <   **38.** [1.9a] 12^4 **39.** [1.9b] 343
40. [1.9b] 100,000   **41.** [1.9b] 625   **42.** [1.9c] 31
43. [1.9c] 98   **44.** [1.9c] 2   **45.** [1.9d] 216
46. [1.9c] 18   **47.** [1.9c] 92   **48.** [1.5c], [1.8a] 336 in^2
49. [1.9c] 9   **50.** [1.8a] 80 payments

CHAPTER 2

Margin Exercises, Section 2.1, pp. 91–95

1. Yes   **2.** No   **3.** 1, 2, 5, 10   **4.** 1, 3, 5, 9, 15, 45
5. 1, 2, 31, 62   **6.** 1, 2, 3, 4, 6, 8, 12, 24   **7.** 5 = 1 · 5;
45 = 9 · 5; 100 = 20 · 5   **8.** 10 = 1 · 10; 60 = 6 · 10;
110 = 11 · 10   **9.** 5, 10, 15, 20, 25, 30, 35, 40, 45, 50
10. Yes   **11.** Yes   **12.** No   **13.** 2, 13, 19, 41, 73 are
prime; 6, 12, 65, 99 are composite; 1 is neither   **14.** 2 · 3
15. 2 · 2 · 3   **16.** 3 · 3 · 5   **17.** 2 · 7 · 7
18. 2 · 3 · 3 · 7   **19.** 2 · 2 · 2 · 2 · 3 · 3
20. 2 · 2 · 2 · 5 · 7 · 7   **21.** 5 · 5 · 7 · 11

Calculator Corner, p. 92

1. Yes   **2.** No   **3.** No   **4.** Yes   **5.** No   **6.** Yes
7. Yes   **8.** No   **9.** Yes   **10.** No

Exercise Set 2.1, p. 96

1. No   **3.** Yes   **5.** 1, 2, 3, 6, 9, 18
7. 1, 2, 3, 6, 9, 18, 27, 54   **9.** 1, 2, 4   **11.** 1
13. 1, 2, 7, 14, 49, 98   **15.** 1, 3, 5, 15, 17, 51, 85, 255
17. 4, 8, 12, 16, 20, 24, 28, 32, 36, 40
19. 20, 40, 60, 80, 100, 120, 140, 160, 180, 200
21. 3, 6, 9, 12, 15, 18, 21, 24, 27, 30
23. 12, 24, 36, 48, 60, 72, 84, 96, 108, 120
25. 10, 20, 30, 40, 50, 60, 70, 80, 90, 100
27. 9, 18, 27, 36, 45, 54, 63, 72, 81, 90   **29.** No   **31.** Yes
33. Yes   **35.** No   **37.** No   **39.** Neither
41. Composite   **43.** Prime   **45.** Prime   **47.** 2 · 2 · 2
49. 2 · 7   **51.** 2 · 3 · 7   **53.** 5 · 5   **55.** 2 · 5 · 5
57. 13 · 13   **59.** 2 · 2 · 5 · 5   **61.** 5 · 7
63. 2 · 2 · 2 · 3 · 3   **65.** 7 · 11   **67.** 2 · 2 · 7 · 103
69. 3 · 17   **71.** 2 · 2 · 2 · 2 · 3 · 5 · 5   **73.** 3 · 7 · 13

75. 2 · 3 · 11 · 17   **77.** D$_W$ **79.** 26   **80.** 256
81. 425   **82.** 4200   **83.** 0   **84.** 22   **85.** 1   **86.** 3
87. $612   **88.** 201 min, or 3 hr 21 min   **89.** Row 1: 48,
90, 432, 63; row 2: 7, 2, 2, 10, 8, 6, 21, 10; row 3: 9, 18, 36,
14, 12, 11, 21; row 4: 29, 19, 42

Margin Exercises, Section 2.2, pp. 98–101

1. Yes   **2.** No   **3.** Yes   **4.** No   **5.** Yes   **6.** No
7. Yes   **8.** No   **9.** Yes   **10.** No   **11.** No   **12.** Yes
13. No   **14.** Yes   **15.** No   **16.** Yes   **17.** No   **18.** Yes
19. No   **20.** Yes   **21.** Yes   **22.** No   **23.** No   **24.** Yes
25. Yes   **26.** No   **27.** No   **28.** Yes   **29.** No   **30.** Yes
31. Yes   **32.** No

Exercise Set 2.2, p. 102

1. 46, 224, 300, 36, 45,270, 4444, 256, 8064, 21,568
3. 224, 300, 36, 4444, 256, 8064, 21,568
5. 300, 36, 45,270, 8064   **7.** 36, 45,270, 711, 8064
9. 324, 42, 501, 3009, 75, 2001, 402, 111,111, 1005
11. 55,555, 200, 75, 2345, 35, 1005   **13.** 324   **15.** 200
17. 313,332, 7624, 111,126, 876, 1110, 5128, 64,000, 9990
19. 313,332, 111,126, 876, 1110, 9990
21. 9990   **23.** 1110, 64,000, 9990   **25.** D$_W$ **27.** 138
28. 139   **29.** 874   **30.** 56   **31.** 26   **32.** 13   **33.** 234
34. 4003   **35.** 45 gal   **36.** 4320 min
37. 2 · 2 · 2 · 3 · 5 · 5 · 13   **39.** 2 · 2 · 3 · 3 · 7 · 11
41. 95,238

Margin Exercises, Section 2.3, pp. 104–109

1. Numerator: 83; denominator: 100
2. Numerator: 27; denominator: 50
3. Numerator: 11; denominator: 25
4. Numerator: 21; denominator: 1000
5. $\frac{1}{2}$ **6.** $\frac{1}{3}$ **7.** $\frac{1}{3}$ **8.** $\frac{2}{3}$ **9.** $\frac{3}{4}$ **10.** $\frac{15}{16}$ **11.** $\frac{15}{15}$ **12.** $\frac{5}{4}$
13. $\frac{7}{4}$ **14.** Clocks: $\frac{3}{5}$; thermometers: $\frac{2}{5}$ **15.** $\frac{98}{64}, \frac{98}{162}, \frac{64}{162}$
16. 1   **17.** 1   **18.** 1   **19.** 1   **20.** 1   **21.** 1   **22.** 0
23. 0   **24.** 0   **25.** 0   **26.** Not defined
27. Not defined   **28.** 8   **29.** 10   **30.** 346   **31.** 23

Exercise Set 2.3, p. 110

1. Numerator: 3; denominator: 4
3. Numerator: 11; denominator: 2
5. Numerator: 0; denominator: 7
7. $\frac{2}{4}$ **9.** $\frac{1}{8}$ **11.** $\frac{4}{3}$ **13.** $\frac{12}{16}$ **15.** $\frac{38}{16}$ **17.** $\frac{3}{4}$ **19.** $\frac{12}{12}$
21. $\frac{4}{8}$ **23.** $\frac{6}{12}$ **25.** $\frac{5}{8}$ **27.** $\frac{4}{7}$ **29. (a)** $\frac{2}{8}$; **(b)** $\frac{6}{8}$
31. (a) $\frac{3}{8}$; **(b)** $\frac{5}{8}$ **33. (a)** $\frac{3}{7}$; **(b)** $\frac{3}{4}$; **(c)** $\frac{4}{7}$; **(d)** $\frac{4}{3}$
35. (a) $\frac{35}{10,000}$; **(b)** $\frac{50}{10,000}$; **(c)** $\frac{44}{10,000}$; **(d)** $\frac{63}{10,000}$; **(e)** $\frac{43}{10,000}$; **(f)** $\frac{21}{10,000}$
37. (a) $\frac{390}{13}$; **(b)** $\frac{13}{390}$ **39.** $\frac{850}{1000}$ **41.** 0   **43.** 7   **45.** 1
47. 1   **49.** 0   **51.** 1   **53.** 1   **55.** 1   **57.** 1   **59.** 18
61. Not defined   **63.** Not defined   **65.** D$_W$
67. 34,560   **68.** 34,600   **69.** 35,000   **70.** 30,000
71. $15,691   **72.** 96 gal   **73.** 2203

74. 848 **75.** 37,239 **76.** 11,851 **77.** $\frac{1}{6}$ **79.** $\frac{2}{16}$, or $\frac{1}{8}$
81. ⬡⬡⬢⬢⬢ **83.**

Margin Exercises, Section 2.4, pp. 115–119

1. $\frac{2}{3}$ **2.** $\frac{5}{8}$ **3.** $\frac{10}{3}$ **4.** $\frac{33}{5}$ **5.** $\frac{46}{5}$
6.   **7.** $\frac{15}{56}$ **8.** $\frac{32}{15}$ **9.** $\frac{3}{100}$ **10.** $\frac{14}{3}$

11. $\frac{3}{8}$ **12.** $\frac{8}{81}$ ft² **13.** $\frac{3}{40}$

Exercise Set 2.4, p. 120

1. $\frac{3}{5}$ **3.** $\frac{5}{8}$ **5.** $\frac{8}{11}$ **7.** $\frac{70}{9}$ **9.** $\frac{2}{5}$ **11.** $\frac{6}{5}$ **13.** $\frac{21}{4}$
15. $\frac{85}{6}$ **17.** $\frac{1}{6}$ **19.** $\frac{1}{40}$ **21.** $\frac{2}{15}$ **23.** $\frac{4}{15}$ **25.** $\frac{9}{16}$ **27.** $\frac{14}{39}$
29. $\frac{7}{100}$ **31.** $\frac{49}{64}$ **33.** $\frac{1}{1000}$ **35.** $\frac{182}{285}$ **37.** $\frac{3}{8}$ cup **39.** $\frac{21}{32}$
41. $\frac{12}{25}$ m² **43.** $\frac{7}{16}$ L **45.** $\mathbf{D_W}$ **47.** 204 **48.** 700
49. 3001 **50.** 204 R 8 **51.** 8 thousands
52. 8 millions **53.** 8 ones **54.** 8 hundreds **55.** 3
56. 81 **57.** 50 **58.** 6399 **59.** $\frac{71,269}{180,433}$ **61.** $\frac{56}{1125}$

Margin Exercises, Section 2.5, pp. 122–126

1. $\frac{8}{16}$ **2.** $\frac{30}{50}$ **3.** $\frac{52}{100}$ **4.** $\frac{200}{75}$ **5.** $\frac{12}{9}$ **6.** $\frac{18}{24}$ **7.** $\frac{90}{100}$
8. $\frac{9}{45}$ **9.** $\frac{56}{49}$ **10.** $\frac{1}{4}$ **11.** $\frac{5}{6}$ **12.** 5 **13.** $\frac{4}{3}$ **14.** $\frac{7}{8}$
15. $\frac{89}{78}$ **16.** $\frac{8}{7}$ **17.** $\frac{1}{4}$ **18.** $\frac{7}{24}$ **19.** $\frac{37}{80}$
20. $\frac{20}{100} = \frac{1}{5}$; $\frac{42}{100} = \frac{21}{50}$; $\frac{8}{100} = \frac{2}{25}$; $\frac{16}{100} = \frac{4}{25}$; $\frac{14}{100} = \frac{7}{50}$
21. = **22.** ≠

Calculator Corner, p. 125

1. $\frac{14}{15}$ **2.** $\frac{7}{8}$ **3.** $\frac{138}{167}$ **4.** $\frac{7}{25}$

Exercise Set 2.5, p. 127

1. $\frac{5}{10}$ **3.** $\frac{20}{32}$ **5.** $\frac{27}{30}$ **7.** $\frac{28}{32}$ **9.** $\frac{20}{48}$ **11.** $\frac{51}{54}$ **13.** $\frac{75}{45}$
15. $\frac{42}{132}$ **17.** $\frac{1}{2}$ **19.** $\frac{3}{4}$ **21.** $\frac{1}{5}$ **23.** 3 **25.** $\frac{3}{4}$ **27.** $\frac{7}{8}$
29. $\frac{6}{5}$ **31.** $\frac{1}{3}$ **33.** 6 **35.** $\frac{1}{3}$ **37.** $\frac{4}{75}$ **39.** $\frac{45}{112}$ **41.** =
43. ≠ **45.** = **47.** ≠ **49.** = **51.** ≠ **53.** =
55. ≠ **57.** $\mathbf{D_W}$ **59.** 4992 ft²; 284 ft **60.** $928
61. 11 **62.** 32 **63.** 186 **64.** 2737 **65.** 5 **66.** 89
67. 3520 **68.** 9001 **69.** $\frac{137}{149}$ **71.** $\frac{2}{5}$; $\frac{3}{5}$
73. No. $\frac{262}{704} \neq \frac{135}{373}$ because $262 \cdot 373 \neq 704 \cdot 135$.

Margin Exercises, Section 2.6, p. 130

1. $\frac{7}{12}$ **2.** $\frac{1}{3}$ **3.** 6 **4.** $\frac{5}{2}$ **5.** 14 lb

Exercise Set 2.6, p. 131

1. $\frac{1}{3}$ **3.** $\frac{1}{8}$ **5.** $\frac{1}{10}$ **7.** $\frac{1}{6}$ **9.** $\frac{27}{10}$ **11.** $\frac{14}{9}$ **13.** 1
15. 1 **17.** 1 **19.** 1 **21.** 2 **23.** 4 **25.** 9 **27.** 9
29. $\frac{26}{5}$ **31.** $\frac{98}{5}$ **33.** 60 **35.** 30 **37.** $\frac{1}{5}$ **39.** $\frac{9}{25}$
41. $\frac{11}{40}$ **43.** $\frac{5}{14}$ **45.** $\frac{5}{8}$ in. **47.** 30 mph
49. 625 addresses **51.** $\frac{1}{3}$ cup **53.** $115,500
55. 160 mi **57.** Food: $9000; housing: $7200; clothing: $3600; savings: $4000; taxes: $9000; other expenses: $3200
59. $\mathbf{D_W}$ **61.** 35 **62.** 85 **63.** 125 **64.** 120
65. 4989 **66.** 8546 **67.** 6498 **68.** 6407 **69.** 4673
70. 5338 **71.** $\frac{129}{485}$ **73.** $\frac{1}{12}$ **75.** $\frac{1}{168}$

Margin Exercises, Section 2.7, pp. 135–139

1. $\frac{5}{2}$ **2.** $\frac{7}{10}$ **3.** $\frac{1}{9}$ **4.** 5 **5.** $\frac{8}{7}$ **6.** $\frac{8}{3}$ **7.** $\frac{1}{10}$ **8.** 100
9. 1 **10.** $\frac{14}{15}$ **11.** $\frac{4}{5}$ **12.** 32 **13.** 320 loops
14. 252 mi

Translating for Success, p. 140

1. C **2.** H **3.** A **4.** N **5.** O **6.** F **7.** I **8.** L
9. D **10.** M

Exercise Set 2.7, p. 141

1. $\frac{6}{5}$ **3.** $\frac{1}{6}$ **5.** 6 **7.** $\frac{3}{10}$ **9.** $\frac{4}{5}$ **11.** $\frac{4}{15}$ **13.** 4 **15.** 2
17. $\frac{1}{8}$ **19.** $\frac{3}{7}$ **21.** 8 **23.** 35 **25.** 1 **27.** $\frac{2}{3}$ **29.** $\frac{9}{4}$
31. 144 **33.** 75 **35.** 2 **37.** $\frac{3}{5}$ **39.** 315
41. 75 times **43.** 32 pairs **45.** 24 bowls **47.** 16 L
49. 288 km; 108 km **51.** $\frac{1}{16}$ in. **53.** $\mathbf{D_W}$
55. associative **56.** factors **57.** prime
58. denominator **59.** additive **60.** reciprocals
61. whole **62.** equation **63.** $\frac{9}{19}$ **65.** 36 **67.** $\frac{3}{8}$

Concept Reinforcement, p. 144

1. True **2.** False **3.** True **4.** True **5.** False
6. True **7.** True **8.** False

Summary and Review: Chapter 2, p. 144

1. 1, 2, 3, 4, 5, 6, 10, 12, 15, 20, 30, 60
2. 1, 2, 4, 8, 11, 16, 22, 44, 88, 176
3. 8, 16, 24, 32, 40, 48, 56, 64, 72, 80
4. Yes **5.** No **6.** Prime **7.** Neither **8.** Composite
9. $2 \cdot 5 \cdot 7$ **10.** $2 \cdot 3 \cdot 5$ **11.** $3 \cdot 3 \cdot 5$ **12.** $2 \cdot 3 \cdot 5 \cdot 5$
13. $2 \cdot 2 \cdot 2 \cdot 3 \cdot 3 \cdot 3 \cdot 3$ **14.** $2 \cdot 3 \cdot 5 \cdot 5 \cdot 5 \cdot 7$
15. 4344, 600, 93, 330, 255,555, 780, 2802, 711
16. 140, 182, 716, 2432, 4344, 600, 330, 780, 2802
17. 140, 716, 2432, 4344, 600, 780 **18.** 2432, 4344, 600
19. 140, 95, 475, 600, 330, 255,555, 780
20. 4344, 600, 330, 780, 2802 **21.** 255,555, 711
22. 140, 600, 330, 780 **23.** Numerator: 2; denominator: 7
24. $\frac{3}{5}$ **25.** $\frac{7}{6}$ **26.** $\frac{2}{7}$ **27.** (a) $\frac{3}{5}$; (b) $\frac{5}{3}$; (c) $\frac{3}{8}$ **28.** 0
29. 1 **30.** 48 **31.** 6 **32.** $\frac{2}{3}$ **33.** $\frac{1}{4}$ **34.** 1 **35.** 0
36. $\frac{2}{5}$ **37.** 18 **38.** 4 **39.** $\frac{1}{3}$ **40.** Not defined

41. Not defined 42. $\frac{11}{23}$ 43. $\frac{2}{7}$ 44. $\frac{39}{40}$ 45. $\frac{32}{225}$
46. $\frac{3}{100}$; $\frac{8}{100}=\frac{2}{25}$; $\frac{10}{100}=\frac{1}{10}$; $\frac{15}{100}=\frac{3}{20}$; $\frac{21}{100}$; $\frac{43}{100}$ 47. ≠ 48. =
49. ≠ 50. = 51. $\frac{3}{2}$ 52. 56 53. $\frac{5}{2}$ 54. 24 55. $\frac{2}{3}$
56. $\frac{1}{14}$ 57. $\frac{2}{3}$ 58. $\frac{1}{22}$ 59. $\frac{3}{20}$ 60. $\frac{10}{7}$ 61. $\frac{5}{4}$ 62. $\frac{1}{3}$
63. 9 64. $\frac{36}{47}$ 65. $\frac{9}{2}$ 66. 2 67. $\frac{11}{6}$ 68. $\frac{1}{4}$ 69. $\frac{9}{4}$
70. 300 71. 1 72. $\frac{4}{9}$ 73. $\frac{3}{10}$ 74. 240 75. 9 days
76. 1000 km 77. $\frac{1}{3}$ cup; 2 cups 78. $15 79. 60 bags
80. 256,000,000 metric tons
81. **DW** To simplify fraction notation, first factor the numerator and the denominator into prime numbers. Examine the factorizations for factors common to both the numerator and the denominator. Factor the fraction, with each pair of like factors forming a factor of 1. Remove the factors of 1, and multiply the remaining factors in the numerator and in the denominator, if necessary.
82. **DW** Taking $\frac{1}{2}$ of a number is equivalent to multiplying the number by $\frac{1}{2}$. Dividing by $\frac{1}{2}$ is equivalent to multiplying by the reciprocal of $\frac{1}{2}$, or 2. Thus taking $\frac{1}{2}$ of a number is not the same as dividing by $\frac{1}{2}$.
83. **DW** $9432 = 9 \cdot 1000 + 4 \cdot 100 + 3 \cdot 10 + 2 \cdot 1 = 9(999 + 1) + 4(99 + 1) + 3(9 + 1) + 2 \cdot 1 = 9 \cdot 999 + 9 \cdot 1 + 4 \cdot 99 + 4 \cdot 1 + 3 \cdot 9 + 3 \cdot 1 + 2 \cdot 1$. Since 999, 99, and 9 are each a multiple of 9, $9 \cdot 999$, $4 \cdot 99$, and $3 \cdot 9$ are multiples of 9. This leaves $9 \cdot 1 + 4 \cdot 1 + 3 \cdot 1 + 2 \cdot 1$, or $9 + 4 + 3 + 2$. If $9 + 4 + 3 + 2$, the sum of the digits, is divisible by 9, then 9432 is divisible by 9.
84. $a = 11,176$; $b = 9887$ 85. 13, 11, 101, 37

Test: Chapter 2, p. 147

1. [2.1a] 1, 2, 3, 4, 5, 6, 10, 12, 15, 20, 25, 30, 50, 60, 75, 100, 150, 300 2. [2.1c] Prime 3. [2.1c] Composite
4. [2.1d] $2 \cdot 3 \cdot 3$ 5. [2.1d] $2 \cdot 2 \cdot 3 \cdot 5$ 6. [2.2a] Yes
7. [2.2a] No 8. [2.2a] No 9. [2.2a] Yes
10. [2.3a] Numerator: 4; denominator: 5 11. [2.3a] $\frac{3}{4}$
12. [2.3a] $\frac{3}{7}$ 13. [2.3a] (a) $\frac{336}{497}$; (b) $\frac{161}{497}$ 14. [2.3b] 26
15. [2.3b] 1 16. [2.3b] 0 17. [2.5b] $\frac{1}{2}$ 18. [2.5b] 6
19. [2.5b] $\frac{1}{14}$ 20. [2.3b] Not defined
21. [2.3b] Not defined 22. [2.5b] $\frac{1}{4}$ 23. [2.5b] $\frac{2}{3}$
24. [2.5c] = 25. [2.5c] ≠ 26. [2.6a] 32 27. [2.6a] $\frac{3}{2}$
28. [2.6a] $\frac{5}{2}$ 29. [2.6a] $\frac{1}{10}$ 30. [2.6a] $\frac{2}{9}$ 31. [2.7a] $\frac{8}{5}$
32. [2.7a] 4 33. [2.7a] $\frac{1}{18}$ 34. [2.7b] $\frac{3}{10}$ 35. [2.7b] $\frac{8}{5}$
36. [2.7b] 18 37. [2.7b] $\frac{18}{7}$ 38. [2.7c] 64 39. [2.7c] $\frac{7}{4}$
40. [2.6b] 4375 students 41. [2.7d] $\frac{3}{40}$ m 42. [2.7d] 5 qt
43. [2.6b] $\frac{3}{4}$ in. 44. [2.6b] $\frac{15}{8}$ tsp 45. [2.6b] $\frac{7}{48}$ acre
46. [2.6a], [2.7b] $\frac{7}{960}$ 47. [2.7c] $\frac{7}{5}$

CHAPTER 3

Margin Exercises, Section 3.1, pp. 150–155

1. 45 2. 40 3. 30 4. 24 5. 10 6. 80 7. 40
8. 360 9. 864 10. 2520 11. $2^3 \cdot 3^2 \cdot 5 \cdot 7$, or 2520
12. $2^3 \cdot 5$, or 40; $2^3 \cdot 3^2 \cdot 5$, or 360; $2^5 \cdot 3^3$, or 864 13. 18
14. 24 15. 36 16. 210 17. 600 18. 3780
19. 600

Exercise Set 3.1, p. 156

1. 4 3. 50 5. 40 7. 54 9. 150 11. 120
13. 72 15. 420 17. 144 19. 288 21. 30
23. 105 25. 72 27. 60 29. 36 31. 900 33. 48
35. 50 37. 143 39. 420 41. 378 43. 810
45. 2160 47. 9828 49. Every 60 yr 51. Every 420 yr
53. **DW** 55. 90 days 56. 704 performances
57. 33,135 58. 6939 59. $\frac{2}{3}$ 60. $\frac{8}{7}$ 61. 18,900
63. 5 in. by 24 in.

Margin Exercises, Section 3.2, pp. 158–161

1. $\frac{4}{5}$ 2. 1 3. $\frac{1}{2}$ 4. $\frac{3}{4}$ 5. $\frac{5}{6}$ 6. $\frac{29}{24}$ 7. $\frac{5}{9}$ 8. $\frac{413}{1000}$
9. $\frac{197}{210}$ 10. $\frac{65}{72}$ 11. $\frac{9}{10}$ mi

Exercise Set 3.2, p. 162

1. 1 3. $\frac{3}{4}$ 5. $\frac{3}{2}$ 7. $\frac{7}{24}$ 9. $\frac{3}{2}$ 11. $\frac{19}{24}$ 13. $\frac{9}{10}$
15. $\frac{29}{18}$ 17. $\frac{31}{100}$ 19. $\frac{41}{60}$ 21. $\frac{189}{100}$ 23. $\frac{7}{8}$ 25. $\frac{13}{24}$
27. $\frac{17}{24}$ 29. $\frac{3}{4}$ 31. $\frac{437}{500}$ 33. $\frac{53}{40}$ 35. $\frac{391}{144}$ 37. $\frac{5}{6}$ lb
39. $\frac{23}{12}$ mi 41. 690 kg; $\frac{14}{23}$ cement; $\frac{5}{23}$ stone; $\frac{4}{23}$ sand; 1
43. $\frac{5}{8}$" 45. $\frac{13}{12}$ lb 47. **DW** 49. 210,528
50. 4,194,000 51. 3,387,807 52. 352,350
53. 537,179 54. 5 55. 84 56. 510,314
57. 21,468,755 58. 33,112,603 59. 12 tickets
60. $\frac{3}{64}$ acre 61. $\frac{4}{15}$; $320

Margin Exercises, Section 3.3, pp. 165–168

1. $\frac{1}{2}$ 2. $\frac{3}{8}$ 3. $\frac{1}{2}$ 4. $\frac{1}{12}$ 5. $\frac{13}{18}$ 6. $\frac{1}{2}$ 7. $\frac{9}{112}$ 8. <
9. > 10. > 11. > 12. < 13. $\frac{1}{6}$ 14. $\frac{11}{40}$
15. $\frac{5}{24}$ mi

Translating for Success, p. 169

1. J 2. E 3. D 4. B 5. I 6. N 7. A 8. C
9. L 10. F

Exercise Set 3.3, p. 170

1. $\frac{2}{3}$ 3. $\frac{3}{4}$ 5. $\frac{5}{8}$ 7. $\frac{1}{24}$ 9. $\frac{1}{2}$ 11. $\frac{9}{14}$ 13. $\frac{3}{5}$ 15. $\frac{7}{10}$
17. $\frac{17}{60}$ 19. $\frac{53}{100}$ 21. $\frac{26}{75}$ 23. $\frac{9}{100}$ 25. $\frac{13}{24}$ 27. $\frac{1}{10}$
29. $\frac{1}{24}$ 31. $\frac{13}{16}$ 33. $\frac{31}{75}$ 35. $\frac{13}{75}$ 37. < 39. >
41. < 43. < 45. > 47. > 49. < 51. $\frac{1}{15}$
53. $\frac{2}{15}$ 55. $\frac{1}{2}$ 57. $\frac{5}{12}$ hr 59. $\frac{19}{24}$ cup
61. $\frac{1}{4}$ of the business 63. $\frac{11}{20}$ lb 65. **DW** 67. 1
68. Not defined 69. Not defined 70. 4 71. $\frac{4}{21}$
72. $\frac{3}{2}$ 73. 21 74. $\frac{1}{32}$ 75. 17 days 76. 9 cups
77. $\frac{14}{3553}$ 79. $\frac{21}{40}$ km 81. $\frac{19}{24}$ 83. $\frac{145}{144}$ 85. >
87. *Day 1*: Cut off $\frac{1}{7}$ of bar and pay him. *Day 2*: Cut off $\frac{2}{7}$ of the bar. Trade him for the $\frac{1}{7}$. *Day 3*: Give him back the $\frac{1}{7}$.

Day 4: Trade him the $\frac{4}{7}$ for his $\frac{3}{7}$. *Day 5*: Give him the $\frac{1}{7}$ again. *Day 6*: Trade him the $\frac{2}{7}$ for the $\frac{1}{7}$. *Day 7*: Give him the $\frac{1}{7}$ again. This assumes that he does not spend parts of the gold bar immediately.

Margin Exercises, Section 3.4, pp. 173–176

1. $1\frac{2}{3}$ **2.** $2\frac{3}{4}$ **3.** $8\frac{3}{4}$ **4.** $12\frac{2}{3}$ **5.** $\frac{22}{5}$ **6.** $\frac{61}{10}$ **7.** $\frac{29}{6}$ **8.** $\frac{37}{4}$ **9.** $\frac{62}{9}$ **10.** $2\frac{1}{3}$ **11.** $1\frac{1}{10}$ **12.** $18\frac{1}{3}$ **13.** $807\frac{2}{3}$ **14.** $134\frac{23}{45}$

Calculator Corner, p. 176

1. $1476\frac{1}{6}$ **2.** $676\frac{4}{9}$ **3.** $800\frac{51}{56}$ **4.** $13{,}031\frac{1}{2}$ **5.** $51{,}626\frac{9}{11}$ **6.** $7330\frac{7}{32}$ **7.** $134\frac{1}{15}$ **8.** $2666\frac{130}{213}$ **9.** $3571\frac{51}{112}$ **10.** $12\frac{169}{454}$

Exercise Set 3.4, p. 177

1. $\frac{57}{4}, \frac{27}{4}, \frac{9}{4}$ **3.** $3\frac{5}{8}, 2\frac{3}{4}$ **5.** $\frac{17}{3}$ **7.** $\frac{13}{4}$ **9.** $\frac{81}{8}$ **11.** $\frac{51}{10}$ **13.** $\frac{103}{5}$ **15.** $\frac{59}{6}$ **17.** $\frac{73}{10}$ **19.** $\frac{13}{8}$ **21.** $\frac{51}{4}$ **23.** $\frac{43}{10}$ **25.** $\frac{203}{100}$ **27.** $\frac{200}{3}$ **29.** $\frac{279}{50}$ **31.** $3\frac{3}{5}$ **33.** $4\frac{2}{3}$ **35.** $4\frac{1}{2}$ **37.** $5\frac{7}{10}$ **39.** $7\frac{4}{7}$ **41.** $7\frac{1}{2}$ **43.** $11\frac{1}{2}$ **45.** $1\frac{1}{2}$ **47.** $7\frac{57}{100}$ **49.** $43\frac{1}{8}$ **51.** $108\frac{5}{8}$ **53.** $618\frac{1}{5}$ **55.** $40\frac{4}{7}$ **57.** $55\frac{1}{51}$ **59.** $\mathbf{D_W}$ **61.** $45{,}800$ **62.** $45{,}770$ **63.** $\frac{8}{15}$ **64.** $\frac{21}{25}$ **65.** $\frac{16}{27}$ **66.** $\frac{583}{669}$ **67.** 18 **68.** $\frac{5}{2}$ **69.** $\frac{1}{4}$ **70.** $\frac{2}{5}$ **71.** 24 **72.** 49 **73.** $\frac{2560}{3}$ **74.** $\frac{4}{3}$ **75.** $237\frac{19}{541}$ **77.** $8\frac{2}{3}$ **79.** $52\frac{2}{7}$

Margin Exercises, Section 3.5, pp. 180–184

1. $7\frac{2}{5}$ **2.** $12\frac{1}{10}$ **3.** $13\frac{7}{12}$ **4.** $1\frac{1}{2}$ **5.** $3\frac{1}{6}$ **6.** $3\frac{5}{18}$ **7.** $3\frac{2}{3}$ **8.** $232\frac{3}{20}$ mi **9.** $4\frac{5}{6}$ ft **10.** $354\frac{23}{24}$ gal

Exercise Set 3.5, p. 185

1. $28\frac{3}{4}$ **3.** $185\frac{7}{8}$ **5.** $6\frac{1}{2}$ **7.** $2\frac{11}{12}$ **9.** $14\frac{7}{12}$ **11.** $12\frac{1}{10}$ **13.** $16\frac{5}{24}$ **15.** $21\frac{1}{2}$ **17.** $27\frac{7}{8}$ **19.** $27\frac{13}{24}$ **21.** $1\frac{3}{5}$ **23.** $4\frac{1}{10}$ **25.** $21\frac{17}{24}$ **27.** $12\frac{1}{4}$ **29.** $15\frac{3}{8}$ **31.** $7\frac{5}{12}$ **33.** $13\frac{3}{8}$ **35.** $11\frac{5}{18}$ **37.** $5\frac{3}{8}$ yd **39.** $7\frac{5}{12}$ lb **41.** $6\frac{5}{12}$ in. **43.** $2\frac{7}{8}$ in., $4\frac{7}{8}$ in. **45.** $19\frac{1}{16}$ in. **47.** $95\frac{1}{5}$ mi **49.** $36\frac{1}{2}$ in. **51.** $20\frac{1}{8}$ in. **53.** $78\frac{1}{2}$ in. **55.** $3\frac{4}{5}$ hr **57.** $28\frac{3}{4}$ yd **59.** $7\frac{3}{8}$ ft **61.** $1\frac{9}{16}$ in. **63.** $\mathbf{D_W}$ **65.** 16 packages **66.** 286 cartons; 2 oz left over **67.** Yes **68.** No **69.** No **70.** Yes **71.** No **72.** Yes **73.** Yes **74.** Yes **75.** $\frac{10}{13}$ **76.** $\frac{1}{10}$ **77.** $8568\frac{786}{1189}$ **79.** $5\frac{3}{4}$ ft

Margin Exercises, Section 3.6, pp. 190–193

1. 20 **2.** $1\frac{7}{8}$ **3.** $12\frac{4}{5}$ **4.** $8\frac{1}{3}$ **5.** 16 **6.** $7\frac{3}{7}$ **7.** $1\frac{7}{8}$ **8.** $\frac{7}{10}$ **9.** $227\frac{1}{2}$ mi **10.** 20 mpg **11.** $240\frac{3}{4}$ ft^2

Calculator Corner, p. 194

1. $\frac{7}{12}$ **2.** $\frac{11}{10}$ **3.** $\frac{35}{16}$ **4.** $\frac{3}{10}$ **5.** $10\frac{2}{15}$ **6.** $1\frac{1}{28}$ **7.** $10\frac{11}{15}$ **8.** $2\frac{91}{115}$

Translating for Success, p. 195

1. O **2.** K **3.** F **4.** D **5.** H **6.** G **7.** L **8.** E **9.** M **10.** J

Exercise Set 3.6, p. 196

1. $22\frac{2}{3}$ **3.** $2\frac{5}{12}$ **5.** $8\frac{1}{6}$ **7.** $9\frac{31}{40}$ **9.** $24\frac{91}{100}$ **11.** $975\frac{4}{5}$ **13.** $6\frac{1}{4}$ **15.** $1\frac{1}{5}$ **17.** $3\frac{9}{16}$ **19.** $1\frac{1}{8}$ **21.** $1\frac{8}{43}$ **23.** $\frac{9}{40}$ **25.** 45,000 beagles **27.** About 4,500,000 **29.** $62\frac{1}{2}$ ft^2 **31.** $13\frac{1}{3}$ tsp **33.** $343\frac{3}{4}$ lb **35.** 68°F **37.** About 1,800,750 **39.** *Cake*: $1\frac{1}{8}$ cups all-purpose flour, $\frac{3}{8}$ cup sugar, $\frac{3}{8}$ cup cold butter, $\frac{3}{8}$ cup sour cream, $\frac{1}{4}$ teaspoon baking powder, $\frac{1}{4}$ teaspoon baking soda, $\frac{1}{2}$ egg, $\frac{1}{2}$ teaspoon almond extract; *filling*: $\frac{1}{2}$ package (4 ounces) cream cheese, $\frac{1}{8}$ cup sugar, $\frac{1}{2}$ egg, $\frac{3}{8}$ cup peach preserves, $\frac{1}{4}$ cup sliced almonds; *cake*: 9 cups all-purpose flour, 3 cups sugar, 3 cups cold butter, 3 cups sour cream, 2 teaspoons baking powder, 2 teaspoons baking soda, 4 eggs, 4 teaspoons almond extract; *filling*: 4 packages (8 ounces each) cream cheese, 1 cup sugar 4 eggs, 3 cups peach preserves, 2 cups sliced almonds **41.** 15 mpg **43.** 4 cu ft **45.** $16\frac{1}{2}$ servings **47.** $35\frac{115}{256}$ sq in. **49.** $59{,}538\frac{1}{8}$ sq ft **51.** $\mathbf{D_W}$ **53.** divisor, quotient, dividend **54.** common **55.** composite **56.** divisible, divisible **57.** multiplications, divisions, additions, subtractions **58.** addends **59.** numerator **60.** reciprocal **61.** $360\frac{60}{473}$ **63.** $35\frac{57}{64}$ **65.** $\frac{4}{9}$ **67.** $1\frac{4}{5}$

Margin Exercises, Section 3.7, pp. 201–204

1. $\frac{1}{2}$ **2.** $\frac{3}{10}$ **3.** $20\frac{2}{3}$, or $\frac{62}{3}$ **4.** $9\frac{7}{8}$ in. **5.** $\frac{5}{9}$ **6.** $\frac{31}{40}$ **7.** $\frac{27}{56}$ **8.** 0 **9.** 1 **10.** $\frac{1}{2}$ **11.** 1 **12.** 12; answers may vary **13.** 32; answers may vary **14.** 27; answers may vary **15.** 15; answers may vary **16.** 1; answers may vary **17.** 1,000,000; answers may vary **18.** $22\frac{1}{2}$ **19.** 132 **20.** 37

Exercise Set 3.7, p. 205

1. $\frac{1}{24}$ **3.** $\frac{2}{5}$ **5.** $\frac{4}{7}$ **7.** $\frac{59}{30}$, or $1\frac{29}{30}$ **9.** $\frac{3}{20}$ **11.** $\frac{211}{8}$, or $26\frac{3}{8}$ **13.** $\frac{7}{16}$ **15.** $\frac{1}{36}$ **17.** $\frac{3}{8}$ **19.** $\frac{37}{48}$ **21.** $\frac{25}{72}$ **23.** $\frac{103}{16}$, or $6\frac{7}{16}$ **25.** $2\frac{41}{128}$ lb **27.** $7\frac{23}{50}$ sec **29.** $\frac{17}{6}$, or $2\frac{5}{6}$ **31.** $\frac{8395}{84}$, or $99\frac{79}{84}$ **33.** 0 **35.** 0 **37.** $\frac{1}{2}$ **39.** $\frac{1}{2}$ **41.** 0 **43.** 1 **45.** 6 **47.** 12 **49.** 19 **51.** 6 **53.** 12 **55.** 16 **57.** 3 **59.** 13 **61.** 2 **63.** $1\frac{1}{2}$ **65.** $\frac{1}{2}$ **67.** $271\frac{1}{2}$ **69.** 3 **71.** 100 **73.** $29\frac{1}{2}$ **75.** $\mathbf{D_W}$ **77.** 3402 **78.** 1,038,180 **79.** 59 R 77 **80.** 348 **81.** 783 **82.** $\frac{8}{3}$ **83.** $\frac{3}{8}$ **84.** Prime: 5, 7, 23, 43; composite: 9, 14; neither: 1 **85.** 16 people **86.** 43 mg

87. (a) $13 \cdot 9\frac{1}{4} + 8\frac{1}{4} \cdot 7\frac{1}{4}$; **(b)** $\frac{2881}{16}$, or $180\frac{1}{16}$ in²;
(c) Multiply before adding. **89.** $a = 2, b = 8$
91. The largest is $\frac{4}{3} + \frac{5}{2} = \frac{23}{6}$.

Concept Reinforcement, p. 209

1. True **2.** True **3.** False **4.** True **5.** True
6. False

Summary and Review: Chapter 3, p. 209

1. 36 **2.** 90 **3.** 30 **4.** 1404 **5.** $\frac{63}{40}$ **6.** $\frac{19}{48}$ **7.** $\frac{25}{12}$
8. $\frac{891}{1000}$ **9.** $\frac{1}{3}$ **10.** $\frac{1}{8}$ **11.** $\frac{5}{27}$ **12.** $\frac{11}{18}$ **13.** > **14.** >
15. $\frac{19}{40}$ **16.** $\frac{2}{5}$ **17.** $\frac{15}{2}$ **18.** $\frac{67}{8}$ **19.** $\frac{13}{3}$ **20.** $\frac{75}{7}$
21. $2\frac{1}{3}$ **22.** $6\frac{3}{4}$ **23.** $12\frac{3}{5}$ **24.** $3\frac{1}{2}$ **25.** $877\frac{1}{3}$
26. $456\frac{5}{23}$ **27.** $10\frac{2}{5}$ **28.** $11\frac{11}{15}$ **29.** $10\frac{2}{3}$ **30.** $8\frac{1}{4}$
31. $7\frac{7}{9}$ **32.** $4\frac{11}{15}$ **33.** $4\frac{3}{20}$ **34.** $13\frac{3}{8}$ **35.** 16 **36.** $3\frac{1}{2}$
37. $2\frac{21}{50}$ **38.** 6 **39.** 12 **40.** $1\frac{7}{17}$ **41.** $\frac{1}{8}$ **42.** $\frac{9}{10}$
43. $4\frac{1}{4}$ yd **44.** $177\frac{3}{4}$ in² **45.** $50\frac{1}{4}$ in²
46. *Serving 2*: $\frac{1}{8}$ cup extra-virgin olive oil, $\frac{3}{4}$ pound fresh red snapper fillets, $\frac{1}{6}$ cup kalamata olives, $1\frac{1}{4}$ tablespoons capers, $\frac{1}{2}$ cup canned tomatoes, $1\frac{1}{2}$ tablespoons chopped shallots, $\frac{1}{4}$ tablespoon fresh rosemary leaves, $\frac{1}{4}$ tablespoon minced garlic, $\frac{1}{6}$ cup white wine; *serving 12*: $\frac{3}{4}$ cup extra-virgin olive oil, $4\frac{1}{2}$ pounds fresh red snapper fillets, 1 cup kalamata olives, $7\frac{1}{2}$ tablespoons capers, 3 cups canned tomatoes, 9 tablespoons chopped shallots, $1\frac{1}{2}$ tablespoons fresh rosemary leaves, $1\frac{1}{2}$ tablespoons minced garlic, 1 cup white wine
47. $1\frac{73}{100}$ in. **48.** 24 lb **49.** About $69\frac{3}{8}$ kg **50.** $8\frac{3}{8}$ cups
51. $63\frac{2}{3}$ pies; $19\frac{1}{3}$ pies **52.** 1 **53.** $\frac{7}{40}$ **54.** 3
55. $\frac{77}{240}$ **56.** $\frac{1}{2}$ **57.** 0 **58.** 1 **59.** 7 **60.** 10
61. 2 **62.** $28\frac{1}{2}$
63. **D$_W$** It might be necessary to find the least common denominator before adding or subtracting. The least common denominator is the least common multiple of the denominators.
64. **D$_W$** Suppose that a room has dimensions $15\frac{3}{4}$ ft by $28\frac{5}{8}$ ft. The equation $2 \cdot 15\frac{3}{4} + 2 \cdot 28\frac{5}{8} = 88\frac{3}{4}$ gives the perimeter of the room, in feet. Answers may vary.
65. 12 min **66.** $\frac{6}{3} + \frac{5}{4} = 3\frac{1}{4}$

Test: Chapter 3, p. 212

1. [3.1a] 48 **2.** [3.1a] 600 **3.** [3.2a] 3 **4.** [3.2a] $\frac{37}{24}$
5. [3.2a] $\frac{921}{1000}$ **6.** [3.3a] $\frac{1}{3}$ **7.** [3.3a] $\frac{1}{12}$ **8.** [3.3a] $\frac{77}{120}$
9. [3.3c] $\frac{15}{4}$ **10.** [3.3c] $\frac{1}{4}$ **11.** [3.3b] > **12.** [3.4a] $\frac{7}{2}$
13. [3.4a] $\frac{79}{8}$ **14.** [3.4a] $4\frac{1}{2}$ **15.** [3.4a] $8\frac{2}{9}$
16. [3.4b] $162\frac{7}{11}$ **17.** [3.5a] $14\frac{1}{5}$ **18.** [3.5a] $14\frac{5}{12}$
19. [3.5b] $4\frac{7}{24}$ **20.** [3.5b] $6\frac{1}{6}$ **21.** [3.6a] 39
22. [3.6a] $4\frac{1}{2}$ **23.** [3.6b] 2 **24.** [3.6b] $\frac{1}{36}$
25. [3.6c] About 105 kg **26.** [3.6c] 80 books
27. [3.5c] **(a)** 3 in.; **(b)** $4\frac{1}{2}$ in. **28.** [3.3d] $\frac{1}{16}$ in.
29. [3.7a] $6\frac{11}{36}$ ft **30.** [3.7a] $3\frac{1}{2}$ **31.** [3.7a] $\frac{3}{4}$

32. [3.7b] 0 **33.** [3.7b] 1 **34.** [3.7b] 4 **35.** [3.7b] $18\frac{1}{2}$
36. [3.7b] 16 **37.** [3.7b] $1214\frac{1}{2}$
38. [3.1a] **(a)** 24, 48, 72; **(b)** 24
39. [3.3b], [3.5c] Rebecca walks $\frac{17}{56}$ mi farther.

Cumulative Review: Chapters 1–3, p. 214

1. [3.3d] **(a)** $\frac{1}{48}$ in.; **(b)** $\frac{1}{12}$ in. **2.** [2.7d] 61
3. (a) [3.5c] $14\frac{13}{24}$ mi; **(b)** [3.6c] $4\frac{61}{72}$ mi
4. [3.6c], [3.5c] **(a)** $142\frac{1}{4}$ ft²; **(b)** 54 ft **5.** [1.8a] 31 people
6. [1.8a] $108 **7.** [2.6b] $\frac{2}{5}$ tsp; 4 tsp **8.** [3.6c] 39 lb
9. [3.6c] 16 pieces **10.** [3.2b] $\frac{33}{20}$ mi **11.** [1.1a] 5
12. [1.1b] 6 thousands + 7 tens + 5 ones
13. [1.1c] Twenty-nine thousand, five hundred
14. [2.3a] $\frac{5}{16}$ **15.** [1.2a] 899 **16.** [1.2a] 8982
17. [3.2a] $\frac{5}{12}$ **18.** [3.5a] $8\frac{1}{4}$ **19.** [1.3b] 5124
20. [1.3b] 4518 **21.** [3.3a] $\frac{5}{12}$ **22.** [3.5b] $1\frac{1}{6}$
23. [1.5a] 5004 **24.** [1.5a] 293,232 **25.** [2.6a] $\frac{3}{2}$
26. [2.6a] 15 **27.** [3.6a] $7\frac{1}{3}$ **28.** [1.6b] 715
29. [1.6b] 56 R 11 **30.** [3.4b] $56\frac{11}{45}$
31. [3.7a] $\frac{1377}{100}$, or $13\frac{77}{100}$ **32.** [2.7b] $\frac{4}{7}$ **33.** [3.6b] $7\frac{1}{3}$
34. [1.4a] 38,500 **35.** [3.1a] 72 **36.** [2.2a] No
37. [2.1a] 1, 2, 4, 8, 16 **38.** [3.3b] > **39.** [2.5c] =
40. [3.3b] < **41.** [2.5b] $\frac{4}{5}$ **42.** [2.3b] 0 **43.** [2.5b] 32
44. [3.4a] $\frac{37}{8}$ **45.** [3.4a] $5\frac{2}{3}$ **46.** [1.7b] 93 **47.** [3.3c] $\frac{5}{9}$
48. [2.7c] $\frac{12}{7}$ **49.** [1.7b] 905 **50.** [3.7b] 1 **51.** [3.7b] $\frac{1}{2}$
52. [3.7b] 0 **53.** [3.7b] 30 **54.** [3.7b] 1 **55.** [3.7b] 42
56. [2.1a, b, c, d], [2.2a]
Factors of 68: 1, 2, 4, 17, 34, 68
Factorization of 68: 2 · 34, or 2 · 2 · 17
Prime factorization of 68: 2 · 2 · 17
Numbers divisible by 6: 12, 54, 72, 300
Numbers divisible by 8: 8, 16, 24, 32, 40, 48, 64, 864
Numbers divisible by 5: 70, 95, 215
Prime numbers: 2, 3, 17, 19, 23, 31, 47, 101
57. [2.6b] (d) **58.** [2.7d] (a) **59.** [3.6c] (a)
60. [3.6c] (a) **61.** [3.2a] **(a)** $\frac{1}{2}, \frac{2}{3}, \frac{3}{4}, \frac{4}{5}$; **(b)** $\frac{9}{10}$
62. [2.1c] 2003

CHAPTER 4

Margin Exercises, Section 4.1, pp. 219–224

1. Eighty and thirty-one hundredths; seventy-seven and forty-three hundredths **2.** Two and six thousand seven hundred sixty-seven hundred-thousandths
3. Two hundred forty-five and eighty-nine hundredths
4. Thirty-four and sixty-four ten-thousandths
5. Thirty-one thousand, seventy-nine and seven hundred sixty-four thousandths
6. $\frac{896}{1000}$ **7.** $\frac{2378}{100}$ **8.** $\frac{56,789}{10,000}$ **9.** $\frac{19}{10}$ **10.** 7.43 **11.** 0.406
12. 6.7089 **13.** 0.9 **14.** 0.057 **15.** 0.083 **16.** 4.3
17. 283.71 **18.** 456.013 **19.** 2.04 **20.** 0.06
21. 0.58 **22.** 1 **23.** 0.8989 **24.** 21.05

25. 2.8 **26.** 13.9 **27.** 234.4 **28.** 7.0 **29.** 0.64
30. 7.83 **31.** 34.68 **32.** 0.03 **33.** 0.943 **34.** 8.004
35. 43.112 **36.** 37.401 **37.** 7459.355 **38.** 7459.35
39. 7459.4 **40.** 7459 **41.** 7460 **42.** 7500 **43.** 7000

Exercise Set 4.1, p. 225

1. Sixty-three and five hundredths **3.** Twenty-six and
fifty-nine hundredths **5.** Eight and thirty-five hundredths
7. Eighty-six and eighty-nine hundredths **9.** Thirty-four
and eight hundred ninety-one thousandths
11. $\frac{83}{10}$ **13.** $\frac{356}{100}$ **15.** $\frac{4603}{100}$ **17.** $\frac{13}{100,000}$ **19.** $\frac{10,008}{10,000}$
21. $\frac{20,003}{1000}$ **23.** 0.8 **25.** 8.89 **27.** 3.798 **29.** 0.0078
31. 0.00019 **33.** 0.376193 **35.** 99.44 **37.** 3.798
39. 2.1739 **41.** 8.953073 **43.** 0.58 **45.** 0.91
47. 0.001 **49.** 235.07 **51.** $\frac{4}{100}$ **53.** 0.4325 **55.** 0.1
57. 0.5 **59.** 2.7 **61.** 123.7 **63.** 0.89 **65.** 0.67
67. 1.00 **69.** 0.09 **71.** 0.325 **73.** 17.002
75. 10.101 **77.** 9.999 **79.** 800 **81.** 809.473
83. 809 **85.** 34.5439 **87.** 34.54 **89.** 35 **91.** **D**w
93. 6170 **94.** 6200 **95.** 6000
96. $2 \cdot 2 \cdot 2 \cdot 2 \cdot 5 \cdot 5 \cdot 5$, or $2^4 \cdot 5^3$
97. $2 \cdot 3 \cdot 3 \cdot 5 \cdot 17$, or $2 \cdot 3^2 \cdot 5 \cdot 17$ **98.** $2 \cdot 7 \cdot 11 \cdot 13$
99. $2 \cdot 2 \cdot 2 \cdot 7 \cdot 7 \cdot 11$, or $2^3 \cdot 7^2 \cdot 11$
101. 2.000001, 2.0119, 2.018, 2.0302, 2.1, 2.108, 2.109
103. 6.78346 **105.** 0.03030

Margin Exercises, Section 4.2, pp. 228–232

1. 10.917 **2.** 34.2079 **3.** 4.969 **4.** 3.5617
5. 9.40544 **6.** 912.67 **7.** 2514.773 **8.** 10.754
9. 0.339 **10.** 0.5345 **11.** 0.5172 **12.** 7.36992
13. 1194.22 **14.** 4.9911 **15.** 38.534 **16.** 14.164
17. 2133.5
18. The "balance forward" column should read:
$3078.92
  2738.23
  2659.67
  2890.47
  2877.33
  2829.33
  2868.91
  2766.04
  2697.45
  2597.45

Calculator Corner, p. 230

1. 317.645 **2.** 506.553 **3.** 17.15 **4.** 49.08 **5.** 4.4
6. 33.83 **7.** 454.74 **8.** 0.99

Exercise Set 4.2, p. 233

1. 334.37 **3.** 1576.215 **5.** 132.560 **7.** 50.0248
9. 40.007 **11.** 977.955 **13.** 771.967 **15.** 8754.8221
17. 49.02 **19.** 85.921 **21.** 2.4975 **23.** 3.397
25. 8.85 **27.** 3.37 **29.** 1.045 **31.** 3.703 **33.** 0.9902
35. 99.66 **37.** 4.88 **39.** 0.994 **41.** 17.802

43. 51.13 **45.** 32.7386 **47.** 4.0622 **49.** 11.65
51. 384.68 **53.** 582.97 **55.** 15,335.3
57. The balance forward should read:
$ 9704.56
  9677.12
  10,677.12
  10,553.17
  10,429.15
  10,416.72
  12,916.72
  12,778.94
  12,797.82
  9997.82
59. **D**w **61.** 35,000 **62.** 34,000 **63.** $\frac{1}{6}$ **64.** $\frac{34}{45}$
65. 6166 **66.** 5366 **67.** $16\frac{1}{2}$ servings **68.** $60\frac{1}{5}$ mi
69. 345.8

Margin Exercises, Section 4.3, pp. 237–240

1. 529.48 **2.** 5.0594 **3.** 34.2906 **4.** 0.348 **5.** 0.0348
6. 0.00348 **7.** 0.000348 **8.** 34.8 **9.** 348 **10.** 3480
11. 34,800 **12.** 3,700,000 **13.** 6,100,000,000
14. 2,700,000,000 **15.** 1569¢ **16.** 17¢ **17.** $0.35
18. $5.77

Calculator Corner, p. 239

1. 48.6 **2.** 6930.5 **3.** 142.803 **4.** 0.5076 **5.** 7916.4
6. 20.4153

Exercise Set 4.3, p. 241

1. 60.2 **3.** 6.72 **5.** 0.252 **7.** 0.522 **9.** 237.6
11. 583,686.852 **13.** 780 **15.** 8.923 **17.** 0.09768
19. 0.782 **21.** 521.6 **23.** 3.2472 **25.** 897.6
27. 322.07 **29.** 55.68 **31.** 3487.5 **33.** 50.0004
35. 114.42902 **37.** 13.284 **39.** 90.72 **41.** 0.0028728
43. 0.72523 **45.** 1.872115 **47.** 45,678 **49.** 2888¢
51. 66¢ **53.** $0.34 **55.** $34.45 **57.** 258,700,000,000
59. 748,900,000 **61.** **D**w **63.** $11\frac{1}{5}$ **64.** $\frac{35}{72}$ **65.** $2\frac{7}{15}$
66. $7\frac{2}{15}$ **67.** 342 **68.** 87 **69.** 4566 **70.** 1257
71. 87 **72.** 1176 R 14 **73.** $10^{21} = 1$ sextillion
75. $10^{24} = 1$ septillion

Margin Exercises, Section 4.4, pp. 244–250

1. 0.6 **2.** 1.5 **3.** 0.47 **4.** 0.32 **5.** 3.75 **6.** 0.25
7. (a) 375; (b) 15 **8.** 4.9 **9.** 12.8 **10.** 15.625
11. 12.78 **12.** 0.001278 **13.** 0.09847 **14.** 67.832
15. 0.78314 **16.** 1105.6 **17.** 0.04 **18.** 0.2426
19. 593.44 **20.** 5967.5 m

Calculator Corner, p. 247

1. 14.3 **2.** 2.56 **3.** 200 **4.** 0.75 **5.** 20 **6.** 0.064
7. 15.7 **8.** 75.8

Exercise Set 4.4, p. 251

1. 2.99　**3.** 23.78　**5.** 7.48　**7.** 7.2　**9.** 1.143
11. 4.041　**13.** 0.07　**15.** 70　**17.** 20　**19.** 0.4
21. 0.41　**23.** 8.5　**25.** 9.3　**27.** 0.625　**29.** 0.26
31. 15.625　**33.** 2.34　**35.** 0.47　**37.** 0.2134567
39. 21.34567　**41.** 1023.7　**43.** 9.3　**45.** 0.0090678
47. 45.6　**49.** 2107　**51.** 303.003　**53.** 446.208
55. 24.14　**57.** 13.0072　**59.** 19.3204　**61.** 473.188278
63. 10.49　**65.** 911.13　**67.** 205　**69.** $1288.36
71. $206.34 billion　**73.** $\mathbf{D_W}$　**75.** $\frac{6}{7}$　**76.** $\frac{7}{8}$　**77.** $\frac{19}{73}$
78. $\frac{19}{73}$　**79.** $2 \cdot 2 \cdot 3 \cdot 3 \cdot 19$, or $2^2 \cdot 3^2 \cdot 19$
80. $2 \cdot 3 \cdot 3 \cdot 3 \cdot 3$, or $2 \cdot 3^4$　**81.** $3 \cdot 3 \cdot 223$, or $3^2 \cdot 223$
82. $5 \cdot 401$　**83.** $15\frac{1}{8}$　**84.** $5\frac{7}{8}$　**85.** 6.254194585
87. 1000　**89.** 100

Margin Exercises, Section 4.5, pp. 255–259

1. 0.8　**2.** 0.45　**3.** 0.275　**4.** 1.32　**5.** 0.4　**6.** 0.375
7. $0.1\overline{6}$　**8.** $0.\overline{6}$　**9.** $0.\overline{45}$　**10.** $1.\overline{09}$　**11.** $0.\overline{428571}$
12. 0.7; 0.67; 0.667　**13.** 0.8; 0.81; 0.808　**14.** 6.2; 6.25;
6.245　**15.** 0.510　**16.** 24.2 mpg　**17.** 12.1 million
digital cameras　**18.** 0.72　**19.** 0.552　**20.** 9.6575

Exercise Set 4.5, p. 260

1. 0.23　**3.** 0.6　**5.** 0.325　**7.** 0.2　**9.** 0.85　**11.** 0.375
13. 0.975　**15.** 0.52　**17.** 20.016　**19.** 0.25　**21.** 1.16
23. 1.1875　**25.** $0.2\overline{6}$　**27.** $0.\overline{3}$　**29.** $1.\overline{3}$　**31.** $1.1\overline{6}$
33. $0.\overline{571428}$　**35.** $0.91\overline{6}$　**37.** 0.3; 0.27; 0.267
39. 0.3; 0.33; 0.333　**41.** 1.3; 1.33; 1.333　**43.** 1.2; 1.17;
1.167　**45.** 0.6; 0.57; 0.571　**47.** 0.9; 0.92; 0.917
49. 0.2; 0.18; 0.182　**51.** 0.3; 0.28; 0.278　**53.** **(a)** 0.429;
(b) 0.75; **(c)** 0.571; **(d)** 1.333　**55.** 15.8 mpg
57. 17.8 mpg　**59.** 15.2 mph　**61.** $24.5625; $24.56
63. $3.734375; $3.73　**65.** $59.875; $59.88　**67.** 11.06
69. 8.4　**71.** $417.51\overline{6}$　**73.** 0　**75.** 2.8125　**77.** 0.20425
79. 317.14　**81.** 0.1825　**83.** 18　**85.** 2.736
87. $\mathbf{D_W}$　**89.** 21　**90.** $238\frac{7}{8}$　**91.** 10　**92.** $\frac{43}{52}$
93. $50\frac{5}{24}$　**94.** $30\frac{7}{10}$　**95.** $1\frac{1}{2}$　**96.** $14\frac{13}{24}$　**97.** $1\frac{1}{24}$ cups
98. $1\frac{33}{100}$ in.　**99.** $0.\overline{142857}$　**101.** $0.\overline{428571}$
103. $0.\overline{714285}$　**105.** $0.\overline{1}$　**107.** $0.\overline{001}$

Margin Exercises, Section 4.6, pp. 264–266

1. (b)　**2.** (a)　**3.** (d)　**4.** (b)　**5.** (a)　**6.** (d)　**7.** (b)
8. (c)　**9.** (b)　**10.** (b)　**11.** (c)　**12.** (a)　**13.** (c)
14. (c)

Exercise Set 4.6, p. 267

1. (d)　**3.** (c)　**5.** (a)　**7.** (c)　**9.** 1.6　**11.** 6　**13.** 60
15. 2.3　**17.** 180　**19.** (a)　**21.** (c)　**23.** (b)　**25.** (b)
27. $1800 \div 9 = 200$ posts; answers may vary　**29.** $\mathbf{D_W}$
31. repeating　**32.** multiple　**33.** distributive
34. solution　**35.** multiplicative　**36.** commutative
37. denominator; multiple　**38.** divisible; divisible
39. Yes　**41.** No　**43.** (a) +, ×; (b) +, ×, −

Margin Exercises, Section 4.7, pp. 270–276

1. $2.5 billion　**2.** 199.1 lb　**3.** $51.26　**4.** $368.75
5. 96.52 cm^2　**6.** $1.33　**7.** 28.6 mpg　**8.** $716,667
9. $158,760

Translating for Success, p. 277

1. I　**2.** C　**3.** N　**4.** A　**5.** G　**6.** B　**7.** D　**8.** O
9. F　**10.** M

Exercise Set 4.7, p. 278

1. $17.8 billion　**3.** $19.15　**5.** 102.8°F
7. $64,333,333.33　**9.** Area: 8.125 cm^2; perimeter: 11.5 cm
11. 22,691.5 mi　**13.** 20.2 mpg　**15.** $30
17. 11.9752 ft^3　**19.** 78.1 cm　**21.** 28.5 cm　**23.** 2.31 cm
25. 876 calories　**27.** $1171.74　**29.** 227.75 ft^2
31. 0.362　**33.** $53.04　**35.** 2152.56 yd^2　**37.** 10.8¢
39. $906.50　**41.** 5.665 billion　**43.** 1.4°F
45. $266,791　**47.** $165,565　**49.** $1,131,429
51. $\mathbf{D_W}$　**53.** 6335　**54.** $\frac{31}{52}$　**55.** $6\frac{5}{6}$　**56.** $\frac{23}{15}$　**57.** $\frac{1}{24}$
58. 2803　**59.** $\frac{2}{15}$　**60.** $1\frac{5}{6}$　**61.** $\frac{129}{251}$　**62.** $\frac{5}{16}$　**63.** $\frac{13}{25}$
64. $\frac{25}{19}$　**65.** 28 min　**66.** $7\frac{1}{5}$ min　**67.** 186 calories
68. 30 calories　**69.** $17.28

Concept Reinforcement, p. 285

1. False　**2.** True　**3.** True　**4.** False　**5.** True

Summary and Review: Chapter 4, p. 285

1. 6,590,000　**2.** 6,900,000　**3.** Three and
forty-seven hundredths　**4.** Thirty-one thousandths
5. Twenty-seven and eleven hundred-thousandths
6. Seven millionths　**7.** $\frac{9}{100}$　**8.** $\frac{4561}{1000}$　**9.** $\frac{89}{1000}$　**10.** $\frac{30,227}{10,000}$
11. 0.034　**12.** 4.2603　**13.** 27.91　**14.** 867.006
15. 0.034　**16.** 0.91　**17.** 0.741　**18.** 1.041　**19.** 17.4
20. 17.43　**21.** 17.429　**22.** 17　**23.** 574.519
24. 0.6838　**25.** 229.1　**26.** 45.551　**27.** 29.2092
28. 790.29　**29.** 29.148　**30.** 70.7891　**31.** 12.96
32. 0.14442　**33.** 4.3　**34.** 0.02468　**35.** 7.5　**36.** 0.45
37. 45.2　**38.** 1.022　**39.** 0.2763　**40.** 1389.2
41. 496.2795　**42.** 6.95　**43.** 42.54　**44.** 4.9911
45. $15.52　**46.** 1.9 lb　**47.** $912.68　**48.** $307.49
49. 14.5 mpg　**50. (a)** 102.6 lb; **(b)** 14.7 lb　**51.** 272
52. 216　**53.** 4　**54.** $125　**55.** 2.6　**56.** 1.28
57. 2.75　**58.** 3.25　**59.** $1.1\overline{6}$　**60.** $1.\overline{54}$　**61.** 1.5
62. 1.55　**63.** 1.545　**64.** $82.73　**65.** $4.87
66. 2493¢　**67.** 986¢　**68.** 1.8045　**69.** 57.1449
70. 15.6375　**71.** $41.537\overline{3}$
72. $\mathbf{D_W}$ Multiply by 1 to get a denominator that is a power
of 10:

$$\frac{44}{125} = \frac{44}{125} \cdot \frac{8}{8} = \frac{352}{1000} = 0.352.$$

We can also divide to find that $\frac{44}{125} = 0.352$.
73. $\mathbf{D_W}$ Each decimal place in the decimal notation
corresponds to one zero in the power of ten in the fraction

notation. When the fractions are multiplied, the number of zeros in the denominator of the product is the sum of the number of zeros in the denominators of the factors. So the number of decimal places in the product is the sum of the number of decimal places in the factors.
74. (a) $2.56 \times 6.4 \div 51.2 - 17.4 + 89.7 = 72.62;$
(b) $(11.12 - 0.29) \times 3^4 = 877.23$
75. $\frac{1}{3} + \frac{2}{3} = 0.33333333\ldots + 0.66666666\ldots$
$= 0.99999999\ldots.$
Therefore, $1 = 0.99999999\ldots$ because $\frac{1}{3} + \frac{2}{3} = 1.$
76. $2 = 1.\overline{9}$ **77.** $a = 5, b = 9$

Test: Chapter 4, p. 288

1. [4.3b] 8,900,000,000   **2.** [4.3b] 3,756,000
3. [4.1a] Two and thirty-four hundredths
4. [4.1a] One hundred five and five ten-thousandths
5. [4.1b] $\frac{91}{100}$ **6.** [4.1b] $\frac{2769}{1000}$ **7.** [4.1b] 0.074
8. [4.1b] 3.7047   **9.** [4.1b] 756.09   **10.** [4.1b] 91.703
11. [4.1c] 0.162   **12.** [4.1c] 0.078   **13.** [4.1c] 0.9
14. [4.1d] 6   **15.** [4.1d] 5.68   **16.** [4.1d] 5.678
17. [4.1d] 5.7   **18.** [4.2a] 0.7902   **19.** [4.2a] 186.5
20. [4.2a] 1033.23   **21.** [4.2b] 48.357   **22.** [4.2b] 19.0901
23. [4.2b] 152.8934   **24.** [4.3a] 0.03   **25.** [4.3a] 0.21345
26. [4.3a] 73,962   **27.** [4.4a] 4.75   **28.** [4.4a] 30.4
29. [4.4a] 0.19   **30.** [4.4a] 0.34689   **31.** [4.4a] 34,689
32. [4.4b] 84.26   **33.** [4.2c] 8.982   **34.** [4.7a] $314.99
35. [4.7a] 28.3 mpg   **36.** [4.7a] $6572.45
37. [4.7a] $181.93   **38.** [4.7a] 58.24 million passengers
39. [4.6a] 198   **40.** [4.6a] 4   **41.** [4.5a] 1.6
42. [4.5a] 0.88   **43.** [4.5a] 5.25   **44.** [4.5a] 0.75
45. [4.5a] $1.\overline{2}$ **46.** [4.5a] $2.\overline{142857}$ **47.** [4.5b] 2.1
48. [4.5b] 2.14   **49.** [4.5b] 2.143   **50.** [4.3b] $9.49
51. [4.4c] 40.0065   **52.** [4.4c] 384.8464   **53.** [4.5c] 302.4
54. [4.5c] $52.339\overline{4}$ **55.** [4.7a] $35
56. [4.1b, c] $\frac{2}{3}, \frac{5}{7}, \frac{15}{19}, \frac{11}{13}, \frac{17}{20}, \frac{13}{15}$

CHAPTER 5

Margin Exercises, Section 5.1, pp. 292–295

1. $\frac{5}{11}$, or $5:11$ **2.** $\frac{57.3}{86.1}$, or $57.3:86.1$ **3.** $\frac{6\frac{3}{4}}{7\frac{2}{5}}$, or $6\frac{3}{4}:7\frac{2}{5}$
4. $\frac{739}{12}$ **5.** $\frac{12}{14}$ **6.** $\frac{71}{214.1}; \frac{214.1}{71}$ **7.** $\frac{38.2}{56.1}$ **8.** $\frac{205}{278}; \frac{278}{205}; \frac{278}{483}$
9. 18 is to 27 as 2 is to 3   **10.** 3.6 is to 12 as 3 is to 10
11. 1.2 is to 1.5 as 4 is to 5   **12.** $\frac{9}{16}$

Exercise Set 5.1, p. 296

1. $\frac{4}{5}$ **3.** $\frac{178}{572}$ **5.** $\frac{0.4}{12}$ **7.** $\frac{3.8}{7.4}$ **9.** $\frac{56.78}{98.35}$ **11.** $\frac{8\frac{3}{4}}{9\frac{5}{6}}$ **13.** $\frac{4}{1}$
15. $\frac{356}{100,000}; \frac{173}{100,000}$ **17.** $\frac{93.2}{1000}$ **19.** $\frac{190}{547}; \frac{547}{190}$ **21.** $\frac{60}{100}; \frac{100}{60}$
23. $\frac{2}{3}$ **25.** $\frac{3}{4}$ **27.** $\frac{12}{25}$ **29.** $\frac{7}{9}$ **31.** $\frac{2}{3}$ **33.** $\frac{14}{25}$ **35.** $\frac{1}{2}$
37. $\frac{3}{4}$ **39.** $\frac{478}{213}; \frac{213}{478}$ **41.** D_W **43.** $=$ **44.** $\neq$ **45.** $\neq$
46. $=$ **47.** 50   **48.** 9.5   **49.** 14.5   **50.** 152
51. $6\frac{7}{20}$ cm   **52.** $17\frac{11}{20}$ cm   **53.** $\frac{30}{47}$ **55.** $1:2:3$

Margin Exercises, Section 5.2, pp. 300–301

1. 5 mi/hr, or 5 mph   **2.** 12 mi/hr, or 12 mph
3. $\frac{89}{13}$ km/h, or 6.85 km/h   **4.** 1100 ft/sec   **5.** 4 ft/sec
6. $\frac{121}{8}$ ft/sec, or 15.125 ft/sec   **7.** $\frac{34 \text{ home runs}}{92 \text{ strikeouts}} \approx$
0.370 home run per strikeout   **8.** $\frac{714 \text{ home runs}}{1330 \text{ strikeouts}} \approx$
0.537 home run per strikeout; Babe Ruth's rate is approximately 0.167 higher   **9.** 7.45¢/oz
10. 24.143¢/oz; 25.9¢/oz; 24.17¢/oz; the 7-oz package has the lowest unit price

Exercise Set 5.2, p. 302

1. 40 km/h   **3.** 7.48 mi/sec   **5.** 24 mpg   **7.** 23 mpg
9. $\frac{32,270 \text{ people}}{0.75 \text{ sq mi}}$; about 43,027 people/sq mi
11. 25 mph; 0.04 hr/mi   **13.** About 18.3 points/game
15. 0.623 gal/ft²   **17.** 186,000 mi/sec   **19.** 124 km/h
21. 25 beats/min   **23.** 19.185¢/oz; 15.709¢/oz; 25.4 oz
25. 11.5¢/oz; 13.833¢/oz; 16 oz
27. 18.174¢/oz; 15.275¢/oz; 34.5 oz
29. 10.5¢/oz; 11.607¢/oz; 12.475¢/oz; 12.484¢/oz; 18 oz
31. 8.58¢/oz; 5.29¢/oz; 5.245¢/oz; 5.263¢/oz; 200 fl oz
33. D_W **35.** 1.7 million   **36.** $37\frac{1}{2}$ servings
37. 30 tests   **38.** 8.9 billion   **39.** 109.608   **40.** 67,819
41. 5833.56   **42.** 466,190.4
43. 6-oz: 10.83¢/oz; 5.5-oz: 10.91¢/oz

Margin Exercises, Section 5.3, pp. 307–310

1. Yes   **2.** No   **3.** No   **4.** Yes   **5.** No   **6.** Yes
7. 14   **8.** $11\frac{1}{4}$ **9.** 10.5   **10.** 2.64   **11.** 10.8
12. $\frac{125}{42}$, or $2\frac{41}{42}$

Calculator Corner, p. 310

1. Left to the student   **2.** Left to the student   **3.** 27.5625
4. 25.6   **5.** 15.140625   **6.** 40.03952941
7. 39.74857143   **8.** 119

Exercise Set 5.3, p. 311

1. No   **3.** Yes   **5.** Yes   **7.** No   **9.** 0.61; 0.66; 0.69; 0.66; the completion rates (rounded to the nearest hundredth) are the same for Brees and Roethlisberger
11. 45   **13.** 12   **15.** 10   **17.** 20   **19.** 5   **21.** 18
23. 22   **25.** 28   **27.** $9\frac{1}{3}$ **29.** $2\frac{8}{9}$ **31.** 0.06   **33.** 5
35. 1   **37.** 1   **39.** 14   **41.** $2\frac{3}{16}$ **43.** $\frac{51}{16}$, or $3\frac{3}{16}$
45. 12.5725   **47.** $\frac{1748}{249}$, or $7\frac{5}{249}$ **49.** D_W **51.** quotient
52. sum   **53.** average   **54.** dollars, cents
55. subtrahend   **56.** terminating   **57.** commutative
58. cross products   **59.** Approximately 2731.4
61. (a) Ruth: 1.863 strikeouts per home run; Schmidt: 3.436 strikeouts per home run; **(b)** Schmidt

Margin Exercises, Section 5.4, pp. 314–318

1. 445 calories   **2.** 15 gal   **3.** 8 shirts   **4.** 38 in. or less
5. 9.5 in.   **6.** 2074 deer

Translating for Success, p. 319

1. N   **2.** I   **3.** A   **4.** K   **5.** J   **6.** F   **7.** M   **8.** B
9. G   **10.** E

Exercise Set 5.4, p. 320

1. 11.04 hr   **3.** 880 calories   **5.** 177 million, or
177,000,000   **7.** 9.75 gal   **9.** 175 bulbs   **11.** 2975 ft^2
13. 450 pages   **15.** (a) 23.63445 British pounds;
(b) $16,450.56   **17.** (a) 21,206 Japanese yen; (b) $29.99
19. (a) About 112 gal; (b) 3360 mi   **21.** 60 students
23. 13,500 mi   **25.** 120 lb   **27.** 64 gal   **29.** 100 oz
31. 954 deer   **33.** 58.1 mi   **35.** (a) 56 games; (b) about
2197 points   **37.** D_W **39.** Neither   **40.** Composite
41. Prime   **42.** Composite   **43.** Prime
44. $2 \cdot 2 \cdot 2 \cdot 101$, or $2^3 \cdot 101$ **45.** $2 \cdot 2 \cdot 7$, or $2^2 \cdot 7$
46. $2 \cdot 433$ **47.** $3 \cdot 31$ **48.** $2 \cdot 2 \cdot 5 \cdot 101$, or $2^2 \cdot 5 \cdot 101$
49. 17 positions   **51.** 2150 earned runs   **53.** CD player:
$150; receiver: $450; speakers: $300

Margin Exercises, Section 5.5, pp. 325–328

1. 15   **2.** 24.75 ft   **3.** 7.5 ft   **4.** 21 cm   **5.** 29 ft

Exercise Set 5.5, p. 329

1. 25   **3.** $\frac{4}{3}$, or $1\frac{1}{3}$ **5.** $x = \frac{27}{4}$, or $6\frac{3}{4}$; $y = 9$ **7.** $x = 7.5$;
$y = 7.2$ **9.** 1.25 m   **11.** 36 ft   **13.** 7 ft   **15.** 100 ft
17. 4   **19.** $10\frac{1}{2}$ **21.** $x = 6$; $y = 5.25$; $z = 3$ **23.** $x = 5\frac{1}{3}$,
or $5.\overline{3}$; $y = 4\frac{2}{3}$, or $4.\overline{6}$; $z = 5\frac{1}{3}$, or $5.\overline{3}$ **25.** 20 ft   **27.** 152 ft
29. D_W **31.** $59.81   **32.** 9.63   **33.** 679.4928
34. 2.74568   **35.** 27,456.8   **36.** 0.549136   **37.** 0.85
38. 1.825   **39.** 0.909   **40.** 0.843   **41.** 13.75 ft
43. 1.25 cm   **45.** 3681.437   **47.** $x = 0.4$; $y \approx 0.35$

Concept Reinforcement, p. 333

1. True   **2.** True   **3.** False   **4.** False   **5.** True

Summary and Review: Chapter 5, p. 333

1. $\frac{47}{84}$ **2.** $\frac{46}{1.27}$ **3.** $\frac{83}{100}$ **4.** $\frac{0.72}{197}$ **5.** (a) $\frac{12,480}{16,640}$, or $\frac{3}{4}$;
(b) $\frac{16,640}{29,120}$, or $\frac{4}{7}$ **6.** $\frac{3}{4}$ **7.** $\frac{9}{16}$ **8.** 26 mpg   **9.** 6300 rpm
10. 0.638 gal/ft^2 **11.** 0.72 serving/lb   **12.** 4.33¢/tablet
13. 14.173¢/oz   **14.** 1.329¢/sheet; 1.554¢/sheet;
1.110¢/sheet; 6 big rolls   **15.** 6.844¢/oz; 5.188¢/oz;
5.609¢/oz; 5.539¢/oz; 48 oz   **16.** No   **17.** No   **18.** 32
19. 7   **20.** $\frac{1}{40}$ **21.** 24   **22.** $4.45   **23.** 351 circuits
24. (a) 202 Euros; (b) $61.88   **25.** 832 mi   **26.** 27 acres
27. Approximately 3,293,558 kg   **28.** 6 in.
29. Approximately 2096 lawyers   **30.** $x = \frac{14}{3}$, or $4\frac{2}{3}$

31. $x = \frac{56}{5}$, or $11\frac{1}{5}$; $y = \frac{63}{5}$, or $12\frac{3}{5}$ **32.** 40 ft
33. $x = 3$; $y = 9$; $z = \frac{15}{2}$, or $7\frac{1}{2}$
34. D_W In terms of cost, a low faculty-to-student ratio is
less expensive than a high faculty-to-student ratio. In terms
of quality of education and student satisfaction, a high
faculty-to-student ratio is more desirable. A college
president must balance the cost and quality issues.
35. D_W Leslie used 4 gal of gasoline to drive 92 mi. At the
same rate, how many gallons would be needed to travel
368 mi?   **36.** 105 min, or 1 hr 45 min
37. $x = 4258.5$; $z \approx 10,094.3$ **38.** Finishing paint: 11 gal;
primer: 16.5 gal

Test: Chapter 5, p. 336

1. [5.1a] $\frac{85}{97}$ **2.** [5.1a] $\frac{0.34}{124}$ **3.** [5.1b] $\frac{9}{10}$ **4.** [5.1b] $\frac{25}{32}$
5. [5.2a] 0.625 ft/sec   **6.** [5.2a] $1\frac{1}{3}$ servings/lb
7. [5.2a] 22 mpg   **8.** [5.2b] About 7.765¢/oz
9. [5.2b] 11.182¢/oz; 7.149¢/oz; 8.389¢/oz; 6.840¢/oz;
263 oz   **10.** [5.3a] Yes   **11.** [5.3a] No   **12.** [5.3b] 12
13. [5.3b] 360   **14.** [5.3b] 42.1875   **15.** [5.3b] 100
16. [5.4a] 1512 km   **17.** [5.4a] 4.8 min   **18.** [5.4a] 525 mi
19. [5.5a] 66 m   **20.** [5.4a] (a) 3498.75 Hong Kong dollars;
(b) $102.25   **21.** [5.4a] About 86,151 arrests
22. [5.5a] $x = 8$; $y = 8.8$ **23.** [5.5b] $x = 8$; $y = 8$; $z = 12$
24. [5.4a] 5888

Cumulative Review: Chapters 1–5, p. 338

1. [1.6b], [4.3b], [4.4a] (a) $252,000,000; (b) $0.252 billion;
(c) $25,200,000; (d) $45,487.36   **2.** [5.2a] 22 mpg
3. [4.2a] 513.996   **4.** [3.5a] $6\frac{3}{4}$ **5.** [3.2a] $\frac{7}{20}$
6. [4.2b] 30.491   **7.** [4.2b] 72.912   **8.** [3.3a] $\frac{7}{60}$
9. [4.3a] 222.076   **10.** [4.3a] 567.8   **11.** [3.6a] 3
12. [4.4a] 43   **13.** [1.6b] 899   **14.** [2.7b] $\frac{3}{2}$
15. [1.1b] 3 ten thousands + 7 tens + 4 ones
16. [4.1a] One hundred twenty and seven hundredths
17. [4.1c] 0.7   **18.** [4.1c] 0.8
19. [2.1d] $2 \cdot 2 \cdot 2 \cdot 2 \cdot 3 \cdot 3$, or $2^4 \cdot 3^2$ **20.** [3.1a] 140
21. [2.3a] $\frac{5}{8}$ **22.** [2.5b] $\frac{5}{8}$ **23.** [4.5c] 5.718
24. [4.5c] 0.179   **25.** [5.1a] $\frac{0.3}{15}$ **26.** [5.3a] Yes
27. [5.2a] 55 m/sec   **28.** [5.2b] 11.769¢/oz; 9.344¢/oz;
9.728¢/oz; 6.357¢/oz; 7.002¢/oz; 28-oz Joy
29. [5.3b] 30.24   **30.** [4.4b] 26.4375   **31.** [2.7c] $\frac{8}{9}$
32. [5.3b] 128   **33.** [4.2c] 33.34   **34.** [3.3c] $\frac{76}{175}$
35. [2.6b] 390 cal   **36.** [5.4a] (a) 1338.7 Rand; (b) $336.89
37. [4.7a] 976.9 mi   **38.** [5.4a] 7 min   **39.** [3.5c] $2\frac{1}{4}$ cups
40. [2.7d] 12 doors   **41.** [4.2b], [2.7d] (a) 40,200 yr;
(b) answers may vary   **42.** [1.8a] (a) $360,000;
(b) $144,000,000; (c) $1,728,000,000
43. [4.7a], [5.2a] 42.2025 mi   **44.** [4.7a] 132 orbits
45. [2.1c] (d)   **46.** [5.2a] (b)   **47.** [1.2b] (b)
48. [5.5a] $10\frac{1}{2}$ ft

CHAPTER 6

Margin Exercises, Section 6.1, pp. 342–345

1. $\frac{70}{100}$; $70 \times \frac{1}{100}$; 70×0.01 **2.** $\frac{23.4}{100}$; $23.4 \times \frac{1}{100}$; 23.4×0.01
3. $\frac{100}{100}$; $100 \times \frac{1}{100}$; 100×0.01 **4.** 0.34   **5.** 0.789
6. 0.06625   **7.** 0.18   **8.** 0.0008   **9.** 24%   **10.** 347%
11. 100%   **12.** 32.1%   **13.** 25.3%

Calculator Corner, p. 343

1. 0.14   **2.** 0.00069   **3.** 0.438   **4.** 1.25

Exercise Set 6.1, p. 346

1. $\frac{90}{100}$; $90 \times \frac{1}{100}$; 90×0.01 **3.** $\frac{12.5}{100}$; $12.5 \times \frac{1}{100}$; 12.5×0.01
5. 0.67   **7.** 0.456   **9.** 0.5901   **11.** 0.1   **13.** 0.01
15. 2   **17.** 0.001   **19.** 0.0009   **21.** 0.0018   **23.** 0.2319
25. 0.14875   **27.** 0.565   **29.** 0.09; 0.58   **31.** 0.44
33. 0.36   **35.** 47%   **37.** 3%   **39.** 870%   **41.** 33.4%
43. 75%   **45.** 40%   **47.** 0.6%   **49.** 1.7%   **51.** 27.18%
53. 2.39%   **55.** 26%; 38%   **57.** 17.7%   **59.** 21.5%
61. $\mathbf{D_W}$ **63.** $33\frac{1}{3}$ **64.** $37\frac{1}{2}$ **65.** $9\frac{3}{8}$ **66.** $18\frac{9}{16}$
67. $5\frac{11}{14}$ **68.** $111\frac{2}{3}$ **69.** $0.\overline{6}$ **70.** $0.\overline{3}$ **71.** $0.8\overline{3}$
72. $1.41\overline{6}$ **73.** $2.\overline{6}$ **74.** 0.9375

Margin Exercises, Section 6.2, pp. 350–352

1. 25%   **2.** 62.5%, or $62\frac{1}{2}$%   **3.** $66.\overline{6}$%, or $66\frac{2}{3}$%
4. $83.\overline{3}$%, or $83\frac{1}{3}$%   **5.** 57%   **6.** 76%   **7.** $\frac{3}{5}$ **8.** $\frac{13}{400}$
9. $\frac{2}{3}$
10.

FRACTION NOTATION	$\frac{1}{5}$	$\frac{5}{6}$	$\frac{3}{8}$
DECIMAL NOTATION	0.2	$0.83\overline{3}$	0.375
PERCENT NOTATION	20%	$83.\overline{3}$%, or $83\frac{1}{3}$%	$37\frac{1}{2}$%

Calculator Corner, p. 350

1. 52%   **2.** 38.46%   **3.** 110.26%   **4.** 171.43%
5. 59.62%   **6.** 28.31%

Calculator Corner, p. 353

1. 30.54; 1.31%   **2.** 32.05; 1.20%   **3.** 34.47; 1.19%
4. 26.47; 1.00%   **5.** 11.98; 4.32%   **6.** 17.52; 0.89%

Exercise Set 6.2, p. 354

1. 41%   **3.** 5%   **5.** 20%   **7.** 30%   **9.** 50%
11. 87.5%, or $87\frac{1}{2}$%   **13.** 80%   **15.** $66.\overline{6}$%, or $66\frac{2}{3}$%
17. $16.\overline{6}$%, or $16\frac{2}{3}$%   **19.** 18.75%, or $18\frac{3}{4}$%

21. 81.25%, or $81\frac{1}{4}$%   **23.** 16%   **25.** 5%   **27.** 34%
29. 40%; 18%   **31.** 22%   **33.** 5%   **35.** 9%   **37.** $\frac{17}{20}$
39. $\frac{5}{8}$ **41.** $\frac{1}{3}$ **43.** $\frac{1}{6}$ **45.** $\frac{29}{400}$ **47.** $\frac{1}{125}$ **49.** $\frac{203}{800}$
51. $\frac{176}{225}$ **53.** $\frac{711}{1100}$ **55.** $\frac{3}{2}$ **57.** $\frac{13}{40,000}$ **59.** $\frac{1}{3}$ **61.** $\frac{2}{25}$
63. $\frac{3}{5}$ **65.** $\frac{1}{50}$ **67.** $\frac{7}{20}$ **69.** $\frac{47}{100}$
71.

FRACTION NOTATION	DECIMAL NOTATION	PERCENT NOTATION
$\frac{1}{8}$	0.125	12.5%, or $12\frac{1}{2}$%
$\frac{1}{6}$	$0.1\overline{6}$	$16.\overline{6}$%, or $16\frac{2}{3}$%
$\frac{1}{5}$	0.2	20%
$\frac{1}{4}$	0.25	25%
$\frac{1}{3}$	$0.\overline{3}$	$33.\overline{3}$%, or $33\frac{1}{3}$%
$\frac{3}{8}$	0.375	37.5%, or $37\frac{1}{2}$%
$\frac{2}{5}$	0.4	40%
$\frac{1}{2}$	0.5	50%

73.

FRACTION NOTATION	DECIMAL NOTATION	PERCENT NOTATION
$\frac{1}{2}$	0.5	50%
$\frac{1}{3}$	$0.\overline{3}$	$33.\overline{3}\%$, or $33\frac{1}{3}\%$
$\frac{1}{4}$	0.25	25%
$\frac{1}{6}$	$0.1\overline{6}$	$16.\overline{6}\%$, or $16\frac{2}{3}\%$
$\frac{1}{8}$	0.125	12.5%, or $12\frac{1}{2}\%$
$\frac{3}{4}$	0.75	75%
$\frac{5}{6}$	$0.8\overline{3}$	$83.\overline{3}\%$, or $83\frac{1}{3}\%$
$\frac{3}{8}$	0.375	37.5%, or $37\frac{1}{2}\%$

75. D_W **77.** 70 **78.** 5 **79.** 400 **80.** 18.75
81. 23.125 **82.** 25.5 **83.** 4.5 **84.** $8\frac{3}{4}$ **85.** $33\frac{1}{3}$
86. $37\frac{1}{2}$ **87.** $83\frac{1}{3}$ **88.** $20\frac{1}{2}$ **89.** $43\frac{1}{8}$ **90.** $62\frac{1}{6}$
91. $18\frac{3}{4}$ **92.** $7\frac{4}{9}$ **93.** $\frac{18}{17}$ **94.** $\frac{209}{10}$ **95.** $\frac{203}{2}$ **96.** $\frac{259}{8}$
97. $11.\overline{1}\%$ **99.** $257.\overline{46317}\%$ **101.** $0.01\overline{5}$
103. $1.04\overline{142857}$

Margin Exercises, Section 6.3, pp. 358–361

1. $12\% \times 50 = a$ **2.** $a = 40\% \times 60$ **3.** $45 = 20\% \times t$
4. $120\% \times y = 60$ **5.** $16 = n \times 40$ **6.** $b \times 84 = 10.5$
7. 6 **8.** $35.20 **9.** 225 **10.** $50 **11.** 40%
12. 12.5%

Calculator Corner, p. 362

1. 1.2 **2.** $5.04 **3.** 48.64 **4.** $22.40 **5.** 0.0112
6. $29.70 **7.** 450 **8.** $1000 **9.** 2.5% **10.** 12%

Exercise Set 6.3, p. 363

1. $y = 32\% \times 78$ **3.** $89 = a \times 99$ **5.** $13 = 25\% \times y$
7. 234.6 **9.** 45 **11.** $18 **13.** 1.9 **15.** 78%
17. 200% **19.** 50% **21.** 125% **23.** 40 **25.** $40
27. 88 **29.** 20 **31.** 6.25 **33.** $846.60 **35. D_W**

37. $\frac{9}{100}$ **38.** $\frac{179}{100}$ **39.** $\frac{875}{1000}$, or $\frac{7}{8}$ **40.** $\frac{125}{1000}$, or $\frac{1}{8}$
41. $\frac{9375}{10,000}$, or $\frac{15}{16}$ **42.** $\frac{6875}{10,000}$, or $\frac{11}{16}$ **43.** 0.89 **44.** 0.07
45. 0.3 **46.** 0.017 **47.** $800 (can vary); $843.20
49. $10,000 (can vary); $10,400 **51.** $1875

Margin Exercises, Section 6.4, pp. 366–368

1. $\frac{12}{100} = \frac{a}{50}$ **2.** $\frac{40}{100} = \frac{a}{60}$ **3.** $\frac{130}{100} = \frac{a}{72}$ **4.** $\frac{20}{100} = \frac{45}{b}$
5. $\frac{120}{100} = \frac{60}{b}$ **6.** $\frac{N}{100} = \frac{16}{40}$ **7.** $\frac{N}{100} = \frac{10.5}{84}$ **8.** $225
9. 35.2 **10.** 6 **11.** 50 **12.** 30% **13.** 12.5%

Exercise Set 6.4, p. 369

1. $\frac{37}{100} = \frac{a}{74}$ **3.** $\frac{N}{100} = \frac{4.3}{5.9}$ **5.** $\frac{25}{100} = \frac{14}{b}$ **7.** 68.4
9. 462 **11.** 40 **13.** 2.88 **15.** 25% **17.** 102%
19. 25% **21.** 93.75% **23.** $72 **25.** 90 **27.** 88
29. 20 **31.** 25 **33.** $780.20 **35. D_W** **37.** 8
38. 4000 **39.** 8 **40.** 2074 **41.** 100 **42.** 15
43. $8.0\overline{4}$ **44.** $\frac{3}{16}$, or 0.1875 **45.** $\frac{43}{48}$ qt **46.** $\frac{1}{8}$ T
47. $1134 (can vary); $1118.64

Margin Exercises, Section 6.5, pp. 372–377

1. About 9.5% **2.** 14,560,000 workers **3.** (a) $1475;
(b) $38,350 **4.** (a) $9218.75; (b) $27,656.25
5. About 2.9% **6.** About 76.6%

Translating for Success, p. 378

1. J **2.** M **3.** N **4.** E **5.** G **6.** H **7.** O **8.** C
9. D **10.** B

Exercise Set 6.5, p. 379

1. About 13,247 wild horses **3.** 3 years: $21,080;
5 years: $17,680 **5.** Overweight: 176.4 million people;
obese: 73.5 million people **7.** Acid: 20.4 mL;
water: 659.6 mL **9.** About 1808 miles
11. About 39,867,000 people **13.** 36.4 correct;
3.6 incorrect **15.** 95 items **17.** 25%
19. 166; 156; 146; 140; 122 **21.** 8% **23.** 20%
25. About 86.5% **27.** $30,030 **29.** $16,174.50;
$12,130.88 **31.** About 27% **33.** 34.375%, or $34\frac{3}{8}\%$
35. (a) 7.5%; (b) 879,675 **37.** 71% **39.** $1560
41. 80% **43.** 98,775; 18.0% **45.** 799,065; 14.8%
47. 4,550,688; 38.1% **49.** $36,400 **51.** 40% **53. D_W**
55. $2.\overline{27}$ **56.** 0.44 **57.** 3.375 **58.** $4.\overline{7}$ **59.** 0.92
60. $0.8\overline{3}$ **61.** 0.4375 **62.** 2.317 **63.** 3.4809
64. 0.675 **65.** About 5 ft 6 in. **67.** $83\frac{1}{3}\%$

Margin Exercises, Section 6.6, pp. 385–389

1. $48.50; $717.45 **2.** $5.39; $140.14 **3.** 6% **4.** $999
5. $5628 **6.** 12.5%, or $12\frac{1}{2}\%$ **7.** $1675 **8.** $180; $360
9. 20%

Exercise Set 6.6, p. 390

1. $19.53 **3.** $2.65 **5.** $16.39; $361.39 **7.** 5%
9. 4% **11.** $2000 **13.** $800 **15.** $719.86 **17.** 5.6%
19. $2700 **21.** 5% **23.** $980 **25.** $5880 **27.** 12%
29. $420 **31.** $30; $270 **33.** $2.55; $14.45
35. $125; $112.50 **37.** 40%; $360 **39.** $30; 16.7%
41. $549; 36.4% **43.** $^{D}_{W}$ **45.** $^{D}_{W}$ **47.** 18 **48.** $\frac{22}{7}$
49. 265.625 **50.** 1.1\overline{13} **51.** 0.\overline{5} **52.** 2.0\overline{9} **53.** 0.91\overline{6}
54. 1.85\overline{7142} **55.** 2.\overline{142857} **56.** 1.58\overline{3}
57. 4,030,000,000,000 **58.** 5,800,000 **59.** 42,700,000
60. 6,090,000,000,000 **61.** $2.69
63. He bought the plaques for 166\frac{2}{3}$ + $250, or 416\frac{2}{3}$, and sold them for $400, so he lost money.

Margin Exercises, Section 6.7, pp. 394–397

1. $301 **2.** $225.75 **3.** (a) $33.53; (b) $4833.53
4. $2376.20 **5.** $7690.94

Calculator Corner, p. 397

1. $16,357.18 **2.** $12,764.72

Exercise Set 6.7, p. 398

1. $8 **3.** $84 **5.** $113.52 **7.** $925 **9.** $671.88
11. (a) $147.95; (b) $10,147.95 **13.** (a) $80.14;
(b) $6580.14 **15.** (a) $46.03; (b) $5646.03 **17.** $441
19. $2802.50 **21.** $7853.38 **23.** $99,427.40
25. $4243.60 **27.** $28,225.00 **29.** $9270.87
31. $129,871.09 **33.** $4101.01 **35.** $1324.58 **37.** $^{D}_{W}$
39. reciprocals **40.** divisible by 6 **41.** additive
42. unit rate **43.** perimeter **44.** divisible by 3
45. prime **46.** proportional **47.** 9.38%

Margin Exercises, Section 6.8, pp. 402–406

1. (a) $97; (b) interest: $86.40; amount applied to principal: $10.60; (c) interest: $55.17; amount applied to principal: $41.83; (d) At 13.6%, the principal was decreased by $31.23 more than at the 21.3% rate. The interest at 13.6% is $31.23 less than at 21.3%. **2.** (a) Interest: $91.78; amount applied to principal: $229.22; (b) $58; (c) $4080
3. Interest: $843.94; amount applied to principal: $135.74
4. (a) Interest: $844.69; amount applied to principal: $498.64; (b) $88,799.40; (c) The Sawyers will pay $110,885.40 less in interest with the 15-yr loan than with the 30-yr loan.
5. (a) Interest: $701.25; amount applied to principal: $548.89; (b) $72,025.20; (c) The Sawyers will pay $16,774.20 less in interest with the 15-yr loan at 5$\frac{1}{2}$% than with the 15-yr loan at 6$\frac{5}{8}$%.

Exercise Set 6.8, p. 407

1. (a) $98; (b) interest: $86.56; amount applied to principal: $11.44; (c) interest: $51.20; amount applied to principal: $46.80; (d) At 12.6%, the principal is decreased by $35.36 more than at the 21.3% rate. The interest at 12.6% is $35.36

less than at 21.3%. **3.** (a) Interest: $125.14; amount applied to principal: $312.79; (b) $51.24; (c) $7991.60, $11,504, $3512.40 **5.** (a) Interest: $854.17; amount applied to principal: $155.61; (b) $199,520.80; (c) new principal: $163,844.39; interest: $853.36; amount applied to principal: $156.42 **7.** (a) Interest: $854.17; amount applied to principal: $552; (b) $89,110.60; (c) The Martinez family will pay $110,410.20 less in interest with the 15-yr loan than with the 30-yr loan. **9.** $99,917.71; $99,834.94 **11.** $99,712.04; $99,422.15 **13.** $149,882.75; $149,764.79 **15.** $199,382.07; $198,760.41
17. (a) $2395, $21,555; (b) $52.09, $401.97; (c) $239.88
19. (a) $595; $11,305; (b) interest: $87.61; amount applied to principal: $273.47; (c) $1693.88 **21.** $^{D}_{W}$ **23.** $^{D}_{W}$
25. 17.5, or $\frac{35}{2}$ **26.** 28 **27.** 100 **28.** 56.25, or $\frac{225}{4}$
29. $\frac{66,875}{36}$, or approximately 1857.64 **30.** 787.69
31. 71.81 **32.** 17.5, or $\frac{35}{2}$ **33.** 56.25, or $\frac{225}{4}$
34. $\frac{66,875}{36}$, or approximately 1857.64

Concept Reinforcement, p. 410

1. True **2.** False **3.** True **4.** False

Summary and Review: Chapter 6, p. 410

1. 56% **2.** 1.7% **3.** 37.5% **4.** 33.$\overline{3}$%, or 33$\frac{1}{3}$%
5. 0.735 **6.** 0.065 **7.** $\frac{6}{25}$ **8.** $\frac{63}{1000}$
9. $30.6 = p \times 90$; 34% **10.** $63 = 84\% \times n$; 75
11. $y = 38\frac{1}{2}\% \times 168$; 64.68 **12.** $\frac{24}{100} = \frac{16.8}{b}$; 70
13. $\frac{42}{30} = \frac{N}{100}$; 140% **14.** $\frac{10.5}{100} = \frac{a}{84}$; 8.82
15. 223 students; 105 students **16.** 42% **17.** 2500 mL
18. 12% **19.** 93.15 **20.** $14.40 **21.** 5% **22.** 11%
23. $42; $308 **24.** $42.70; $262.30 **25.** $2940
26. Approximately 18.4% **27.** $36 **28.** (a) $394.52;
(b) $24,394.52 **29.** $121 **30.** $7727.26 **31.** $9504.80
32. (a) $129; (b) interest: $100.18; amount applied to principal: $28.82; (c) interest: $70.72; amount applied to principal: $58.28; (d) At 13.2%, the principal is decreased by $29.46 more than at the 18.7% rate. The interest at 13.2% is $29.46 less than at 18.7%.
33. $^{D}_{W}$ No; the 10% discount was based on the original price rather than on the sale price.
34. $^{D}_{W}$ A 40% discount is better. When successive discounts are taken, each is based on the previous discounted price rather than on the original price. A 20% discount followed by a 22% discount is the same as a 37.6% discount off the original price.
35. 19.5% increase **36.** 66$\frac{2}{3}$% **37.** $168

Test: Chapter 6, p. 413

1. [6.1b] 0.064 **2.** [6.1b] 38% **3.** [6.2a] 137.5%
4. [6.2b] $\frac{13}{20}$ **5.** [6.3a, b] $a = 40\% \cdot 55$; 22
6. [6.4a, b] $\frac{N}{100} = \frac{65}{80}$; 81.25% **7.** [6.5a] 400 passengers;
575 passengers **8.** [6.5a] About 539 at-bats

9. [6.5b] 50.$\overline{90}$%   **10.** [6.5a] 5.5%   **11.** [6.6a] $16.20; $340.20   **12.** [6.6b] $630   **13.** [6.6c] $40; $160
14. [6.7a] $8.52   **15.** [6.7a] $5356   **16.** [6.7b] $1110.39
17. [6.7b] $11,580.07   **18.** [6.5b] Registered nurses: 2.9, 26.1%; post-secondary teachers: 0.6, 37.5%; food preparation and service workers: 2.0, 20%; restaurant servers: 2.1, 0.4   **19.** [6.6c] $275, about 14.1%
20. [6.8a] $119,909.14; $119,817.72
21. [6.6b] $194,600   **22.** [6.6b], [6.7b] $2546.16

CHAPTER 7

Margin Exercises, Section 7.1, pp. 418–423

1. 75   **2.** 54.9   **3.** 81   **4.** 19.4   **5.** 37   **6.** 34 mpg
7. 2.5   **8.** 94   **9.** 17   **10.** 17   **11.** 91   **12.** $3700
13. 67.5   **14.** 45   **15.** 34, 67   **16.** No mode exists.
17. (a) 17 g; (b) 18 g; (c) 19 g
18. Wheat A: average stalk height $\approx$ 25.21 in.; wheat B: average stalk height $\approx$ 22.54 in.; wheat B is better.

Calculator Corner, p. 420

1. 285.5   **2.** 75; 54.9; 81; 19.4   **3.** 202.$\overline{3}$

Exercise Set 7.1, p. 424

1. Average: 21; median: 18.5; mode: 29
3. Average: 21; median: 20; modes: 5, 20
5. Average: 5.2; median: 5.7; mode: 7.4
7. Average: 239.5; median: 234; mode: 234
9. Average: 23.$\overline{8}$; median: 15; mode: 1   **11.** 31 mpg
13. 2.7   **15.** Average: $8.19; median: $8.49; mode: $6.99
17. 90   **19.** 263 days
21. Bulb A: average time = 1171.25 hr; bulb B: average time $\approx$ 1251.58 hr; bulb B is better   **23.** $\mathbf{D_W}$ **25.** 196
26. $\frac{4}{9}$ **27.** 1.96   **28.** 1.999396   **29.** 225.05
30. 126.0516   **31.** $\frac{3}{35}$ **32.** $\frac{14}{15}$ **33.** 118.75%
34. 68.75%   **35.** 51.2%   **36.** 97.81%   **37.** 182
39. 10 home runs   **41.** 58

Margin Exercises, Section 7.2, pp. 427–430

1. Zone (Men's Menu)   **2.** Both Ornish (Eat More, Weigh Less) and Zone (Men's Menu) have 17 servings of fruits and vegetables   **3.** Ornish (Eat More, Weigh Less)
4. About 43 g   **5.** About 24.5 g   **6.** 22 g
7. Average: about 9.3 g; median: 7 g; mode: 7 g
8. 20 calories   **9.** Yes   **10.** 560 mg   **11.** 24%
12. 14 g   **13.** 2100 rhinos   **14.** There are twelve times as many black rhinos as Sumatran rhinos, or 3300 more.
15. 3537 rhinos; answers may vary
16. 795 cups; answers may vary
17. 750 cups; answers may vary
18. 1830 cups; answers may vary

Exercise Set 7.2, p. 431

1. 483,612,200 mi   **3.** Neptune   **5.** All
7. 11 Earth diameters   **9.** Average: 27,884.$\overline{1}$ mi; median: 7926 mi; no mode exists   **11.** 92°   **13.** 108°
15. 3   **17.** 90° and higher   **19.** 30% and higher
21. 50%   **23.** 1940: 1976; 1980: 3849; 94.8%
25. 1920; 3186; 1266   **27.** 1.0 billion   **29.** 2070
31. 1650 and 1850   **33.** 3 billion; 75%   **35.** Africa
37. 475,000 gal   **39.** 325,000 gal   **41.** $\mathbf{D_W}$
43. Cabinets: $13,444; countertops: $4033.20; appliances: $2151.04; fixtures: $806.64
44. Cabinets: $3597.55; countertops: $1276.55; labor: $2901.25; flooring: $696.30   **45.** $\frac{6}{25}$ **46.** $\frac{9}{20}$
47. $\frac{6}{125}$ **48.** $\frac{8}{125}$ **49.** $\frac{531}{1000}$ **50.** $\frac{873}{1000}$ **51.** 1   **52.** $\frac{1}{50}$
53.

Coffee Consumption

Germany / United States / Switzerland / France / Italy   = 150 cups

Margin Exercises, Section 7.3, pp. 436–441

1. About 77 million   **2.** NASCAR   **3.** NFL, NBA, and MLB
4. 60   **5.** 85+   **6.** 60–64   **7.** Yes
8.

9. Month 7, June   **10.** Months 1 and 2, December and January, months 4 and 5, March and April, months 6 and 7, May and June, and months 11 and 12, October and November
11. Month 2, January, and month 8, July   **12.** $900
13. 40 yr   **14.** $1300
15.

Exercise Set 7.3, p. 442

1. 190 calories **3.** 1 slice of chocolate cake with fudge frosting **5.** 1 cup of premium chocolate ice cream **7.** About 120 calories **9.** About 920 calories **11.** About 28 lb **13.** 1970: $11,000; 2002: $58,000; $47,000; 427% **15.** 1970: $6000; 2002: $25,000; $19,000; 317% **17.** $4000 **19.** $11,000
21.

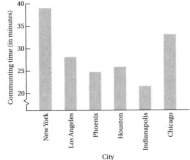

23. Indianapolis **25.** 28.73 min **27.** 1995 and 1996, 1996 and 1997, 1998 and 1999 **29.** 51% **31.** $3.9 billion **33.** 30.4 yd **35.** 1988 and 1995
37.

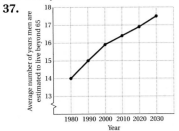

39. 25%0 **41.** 10.1% **43.** 1995 and 1996 **45.** About 308 murders **47.** 14.1% **49.** **D**w **51.** natural **52.** add; divide **53.** simple **54.** marked price; rate of discount; discount; sale price **55.** compound **56.** repeating **57.** distributive **58.** commutative
59.

Using the line through the points for 1996 and 1998 (instead of a line through the point for 1999) probably gives more realistic estimates of the number of movie screens for the years 2000, 2001, and 2003. 2000: 37,500; 2001: 40,000; 2003: 44,500. Answers may vary.

Margin Exercises, Section 7.4, pp. 448–449

1. Spaying or neutering **2.** 87% **3.** $1122 **4.** 8% + 3%, or 11%

5. **Times of Engagement of Married Couples**

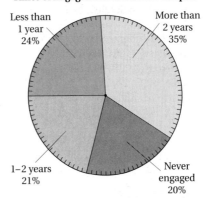

Translating for Success, p. 450

1. D **2.** B **3.** J **4.** K **5.** I **6.** F **7.** N **8.** E **9.** L **10.** M

Exercise Set 7.4, p. 451

1. 17.1% **3.** About 613 students **5.** 86.1% **7.** Food **9.** 14% **11.**

13.

15.

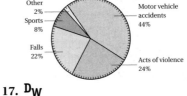

17. **D**w

Concept Reinforcement, p. 454

1. True **2.** False **3.** True **4.** True

Summary and Review: Chapter 7, p. 454

1. $32.25 **2.** $47.00 **3.** $20.95 **4.** $25.40 **5.** No **6.** $19.75 **7.** 14,000 officers **8.** Los Angeles **9.** Houston **10.** 12,500 **11.** 26 **12.** 11 and 17 **13.** 0.2 **14.** 700 and 800 **15.** $17 **16.** 20 **17.** $110.50; $107 **18.** 33 mpg **19.** 96

20. About 420 calories   **21.** About 440 calories
22. Big Bacon Classic   **23.** Plain Single   **24.** Plain Single
25. Chicken Club   **26.** About 50 calories
27. About 220 calories   **28.** Under 20   **29.** 12 accidents
30. 13 accidents per 100 drivers   **31.** 45–74
32. 11 accidents per 100 drivers   **33.** Under 20
34. 22%   **35.** 11%   **36.** 1600 travelers   **37.** 25%
38.

39.

40. Battery A: average ≈ 43.04 hr; battery B:
average = 41.55 hr; battery A is better.   **41.** 38.5
42. 13.4   **43.** 1.55   **44.** 1840   **45.** $16.\overline{6}$ **46.** $321.\overline{6}$
47. 38.5   **48.** 14   **49.** 1.8   **50.** 1900   **51.** $17
52. 375   **53.** 3.1
54. ^D**W** The average, the median, and the mode are
"center points" that characterize a set of data. You might use
the average to find a center point that is midway between
the extreme values of the data. The median is a center point
that is in the middle of all the data. That is, there are as
many values less than the median than there are values
greater than the median. The mode is a center point that
represents the value or values that occur most frequently.
55. ^D**W** The equation could represent a person's average
income during a 4-yr period. Answers may vary.
56. $a = 316, b = 349$

Test: Chapter 7, p. 458

1. [7.2a] 179 lb   **2.** [7.2a] 111 lb   **3.** [7.2a] 5 ft, 3 in.;
medium frame   **4.** [7.2a] 5 ft, 11 in.; medium frame
5. [7.2b] Japan   **6.** [7.2b] United States
7. [7.2b] 800 pounds   **8.** [7.2b] 400 pounds
9. [7.1a] 49.5   **10.** [7.1a] 2.6   **11.** [7.1a] 15.5
12. [7.1b, c] 50.5; 52   **13.** [7.1b, c] 3; 1 and 3
14. [7.1b, c] 17.5; 17 and 18   **15.** [7.1a] 27 mpg
16. [7.1a] 76   **17.** [7.3c] 53%   **18.** [7.3c] 41%
19. [7.3c] 1967   **20.** [7.3c] 2006

21. [7.3b]

22. [7.2a], [7.3a] 197 mph   **23.** [7.2a], [7.3a] No; the
greyhound can run 14 mph faster than a human.
24. [7.2a], [7.3a] 89 mph   **25.** [7.2a], [7.3a] 61 mph
26. [7.4b]

27. [7.2a], [7.4a] Employee theft: $10.12 billion; shoplifting:
$7.521 billion; administrative error: $4.025 billion; vendor
fraud: $1.173 billion; other: $0.161 billion
28. [7.3b]

29. [7.3d]

30. [7.1d] Bar A: average ≈ 8.417; bar B: average ≈ 8.417;
equal quality
31. [7.1a] 2.9   **32.** [7.1a, b] $a = 74, b = 111$

Cumulative Review: Chapters 1–7, p. 462

1. [4.3b] $46,600,000,000   **2.** [7.1a] 27 mpg
3. [1.1a] 5 hundreds   **4.** [1.9c] 128   **5.** [2.1a] 1, 2, 3, 4, 5,
6, 10, 12, 15, 20, 30, 60   **6.** [4.1d] 52.0   **7.** [3.4a] $\frac{33}{10}$
8. [4.3b] $2.10   **9.** [4.3b] $3,250,000,000,000
10. [5.3a] No   **11.** [3.5a] $6\frac{7}{10}$ **12.** [4.2a] 44.6351
13. [3.3a] $\frac{1}{3}$ **14.** [4.2b] 325.43   **15.** [3.6a] 15
16. [1.5a] 2,740,320   **17.** [2.7b] $\frac{9}{10}$ **18.** [3.4b] $4361\frac{1}{2}$
19. [5.3b] $9\frac{3}{5}$ **20.** [2.7c] $\frac{3}{4}$ **21.** [4.4b] 6.8
22. [1.7b] 15,312   **23.** [5.2b] 20.6 ¢/oz
24. [6.5a] 3324 students   **25.** [3.6c] $\frac{1}{4}$ yd   **26.** [2.6b] $\frac{3}{8}$ cup
27. [4.7a] 6.2 lb   **28.** [1.8a] 2572 billion kWh
29. [6.5a] 2.5%; 0.2%   **30.** [6.5b] About 4.1%

31. (a) [3.6c] 1870 sq yd; **(b)** [3.6c] 6400 sq yd; **(c)** [1.8a] 4530 sq yd **32.** [7.1a, b] $38.25; $38.25
33. [7.3b]

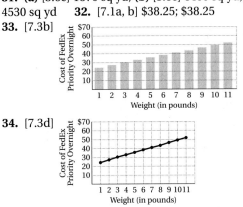

34. [7.3d]

35. [3.2b], [3.3d] $\frac{1}{4}$ **36.** [5.4a] 1122 defective valves
37. [4.7a] $9.55 **38.** [6.6b] 7% **39.** [7.2a], [6.5a]
Clothing: $129.54; entertainment: $83.82; food: $83.82;
other: $83.82 **40.** [2.6b] $254; $127
41. [7.2a], [6.5a] Clothing: $44.064 billion;
entertainment: $28.512 billion;
food: $28.512 billion; other: $28.512 billion
42. [6.5b] 73% **43.** [6.5b], [7.1a] 12% decrease

CHAPTER 8

Margin Exercises, Section 8.1, pp. 466–469

1. 2 **2.** 3 **3.** $1\frac{1}{2}$ **4.** $2\frac{1}{2}$ **5.** 288 **6.** $8\frac{1}{2}$
7. 240,768 **8.** 6 **9.** $1\frac{5}{12}$ **10.** 8
11. $11\frac{2}{3}$, or $11.\overline{6}$ **12.** 5 **13.** 0.5
14. 288 **15.** 40,480 yd

Exercise Set 8.1, p. 470

1. 12 **3.** $\frac{1}{12}$ **5.** 5280 **7.** 108 **9.** 7 **11.** $1\frac{1}{2}$
13. 26,400 **15.** $5\frac{1}{4}$, or 5.25 **17.** $3\frac{1}{3}$ **19.** 37,488
21. $1\frac{1}{2}$ **23.** $1\frac{1}{4}$, or 1.25 **25.** 110 **27.** 2 **29.** 300
31. 30 **33.** $\frac{1}{36}$ **35.** 126,720 **37.** **D**w **39.** $\frac{37}{400}$
40. $\frac{7}{8}$ **41.** $\frac{11}{40}$ **42.** $\frac{57}{100}$ **43.** 137.5%
44. $66.\overline{6}$%, or $66\frac{2}{3}$% **45.** 25% **46.** 43.75%
47. $\frac{373}{100,000}$; $\frac{100,000}{373}$ **48.** $\frac{29}{2000}$; $\frac{2000}{29}$ **49.** Length: 5400 in., or
450 ft; breadth: 900 in., or 75 ft; height: 540 in., or 45 ft

Margin Exercises, Section 8.2, pp. 473–477

1. 2 cm, or 20 mm **2.** 2.3 cm, or 23 mm **3.** 4.4 cm, or
44 mm **4.** cm **5.** km **6.** mm **7.** m **8.** cm
9. m **10.** 23,000 **11.** 400 **12.** 178 **13.** 9040
14. 7.814 **15.** 781.4 **16.** 0.967 **17.** 8,900,000
18. 6.78 **19.** 97.4 **20.** 0.1 **21.** 8.451

Exercise Set 8.2, p. 478

1. (a) 1000; **(b)** 0.001 **3. (a)** 10; **(b)** 0.1 **5. (a)** 0.01;
(b) 100 **7. (a)** 6700; **(b)** Substitute **9. (a)** 0.98;
(b) Multiply by 1 **11.** 8.921 **13.** 0.05666 **15.** 566,600

17. 4.77 **19.** 688 **21.** 0.1 **23.** 100,000 **25.** 142
27. 0.82 **29.** 450 **31.** 0.000024 **33.** 0.688 **35.** 230
37. 3.92 **39.** 180; 0.18 **41.** 278; 27.8 **43.** 48,440;
48.44 **45.** 4000; 400 **47.** 0.027; 0.00027 **49.** 442,000;
44,200 **51.** **D**w **53.** 0.234 **54.** 0.0234 **55.** 13.8474
56. $80\frac{1}{2}$ **57.** 37.5% **58.** 62.5% **59.** $66.\overline{6}$%, or $66\frac{2}{3}$%
60. 20% **61.** 0.9 **62.** 0.108 **63.** 1.0 m **65.** 1.4 cm

Margin Exercises, Section 8.3, pp. 480–481

1. 91.4 **2.** 804.5 **3.** 1479.843 **4.** 913.912
5. 343.735 m **6.** 0.125

Exercise Set 8.3, p. 482

1. 100.65 **3.** 727.4394 **5.** 104.585 **7.** 289.62
9. 112.63 **11.** 9.14 **13.** 81.8896 **15.** 1250.061
17. 19.05 **19.** 1376.136 **21.** 0.51181
23.–33. Answers may vary, depending on the conversion
factor used.

	yd	cm	in.	m	mm
23.	0.2604	23.8	9.37006	0.238	238
25.	0.2361	21.59	$8\frac{1}{2}$	0.2159	215.9
27.	52.9934	4844	1907.0828	48.44	48,440
29.	4	365.6	144	3.656	3656
31.	0.000295	0.027	0.0106299	0.00027	0.27
33.	483.548	44,200	17,401.54	442	442,000

35. **D**w **37.** 23,072,000,000,000 **38.** 25,100,000
39. 366,000,000 **40.** 585,000,000,000
41. 1 in. = 25.4 mm **43.** 21.3 mph

Margin Exercises, Section 8.4, pp. 484–488

1. 80 **2.** 4.32 **3.** 32,000 **4.** kg **5.** kg **6.** mg
7. g **8.** t **9.** 6200 **10.** 3.048 **11.** 77 **12.** 234.4
13. 6700 **14.** 0.001 **15.** 0.5 mg

Exercise Set 8.4, p. 489

1. 2000 **3.** 3 **5.** 64 **7.** 12,640 **9.** 0.1 **11.** 5
13. 26,000,000,000 lb **15.** 1000 **17.** 10
19. $\frac{1}{100}$, or 0.01 **21.** 1000 **23.** 10 **25.** 234,000
27. 5.2 **29.** 6.7 **31.** 0.0502 **33.** 8.492 **35.** 58.5
37. 800,000 **39.** 1000 **41.** 0.0034 **43.** 1000
45. 0.325 **47.** 125 mcg **49.** 0.875 mg; 875 mcg
51. 4 tablets **53.** 8 cc **55.** **D**w **57.** $\frac{7}{20}$ **58.** $\frac{99}{100}$
59. $\frac{171}{200}$ **60.** $\frac{171}{500}$ **61.** $\frac{3}{8}$ **62.** $\frac{2}{3}$ **63.** $\frac{5}{6}$ **64.** $\frac{1}{6}$
65. 187,200 **66.** About 324 students **67.** $24.30
68. 12% **69.** 7500 sheets **70.** 46 test tubes, 2 cc left
over **71.** 144 packages **73. (a)** 109.134 g; **(b)** 9.104 g;
(c) Golden Jubilee: 3.85 oz; Hope: 0.321 oz

Margin Exercises, Section 8.5, pp. 493–495

1. 40 **2.** 20 **3.** mL **4.** mL **5.** L **6.** L **7.** 970
8. 8.99 **9.** 2.4 L **10.** (a) About 59.14 mL;
(b) about 0.059 L

Exercise Set 8.5, p. 496

1. 1000; 1000 **3.** 87,000 **5.** 0.049 **7.** 0.000401
9. 78,100 **11.** 320 **13.** 10 **15.** 32 **17.** 20
19. 14 **21.** 88

	gal	qt	pt	cups	oz
23.	1.125	4.5	9	18	144
25.	16	64	128	256	2048
27.	0.25	1	2	4	32
29.	0.1171875	0.46875	0.9375	1.875	15

	L	mL	cc	cm^3
31.	2	2000	2000	2000
33.	64	64,000	64,000	64,000
35.	0.443	443	443	443

37. 2000 mL **39.** 0.32 L **41.** 59.14 mL **43.** 500 mL
45. 125 mL/hr **47.** 9 **49.** $\frac{1}{5}$ **51.** 6 **53.** 15
55. ^{D}W **57.** 45.2% **58.** 99.9% **59.** 33.$\overline{3}$%, or 33$\frac{1}{3}$%
60. 66.$\overline{6}$%, or 66$\frac{2}{3}$% **61.** 55% **62.** 105% **63.** 88%
64. 8% **65.** $\frac{18.2}{79.3}; \frac{79.3}{18.2}$ **66.** $\frac{1}{4}; \frac{3}{4}$
67. 1.75 gal/week; 7.5 gal/month; 91.25 gal/year;
73,500,000 gal/day; 26,827,500,000 gal/year **69.** 0.21$\overline{3}$ oz

Margin Exercises, Section 8.6, pp. 499–501

1. 120 **2.** 1461 **3.** 1440 **4.** 1 **5.** 80°C **6.** 0°C
7. −20°C **8.** 80°F **9.** 100°F **10.** 50°F **11.** 176°F
12. 95°F **13.** 35°C **14.** 45°C

Calculator Corner, p. 501

1. 41°F **2.** 122°F **3.** 20°C **4.** 45°C

Exercise Set 8.6, p. 502

1. 24 **3.** 60 **5.** 365$\frac{1}{4}$ **7.** 0.05 **9.** 8.2 **11.** 6.5
13. 10.75 **15.** 336 **17.** 4.5 **19.** 56 **21.** 86,164.2 sec
23. 77°F **25.** 104°F **27.** 186.8°F **29.** 136.4°F
31. 35.6°F **33.** 41°F **35.** 5432°F **37.** 30°C
39. 55°C **41.** 81.$\overline{1}$°C **43.** 60°C **45.** 20°C **47.** 6.$\overline{6}$°C
49. 37°C **51.** (a) 136°F = 57.$\overline{7}$°C, 56$\frac{2}{3}$°C = 134°F;
(b) 2°F **53.** ^{D}W **55.** compound **56.** ratio
57. median **58.** commission **59.** similar
60. mean **61.** composite **62.** subtrahend
63. About 0.03 yr **65.** About 31,688 yr **67.** 0.25

Margin Exercises, Section 8.7, pp. 505–506

1. 9 **2.** 45 **3.** 2880 **4.** 2.5 **5.** 3200 **6.** 1,000,000
7. 1 **8.** 28,800 **9.** 0.043 **10.** 0.678

Translating for Success, p. 507

1. E **2.** H **3.** O **4.** A **5.** G **6.** C **7.** M
8. I **9.** K **10.** N

Exercise Set 8.7, p. 508

1. 144 **3.** 640 **5.** $\frac{1}{144}$ **7.** 198 **9.** 396 **11.** 12,800
13. 27,878,400 **15.** 5 **17.** 1 **19.** $\frac{1}{640}$, or 0.0015625
21. 5,210,000 **23.** 140 **25.** 23.456 **27.** 0.085214
29. 2500 **31.** 0.4728 **33.** ^{D}W **35.** $240 **36.** $212
37. (a) $484.11; (b) $15,984.11 **38.** (a) $209.59;
(b) $8709.59 **39.** (a) $220.93; (b) $6620.93
40. (a) $37.97; (b) $4237.97 **41.** 10.76 **43.** 1.67
45. 1876.8 m^2

Concept Reinforcement, p. 510

1. False **2.** True **3.** True **4.** True **5.** False
6. False

Summary and Review: Chapter 8, p. 511

1. 2$\frac{2}{3}$ **2.** 30 **3.** 0.03 **4.** 0.004 **5.** 72 **6.** 400,000
7. 1$\frac{1}{6}$ **8.** 0.15 **9.** 218.8 **10.** 32.18 **11.** 10; 0.01
12. 305,000; 30,500 **13.** 112 **14.** 0.004 **15.** $\frac{4}{15}$, or 0.2$\overline{6}$
16. 0.464 **17.** 180 **18.** 4700 **19.** 16,140 **20.** 830
21. $\frac{1}{4}$, or 0.25 **22.** 0.04 **23.** 200 **24.** 30 **25.** 0.0007
26. 0.06 **27.** 1600 **28.** 400 **29.** 1.25 **30.** 50
31. 160 **32.** 7.5 **33.** 13.5 **34.** 60 **35.** 0.0302
36. 5.25 **37.** 6 mL **38.** 3000 mL **39.** 250 mcg
40. 80.6°F **41.** 23°F **42.** 20°C **43.** −12.$\overline{2}$°C
44. 36 **45.** 300,000 **46.** 14.375 **47.** 0.06
48. ^{D}W The metric system was adopted by law in France in
about 1790, during the rule of Napoleon I.
49. ^{D}W 1 gal = 128 oz, so 1 oz of water (as capacity) weighs
$\frac{8.3453}{128}$ lb, or about 0.0652 lb. An ounce of pennies weighs $\frac{1}{16}$ lb,
or 0.0625 lb. Thus an ounce of water (as capacity) weighs
more than an ounce of pennies. **50.** 17.66 sec

Test: Chapter 8, p. 513

1. [8.1a] 48 **2.** [8.1a] $\frac{1}{3}$ **3.** [8.2a] 6000 **4.** [8.2a] 0.87
5. [8.3a] 182.8 **6.** [8.3a] 1490.4 **7.** [8.2a] 5; 0.005
8. [8.2a] 1854.2; 185.42 **9.** [8.5a] 3.08 **10.** [8.5a] 240
11. [8.4a] 64 **12.** [8.4a] 8220 **13.** [8.4b] 3800
14. [8.4b] 0.4325 **15.** [8.4b] 2.2 **16.** [8.6a] 300
17. [8.6a] 360 **18.** [8.5a] 32 **19.** [8.5a] 1280
20. [8.5a] 40 **21.** [8.4c] 370 **22.** [8.6b] 35°C
23. [8.6b] 138.2°F **24.** [8.3a] 1.094; 100; 39.370; 1000
25. [8.3a] 36,377.2; 14,328; 363.772; 363,772
26. [8.5b] 2500 mL **27.** [8.4c] 1500 mcg
28. [8.5b] About 118.28 mL **29.** [8.7a] 1728
30. [8.7b] 0.0003 **31.** [8.3a] About 39.47 sec

CHAPTER 9

Margin Exercises, Section 9.1, pp. 516–518

1. 26 cm   **2.** 46 in.   **3.** 12 cm   **4.** 17.5 yd   **5.** $27\frac{5}{6}$ in.
6. 40 km   **7.** 21 yd   **8.** 31.2 km   **9.** 70 ft; $346.50

Exercise Set 9.1, p. 519

1. 17 mm   **3.** 15.25 in.   **5.** 18 km   **7.** 30 ft
9. 79.14 cm   **11.** 88 ft   **13.** 182 mm
15. 826 m; $6021.54   **17.** 122 cm   **19.** (a) 228 ft;
(b) $1046.52   **21.** $\mathbf{D_W}$ **23.** $19.20   **24.** $96
25. 1000   **26.** 1331   **27.** 225   **28.** 484   **29.** 49
30. 64   **31.** 5%   **32.** 11%   **33.** 9 ft

Margin Exercises, Section 9.2, pp. 521–526

1. 8 cm^2 **2.** 56 km^2 **3.** $18\frac{3}{8}$ yd^2 **4.** 144 km^2
5. 118.81 m^2 **6.** $12\frac{1}{4}$ yd^2 **7.** 43.8 cm^2 **8.** 12.375 km^2
9. 96 m^2 **10.** 18.7 cm^2 **11.** 100 m^2 **12.** 88 cm^2
13. 228 in^2

Exercise Set 9.2, p. 527

1. 15 km^2 **3.** 1.4 in^2 **5.** $6\frac{1}{4}$ yd^2 **7.** 8100 ft^2 **9.** 50 ft^2
11. 169.883 cm^2 **13.** $41\frac{2}{9}$ in^2 **15.** 484 ft^2
17. 3237.61 km^2 **19.** $28\frac{57}{64}$ yd^2 **21.** 32 cm^2 **23.** 60 in^2
25. 104 ft^2 **27.** 45.5 in^2 **29.** 8.05 cm^2 **31.** 297 cm^2
33. 7 m^2 **35.** 1197 m^2 **37.** (a) About 8473 ft^2;
(b) about $102   **39.** 630.36 ft^2 **41.** (a) 819.75 ft^2;
(b) 10 gal; **(c)** $179.50   **43.** 80 cm^2 **45.** 675 cm^2
47. 21 cm^2 **49.** 852.04 ft^2 **51.** $\mathbf{D_W}$ **53.** 234
54. 230   **55.** 336   **56.** 24   **57.** 0.724   **58.** 0.0724
59. 2520   **60.** 6   **61.** 28   **62.** $22\frac{1}{2}$ **63.** 12
64. 46,252.8   **65.** $33,597.91   **66.** $413,458.31
67. $641,566.26   **68.** $684,337.34   **69.** 16,914 in^2

Margin Exercises, Section 9.3, pp. 532–536

1. 9 in.   **2.** 5 ft   **3.** 62.8 m   **4.** 88 m   **5.** 34.296 yd
6. $78\frac{4}{7}$ km^2 **7.** 339.62 cm^2 **8.** 12-ft–diameter flower
bed, by about 13.04 ft^2

Exercise Set 9.3, p. 537

1. 14 cm; 44 cm; 154 cm^2 **3.** $1\frac{1}{2}$ in.; $4\frac{5}{7}$ in.; $1\frac{43}{56}$ in^2
5. 16 ft; 100.48 ft; 803.84 ft^2 **7.** 0.7 cm; 4.396 cm;
1.5386 cm^2 **9.** 3 cm; 18.84 cm; 28.26 cm^2 **11.** 153.86 ft^2
13. 2.5 cm; 1.25 cm; 4.90625 cm^2 **15.** 3.454 ft
17. 65.94 yd^2 **19.** 45.68 ft   **21.** 26.84 yd   **23.** 45.7 yd
25. 100.48 m^2 **27.** 6.9972 cm^2 **29.** 64.4214 in^2
31. $\mathbf{D_W}$ **33.** 16   **34.** 289   **35.** 125   **36.** 64
37. 543   **38.** 0.00543   **39.** 41.01%   **40.** $730
41. 10 cans   **42.** 5 lb   **43.** 3.1416
45. $3d$; πd; circumference of one ball, since $\pi > 3$

Margin Exercises, Section 9.4, pp. 541–545

1. 12 cm^3 **2.** 3087 in^3 **3.** 128 ft^3 **4.** 785 ft^3
5. 67,914 m^3 **6.** $91,989\frac{1}{3}$ ft^3 **7.** 38.77272 cm^3
8. 1695.6 m^3 **9.** 528 in^3 **10.** $83.7\overline{3}$ mm^3

Calculator Corner, p. 544

1. Left to the student   **2.** Left to the student

Exercise Set 9.4, p. 546

1. 768 cm^3 **3.** 45 in^3 **5.** 75 m^3 **7.** $357\frac{1}{2}$ yd^3
9. 803.84 in^3 **11.** 353.25 cm^3 **13.** 41,580,000 yd^3
15. $4,186,666\frac{2}{3}$ in^3 **17.** 124.72 m^3 **19.** $1950\frac{101}{168}$ ft^3
21. 113,982 ft^3 **23.** 24.64 cm^3 **25.** 4747.68 cm^3
27. 367.38 m^3 **29.** 143.72 cm^3 **31.** 32,993,441,150 mi^3
33. 646.74 cm^3 **35.** 61,600 cm^3 **37.** 5832 yd^3
39. $\mathbf{D_W}$ **41.** 396   **42.** 3   **43.** 14   **44.** 253,440
45. 12   **46.** 335,808   **47.** 24   **48.** 14   **49.** 0.566
50. 13,400   **51.** About 57,480 in^3 **53.** 9.425 L
55. The diameter of the earth at the equator is about
7930 mi. The diameter of the earth between the north and
south poles is about 7917 mi. If we use the average of these
two diameters (7923.5 mi), the volume of the earth is
$\frac{4}{3} \cdot \pi \cdot \left(\frac{7923.5}{2}\right)^3$, or about 260,000,000,000 mi^3.   **57.** 0.331 m^3

Margin Exercises, Section 9.5, pp. 551–556

1. Angle DEF, angle FED, $\angle DEF$, $\angle FED$, or $\angle E$
2. Angle PQR, angle RQP, $\angle PQR$, $\angle RQP$, or $\angle Q$ **3.** 127°
4. 33°   **5.** Right   **6.** Acute   **7.** Obtuse   **8.** Straight
9. $\angle 1$ and $\angle 2$; $\angle 1$ and $\angle 4$; $\angle 2$ and $\angle 3$; $\angle 3$ and $\angle 4$
10. 45°   **11.** 72°   **12.** 5°
13. $\angle 1$ and $\angle 2$; $\angle 1$ and $\angle 4$; $\angle 2$ and $\angle 3$; $\angle 3$ and $\angle 4$
14. 142°   **15.** 23°   **16.** 90°   **17.** (a) $\triangle ABC$;
(b) $\triangle ABC$, $\triangle MPN$;   **(c)** $\triangle DEF$, $\triangle GHI$, $\triangle JKL$, $\triangle QRS$
18. Yes   **19.** No   **20.** (a) $\triangle DEF$; **(b)** $\triangle GHI$, $\triangle QRS$;
(c) $\triangle ABC$, $\triangle MPN$, $\triangle JKL$ **21.** 180°   **22.** 64°

Exercise Set 9.5, p. 557

1. Angle GHI, angle IHG, $\angle GHI$, $\angle IHG$, or $\angle H$ **3.** 10°
5. 180°   **7.** 130°   **9.** Obtuse   **11.** Acute   **13.** Straight
15. Obtuse   **17.** Acute   **19.** Obtuse   **21.** 79°
23. 23°   **25.** 32°   **27.** 61°   **29.** 177°   **31.** 41°
33. 95°   **35.** 78°   **37.** Scalene; obtuse   **39.** Scalene;
right   **41.** Equilateral; acute   **43.** Scalene; obtuse
45. 46°   **47.** 120°   **49.** $\mathbf{D_W}$ **51.** three   **52.** parallel
53. gram   **54.** prime   **55.** angle   **56.** cents, dollars
57. perimeter   **58.** multiplicative, additive
59. $m\angle 2 = 67.13°$; $m\angle 3 = 33.07°$; $m\angle 4 = 79.8°$;
$m\angle 5 = 67.13°$ **61.** $m\angle ACB = 50°$; $m\angle CAB = 40°$;
$m\angle EBC = 50°$; $m\angle EBA = 40°$; $m\angle AEB = 100°$;
$m\angle ADB = 50°$

Margin Exercises, Section 9.6, pp. 560–563

1. 81 **2.** 100 **3.** 121 **4.** 144 **5.** 169 **6.** 196
7. 225 **8.** 256 **9.** 289 **10.** 324 **11.** 400 **12.** 625
13. 3 **14.** 4 **15.** 11 **16.** 10 **17.** 9 **18.** 8
19. 18 **20.** 20 **21.** 15 **22.** 13 **23.** 1 **24.** 0
25. 2.236 **26.** 8.832 **27.** 12.961 **28.** $c = 13$
29. $a = \sqrt{75}$; $a \approx 8.660$ **30.** $b = \sqrt{120}$; $b \approx 10.954$
31. $a = \sqrt{175}$; $a \approx 13.229$ **32.** $\sqrt{424}$ ft ≈ 20.6 ft

Calculator Corner, p. 561

1. 6.6 **2.** 9.7 **3.** 19.8 **4.** 17.3 **5.** 24.9 **6.** 24.5
7. 121.2 **8.** 115.6 **9.** 16.2 **10.** 85.4

Translating for Success, p. 564

1. K **2.** G **3.** B **4.** H **5.** O **6.** M **7.** E
8. A **9.** D **10.** I

Exercise Set 9.6, p. 565

1. 10 **3.** 21 **5.** 25 **7.** 19 **9.** 23 **11.** 100
13. 6.928 **15.** 2.828 **17.** 4.243 **19.** 2.449
21. 3.162 **23.** 8.660 **25.** 14 **27.** 13.528
29. $c = \sqrt{34}$; $c \approx 5.831$ **31.** $c = \sqrt{98}$; $c \approx 9.899$
33. $a = 5$ **35.** $b = 8$ **37.** $c = 26$ **39.** $b = 12$
41. $b = \sqrt{1023}$; $b \approx 31.984$ **43.** $c = 5$
45. $\sqrt{250}$ m ≈ 15.8 m **47.** $\sqrt{8450}$ ft ≈ 91.9 ft
49. $h = \sqrt{500}$ ft ≈ 22.4 ft
51. $\sqrt{211{,}200{,}000}$ ft $\approx 14{,}532.7$ ft **53.** $\sqrt{832}$ ft ≈ 28.8 ft
55. D_W **57.** 1000 **58.** 100 **59.** 100,000 **60.** 10,000
61. $208 **62.** $11.25 **63.** $115.02 **64.** $994,274.04
65. $1,686,435.46 **67.** The areas are the same.
69. Length: 36.6 in.; width: 20.6 in.

Concept Reinforcement, p. 569

1. False **2.** True **3.** True **4.** False **5.** True
6. False

Summary and Review: Chapter 9, p. 570

1. 23 m **2.** 4.4 m **3.** 228 ft; 2808 ft^2 **4.** 36 ft; 81 ft^2
5. 17.6 cm; 12.6 cm^2 **6.** 60 cm^2 **7.** 35 mm^2
8. 22.5 m^2 **9.** 29.64 cm^2 **10.** 88 m^2 **11.** $145\frac{5}{9}$ in^2
12. 840 ft^2 **13.** 8 m **14.** $\frac{14}{11}$ in., or $1\frac{3}{11}$ in. **15.** 14 ft
16. 20 cm **17.** 50.24 m **18.** 8 in. **19.** 200.96 m^2
20. $5\frac{1}{11}$ in^2 **21.** 1038.555 ft^2 **22.** 93.6 m^3
23. 193.2 cm^3 **24.** 31,400 ft^3 **25.** $33.49\overline{3}$ cm^3
26. 4.71 in^3 **27.** 942 cm^3 **28.** 26.28 ft^2; 20.28 ft
29. 54° **30.** 180° **31.** 140° **32.** 90° **33.** Acute
34. Straight **35.** Obtuse **36.** Right **37.** 49°
38. 136° **39.** 60° **40.** Scalene **41.** Right **42.** 8
43. 9.110 **44.** $c = \sqrt{850}$; $c \approx 29.155$
45. $b = \sqrt{51}$; $b \approx 7.141$ **46.** $c = \sqrt{89}$ ft; $c \approx 9.434$ ft
47. $a = \sqrt{76}$ cm; $a \approx 8.718$ cm **48.** About 28.8 ft
49. About 44.7 ft **50.** About 85.9 ft

51. D_W See the volume formulas listed at the beginning of the Summary and Review Exercises for Chapter 9.
52. D_W Volume of two spheres, each with radius r:
$2\left(\frac{4}{3}\pi r^3\right) = \frac{8}{3}\pi r^3$; volume of one sphere with radius $2r$:
$\frac{4}{3}\pi(2r)^3 = \frac{32}{3}\pi r^3$. The volume of the sphere with radius $2r$ is four times the volume of the two spheres, each with radius r: $\frac{32}{3}\pi r^3 = 4 \cdot \frac{8}{3}\pi r^3$. **53.** 100 ft^2
54. 7.83998704 m^2 **55.** 42.05915 cm^2

Test: Chapter 9, p. 574

1. [9.1a], [9.2a] 32.82 cm; 65.894 cm^2
2. [9.1a], [9.2a] $19\frac{1}{2}$ in.; $23\frac{49}{64}$ in^2 **3.** [9.2b] 25 cm^2
4. [9.2b] 12 m^2 **5.** [9.2b] 18 ft^2 **6.** [9.3a] $\frac{1}{4}$ in.
7. [9.3a] 9 cm **8.** [9.3b] $\frac{11}{14}$ in. **9.** [9.3c] 254.34 cm^2
10. [9.3d] 65.46 km; 103.815 km^2 **11.** [9.4a] 84 cm^3
12. [9.4e] 420 in^3 **13.** [9.4b] 1177.5 ft^3
14. [9.4c] $4186.\overline{6}$ yd^3 **15.** [9.4d] 113.04 cm^3
16. [9.5a] 90° **17.** [9.5a] 35° **18.** [9.5a] 180°
19. [9.5a] 113° **20.** [9.5b] Right **21.** [9.5b] Acute
22. [9.5b] Straight **23.** [9.5b] Obtuse **24.** [9.5e] 35°
25. [9.5d] Isosceles **26.** [9.5d] Obtuse
27. [9.5c] Complement: 25°; supplement: 115°
28. [9.6a] 15 **29.** [9.6b] 9.327 **30.** [9.6c] $c = 40$
31. [9.6c] $b = \sqrt{60}$; $b \approx 7.746$ **32.** [9.6c] $c = \sqrt{2}$;
$c \approx 1.414$ **33.** [9.6c] $b = \sqrt{51}$; $b \approx 7.141$
34. [9.6d] About 15.8 m **35.** [8.1a], [9.2a] 2 ft^2
36. [8.1a], [9.2b] 1.875 ft^2 **37.** [8.1a], [9.4a] 0.65 ft^3
38. [8.1a], [9.4d] 0.033 ft^3 **39.** [8.1a], [9.4b] 0.055 ft^3

Cumulative Review: Chapters 1–9, p. 577

1. [4.3b] 1,500,000 **2.** [8.1a], [8.3a] $437\frac{1}{3}$ yd; about 400 m
3. [7.3c] 236 in 1993 and 1995 **4.** [7.3c] 258 in 2000
5. [7.1a, b, c] Average: 243.375; median: 239.5; mode: 236
6. [7.1a] 249.5 **7.** [7.1a] 237.25; the average over the years 1993–1996 is 12.25 lower than the average over the years 1997–2000. **8.** [6.5b] About 8.4%
9. [8.5b] 56 mg **10.** [9.4e] 93,750 lb **11.** [3.5a] $4\frac{1}{6}$
12. [4.5c] 49.2 **13.** [4.2b] 87.52 **14.** [1.6b] 1234
15. [1.9d] 2 **16.** [1.9c] 1565 **17.** [4.1b] $\frac{1209}{1000}$
18. [6.2b] $\frac{17}{100}$ **19.** [3.3b] $<$ **20.** [2.5c] $=$ **21.** [8.4a] $\frac{3}{8}$
22. [8.6b] 59 **23.** [8.5a] 87 **24.** [8.6a] $\frac{3}{20}$ **25.** [8.7a] 27
26. [8.2a] 0.17 **27.** [8.4c] 2.437 **28.** [8.3a] 8870.01
29. [8.5a] 9.51 **30.** [5.3b] $14\frac{2}{5}$ **31.** [5.3b] $4\frac{2}{7}$
32. [4.4b] 113.4 **33.** [3.3c] $\frac{1}{8}$ **34.** [9.1a], [9.2b] 380 cm; 5500 cm^2 **35.** [9.1a], [9.2b] 32.3 ft; 56.55 ft^2
36. [9.3a, c] 70 in.; 3850 in^2
37. [9.4c] $179{,}666\frac{2}{3}$ in^3 **38.** [7.1a] 99
39. [1.8a] 528 million pets **40.** [6.7a] $84
41. [6.7b] $22,376.03 **42.** [9.6d] 17 m **43.** [6.6a] 6%
44. [3.5c] $2\frac{1}{8}$ yd **45.** [4.7a] $37.42
46. [5.2b] The 8-qt box **47.** [2.6b] $\frac{7}{20}$ km **48.** [9.5e] 30°
49. [9.5d] Isosceles **50.** [9.5d] Obtuse
51. [9.3c] Area of a circle of radius 4 ft: $16 \cdot \pi$ ft^2;
[9.2a] Area of a square of side 4 ft: 16 ft^2;
[9.3b] Circumference of a circle of radius 4 ft: $8 \cdot \pi$ ft;
[9.4d] Volume of a cone with radius of base 4 ft and height

8 ft: $\frac{128}{3} \cdot \pi$ ft^3; [9.2a] Area of a rectangle of length 8 ft and width 4 ft: 32 ft^2; [9.2b] Area of a triangle with base 4 ft and height 8 ft: 16 ft^2; [9.4c] Volume of a sphere of radius 4 ft: $\frac{4^4}{3} \cdot \pi$ ft^3; [9.4b] Volume of a right circular cylinder with radius of base 4 ft and height 8 ft: 128 $\cdot \pi$ ft^3; [9.1a] Perimeter of a square of side 4 ft: 16 ft; [9.1a] Perimeter of a rectangle of length 8 ft and width 4 ft: 24 ft   **52.** [8.1a], [9.4e] 272 ft^3 **53.** [8.1a], [9.4b] 94,200 ft^3 **54.** [8.1a], [9.4a] 1.342 ft^3 **55.** [9.4b, c] 1204.260429 ft^3; about 22,079,692.8 ft, or 4181.76 mi

CHAPTER 10

Margin Exercises, Section 10.1, pp. 583–587

1. 8; −5   **2.** 125; −50   **3.** The integer −3 corresponds to the decrease in the stock value.
4. −10; 148   **5.** −120; 50; −80
6.

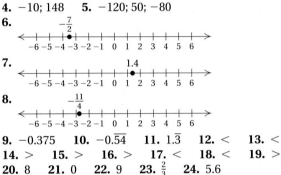

7.

8.

9. −0.375   **10.** −0.5$\overline{4}$ **11.** 1.$\overline{3}$ **12.** <   **13.** <
14. >   **15.** >   **16.** >   **17.** <   **18.** <   **19.** >
20. 8   **21.** 0   **22.** 9   **23.** $\frac{2}{3}$ **24.** 5.6

Exercise Set 10.1, p. 588

1. −34,000,000   **3.** 24; −2   **5.** 950,000,000; −460
7. −34; 15   **9.**

11.

13. −0.875   **15.** 0.8$\overline{3}$ **17.** 1.1$\overline{6}$ **19.** 0.$\overline{6}$ **21.** −0.5
23. −8.28   **25.** >   **27.** <   **29.** <   **31.** <   **33.** >
35. <   **37.** >   **39.** <   **41.** <   **43.** <   **45.** 3
47. 18   **49.** 325   **51.** 3.625   **53.** $\frac{2}{3}$ **55.** $\frac{0}{4}$, or 0
57. $\mathbf{D_W}$ **59.** $2 \cdot 3 \cdot 3 \cdot 3$, or $2 \cdot 3^3$
60. $2 \cdot 2 \cdot 2 \cdot 2 \cdot 2 \cdot 2 \cdot 3$, or $2^6 \cdot 3$ **61.** $2 \cdot 3 \cdot 17$
62. $2 \cdot 2 \cdot 5 \cdot 13$, or $2^2 \cdot 5 \cdot 13$
63. $2 \cdot 2 \cdot 2 \cdot 2 \cdot 3 \cdot 3 \cdot 3$, or $2^5 \cdot 3^3$
64. $2 \cdot 2 \cdot 3 \cdot 3 \cdot 13$, or $2^2 \cdot 3^2 \cdot 13$ **65.** 18   **66.** 72
67. 96   **68.** 72   **69.** 1344   **70.** 252   **71.** >   **73.** =
75. $-100, -8\frac{7}{8}, -8\frac{5}{8}, -\frac{67}{8}, -5, 0, 1^7, |3|, \frac{14}{4}, 4, |-6|, 7$

Margin Exercises, Section 10.2, pp. 590–593

1. −3   **2.** −3   **3.** −5   **4.** 4   **5.** 0   **6.** −2   **7.** −11
8. −12   **9.** 2   **10.** −4   **11.** −2   **12.** 0   **13.** −22
14. 3   **15.** 0.53   **16.** 2.3   **17.** −7.7   **18.** −6.2
19. $-\frac{2}{9}$ **20.** $-\frac{19}{20}$ **21.** −58   **22.** −56   **23.** −14
24. −12   **25.** 4   **26.** −8.7   **27.** 7.74   **28.** $\frac{8}{9}$ **29.** 0

30. −12   **31.** −14; 14   **32.** −1; 1   **33.** 19; −19
34. 1.6; −1.6   **35.** $-\frac{2}{3}; \frac{2}{3}$ **36.** $\frac{9}{8}; -\frac{9}{8}$ **37.** 4   **38.** 13.4
39. 0   **40.** $-\frac{1}{4}$

Calculator Corner, p. 593

1. 2   **2.** 13   **3.** −15   **4.** −8   **5.** −3.3   **6.** −13.4

Exercise Set 10.2, p. 594

1. −7   **3.** −4   **5.** 0   **7.** −8   **9.** −7   **11.** −27
13. 0   **15.** −42   **17.** 0   **19.** 0   **21.** 3   **23.** −9
25. 7   **27.** 0   **29.** 45   **31.** −1.8   **33.** −8.1   **35.** $-\frac{1}{5}$
37. $-\frac{8}{7}$ **39.** $-\frac{3}{8}$ **41.** $-\frac{29}{35}$ **43.** $-\frac{11}{15}$ **45.** −6.3
47. $\frac{7}{16}$ **49.** 39   **51.** 50   **53.** −1093   **55.** −24
57. 26.9   **59.** −9   **61.** $\frac{14}{3}$ **63.** −65   **65.** $\frac{5}{3}$ **67.** 14
69. −10   **71.** $\mathbf{D_W}$ **73.** 357.5 ft^2 **74.** 45.51 ft^2
75. 54,756 mi^2 **76.** 1.4196 mm^2 **77.** 66.97 ft^2
78. 7976.1 m^2 **79.** All positive   **81.** −640,979
83. Negative

Margin Exercises, Section 10.3, pp. 596–598

1. −10   **2.** 3   **3.** −5   **4.** −1   **5.** 2   **6.** −4   **7.** −2
8. −11   **9.** 4   **10.** −2   **11.** −6   **12.** −16   **13.** 7.1
14. 3   **15.** 0   **16.** $\frac{3}{2}$ **17.** Three minus eleven; −8
18. Twelve minus five; 7   **19.** Negative twelve minus negative nine; −3   **20.** Negative twelve point four minus ten point nine; −23.3   **21.** Negative four-fifths minus negative four-fifths; 0   **22.** −9   **23.** $-\frac{6}{5}$ **24.** 12.7
25. 214°F higher

Exercise Set 10.3, p. 599

1. −4   **3.** −7   **5.** −6   **7.** 0   **9.** −4   **11.** −7
13. −6   **15.** 0   **17.** 11   **19.** −14   **21.** 5   **23.** −7
25. −5   **27.** −3   **29.** −23   **31.** −68   **33.** −73
35. 116   **37.** −2.8   **39.** $-\frac{1}{4}$ **41.** $\frac{1}{12}$ **43.** $-\frac{17}{12}$ **45.** $\frac{1}{8}$
47. 19.9   **49.** −9   **51.** −0.01   **53.** −2.7   **55.** −3.53
57. $-\frac{1}{2}$ **59.** $\frac{6}{7}$ **61.** $-\frac{41}{30}$ **63.** $-\frac{1}{156}$ **65.** 37
67. −62   **69.** 6   **71.** 107   **73.** 219   **75.** Down 3 ft
77. −3°   **79.** 20,602 ft   **81.** −$85   **83.** 100°F
85. $\mathbf{D_W}$ **87.** 96.6 cm^2 **88.** $2 \cdot 3 \cdot 5 \cdot 5 \cdot 5$, or $2 \cdot 3 \cdot 5^3$
89. 108   **90.** 125.44 km^2 **91.** 64   **92.** 125
93. 8 cans   **94.** 288 oz   **95.** True   **97.** True
99. True   **101.** True

Margin Exercises, Section 10.4, pp. 602–603

1. 20; 10; 0; −10; −20; −30   **2.** −18   **3.** −100   **4.** −80
5. $-\frac{5}{9}$ **6.** −30.033   **7.** $-\frac{7}{10}$ **8.** −10; 0; 10; 20; 30
9. 12   **10.** 32   **11.** 35   **12.** $\frac{20}{63}$ **13.** $\frac{2}{3}$ **14.** 13.455
15. −30   **16.** 30   **17.** −32   **18.** $-\frac{8}{3}$ **19.** −30
20. −30.75   **21.** $-\frac{5}{3}$ **22.** 120   **23.** −120   **24.** 6

Exercise Set 10.4, p. 604

1. -16 **3.** -24 **5.** -72 **7.** 16 **9.** 42 **11.** -120
13. -238 **15.** 1200 **17.** 98 **19.** -12.4 **21.** 24
23. 21.7 **25.** $-\frac{2}{5}$ **27.** $\frac{1}{12}$ **29.** -17.01 **31.** $-\frac{5}{12}$
33. 420 **35.** $\frac{2}{7}$ **37.** -60 **39.** 150 **41.** $-\frac{2}{45}$
43. 1911 **45.** 50.4 **47.** $\frac{10}{189}$ **49.** -960 **51.** 17.64
53. $-\frac{5}{784}$ **55.** -756 **57.** -720 **59.** $-30{,}240$
61. $\mathbf{D_W}$ **63.** isosceles **64.** acute
65. complementary **66.** simple **67.** unit
68. mode **69.** rectangle **70.** square root
71. **(a)** One must be negative and one must be positive.
(b) Either or both must be zero. **(c)** Both must be negative or both must be positive.

Margin Exercises, Section 10.5, pp. 606–611

1. -2 **2.** 5 **3.** -3 **4.** 9 **5.** -6 **6.** $-\frac{30}{7}$ **7.** Not defined **8.** 0 **9.** $\frac{3}{2}$ **10.** $-\frac{4}{5}$ **11.** $-\frac{1}{3}$ **12.** -5
13. $\frac{1}{5.78}$ **14.** $-\frac{7}{2}$
15.

NUMBER	OPPOSITE	RECIPROCAL
$\frac{2}{3}$	$-\frac{2}{3}$	$\frac{3}{2}$
$-\frac{5}{4}$	$\frac{5}{4}$	$-\frac{4}{5}$
0	0	Undefined
1	-1	1
-4.5	4.5	$-\frac{1}{4.5}$

16. $\frac{4}{5} \cdot \left(-\frac{5}{3}\right)$ **17.** $5 \cdot \left(-\frac{1}{8}\right)$ **18.** $-10 \cdot \left(\frac{1}{7}\right)$ **19.** $-\frac{2}{3} \cdot \frac{7}{4}$
20. $-5 \cdot \left(\frac{1}{7}\right)$ **21.** $-\frac{20}{21}$ **22.** $-\frac{12}{5}$ **23.** $\frac{16}{7}$ **24.** -7
25. $-3.4°F$ per minute **26.** -1237 **27.** 8 **28.** 381
29. -12

Translating for Success, p. 612

1. I **2.** C **3.** F **4.** L **5.** B **6.** O **7.** A **8.** D
9. H **10.** M

Exercise Set 10.5, p. 613

1. -6 **3.** -13 **5.** -2 **7.** 4 **9.** -8 **11.** 2
13. -12 **15.** -8 **17.** Not defined **19.** -9 **21.** $-\frac{7}{15}$
23. $\frac{1}{13}$ **25.** $-\frac{9}{8}$ **27.** $\frac{5}{3}$ **29.** $\frac{9}{14}$ **31.** $\frac{9}{64}$ **33.** -2
35. $\frac{11}{13}$ **37.** -16.2 **39.** Not defined **41.** $-54°C$
43. $\$12.71$ **45.** 32 m below sea level **47.** 44.3%
49. -5.1% **51.** -7 **53.** -7 **55.** -334 **57.** 14
59. 1880 **61.** 12 **63.** 8 **65.** -86 **67.** 37 **69.** -1
71. -10 **73.** -67 **75.** -7988 **77.** -3000 **79.** 60
81. 1 **83.** 10 **85.** $-\frac{13}{45}$ **87.** $-\frac{4}{3}$ **89.** $\mathbf{D_W}$
91. $2 \cdot 2 \cdot 2 \cdot 2 \cdot 2 \cdot 2 \cdot 3 \cdot 5$, or $2^6 \cdot 3 \cdot 5$
92. $5 \cdot 5 \cdot 41$, or $5^2 \cdot 41$ **93.** 665 **94.** 320 **95.** 1710

96. 4%
97.

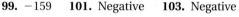

Light whisper	
Normal conversation	
Noisy office	
Thunder	
Moving car	
Chain saw	
Jet takeoff	

Loudness (in decibels)
Sound

99. -159 **101.** Negative **103.** Negative

Concept Reinforcement, p. 616

1. True **2.** False **3.** True **4.** True **5.** False
6. True

Summary and Review: Chapter 10, p. 616

1. $-45; 72$ **2.** 38 **3.** 7.3 **4.** $\frac{5}{2}$ **5.** -0.2 **6.** -1.25
7. $-0.8\overline{3}$ **8.** $-0.41\overline{6}$ **9.** $-0.\overline{27}$
10.

$$-2.5$$
$$\xleftarrow{\hspace{1cm}} \underset{-6\ -5\ -4\ -3\ -2\ -1\ \ 0\ \ 1\ \ 2\ \ 3\ \ 4\ \ 5\ \ 6}{\bullet} \xrightarrow{\hspace{1cm}}$$

11.

$$\frac{8}{9}$$
$$\xleftarrow{\hspace{1cm}} \underset{-6\ -5\ -4\ -3\ -2\ -1\ \ 0\ \ 1\ \ 2\ \ 3\ \ 4\ \ 5\ \ 6}{\bullet} \xrightarrow{\hspace{1cm}}$$

12. $<$ **13.** $>$ **14.** $>$ **15.** $<$ **16.** -3.8 **17.** $\frac{3}{4}$
18. 34 **19.** 5 **20.** $\frac{8}{3}$ **21.** $-\frac{1}{7}$ **22.** -10 **23.** -3
24. $-\frac{7}{12}$ **25.** -4 **26.** -5 **27.** 4 **28.** $-\frac{7}{5}$
29. -7.9 **30.** 54 **31.** -9.18 **32.** $-\frac{2}{7}$ **33.** -210
34. -7 **35.** -3 **36.** $\frac{3}{4}$ **37.** 40.4 **38.** -62 **39.** -5
40. 2 **41.** -180 **42.** 8-yd gain **43.** $-\$130$
44. $\$4.64$ per share **45.** $\$18.95$
46. $\mathbf{D_W}$ Yes; the numbers 1 and -1 are their own reciprocals: $1 \cdot 1 = 1$ and $-1(-1) = 1$.
47. $\mathbf{D_W}$ We know that $a + (-a) = 0$, so the opposite of $-a$ is a. That is, $-(-a) = a$.
48. 403 and 397 **49.** 12 and -7
50. **(a)** $-7 + (-6) + (-5) + (-4) + (-3) + (-2) + (-1) + 0 + 1 + 2 + 3 + 4 + 5 + 6 + 7 + 8 = 8$; **(b)** 0
51. -1; Consider reciprocals and pairs of products of negative numbers.
52. $-\frac{5}{8}$ **53.** -2.1

Test: Chapter 10, p. 619

1. [10.1d] $<$ **2.** [10.1d] $>$ **3.** [10.1d] $>$ **4.** [10.1d] $<$
5. [10.1c] -0.125 **6.** [10.1c] $-0.\overline{4}$ **7.** [10.1c] $-0.\overline{18}$
8. [10.1e] 7 **9.** [10.1e] $\frac{9}{4}$ **10.** [10.1e] -2.7
11. [10.2b] $-\frac{2}{3}$ **12.** [10.2b] 1.4 **13.** [10.2b] 8
14. [10.1b]

$$-0.2$$
$$\xleftarrow{\hspace{1cm}} \underset{-6\ -5\ -4\ -3\ -2\ -1\ \ 0\ \ 1\ \ 2\ \ 3\ \ 4\ \ 5\ \ 6}{\bullet} \xrightarrow{\hspace{1cm}}$$

15. [10.5b] $-\frac{1}{2}$ **16.** [10.5b] $\frac{7}{4}$ **17.** [10.3a] 7.8
18. [10.2a] -8 **19.** [10.2a] $\frac{7}{40}$ **20.** [10.3a] 10
21. [10.3a] -2.5 **22.** [10.3a] $\frac{7}{8}$ **23.** [10.4a] -48

24. [10.4a] $\frac{3}{16}$ **25.** [10.5a] -9 **26.** [10.5c] $\frac{3}{4}$
27. [10.5c] -9.728 **28.** [10.5e] 109 **29.** [10.3b] $14°F$
higher   **30.** [10.3b] Up 15 pts   **31.** [10.5d] 16,080
32. [10.5d] About $0.9429°C$ each minute
33. [10.1e], [10.3a] 15   **34.** [10.3b] 2385 m
35. (a) [10.3a] $-4, -9, -15$; (b) [10.3a] $-2, -6, -10$;
(c) [10.3a] $-18, -24, -31$; (d) [10.5c] $-0.25, 0.125, -0.0625$

CHAPTER 11

Margin Exercises, Section 11.1, pp. 622–628

1. 64   **2.** 28   **3.** $60; -27.2$ **4.** 192 ft^2 **5.** -25
6. 16   **7.**

Value	$1 \cdot x$	x
$x = 3$	3	3
$x = -6$	-6	-6
$x = 4.8$	4.8	4.8

8.

Value	$2x$	$5x$
$x = 2$	4	10
$x = -6$	-12	-30
$x = 4.8$	9.6	24

9.

Value	$5(x + y)$	$5x + 5y$
$x = 6, y = 7$	65	65
$x = -3, y = 4$	5	5
$x = -10, y = 5$	-25	-25

10. $90; 90$ **11.** $64; 64$ **12.** $14; 14$ **13.** $30; 30$
14. $12; 12$ **15.** $5x; -4y; 3$ **16.** $-4y; -2x; 3z$
17. $3x - 15$ **18.** $5x + 5$ **19.** $\frac{5}{4}x - \frac{5}{4}y + 5$
20. $-2x + 6$ **21.** $-5x + 10y - 20z$ **22.** $6(z - 2)$
23. $3(x - 2y + 3)$ **24.** $2(8a - 18b + 21)$
25. $-4(3x - 8y + 4z)$ **26.** $3x$ **27.** $6x$ **28.** $-8x$
29. $0.59x$ **30.** $3x + 3y$ **31.** $-4x - 5y - 7$
32. $\frac{1}{10}x + \frac{7}{9}y - \frac{2}{3}$

Calculator Corner, p. 623

1. 51   **2.** 98   **3.** -6 **4.** -19 **5.** 9   **6.** 12
7. $52; 52$ **8.** $156; 156$ **9.** $228; 228$ **10.** $84; 84$

Exercise Set 11.1, p. 629

1. 42   **3.** 3   **5.** -1 **7.** 6   **9.** -20 **11.** $240; 240$
13. $160; 160$ **15.** $2b + 10$ **17.** $7 - 7t$ **19.** $30x + 12$
21. $7x + 28 + 42y$ **23.** $-7y + 14$ **25.** $45x + 54y - 72$
27. $\frac{3}{4}x - \frac{9}{4}y - \frac{3}{2}z$ **29.** $-3.72x + 9.92y - 3.41$
31. $2(x + 2)$ **33.** $5(6 + y)$ **35.** $7(2x + 3y)$

37. $5(x + 2 + 3y)$ **39.** $8(x - 3)$ **41.** $4(8 - y)$
43. $2(4x + 5y - 11)$ **45.** $6(-3x - 2y + 1)$, or
$-6(3x + 2y - 1)$ **47.** $19a$ **49.** $9a$ **51.** $8x + 9z$
53. $-19a + 88$ **55.** $4t + 6y - 4$ **57.** $8x$ **59.** $5n$
61. $-16y$ **63.** $17a - 12b - 1$ **65.** $4x + 2y$
67. $\frac{39}{20}x + \frac{1}{2}y + 12$ **69.** $0.8x + 0.5y$ **71.** $\mathbf{D_W}$
73. 30 yd; 94.2 yd; 706.5 yd^2 **74.** 16.4 m; 51.496 m;
211.1336 m^2 **75.** 19 mi; 59.66 mi; 283.385 mi^2
76. 4800 cm; 15,072 cm; $18,086,400 \text{ cm}^2$ **77.** 10 mm;
62.8 mm; 314 mm^2 **78.** 132 km; 828.96 km; $54,711.36 \text{ km}^2$
79. 2.3 ft; 14.444 ft; 16.6106 ft^2 **80.** 5.15 m; 32.342 m;
83.28065 m^2 **81.** $q(1 + r + rs + rst)$

Margin Exercises, Section 11.2, pp. 631–632

1. (a) $2 + \boxed{-2} = 0$; (b) 9   **2.** -5 **3.** 13.2 **4.** -6.5

Exercise Set 11.2, p. 633

1. 4   **3.** -20 **5.** -14 **7.** 7   **9.** -26 **11.** -18
13. 15   **15.** -14 **17.** 2   **19.** 20   **21.** -6 **23.** $\frac{7}{3}$
25. $-\frac{7}{4}$ **27.** $\frac{41}{24}$ **29.** $-\frac{1}{20}$ **31.** 5.1   **33.** 12.4   **35.** -5
37. $1\frac{5}{6}$ **39.** $-\frac{10}{21}$ **41.** $\mathbf{D_W}$ **43.** -11 **44.** $-\frac{1}{24}$
45. -34.1 **46.** -1.7 **47.** 5   **48.** $-\frac{31}{24}$ **49.** 5.5
50. 8.1   **51.** 24   **52.** $-\frac{5}{12}$ **53.** 283.14   **54.** -15.68
55. 8   **56.** $-\frac{16}{15}$ **57.** -14.3 **58.** -4.9 **59.** 342.246
61. $-\frac{26}{15}$ **63.** -10 **65.** $-\frac{5}{17}$

Margin Exercises, Section 11.3, pp. 635–637

1. 15   **2.** $-\frac{7}{4}$ **3.** -18 **4.** 10   **5.** $-\frac{4}{5}$ **6.** 7800
7. -3 **8.** 28

Exercise Set 11.3, p. 638

1. 6   **3.** 9   **5.** 12   **7.** -40 **9.** 5   **11.** -7 **13.** -6
15. 6   **17.** -63 **19.** 36   **21.** -21 **23.** $-\frac{3}{5}$ **25.** $-\frac{3}{2}$
27. $\frac{9}{2}$ **29.** 7   **31.** -7 **33.** 8   **35.** 15.9   **37.** -50
39. -14 **41.** $\mathbf{D_W}$ **43.** 62.8 ft; 20 ft; 314 ft^2
44. 75.36 cm; 12 cm; 452.16 cm^2 **45.** 8000 ft^3
46. 31.2 cm^3 **47.** 53.55 cm^2 **48.** 64 mm^2 **49.** 68 in^2
50. 38.25 m^2 **51.** -8655 **53.** No solution
55. No solution

Margin Exercises, Section 11.4, pp. 640–646

1. 5   **2.** 4   **3.** 4   **4.** 39   **5.** $-\frac{3}{2}$ **6.** -4.3 **7.** -3
8. 800   **9.** 1   **10.** 2   **11.** 2   **12.** $\frac{17}{2}$ **13.** $\frac{8}{3}$
14. $-\frac{43}{10}$, or -4.3 **15.** 2   **16.** 3   **17.** -2 **18.** $-\frac{1}{2}$

Calculator Corner, p. 646

1. Left to the student   **2.** Left to the student

Exercise Set 11.4, p. 647

1. 5 **3.** 8 **5.** 10 **7.** 14 **9.** -8 **11.** -8 **13.** -7
15. 15 **17.** 6 **19.** 4 **21.** 6 **23.** -3 **25.** 1
27. -20 **29.** 6 **31.** 7 **33.** 2 **35.** 5 **37.** 2
39. 10 **41.** 4 **43.** 0 **45.** -1 **47.** $-\frac{4}{3}$ **49.** $\frac{2}{5}$
51. -2 **53.** -4 **55.** $\frac{4}{5}$ **57.** $-\frac{28}{27}$ **59.** 6 **61.** 2
63. 6 **65.** 8 **67.** 1 **69.** 17 **71.** $-\frac{5}{3}$ **73.** -3
75. 2 **77.** $-\frac{51}{31}$ **79.** 2 **81.** $\mathbf{D_W}$ **83.** rational, zero
84. is greater than **85.** one **86.** absolute value
87. obtuse **88.** equilateral **89.** meter **90.** factor
91. 2 **93.** -2 **95.** 8 **97.** 2 cm

Margin Exercises, Section 11.5, pp. 652–659

1. $x - 12$ **2.** $y + 12$, or $12 + y$ **3.** $m - 4$
4. $\frac{1}{2} \cdot p$ or $\frac{p}{2}$ **5.** $6 + 8x$, or $8x + 6$ **6.** $a - b$
7. $59\%x$, or $0.59x$ **8.** $xy - 200$ **9.** $p + q$ **10.** $62\frac{2}{3}$ mi
11. Jenny: 8 in.; Emma: 4 in.; Sarah: 6 in. **12.** 587.8 mi
13. $30°, 90°, 60°$ **14.** Second: 80; third: 89 **15.** $8400

Translating for Success, p. 660

1. B **2.** H **3.** G **4.** N **5.** J **6.** C **7.** L
8. E **9.** F **10.** D

Exercise Set 11.5, p. 661

1. $2x - 3$ **3.** $97\%y$, or $0.97y$ **5.** $5x + 4$, or $4 + 5x$
7. $\frac{1}{3}x$, or $240 - x$ **9.** 32 **11.** 305 ft **13.** $4.29
15. 57 **17.** -12 **19.** $699\frac{1}{3}$ mi **21.** First: 30 m;
second: 90 m; third: 360 m **23.** Length: 94 ft; width: 50 ft
25. Length: 265 ft; width: 165 ft; area: 43,725 ft^2
27. About 412.6 mi **29.** First: 22.5°; second: 90°;
third: 67.5° **31.** $350 **33.** $852.94 **35.** 12 mi
37. $36 **39.** Length: 48 ft; width: 14 ft **41.** $\mathbf{D_W}$
43. $-\frac{47}{40}$ **44.** $-\frac{17}{40}$ **45.** $-\frac{3}{10}$ **46.** $-\frac{32}{15}$ **47.** 1.6
48. 409.6 **49.** -9.6 **50.** -41.6
51. $1863 = 1776 + (4s + 7)$ **53.** Length: 12 cm;
width: 9 cm **55.** Half dollars: 5; quarters: 10; dimes: 20;
nickels: 60

Concept Reinforcement, p. 666

1. False **2.** True **3.** True **4.** False **5.** True

Summary and Review: Chapter 11, p. 666

1. 4 **2.** $15x - 35$ **3.** $-8x + 10$ **4.** $4x + 15$
5. $-24 + 48x$ **6.** $2(x - 7)$ **7.** $6(x - 1)$ **8.** $5(x + 2)$
9. $3(4 - x)$ **10.** $7a - 3b$ **11.** $-2x + 5y$ **12.** $5x - y$
13. $-a + 8b$ **14.** -22 **15.** 7 **16.** -192 **17.** 1
18. $-\frac{7}{3}$ **19.** 25 **20.** $\frac{1}{2}$ **21.** $-\frac{15}{64}$ **22.** 9.99 **23.** -8
24. -5 **25.** $-\frac{1}{3}$ **26.** 4 **27.** 3 **28.** 4 **29.** 16
30. 6 **31.** -3 **32.** -2 **33.** 4 **34.** $19\%x$, or $0.19x$
35. Length: 365 mi; width: 275 mi **36.** $2117

37. 27 appliances **38.** $35°, 85°, 60°$ **39.** $220
40. $75,000 **41.** $138.95 **42.** Amazon: 6437 km;
Nile: 6671 km **43.** $791 **44.** Width: 11 cm;
length: 17 cm **45.** $\mathbf{D_W}$ The distributive laws are used to
multiply, factor, and collect like terms in this chapter.
46. $\mathbf{D_W}$ **(a)** $4 - 3x = 9$
$$3x = 9 \quad (1)$$
$$x = 3 \quad (2)$$

    1. 4 was subtracted on the left side but not on
the right side. Also, the minus sign preceding
$3x$ has been dropped.
    2. This step would give the correct result if the
preceding step were correct.
    The correct steps are
$$4 - 3x = 9$$
$$-3x = 5 \quad (1)$$
$$x = -\tfrac{5}{3}. \quad (2)$$

    (b) $2(x - 5) = 7$
$$2x - 5 = 7 \quad (1)$$
$$2x = 12 \quad (2)$$
$$x = 6 \quad (3)$$

    1. When a distributive law was used to remove
parentheses, x was multiplied by 2 but 5
was not.
    2. and 3. These steps would give the correct
result if the preceding step were correct.
    The correct steps are
$$2(x - 5) = 7$$
$$2x - 10 = 7 \quad (1)$$
$$2x = 17 \quad (2)$$
$$x = \tfrac{17}{2}. \quad (3)$$

47. $23, -23$ **48.** $20, -20$

Test: Chapter 11, p. 669

1. [11.1a] 6 **2.** [11.1b] $18 - 3x$ **3.** [11.1b] $-5y + 5$
4. [11.1c] $2(6 - 11x)$ **5.** [11.1c] $7(x + 3 + 2y)$
6. [11.1d] $-5x - y$ **7.** [11.1d] $4a + 5b$ **8.** [11.2a] 8
9. [11.2a] 26 **10.** [11.3a] -6 **11.** [11.3a] 49
12. [11.4b] -12 **13.** [11.4a] 2 **14.** [11.4a] -8
15. [11.2a] $-\frac{7}{20}$ **16.** [11.4b] 2.5 **17.** [11.4c] 7
18. [11.4c] $\frac{5}{3}$ **19.** [11.5a] $x - 9$ **20.** [11.5b] Width: 7 cm;
length: 11 cm **21.** [11.5b] About $240.7 billion
22. [11.5b] 3 m, 5 m **23.** [11.5b] $880 **24.** [11.5b] 6
25. [11.5b] 25.625° **26.** [10.1e], [11.4a] $15, -15$
27. [11.5b] 60 tickets

Cumulative Review/Final Examination:
Chapters 1–11, p. 671

1. [5.2b] Unit prices: 7.647¢/oz, 7.133¢/oz, 7.016¢/oz,
5.828¢/oz, 4.552¢/oz; brand E has the lowest unit price.
2. [4.1a] Seven and four hundred sixty-three thousandths
3. [9.3b], [9.4c] **(a)** 4400 mi; **(b)** about 1,437,333,333 mi^3
4. [6.5a] $13,444 **5.** [6.5a] 15% **6.** [6.5a] $2151.04
7. [6.5a] 30% **8.** [6.5a] $537.76 **9.** [1.1a] 7

10. [1.1b] 7 thousands + 4 hundreds + 5 ones
11. [1.2a] 1012   **12.** [1.2a] 21,085   **13.** [3.2a] $\frac{5}{26}$
14. [3.5a] $5\frac{7}{9}$ **15.** [4.2a] 493.971   **16.** [4.2a] 802.876
17. [1.3b] 152   **18.** [1.3b] 674   **19.** [3.3a] $\frac{5}{24}$
20. [3.5b] $2\frac{17}{24}$ **21.** [4.2b] 19.9973   **22.** [4.2b] 34.241
23. [1.5a] 4752   **24.** [1.5a] 266,287   **25.** [3.6a] $4\frac{1}{12}$
26. [2.6a] $\frac{6}{5}$ **27.** [2.4a] 10   **28.** [4.3a] 259.084
29. [1.6b] 573   **30.** [1.6b] 56 R 10   **31.** [3.4b] $56\frac{5}{17}$
32. [2.7b] $\frac{3}{2}$ **33.** [3.6b] $\frac{7}{90}$ **34.** [4.4a] 39   **35.** [1.4a] 68,000
36. [4.1d] 0.428   **37.** [4.5b] 21.84   **38.** [2.2a] Yes
39. [2.1a] 1, 3, 5, 15   **40.** [3.1a] 800   **41.** [2.5b] $\frac{7}{10}$
42. [2.5b] 55   **43.** [3.4a] $3\frac{3}{5}$ **44.** [2.5c] $\neq$ **45.** [3.3b] $<$
46. [4.1c] 1.001   **47.** [1.4c] $>$ **48.** [2.3a] $\frac{3}{5}$
49. [4.1b] 0.037   **50.** [4.5a] 0.52   **51.** [4.5a] $0.\overline{8}$
52. [6.1b] 0.07   **53.** [4.1b] $\frac{463}{100}$ **54.** [3.4a] $\frac{29}{4}$ **55.** [6.2b] $\frac{2}{5}$
56. [6.2a] 85%   **57.** [6.1b] 150%   **58.** [1.7b] 555
59. [4.4b] 64   **60.** [2.7c] $\frac{5}{4}$ **61.** [5.3b] $76\frac{1}{2}$, or 76.5
62. [7.4b]

Number of Rounds of Golf Per Year

25 – 49 25.3%
8 – 24 14.9%
1 – 7 2.2%
100 or more 18.6%
50 – 99 39.0%

63. [7.3b]

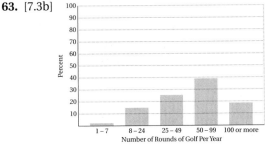

Number of Rounds of Golf Per Year

64. [9.5e] 118°   **65.** [9.5d] Obtuse   **66.** [1.8a] $675
67. [1.8a] 65 min   **68.** [4.7a] $25.75   **69.** [4.7a] 485.9 mi
70. [1.8a] $8100   **71.** [1.8a] $783   **72.** [2.4c] $\frac{3}{10}$ km
73. [4.7a] $84.96   **74.** [5.4a] 13 gal   **75.** [6.7a] $150
76. [6.6b] 7%   **77.** [6.5b] 30,160   **78.** [7.1a, b, c] 28;
26; 18   **79.** [1.9b] 324   **80.** [1.9b] 400   **81.** [9.6a] 3
82. [9.6a] 11   **83.** [9.6b] 4.472   **84.** [8.1a] 12
85. [8.2a] 428   **86.** [8.6a] 72   **87.** [8.4b] 20
88. [8.4a] 80   **89.** [8.4b] 0.08   **90.** [8.5a] 8.19
91. [8.5a] 5   **92.** [9.6c] $c = \sqrt{50}$ ft; $c \approx 7.071$ ft
93. [9.1a], [9.2a] 25.6 m; 25.75 m²   **94.** [9.2b] 25 in²
95. [9.2b] 128.65 yd²   **96.** [9.2b] 61.6 cm²
97. [9.3a, b, c] 20.8 in.; 65.312 in.; 339.6224 in²
98. [9.4a] 52.9 m³   **99.** [9.4b] 803.84 ft³
100. [9.4d] 267.94$\overline{6}$ cm³   **101.** [1.9c] 238   **102.** [1.9c] 172
103. [10.1e], [10.4a] 3   **104.** [10.2a] 14   **105.** [10.3a] $\frac{1}{3}$
106. [10.4a] 30   **107.** [10.4a] $-\frac{2}{7}$ **108.** [10.5a] -8
109. [11.4a] -5 **110.** [11.3a] 4   **111.** [11.4b] -8
112. [11.4c] $\frac{2}{3}$ **113.** [11.5a] $y + 17$, or $17 + y$
114. [11.5a] 38%x, or 0.38x **115.** [11.5b] $45
116. [11.5b] $1050   **117.** [11.5a] 50 m; 53 m; 40 m
118. [11.4b] 0   **119.** [11.4b] 2   **120.** [11.1d] (c)
121. [11.1c] (e)   **122.** [10.5c] (d)   **123.** [10.2a] (a)

Glossary

A

Absolute value   The distance that a number is from 0 on the number line

Acute angle   An angle whose measure is greater than $0°$ and less than $90°$

Acute triangle   A triangle in which all three angles are acute

Addend   In addition, a number being added

Additive identity   The number 0

Additive inverse   A number's opposite; two numbers are additive inverses of each other if their sum is 0.

Algebraic expression   An expression consisting of variables, constants, numerals, and operation signs

Angle   A set of points consisting of two rays (half-lines) with a common endpoint (vertex)

Area   The number of square units that fill a plane region

Arithmetic mean   A center point of a set of numbers found by adding the numbers and dividing by the number of items of data; also called mean or average

Arithmetic numbers   The whole numbers and the positive fractions

Associative law of addition   The statement that when three numbers are added, regrouping the addends gives the same sum

Associative law of multiplication   The statement that when three numbers are multiplied, regrouping the factors gives the same product

Average   A center point of a set of numbers found by adding the numbers and dividing by the number of items of data; also called the arithmetic mean or mean

B

Bar graph   A graphic display of data using bars proportional in length to the numbers represented

Base   In exponential notation, the number being raised to a power

C

Celsius   A temperature scale for metric measure

Circle graph   A graphic means of displaying data using sectors of a circle; often used to show the percent of a quantity in different categories or to show visually the ratio of one category to another; also called a pie chart

Circumference   The distance around a circle

Coefficient   The numerical multiplier of a variable

Commutative law of addition   The statement that when two numbers are added, changing the order in which the numbers are added does not affect the sum

Commutative law of multiplication   The statement that when two numbers are multiplied, changing the order in which the numbers are multiplied does not affect the product

Commission   A percent of total sales paid to a salesperson

Complementary angles   Two angles for which the sum of their measures is $90°$

Composite number   A natural number, other than 1, that is not prime

Compound interest   Interest paid on interest

Constant   A known number

Cross products   Given an equation with a single fraction on each side, the products formed by multiplying the left numerator and the right denominator, and the left denominator and the right numerator

D

Decimal notation   A representation of a number containing a decimal point

Denominator   The number below the fraction bar in a fraction

Diameter   A segment that passes through the center of a circle and has its endpoints on the circle

Difference   The result of subtracting one number from another

Digit   A number 0, 1, 2, 3, 4, 5, 6, 7, 8, or 9 that names a place-value location

Discount   The amount subtracted from the original price of an item to find the sales price

Distributive law of multiplication over addition   The statement that multiplying a factor by the sum of two numbers gives the same result as multiplying the factor by each of the two numbers and then adding

Distributive law of multiplication over subtraction   The statement that multiplying a factor by the difference of two numbers gives the same result as multiplying the factor by each of the two numbers and then subtracting

Dividend   In division, the number being divided

Divisible The number a is divisible by another number b if there exists a number c such that $a = b \cdot c$.

Divisor In division, the number dividing another number

E

Equation A number sentence that says that the expressions on either side of the equals sign, $=$, represent the same number

Equilateral triangle A triangle in which all sides are the same length

Equivalent equations Equations with the same solution set

Equivalent expressions Expressions that have the same value for all allowable replacements

Even number A number that is divisible by 2; that is, it has an even ones digit

Exponential notation A representation of a number using a base raised to a power

Exponent In expressions of the form a^n, the number n is an exponent. For n a natural number, a^n represents n factors of a.

F

Factor *Verb:* to write an equivalent expression that is a product. *Noun:* a multiplier

Factorization A number expressed as a product of natural numbers

Fahrenheit A temperature scale for American measure

Fraction notation A number written using a numerator and a denominator

H

Hypotenuse In a right triangle, the side opposite the right angle

I

Identity property of 1 The statement that the product of a number and 1 is always the original number

Identity property of 0 The statement that the sum of a number and 0 is always the original number

Inequality A mathematical sentence using $<, >, \leq, \geq$, or $\neq$.

Integers The whole numbers and their opposites

Interest A percentage of an amount invested or borrowed

Irrational number A real number that cannot be named as a ratio of two integers

Isosceles triangle A triangle in which two or more sides are the same length

L

Least common denominator (LCD) The least common multiple of the denominators of two or more fractions

Least common multiple (LCM) The smallest number that is a multiple of two or more numbers

Legs In a right triangle, the two sides that form the right angle

Like terms Terms that have exactly the same variable factors

Line graph A graphic means of displaying data by connecting adjacent data points with line segments

M

Marked price The original price of an item

Mean A center point of a set of numbers found by adding the numbers and dividing by the number of items of data; also called the arithmetic mean or average

Median In a set of data listed in order from smallest to largest, the middle number if there is an odd number of data items, or the average of the two middle numbers if there is an even number of data items

Minuend The number from which another number is being subtracted

Mixed numeral A number represented by a whole number and a fraction less than 1

Mode The number or numbers that occur most often in a set of data

Multiple of a number A product of the number and some natural number

Multiplicative identity The number 1

Multiplicative inverses Reciprocals; two numbers whose product is 1

N

Natural numbers The counting numbers: 1, 2, 3, 4, 5, …

Numerator The number above the fraction bar in a fraction

O

Obtuse angle An angle whose measure is greater than $90°$ and less than $180°$

Obtuse triangle A triangle in which one angle is an obtuse angle

Opposite The opposite, or additive inverse, of a number a is written $-a$. Opposites are the same distance from 0 on the number line but on different sides of 0.

Original price The price of an item before a discount is deducted

P

Parallelogram A four-sided polygon with two pairs of parallel sides

Percent notation A representation of a number as parts per 100

Perimeter The sum of the lengths of the sides of a polygon

Pi (π) The number that results when the circumference of a circle is divided by its diameter; $\pi \approx 3.14$, or $22/7$

Pictograph A graphic means of displaying data using pictorial symbols

Pie chart   A graphic means of displaying data using sectors of a circle; often used to show the percent of a quantity used in different categories or to show visually the ratio of one category to another; also called a circle graph

Polygon   A closed geometric figure with three or more sides

Prime factorization   A factorization of a composite number as a product of prime numbers

Prime number   A natural number that has exactly two different factors: itself and 1

Protractor   A device used to measure angles

Purchase price   The price of an item before sales tax is added

Pythagorean equation   The equation $a^2 + b^2 = c^2$, where a and b are the lengths of the legs of a right triangle and c is the length of the hypotenuse

Q

Quotient   The result when one number is divided by another

R

Radical sign   The symbol $\sqrt{}$

Radius   A segment with one endpoint on the center of a circle and the other endpoint on the circle

Rate   A ratio used to compare two different kinds of measure

Ratio   The quotient of two quantities

Rational number   A number that can be written in the form a/b, where a and b are integers and $b \neq 0$

Real numbers   All rational and irrational numbers

Reciprocal   A multiplicative inverse; two numbers are reciprocals if their product is 1

Rectangle   A four-sided polygon with four right angles

Right angle   An angle whose measure is 90°

Right triangle   A triangle that includes a right angle

Rounding   Approximating the value of a number; used when estimating

S

Sale price   The price of an item after a discount has been deducted

Sales tax   A tax added to the purchase price of an item

Scalene triangle   A triangle in which each side is a different length

Similar triangles   Triangles in which corresponding sides are proportional

Simple interest   A percentage of an amount P invested or borrowed for t years, computed by calculating principal × interest rate × time

Simplify   To rewrite an expression in an equivalent, abbreviated form

Solution of an equation   A replacement or substitution that makes an equation true

Sphere   The set of all points in space that are a given distance from a given point

Square   A four-sided polygon with four right angles and all sides of equal length

Square of a number   A number multiplied by itself

Square root of a number   A number that when multiplied by itself yields the given original number

Straight angle   An angle whose measure is 180°

Substitute   To replace a variable with a number

Subtrahend   In subtraction, the number being subtracted

Sum   The result in addition

Supplementary angles   Two angles for which the sum of their measures is 180°

T

Term   A number, a variable, or a product or a quotient of numbers and/or variables

Terminating decimal   A decimal that can be written using a finite number of decimal places

Total price   The sum of the purchase price of an item and the sales tax on the item

Trapezoid   A four-sided polygon with two parallel sides

Triangle   A three-sided polygon

U

Unit price   The ratio of price to the number of units; also called unit rate

Unit rate   The ratio of price to the number of units; also called unit price

V

Variable   A letter that represents an unknown number

Vertex   The common endpoint of the two rays that form an angle

W

Whole numbers   The natural numbers and 0: 0, 1, 2, 3, …

Photo Credits

xxxi, Michael St.–Andre **6,** © Ed Quinn, CORBIS **7,** Tony Stone/Getty **8 (left),** © Jeff Greenberg/The Image Works; **(right),** © CLARO CORTES IV/Reuters/Corbis **36,** © Kin Cheung/Reuters/Corbis **38,** © Michael S. Yamshita/CORBIS **69,** © Pete Saloutos/CORBIS **72,** CORBIS **84,** Tony Stone/Getty **104 (top),** © Keith Dannemiller/CORBIS; **(bottom),** CORBIS **107 (Kidman),** © Rufus F. Folkks/CORBIS; **(Berry),** © Eric Robert/CORBIS SYGMA; **(Dole),** © Reuters/CORBIS; **(Swank),** © Christy Bowe/CORBIS; **(Paltrow),** © Dusko Despotovic/Corbis; **(Roberts),** © Reuters/CORBIS; **(Feinstein),** © Roger Ressmeyer/CORBIS **112,** © Ariel Skelley/ Corbis Stock Market **113,** © LWA-JDC/CORBIS **119,** © Laura Dwight/CORBIS **128,** The Image Bank **138,** © Bill Varie/CORBIS **146 (left),** Greg Probst/Stone/ Getty; **(right),** © Craig Aurness/CORBIS **163,** photodisc **164,** © Ted Horowitz/ CORBIS **171,** © Royalty-Free/Corbis **177,** photodisc **182,** J. P. Jenkins **183,** © Joseph Sohm; ChromoSohm Inc./CORBIS **184,** John Deere **186,** © Anthony Redpath/CORBIS **197 (left),** © Ted Spiegel/CORBIS; **(right),** © Royalty-Free/Corbis **199,** photodisc **202,** © Annie Griffiths Belt/CORBIS **211,** CORBIS **214 (left),** photodisc; **(right),** Karen Bittinger **219 (left),** Plush Studios/Getty; **(right),** © Rick Rickman/NewSport **225 (left),** Digital Vision/Getty; **(right),** © Sion Touhig/Corbis **239,** Bettman/CORBIS **257,** Mark Gibson, Digital Vision/Getty **261,** Chuck Savage/ CORBIS **262, 265, 268,** © Reuters/CORBIS **276,** Think-stock/Getty **285,** © Tom Nebbia/Corbis Stock Market **286,** © Royalty-Free/Corbis **292,** David Muench, CORBIS **293,** © Matt A. Brown/NewSport **294,** apwire photos **296 (left),** © Royalty-Free/Corbis; **(right),** Albert J. Copley, photodisc **297,** © Reuters/CORBIS **300 (top),** Eyewire Collection; **(bottom),** © Reuters/CORBIS **301,** Dick Morton **303 (top),** © JOE GIZA/Reuters/Corbis; **(left),** © Royalty-Free/Corbis; **(right),** © Joe McDonald/CORBIS **304,** Jeremy Woodhouse, Photodisc **306 (top),** © Reuters/ CORBIS; **(bottom),** © Bil Janscha/Reuters/Corbis **314,** Brian Spurlock **315,** Buccina Studios, photodisc **318,** Nicholas Devore III/Photographers Aspen **321 (left),** Barry Rosenthal, FPG International; **(right),** Keith Brofsky, photodisc **322 (left),** © Rob Lewine/CORBIS; **(right),** © Kevin Fleming, Corbis **323 (left),** © Royalty-Free/Corbis; **(right),** © Chase Jarvis/CORBIS **333,** Karen Bittinger **334,** Dick Morton **335 (top),** Patrick Clark, photodisc; **(bottom),** Phil Schermeister, Corbis **339,** © Carl & Ann Purcell/CORBIS **342,** photodisc **343,** © Turbo/zefa/ CORBIS **347, 348,** photodisc **350,** © Guzelian Asadour/CORBIS **359, 360, 361, 365, 372,** © Royalty-Free/CORBIS **376,** Digital Vision **379,** photodisc **380,** © David H. Wells/CORBIS **391,** Christopher Bissell, Stone/Getty Images **401,** photodisc **403, 404, 413,** © Royalty-Free/CORBIS **418,** Hulton Archive/Getty Images **419,** Reportage/Getty Images **423,** Tom Stratmen **425,** © Royalty-Free/ Corbis **429,** Cincinnati Zoo & Botanical Garden, Photo, S. David Jenike **430,** Stone/ Getty Images **448,** © Larry Williams/CORBIS **462,** © Reuters/CORBIS **464,** FPG International **480,** © Owen Franken/CORBIS **488,** © Tom and Dee Ann McCarthy/ Corbis Stock Market **489,** Digital Vision/photodisc **491,** David Young-Wolff/ PhotoEdit **492,** © Richard T. Nowitz/CORBIS **498,** © Jonathan Blair/CORBIS **503,** © Galen Rowell/Corbis **509,** © Jeremy Woodhouse/photodisc **512,** © Warren Morgan/CORBIS **519,** © Michael Prince/CORBIS **538,** © Nancy Dudley/Stock Boston **548,** © Richard Cummins/CORBIS **549,** Scott T. Baxer/photodisc **550,** © 1988 Al Satterwhite **577 (top),** © National Geographic Image Collection; **(bottom),** photodisc **588,** © Buddy Mays/CORBIS **600,** © Douglas Peebles/ CORBIS **618,** Comstock Images **653,** Appalachian Trail Conference **662 (top),** EyeWire Collection; **(bottom left),** Kevin Horan, Tony Stone Images; **(bottom right),** © Ed Bock/CORBIS **668,** © Warren Morgan/CORBIS **676,** © ML Sinibaldi/The Stock Market

Index

H

Hectogram, 485
Hectometer, 472
Height of a parallelogram, 522
Home loans, 404
Hour, 499
Hundred, 239
Hundred-thousandths, 218
Hundredths, 218
   dividing by, 248
   multiplying by, 238
Hypotenuse, 561

I

Identity
   additive, 9
   multiplicative, 32
Identity property
   of one, 624
   of zero, 591
Improper fraction, 106, 176
Inch, 466
Increase, percent of, 374, 376
Inequality, 26
Inequality symbols, 26
Integers, 582
   division of, 606
Interest
   compound, 395–397, 410
   rate, 394
   simple, 394, 410
Inverse
   additive, 592
   multiplicative, 635
Irrational numbers, 585, 586
Isosceles triangle, 556

K

Karat, 492
Key for a pictograph, 429
Kilogram, 485
Kilometer, 472

L

Large numbers, naming, 239
Laws
   associative, 9, 34
   commutative, 9, 34
   distributive, 33, 625, 626, 666
Leap year, 499
Least common denominators, 159
Least common multiple (LCM), 150
   of denominators, 159, 643
Legs of a right triangle, 561
Length, *see* American system of
     measures; Metric system of
     measures

Less than ($<$), 26, 586
Liebniz, Gottfried Wilhelm von, 31
Like terms, 628
Line graph, 417, 438
Linear measure, see American system
     of measures; Metric system of
     measures
List of multiples, 150, 151
Liter, 493, 494
Loans, 403–406

M

Marked price, 388
Mass, 484, 487, 510. *See also* Weight.
Mean, 418
Measures of central tendency, 417,
     421, 422
Measures, *see* American system of
     measures; Metric system of
     measures
Measuring angles, 551–553
Median, 421
Medical applications, 487, 495
Meter, 472
Metric system of measures, 472
   area, 505
   capacity, 493, 494, 510
   changing units, 475, 486, 494, 500,
     501, 505, 506
     to American system, 480
     mentally, 476, 486, 494, 506
   length, 472, 510
   mass, 484, 487, 510
   temperature, 500
Metric ton, 485
Microgram, 487
Mile, 466
Milligram, 485
Milliliter, 494
Millimeter, 472
Million, 239
Millions period, 2
Minimum payment, 401
Minuend, 15
Minute, 499
Missing addend, 16
Mixed numerals, 173
   and addition, 180, 194
   converting from/to fraction
     notation, 173, 174
   converting to decimal notation, 222
   and division, 191, 194
   estimation with, 204
   and multiplication, 190, 194
   order of operations, 201
   as quotients, 175, 176
   and subtraction, 181, 194
Mode, 422
Multiples, 91
   and divisibility, 92

   least common, 150
   list of, 150, 151
Multiples method, 151, 154
Multiplication
   of algebraic expressions, 626
   associative law of, 34
   on a calculator, 35
   commutative law of, 34
   with decimal notation, 236–239
   using the distributive laws, 33, 626
   estimating products, 35, 265
   factors, 31
   of a fraction by a whole number,
     115
   using fraction notation, 115, 116,
     129, 194
   by a hundredth, 238
   using mixed numerals, 190, 194
   by one, 122, 159, 467, 476, 484, 493,
     499, 505
   by a power of ten, 238
   product, 31
   of real numbers, 602, 603
   as a rectangular array, 31
   related division sentence, 41
   as repeated addition, 31
   and simplifying, 129
   by a tenth, 238
   by a thousandth, 238
   of whole numbers, 31, 35
     and fractions, 115
Multiplication principle, 635, 666
Multiplicative identity, 32, 122
Multiplicative inverse, 635. *See also*
     Reciprocal.
Multiplying, *see* Multiplication

N

Naming angles, 551
Naming large numbers, 239
Natural numbers, 3, 90
   composite, 93
   divisibility, 92. *See also* Divisibility.
   factoring, 91, 93
   least common multiple, 150
   multiples, 91
   prime, 93
Negative integers, 582
Nine, divisibility by, 100
Nonnegative rational numbers, 218
Not equal to ($\neq$), 126
Notation
   decimal, 218
   expanded, 3
   exponential, 75
   fraction, 104
   percent, 342, 343
   ratio, 292
   for rational numbers, 584
   standard, for numbers, 3

Number line, addition on, 590
Numbers
  arithmetic, 218
  composite, 93
  digits, 2
  equivalent, 122
  even, 98
  expanded notation, 3
  graphing, 584
  integers, 582
  irrational, 585
  natural, 3, 90
  nonnegative rational, 218
  palindrome, 146
  periods, 2, 4
  place-value, 2
  prime, 93
  ratio, 292
  rational, 583, 584
    nonnegative, 218
  real, 586
  signs of, 593
  standard notation, 3
  whole, 3
  word names, 4
Numerals, mixed, 173. *See also* Mixed
    numerals.
Numerator, 104

O

Obtuse angle, 553
Obtuse triangle, 556
One
  division by, 42
  fraction notation for, 108
  identity property, 624
  multiplying by, 467, 476, 484, 493,
    499, 505
Ones period, 2
Operations, order of, 76, 78, 201, 249
Opposite, 582, 592, 593
  and subtraction, 597
Order
  in decimal notation, 223
  in fraction notation, 167
  of operations, 76, 78, 201, 249, 610
  of real numbers, 586
  of whole numbers, 26
Original price, 388
Ounce, 493

P

Palindrome prime, 146
Paragraph presentation of data, 416
Parallelogram, 522
  area, 523, 569
Parentheses, 76
  equations containing, 645
  in multiplication, 31

  within parentheses, 79
Percent, *see* Percent notation
Percent equation, 358
Percent equivalent, 352
Percent notation, 342, 343
  converting
    from/to decimal notation,
      343–345
    from/to fraction notation,
      349–352
  in pie charts, 342, 448
  solving problems involving,
    359–362, 365–368, 371–377
Percent of decrease, 374, 375
Percent of increase, 374, 376
Perimeter, 11, 516
  rectangle, 517, 569
  square, 517, 569
Periods, 2, 4
Pi (π), 533
  on a calculator, 544
Pictographs, 416, 429
Picture graphs, 416, 429
Pie chart, 448
Pint, 493
Pitch of a screw, 132
Pixels, 70
Place value, 2, 218
Polygon, 516
Positive integers, 582
Pound, 484
Power, 75
Power of ten
  dividing by, 247, 248
  multiplying by, 238
Price
  marked, 388
  original, 388
  purchase, 385
  sale, 388
  total, 385
  unit, 300
Price-earnings ratio, 353
Prime factorization, 93
  in finding LCMs, 151, 152
  in simplifying fractions, 124
Prime numbers, 93
  division by, 155
  palindrome, 146
Principal, 394
Principles
  addition, 631, 666
  multiplication, 635, 666
Problem solving, 58, 653. *See also*
    Index of Applications.
Products, 31. *See also* Multiplication.
  cross, 307, 308
  estimating, 35, 265
Proportional, 306. *See also*
    Proportions.
Proportions, 306

  and geometric shapes, 325, 327
  and similar triangles, 325
  solving, 307, 308, 310
  used in solving percent problems,
    365
Protractor, 552
Purchase price, 385
Pythagoras, 562
Pythagorean equation, 561, 569
Pythagorean theorem, 561

Q

Quadrillion, 243
Quality points, 419
Quart, 493
Quintillion, 243
Quotient, 40, 41
  estimating, 265
  as a mixed numeral, 175
  zeros in, 46

R

Radical sign ($\sqrt{}$), 560
Radius, 532, 569
Rate, 299
  commission, 387
  of discount, 388
  interest, 394
  sales tax, 385
  unit, 300
Ratio, 107, 292
  and circle graphs, 448
  notation, 292
  percent as, 343
  price-earnings, 353
  and proportion, 309
  as a rate, 299
  simplifying, 294
Rational numbers, 583, 584
  and decimal notation, 585
  and fraction notation, 584
  graphing, 584
  nonnegative, 218
Rays, 551
Real-number system, 586
  and absolute value, 587
  and addition, 590, 591
  additive inverses, 592
  and division, 606–609
  integers, 582
  and multiplication, 602, 603
  opposites, 592, 593
  and order, 586
  and order of operations, 610
  rational numbers, 583, 584
  and subtraction, 596
Reciprocal, 135, 607
  and division, 136, 608
  of a fraction, 135

equilateral, 556
isosceles, 556
obtuse, 556
right, 556, 561
scalene, 556
similar, 625
sum of angle measures, 556, 569
Trillion, 239
Trillions period, 2
True equation, 26
True inequality, 26
Truncating, 227
Two, divisibility by, 98

U

Unit angle, 551
Unit cube, 541
Unit price, 300
Unit rate, 300
Unit segment, 466

V

Value of an expression, 622
Variable, 622
    substituting for, 622

Vertex, 551
Volume
    of a circular cone, 544, 569
    of a circular cylinder, 543, 569
    of a rectangular solid, 541, 569
    of a sphere, 543, 569

W

Week, 499
Weight, 484, 510. *See also* Mass.
Whole numbers, 3
    addition, 9, 12
    division, 40, 47
    expanded notation, 3
    fraction notation for, 108, 109
    multiplication, 31, 35
        by fractions, 115
    order, 26
    rounding, 22
    standard notation, 3
    subtraction, 15, 17, 18
    word names for, 4
Wildlife population, estimating, 318
Word names
    for decimal notation, 219
    for whole numbers, 4

Y

Yard, 466
Year, 499
Yield
    effective, 400
    stock, 353

Z

Zero
    addition of, 9, 591
    divided by a nonzero number, 42,
        607
    division by, 42, 607
    fraction notation for, 108
    identity property, 591
    in quotients, 46
    reciprocal, 135